NUTRITIONAL AND TOXICOLOGICAL SIGNIFICANCE OF ENZYME INHIBITORS IN FOODS

ADVANCES IN EXPERIMENTAL MEDICINE AND BIOLOGY

Recent Volumes in this Series

Volume 195A
PURINE AND PYRIMIDINE METABOLISM IN MAN V, Part A: Clinical Aspects Including Molecular Genetics
Edited by W. L. Nyhan, L. F. Thompson, and R. W. E. Watts

Volume 195B
PURINE AND PYRIMIDINE METABOLISM IN MAN V, Part B: Basic Science Aspects
Edited by W. L. Nyhan, L. F. Thompson, and R. W. E. Watts

Volume 196
STEROID HORMONE RESISTANCE: Mechanisms and Clinical Aspects
Edited by George P. Chrousos, D. Lynn Loriaux, and Mortimer B. Lipsett

Volume 197
BIOLOGICAL REACTIVE INTERMEDIATES III: Mechanisms of Action in Animal Models and Human Disease
Edited by James J. Kocsis, David J. Jollow, Charlotte M. Witmer, Judd O. Nelson, and Robert Snyder

Volume 198A
KININS IV, Part A
Edited by Lowell M. Greenbaum and Harry S. Margolius

Volume 198B
KININS IV, Part B
Edited by Lowell M. Greenbaum and Harry S. Margolius

Volume 199
NUTRITIONAL AND TOXICOLOGICAL SIGNIFICANCE OF ENZYME INHIBITORS IN FOODS
Edited by Mendel Friedman

Volume 200
OXYGEN TRANSPORT TO TISSUE VIII
Edited by Ian S. Longmuir

Volume 201
LIPOPROTEIN DEFICIENCY SYNDROMES
Edited by Aubie Angel and Jiri Frohlich

NUTRITIONAL AND TOXICOLOGICAL SIGNIFICANCE OF ENZYME INHIBITORS IN FOODS

Edited by

Mendel Friedman

U.S. Department of Agriculture
Berkeley, California

PLENUM PRESS • NEW YORK AND LONDON

Library of Congress Cataloging in Publication Data

American Institute of Nutrition FASEB Symposium on Nutritional and Toxicological Significance of Enzyme Inhibitors in Foods (1985: Anaheim, Calif.)
Nutritional and toxicological significance of enzyme inhibitors in foods.

(Advances in experimental medicine and biology; v. 199)
"Proceedings of the American Institute of Nutrition FASEB Symposium on Nutritional and Toxicological Significance of Enzyme Inhibitors in Foods, held April 21–26, 1985, in Anaheim, California"—T.p. verso.
Includes bibliographies and indexes.
1. Digestive enzymes—Inhibitors—Congresses. 2. Food—Protein content—Congresses. 3. Nutritionally induced diseases—Congresses. I. Friedman, Mendel. II. American Institute of Nutrition. III. Federation of American Societies for Experimental Biology. IV. Title. V. Series.
QP609.D52A44 1985 612′.398 86-15151

Proceedings of the American Institute of Nutrition FASEB Symposium on Nutritional and Toxicological Significance of Enzyme Inhibitors in Foods, held April 21–26, 1985, in Anaheim, California, with additional invited contributions

Softcover reprint of the hardcover 1st edition 1986

A Division of Plenum Publishing Corporation
233 Spring Street, New York, N.Y. 10013

ISBN 978-1-4757-0024-4 ISBN 978-1-4757-0022-0 (eBook)
DOI 10.1007/978-1-4757-0022-0

PREFACE

Soybean proteins are widely used in human foods in a variety of forms, including baby formulas, flour, soy protein concentrates, soy protein isolates, soy sauces, textured soy fibers, and tofu. The presence of inhibitors of digestive enzymes in soy proteins impairs nutritional quality and possible safety of this impportant legume. Normal processing conditions based on the use of heat do not completely inactivate these inhibitors, so that residual amounts of plant protease inhibitors are consumed by animals and man.

Inhibitors of digestive enzymes are present not only in legumes, such as soybeans, lima beans, and kidney beans, but also in nearly all plant foods, including cereals and potatoes, albeit in much smaller amounts. The antinutritional effects of inhibitors of proteolytic enzymes have been widely studied and can be ameliorated by processing and/or sulfur amino acid fortification. A more urgent concern is reports that rats fed diets containing even low levels of soybean-derived inhibitors, which are found in foods such as soy-based baby formulas, may develop over their lifespan pancreatic lesions leading eventually to neoplasia or tumor formation. On the other hand, recent studies suggest that certain enzyme inhibitors from plant foods may prevent cancer formation in other tissues. A key question, therefore, is whether inhibitors from plant foods constitute a human health hazard.

Since the interaction of plant-derived inhibitors with proteases and other enzymes including trypsin, chymotrypsin, elastase, amylase, and lipase is a subject of increasing importance in animal and human nutrition, food chemistry, and food safety, I organized a symposium broadly concerned with the nutritional and toxicologial significance of enzyme inhibitors in foods. The symposium was sponsered by the American Institute of Nutrition - Federation of American Societies for Experimental Biology (FASEB) for its annual meeting in Anaheim, California, April 21-26, 1985. Invited speakers were asked to develop at least one of the following topics:

1. Mechanism of action of inhibitors - biofeedback mechanisms, role of hormones, differentiation of antinutritional from pancreatic effects, etc.

2. Inhibitors as inducers or promoters of pancreatic lesions.

3. Pancreatic response to inhibitors in different animal species.

4. Beneficial effects of inhibitors.

5. The importance of inhibitors in the human diet.

6. Inactivation and removal of inhibitors from plant foods.

The most important function of a symposium, I believe, is dissemination of insights and exchange of ideas so as to catalyze progress by permitting synergistic interaction among related disciplines. I hope that the reports presented at the symposium fulfilled this purpose. In addition, a number of scientists who could not participate in the symposium accepted invitations to contribute papers to this volume on the theme of the symposium. Indeed, about half the papers are especially written invited contributions. This book is, therefore, a hybrid between symposium proceedings and a collection of invited papers.

Brought together here are outstanding international authors from nine countries, who discuss the multidisciplinary aspects of the nutritional and toxicological significance and related agronomic, biochemical, and food processing aspects of naturally occurring inhibitors of digestive enzymes. Although a comprehensive summary is not possible in a preface, I wish to call special attention to several manuscripts. First, Weder's comprehensive review of the comparative potencies of plant inhibitors against digestive enzymes from animal and human sources is especially welcome, since it directly relates to the importance of enzyme inhibitors in human nutrition. The related report by Toskes on biofeedback effects in human patients should stimulate needed studies on the role of inhibitors in human health. The manuscripts by Green, Howatson, Morgan, and Roebuck describe pioneering studies on the relationship between the amounts of soybean inhibitors in the diet and the induction of pancreatic hypertrophy, hyperplasia, and adenoma. These nicely complement the comprehensive paper by Gumbmann and colleagues on long-term effects of soybean inhibitors and dietary protein levels on pancreatic histology and pathology. The papers by Schneeman, Liener, Nitsan, and Harwood and their colleagues on the observed species-variation of the pancreatic effects suggest that the rat may be especially sensitive and the monkey much less so or not at all. If these observations turn out to be generally true, then primates including humans may be more tolerant of inhibitors than rodents. The papers by Green and colleagues, Temler and Mettraux, Howatson, and Toskes strikingly demonstrate the key role of cholecystokinin and other hormones in the mechanism of action of inhibitors. The comprehensive paper by Rackis and collaborators on the widespread occurrence of inhibitors in the plant kingdom and on possible approaches to inactivate or remove them from foods offers an authoritative and critical assessment of the subject. This is complemented by the paper authored by Hymowitz on the development of new genetic varieties of soybeans with low trypsin inhibitor content. These studies, and related aspects described in the paper by Friedman and Gumbmann, could serve as the basis for possible new or improved ways to completely eliminate inhibitors from our food supply. The invited paper by Buonocore and Silano on amylase inhibitors and the two papers on effects of processed foods on digestive and metalloenzymes broaden the scope of the book from dealing with purely protease inhibitors to enzyme inhibitors in general. Other papers provide valuable insight into biochemical, immunochemical, and nutritional aspects of enzyme inhibitors from plant sources.

The described multidisciplinary studies reveal a complex and fascinating interplay between evolution, chemistry, nutrition, and toxicology of plant foods.

I am particularly grateful to all contributors for excellent cooperation, to Prof. James A. Olson of the American Institute of

Nutrition for helpful correspondence on the theme and contents of the symposium, to Prof. Barbara O. Schneeman and Dr. Michael R. Gumbmann for co-chairing the symposium sessions, and to Glenda M. Dugan for editorial assistance.

Plenum Press is publishing the papers under the title NUTRITIONAL AND TOXICOLOGICAL SIGNIFICANCE OF ENZYME INHIBITORS IN FOODS, as a volume in the series Advances in Experimental Medicine and Biology. This book is intended to complement the following published volumes which I edited for the same series: PROTEIN-METAL INTERACTIONS (1974); PROTEIN CROSSLINKING: BIOCHEMICAL AND MOLECULAR ASPECTS (1977); PROTEIN CROSSLINKING: NUTRITIONAL AND MEDICAL CONSEQUENCES (1977); NUTRITIONAL IMPROVEMENT OF FOOD AND FEED PROTEINS (1978); and NUTRITIONAL AND TOXICOLOGICAL ASPECTS OF FOOD SAFETY (1984).

I very much hope that these volumes will be a valuable resource for further progress in agriculture, protein chemistry, food chemistry, food safety, animal and human nutrition, pharmacology, physiology, toxicology, and medicine; all areas of world-wide urgency.

Mendel Friedman
April, 1986

The eyes of all look to you expectantly,
and you give them their food when it is due,
You give it openhandedly, feeding every
creature to its heart's content.

Psalm 145: 15-16

Eat nothing that will prevent you from eating.

Ibn Tibbon, c. 1190

CONTENTS

1. PROTEIN INHIBITORS OF SERINE PROTEINASES - MECHANISM AND CLASSIFICATION.......... 1
Michael Laskowski, Jr.

2. THE BIOLOGICAL FUNCTIONS AND PHYSIOLOGICAL EFFECTS OF INGESTED INHIBITORS OF DIGESTIVE REACTIONS.......... 19
S. S. Rothman

3. SAFETY OF TRYPSIN INHIBITORS IN THE DIET: EFFECTS ON THE RAT PANCREAS OF LONG-TERM FEEDING OF SOY FLOUR AND SOY PROTEIN ISOLATE...... 33
Michael R. Gumbmann, William L. Spangler, Glenda M. Dugan, and Joseph J. Rackis

4. DOSE EFFECTS OF RAW SOYABEAN FLOUR ON PANCREATIC GROWTH.......... 81
R. G. H. Morgan, R. A. Crass, and P. S. Oates

5. ENHANCEMENT OF PANCREATIC CARCINOGENESIS OF RAW SOY PROTEIN ISOLATE QUANTITATIVE RAT MODEL AND NUTRITIONAL CONSIDERATIONS...... 91
B. D. Roebuck

6. PANCREATIC CARCINOGENESIS - THE POTENTIAL OF CHOLECYSTOKININ AS A COCARCINOGEN IN THE HAMSTER - NITROSAMINE MODEL..........109
Allan G. Howatson

7. INTERACTION OF DIETARY PROTEIN AND TRYPSIN INHIBITOR ON PLASMA CHOLECYSTOKININ AND PANCREATIC GROWTH IN RATS..........123
Gary M. Green, Van H. Levan, and Rodger A. Liddle

8. GASTRIN AND CHOLECYSTOKININ LEVELS IN RATS FED SOYA BEAN TRYPSIN INHIBITOR..........133
Roman S. Temler and Christine Mettraux

9. NEGATIVE FEEDBACK INHIBITION OF PANCREATIC EXOCRINE SECRETION IN HUMANS..........143
P. P. Toskes

10. PROTEASE INHIBITORS: THEIR ROLE AS MODIFIERS OF CARCINOGENIC PROCESSES....................153
Walter Troll, Krystyna Frenkel, and Rakoma Wiesner

11. NUTRITIONAL AND METABOLIC RESPONSE TO PLANT INHIBITORS OF DIGESTIVE ENZYMES....................167
Daniel Gallaher and Barbara O. Schneeman

12. AMONG SPECIES SPONSE TO DIETARY TRYPSIN INHIBITOR: VARIATION SPECIES....................185
Barbara O. Schneeman and Daniel Gallaher

13. THE EFFECT OF THE LONG-TERM FEEDING OF RAW SOYFLOUR ON THE PANCREAS OF THE MOUSE AND HAMSTER....................189
Irvin E. Liener and Aharon Hasdai

14. ACCENTUATED RESPONSE TO DIETARY RAW SOYFLOUR BY MEAL-FEEDING IN VARIOUS SPECIES....................199
Zafrira Nitsan and Israel Nir

15. EFFECT OF LONG-TERM FEEDING OF SOY-BASED DIETS ON THE PANCREAS OF CEBUS MONKEYS....................223
J. P. Harwood, L. M. Ausman, N. W. King, P. K. Sehgal, R.J. Nicolosi, I.E. Liener, D. Donatucci, and J. Tarcza

16. INHIBITION OF HUMAN PROTEINASES BY GRAIN LEGUMES....................239
Juergen K. P. Weder

17. REGULATION OF PROTEINASE INHIBITOR GENES IN FOOD PLANTS....................281
Willis E. Brown, John S. Graham, Jong S. Lee, and Clarence A. Ryan

18. GENETICS AND BREEDING OF SOYBEANS LACKING THE KUNITZ TRYPSIN INHIBITOR....................291
Theodore Hymowitz

19. PROTEASE INHIBITORS IN PLANT FOODS: CONTENT AND INACTIVATION......299
J. J. Rackis, W. J. Wolf, and E. C. Baker

20. PREPARATION OF UNHEATED SOY PROTEIN ISOLATE WITH LOW TRYPSIN INHIBITOR CONTENT....................349
E. C. Baker and J. J. Rackis

21. NUTRITIONAL IMPROVEMENT OF LEGUME PROTEINS THROUGH DISULFIDE INTERCHANGE....................357
Mendel Friedman and Michael R. Gumbmann

22. RELEVANCE OF MULTIPLE SOYBEAN TRYPSIN INHIBITOR FORMS TO NUTRITIONAL QUALITY..391
Anna L. Tan-Wilson and Karl A. Wilson

23. ANTINUTRITIONAL AND BIOCHEMICAL PROPERTIES OF WINGED BEAN TRYPSIN INHIBITORS..413
Benito O. de Lumen and Jones Chan

24. PROTEASE INHIBITORS OF THE MARAMA BEAN............................429
Barry Starcher, Leah Bryant, and Michael Elfant

25. TRYPSIN/CHYMOTRYPSIN INHIBITORS FROM MILLETS......................439
Thillaisthanam N. Pattabiraman

26. ANTIGENICITY OF NATIVE AND MODIFIED KUNITZ SOYBEAN TRYPSIN INHIBITORS..449
David L. Brandon, Sakhina Haque, and Mendel Friedman

27. PHOTOREACTIVE, ACTIVE DERIVATIVES OF TRYPSIN- AND CHYMOTRYPSIN-INHIBITORS FOMR SOYBEANS AND CHICK PEAS..............469
Yehudith Birk, Patricia Smirnoff, and J. Ramachandran

28. BIOCHEMICAL, NUTRITIONAL, AND TOXICOLOGICAL ASPECTS OF ALPHA-AMYLASE INHIBITORS FROM PLANT FOODS.........................483
Vincenzo Buonocore and Vittorio Silano

29. THE INHIBITION OF DIGESTIVE ENZYMES BY POLYPHENOLIC COMPOUNDS.....509
David Wynne Griffiths

30. EFFECT OF SEVERELY ALKALI-TREATED CASEIN ON GASTROINTESTINAL TRANSIT AND SELECTED INTESTINAL ENZYME ACTIVITY..................517
Bernard Possompes and Jacques Berger

31. INACTIVATION OF METALLOENZYMES BY LYSINOALANINE, PHENYLETHYLAMINOALANINE, ALKALI-TREATED FOOD PROTEINS, AND SULFUR AMINO ACIDS...531
Mendel Friedman, Ok-Koo K. Grosjean, and James C. Zahnley

INDEX...561

1

PROTEIN INHIBITORS OF SERINE PROTEINASES - MECHANISM AND CLASSIFICATION

Michael Laskowski, Jr.

Department of Chemistry
Purdue University
West Lafayette, IN 47907

ABSTRACT

Protein proteinase inhibitors are widely distributed in plants, animals and microorganisms. They can be conveniently grouped since most frequently they inhibit proteinases belonging only to a single mechanistic class. Protein inhibitors of serine proteinases have been most extensively studied. They are strictly competitive inhibitors forming 1:1 complexes with the enzymes they inhibit. In these complexes, all activities of the enzyme are completely abolished. The inhibitors are substrates for the enzyme they inhibit at a unique peptide bond called the reactive site peptide bond (one for each inhibitory domain). However, compared to normal substrates where the enzyme-substrate and enzyme-product complexes dissociate very readily here, the complexes are very stable. Serine proteinase inhibitors can be divided into at least 13 families. Within each family the position of the reactive site and the closure of disulfide bridges can be inferred by homology. In enzyme-inhibitor complexes, about 10-15 residues of the inhibitor are in contact with the enzyme. Their specific nature strongly affects both the strength and the specificity of enzyme-inhibitor interaction. In all cases where the sequences of many inhibitors from the same family can be compared, the contact residues are not strongly conserved - instead, they are hypervariable. This raises major problems but also offers huge opportunities to those concerned with the role of inhibitors in biology and in medicine.

INTRODUCTION

The objective of this introductory chapter is to review the mechanism of action of protein inhibitors of serine proteinases, to update their system of classification and to comment on two major evolutionary trends exhibited by proteinase inhibitors - the tendency to acquire several inhibitory domains in the same molecule and to fix mutations more rapidly in those positions of the inhibitor molecule which are in contact with the enzyme. This chapter is an update of the review by Laskowski and Kato (1980). However, the selection of material is somewhat different. Aside from providing a narrative description, I focus especially strongly on those aspects of inhibitors which are most often misunderstood by workers who are only tangently acquainted with inhibitor chemistry. I ask the readers for forgiveness in advance for this being only in part a review and in part a sermon. I hope that those who do not need the latter can still go on reading it and feeling superior to those who do.

Proteolysis is one of the most important of biological functions (Neurath, 1984). It is almost always involved in biological protein synthesis, in control of protein destination (signal peptide removal), in control of protein activity (zymogen activation) in control of protein death (protein turnover). Aside from this cradle to grave involvement with the functioning of proteins, it is also often involved with facilitating the transport of large systems (acrosin in sperm penetration of ova) or of an entire organism (cocoonase and emergence of a silk moth). Most importantly from the point of view of this volume, proteolysis is essential for ingestion of foreign protein foodstuffs.

It is clear that because of the great importance of proteolysis it must be controlled. We wish to eat the other guy's pancreas or intestine - not our own. Similarly, we wish our blood to clot at specific locations but not in the entire circulatory system. One of the steps in the control of proteolysis is the existance of proteinase inhibitors. These are ubiquitious. They were found in almost all tissues where they were looked for - animal, plant and microbial. There are very many of them. It is still an open question of whether there are in nature more proteinases or protein-proteinase inhibitors. Some of the inhibitors are highly specific and inhibit only a small fraction of proteinases - others show a broad inhibitory spectrum. For the latter, we talk about two types of enzymes - cognate enzymes, i.e., enzymes which recognize the inhibitor's reactive site and combine with it to form the inactive enzyme-inhibitor complex and target enzymes - enzymes that the inhibitor has evolved to control. Cognate enzyme-inhibitor combinations often are very arbitrary from the point of view of biology. Thus, bovine cationic β trypsin is a frequently employed cognate enzyme for a great number of "trypsin" inhibitors even though in the natural order of things these inhibitors are not likely to have any contact with bovine β trypsin. Finding the target enzyme for an already characterized inhibitor is one of the major challenges of inhibitor research. It is, however, largely although not entirely beyond the scope of this review.

Protein proteinase inhibitors can be broadly classified as shown in Table I.

TABLE I

PROTEIN PROTEINASE INHIBITORS

Macroglobulins

Inhibitors of Single Mechanistic Classes

1. Serine Proteinase Inhibitors
2. Cysteine Proteinase Inhibitors
3. Aspartic Acid Proteinase Inhibitors
4. Metallo Proteinase Inhibitors

Macroglobulins (Barrett and Starkey, 1973; Harpel, 1973; Sottrup-Jensen et al., 1984) tend to "inhibit" proteinases of all mechanistic classes - they are not quite universal proteinase inhibitors but the exceptions are moderately few. Thus, they are the best thing to suggest to someone who wishes to "inhibit" excessive proteolysis without specifying what enzyme is responsible for it. On the other hand, macroglobulins do not inhibit their cognate enzymes at all. Instead, they trap them leaving the active site of the enzyme open and fully operational toward small substrates. The enzyme-macroglobulin complexes can even be inhibited by small protein proteinase inhibitors. The observed "inhibition" by macroglobulins is a consequence of entrapment - large substrates and large inhibitors are sterically excluded. In my opinion, it is a great pity that macroglobulins were called inhibitors before their mechanism of action was understood. The huge distinction between them and the other inhibitors would be better appreciated if they were not.

The remainder of inhibitors listed in Table I have an unusual characteristic. They appear to be specific for enzymes from single mechanistic classes (Laskowski and Sealock, 1971; Laskowski and Kato, 1980). In spite of numerous claims to the contrary, I am not aware of any documented example where enzymes of two different mechanistic classes were shown to be inhibited at the same reactive site of a protein inhibitor. This is quite surprising since cysteine proteinases and serine proteinases frequently hydrolyze the same bond in protein substrates and, more importantly, are inhibited by the same peptide aldehyde inhibitors such as leupeptin and antipain. Furthermore, it is surprising that inhibitors are not restricted to a single homologous family of proteinases as several inhibitors inhibit both members of the chymotrypsin and of the subtilisin family of serine proteinases.

For all four groups of inhibitors listed in Table I, there are known protein inhibitors although the number of well-characterized inhibitors of serine proteinases far exceeds the number known for other mechanistic classes. This is not likely to be a reflection of the situation that exists in nature but simply a consequence of the search made by the isolators. Three decades ago my (late) father and I wrote a review on "naturally occurring trypsin inhibitors" (Laskowski and Laskowski, 1954). Chymotrypsin inhibition was considered but only as an unusual property of some trypsin inhibitors. Later it was realized that chymotrypsin inhibitors occur in nature - this was followed by an outpouring of findings of new chymotrypsin inhibitors. Another miniboomlet followed the discovery of specific elastase inhibitors. At present, we are living in a renaissance of cysteine proteinase inhibitors which were not pursued since the initial discovery by Fossum and Whittaker (1968) but now are abundantly discovered in numerous sources by the groups of Barrett, Turk, Katunuma and Fritz (for example, see Mueller-Esterl, 1985). Similar explosions undoubtedly await the two other groups of inhibitors listed in Table I.

Our understanding of serine proteinase inhibitors is greater than that of other groups of inhibitors because the research on them stems from the tradition of the brilliant beginnings of Moses Kunitz who described and crystallized bovine pancreatic trypsin inhibitor (Kunitz) in 1936 (Kunitz and Northrop, 1936) and soybean trypsin inhibitor (Kunitz) in 1947 (Kunitz, 1947a,b). Kunitz realized that enzyme and inhibitor combined with 1:1 stoichiometry to form a totally inactive complex according to the equation

$$E + I \underset{k_{off}}{\overset{k_{on}}{\rightleftharpoons}} C \qquad (1)$$

where E is the enzyme, I the inhibitor and C the enzyme-inhibitor complex. It is a common error to claim that the complex poses some residual activity (sometimes toward some substrates but not others). There are two reasons for this mistaken view - one is that, under some assay conditions, part of the complexes are allowed to dissociate and, thus, exhibit activity; the other is that occasionally a complex is made from an enzyme slightly contaminated by another enzyme which is not cognate (not efficiently inhibited) and, thus, the contaminant enzyme's activity is seen in the "complex."

Equation (1) yields

$$K_a = \frac{1}{K_I} = \frac{(C)}{(I)\ (E)} = \frac{k_{on}}{k_{off}} \tag{2}$$

For many protein proteinase inhibitors-proteinase pairs, the values of K_a are very large (the value for bovine β trypsin-bovine trypsin inhibitor (Kunitz) is probably the largest reported and is $1.8 \times 10^{13}\ M^{-1}$) (Vincent and Lazdunski, 1972). However, values between 1×10^8 and $1 \times 10^{12}\ M^{-1}$ are quite common. This is in sharp contrast to the typical values for small synthetic competitive inhibitors of many enzymes discussed in elementary biochemistry texts under the heading of competitive inhibition. These values tend to range between 10^3 and $10^5\ M^{-1}$. The values of k_{on} for most, but not all inhibitors, tend to be of the order of $10^6\ M^{-1}\ s^{-1}$ (Finkenstadt et al., 1974; Empie and Laskowski, 1982; Ardelt and Laskowski, 1985; Hiromi et al., 1985). It is, therefore, a consequence of Eq. (2) that for the small, textbook inhibitors dissociation half lives are very short (1 millisecond - 1 second) and, therefore, on laboratory time scales the enzyme-inhibitor-substrate system is in pseudo equilibrium. For protein inhibitors of serine proteinases, the dissociation half lives range from minutes to weeks and, therefore, it is frequently observed that addition of even concentrated substrate does not dissociate the enzyme-inhibitor complex on laboratory time scales. As a result, it appears that the substrate and the inhibitor do not compete because they are not given enough time to do it and, therefore, people frequently reach the grossly incorrect conclusion that inhibition is non-competitive. This and other aspects of differences between strong and weak competitive inhibitors were analyzed by N. M. Green in 1953 (Green, 1953) and frequently thereafter; but, the repeated non-original (but obviously needed) papers appear to fall on deaf ears. This is a pity because, while the distinction between competitive and non-competitive inhibition may be dismissed as a semantic nicety, taking an incorrect view of the situation leads to grossly (many orders of magnitude) incorrect K_I or K_a values. On the other hand, a hard-headed view that addition of substrate will <u>never</u> cause dissociation of protein inhibitor-proteinase complexes has also caused considerable trouble. Several authors reported that inhibitors with K_a of $10^8\ M^{-1}$ are "ineffective" even though this is a quite respectable association equilibrium constant. The reason for this was that, since complexes of such inhibitors have dissociation half lives of about 1 min., the complexes often dissociate completely during the assay and, thus, are overlooked.

While the "mechanism" embodied in equations (1) and (2) is very helpful to those who wish to describe the enzyme-inhibitor association, they provide no clue or guess about the molecular details of complex formation. We were fortunate in our laboratory at Purdue (Lebowitz and Laskowski, 1962; Finkenstadt and Laskowski, 1965, 1967; Ozawa and Laskowski, 1966) in getting such a clue by finding that the mechanism can be more correctly stated as

$$E + I \underset{k_{off}}{\overset{k_{on}}{\rightleftharpoons}} C \underset{k^*_{on}}{\overset{k^*_{off}}{\rightleftharpoons}} E + I^* \tag{3}$$

where the symbols have the same meaning as before except for the newly introduced I^*, which is called modified inhibitor and for I which is rechristened virgin inhibitor. Virgin inhibitor has all of its peptide bonds intact - modified inhibitor has one peptide bond - the reactive site peptide bond - hydrolyzed. This is illustrated for soybean trypsin inhibitor (Kunitz) in Fig. 1.

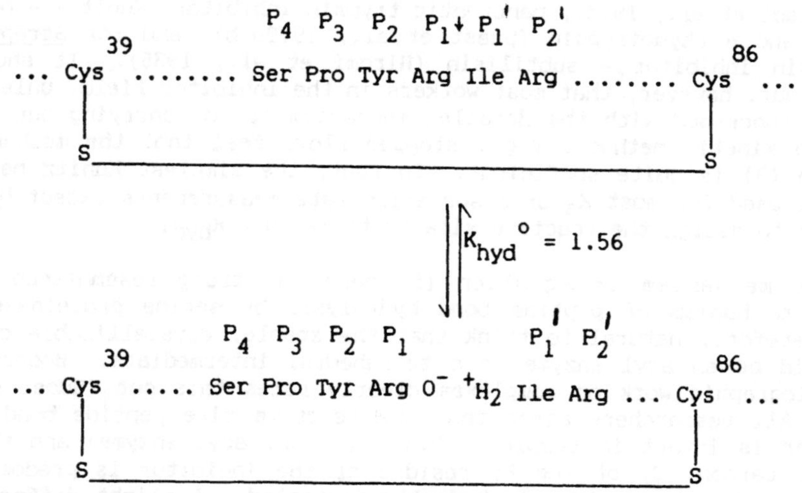

Fig. 1. Schematic diagram of the reactive site of soybean trypsin inhibitor (Kunitz) as identified by Ozawa and Laskowski (1966). The inhibitor sequence was determined by Koide and Ikenaka (1973). Aside from the conventional numbering system, the reactive site numbering system based on Schechter and Berger (1967) is also shown and later used throughout this chapter. The value of K_{hyd}° was determined by Mattis and Laskowski (1973).

Note that complicating the mechanism from (1) to (3) allows us to introduce another equilibrium constant in addition to K_a. We choose for this constant to be K_{hyd}

$$K_{hyd} = \frac{(I^*)}{(I)} = \frac{k_{on}\ k^*_{off}}{k^*_{on}\ k_{off}} \tag{4}$$

Many people find it surprising that the value of K_{hyd} is (for most good inhibitors) rather close to unity. Most people expect proteolysis to be "irreversible", i.e., for K_{hyd} to be very high. Another common misconception about K_{hyd} is that its value depends not only on the inhibitor but also on the enzyme used to carry out the hydrolysis. A moment's reflection convinces one about the fallacy of this view.

The work on the mechanism was extended to complicate it further. The mechanism as given by equation (3) demands that the rate of formation of complex from equimolar enzyme and inhibitor mixtures (E + I or E + I*) should increase as the square of the starting concentration. This is, indeed, true for most systems that were studied but it does not go on without a limit. At concentrations of about 10^{-5} M, the association reactions (E + I and E + I*) switch from the simple second order to first order indicating the presence of intermediate "loose" complexes L and L* (Luthy et al., 1973; Quast and Engel, 1978a,b; Hiromi et al., 1985). An even more penetrating study allowed Quast and Engel to postulate another intermediate, X, only on the side of the modified inhibitor. Thus, the "full" mechanism now reads

$$E + I \rightleftharpoons L \rightleftharpoons C \rightleftharpoons X \rightleftharpoons L^* \rightleftharpoons E + I^* \tag{5}$$

Laskowski and Kato (1980) referred to this as the "standard mechanism." Many of the rate constants (and in some cases, their pH dependence) involved in this mechanism have been measured for the systems - soybean trypsin inhibitor - (Kunitz) bovine β trypsin (Luthy et al., 1973; Finkenstadt et al., 1974), pancreatic trypsin inhibitor (Kunitz) - bovine β trypsin and α chymotrypsin (Quast et al., 1978a,b), and for Streptomyces subtilisin inhibitor - subtilisin (Hiromi et al., 1985). It should be pointed out, however, that most workers in the inhibitor field, unless they are (a) concerned with the detailed mechanism or (b) carrying out studies by rapid kinetic methods, e.g., stopped flow, feel that the mechanism of equation (3) is quite sufficient. In fact, the simplest Kunitz mechanism is still used for most K_a or association rate measurements except by those who wish to assign the reactive site or to measure K_{hyd}.

The mechanism of equation (5) bears a strong resemblance to the general mechanism of peptide bond hydrolysis by serine proteinases. It was, therefore, natural to think that the stable, crystallizable complex, C, should be an acyl enzyme or a tetrahedral intermediate. Modern X-ray crystallographic work on complexes clearly rules this out (Huber et al., 1974). All researchers agree that the reactive site peptide bond of the inhibitor is intact in complex (this rules out acyl enzyme) and that the carbonyl carbon, C, of the P_1 residue of the inhibitor is predominantly trigonal and not greatly tetrahedrally distorted. A slight difference of opinion exists between fully trigonal (Read and James, 1986) and partially tetrahedral C (Marquart et al., 1983). All structures agree in showing that the oxygen O^{γ} of the catalytic Ser residue of the enzyme approaches the C atom of the P_1 residue of the inhibitor closer than Van der Waal's contact but not nearly as close as a covalent bond. Therefore, some partial covalent bonding is involved but it is not an important consideration in the energetics of the complex formation as complexes of similar stability form with anhydrotrypsin, where the O^{γ} atom of catalytic Ser is removed (Huber et al., 1975).

Nuclear magnetic resonance measurements were also carried out on complexes in which the C atom of P_1 was specifically enriched by ^{13}C (and in the SSI case, the P_1' N atom by ^{15}N). All of these studies confirm intact peptide bond in complex and absence of large tetrahedral distortion around the C atom. Small distortions cannot be excluded by NMR data.

This highly detailed view of the nature of the stable enzyme-inhibitor complex somewhat obstructs what I believe is the principal lesson of all of these detailed studies. In every case where the experiment was done, the X-ray results confirmed the previously chemically assigned reactive site placement. In every case, it is the P_1 of the inhibitor with which the catalytic O^{γ} interacts. Furthermore, the side chain of P_1 is always imbedded in the primary specificity pocket of the enzyme, the O atom of P_1 forms the two expected hydrogen bonds with the oxyanion hole of the enzyme, residues P_3, P_2, P_1 form an extended β-type interaction with residues of the enzyme, etc. In summary, they bound to the enzyme in a substrate-like manner. The lesson of both the crystallographic and NMR studies and of all mechanistic studies embodied in equations (3) and (5) is - INHIBITORS ARE SUBSTRATES.

Another way of buttressing this conclusion is to recall that an "ideal" enzyme has the highest possible k_{cat}/K_m value (Fersht, 1985). Since k_{cat}/K_m has units of a second order reaction ($M^{-1}S^{-1}$), it is restricted by the diffusion upper limit, i.e., it can be no greater (generally, it is much less) than the rate of first encounters between enzyme and inhibitor molecules. We could turn this reasoning around for substrates, i.e., the substrates with the highest possible k_{cat}/K_m is the "ideal" substrate. It turns out that many protein inhibitors of serine proteinases have very large k_{cat}/K_m values, values that rival many peptide or even p-nitroanilide substrates which were expressly designed by chemists to be "ideal" substrates for the proteinases.

This conclusion seems absurd as inhibitors differ strikingly from superb substrates. The reason is that we did not as yet complete the definition of an "ideal" substrate. Subject to the value of k_{cat}/K_m being the largest possible, i.e., approaching the diffusion upper limit, both k_{cat} and K_m should be as large as possible, i.e., weak binding and rapid, productive breakdown of complex. "Ideal" inhibitors are just the opposite. They partition the highest possible k_{cat}/K_m into the lowest possible k_{cat} and K_m. Typically, the values of k_{cat} for some very good inhibitors are at optimal pH 10^6 - 10^{12} times slower than those for the best peptide and p-nitroanilide substrates. It is not yet known what in detailed molecular terms controls the k_{cat}/K_m partitioning and, thus, is responsible for making some enzyme-(substrate or inhibitor) pairs enzyme-substrate pairs while others are enzyme-inhibitor pairs. It is clear that the explanation lies in detailed understanding of contacts between both partners of the complex. However, it is interesting to note that some enzymes treat the reactive sites of inhibitors almost as normal substrates (Estell and Laskowski, 1980; Ardelt and Laskowski, 1983).

Once it was realized that inhibitors are substrates and that they possess reactive sites, numerous workers found that a very large number of inhibitors from different sources obeyed the mechanism shown in equation (5). This was quite surprising since many of the inhibitors obeying this common mechanism did not seem at all similar in molecular weight, amino acid composition and in physico-chemical properties. The huge effort devoted to sequencing of many inhibitors and to the determination of their three-dimensional structure proved the point beyond any reasonable doubt - inhibitors which obey the standard mechanism are not all homologous. However, X-ray crystallography suggests that those portions of the inhibitor molecule which are in enzyme-inhibitor contact in complexes have very similar conformations (Read and James, 1986). We are, thus, dealing here with manifestation of convergent evolution. Inhibitors are so important that essentially the same inhibitory mechanism developed independently several times during the evolutionary history of life. The classical example of this situation is the serine proteinase group where the chymotrypsin family and the subtilisin family acquired their activity by convergence.

We now wish to ask the more detailed question. How many times has the inhibitory mechanism been independently invented? This question is essentially equivalent to asking - how many families of inhibitors exist? It is clear that a detailed answer to this question is not straightforward. The most obvious problem is that we are still at an early stage of serine proteinase inhibitor research and new families are still being found, although, the frequency of such discoveries is decreasing. The other problem is opposite. It cannot be rigorously excluded that some of the listed families should not be lumped as it is very hard to prove that two proteins are not divergently related.

Laskowski and Kato (1980) have proposed a list of families of protein inhibitors of serine proteinases which met with reasonably warm reception. Table II is an update obtained by consultation with several workers in the inhibitor sequencing field early in 1985. The division of inhibitors into bacterial, plant and animal, though quite attractive for many workers, e.g., nutritionists, is probably flawed as one plant family - the potato I family - already contains one animal member, eglin, an inhibitor from leeches. The serpin family both belongs and does not belong on this list

TABLE II

FAMILIES OF PROTEIN INHIBITORS OF SERINE PROTEINASES

	Animals
1.	Bovine pancreatic trypsin inhibitor (Kunitz) family[a]
2.	Pancreatic secretory trypsin inhibitor (Kazal) family[b]
3.	*Ascaris* inhibitor family[c]
4.	Serpin family (mechanistically distinct)[d]
5.	Hirudin family[e]
	Plants
6.	Soybean trypsin inhibitor (Kunitz) family[f]
7.	Soybean proteinase inhibitor (Bowman-Birk) family[g]
8.	Potato 1 family[h]
9.	Potato 2 family[i]
10.	Barley trypsin inhibitor family[j]
11.	Squash inhibitor family[k]
	Microbial
12.	Streptomyces subtilisin inhibitor (SSI) family[l]
13.	OTHER FAMILIES

[a]Gebhardt and Fritz (1986); Gebhardt and Hochstrasser (1986).
[b]Laskowski et al. (1980); a list of all known Kazal sequences is maintained at Purdue. Currently, it contains 300 domain sequences.
[c]Babin, Peanasky and Goos (1984).
[d]Carrell and Travis (1985).
[e]Seemueller et al. (1986).
[f]Hejgaard, Svendsen and Mundy (1983).
[g]Ikenaka and Norioka (1986).
[h]Graham et al. (1985a).
[i]Graham et al. (1985b).
[j]Odani et al. (1983), Campos and Richardson (1983), Mahoney et al. (1984).
[k]Wieczorek et al. (1985), Joubert (1984).
[l]Hiromi et al. (1985).

as the mechanism of serpin-serine proteinase interaction is closely similar but not identical to that given by equation (5). The differences between the serpins and the other inhibitors listed in Table II are not as yet very clearly delineated. An additional objection to Table II is that some new proposed families are not listed. This is purely a conservative approach on my part - I am waiting for more sequences to clarify the situation.

An important, recent development from the laboratories of Odani, Richardson and Svendsen is that several serine proteinase inhibitors are also amylase inhibitors. This is likely to lead to considerable complications in future classification efforts.

Aside from purely intellectual satisfaction, why is it worthwhile to identify any newly found inhibitor with a family? It is clear that, if after sequencing you can make a clear cut family identification, a great deal of work has been saved for you. Unless you are a purist, you no longer need to determine the disulfide bridge pairing and the position of the reactive site. If in addition a three-dimensional structure of one family member is known, you can reasonably and safely assume that your inhibitor has a closely similar structure. It is even better if the structure of enzyme-inhibitor complex is known for one of the family members since, in such a case, you can infer those residues of your inhibitor which are in enzyme-inhibitor contact. From these, given a sequence to reactivity algorithm, you could infer the reactivity of your new inhibitor toward many enzymes. There is, however, a false bonus that one should strongly guard against. This is assuming that inhibitory properties of a new inhibitor in a family will be closely similar to the properties of the other members. Yet, it is an easy error to make; and, I have heard it often - "as my inhibitor is a member of the Bowman-Birk family, its property should be similar to that of the classical Bowman-Birk inhibitor." If for property we read strength of inhibition or inhibitory specificity and by inference nutritional or toxic effect, then the statement is likely to be wrong. Yet, the reasoning seems sound. This is because in most protein families highly homologous members have closely similar activities and specificities. This is frequently not the case for protein inhibitors of serine proteinases.

Hypervariability of Enzyme-Inhibitor Contact Residues

Soon after the development of the reactive site concept, it became clear that the specificity of the inhibitor is predominantly governed by the residues comprising the reactive site. In a more modern version, the specificity is predominantly governed by those residues which are in contact with the enzyme in enzyme-inhibitor complex; changing of the residues which are not in contact seldom has an effect on inhibitory activity (Empie and Laskowski, 1982; Laskowski et al., 1983; S. J. Park, 1985). In my laboratory, Sealock succeeded in replacing P_1 Arg^{63} in soybean trypsin inhibitor (Kunitz) by Lys^{63} (Sealock and Laskowski, 1969). The product retained full trypsin inhibitory activity. Later, Leary (Kowalski et al., 1974) replaced Arg^{63} by Trp^{63} and converted a trypsin inhibitor to a chymotrypsin inhibitor. The reactions are illustrated in Fig. 2. What people can do in the laboratory, nature can do better. Stevens (1971) has provided the preliminary sequence of lima bean inhibitor (a Bowman-Birk family inhibitor) and showed that this inhibitor consists of two homology regions each with a reactive site. The first reactive site has a P_1 Lys and inhibits trypsin, the second has P_1 Leu and inhibits chymotrypsin. The work on Bowman-Birk inhibitors has been greatly extended in the laboratory of Ikenaka (Ikenaka and Narioka, 1986). When we examine his alignments of Bowman-Birk inhibitor sequences, we note

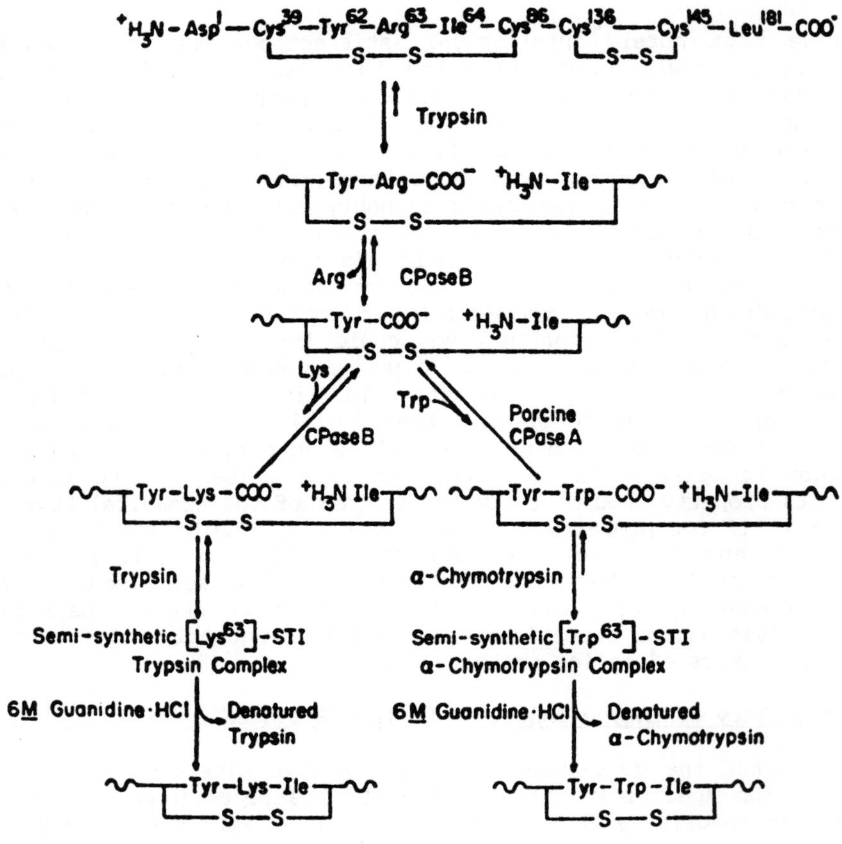

Fig. 2. Detailed procedure for replacing the P_1 Arg^{63} of soybean trypsin inhibitor (Kunitz) by Lys^{63} on the left and by Trp^{63} on the right (adapted from Kowalski et al. (1974)). The Lys^{63} inhibitor is a powerful inhibitor of trypsin, the Trp^{63} inhibitor is a powerful inhibitor of chymotrypsin but is ineffective against trypsin.

TABLE III

AMINO ACID RESIDUES IN P_1 POSITION

	Found		Not Found Yet[a]
Overall[b]	In Kazal Family	Inhibition Demonstrated[c]	
Arg	+	+	Trp
Lys	+	+	His
Tyr	+	+	Ile
Phe	+	+	Cys
Leu	+	+	
Met	+	+	
Val	+	+	
Ala	+	+	
Ser	+	+	
Thr	+	–	
Asp	+	–	
Asn	–	+	
Glu	+	–	
Gln	+	+	
Gly	+	+	
Pro	+	–	

[a]Several of these were artificially introduced, e.g., see Fig. 2 for Trp. Here, I list only those found naturally. The general impression is that, as research continues, the Not Found Yet column will disappear.

[b]This is a composite of all families listed in Table II.

[c]This is a listing of residues found by homology. In some cases, these substances may not serve as inhibitors. In other cases, the (-) answer should not be taken too seriously as testing may not have been extensive enough.

great variation of P_1 residues and of several other residues surrounding the reactive site. The specificity of the various inhibitors also differs strikingly from one to another. The same phenomenon is evident whenever sequences of many inhibitors from any of the families listed in Table II are aligned or when the inhibitory activities are compared. This behavior was a part of the reason why Kato and I (Laskowski and Kato, 1980) started speaking about P_1 hypervariability or about hypervariability of the reactive site region. Table III is a list of all residues thus far found in the P_1 position of serine proteinase inhibitors or of their close homologs. It appears that soon all amino acid residues will be represented at this position.

However, such conclusions had to be guarded. First, in most inhibitor sequence sets that are now available, many members are paralogous*. In any case, even when we suspect that they are orthologous, it is difficult to prove it. This is particularly so in plants where a multiplicity of related inhibitors are often present in the same plant - therefore, orthologous relationship to an inhibitor in another plant is difficult to establish. In order to better define hypervariability, it is best to deal with an orthologous set.

The second problem associated with the definition of hypervariability on the basis of the various sets of inhibitors from the families of Table II is that the three-dimensional structures of enzyme-inhibitor complexes are available for only a few families. Thus, the residues in contact cannot be precisely inferred and we are reduced to guessing which residues other than P_1 are in contact.

We have been fortunate to get a large set (106 sequences) which has neither of these two objections (Table IV). They are ovomucoid third domains from various avian species. Ovomucoid is one of the major proteins in avian egg white. It is frequently a potent inhibitor of serine proteinases. A comparative study of ovomucoids from various species was pioneered by Rhodes, Bennett and Feeney (1960). We have sequenced several ovomucoids (Kato et al., 1976, 1978, 1986) and showed that they consist of three tandem homologous domains. The domains belong to the Kazal inhibitor family (Table II). I digress here briefly to point out that the acquisition of several inhibitory domains on a single polypeptide chain is a frequent phenomenon in inhibitor evolution - it is manifested in most of the well-studied families. This phenomenon is dealt with in greater detail in Laskowski and Kato (1980) and for Kazal family in Laskowski et al. (1980).

Now returning to ovomucoid. There appears to be strong evidence that there is only one ovomucoid (Bogard, Kato and Laskowski, 1980) and one ovomucoid gene (Stein et al., 1980) in each bird and, therefore, that ovomucoids from various birds are orthologous. Limited proteolysis of ovomucoid isolated from 101 avian species allowed us to isolate intact ovvomucoid third domain (51-56 amino acids depending on species). We have sequenced them and composite results are shown in Table IV. As with any other protein when many sequences are examined, it is seen that some

---

*Some proteins are homologous to one another as a consequence of a gene duplication followed by divergence, e.g., α and β globin chains. Such proteins are called paralogous to one another. Trypsin, chymotrypsin and elastase are paralogous. The more obvious way for proteins to be homologous is to be products of the same gene in different species. Such proteins are called orthologous. The distinction is very important to protein evolutionists, but, in fact, it is difficult to make.

TABLE IV

THE AMINO ACID SEQUENCE OF TURKEY OVOMUCOID THIRD DOMAIN AND THE ALTERNATIVES FROM 100 ADDITIONAL AVIAN SPECIES[a]

Val Ile	6	136	13	
Asp Asn	7	137	12	
Cys	8	138	11	
Ser	9	139	10	
Glu Asp Gly	10	140	9	
Tyr His	11	141	8	
Pro	12	142	7	
*****Lys Arg Thr Met Gln**	13	143	6	P_n
*****Pro His**	14	144	5	
*****Ala Val Asp Ser Gly Glu Thr**	15	145	4	
*****Cys**	16	146	3	
*****Thr Pro Ser Arg Leu Met**	17	147	2	
*****Leu Pro Met Val Gln Ser Ala Thr Lys**	18	148	1	
*****Glu Asp Leu**	19	149	1	
*****Tyr Phe Asn Asp Leu Gln Glu His Arg**	20	150	2	
*****Arg Met Phe Lys Val Thr Leu**	21	151	3	
Pro	22	152	4	
Leu Val Ile Phe	23	153	5	
Cys	24	154	6	
Gly	25	155	7	
Ser	26	156	8	
Asp Asn	27	157	9	
Asn Ser	28	158	10	
Lys Gln Ile Glu	29	159	11	
Thr Ser Ile	30	160	12	
Tyr	31	161	13	
*****Gly Ser Asn Ala Asp Val Arg His**	32	162	14	
Asn Ser Asp	33	163	15	
Lys Arg	34	164	16	P_n'
Cys	35	165	17	
*****Asn Asp Ala Ser Gly Tyr**	36	166	18	
Phe	37	167	19	
Cys	38	168	20	
Asn Ser	39	169	21	
Ala	40	170	22	
Val Ala Phe	41	171	23	
Val Ala Leu Met	42	172	24	
Glu Asp Gln Lys His	43	173	25	
Ser Lys	44	174	26	
Asn Ser	45	175	27	
Gly Val	46	176	28	
Thr -	47	177	29	
Leu -	48	178	30	
Thr Ile Asn Ser	49	179	31	
Leu Val Phe	50	180	32	
Ser Arg Gly Asn	51	181	33	
His Arg	52	182	34	
Phe Leu Ile	53	183	35	
Gly Gly	54	184	36	
Lys Glu Thr Gln	55	185	37	
Cys	56	186	38	

[a]Three numbering systems are presented: turkey third domain, entire chicken ovomucoid, and Schechter-Berger (1967). The residues which are in contact with the enzyme in the turkey ovomucoid third domain - Streptomyces griseus proteinase B (Read et al., 1983), are starred. (Taken from Kato et al., 1986)

positions are unvaried, others fix but a few replacements and yet others fix very many. It is the nature of the positions that change rapidly that is surprising - they are the residues in enzyme-inhibitor contact. These changes matter. The enzyme-inhibitor equilibrium constants for virtually all of these inhibitors were measured (Empie and Laskowski, 1982; Laskowski et al., 1983; Park, 1985) for bovine chymotrypsin α, porcine pancreatic elastase, and for three bacterial enzymes - subtilisin, Streptomyces griseus proteinase A and B. They range from 10^3 M^{-1} (very weak inhibition or no inhibition at all) to 10^{12} M^{-1} (extremely strong inhibition). Generally, those inhibitors which are very strong for one enzyme are strong for all five and those which are weak for one are weak for all.

The very large set of variants with different sequences allows us to work on a sequence to reactivity algorithm for protein inhibitors of serine proteinases. First conclusion is relatively simple. Changes of residues which are not in contact (no stars in Table IV) have essentially no effect on the enzyme-inhibitor residues. There are some exceptions involving changes of residues which affect the inhibitor conformation.

In contrast to the relatively small effects of changing residues not in contact, changing of the 11 contact residues has very large effects. While these effects are especially large at P_1, they are by no means limited to that position alone. Some very large effects are caused by changes at P_4, P_2, P_2', P_3' and P_{14}'. Furthermore, several changes at these positions cause large differential effects - they affect interaction with one enzyme much more than interaction with the others. By combining such differential effects on paper, one can generate sequences of inhibitors which are expected to be highly specific for only one enzyme within the set (Park, 1985). Such highly specific inhibitors seem rare in nature because the choices of various differential residues are seldom made all in one direction.

Inhibitor hypervariability has two major consequences for nutritionists and toxicologists. The first, already pointed out, is that it is very dangerous to base any conclusions on plant "trypsin" or plant "chymotrypsin" inhibitors. One probably has to be far more specific. It probably does not suffice to base such conclusions on the behavior of inhibitors in the various plant inhibitor families, without specifying the exact inhibitor. This presents the nutritionist with an awesome dilemma as most plant foodstuffs simultaneously contain numerous inhibitors, not just one.

However, there is a hope, albeit far off hope, of a great future progress. If the advantages to the plant of containing inhibitors can be made very precise - including the specification of the target enzymes - and if the same can be done for their various nutritional advantages and disadvantages - genetic engineering might be employed to get the best of both worlds. What will be needed is the ability to affect the sequences of plant inhibitors at will. It is then likely that we could design inhibitors which will, say, interact with insect enzymes very well but not interact with human enzymes at all. However, before such interventions are even worthwhile, the bill of particulars - which properties of the inhibitor we wish to retain and which we wish to abolish - should be very well understood. Many subsequent chapters in this book address this question implicitly or explicitly but the problems of understanding the biological function of proteins in complex organisms are very great. It is likely that the ability to make the genetic manipulations and even the ability to predict the consequences of such manipulations on well-defined in vitro systems will be acquired faster than the ability to specify the enzymes that we wish to have inhibited and that we wish to have left alone.

It is a pity that this book does not contain a chapter on X-ray crystallography of inhibitors and of their complexes with enzymes as the interplay of chemical and X-ray crystallographic studies was particularly strong in this field. The concept of the reactive site was developed solely by chemical studies but was supported strongly by all the available crystallographic studies. The chemical studies did not yield unambiguous results about the nature of the enzyme-inhibitor complex; this was provided mainly by X-ray crystallography and somewhat by NMR studies. The concept of many inhibitor families sharing the same "standard mechanism" originated in sequencing and in mechanistic studies. But, it receives very strong support from crystallographers who found that members of different families have very different three-dimensional structures yet very similar geometry of the reactive site regions. In most current applications, X-ray crystallography allows for the identification of the enzyme-inhibitor contact residues and, thus, allows for making the concept of hypervariability quantitative. Most importantly, in the field of sequence to reactivity algorithm, X-ray crystallography allows for checking inferences on why the values of K_a change on residue substitution and for suggesting in a few cases what such changes might be even without a substitution experiment. It has been the very good fortune of the proteinase inhibitor field that the laboratories of some of the foremost X-ray crystallographers in the world (Bode and Huber, Blow, Mitsui, and James) worked, and are still working, in this field.

ACKNOWLEDGMENT

In revising the list of families (Table II), I had helpful discussions with H. Fritz, T. Ikenaka, S. Odani, M. Richardson, C. A. Ryan and I. Svendsen. However, the final list is my own - they should not be blamed. Whatever I know about inhibitors has been taught to me over the years by my late father, M. Laskowski, Sr. and by various collaborators who, for various periods of time, worked in my laboratory at Purdue. They are: W. Ardelt, M. Welch Baillargeon, U. DeVonis Biddlingmeyer, W. C. Bogard, C.-W. Chi, J. A. Cook, S. Crema, H. Crose, A. Denton, M. Empie, D. A. Estell, P. Fankhauser, W. R. Finkenstadt, M. Hall, M. A. Hamid, S. Herbert, H. F. Hixson, Jr., G. A. Homandberg, I. Kato, C. A. Kelly, R. Duran Klopp, W. J. Kohr, D. Kowalski, T. R. Leary, J. Lebowitz, T.-Y. Lin, J. A. Luthy, C. J. March, J. A. Mattis, R. E. McKee, J. E. McKie, Jr., J. Otlewski, S. J. Park, K. Parks, B. L. Schatzley, Oe. Schoenberger, J. Schrode, R. W. Sealock, M. Tashiro, G. Vichot, D. Wang, N. Warne, H. Whatley, G. H. Widener, Jr., A. Wieczorek, M. Wieczorek, K. A. Wilson and H. Wojciechowska.

During this long time, our work was supported by National Institutes of Health Grant GM10831 from the Institute of General Medical Sciences.

REFERENCES

Ardelt, W. and Laskowski, M. Jr. (1983), Acta Biochim. Pol., 30:115-126.
Ardelt, W. and Laskowski, M. Jr. (1985), Biochemistry, 24:5313-5320.
Babin, D. R., Peanasky, R. J. and Goos, S. M. (1984), Arch. Biochem. Biophys., 232:143-161.
Barrett, A. J. and Starkey, P. M. (1973), Biochem. J., 133:709-724.
Bogard, W. C. Jr., Kato, I. and Laskowski, M. Jr. (1980), J. Biol. Chem., 255:6569-6574.
Campos, F. A. P. and Richardson, M. (1983), FEBS LETTERS, 152:300-304.
Carrell, R. and Travis, J. (1985), Trends in Biochem. Sci., 10:20-24.
Empie, M. W. and Laskowski, M. Jr. (1982), Biochemistry, 21:2274-2284.
Estell, D. A. and Laskowski, M. Jr. (1980), Biochemistry, 19:124-131.

Fersht, A. (1985), "Enzyme Structure and Mechanism," 2nd ed., W. H. Freeman and Co., New York.
Finkenstadt, W. R. and Laskowski, M. Jr. (1965), J. Biol. Chem., 240:962-963.
Finkenstadt, W. R. and Laskowski, M. Jr. (1967), J. Biol. Chem., 242:771-773.
Finkenstadt, W. R., Hamid, M. A., Mattis, J. A., Schrode, J., Sealock, R. W., Wang, D. and Laskowski, M. Jr. (1974), Bayer Symp., 5:389-411.
Fossum, K. and Whitaker, J. R. (1968) Arch. Biochem. Biophys., 125:367-375.
Gebhardt, W. and Fritz, H. (1986), in "Proteinase Inhibitors," A. J. Barrett and G. S. Salvesen, eds., Elsevier, Amsterdam/Oxford/New York (in press).
Gebhardt, W. and Hochstrasser, K. (1986), in "Proteinase Inhibitors," A. J. Barrett and G. S. Salvesen, eds., Elsevier, Amsterdam/Oxford/New York (in press).
Graham, J. S., Pearce, G., Merryweather, J., Titani, K., Ericsson, L. and Ryan, C. A. (1985a), J. Biol. Chem., 260:6555-6560.
Graham, J. S., Pearce, G., Merryweather, J., Titani, K., Ericsson, L. and Ryan, C. A. (1985b), J. Biol. Chem., 260:6561-6564.
Green, N. M. (1953), J. Biol. Chem., 205:535.
Harpel, P. C. (1973), J. Exp. Med., 138:508-521.
Hejgaard, J., Svendsen, I. and Mundy, J. (1983), Carlsberg Res. Commun., 48:91-94.
Hiromi, K., Akasaka, K., Mitsui, Y., Tonumura, B. and Murao, S. (1985), "Protein Protease Inhibitor - The Case of Streptomyces Subtilisin Inhibitor (SSI)," Elsevier, Amsterdam/Oxford/New York.
Huber, R., Kukla, D., Bode, W., Schwager, P., Bartels, K., Deisenhofer, J. and Steigman, W. (1974), J. Mol. Biol., 89:73-101.
Huber, R., Bode, W., Kukla, D., Kohl, U. and Ryan, C. A. (1975), Biophys. Struct. Mechanism, 1:189-201.
Ikenaka, T. and Norioka, S. (1986), in "Proteinase Inhibitors," A. J. Barrett and G. S. Salvesen, eds., Elsevier, Amsterdam/Oxford/New York (in press).
Joubert, F. J. (1984), Phytochemistry, 23:1401-1406.
Kato, I., Schrode, J., Wilson, K. A. and Laskowski, M. Jr. (1976), Protides Biol. Fluids, 23:235-243.
Kato, I., Kohr, W. J. and Laskowski, M.Jr. (1978), Proc. FEBS Mtg., 47:197-206.
Kato, I., Ardelt, W., Cook, J., Denton, A., Empie, M. W., Kohr, W. J., Park, S. J., Parks, K., Schatzley, B. L., Schoenberger, Oe. L., Tashiro, M., Vichot, G., Whatley, H. E., Wieczorek, A., Wieczorek, M. and Laskowski, M. Jr. (1986), Biochemistry, submitted.
Koide, T. and Ikenaka, T. (1973), Eur. J. Biochem., 32:417-431.
Kowalski, D., Leary, T. R., McKee, R. E., Sealock, R.W., Wang, D. and Laskowski, M. Jr. (1974), Bayer Symp., 5:311-324.
Kunitz, M. and Northrop, J. H. (1936), J. Gen. Physiol., 19:991.
Kunitz, M. (1947a), J. Gen. Physiol., 30:291.
Kunitz, M. (1947b), J. Gen. Physiol., 30:311.
Laskowski, M. and Laskowski, M. Jr. (1954), Adv. Protein Chem., 9:203-242.
Laskowski, M. Jr. and Sealock, R. W. (1971), The Enzymes, 3rd. ed., 3:375-473.
Laskowski, M. Jr. and Kato, I. (1980), Ann. Rev. Biochem., 49:593-626.
Laskowski, M. Jr., Kato, I., Kohr, W. J., March, C. J. and Bogard, W. C. (1980), Protides Biol. Fluids, 28:123-128.
Laskowski, M. Jr., Tashiro, M., Empie, M. W., Park, S. J., Kato, I., Ardelt, W. and Wieczorek, M. (1983), in: "Proteinase Inhibitors: Medical and Biological Aspects," N. Katanuma, H. Umezawa and H. Holtzer, eds., Japan Scientific Societies Press, Tokyo/Springer Verlag, Berlin.
Lebowitz, J. and Laskowski, M. Jr. (1962), Biochemistry, 1:1044-1055.

Luthy, J. A., Praissman, M., Finkenstadt, W. R. and Laskowski,M. Jr. (1973) J. Biol. Chem., 248:1760-1771.

Mahoney, W. C., Hermodson, M. A., Jones, B., Powers, D. D., Corfman, R. S. and Reeck, G. R. (1984), J. Biol. Chem., 259:8412-8416.

Marquart, M., Walter, J., Deisenhofer, J., Bode, W. and Huber, R. (1983), Acta Cryst., B39:480-490.

Mattis, J. A. and Laskowski, M. Jr. (1973), Biochemistry, 12:2239-2245.

Mueller-Esterl, W., Fritz, H., Kellermann, J., Lottspeich, F., Machleidt, W. and Turk, V. (1985), FEBS LETTERS, 191:221.

Neurath, H. (1984), Science, 224:350-357.

Odani, S., Koide, T., Ono, T. and Ohnishi, K. (1983), Biochem. J., 213:543-545.

Ozawa, K. and Laskowski, M. Jr. (1966), J. Biol. Chem., 241:3955-3961.

Park, S. J. (1985), Ph.D. Thesis, Purdue University, West Lafayette, IN.

Quast, U., Engel, J., Steffen, E., Tschesche, H. and Kupfer, S. (1978a), Eur. J. Biochem., 86:353-360.

Quast, U., Engel, J., Steffen, E., Tschesche, H. and Kupfer, S. (1978b), Biochemistry, 17:1675-1682.

Read, R. J., Fujinaga, M., Sielecki, A. R. and James, M.N.G. (1983), Biochemistry, 22:4420-4433.

Read, R. J. and James, M.N.G. (1986), in "Proteinase Inhibitors," A. J. Barrett and G. S. Salvesen, eds., Elsevier, Amsterdam/Oxford/New York (in press).

Rhodes, M. B., Bennett, N. and Feeney, R. E. (1960), J. Biol. Chem., 235:1686-1693.

Schechter, I. and Berger, M. (1967), Biochem. Biophys. Res. Comm., 27:157-162.

Sealock, R. W. and Laskowski, M. Jr. (1969), Biochemistry, 8:3703-3710.

Seemueller, U., Dodt, J., Fink, E. and Fritz, H. (1986), in "Proteinase Inhibitors," A. J. Barrett and G. S. Salvesen, eds., Elsevier, Amsterdam/Oxford/New York (in press).

Sottrup-Jensen, L., Stepanik, T. M., Kristensen, T., Wierzbicki, D., Jones, C. M., Lonblad, P. B., Magnusson, S. and Petersen, T. E. (1984), J. Biol. Chem., 259:8318-8327.

Stein, J. P., Catterall, J. F., Kristo, P., Means, A. R. and O'Malley, B. W. (1980), Cell, 21:681-687.

Stevens, F. C. (1971), Proc. Int. Res. Conf., Proteinase Inhibitors, 1:149-155.

Vincent, J.-P. and Lazdunski, M. (1972), Biochemistry, 11:2967-2977.

Wieczorek, M., Otlewski, J., Cook, J., Parks, K., Leluk, K., Wilmowska-Pelc, A., Polanowski, A., Wilusz, T. and Laskowski, M. Jr. (1985), Biochem. Biophys. Res. Comm., 126:646-652.

2

THE BIOLOGICAL FUNCTIONS AND PHYSIOLOGICAL EFFECTS OF INGESTED INHIBITORS OF DIGESTIVE REACTIONS

S. S. Rothman

Department of Physiology, University of California
San Francisco, California 94143

INTRODUCTION

In terms of everyday experience, we think of disease and illness as something visited upon a healthy body from the outside; analogous to an unfortunate accident. Microbes infect us, and plagues and epidemics come upon us. In this way, man has come to view illness as an event that is discontinuous with normal function. However, normal processes can go awry without particular help from outside agents and produce disease and illness. In this circumstance, the line between the normal and abnormal can often be hard to discern, and the old adage that there can be too much (or too little) of a good thing may apply. Thus, we have come to appreciate that disease is often continuous with, rather than a break and something different from, normal function.

Similarly, when a harmful substance is ingested,we are likely to consider it as separate from those healthful things that are normally eaten, namely food. In this context, non-nutrient substances in our food that, when eaten in sufficient quantity, cause pathologic sequelae are often considered akin to "foreign" invaders; "toxic" substances whose untoward ingestion produces pathology.

In this way, the abnormal, as well as the frankly pathological, effects of ingested trypsin inhibitors, the subject of this volume, may be viewed as toxicological disease artifacts wholly unrelated to, rather than extreme distortions of, normal biological events. In addition, the physiological purpose of endogenous trypsin inhibitors, those normally present in procaryotic and eucaryotic cells and blood, has sometimes been thought to be solely security devices to prevent uncontrolled proteolysis. For these reasons, our perspective of the actions of trypsin inhibitors has a strongly pathologic flavor. Their ingestion may be viewed as unnatural, producing toxic pathologic effects, and their physiological purpose in the cells and organisms that produce them may be thought to be to prevent or combat pathologic occurrences; namely, uncontrolled proteolysis.

WHAT ARE TRYPSIN INHIBITORS?

It is in this seemingly "unphysiologic" context that we approach the question of the biological functions and physiological effects of trypsin inhibitors. Although it is clear that the ingestion of trypsin inhibitors can cause pathologic sequelae, some of which are discussed by others in this volume, are these discontinuous pathologic aberrations unrelated to normal physiological functions, or do they reflect normal processes moved by the presence of the trypsin inhibitor into an abnormal part of a continuum? Similarly, although it is clear that the presence of protease inhibitors in cells and blood can help prevent or ameliorate potentially explosive proteolysis, is this their sole or even their normal every day function in cells? Indeed, can we say with any assurance that this is really their function at all? Such questions make it apparent how little we know about the biological effects of the numerous molecules that are classified as tryspin inhibitors.

This difficulty derives in great part from the fact that trypsin inhibitors and other protease inhibitors are for the most part simply molecules purified by chemists that have been found to inhibit proteolytic activity of one sort or another in the test tube. It is in the main from this knowledge that inferences have been made about their "natural", "physiological" activities. Although a great many different molecules have been identified and, to one extent or another, chemically purified and characterized, the reason for purification and characterization has most often been chemical (i. e., interest in identifying protease inhibitors, or merely finding them accidentally), and not the result of the need to identify the reactive elements in particular physiological processes that have been elucidated previously. That is, although we have a great deal of knowledge of protease inhibitor molecules from a chemical perspective, we have little knowledge of what they do in situ; indeed, often, if not usually, we cannot even presume that inhibition of trypsin, or other proteases, is what they are up to, although it is widely assumed. We merely know that they are capable of such behavior; and this knowledge is ours often by dint of the investigator's choice of which chemical to take off the shelf and test. Thus, this is an area of research where many carts have been put before who knows how many horses.

Thus, even looking for the physiological actions of these proteins as trypsin inhibitors carries with it the risk that they may not really be trypsin inhibitors at all in nature, or perhaps more likely, some are and some are not; or this may be one thing that they are capable of doing, but they may peform other functions as well. The point is simply that frequently we cannot say.

THE INGESTION OF TRYPSIN INHIBITORS

With this caveat in mind, substances that can be classified clearly as trypsin inhibitors form several relatively specific structural and evolutionary subclasses, and are molecules found widely throughout bacterial, animal, and plant kingdoms (Laskkowski and Kato, 1980). Thus, they are a seemingly ubiquitous group of proteins that have evolved, and this suggests that they serve some useful function or functions, as inhibitors or otherwise.

It also seems likely, and is certainly a reasonable hypothesis, that among these functions trypsin inhibitors in beans and other seed-containing foods act as protease inhibitors to protect the seed. As seed-containing structures, beans are of course crucial to the survival of their plant species. For the species to propagate across generations, the bean must be sown successfully. The cotyledonous mass provides food for the seed, but may also be a source of nutrient for other organisms. The inhibitors would protect both the food supply and the seed itself from insects (Green and Ryan, 1972), and from proteolytic digestion by surface bacteria or fungi.

In addition, vertebrate species for whom beans represent a source of food, notably man for whom they are a major cultivated food, may display analogous interactions with ingested beans. The presence of trypsin inhibitors in beans, along with other protective devices such as cellulose shells, may permit a few bean seeds to survive ingestion. That is, if digestion can be prevented, or at least sufficiently retarded or delayed by the inhibitors, the seed will pass through the gut intact, and be deposited on the earth within the high nitrogen mass of feces. Indeed, if successful, passage out of the animal embedded in fertilizer might enhance survival of the seed and its species, and might properly be considered part of a natural life cycle, analogous to that of animal parasites.

Despite the modern penchant for thinking at the molecular level, when we think of the survival of species, we are likely to be drawn both by common experience and the observations of biologists, to conceptualize the conflict between species as involving the outright life and death struggle between organisms in the same ecological niche or competition between organisms for territory and food resources. In this view of things, once ingestion occurs the battle is over; the endpoint of victory or defeat for the individual. However, as I have suggested, ingestion does not necessarily signal the end of the battle. In the case at hand, the molecular level battle between trypsin and inhibitors that help provide safe passage for seed through the intestine into the haven of the fecal mass continues.

THE INHIBITOR FROM THE INGESTING ORGANISM'S VIEWPOINT

Although substances that inhibit digestive reactions are probably contained in many foods, only when the germ of a species is involved, such as in eggs and beans, does their presence suggest a biological purpose for the organism that is the "food." But whether or not a particular inhibitor has a biological purpose for an ingested species, it, in any event, can present a problem, a physiological problem if you will, for the ingesting species, decreasing the efficiency, and perhaps even preventing the accomplishment, of the digestive process.

Therefore, trypsin inhibitors, as well as other potential inhibitory substances normally present in food, must be viewed as "natural" inhibitors; not only in the sense that they are naturally present in foods, but in the sense that they are therefore inhibitors whose presence is to be expected from the assimilating organism's point of view. That is, in this light, it would not be surprising if the "host" evolved functional adaptations to overcome digestive difficulties that such substances might produce and that might normally be encountered during the digestion of food. Though one might expect there to be many other substances in food, as well as physical circumstances associated with particular foods, that impair digestive efficiency, the actions and effects of dietary trypsin inhibitors represent the only case for which we have substantive knowledge, however incomplete.

THE STOMACH

The stomach is sometimes viewed today as more of a detoxifying antechamber, than a digestive organ. It is argued that acidification and peptic digestion, the actions of its two major secretory products in non-ruminant mammals, hydrochloric acid and pepsinogen, play a relative insignificant role in the digestion of protein, but are important as a means of making safe ingested substances that have the potential to cause harm. From this perspective, acidification and peptic digestion of ingested trypsin inhibitors, could make the inhibitors ineffective by the time they reach the major digestive chamber; no longer a problem for the adequate digestion of food. Though some trypsin inhibitors are acid sensitive and susceptible to peptic digestion and detoxification no doubt occurs to some extent in these cases, not all trypsin inhibitors are, and in any event we cannot assume that simply because the possibility exists for "detoxification" in simple pepsin-containing acid solution, that it would occur in gastric contents, or, if it did, that it would be effective.

During the digestion of a meal, gastric contents are a complex suspension of substances and particles that cannot be considered a simple solution by any means. Substances within the neutral pH, relatively pepsin-free bolus of food, as trypsin inhibitors would be, would be protected from damage even if the contents of the bulk fluid phase in the stomach could destroy the particular substance. In any event, the issue is quantitative, not qualitative. That is, the effectivness of protection provided by the stomach must be evaluated in situ by measuring the degree to which material entering the lower gastrointestinal tract is indeed free of active inhibitors after their ingestion.

THE PANCREAS AND INGESTED TRYPSIN INHIBITORS

Whatever notions we might harbor about the effectivness and generality of gastric detoxification, we know that ingested trypsin inhibitors, when present in sufficient quantity in the diet, are not detoxified, or at least not sufficiently so, in the sense that the organism responds to their presence. That is, they have biological effects on the "host" organism. This brings us to the question of the nature of the effects of ingested trypsin inhibitors.

If trypsin inhibitors produce their biological effects by impairing, or at least "challenging" the efficiency of proteolytic digestion, then as a result of feeding trypsin inhibitor-containing foods we might see wasting or other signs of inadequate nutrition (if adequate proteolytic digestion cannot be maintained) or adaptation (if the animal can compensate, and maintain adequate protein digestion in its presence). In the latter case, we should observe signs of adaptation, most obviously the increased secretion of trypsin by the pancreas. Of course, both wasting and attempts at adaptation might occur simultaneously.

These were indeed the effects that were observed when trypsin inhibitor-containing soy bean meal was fed to chickens (Chernick et al., 1948). Animals grew poorly on the diet, and the pancreas was enlarged, primarily as the result of cellular hypertrophy, and contained increased stores of digestive enzymes, suggesting increased secretory activity. Thus, the hypothesis that trypsin inhibitors acted by impairing digestion seemed to be affirmed by its observed effects on treated animals.

However, this straightforward interpretation was incorrect in part. The failure of animals to thrive on a soybean diet did not appear to be due to the presence of trypsin inhibitors (Lyman and Lepkovsky, 1957), but to other causes,probably the limited availability of certain essential amino acids, such as methionine, in raw meal (Khayambashi and Lyman, 1966; Rackis, 1974). On the other hand, the trypsin inhibitor was responsible for the pancreatic sequelae; and assuming that the inhibitor was not absorbed from the intestine in any quantity, its actions were related to effects that it produced within the gut. For example, in one study, Wells and the author (Rothman and Wells, 1969) added purified egg white trypsin inhibitor to otherwise identical diets fed to rats. Classical pancreatic effects were seen. The pancreas enlarged, by one third within six days, and contained elevated levels of digestive enzymes. Since only the inhibitor protein was added, the effects were clearly attributable to it.

The hypothesis that the pancreas increased its production of enzymes as a result of the inhibition of digestive reactions in the intestine caused by trypsin inhibitor, was supported by the observation that the secretion of trypsinongen, and other digestive enzymes, was increased (Lyman and Lepkovsky, 1957; Rothman and Wells, 1969; Applegarth et al., 1964). Indeed, a remarkable hypersecretion could be observed; a five-fold augmentation in the maximal response to a cholinergic stimulant in the cited study by Wells and the author (Figure 1). In addition, in the early 1970s, Lyman, Green and coworkers (Green and Lyman, 1972; Schneeman and Lyman, 1975) at Berkeley found that if they reduced the concentration of proteolytic enzymes in the intestine acutely, either by diverting pancreatic juice, or infusing trypsin inhibitors, the pancreas began to hypersecrete, in seeming compensation for the lack (Figures 2 and 3).

THE REGULATION OF DIGESTIVE REACTIONS

Thus, wasting or no wasting, it seems that trypsin inhibitors, ingested in sufficient quantity, impair proteolytic digestion, and that the pancreatic response reflects an attempt on the part of the organism to overcome this impairment. Assuming this to be the case, then there must normally be some means of monitoring the rate of digestive reactions, at least proteolytic reactions, in the intestine; that is, digestive reactions must be regulated in one fashion or another. And of course, whatever their biological purpose or lack of purpose, the ingestion of trypsin inhibitors and other naturally occurring substances in food that can inhibit particular digestive reactions, provide one important reason why the body has need to continuously monitor the ongoing rate of these reactions. Therefore, the body's response to the ingestion of trypsin inhibitors tells us that it is neither ignorant of, nor uncaring in regard to, the rates at which digestive processes are carried out in the intestine. There are a variety of observations, from different experimental and theoretical perspectives, at this point that demonstrate that digestive reactions are regulated (Grossman et al., 1943; Rothman, 1974: Adelson and Rothman, 1974; Felber et al., 1974; Dick and Felber, 1975; Grendell and Rothman, 1981; Grendell et al., 1984; Tseng et al., 1982; 1984; Reisenauer and Gray, 1985; Adelson and Miller, 1985). The view that digestion is accomplished by means of a series of unregulated reactions whose ongoing rate, or even success, goes unnoticed by the organism is not supported by the available evidence.

Regulation can be accomplished by monitoring the concentrations of the enzymes themselves in the gut, as suggested by Lyman and Green (1972), or by monitoring the substrates or end products of digestive reactions, or all three relevant "Modalities" may be involved, acting in the intestine, at the pancreas, and at other sites. The evidence supports the view that indeed all three elements of digestive reactions, substrates, end products and enzymes, are involved in one fashion or another (Green and Lyman, 1972;

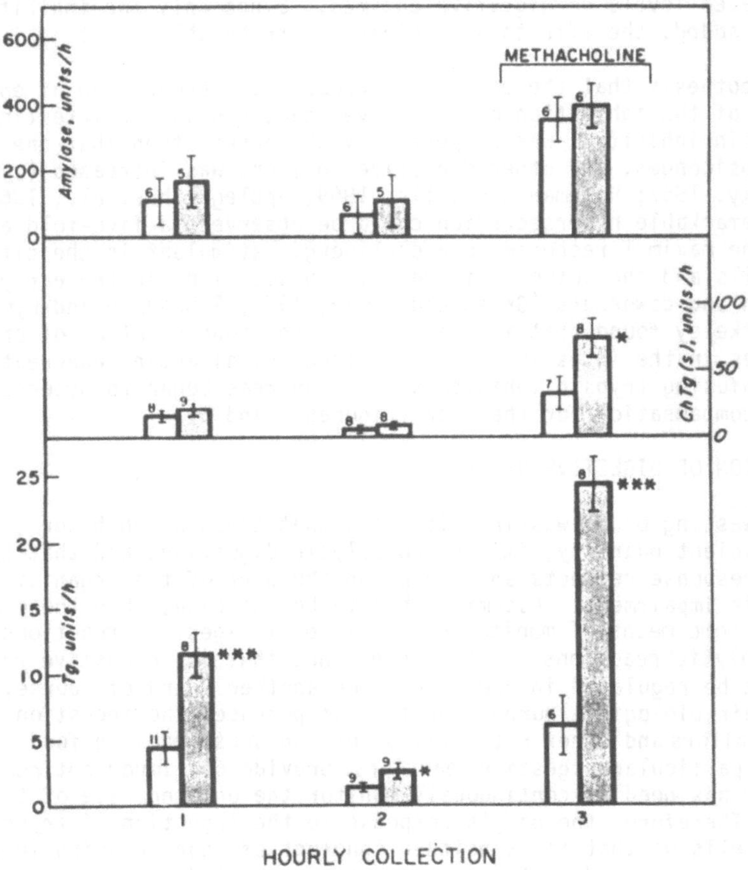

Figure 1. Effect of dietary egg white trypsin inhibitor(EWTI)on pancreatic enzyme secretion in anesthesized rats. Enzyme output is expressed as units per hour. Clear bar (left) = control; shaded bar (right) = EWTI fed. Tg = trypsinogen, ChTg(s) = chymotrypsinogens. Values are means ± SEM. Numbers on top of bar refer to the number of animals. Collections were made at 1-hr intervals and after 2 hr of unstimulated secretion, methacholine chloride was given sc (0.8 mg/Kg body weight). Control vs. EWTI fed, * $P < 0.05$, *** $P < 0.001$. (Reproduced with permission from Rothman and Wells, 1969).

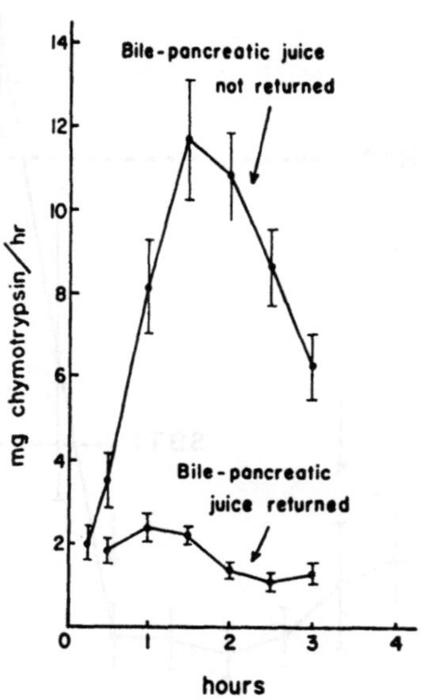

Figure 2. Spontaneous pancreatic enzyme secretion with and without bile-pancreatic juice returned. Time scale starts with cannulation of common bile duct. Chymotrypsin activity of the secreted bile-pancreatic juice served as an indication of enzyme secretion. Each point is the mean ± SEM from 6 animals. (Reproduced with permission from Green and Lyman, 1972).

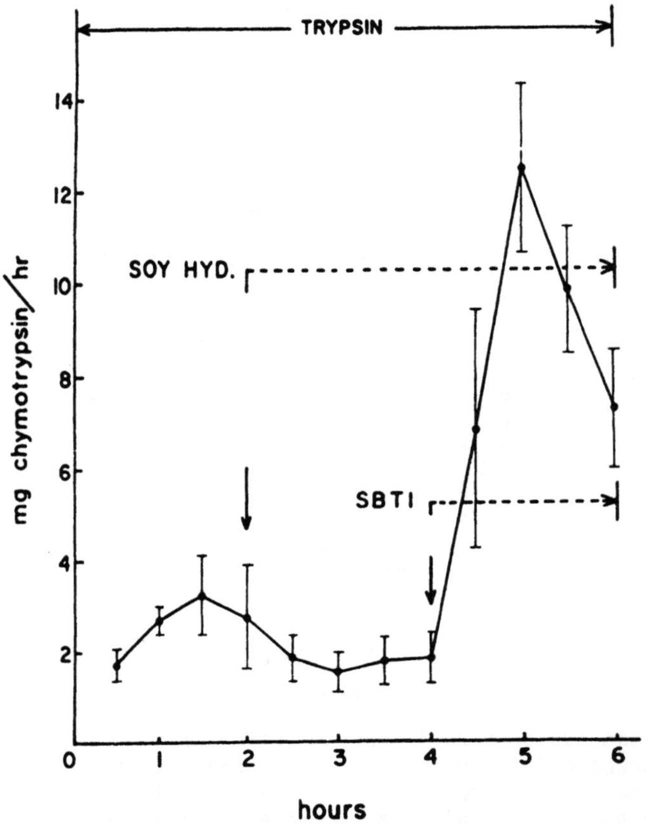

Figure 3. Effect of trypsin inhibitor on pancreatic enzyme secrection during intestinal infusion of trypsin. Trypsin was infused at 2 mg/hr in 0.05 N $NaHCO_3$. Soybean trypsin inhibitor (SBTI) and soy hydrolysate (in saline) were each infused at 30 mg/hr. Enzyme secretion is represented by total chymotrypsin activity of the secreted bile-pancreatic juice. Each point is the mean $\pm$ SEM for 6 rats. (Reproduced with permission from Green and Lyman, 1972).

Schneeman and Lyman, 1975; Grossman et al., 1943; Rothman, 1974; Adelson and Rothman, 1974; Felber et al., 1974; Dick and Felber, 1975; Grendell and and Rothman, 1981; Grendell et al., 1984; Tseng et al., 1982; 1984; Reisenauer and Gray, 1985).

In addition, it has become clear that digestive reactions are regulated by type, and not merely en masse by increasing or decreasing the overall amount of digestive enzyme in the gut regardless of the contents of the meal and the rates at which particular reactions progress during digestion. Various digestive end products, hormones of internal metabolism, and known and putative gastrointestinal hormones have been shown to modify the mixture of digestive enzymes secreted into the intestine (Grossman et al., 1943; Rothamn, 1974; Felber et al., 1974; Dick and Felber, 1975; Grendell and Rothman, 1981; Grendell et al., 1984; Tseng et al., 1982; 1984; Reisenauer and Gray, 1985; Adelson and Miller, 1985).

For example, in the trypsin inhibitor study by Wells and myself mentioned above (Rothman and Wells, 1969), we found that there were variations in the relative rates of secretion of the three enzymes that we measured, trypsinogen, chymotrypsinogen, and amylase (Figure 1). Trypsinogen secretion was enhanced most by trypsin-inhibitor feeding, a lesser enhancement was observed for chymotrypsinogen, and amylase secretion was unaltered by the diet (Figure 1; bars on right showing the response to cholinergic agonist). That is, the animal's response not only took into consideration the fact that digestion was being impaired, but proteolytic digestion in particular, and trypsin digestion even more specifically.

The occurrence of the regulation of digestion at the specific reaction or molecular level accounts for certain paradoxes that <u>en masse</u> regulation, as the sole means, would introduce. For example, <u>if regulation</u> occurred solely as a function of the amount of protein in the meal, as an indicator of digestive mass, with the amount of free trypsin in the intestine indicating the ongoing need for digestion, and with the secretion of some 20 different digestive enzymes by the pancreas occurring <u>en masse</u> in unvarying proportions in accordance with the magnitude of that <u>signal</u>, then a high protein, low starch meal with, let us say for the sake of discussion, a 10: 1 protein to starch ratio, would provide much more efficient starch digestion, than an equivalent meal with an inverse 10: 1 ratio of starch to protein. Indeed, if there was a linear relationship, then starch digestion would be accomplished 100 times more efficiently for a diet containing ten times as much protein in relative terms, as one containing ten times as much starch. We would have the curious fact that the efficiency of starch digestion would be a function of the protein content of the meal. Of course, things can be designed to help overcome such difficulties. However, from what we know of the regulation of digestion such a problem would exist in one form or another if the regulation of the digestion of food occurred only as a function of its mass or bulk, however this might be monitored. This is not to say that bulk regulation does not occur. It may well. Nor that free trypsin concentration might not be an important means of assessing bulk in need of digestion, but that as a means of regulation unto itself it would seem inadequate.

It only recently has been generally appreciated that pancreatic enzymes can be secreted in a non-parallel fashion, one in which the rate of secretion of one enzyme can vary relative to another. After prolonged controversy, the wide variety of published observations of non-parallel transport leave little doubt of its existence today. I and others have written about these issues elsewhere (for example, Rothman, 1977; 1985; Grendell, 1985). In the past, the central reason fur concluding that digestive reactions, unique among biological reaction sequences, were not regulated at the molecular or reaction level was the belief that the pancreas was only capable of secreting its enzymes en masse in fixed proportions, or in parallel, in response to any and all food stimuli. If this were the case, then, although the overall rate of digestion might be varied, the rate of digestion of particular foodstuffs could not be. This might pose a problem; too much or too little of a particular enzyme being secreted relative to need, if, unlike that of a laboratory rat, meals in nature were not invariant in their nutritional content. However, the nature of the secretory mechanism would make such difficulties, if they occurred, unavoidable; a simple fact of life.

THE FUNCTION OF ENDOGENOUS TRYPSIN INHIBITORS

Trypsin is one of a group of enzymes of common ancestry and substantial structural homology, the serine proteases, found widely throughout animal species. It is designed to hydrolyze peptide bonds containing basic amino acids, in particular lysine or arginine. The term "trypsin" is often used to refer to the enzyme produced and secreted by the pancreas of vertebrates, the first enzyme to be purified, but it is also often used today to describe a variety of other enzymes of like function, "trypsin-like" enzymes, that may or may not be structurally homologous with pancreatic trypsins. Such trypsin-like enzymes are also found widely throughout plant, animal, and bacterial kingdoms.

There appear to be two general functions of trypsins, and their relatives. As we have discussed, a digestive function, as exemplified by cavital digestion in vertebrates, where as an endopeptidase trypsin splits peptide chains internally and is central to the digestive process that leads to the liberation of amino acids and small peptides that are then assimilated. It also serves another important digestive function as the enzyme that initiates the activation of other proteolytic enzymes. Digestive enzymes, particularly proteases, are in the main secreted as inactive precursors, zymogens or proenzymes, that are relatively, although not absolutely, inactive. Trypsin can activate these proenzymes, which in turn also leads to their autocatalytic activation.

The second general function of trypsins is not digestive. Acting within cells and in extracellular fluids such as blood, "trypsin-like" enzymes split basic amino acid-containing peptide bonds in many different protein molecules, in many different cells, in pursuit of many different functions. Such tryptic hydrolysis is often central to the initiation of physiological actions, such as those of some peptide hormones, and probably the prevention of others. Perhaps most well known are the trypsin-like enzymes of plasma some of which play a central role in the reactions involved in the clotting of blood.

Thus, numerous biological reactions and processes involve and are perhaps controlled, at least in part, by the cleavage of a lysine or arginine peptide bond in a protein; whether we are referring to digestion in the intestine, or the modulation of processes in blood and within cells. Indeed, trypsinogen itself is activated by a trypsin-like enzyme, enteropeptidase, produced by the duodenum.

The view that trypsin inhibitors are there to protect against an untoward eventuality, uncontrolled proteolytic digestion, must be evaluated in this light; that is, in light of knowledge that tryptic digestion appears to be constantly occurring within organisms, not only within the gut, but in cells and extracellular fluid, seemingly at modulated, not uncontrolled rates. Hydrolysis of lysine and arginine-containing peptide bonds appear to serve important functions in normal processes of various kinds. If this is the case, then inhibitor molecules may provide one means of regulating the rates of these tryptic reactions; keeping the amount of a particular active enzyme in a given compartment at "physiological" levels.

Of course, this is really not different in kind than preventing uncontrolled proteolysis; only the place on the "continuum of function" differs. In this view, under normal physiological conditions, inhibitors would help maintain the rates of reactions at appropriate levels, and only if tryptic or other proteolytic hydrolysis were to get out of hand, that is, occurred at abnormal rates, would their function become "protective". Of course, in or out of the normal range, the function of such inhibitors would be in actuality simultaneously regulatory and protective.

The important point is that proteolytic activity is not only an ongoing process in cells and organisms, beyond digestion of food, but is a central aspect of biological systems. That is, the occurrence of potentially dangerous chemical reactions is part and parcel of an organism's normal physiology. Trypsin, and other protease inhibitors of various structural varieties and modes of action, may represent important means of controlling such reactions; helping, along with other regulatory modalities, make sure that they occur at rates appropriate to normal organismal function, as well as helping to prevent untoward proteolysis.

It seems a fair and true statement that the body is continuously "playing with poison;" even poison of its own construction. It has no choice. To maintain normal function, regulate reactions, turn over its protein mixture, deal with foreign substances, the potential always exists that enzymes manufactured by an organism for its own use may pose a danger to the body's own internal processes, occurring at abnormal rates and producing self-pathology, pathology that is derived endogenously.

CONCLUDING COMMENT

As I said at the outset, one central problem with our understanding of the biological and physiological role or roles of trypsin inhibitors is that our knowledge is primarily, and often in the first instance, chemical. Much of it is not derived from knowledge of physiological processes in which these substances have been shown to be involved. But how can we overcome this difficulty? How can we learn more about their role in situ? I believe that in order to develop this understanding we must determine what effects variation in the concentrations of the putative inhibitors, and their "substrates", have on physiological functions related to trypsin-like proteolysis. That is, what effect does varying their concentration, or the concentration of their "substrates", have on physiological activities?

To carry out such a program of investigation we must accept, at least for purposes of carrying out experimental work, the hypothesis that a continuum exists for the function of these molecules, and that we can properly attempt to vary the relevant parameters, thereby introducing abnormal behavior as a means of evaluating the normal. Such an approach has a valued history in biology. Our knowledge of endocrine systems has come about in great measure by appreciating the utility of producing experimental states of mal-(hyper or hypo) function. Thus, by accepting, at least as appropriate hypothesis, the notion that abnormal conditions involving trypsin and other protease inhibitors in the gut, and in the organism more generally, reflect distortions of normal, regulated processes, not activities discontinuous with them, we allow ourselves the means of probing for potential functional modalities in which they may be involved.

REFERENCES

Adelson, J. W. and Miller, P. E. (1985), Pancreatic secretion by nonparallel exocytosis: potential resolution of a long controversy. Science, 228, 993-996.

Adelson, J. W. and Rothman, S. S. (1974). Selective pancreatic enzyme secretion due to a new peptide called chymodenin. Science, 183, 1087-1089.

Applegarth, A., Furuta, F. and Lepkovsky, S. (1964). Response of the chicken pancreas to raw soybean. Poultry Sci., 43, 733-739.

Chernick, S. S., Lepkovsky, S., and Chaikoff, I. L. (1948). A dietary factor regulating the enzyme content of the pancreas: changes induced in size and proteolytic activity of the chick pancreas by the ingestion of raw soybean meal. Am. J. Physiol., 155: 33-41.

Dick, J. and Felber, J.P. (1975).Specific hormonal regulation, by food, of the pancreas enzymatic (amylase and trypsin) secretion. Horm. Metab. Res., 7, 161-166.

Felber, J. P., Zermatten, A. and Dick, J. (1974). Modulation, by food, of hormonal system regulating rat pancreatic secretion. Lancet, 2, 185-188.

Green, G. M. and Lyman, R. L. (1972). Feedback regulation of pancreatic enzyme secretion as a mechanism for trypsin inhibitor-induced hypersecretion in rats. Proc. Soc. Exp. Biol. Med., 140, 6-12.

Green, T. R. and Ryan, C. A. (1972). Wound-induced proteinase inhibitor in plant leaves: a possible defense mechanism against insects. Science, 175, 776-777. See also, Brown et al., this volume.

Grendell, J. H. (1985). Nonparallel secretion of digestive enzyme by the pancreas - implications for models for protein secretion. In "Nonvesicular Transport," S. S. Rothman and J. J. J. Ho, eds., John Wiley & Sons, New York, pp. 347-357.

Grendell, J. H. and Rothman, S. S. (1981). Digestive end products mobilize secretory proteins from subcellular stores in the pancreas. Am. J. Physiol., 241, G67-G73.

Grendell, J. H., Tseng, H. C. and Rothman, S. S. (1984). Regulation of digestion. I. Effects of glucose and lysine on pancreatic secretion. Am. J. Physiol., 246, G445-G450.

Grossman, M. I., Greengard, H. and Ivy, A. C. (1943). The effect of dietary composition on pancreatic enzymes. Am. J. Physiol., 138, 676-682.

Khayambashi, H. and Lyman, R. L. (1966). Growth depression and pancreatic and intestinal changes in rats force-fed amino acid diets containing soybean trypsin inhibitor. J. Nutr., 89, 455-464.

Laskowski, M., Jr., and Kato, I. (1980). Protein inhibitors of proteinases. Ann. Rev. Biochem., 49, 593-626.

Lyman, R. L. and Lepkovsky, S. (1957). The effect of raw soybean meal and trypsin inhibitor diets on pancreatic enzyme secretion in the rat. J. Nutr., 62, 269-284.

Rackis, J. J. (1974). Biological and physiological factors in soybeans. J. Am. Oil Chem. Soc., 51, 161A-174A.

Reisenauer A. and Gray, G. M. (1985). Abrupt induction of a membrane digestive enzyme by its intraintestinal substrate. Science, 227, 70-72.

Rothman, S. S. (1974). Molecular regulation of digestion: short-term and bond specific. Am. J. Physiol., 226, 77-83

Rothman, S. S. (1977). The digestive enzymes of the pancreas: a mixture of inconstant proportions. Ann. Rev. Physiol., 39, 373-389.

Rothman, S. S. (1985). "Protein Secretion: A Critical Analysis of the Vesicle Model". John Wiley & Sons, New York.

Rothman, S. S. and Wells, H. (1969). Selective effects of dietary egg white trypsin inhibitor on pancreatic enzyme secretion. Am. J. Physiol., 216, 504-507.

Schneeman, B. O. and Lyman, R. L. (1975). Factors involved in the intestinal feedback regulation of pancreatic enzyme secretion in the rat. Proc. Soc. Exp. Biol. Med., 148, 897-903.

Tseng, H. C., Grendell, J. H. and Rothman, S. S. (1984). Regulation of digestion. II. Effects of insulin and glucagon on pancreatic secretion. Am. J. Physiol., 246, G451-G456.

3

SAFETY OF TRYPSIN INHIBITORS IN THE DIET: EFFECTS ON THE RAT PANCREAS OF LONG-TERM FEEDING OF SOY FLOUR AND SOY PROTEIN ISOLATE

Michael R. Gumbmann[1], William L. Spangler[2], Glenda M. Dugan[1], and Joseph J. Rackis[3]

[1]Western Regional Research Center, ARS, USDA, Berkeley California 94710

[2]Veterinary Pathology Consultants, Inc., West Sacramento California 95691

[3]3411 North Elmcroft Terrace, Peoria, Illinois 61604

ABSTRACT

The effects on the pancreas of chronic dietary exposure to defatted soy flour and soy protein isolate have been studied in two two-year feeding trials in rats. Emphasis was placed on detecting changes that might accompany low levels of dietary trypsin inhibitor (TI) as might be found in edible grade soy products and on studying the influence of protein nutrition. The major pathological findings in the pancreas were nodular hyperplasia (NH), consisting of foci of hyperplastic acinar cells often grossly visible by six months, and the benign neoplastic lesion, acinar adenoma (AA), which developed more slowly.

In the first feeding trial, the objectives were to obtain the dose-response relationship of pancreatic pathology to dietary TI provided by raw and heated soy flour and to study the nutritional interaction of protein level which was varied from 10% to 30% using casein supplementation. Also, the responses to raw and heated soy protein isolate were compared to determine whether the removal of more than 50% of the constituents found in soy flour would alter the development of pancreatic lesions.

In the second trial, the effect of unusually low levels of TI in raw and heat-treated soy protein isolate, prepared through a salt extraction process and fed at 10% and 30% protein in the diet, was investigated.

The incidence of both NH and AA was positively related to the TI content of the diet. The probit transformation of the percent incidence of AA was linearly related to the log of TI/g protein in the diet. A single curve best described the response to 20% and 30% protein, with a slope that was distinctly greater than that for 10% protein. The intersection of the two curves near the TI concentration of edible grade soy flour predicts that protein level in the diet can be expected to have essentially no effect on the incidence of AA when TI activity is in this range. But, for proteins containing greater concentrations of TI, increasing the level of protein in the diet will increase the incidence of pancreatic pathology, while for proteins with quite low levels of TI, increasing the protein in the diet above 10% will have a protective effect. The basis for this interaction between dietary protein and TI is to be found in the nutritional demands for increased protein synthesis accompanying the adaptive response of the pancreas to TI and in the reported influence of low protein levels in sustaining elevated plasma cholecystokinin concentrations.

No significant difference between the response to soy protein isolate, providing 30% protein and graded levels of TI, and that of diets containing soy flour was obtained. This indicates that the non-proteinaceous components removed from soy flour were not factors in the development of pancreatic pathology. The response to the low TI, extracted soy protein isolate (both raw and heated) fed at 30% protein was also equivalent to that of diets containing heated soy flour with casein. However, with 10% protein, the incidence of AA was greater than might be predicted from the probit model, particularly for the raw, low TI soy protein isolate. Thus, with restricted protein in the diet, proteins of lower quality can be expected to stimulate greater development of pancreatic lesions than would be predicted from TI content alone.

The results of the two feeding studies reported here with defatted soy flour and soy protein isolate indicate that low levels of residual TI equal to those encountered in commercially prepared soy products can be expected to increase the probability of pancreatic adenoma formation in the rat. However, the near exclusive use of the rat as a model has not supplied the breadth of information required to make a useful safety assessment of residual TI in man's diet.

The appearance of pre-neoplastic and neoplastic lesions of the pancreas in response to chronic dietary exposure of TI in the rat appears to be mediated by the endocrine system of the gastro-intestinal tract which regulates pancreatic function. There is no evidence that TI or other factors in soy are directly involved in the production of pancreatic tumors. Cholecystokinin, a key hormone, can essentially reproduce the early pancreatic alterations brought about by raw soy in the diet, including the development of hyperplasia. Whether the pancreas of the rat is unusually susceptible to the induction of neoplasia when stimulated by the normal regulatory mechanism for pancreatic adaptation to diet remains to be determined. With only limited information as to the generality of the long-term effects of TI on the pancreas in other species, it would be prudent to minimize the intake of dietary factors, such as TI, which may stimulate an undue proliferative response.

INTRODUCTION

Substances that inhibit digestive proteinases are well known to be widely distributed in man's food supply, especially among the legumes (Liener and Kakade, 1980). Such inhibitors, commonly termed protease or trypsin inhibitors (TI), are particularly noted for their occurrence in soybeans which are an increasingly important nutritional source of protein. It has been estimated that the daily intake of TI for Western man is 4.1 mg per gram of protein (Doell et al., 1981), even though conventional cooking and food processing destroys most TI activity.

The ability of dietary TI and raw soy protein to inhibit growth and alter pancreatic function of various animals is also well documented as evidenced in the extensive literature on the subject (Rackis and Gumbmann, 1981; Senti, 1982). Pancreatic effects include increased organ weight or hypertrophy (Booth et al., 1960) and increased enzyme secretion and synthesis (Lyman, 1957; Green and Lyman, 1972). Histologically and biochemically, evidence of hypertrophy and hyperplasia at the cellular level of the exocrine pancreas has been obtained in rats fed raw soy or purified soybean TI (Yanatori and Fujita, 1976; Melmed et al., 1976).

The safety of soybeans and TI was questioned when rats developed adenomatous nodules and invasive cancer of the pancreas after being fed full-fat, raw soy flour for two years (McGuinness et al., 1980). Heating the soy flour markedly reduced the TI content as well as the incidence of pancreatic pathology. The formulation of the test diets (Fölsch and Wormsley, 1974) which contained 40% protein and 19% polyunsaturated fat may have contributed to the development of the observed pathology, since both high protein and high unsaturated fat are dietary factors shown to promote hyperplastic nodules and neoplastic lesions of the pancreas (Temler, 1980; Roebuck et al., 1981, 1985).

Much evidence supports the participation of gastro-intestinal hormones in the regulation of pancreatic function (Harper and Scratcherd 1979), and the mechanism whereby TI exerts its effects on the pancreas appears to involve a negative feedback system largely mediated by cholecystokinin (CCK) (Green and Lyman, 1972; Green et al., 1973; Schneeman and Lyman, 1975), a hormone produced in endocrine cells of the intestinal mucosa. Proteases secreted by the pancreas (trypsin and chymotrypsin activities) suppress the release of CCK from the endocrine cells, thereby providing a monitoring system for sensing enzyme levels in the intestinal tract. When the concentration of free enzyme falls at the endocrine cell sites as a result of the formation of complexes between the proteases and either ingested protein or, especially, ingested TI, which binds nearly irreversibly, suppression is relieved. This, in turn, stimulates the release of CCK resulting in increased hormone levels in the blood.

CCK, which has a trophic effect on the pancreas, has been shown to induce the observed adaptive compensatory response to TI, with increased protease synthesis and secretion. This feedback system controlling the regulation of pancreatic function is undoubtably more complex than presently understood, with the involvement of additional gastro-intestinal hormones and nonparallel induction of digestive enzymes (Temler et al., 1984; Adelson and Miller, 1985).

Presented here are the results of two feeding trials in rats investigating the effects on the pancreas of feeding raw and edible grade soy preparations for two years. In contrast to work cited above, in which pancreatic neoplasia was potentiated by full-fat, raw soy flour in diets containing high TI and high fat, emphasis has been placed on the effects of raw and heat-treated defatted soy flour and soy protein isolate possessing low levels of TI. The pancreatic regulatory responses mediated through the negative feedback mechanism are believed to lead to the pancreatic pathology associated with chronic dietary exposure to raw soy and TI. Since both protein (Giorgi et al., 1985) and TI influence pancreatic adaptation, it was of interest to determine the interaction of these two dietary constituents on the development of pancreatic lesions.

In Trial I, graded concentrations of TI provided by defatted, raw and heated soy flour and by soy protein isolate with partially reduced TI content were investigated. Added casein was used to vary protein quality and content of the diets. Changes in the pancreas, including hyperplastic lesions, were observed as early as six months (Liener et al., 1985; Spangler et al., 1985). After two years, there was a positive dose-response relationship between TI level and the incidence of pancreatic pathology (Gumbmann et al., 1985). In Trial II, the question of whether soy protein, per se, in addition to TI might contribute to the pathology of the pancreas was addressed by feeding raw and heated soy protein isolate from which most of the TI had been removed by an extraction process. The dose related responses to the graded level of TI and protein provided by soy flour, casein, and soy protein isolates were quantitated and compared. Of particular interest was the response to low dietary levels of TI equivalent to those found in edible grade soy products and the possible detection of a threshold for the development of pancreatic neoplasia.

The neoplastic incidence in these studies has been related to an interaction of protein nutrition with the hormonal feedback mechanism for the regulation of pancreatic function, and food safety considerations based on the development of neoplasia in the rat model are discussed. Histologic criteria for diagnosing the observed pancreatic lesions are also presented.

METHODS

Animals

Outbred albino male Wistar rats, Crl:COBS[R](WI) Br, were obtained from Charles River Breeding Laboratories, Inc., Wilmington, Massachusetts when 21 days old for the chronic feeding studies. In Trial I, the rats were housed four per cage in stainless steel, solid bottom cages using hardwood chips (Absorb-Dri[R], Lab Products, Inc., Maywood, New Jersey) as bedding. In Trial II, stainless-steel auto-flush racks with wire bottom cages were used. The rats were housed two per cage.

Forty rats per diet were randomly assigned to each dietary group. Feed and water were offered <u>ad libitum</u> throughout the study. Body weights, feed consumption, and clinical signs were monitored. Pancreata from dead and moribund animals and those surviving to the end of the study (99 to 104 weeks) were preserved in 10% buffered neutral formalin, embedded in paraffin, and sectioned and stained with hematoxylin and eosin. Histological evaluations and scoring were made "blind" without reference to treatment.

During Trial I, interval sacrifices were performed every three months on the same strain of rats fed the twelve soy flour diets and three casein controls described below, from which sequential data over time were obtained (Liener, et al., 1985).

Short-term (28 day) nutritional assays of the soy preparations utilized in the chronic feeding trials were conducted with Sprague-Dawley rats from Simonsen Laboratories, Inc., Gilroy, California.

Test Substances

See Table 1 for the protein and TI content of the soy and casein test substances used in the two feeding trials. Conditions used in the manufacture of the soy products of Trial I have been described previously by Rackis et al.(1985).

The three soy protein isolates used in Trial II were prepared at the Kraft Pilot Plant, Coshocton, Ohio. The isolates were prepared from raw, dehulled, defatted soy flakes (Nutrisoy-7B, Archer Daniels Midland Co., Decatur, Illinois). A combination of mild alkali and sodium

Table 1. Trypsin inhibitor activity and protein content of soy preparations and casein used in Trials I and II[1]

Test substance[2]	Protein	Trypsin inhibitor content	
	(%)[3]	(mg/g)[4]	(mg/g protein)
Defatted soy flour			
Raw (I)	52.3	30.2	57.7
Toasted (I)	53.3	8.5	16.0
Over-toasted (I)	54.6	5.1	9.3
Soy protein isolate			
Raw, unextracted high TI (II)	83.9	30.0	35.8
Raw, unextracted high TI (I)	82.0	24.1	29.4
Raw, extracted intermediate TI (I)	78.9	16.1	20.4
Raw, extracted low TI (II)	86.4	6.0	7.0
Heated, unextracted low TI (I)	77.4	4.7	6.1
Heated, extracted low TI (II)	84.1	2.7	3.2
Lactic casein			
(I)	86.8	0.6	0.7
(II)	86.4	0.7	0.8

[1]Trial I data from Rackis et al., 1985.

[2]Trial number in parentheses.

[3]Protein equals percent nitrogen X 6.25.

[4]As is basis.

chloride extraction of the proteinaceous components of soy flakes, followed by hydrochloric acid precipitation of the extracted proteins was employed.

Protein, moisture, ash, and microbial profiles were determined. Plate counts of bacteria, molds, and yeast were well below acceptable levels. All test substances were Salmonella-free. The test materials were stored at 32 to 40°F, where changes in TI activity during 30 months storage were within the range of experimental error. Nitrite content was 2.1, 14.0, and 22.6 ppm in raw, high-TI; raw, low-TI; and heated, low-TI soy protein isolates, respectively. Less than 1 ppb of nitrosamine was present in all three isolates.

Diets

Defatted soy flours, soy protein isolates, and lactic casein served as the sources of protein. In addition, the diets contained glucose, 15%; non-nutritive fiber, 5.0%; corn oil, 5.5%; lard, 2.5%; DL-methionine 0.012 to 0.1%; vitamin mixture, 2.0%; mineral mixture, 5.0%; choline chloride, 0.2%; antioxidant Tenox[R] 24, 0.5g/kg diet; and dextrin to make 100%. Diets were freshly prepared at approximately three month intervals and stored at - 10°F until used. TI activity of the protein sources used in these studies was determined by the method of Kakade et al. (1974) as modified by Hamerstrand et al. (1981).

In Trial I, twelve diets were formulated with raw and over-toasted soy flours to explore the response to four graded levels of TI (93, 215, 337, and 577mg TI/100g diet) at three protein concentrations (10, 20, and 30%) (Rackis et al., 1985). The two soy flours alone or in combination established the TI levels while providing 10% protein in the diet. Casein was added to achieve 20% and 30% protein. Three diets, containing only casein as the source of protein at 10%, 20%, and 30%, were included. Additional diets (without casein) compared the response to over-toasted and toasted soy flours at 10% and 30% protein, to graded levels of TI provided by raw and heated soy protein isolates at 30% protein, and to a positive control (1,271mg TI/100g diet) provided by raw soy flour at 22% protein. The TI content of one of the raw soy protein isolate preparations was reduced to intermediate levels by a salt extraction procedure (Rackis et al., 1985). The total number of diets in Trial I was 26.

In Trial II, there were eight treatment groups: extracted soy protein isolate, raw and heated, possessing unusually low TI and providing 10% and 30% protein; raw, unextracted (regular process) soy protein isolate as a positive control furnishing 10% and 30% protein; and two casein based diets at 10% and 30% protein. The isolate with unusually low TI concentration was prepared, as in Trial I, by a salt extraction process (Rackis et al., 1985).

Statistical Analysis

Mean body weight data were subjected to analysis of variance and Duncan's multiple range test (Duncan, 1955). Tumor incidence data were analyzed by Fisher's exact test (Bliss, 1967a), by the method of Peto for "incidental" tumors (1974), and with SAS (SAS Institute, Inc., Gary, North Carolina, as run by the Washington Computer Center, USDA, Beltsville, Maryland) using analysis of covariance (GLM) (SAS, 1982a) and FUNCAT for trend analysis (SAS, 1982b) based on the method of Grizzle et al. (1969). Whenever the direction of the potential difference to be tested is uncertain, two-tailed probabilities are given for Fisher's exact test.

RESULTS AND DISCUSSION

Nutritional Considerations

Protein sources. In these studies, in which pancreatic response to chronic exposure to graded levels of dietary TI was investigated, the concentration of TI in the diet was controlled by feeding mixtures of raw and heated soy preparations. It is well known that heat treatment of soybeans brings about the destruction of TI activity while nutritional quality is improved (Rackis and Gumbmann, 1981). The changes in TI content, protein quality (protein efficiency ratio), and nitrogen digestibility of the protein sources utilized in this study as a result of heating or a special salt extraction procedure are shown in Table 2.

Toasting raw soy flour according to commercial practice for producing edible-grade soy, i.e. heat treatment by steam for approximately 20 minutes, brought about a marked reduction in TI activity, although more than 25% residual activity still remained. This was accompanied by a significant increase in protein quality which can be attributed largely to improved protein digestibility. Longer heating (30 minutes) for the over-toasted sample further reduced the TI content, but with essentially no change in protein efficiency ratio (PER) and digestibility. As shown in Figure 1, heat treatment alone to achieve optimum protein quality results in a significant level of residual TI activity resistent to further destruction, and extensive heating, beyond that used to produce the over-toasted soy flour product fed in Trial I, can only be expected to degrade protein quality below optimum values.

Preparation of raw soy protein isolate, an acid precipitated water soluble protein fraction in which more than half of the mass present in soy flour is removed, resulted in a product with about 62% of the TI activity found in raw soy flour. Although nitrogen digestibility improved, there was no significant change in PER. Raw soy protein isolate with reduced TI content was prepared through a combination of mild alkali and sodium chloride extraction process in which TI was selectively removed (Rackis et al., 1985). By controlling pH, products of intermediate and low TI content were obtained. For each increment of TI removed through the extraction process, there was a corresponding increase in PER and digestibility of the raw isolate (Table 2) similar to the results of Kakade et al. (1973).

Heating the extracted isolate further reduced TI activity and increased nutritive value. The heated product, with 3.2mg TI/g protein (approximately 6% of the TI found in raw soy flour), contained the lowest level of residual soybean TI investigated in these studies. It should be noted that heating the low TI, extracted isolate did not improve protein quality equivalent to that of heated soy flour despite superior nitrogen digestibility. This reflects the greater deficiency of sulfur amino acids in soy protein isolate compared to the protein of soy flour, a deficiency which was augmented through extraction by the removal of TI, known to be especially rich in sulfur amino acids and presumably nutritionally available after heat-denaturation (Liener, 1972; Liener and Kakade, 1980).

The protein quality and digestibility of lactic casein used in these studies was essentially equivalent to that of ANRC casein which served as a standard in the PER assays (Table 2). It is unclear whether the nominal TI activity associated with casein represents any specific moiety analogous to soybean TI, is the result of non-specific binding, or reflects an artifact of the TI assay at its sensitivity limits. In

Table 2. PER and nitrogen digestibility of soy preparations used in Trials I and II[1]

Protein source	Trypsin inhibitor (mg/g protein)	PER[2,3]	Nitrogen digestibility[2,4] %
Defatted soy flour			
Raw (I)	57.7	1.18[e]	77.8[f]
Toasted (I)	16.0	2.10[b]	85.1[d]
Over-toasted (I)	9.3	2.02[b]	85.6[d]
Soy protein isolate			
Raw, unextracted (II)	35.8	1.16[e]	82.1[e]
Raw, extracted intermediate TI (I)	20.4	1.21[e]	84.3[d]
Raw, extracted low TI (II)	7.0	1.51[d]	87.2[c]
Heated, extracted (II)	3.2	1.64[c]	89.9[b]
Casein controls (I and II)	(0.7-0.8)	2.46-2.48[a]	92.9-94.2[a]

[1]Weanling male Sprague-Dawley rats (11 per group) were fed for 28 days standard PER diets (A.O.A.C., 1975) containing 10% protein from the source indicated. (Trial number designated in parentheses).

[2]Means with no letter in common are significantly different, $P < 0.05$ (Duncan's Multiple range test). SE = ±0.04, PER; ±0.4, nitrogen digestibility.

[3]PER values adjusted to that of ANRC casein standardized at 2.50.

[4]Nitrogen digestibility (apparent) = 100 x (nitrogen intake - fecal nitrogen)/nitrogen intake.

Table 3. Body weight gain at eight months for rats fed soy protein isolate in Trial II

Protein source[1]	Trypsin inhibitor (mg/g protein)	Weight gain (grams)[2] 10% protein	30% protein
Soybean protein isolate			
Raw, unextracted	35.8	239[e]	281[cd]
Raw, extracted	7.0	277[d]	290[bc]
Heated, extracted	3.2	285[cd]	283[cd]
Casein	0	298[b]	313[a]

[1]Dietary protein at 10% or 30% from the single source indicated. The range of TI in the diet (mg/100g) for soy protein isolate was: 32 to 358, 10% protein and 96 to 1075, 30% protein.

[2]Means with no letter in common are significantly different, $P < 0.05$ (Duncan's multiple range test). SE = ±4, 37-40 rats per group.

subsequent tables and figures, the apparent TI activity supplied by casein has been designated as zero.

Growth of rats. The growth of rats fed raw and over-toasted soy flour preparations throughout the two year duration of these studies closely reflected the protein quality and protein level of the diets. As shown in Figure 2, Trial I diets containing raw soy flour (577mg TI/100g diet) generally resulted in the lowest body weights, while rats fed over-toasted soy flour and the casein controls (93 and 0 mg TI/100g diet, respectively) emerged in the latter half of the study with the greatest body weights. Mixtures of over-toasted and raw soy flours yielding intermediate levels of TI produced intermediate growth. By combining data over TI concentration to demonstrate the effects of protein level on body weight (Figure 3), it is evident that rats fed 10% protein diets failed to attain maximum growth.

Similar differences in response to the raw versus heated soy protein isolates fed in Trial II were observed. With 10% protein in the diet, growth was restricted for rats fed the regular process, raw soy protein isolate (35.8mg TI/g protein) (Table 3). Extraction or extraction plus heating to reduce TI content were both effective in significantly increasing weight gain. With 30% protein, the nutritional differences among the three soy protein isolates were obscured and similar weight gains were observed.

Pancreatic Histology and Diagnostic Criteria

In rats, the exocrine pancreas rapidly responds, within hours, to ingestion of TI with increased secretion and synthesis of digestive enzymes. The TI may be that found in raw soy preparations, purified TI from various sources, or low molecular weight TI such as synthetic p-aminobenzamidine and Foy 305 (Geratz, 1968; Geratz and Hurt, 1970; Melmed and Bouchier, 1969; Dijkhof and Poort, 1978; Rackis and Gumbmann, 1981; Goebell et al., 1983). Following continued TI intake, pancreatic hypertrophy (increased organ weight) and hyperplasia are detectable within nine days to a few weeks (Booth et al., 1960; Rackis, 1965; Rackis and Gumbmann, 1981; Geratz, 1968; Senti, 1982). Histologically, little or no structural alterations may be readily evident at this time. In the case of cellular hypertrophy, detection is dependent upon quantitative morphometric determination (Yanatori and Fujita, 1976; Spangler et al., 1985). Hyperplasia in the presence of cellular hypertrophy is diagnosed through the detection of an increased frequency of mitotic figures, which may be difficult unless the animals have been treated with colchicine prior to sacrifice (Yanatori and Fugita, 1976; Senti, 1982).

Chronic dietary exposure of rats to raw soy flour, alone or after treatment with azaserine (reported to be a pancreatic carcinogen), has been associated with the development of focal nodular hyperplasia of the acinar pancreas (NH), observed grossly after six months, and eventually the benign neoplastic lesion, pancreatic acinar adenoma (AA) (Morgan et al., 1977; Levison et al., 1979; McGuinness et al., 1980; Crass and Morgan, 1982; Spangler et al., 1985). Carcinomatous change in the rat pancreas was reported after 60 weeks with exposure to a raw soy flour diet (McGuinness et al., 1984).

The response of the exocrine pancreas to TI or raw soy represents a continuous spectrum of histologic changes. Unfortunately, criteria used in the diagnosis of these changes often remain unclear in many published reports. Also, it is not always possible to define absolute morphologic or functional criteria to separate hyperplasia from benign neoplasia, and this becomes a matter of subjective judgment. In the presence of defining criteria, however, a subjective judgment of sustaining or reproducible accuracy can be made. Recent evaluation of the pancreas in rats from two-year studies conducted by the National Toxicology Program of the National Cancer Institute has lead to a more systematic classification of pancreatic lesions, which should provide a better understanding of criteria for the diagnosis of NH and AA in the future (Boorman and Eustis, 1984). The following criteria and definitions were used in classifying the pancreatic lesions observed in the two long-term feeding trials discussed here.

Hypertrophy. An increase in the size of the described component, while normal cellular relationships are maintained. For an organ, hypertrophy indicates a generalized increase in size and weight. At the cellular level, hypertrophy indicates a cell of greater proportions, generally involving both the nucleus and cytoplasm. Thus, when most cells of an organ are involved, the organ is larger with fewer cells per unit volume or cross-sectional area.

Hyperplasia. Changes at the cellular level to denote proliferation of a specific cell type resulting in an increase in the total number of cells in the organ as a whole or a specified portion. Foci of hyperplastic acinar cells, NH, may increase rapidly in size, becoming grossly visible nodules by at least six months, and are easily distinguished from surrounding normal structures (Figure 4). Such nodules have sometimes been termed atypical acinar cell nodules (Longnecker et al., 1981).

Neoplasia. A change in cellular state in which there is an autonomous proliferation of cellular elements not subject to the normal controlling mechanisms of cellular growth or proliferation. Defining criteria used to differentiate the benign neoplastic lesion, AA, from the pre-neoplastic NH were:

1. Relatively more rapid cellular proliferation resulting in a larger mass of tissue in a relatively shorter period of time, as evidenced by compression of surrounding normal tissue in such a way that the architecture of the normal adjacent tissue is distorted (Figure 5). In a more slowly growing tissue mass, the adjacent tissue would have adequate time to remodel and conform to the increased bulk of the proliferating tissue mass. Rapidly proliferating and expansile nodules within the pancreatic parenchyma were frequently surrounded by a connective tissue capsule of varying thickness. This capsule was generally composed of mature fibrocytes with abundant collagenous matrix indicating that the capsule was formed from mature pancreatic stroma which condensed following atrophy or degeneration of resident pancreatic acinar cells.

2. Evidence of rapid growth as manifested by increased frequency of mitotic figures (Figure 6).

3. Substantial alteration of the histologic architecture of the proliferating neoplastic mass relative to the structure normally observed (Figure 7). Thus, in contrast to a hyperplastic nodule in which the basic architecture was maintained allowing the tissue to be easily recognized, the histologic features of neoplastic proliferation were distorted to varying degrees frequently resulting in structural changes which put the histogenesis of the lesion in doubt.

4. Alteration in the cytology of the neoplastic tissue. Altered cytology is the basis for definition and diagnosis of neoplastic tissue and is an important distinguishing feature. These changes were less

easily defined, but generally seen as nuclear enlargement with unusual or abnormal arrangements of chromatin material as well as an alteration in the distribution and character of cytoplasmic organelles with a resultant shift in the ratio of nuclear to cytoplasmic volume.

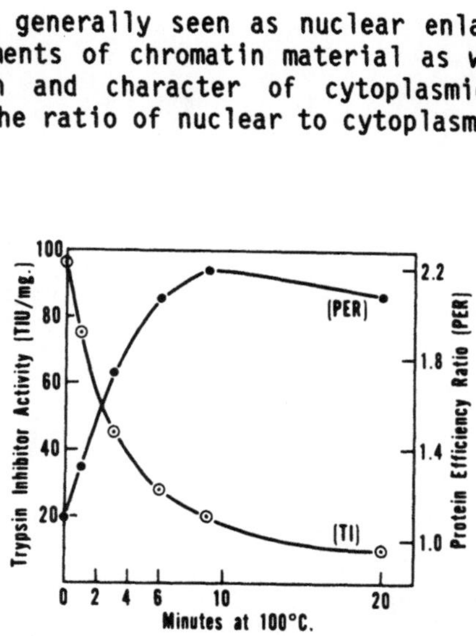

Figure 1. Effect of steam heat treatment on trypsin inhibitor activity and protein efficiency ratio of defatted soy flour. TIu = trypsin inhibitor units as defined by Kadade et al., 1969. Data from Rackis et al., 1975.

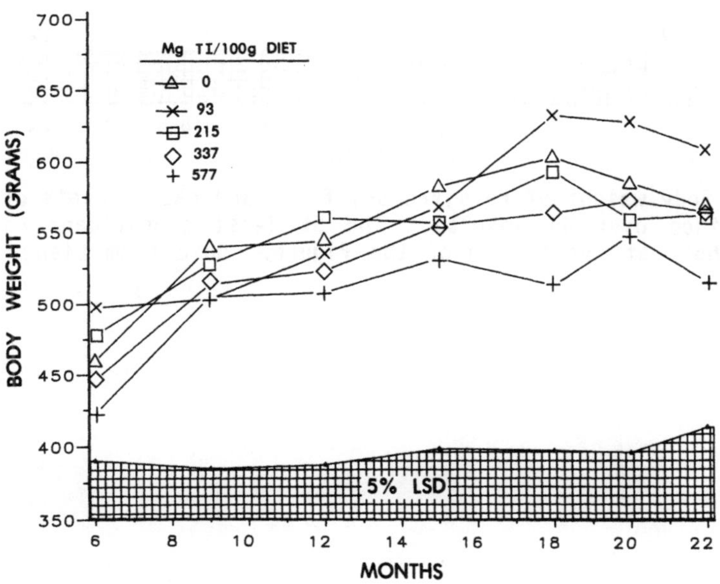

Figure 2. Body weight of rats fed soy flour and casein diets of Trial I (data combined over 10%, 20%, and 30% protein levels). Diets and their TI content in mg/100g were: casein, 0; over-toasted soy flour, 93; blends of over-toasted and raw soy flours, 215 and 337; and raw soy flour, 577. The 5% least significant difference (LSD) is shown at the bottom of the figure. Data from Liener et al., 1985.

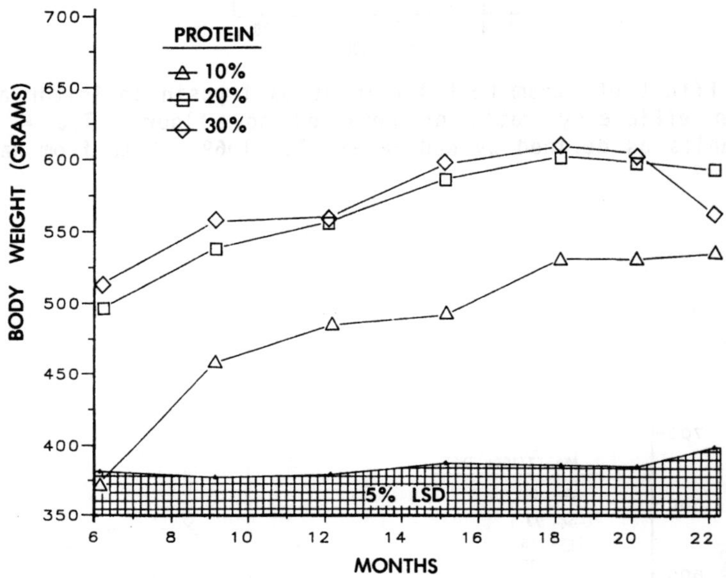

Figure 3. Body weight of rats fed soy flour and casein diets of Trial I (data combined over TI levels). The 5% least significant difference (LSD) is shown at the bottom of the figure. Data from Liener et al., 1985.

Carcinoma. Criteria used for distinguishing carcinoma, which would allow differentiation from benign neoplasia, were:

1. Substantial anaplasia with loss of cellular orientation.
2. Unambiguous indication of invasiveness into surrounding tissue.
3. Direct evidence of metastases to other organs.

Animals from the long-term feeding studies, Trials I and II, developed histologic changes in the pancreas consisting of generalized cellular hypertrophy, NH, and AA as the major departures from normal pancreatic morphology. Pancreatic carcinoma was not observed. When morphometric criteria similar to those described below for generalized pancreatic hypertrophy were applied to the isolated foci of NH, confirmation of hyperplasia was achieved quantitatively, i.e., smaller cell size with an increase in cell density. Grossly, the pancreata of animals fed the higher levels of TI showed massive involvement, consisting of numerous nodules of various size up to 5mm or more in diameter (Figure 8).

Pancreatic islet cell adenoma and mild to moderate fatty infiltration were noted and found to be strongly related to nutritional considerations of the diets. Pancreatic islet cell adenoma was induced a little more readily in animals fed diets containing better quality protein and in diets containing higher levels of protein, while fatty infiltration was promoted by better quality protein at low dietary levels (Gumbmann et al., 1985). Other lesions were of such minor incidence that no association with treatment could be discerned.

Pancreatic Hypertrophy

Soy flour. Chronic feeding of the defatted soy flour diets of Trial I resulted in pancreatic enlargement or organ hypertrophy which was positively related to the concentration of TI as provided by raw soy flour (Figure 9). Pancreas weights associated with diets containing 577mg TI/100g were elevated above all others over a period of 22 months, beginning with the first observation at 6 months. Lowest pancreas weights occurred with diets containing only oven-toasted soy flour or casein as the source of protein. Next to TI and raw soy flour in importance as factors influencing pancreas weight was dietary protein level (Figure 10). Pancreas weight was markedly decreased throughout the study when protein was limited to 10%. This is similar to the finding of Temler (1980) in which pancreatic weight relative to body weight increased with increasing level of dietary protein from either casein or soy flour throughout the range of 10% to 40%.

To determine whether the observed pancreatic hypertrophy was related to changes in cell size, morphometric techniques were used to measure the cross sectional density of acinar cells in a given area on photomicrographs of pancreatic tissue. At each sacrifice period from 6 through 16 months, pancreatic hypertrophy existing at the cellular level, as evidenced by reduced cellular density, was associated with rats consuming raw soy flour providing 577mg TI/100g diet (Figure 11). Differences from the casein control were statistically significant for each period except 12 months.

The effect of dietary protein level on cellular density is shown in Figure 12. Cellular density decreased markedly in the casein fed controls when dietary protein was increased from 10% to 20% or 30%. However, for rats fed diets containing 577mg TI/100g, uniformly low cellular density, significantly less than that for the casein controls, was encountered at each level. Thus, pancreatic hypertrophy at the

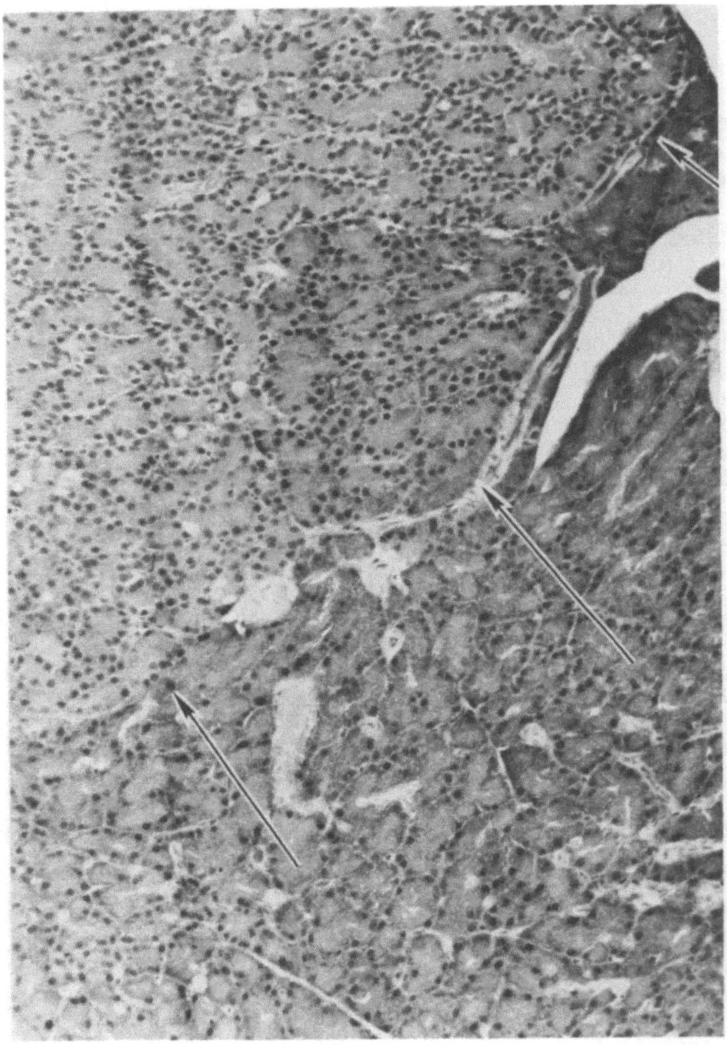

Figure 4. Focal nodular hyperplasia (boundry indicated by arrows). Note that the pancreatic acinar structure, although somewhat distorted, is still evident. Cells in the area of hyperplasia are obviously smaller and more numerous.

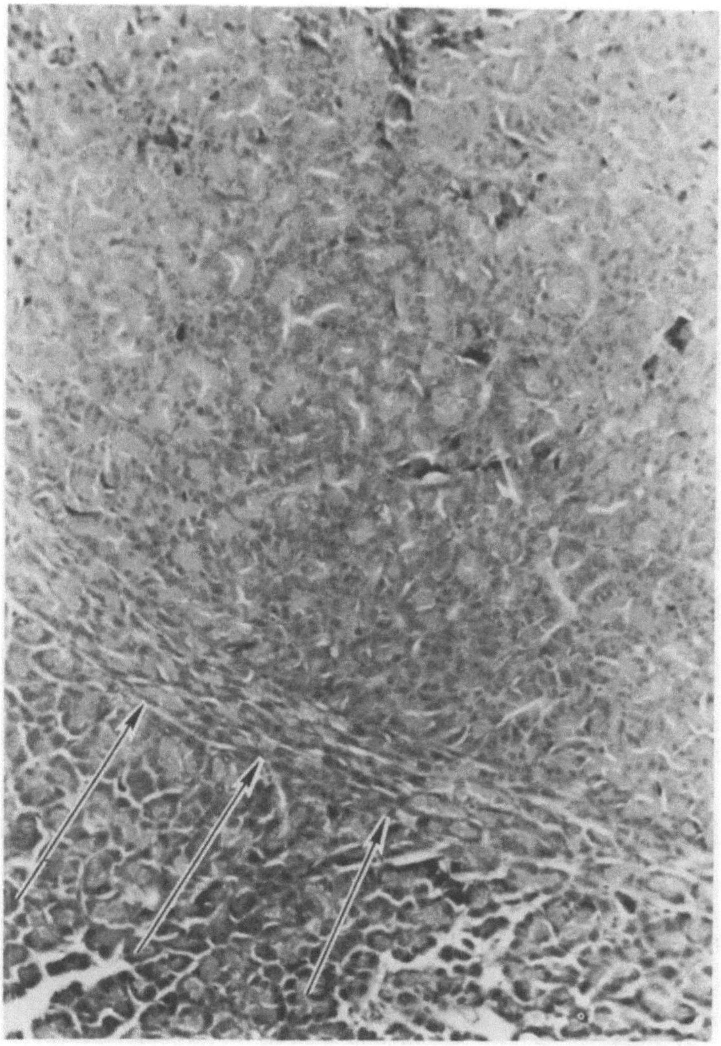

Figure 5. A neoplastic nodule (adenoma) of the pancreas. Acinar architecture is well preserved; however, the nodule is large and rapid growth is indicated by compression (arrows) of surrounding normal pancreas.

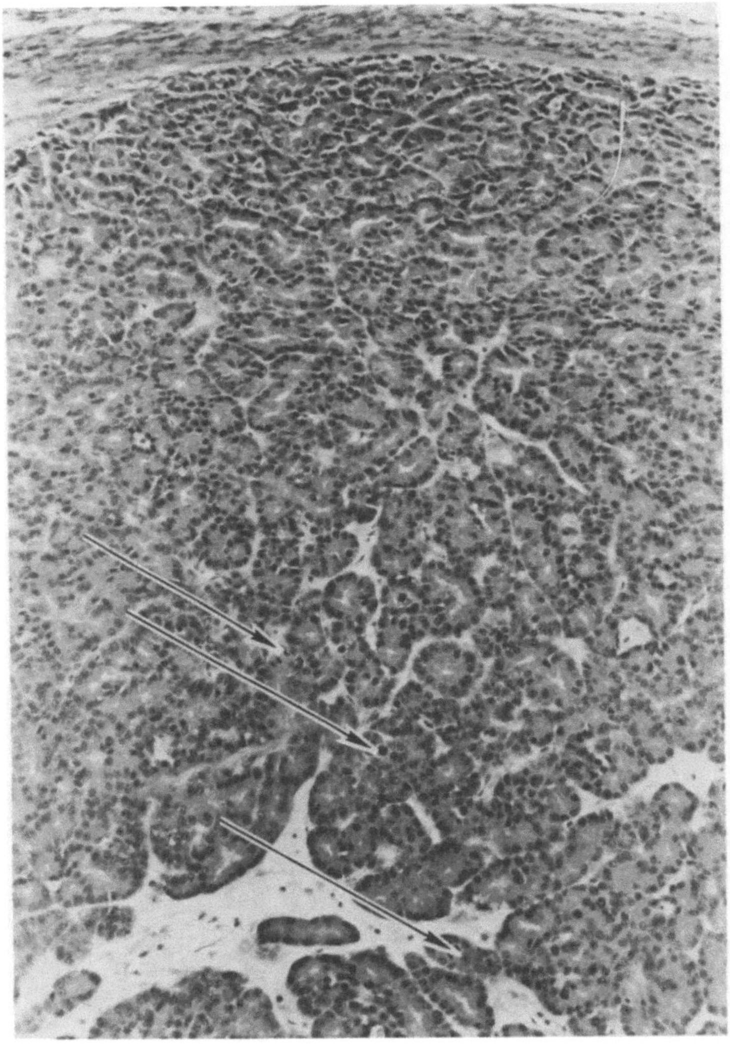

Figure 6. Adenomatous proliferation in the pancreas illustrating increased mitotic figures (arrows) indicative of rapid growth.

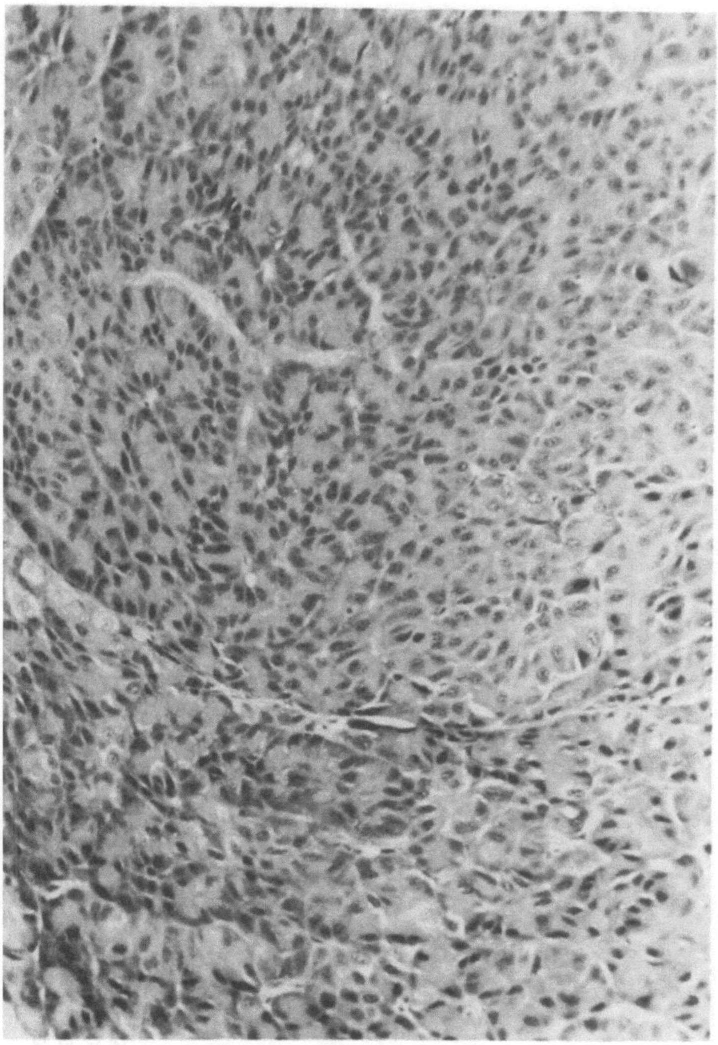

Figure 7. Sèvere distortion of acinar structure in pancreatic adenoma. Histologic recognition of the tissue in this field is compromised due to architectural distortion.

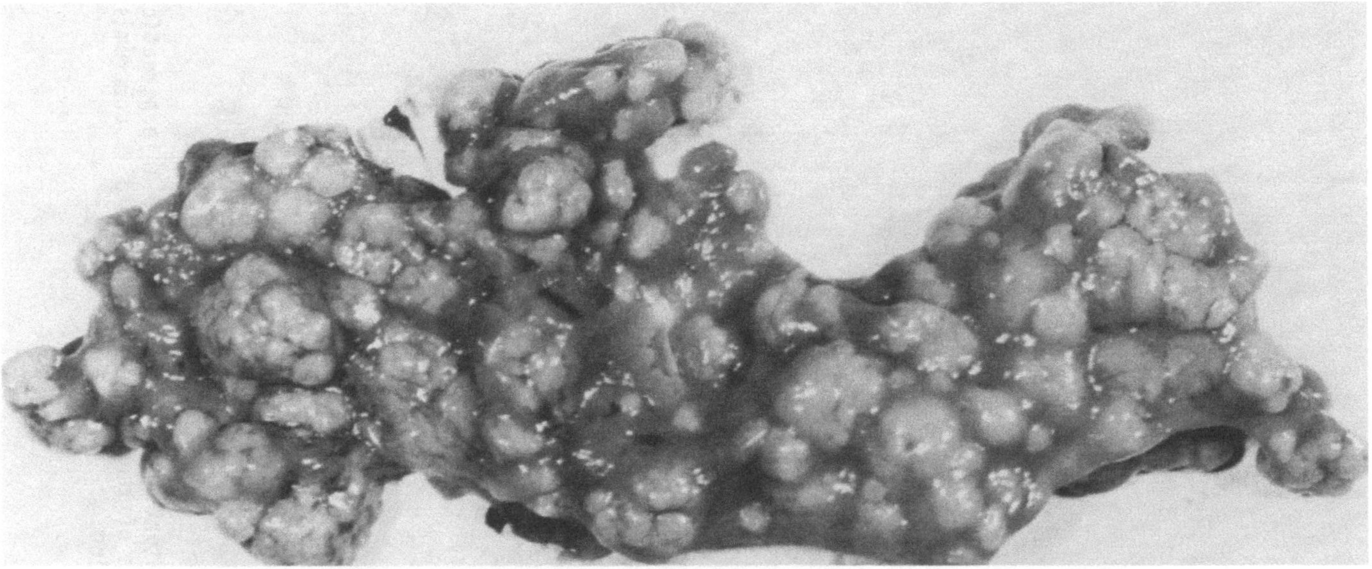

Figure 8. Typical gross appearance of pancreas of rat fed raw soy protein isolate with high TI content for two years. Note nodularity and massiveness. Normal pancreatic tissue is thin and lightly dispersed along the intestinal mesentery.

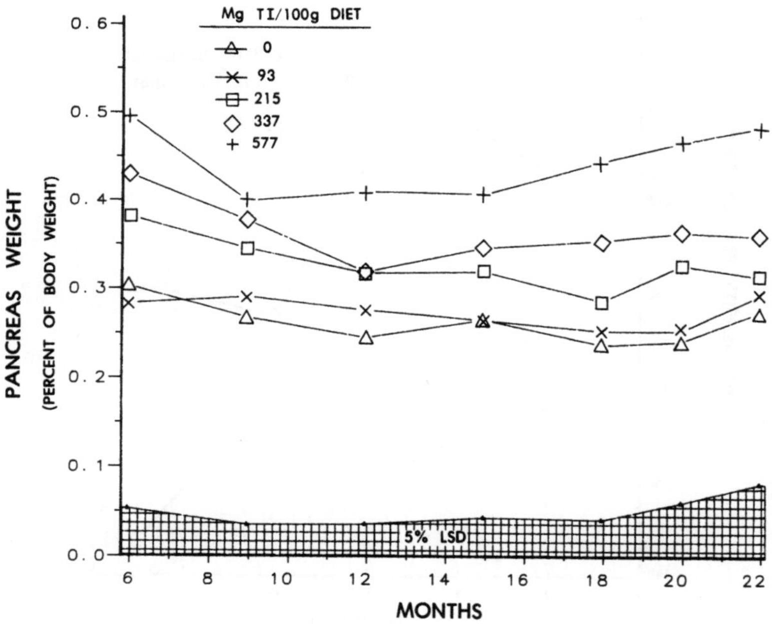

Figure 9. Pancreas weights of rats fed soy flour and casein diets of Trial I (data combined over 10%, 20%, and 30% protein levels). Diets and their TI content in mg/100g were: casein, 0; over-toasted soy flour, 93; blends of over-toasted and raw soy flours, 215 and 337; and raw soy flour, 577. The 5% least significant difference (LSD) is shown at the bottom of the figure. Data from Liener et al., 1985.

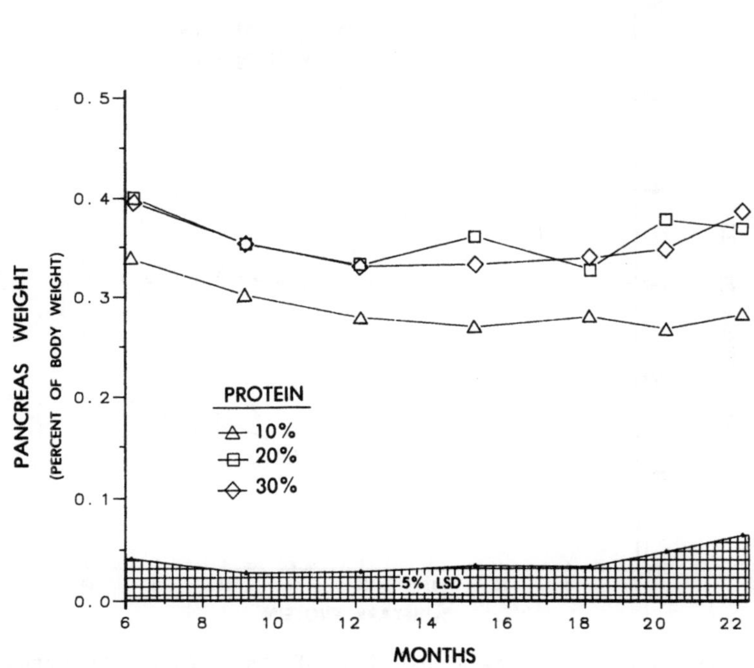

Figure 10. Pancreas weights of rats fed soy flour and casein diets of Trial I (data combined over TI levels). The 5% least significant difference (LSD) is shown at the bottom of the figure. Data from Liener et al., 1985.

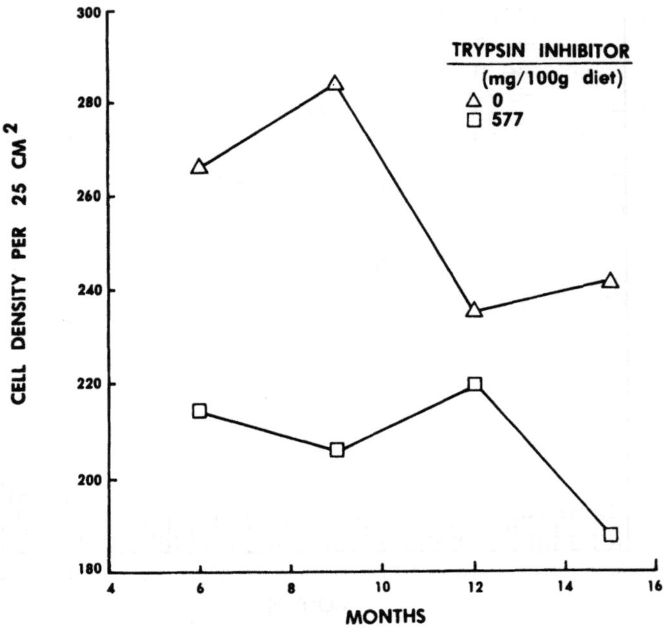

Figure 11. Effect of raw soy flour on cell density of the acinar pancreas. Cell density was determined by counting the number of nuclei in 15 well defined acinar units in each of three rats per dietary group of Trial I; casein control diets, 0 mg TI/g; raw soy flour diets, 577mg TI/g. The difference between means was significant ($P < 0.05$) at each time except 12 months. Data averaged over protein levels (Spangler et al., 1985).

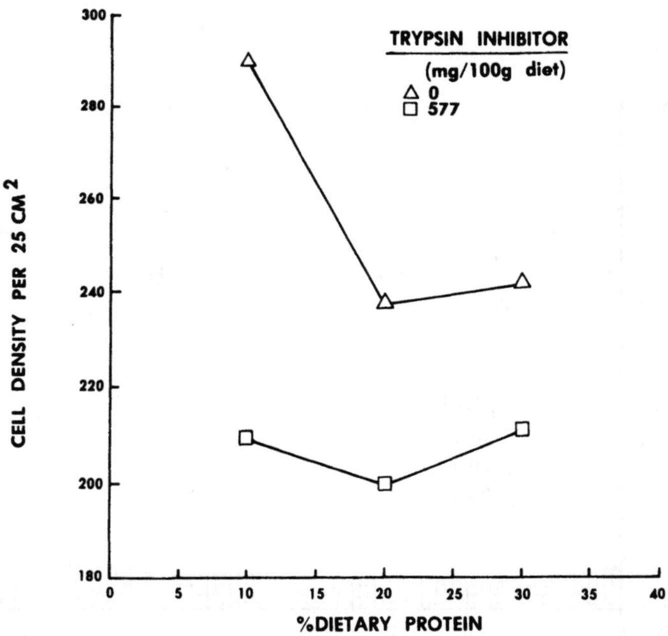

Figure 12. The change in cell density of the acinar pancreas with dietary protein level, determined as described in Figure 11. Casein control diets, 0mg TI/g; raw soy flour diets, 577mg TI/g. At each protein level, the difference between means was significant ($P < 0.05$). Data averaged over time (Spangler et al., 1985).

cellular level occurred when the concentration of casein in the diet was increased above 10%, but never to the extent brought about by raw soy flour. Also, pancreatic organ enlargement associated with TI or protein can be explained, at least in part, by a relatively greater size of pancreatic acinar cells.

Soy protein isolates. The effect on pancreatic weight from feeding soy protein isolates of Trial II was also investigated. In a short term assay (Table 4), elevated pancreas weights were observed when rats were fed the high TI, regular process isolate, but not after the TI content had been lowered through an extraction process. Even though the soy protein was raw in the extracted isolate, pancreas weights were equivalent to those of rats fed the heated isolate or casein.

The mean pancreas weights of rats fed these isolates for two years in Trial II are shown in Table 5. The response was similar to that of the short-term assay, except that pancreas weights were generally greater in these older animals. For either 10% or 30% dietary protein, mean pancreas weights for rats fed extracted raw soy protein isolates were again not significantly different from those fed heated isolate. Although not statistically significant, the data suggest that heating the extracted isolate slightly lowered pancreas weights commensurate with the further reduction in TI content.

The concentration of dietary protein had a marked effect on pancreas weight which was clearly greater in rats fed 30% protein, particularly with the diet containing high TI raw isolate.

Pancreatic Nodular Hyperplasia and Acinar Adenoma

Onset. Focal areas of nodular hyperplasia (NH) were observed histopathologically in the pancreas throughout the duration of Trial I, beginning at six months in the first group of the serially sacrificed animals (Liener et al., 1985). Acinar adenoma (AA), which has a slower onset, was first observed in a scheduled sacrifice of five rats at 18 months, and in one early death animal at 15 months. The incidences of NH and AA in interval sacrifice animals fed raw unextracted and heated extracted soy protein isolate of Trial II are compared in Table 6. NH was present in all animals of the raw soy group at 12 and 15 months, while AA was present in 40% at 12 months, increasing to 60% at 15 months. The reduction in TI content brought about by extraction and heating was associated with a marked decrease in the proportion of rats with these lesions.

The relative frequency with which NH and AA were manifested after two years is shown in Figure 13 for the 26 diets of Trial I. The best-fitting curve is clearly exponential:

$$NH = 100\ (1 - e^{-0.06\ AA})$$

Another model based on a linear regression of log transformed data shows that the incidence rate of AA is roughly equal to 0.3% to 0.4% of that of NH squared. Either model indicates a disproportionately greater risk of AA in populations with a large incidence of NH.

Incidence. Mixtures of proteins from raw and heated soy flour and casein provided diets containing four graded levels of TI repeated at three protein concentrations. The incidences of NH and AA associated with these twelve diets plus three casein controls of Trial I after two years are shown in Tables 7 and 8, respectively. NH was positively

Table 4. Pancreas weight after 28 days in rats fed soy protein isolates used in Trial II[1]

Protein source	Trypsin inhibitor (mg/g protein)	Pancreas weight[2] (% body weight)
Soy protein isolate		
Raw, unextracted	35.8	0.34[a]
Raw, extracted	7.0	0.28[b]
Heated, extracted	3.2	0.27[b]
Casein control	0	0.25[b]

[1]Weanling male Sprague-Dawley rats (10 per group) were fed for 28 days standard PER diets (A.O.A.C., 1975) containing 10% protein provided by the single source indicated. The range of TI in the diet (mg/100g) for soy protein isolate was 32 to 358.

[2]Means with no letter in common are significantly different, $P < 0.05$ (Duncan's multiple range test). SE = ±0.02.

Table 5. Pancreas weights in rats fed soy protein isolates in Trial II for two years

Protein source[1]	Trypsin inhibitor (mg/g protein)	Pancreas weight[2] (% body weight)	
		10% protein	30% protein
Soy protein isolate			
Raw, unextracted	35.8	0.49[c]	1.74[a]
Raw, extracted	7.0	0.39[de]	0.60[b]
Heated, extracted	3.2	0.31[e]	0.52[bc]
Casein	0	0.35[de]	0.43[cd]

[1]Dietary protein at 10% or 30% from the single source indicated. The range of TI in the diet (mg/100g) for soy protein isolate was: 32 to 358, 10% protein and 96 to 1075, 30% protein.

[2]Means with no letter in common are significantly different, $P < 0.05$ (Duncan's multiple range test). SE = ± 0.17, 20-33 rats per group.

Table 6. Incidence of pancreatic nodular hyperplasia and acinar adenoma in rats fed soy protein isolates in Trial II for 12 and 15 months[1]

Soy protein isolate	Trypsin inhibitor (mg/g protein)	Nodular hyperplasia		Acinar adenoma	
		12 months	15 months	12 months	15 months
		(% incidence)			
Raw, unextracted	35.8	100	100	40	60
Heated, extracted	3.2	0	20	0	0

[1]Diets contained 30% protein from the source indicated, five rats per group. TI (mg/100g diet) was 1075 and 96 for raw and heated proteins, respectively.

Table 7. Incidence of pancreatic nodular hyperplasia in rats fed soy flour and casein diets for two years[1,2]

Protein[3] %	Incidence of nodular hyperplasia					TI Combined
	Casein control	Trypsin inhibitor (mg/100g diet)[4]				
		93	215	337	557	
10	17/30[xa] (56.7)	29/35[xbc] (82.9)	29/35[xbc] (82.9)	26/35[xab] (74.3)	30/31[xc] (98.6)	131/166[x] (78.9)
20	6/31[ya] (19.4)	10/29[ya] (34.5)	25/32[xb] (78.1)	34/37[xbc] (91.9)	32/32[xc] (100.0)	107/161[y] (66.5)
30	5/25[ya] (20.0)	7/33[ya] (21.2)	20/29[xb] (69.0)	25/30[xb] (83.3)	31/31[xc] (100.0)	88/148[y] (59.5)
Protein combined	28/86[a] (32.6)	46/97[b] (47.4)	74/96[c] (77.1)	85/102[c] (83.3)	93/94[d] (98.9)	

[1]Number of animals with lesion/number of surviving animals, percent incidence in parentheses.

[2]One-tailed Fisher's exact test: Within each protein level or protein levels combined, elevated incidence at higher TI is significantly greater ($P < 0.05$) than a lower incidence at lower TI if there are no letters (a,b,c,d) in common. Within each TI level or TI levels combined, decreased incidence at higher protein is significantly less ($P < 0.05$) than a higher incidence at lower protein if there are no superscript letters (x,y) in common.

[3]Casein was added to achieve 20% and 30% protein.

[4]Graded TI concentrations (93-577 mg/100g diet) resulted from over-toasted (9.3 mg TI/g protein) and raw (57.7 mg TI/g protein) soy flours and mixtures of the two providing 10% protein in the diet.

Table 8. Incidence of pancreatic acinar adenoma in rats fed soy flour and casein diets for two years[1,2]

Protein[3] %	Incidence of acinar adenoma					
	Casein control	Trypsin inhibitor (mg/100g diet)[4]				TI Combined
		93	215	337	557	
10	$0/30^{xa}$ (0.0)	$8/35^{xb}$ (22.9)	$14/35^{xb}$ (40.0)	$15/35^{xbc}$ (42.9)	$20/31^{xc}$ (64.5)	$57/166^{x}$ (34.3)
20	$0/31^{xa}$ (0.0)	$1/29^{ya}$ (3.4)	$7/32^{xyb}$ (21.9)	$19/37^{xc}$ (51.4)	$24/32^{xd}$ (75.0)	$51/161^{xy}$ (31.7)
30	$1/25^{xa}$ (4.0)	$1/33^{ya}$ (3.0)	$2/29^{ya}$ (6.9)	$12/30^{xb}$ (40.0)	$21/31^{xc}$ (67.7)	$37/148^{y}$ (25.0)
Protein combined	$1/86^{a}$ (1.2)	$10/97^{b}$ (10.3)	$23/96^{c}$ (24.0)	$46/102^{d}$ (45.1)	$65/94^{e}$ (69.1)	

[1]Number of animals with lesion/number of surviving animals, percent incidence in parentheses.

[2]One-tailed Fisher's exact test: Within each protein level or protein levels combined, elevated incidence at higher TI is significantly greater ($P < 0.05$) than a lower incidence at lower TI if there are no letters (a,b,c,d,e) in common. Within each TI level or TI levels combined, decreased incidence at higher protein is significantly less ($P < 0.05$) than a higher incidence at lower protein if there are no superscript letters (x,y) in common.

[3]Casein was added to achieve 20% and 30% protein.

[4]Graded TI concentrations (93-577 mg/100g diet) resulted from over-toasted (9.3 mg TI/g protein) and raw (57.7 mg TI/g protein) soy flours and mixtures of the two providing 10% protein in the diet.

related to the concentration of dietary TI in the 20% and 30% protein groups. With 10% protein, the incidence of NH was relatively high in all five diets, including the casein control, regardless of TI concentration. The protein level of the diets containing low levels of TI, the casein controls (0 mg/100g) and over-toasted soy flour (93mg/100g), had a marked effect on the incidence of NH. NH was significantly decreased when protein in the diet was increased from 10% to 20% or 30%.

The incidence of AA also showed a positive association to TI concentration. For rats fed the diets with over-toasted soy flour (93mg TI/100g) and the lowest level of raw soy flour (215mg TI/100g), increasing protein in the diet through addition of casein greatly reduced the incidence of AA, similar to the results for NH. At higher TI concentrations, casein offered no protection.

Soy flours. The long term effect of substituting edible grade soy flours for casein in the diet on the development of pancreatic pathology is shown in Table 9. At 10% protein, both NH and AA became significantly elevated. The actual P value for the significance of the difference in incidence of AA between casein and over-toasted soy flour was 0.005 (one-tailed). If the time period selected for consideration is 86 weeks, rather than 99 weeks, to include the one occurrence of AA that was observed in the casein control group before the terminal sacrifice period, then the incidences for the casein control and over-toasted soy flour diets change from 0/30 and 8/35 to 1/34 and 8/39, respectively ($P = 0.024$). At the higher protein concentration, both over-toasted and toasted soy flours resulted in increased NH. AA was significantly increased only with toasted soy flour.

The somewhat greater heat treatment used to produce over-toasted soy flour, which further reduced the TI content, decreased the incidence of pancreatic pathology below that associated with regular toasted flour (Table 9). Comparing only these two heated flours to each other shows that the incidences of AA at 10% and both NH and AA at 30% protein levels were significantly elevated in the regularly toasted flour.

Dose-response model. A mathematical model demonstrating the dose-response relationship of AA to dietary TI and the interaction with protein level in the diet is shown in Figure 14. Data from Table 8 were linearized by plotting the probit (Bliss, 1967b) of percent incidence of AA against the log of the concentration of TI expressed per unit of dietary protein. Probit analysis (SAS, 1982c) shows that the data fit this model well, which suggests that the biological response to dietary TI may be, at least in part, a function of the concentration of TI in the protein rather than the absolute quantity of TI ingested. The difference in response to 20% and 30% protein diets was not significantly different. However, the slopes of both were significantly different from that of 10% protein. Similar results were obtained for the incidence of NH, except for the 10% protein diets where divergence from linearity was excessive.

Soy Protein Isolates. Soy protein isolate in Trial I, which was prepared with intermediate TI content through a salt extraction procedure, was fed raw and heated and in mixtures of the two to provide graded levels of TI at 30% dietary protein (Table 10). The incidence of NH was least in the group fed the heated isolate. With 10% or more protein in the diet from raw isolate, the incidence of NH was 90% or more.

The incidence of AA was positively related to TI concentration (P = 0.01) (SAS, 1982b). The probit model for the dose-related response of the soy protein isolates compared to that of the soy flour diets is shown in Figure 15. The response fitted the probit model well and was significantly different from that of the 10% protein curve, but not the 20-30% protein curve (SAS, 1982c). Thus, removal of more than 50% of the mass of soy flour in the production of protein isolate did not substantially alter the effect of soy on the development of pancreatic pathology, which showed clear dependence on TI content.

Whether raw soy protein, per se, may contribute to pancreatic pathology was investigated in Trial II. Raw soy protein isolate possessing an unusually low TI content of only 7.0mg/g protein was also prepared through a salt extraction process and fed both raw and heated. The incidence of NH with the raw extracted isolate was not significantly different from that of raw unextracted isolate when fed at either 10% or 30% protein in the diet; only heating the extracted isolate significantly reduced the incidence of NH (Table 11). Extraction of the raw isolate did, however, bring about a significant reduction in the severity of the lesion when fed at 30% protein but not 10%. From the occurrence of NH in association with diets based on heated soy protein or casein, it is evident that NH is a prominent lesion in the older rat.

With 30% protein in the diet, but not 10%, the incidence of AA was considerably reduced when rats were fed the raw soy protein isolate from which a large portion of TI had been extracted (Table 12). AA was further reduced in both protein levels after the isolate was heated. Figure 16 compares the dose-related responses of rats fed the soy protein isolates of Trial II and soy flour of Trial I. At 30% protein, the three isolate preparations, raw unextracted, raw extracted, and heated, did not diverge significantly from the 20-30% protein curve for flour. Raw, unextracted isolate at 10% protein was also in close correspondence with the 10% flour curve. The responses to low TI from raw and heated extracted isolates at the 10% protein level, however, were greater than would be predicted from TI concentration alone.

Protein-TI Interaction and Pancreatic Adenoma

Probit plot. Two interacting dietary factors, TI concentration and protein level, have been shown by the data of Trials I and II to be involved in the development of pancreatic neoplasia in the rat. A probit plot of the incidence of AA for all diets containing soy preparations in both trials is shown in Figure 17. It is noted that the slope of the neoplastic response is much less with low dietary protein. Of interest is the intersection of the 10% protein curve with that of 20-30% protein which occurs near the region of edible grade, over-toasted soy flour (9.3mg TI/g protein). Exactly at this point, the development of AA would be predicted to be independent of protein level in the diet. Increasing the protein level of the diet while maintaining constant TI concentration per gram of protein (vertical movement on the figure) would be expected to result in a decreased incidence of AA at low TI concentrations (to the left of the intersection), and an increased incidence of AA at high TI concentrations (to the right of the intersection).

Experimentally, this was observed with specific soy preparations investigated in these studies. Examples of response to diets showing this interaction of TI and protein are given in Table 13. Near or to the left of the intersection (low TI region) are over-toasted and toasted soy flour diets of Trial I and raw and heated, extracted soy protein isolate diets of Trial II. In each case, opposite to what might

Table 9. Incidence of pancreatic nodular hyperplasia and acinar adenoma in rats fed heat-treated soy flours and casein for two years

Protein source[1]	Trypsin inhibitor (mg/100g diet)	Lesion Incidence[2,3] Nodular hyperplasia	Acinar adenoma
		10% Protein	
Casein	0	17/30 (56.7)[bcBC]	0/30 (0.0)[dD]
Over-toasted soy flour	93	29/35 (82.9)[aA]	8/35 (22.9)[bcBC]
Toasted soy flour	160	27/33 (81.9)[aAB]	18/33 (54.5)[aA]
		30% Protein	
Casein	0	5/25 (20.0)[dD]	1/25 (4.0)[dCD]
Over-toasted soy flour	281	16/32 (50.0)[cC]	2/32 (6.3)[cdCD]
Toasted soy flour	479	26/34 (76.5)[abAB]	11/34 (32.4)[abAB]

[1]Trial I diets: dietary protein at 10% or 30% from the single source indicated.

[2]Number of animals with lesion/number of animals observed; percent incidence in parentheses.

[3]For each lesion, incidences with no letter in common are significantly different ($P < 0.05$): one-tailed, lower case; two-tailed, upper case (Fisher's exact test).

Table 10. Incidence of pancreatic nodular hyperplasia and acinar adenoma in rats fed raw and heated soy protein isolate for two years

Soy protein isolate[1]	Trypsin inhibitor (mg/g protein)	Lesion incidence[2,3] Nodular hyperplasia	Acinar adenoma
Raw	20.4	27/28 (96.4)[a]	14/28 (50.0)[a]
Raw/heated (20:10)	15.6	28/28 (100.0)[a]	12/28 (42.9)[a]
Raw/heated (10:20)	10.9	27/30 (90.0)[a]	11/30 (36.7)[ab]
Heated	6.1	20/33 (60.6)[b]	6/33 (18.2)[b]

[1]Trial I diets providing 30% protein and 182 to 612 mg TI/100g diet. For raw/heated combinations, the percent of protein in the diet from each source is given in parentheses.

[2]Number of animals with lesion/number of animals observed; percent incidence in parentheses.

[3]For each lesion, incidences with no letter in common are significantly different, one-tailed $P < 0.05$ (Fisher's exact test).

Table 11. Incidence and severity of pancreatic nodular hyperplasia in rats fed soy protein isolates for two years

Protein source[1]	Trypsin inhibitor (mg/100g diet)	Nodular hyperplasia[2,3]	Severity[4]
		10% protein	
Soy protein isolate			
Raw, unextracted	358	25/27 (92.6)[a]	2.6[b]
Raw, extracted	70	29/29 (100.0)[a]	2.7[b]
Heated, extracted	32	23/33 (69.7)[bc]	1.4[d]
Casein	0	12/21 (57.1)[c]	1.1[d]
		30% Protein	
Soy protein isolate			
Raw, unextracted	1075	28/28 (100.0)[a]	4.6[a]
Raw, extracted	206	24/27 (88.9)[ab]	2.0[c]
Heated, extracted	94	17/28 (60.7)[c]	1.2[d]
Casein	0	10/22 (45.5)[c]	1.0[d]

[1]Trial II diets: dietary protein at 10% or 30% from the single source indicated.

[2]Number of animals with lesion/number of animals observed; percent incidence in parentheses.

[3]Incidences with no letter in common are significantly different, one-tailed $P < 0.05$ (Fisher's exact test).

[4]Subjective grading of severity ranged from 1 (minimal) to 5 (extreme). Means with no superscript letter in common are significantly different, $P < 0.05$ (Duncan's multiple range test). SE = ±0.1.

Table 12. Incidence of pancreatic acinar adenoma in rats fed soy protein isolates for two years

Protein source[1]	Trypsin inhibitor (mg/g protein)	Acinar adenoma[2,3] 10% protein	30% protein
Soy protein isolate			
Raw, unextracted	35.8	15/27 (55.6)[bB]	24/28 (85.7)[aA]
Raw, extracted	7.0	14/29 (48.3)[bBC]	6/27 (22.2)[cCD]
Heated, extracted	3.2	6/33 (18.2)[cdD]	1/28 (3.6)[dD]
Casein	0	2/21 (9.5)[cdD]	2/22 (9.1)[cdD]

[1]Trial II diets: dietary protein at 10% or 30% from the single source indicated.

[2]Number of animals with lesion/number of animals observed; percent incidence in parentheses.

[3]Incidences for any two diets with no letter in common are significantly different ($P < 0.05$): one-tailed, lower case; two-tailed, upper case (Fisher's exact test).

Table 13. Effect of dietary protein level at low and high trypsin inhibitor concentration on the incidence of acinar adenoma after two years in rats

Soy preparation[1]	Trypsin inhibitor (mg/g protein)	Acinar adenoma[2,3] 10% protein	30% protein
	Low trypsin inhibitor		
Toasted flour (I)	16.0	18/33 (54.5)[a]	11/34 (32.4)[a]
Over-toasted flour (I)	9.3	8/35 (22.9)[a]	2/32 (6.3)[a]
Raw extracted isolate (II)	7.0	14/29 (48.3)[a]	6/27 (22.2)[b]
Heated extracted isolate (II)	3.2	6/33 (18.2)[a]	1/28 (3.6)[a]
	High trypsin inhibitor		
Raw flour (I)	57.7	20/31 (64.5)[a]	27/33 (81.8)[4a]
Raw unextracted isolate (II)	35.8	15/27 (55.6)[a]	24/28 (85.7)[b]
Raw unextracted isolate (I)	29.4	13/34 (38.2)[a]	23/26 (88.5)[b]

[1]Dietary protein at 10% or 30% from the single source indicated; data from Trial I or II indicated in parentheses.

[2]Number of animals with lesion/number of animals observed; percent incidence in parentheses.

[3]Incidences of AA for a given soy preparation with no letter in common between protein concentrations are significantly different, one-tailed $P < 0.05$ (Fisher's exact test).

[4]Raw flour fed at 22%, not 30%, protein.

Table 14. Pancreas weights and PER of rats fed raw soy flour supplemented with sulfur amino acids[1]

Diet supplementation	Pancreas weight[2]	PER[2]
	(% body weight)	(unadjusted)
Soy flour		
none	0.45[b]	1.79[c]
0.17% L-cystine	0.76[a]	2.03[bc]
0.25% L-cysteine	0.68[a]	2.12[bc]
0.23% N-acetyl-L-cysteine	0.66[a]	2.16[b]
Casein control	0.39[b]	3.40[a]
Standard error	± 0.05	± 0.11

[1]Weanling male Sprague-Dawley rats (five per group) were fed standard PER diets (A.O.A.C., 1975) containing 10% protein provided by raw soy flour (47.2mg TI/g protein) for 28 days (Friedman and Gumbmann, unpublished results).

[2]Means with no letter in common are significantly different, $P < 0.05$ (Duncan's multiple range test).

be expected, increasing the protein content of the diets to 30% resulted in no increase, but tended to decrease the incidence of AA, even though the total TI content of the diet tripled. Examples in the high TI region (to the right of the intersection) include raw soy flour and raw, unextracted soy protein isolates. At these elevated TI concentrations, the incidence of AA increased when the amount of soy preparation in the diet was increased.

Protein nutrition. The basis for the observed TI-protein interaction is to be found in nutritional considerations governing the operation of the hormonal feedback mechanism which regulates normal pancreatic function. It is suggested that the degree to which the pancreas is capable of responding to the hormonal signal resulting from a given TI challenge depends upon the status of protein nutrition, especially with regard to the adequacy of the supply of sulfur amino acids which are limiting in soy.

Early investigators supplemented raw soy diets with methionine and other amino acids to test whether correcting the nutritional deficiency of raw soy might prevent hypertrophy of the pancreas (Booth et al., 1960). Although growth was improved, pancreatic hypertrophy was not prevented. Instead, their data show that pancreas weights increased slightly after supplementation, although not significantly.

Figure 18 shows the profound effect of protein level on pancreas size in a short term study where all rats were fed raw soy flour, containing a high level of TI, as the sole source of protein. Marked increases in pancreatic weight accompanied increases in protein level. Supplementation with sulfur amino acids significantly augmented additional pancreatic growth after sufficient protein for basic needs had been supplied. In a second study, striking changes in pancreatic enlargement occurred when diets containing raw soy flour furnishing 10% protein were supplemented with moderate levels of L-cystine, L-cysteine, and N-acetyl-L-cysteine (Table 14). Without supplementation, pancreas weight was elevated above that for casein fed rats, but not significantly. With added sulfur amino acids, including the N-acetyl derivative of L-cysteine, hypertrophy was marked. As expected, supplementation also improved protein quality (PER). Similar results were obtained when the experiment in Table 14 was repeated with a different preparation of raw soy flour.

In the long term study of Trial II, pancreatic hypertrophy was increased in rats fed 30% protein over those fed 10%, especially at the high TI concentration provided by raw, regular process soy protein isolate (Table 5).

The response of the rat to TI, at least initially, is an adaptive one in which the capacity of the pancreas to produce digestive enzymes to compensate for those lost through TI binding is increased; a response requiring an increased rate of protein synthesis and resulting in pancreatic hypertrophy. The hypothesis is proposed that the extent to which pancreatic hypertrophy, including hypertrophy and hyperplasia at the cellular level, develops as a result of a given exposure to dietary TI may be limited when the protein nutrition required to bring about these changes is less than optimal. Thus, the greater the TI dosage, the greater the need for adequate protein nutrition to achieve the full pancreatic response dictated by the hormonal feed-back, regulatory mechanism. Since the magnitude and duration of these initial pancreatic changes are believed to be directly related to the rate at which pancreatic adenoma is eventually expressed, this provides an explanation for the lesser slope of the dose-response curve for the incidence of AA when protein in the diet is low and TI concentrations are high (Figure 17).

Cholecystokinin. An additional factor is needed to understand why, at low TI levels, the incidence of pancreatic adenoma is greater when dietary protein is low. One possible explanation has been provided by Green et al. (1985, 1986), who have taken the protein-TI interaction one step further to study changes at the hormonal level. These investigators administered a single oral dose of concentrated soybean TI to rats adapted to diets either low (5% casein) or high (24% casein) in protein. In both groups after the TI challenge, plasma cholecystokinin (CCK) became similarly elevated. However, for rats fed a low level of protein, elevated CCK persisted, whereas in rats receiving a high protein diet, plasma CCK soon returned to control levels. Presumably, the pancreas in rats fed a restricted protein diet was unable to respond fully to the TI challenge with sufficiently increased enzyme production and secretion and, thereby, failed to turn off the hormonal feedback control system. This resulted in elevated CCK homeostasis. The limited pancreatic response to CCK was reflected by lower pancreatic weight in rats fed the restricted protein diet. Reduced pancreatic activity from experimentally induced pancreatic atrophy, rather than through protein restriction of dietary protein, also has been shown to result in elevated plasma CCK concentrations (Fölsch et al., 1983).

Thus, with protein containing elevated TI, considerable hormonal stimulation of the pancreas is expected. However, when protein is chronically limited, the pancreas is unable to respond fully to the large hormonal signal to become hyperactive, and the eventual incidence of AA is less than that associated with diets more adequate in protein and containing even greater total TI. At low TI concentrations where hormonal stimulation of the pancreas should be minimal, the factor of CCK homeostasis in rats fed sub-optimal levels of protein would tend to prolong pancreatic stimulation, keeping pancreatic response at the maximum possible comensurate with the low TI challenge. For rats fed higher protein levels, also at low TI concentrations, CCK release should be limited and not prolonged, thereby eliciting little pancreatic response to the low dietary concentration of TI, and hence, a relatively lower incidence of AA.

To what extent raw soy protein devoid of TI may also trigger excessive CCK release and, thus, contribute to the relatively greater production of AA in rats fed a low protein diet is not clear. The data suggest that at low dietary protein levels, the lower protein quality of soy protein isolate, as compared to soy flour, accentuates the effect of low protein in the diet, particularly when raw. However, with 30% protein in the diet, the dose-response relationship demonstrated in Trials I and II (Figure 16) does not support the idea that raw soy protein, per se, was a factor in the development of AA. A comparison of actual incidence data for AA associated with heated, low TI soy protein isolate, which was fed at 30% protein in Trial I, shows remarkable similarity to that of raw, low TI, extracted soy protein isolate in Trial II, also at 30% protein. AA with heated isolate (6.1 mg TI/g protein) was 18.2% (Table 10) and with raw isolate (7.0mg TI/g protein) 22.2% (Table 12).

Food Safety Considerations

Residual TI. Proteins that are potent inhibitors of pancreatic proteinases are widely distributed in man's food supply and include important dietary sources such as legumes, cereals, and potatoes (Liener and Kakade, 1980; Rackis and Gumbmann, 1981; Ryan and Hass, 1981).

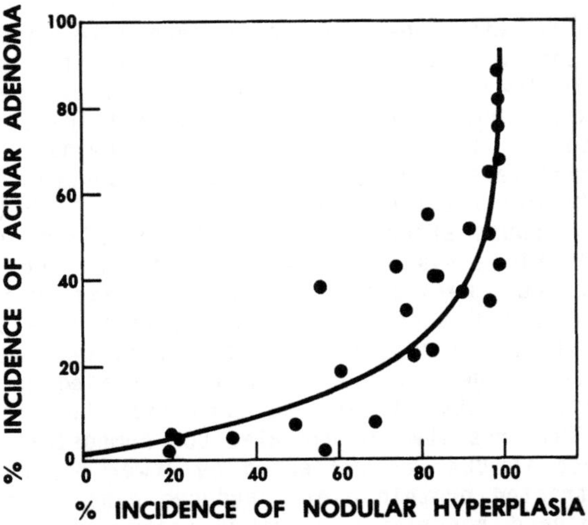

Figure 13. Relationship of the incidence of pancreatic acinar adenoma to that of nodular hyperplasia. Plotted are data for the 26 diets of Trial I.

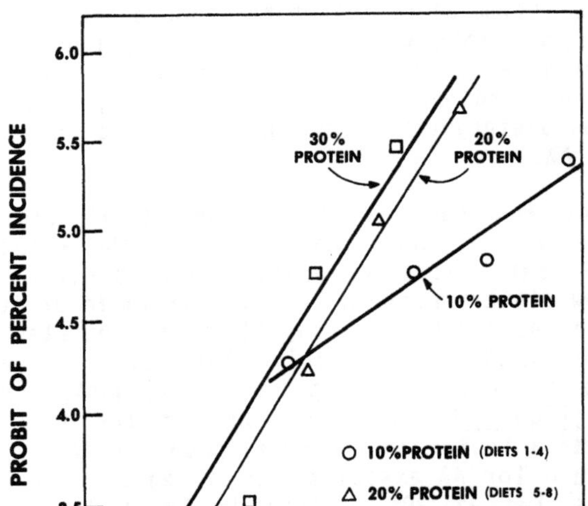

Figure 14. Probit of percent incidence of pancreatic acinar adenoma plotted against log of trypsin inhibitor concentration of soy flour diets expressed as mg/g protein (incidence data from Table 8).

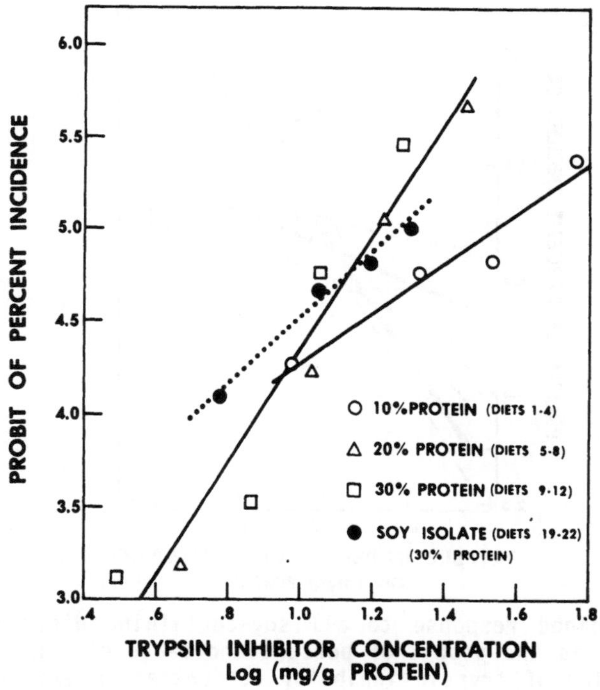

Figure 15. Comparison of the response to soy protein isolate and soy flour (open points) of Trial I. Plotted is the probit of percent incidence of pancreatic acinar adenoma versus log of trypsin inhibitor concentration expressed as mg/g protein in the diet (incidence data from Tables 8 and 10).

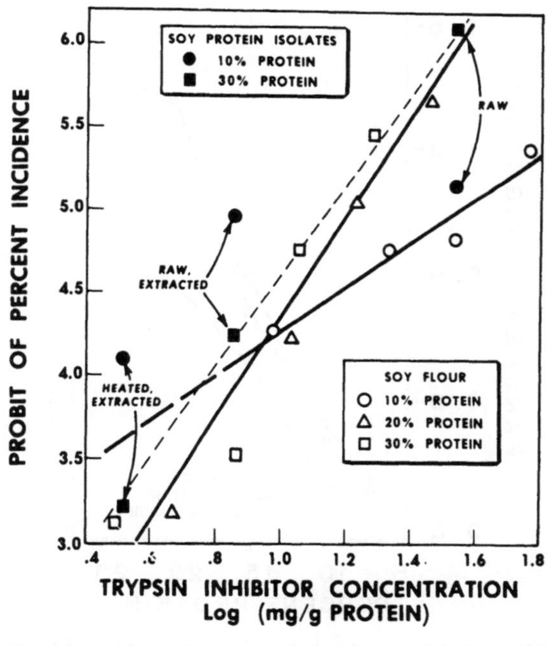

Figure 16. Comparison of the reponse to extracted soy protein isolate of Trial II and soy flour of Trial I. Plotted is the probit of percent incidence of pancreatic acinar adenoma versus log of trypsin inhibitor concentration expressed as mg/g protein in the diet (incidence data from Tables 8 and 12). The best fitting curves are: soy flour diets, solid lines; soy protein isolate diets at 30% protein, dashed line.

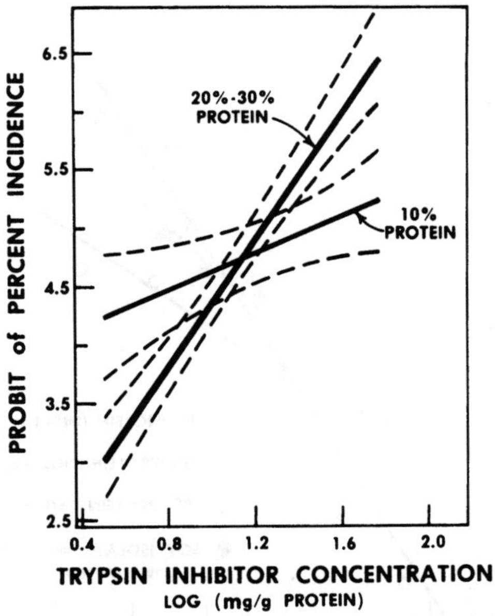

Figure 17. Combined response to all soy-containing diets of Trials I and II. Plotted is the probit of percent incidence of pancreatic acinar adenoma versus log of trypsin inhibitor concentration expressed as mg/g protein in the diet for 23 dietary groups of Trial I and 6 of Trial II. The broken curves indicate 95% confidence intervals.

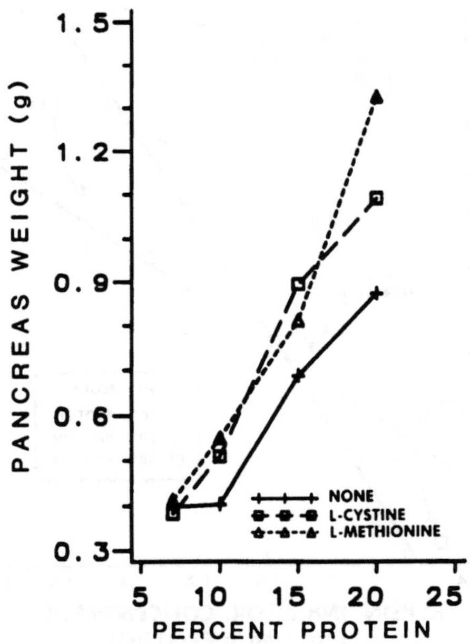

Figure 18. Pancreas weights in rats fed raw soy flour (47.2mg TI/g protein) as a function of dietary protein level and sulfur amino acid supplementation (0.3% in the diet). Standard PER diets (A.O.A.C., 1975), supplemented with protein and amino acids at the expense of cornstarch, were fed to weanling male Sprague-Dawley rats for 28 days.

While a large portion of inhibitor activity in food is heat-labile and destroyed during processing, residual levels of significant physiological importance remain. As shown in Trial I, the incidence of pancreatic adenoma was demonstrated to be greater in rats fed over-toasted soy flour relative to those fed casein in diets containing 10% protein (Table 8). Further, the dose-response relationship to TI indicated no particular threshold, other than that dictated by the experimental sensitivity inherent in the design of the studies. It may be concluded that low levels of residual TI equal to those encountered in commercially prepared soy products can be expected to increase the probability of pancreatic adenoma formation in the rat. With diets of low protein content, inferior protein quality, as found in raw soy, is also a factor.

Hormonal factors. These studies support the view that pancreatic neoplasia results from a progression of histological changes, beginning with those associated with the normal adaptive response of the pancreas to dietary components, including TI. Such response includes the ability of the pancreas to change its rates of both cellular division and enzyme synthesis (Dijkhof and Poort, 1978; Solomon et al., 1981). The feed-back control mechanism, mediated by CCK and other gastro-intestinal hormones, for pancreatic regulation in mammals provides the basis for explaining the observed pancreatic changes in the rat, such as organ enlargement, cellular hypertrophy and hyperplasia, and stimulated production of pancreatic enzymes (Green and Lyman, 1972; Green et al., 1973; Brand and Morgan, 1981; Rackis and Gumbmann, 1981; Fölsch et al., 1983).

Underlying biochemical changes associated with protein synthesis and cell division and used as indices for the effect of TI and raw soy on the pancreas in the rat include increases in DNA synthesis, cellular DNA content, protein-DNA ratio, and rate of thymidine incorporation into pancreatic DNA (Melmed et al., 1976; Oates and Morgan, 1982). Regarding these indices, the trophic action of CCK and its analogs on the pancreas can essentially reproduce the firmly established alterations brought about by raw soy in the diet, including hyperplasia (Mainz et al., 1973; Yanatori and Fujita, 1976; Solomon et al., 1981; Wood et al., 1984; Temler et al., 1982, 1984). In addition to increased DNA, RNA, and protein, other determinants of the trophic effect of CCK on the pancreas are elevations in the pancreatic concentrations of ornithine decarboxylase and putrescine, known to accompany the early stages of cellular proliferation (Haarstad et al., 1985). In man, plasma CCK is significantly increased after ingestion of food, and intravenously infused CCK at physiological concentrations has been shown to exert a trophic effect on the pancreas resulting in increased protease secretion (Jansen et al., 1983; Anagnostides et al., 1985; Variyam et al., 1985). Although such changes represent an adaptive response to diet through normal physiological control mechanisms, the appearance of nodular hyperplasia and the neoplastic lesion, acinar adenoma, after chronic pancreatic stimulation, is distinctly pathological and not necessarily to be predicted by our understanding of pancreatic function alone.

The framework for understanding this phenomenon lies in the field of oncology. The appearance of the lesions NH and AA is consistent with the generally accepted fact that cell populations under constant or repeated stimulation to divide may well be at increased risk of becoming neoplastic. The evidence suggests that tumor production in response to raw soy is mediated by a tumor promoting effect of CCK and perhaps other intestinal hormones, in which CCK produces the cellular environment favoring the growth and transformation of previously initiated cells. The initiated cells represent a background population of unknown origin. Even so, such development of neoplasia cannot be held as inevitable; an animal's average life-span, for example, may be insufficient to permit expression of the lesion, in which case the problem would be of no practical concern for the species. Generalized organ hyperplasia in response to hormonal stimulation is a common, reversible adaptive mechanism in many tissues, the mammary gland during pregnancy for example, and is not expected to lead to neoplasia. The mechanism for the transition from pancreatic hyperplasia to the development of overt neoplasia is unknown, and the ease by which it occurs can be expected to vary among species.

As a matter of definition, TI and, in turn, CCK may be viewed as pancreatic tumor promotors. But even a more immediate factor governing the development of neoplasia, as supported by our data, is the actual amount of pancreatic alteration, hypertrophic and hyperplastic change, that has taken place. In this regard, intermittent and repeated pancreatic stimulation has been shown by McGuinness et al. (1982) to be at least as effective as sustained levels of TI in the diet of rats for the development of pancreatic neoplasia. Thus, the number of times that a cell population is induced to divide is an important factor. It is not surprising, therefore, that pancreatic acinar tumorigenesis (production of hyperplastic nodules and adenocarcinomas) induced by azaserine was found to be enhanced by pancreatic regeneration following partial pancreatectomy in the rat (Konishi et al., 1980).

<u>Species variation</u>. It is of paramont importance to know whether the rat is unusually susceptible to the development of pancreatic NH and AA when the pancreas is stimulated through feedback mechanisms designed to regulate pancreatic function. Spontaneous NH is a common finding in older rats (Altman and Goodman, 1979). Work reported by Liener and Hasdai (1986, this volume) would indicate that rats are, indeed, more susceptible to neoplasia than hamsters and mice when fed raw soy flour for extended periods. A similar species difference between rats and mice was noted in the responsiveness to induction of pancreatic adenocarcinomas by azaserine (Roebuck et al., 1980). It should be noted that the degree and persistence of early DNA damage which resulted from azaserine treatment failed to predict the relative carcinogenic potential among the rat, mouse, Syrian golden hamster, and guinea pig. It would be important to know whether the rat pancreas, which is structurally unstable and capable of marked reversible changes in weight, DNA, RNA, and protein content in response to dietary stimulus (Crass and Morgan, 1981), is typical of that of other species. Perhaps viral or genetic factors predispose the laboratory rat to the development of pancreatic pathology and neoplasia.

The near exclusive use of the rat as a model for the expression of chronic effects of TI containing diets has not supplied the breadth of information required to make a useful safety assessment of residual TI in man's diet. In short-term studies with hamsters, a negative feedback regulation of the pancreas has been demonstrated, in which CCK and orally administered TI were both found to exert trophic effects on the pancreas (Andren-Sanberg and Ihse, 1983). In this species, CCK stimulated pancreatic secretory response, caused hypertrophy and hyperplasia of the pancreas and enhanced pancreatic carcinogenesis after treatment with a tumorigenic nitrosamine (Howatson and Carter, 1985). A negative feedback regulation of pancreatic secretion by gastro-intestinal hormones also has been investigated in pigs (Bellili et al., 1983; Charbonneau et al., 1982; Corring et al., 1985).

Although evidence is less for humans, such a system for the control of pancreatic exocrine secretion apparently exists (Toskes, 1980; Dlugosz et al., 1982; Beglinger et al., 1985), and it must be presumed that all stages of pancreatic change similar to those observed in the rat model can occur, including the development of overt pancreatic pathology. As indicated in a limited series of autopsies (Longnecker et al., 1980), focal acinar cell lesions resembling NH in the rat may be a common occurrence in man; however, if they have pre-neoplastic potential, at most only a small minority of such foci progress to malignancy. In contrast to the neoplasia observed in the rat pancreatic model, which is presumed to be derived from acinar tissue, the histogenesis of human pancreatic cancer, including hyperplastic and preneoplastic lesions of the pancreas, appears to be mainly of ductular cell origin (Rao and Reddy, 1985; Pour et al., 1979, 1982; Parsa et al., 1985).

The likelihood that common dietary levels of TI significantly enhance the frequency of pancreatic neoplasia in man's lifetime, either in the presence or absence of predisposing environmental factors such as exposure to primary pancreatic carcinogens, remains to be determined. The critical factor is the rate at which one stage may progress to the next. The rate of formation of NH in the rat given even mild TI stimulation is considerable, and the probability of developing AA within the life-span of the rat given this incident rate of NH is significant (approximately 0.4% of that of NH squared in two years) <u>(Figure 13)</u>. Whether other species resemble the rat in their propensity to develop pancreatic hyperplasia and neoplasia or whether they show greater pancreatic stability when subjected to dietary or hormonal stimulation is vital knowledge for predicting the safety of TI in man's diet.

<u>Epidemiology</u>. Pancreatic cancer in man is common in western countries and increasing in frequency in the United States (Fraumeni, 1975). Epidemiologic studies have identified many sub-populations showing positive correlations with incidence, including racial differences denoting a complex etiology (Wynder, 1975; Fraumeni, 1975; Kato et al., 1985). Although environmental factors are strongly suggested, genetic predisposition may not be ruled out (Gordis and Gold, 1984). As with rats, a hormonal involvement in the development of pancreatic cancer is evident, in that the incidence of pancreatic cancer in men exceeds that of women. The major nutritional factor which has been associated with increased rates of spontaneous pancreatic cancer in man is elevated dietary fat. Fat in the form of corn oil also has been implicated as a factor in the rat in large, long-term studies conducted by the National Toxicology Program. In these studies, corn oil-treated male animals developed a significantly greater incidence of proliferative acinar cell lesions of the pancreas compared to untreated controls (Food Chem. News, 1983a, 1983b; Haseman et al., 1985). Any correlation of pancreatic cancer in man to TI intake must await reliable information on dietary levels of TI encountered in sub-populations.

<u>Future research</u>. Future research concerned with the food safety implications of protease inhibitors in the diet should emphasize the determination of species variability regarding pancreatic regulation at the hormonal level and promotion of pancreatic pathology by TI. Since man's diet contains protease inhibitors from many sources, the equivalency of action on the pancreas for these inhibitors also needs to be established. Three approaches, in which some work is beginning to be reported, are suggested:

1. Hormonal response to TI. Determine whether there is an equivalent endocrine response to TI among species, including primates and man, after oral challenges of TI by measuring CCK and perhaps other hormonal levels in blood. Determine whether there is a difference in response of the rat to different sources of TI, and whether differences can be related to in vitro studies of protease binding affinity. Differences in the short-term adaptive response of the rat pancreas to dietary egg white and soy TIs have been noted, for example. Although both induced the synthesis of proteases, the overall pattern of enzyme induction was not identical (Snook, 1969). It remains to be shown that soy TI has any unique effects regarding the development of pancreatic pathology compared to other sources of dietary TI.

2. Pancreatic response to CCK and other gastrointestinal hormones. Compare the effects of administered CCK on pancreatic response using early indices of pancreatic changes leading to hyperplasia and hypertrophy for several species, including primates and possibly man. Determine whether the progression of events observed in the rat pancreas can be duplicated in other species. Proglumide, which is a specific gastrin and CCK receptor antagonist, may be a valuable tool in studying the relative pancreatic response to CCK peptides and their binding affinities among species (Fried et al., 1983; Jensen et al., 1985; Yamaguchi et al., 1985). Use of this antagonist has provided evidence that CCK is the major mediator of the intestinal phase of exocrine pancreatic secretion in dogs (Stubbs and Stabile, 1985).

3. TI related pancreatic pathology. Demonstrate the relative propensity among species for the development of overt pancreatic pathology, defined histologically; alterations which lie beyond those generally believed to be part of the normal adaptive response. Determine whether pre-neoplastic lesions similar to NH, which appear in rats after a few months, can be promoted by TI in other common laboratory animals, pigs, and primates.

Conclusion. Without a full understanding of oncogenic mechanisms, caution is necessary in making judgments based upon tumor pathogenesis alone in a single species. Unlike primary carcinogens, much evidence indicates that TI indirectly promotes pancreatic tumor formation in the rat and that the phenomenon is mediated by and under control of the animal's endocrine system. For TI, there is no evidence of absorption from the gastro-intestinal tract, direct neoplastic action or tumor induction, genotoxicity, interaction with cellular genetic material, or epidemiological indication of potential risk in man. Such a profile on chemicals in general would provide little basis for severely restricting exposure.

Most toxicologists would accept that it would be prudent to avoid environmental factors which produce an undue proliferative response in a tissue. Of most importance, then, is whether the normal pancreatic regulation of pancreatic function in man produces the pattern of changes observed in the rat, both qualitatively and quantitatively, especially including those leading to cellular proliferation.

ACKNOWLEDGEMENTS

The authors wish to ackowledge Jim Bowie and Sal Torres for care of the animals and Feli Bautista, MacDonald Calhoun, Sandy Heier, Billie

Reiydelle, and Lucille Rossi for necropsy, histology, and data work-up. We thank Dr. Bruce Mackey for his advice and assistance with the statistical analyses. Special thanks go to Dr. Mendel Friedman for his encouragement and advice during the preparation of this manuscript.

REFERENCES

Adelson, J. W. and Miller, P. E. (1985). Pancreatic secretion by nonparallel exocytosis: Potential resolution of a long controversy. Science, 228, 993-996.

Altman, N. H. and Goodman, D. G. (1979). Neoplastic diseases. In: "The Laboratory Rat, Vol. I, Biology and Diseases", H. J. Baker, J. R. Lindsey, and S. H. Weisbroth, eds., Academic Press, New York, New York, pp. 333-376.

Anagnostides, A. A., Chadwick, V. S., Selden, A. C., Barr, J., and Maton, P. N. (1985). Human pancreatic and biliary responses to physiological concentrations of cholecystokinin octapeptide. Clin. Sci. (Lond.), 69, 259-263.

Andren-Sandberg, A. and Ihse, I. (1983). Regulatory effects on the pancreas of intraduodenal pancreatic juice and trypsin in the Syrian golden hamster. Scand. J. Gastroenterol., 18, 697-706.

A.O.A.C. (1975). "Official Methods of Analysis", 12th ed. Assoc. Offic. Anal. Chem., Washington, D.C., p. 857.

Bellili, C., Ormas, P., Cissokho, S., and Beretta, C. (1983). The effects of caerulein on exocrine pancreatic secretion in pigs. Vet. Res. Commun., 6, 43-50.

Beglinger, C., Fried, M., Whitehouse, I., Jansen, J. B., Lamers, C. B., and Gyr, K. (1985). Pancreatic enzyme response to a liquid meal and to hormonal stimulation. J. Clin. Invest., 75, 1471-1476.

Bliss, C. I. (1967a). Analysis of fourfold tables. In: "Statistics in Biology, Vol. I", McGraw-Hill Book Co., New York, New York, pp. 53-91.

Ibid. (1967b). Graphic Tests for Normality. pp. 101-107.

Boorman, G. A. and Eustis, S. L. (1984). Proliferative lesions of the exocrine pancreas in male F344/N rats. Environ. Health Perspectives, 56, 213-217.

Booth, A. N., Robbins, D. J., Ribelin, W. E., and DeEds, F. (1960). Effect of raw soybean meal and amino acids on pancreatic hypertrophy in rats. Proc. Soc. Exp. Biol. Med., 104, 681-683.

Brand, S. J. and Morgan, R. G. H. (1981). The release of rat intestinal cholecystokinin after oral trypsin inhibitor measured by bio-assay. J. Physiol. (Lond.), 319, 325-343.

Charbonneau, P., Pelletier, G., and Morisset, J. (1982). Development of the pancreas during gestation and lactation in swine. Can. J. Physiol. Pharmacol., 60, 1229-1235.

Corring, T., Chayvialle, J. A., Simoes-Nunes, C., and Abello, J. (1985). Regulation of pancreatic secretion in the pig by negative feedback and plasma gastrointestinal hormones. Reprod. Nutr. Develop., 25, 439-450.

Crass, R. A. and Morgan, R. G. H. (1981). Rapid changes in pancreatic DNA, RNA and protein in the rat during pancreatic enlargement and involution. Internat. J. Vit. Nutr. Res., 51, 85-91.

Crass, R. A. and Morgan, R. G. H. (1982). The effect of long-term feeding of soya-bean flour diets on pancreatic growth in the rat. Br. J. Nutr., 47, 119-129.

Dijkhof, J. and Poort C. (1978). Changes in rat pancreatic protein synthesis after a single feeding with diets containing raw or heated soybeans. J. Nutr., 108, 1222-1228.

Dlugosz, J., Fölsch, U. R., and Creutzfeldt, W. (1982). Effect of intraduodenal inhibition of trypsin on pancreatic exocrine secretion in man. Digestion, 25, 24.

Doell, B. H., Ebden, C. J., and Smith, C. A. (1981). Trypsin inhibitor activity of conventional foods which are part of the British diet and some soya products. Qual. Plant. Plant Foods Hum. Nutr., 31, 139-150.

Duncan, D. B. (1955). Multiple range and multiple F tests. Biometrics, 11, 1-42.

Fölsch, U. R., Schafmayer, A., Becker, H. D., and Creutzfeldt, W. (1983). Elevated plasma CCK concentrations in exocrine pancreatic atrophy in the rat. Digestion, 28, 27.

Fölsch, U. R., Wormsley, K. G. (1974). The pancreatic secretion of enzymes in rats treated with soybean diet. Scand. J. Gastroenterol., 9, 679-683.

Food Chem. News. (1983a). Corn oil gavage bioassay results questioned by NTP scientist. February 28, 27-30.

Food Chem. News. (1983b). NTP approves 4-year studies on role of corn oil in gavage tests. October 3, 39-41.

Fraumeni, J. F. (1975). Cancers of the pancreas and biliary tract: Epidemiological considerations. Cancer Res., 35, 3437-3446.

Fried, M., Beglinger, C., Whitehouse, I., and Gyr, K. (1983). Effect of proglumide, a CCK receptor antagonist, on caerulein-induced pancreatic enzyme secretion and plasma PP release. Digestion, 28, 28.

Geratz, J. D. (1968). Growth retardation and pancreatic enlargment in rats due to p-aminobenzamidine. Am. J. Physiol., 214, 595-600.

Geratz, J. D. and Hurt, J. P. (1970). Regulation of pancreatic enzyme levels by trypsin inhibitors. Am. J. Physiol., 219, 705-711.

Giorgi, D., Renaud, W., Bernard, J. P., and Dagorn, J. C. (1985). Regulation of proteolytic enzyme activities and mRNA concentrations in rat pancreas by food content. Biochem. Biophys. Res. Commun., 127, 937-942.

Goebell, H., Schmitz-Moormann, P., Pfannkuch, U., and Wiechmann, A. (1983). Hypertrophy of the exocrine pancreas: Induction by oral application of the synthetic trypsin inhibitor Foy 305 in rats. Digestion, 28, 31.

Gordis, L. and Gold, E. B. (1984). Epidemiology of pancreatic cancer. World J. Surg., 8, 808-821.

Green, G. M., Levan, V. H., and Liddle, R. A. (1986). Interaction of dietary protein and trypsin inhibitor on plasma cholecystokinin and pancreatic growth in rats. This volume.

Green, G., Levan, V., and Liddle, R. (1985). Homeostasis of plasma cholecystokinin (CCK) in rats. Fed. Amer. Soc. Exp. Biol. Proc., 44, xii.

Green, G. M. and Lyman, R. L. (1972). Feedback regulation of pancreatic enzyme secretion as a mechanism for trypsin inhibitor-induced hypersecretion in rats. Proc. Soc. Exp. Biol. Med., 140, 6-12.

Green, G. M., Olds, B. A., Matthews, G., and Lyman, R. L. (1973). Protein, as a regulator of pancreatic enzyme secretion in the rat. Proc. Soc. Exp. Biol. Med., 142, 1162-1167.

Grizzle, J. E., Starmer, C. F., and Koch, G. G. (1969). Analysis of categorical data by linear models. Biometrics, 25, 489-504.

Gumbmann, M. R., Spangler, W. L., Dugan, G. M., Rackis, J. J., and Liener, I. E. (1985). The USDA trypsin inhibitor study. IV. The chronic effects of soy flour and soy protein isolate on the pancreas in rats after two years. Qual. Plant. Plant Foods Hum. Nutr., 35, 275-314.

Haarstad, H., Winnberg, A., and Petersen, H. (1985). Effects of a cholecystokinin-like peptide on DNA and polyamine synthesis in the rat pancreas. Scand. J. Gastroenterol., 20, 530-538.

Hamerstrand, G. E., Black, L. T., and Glover, J. D. (1981). Trypsin inhibitors in soy products: Modification of the standard analytical procedure. Cereal Chem., 58, 42-45.

Harper, A. A. and Scratcherd, T. (1979). Physiology. In: "The Exocrine Pancreas", H. T. Howat and H. Sarles, eds., W. B. Saunders Company Ltd., London, pp 50-85.

Haseman, J. K., Huff, J. E., Rao, G. N., Arnold, J. E., Boorman, G. A., and McConnell, E. E. (1985). Neoplasms observed in untreated and corn oil gavage control groups of F344/N rats and (C57BL/6N X C3H/HeN) F_1 ($B6C3F_1$) mice. J. Natl. Cancer Inst., 75, 975-984.

Howatson, A. G. and Carter, D. C. (1985). Pancreatic carcinogenesis-enhancement by cholecystokinin in the hamster-nitrosamine model. Br. J. Cancer, 51, 107-114.

Jansen, J., Kerstens, P., Welberg, J., Hessels, M., Hafkenscheid J., and Lamers, C. (1983). Physiological plasma concentrations of cholecystokinin stimulate pancreatic enzyme secretion in man. Digestion, 28, 37.

Jensen, R. T., Murphy, R. B., Trampota, M., Schneider, L. H., Jones, S. W., Howard, J. M., and Gardner, J. D. (1985). Proglumide analogs potent cholecystokinin receptor antagonists. Am. J. Physiol., 249, G214-G220.

Kakade, M. L., Hoffa, D. E., and Liener, I. E. (1973). Contribution of trypsin inhibitors to the deleterious effects of unheated soybeans fed to rats. J. Nutr., 103, 1772-1778.

Kakade, M. L., Rackis, J. J., McGhee, J. E., and Puski, G. (1974). Determination of trypsin inhibitor activity of soy products: A collaborative analysis of an improved procedure. Cereal Chem., 51, 376-382.

Kakade, M. L., Simons, N., and Liener, I. E. (1969). An evaluation of natural vs. synthetic substrates for measuring the antitryptic activity of soybean samples. Cereal Chem., 46, 518-526.

Kato, I., Tajima, K., Kuroishi, T., and Tominaga, S. (1985). Latitude and pancreatic cancer. Jpn. J. Clin. Oncol., 15, 403-414.

Konishi, Y., Denda, A., Maruyama, H., Yoshimura, H., Nobuoka, J., and Sunagawa, M. (1980). Pancreatic tumors induced by a single intraperitoneal injection of azaserine in partial pancreatectomized rats. Cancer Lett., 9, 43-46.

Levison, D. A., Morgan, R. G. H., Brimacombe, J. S., Hopwood, D., Coghill, G., and Wormsley, K. G. (1979). Carcinogenic effects of di(2-hydroxypropyl) nitrosamine (DHPN) in male Wistar rats: Promotion of pancreatic cancer by a raw soya flour diet. Scand. J. Gastroenterol., 14, 217- 224.

Liener, I. E. (1972). Nutritional value of food protein products. In: "Soybeans: Chemistry and Technology", A. K. Smith and S. J. Circle, eds., AVI Publishing Co., Westport, Connecticut, pp. 203.

Liener, I. E. and Hasdai, A. (1986). The effect of long-term feeding of raw soy flour on the pancreas of the mouse and hamster. This volume.

Liener, I. E. and Kakade, M. L. (1980). Protease inhibitors. In: "Toxic Constituents of Plant Foodstuffs", I. E. Liener, ed., Academic Press, New York, New York, pp. 7-71.

Liener, I. E., Nitsan, Z., Srisangnam, C., Rackis, J. J., and Gumbmann, M. R. (1985). The USDA trypsin inhibitor study. II. Timed related biochemical changes in the pancreas of rats. Qual. Plant. Plant Foods Hum. Nutr., 35, 243-257.

Longnecker, D. S., Roebuck, B. D., Yager, J. D., Lilja, H. S., and Siegmund, B. (1981). Pancreatic carcinoma in azaserine-treated rats: Induction, classification and dietary modulation of incidence. Cancer, 47, 1562-1572.

Longnecker, D. S., Shinozuka, H., and Dekker, A. (1980). Focal acinar cell dysplasia in human pancreas. Cancer, 45, 534-540.

Lyman, R. L. (1957). The effect of raw soybean meal and trypsin inhibitor diets on the intestinal and pancreatic nitrogen in the rat. J. Nutr., 62, 285-294.

Mainz, D. L., Black, O., and Webster, P. D. (1973). Hormonal control of pancreatic growth. J. Clin. Invest., 52, 2300-2304.

McGuinness, E. E., Hopwood, D., and Wormsley, K. G. (1982). Further studies of the effects of raw soya flour on the rat pancreas. Scand. J. Gastroenterol., 17, 273-277.

McGuinness, E. E., Morgan, R. G. H., Levison, D. A., Frape, D. L., Hopwood, D., and Wormsley, K. G. (1980). The effects of long-term feeding of soya flour on the rat pancreas. Scand. J. Gastroenterol. 15, 497-502.

McGuinness, E. E., Morgan, R. G. H., and Wormsley, K. G. (1984). Effects of soybean flour on the pancreas of rats. Environ. Health Perspectives, 56, 205-212.

Melmed, R. N. and Bouchier, I. A. D. (1969). A further physiological role for naturally occurring trypsin inhibitors: The evidence for a trophic stimulant of the pancreatic acinar cell. Gut, 10, 973-979.

Melmed, R. N., El-Aaser, A. A. A., and Holt, S. J. (1976). Hypertrophy and hyperplasia of the neonatal rat exocrine pancreas induced by orally administered soybean trypsin inhibitor. Biochim. Biophys. Acta, 421, 280-288.

Morgan, R. G. H., Levinson, D. A., Hopwood, D., Saunders, J. H. B., and Wormsley, K. G. (1977). Potentiation of the action of azaserine on the rat pancreas by raw soya bean flour. Cancer Lett., 3, 87-90.

Oates, P. S. and Morgan, R. G. H. (1982). Pancreatic growth and cell turnover in the rat fed raw soy flour. Am. J. Pathol., 108, 217-224.

Parsa, I., Longnecker, D. S., Scarpelli, D. G., Pour, P., Reddy, J. K., and Lefkowitz, M. (1985). Ductal metaplasia of human exocrine pancreas and its association with carcinoma. Cancer Res., 45, 1285-1290.

Peto, R. (1974). Guidelines on the analysis of tumour rates and death rates in experimental animals. Br. J. Cancer, 29, 101-105.

Pour, P., Salmasi, S. Z., and Runge, R. G. (1979). Ductular origin of pancreatic cancer and its multiplicity in man comparable to experimentally induced tumors. A preliminary study. Cancer Lett., 6, 89-97.

Pour, P. M., Sayed, S., and Sayed, G. (1982). Hyperplastic, preneoplastic and neoplastic lesions found in 83 human pancreases. Am. J. Clin. Pathol., 77, 137-152.

Rackis, J. J. (1965). Physiological properties of soybean trypsin inhibitor and their relationship to pancreatic hypertrophy and growth inhibition of rats. Fed. Proc., 24, 1488-1493.

Rackis, J. J. and Gumbmann, M. R. (1981). Protease inhibitors: Physiological properties and nutritional significance. In: "Antinutrients and Natural Toxicants in Foods", R. L. Ory, ed., Food and Nutrition Press, Inc., Westport, Connecticut, pp. 203-237.

Rackis, J. J., Gumbmann, M. R., and Liener, I. E. (1985). The USDA trypsin inhibitor study. I. Background, objectives, and procedural details. Qual. Plant. Plant Foods Hum. Nutr., 35, 213-242.

Rackis, J. J., McGhee, J. E., and Booth, A. N. (1975). Biological threshold levels of soybean trypsin inhibitors by rat bioassay. Cereal Chem., 52, 85-92.

Rao, M. S. and Reddy, J. K. (1985). Induction and differentiation of exocrine pancreatic tumors in the rat. Exp. Path., 28, 67-87.

Roebuck, B. D., Kaplita, P. V., and Macmillan, D. L. (1985). Interaction of dietary fat and soybean isolate (SBI) on azaserine-induced pancreatic carcinogenesis. Qual. Plant. Plant Foods Hum. Nutr., 35, 323-329.

Roebuck, B. D., Lija, H. S., Curphey, T. J. and Longnecker, D. S. (1980). Pathologic and biochemical effects of azaserine in inbred Wistar/Lewis rats and nonibred CD^R-1 mice. J. Nat. Cancer. Inst., 65, 383-389.

Roebuck, B. D., Yager, J. D., Jr., Longnecker, D. S., and Wilpone, S. A. (1981). Promotion by unsaturated fat of azaserine-induced pancreatic carcinogenesis in the rat. Cancer Res., 41, 3961-3966.

Ryan, C. A. and Hass, G. M. (1981). Structural, evolutionary and nutritional properties of proteinase inhibitors from potatoes. In: "Antinutrients and Natural Toxicants in Foods", R. L. Ory, ed., Food and Nutrition Press, Inc., Westport, Connecticut, pp. 169-185.

SAS Instutite, Inc. (1982a). The GLM procedure. In: "SAS User's Guide: Statistics, 1982 Edition", A. A. Ray, ed., SAS Institute Inc., Gary, North Carolina, pp. 139-199.

Ibid. (1982b). The FUNCAT procedure. In: "SAS User's Guide: Statistics, 1982 Edition", A. A. Ray, ed., SAS Institute Inc., Gary, North Carolina, pp. 257-285.

Ibid. (1982c). The PROBIT procedure. In: "SAS User's Guide: Statistics, 1982 Edition", A. A. Ray, ed., SAS Institute Inc., Gary, North Carolina, pp. 287-292.

Schneeman, B. O. and Lyman, R. L. (1975). Factors involved in the intestinal feedback regulation of pancreatic enzyme secretion in the rat. Proc. Soc. Exp. Biol. Med., 148, 897-903.

Senti, F. R. (1982). Annotated bibliographies on pancreatic changes in experimental animals fed soybeans, processed soybean products, soybean trypsin inhibitor, or cholecystokinin-pancreozymin and antinutritional factors in processed soybean products. Report prepared for Bureau of Foods, Food and Drug Administration, Washington, D. C. Contract no. FDA 223-79-2275, Life Sciences Research Office, Federation of American Societies for Experimental Biology, Bethesda, Maryland.

Snook, J. T. (1969). Factors in whole-egg protein influencing dietary induction of increases in enzyme and RNA levels in rat pancreas. J. Nutr., 97, 286-294.

Solomon, T. E., Grossman, M. I., and Williams, J. A. (1981). Recent advances in pancreatic physiology: Summary of a conference. Fed. Proc., 40, 2105-2110.

Spangler, W. L., Gumbmann, M. R., Liener, I. E., and Rackis, J. J. (1985). The USDA trypsin inhibitor study. III. Sequential development of pancreatic pathology in rats. Qual. Plant. Plant Foods Hum. Nutr., 35, 259-274.

Stubbs, R. S. and Stabile, B. E. (1985). Role of cholecystokinin in pancreatic exocrine response to intraluminal amino acids and fat. Am. J. Physiol., 248, G347-G352.

Temler, R. S. (1980). Alterations in the pancreas of rats fed on different levels of soya flour and casein. Internat. J. Vit. Nutr. Res., 50, 212-214.

Temler, R. S., Dormond, C. A., Simon, E., and Morel, B. (1984). The effect of feeding soybean trypsin inhibitor and repeated injections of cholecystokinin on rat pancreas. J. Nutr., 114, 1083-1091.

Temler, R. S., Simon, E., Morel, B., and Dormond, C. (1982). Comparison between the effects on rat pancreas of repeated injections of cholecystokinin and soya bean trypsin inhibitor. Digestion, 25, 72.

Toskes, P. P. (1980). Does a negative feedback system for the control of pancreatic exocrine secretion exist and is it of any clinical significance? J. Lab. Clin. Med., 95, 11-12.

Variyam, E. P., Fuller, R. K., Brown, F. M., and Quallich, L. G. (1985). Effect of parenteral amino-acids on human pancreatic exocrine secretion. Dig. Dis. Sci., 30, 541-546.

Wood, J. G., Dale, W. E., and Solomon, T. E. (1984). Effects of chronic administration of CCK on pancreas of the dog and rat. Gastroenterol., 86, 1302.

Wynder, E. L. (1975). An epidemiological evaluation of the causes of cancer of the pancreas. Cancer Res., 35, 2228-2233.

Yamaguchi, T., Tabata, K., and Johnson, L. R. (1985). Effect of proglumide on rat pancreatic growth. Am. J. Physiol., 249, G294-G298.

Yanatori, Y. and Fujita, T. (1976). Hypertrophy and hyperplasia in the endocrine and exocrine pancreas of rats fed soybean trypsin inhibitor or repeatedly injected with pancreozymin. Arch. Histol. Jap., 39, 67-78.

4

DOSE EFFECTS OF RAW SOYABEAN FLOUR ON PANCREATIC GROWTH

R.G.H. Morgan, R.A. Crass, and P.S. Oates

Department of Physiology
University of Western Australia
Nedlands, Western Australia, 6009

Raw soya flour (RSF) feeding to rats produces pancreatic hypertrophy and hyperplasia, and, if sufficiently prolonged, the spontaneous development of pancreatic neoplasms and the potentiation of pancreatic carcinogens. With continuous exposure to RSF the threshold dose for pancreatic growth is approximately 20% RSF, but the threshold for the other effects has not been defined. If 100% RSF is fed for less than 24 weeks continuously the effects on the pancreas are completely reversible, but feeding for longer than this leads to irreversible progression to pancreatic adenoma and carcinoma over the next 50-70 weeks. Repeated alternation of the diet between 100% RSF and rat chow for prolonged periods leads to effects on the pancreas at least as marked as those seen with continuous RSF feeding. This occurs with RSF feeding for periods as short as 2 days out of 7. The effects seen in the rat must be applied to human nutrition with caution, but it is suggested that infants fed soya based milk substitutes and fad dieters who frequently radically alter dietary composition may be more susceptible to spontaneous or carcinogen-induced pancreatic neoplasms.

INTRODUCTION

When the rats are fed raw soyabean flour (RSF) pancreatic growth by hypertrophy and hyperplasia occurs (Crass and Morgan, 1982). Although RSF contains many biologically active components which may affect body or organ growth (Rackis, 1974), the major component responsible for the stimulation of pancreatic growth appears to be a heat-labile trypsin inhibitor (TI), which comprises about 6% of the total protein (Rackis and Anderson, 1964). Consequently, when soya flour is appropriately heated (heated soyabean flour, HSF) the effect on pancreatic growth is lost (Rackis, 1965). The mechanism by which TI itself acts is almost certainly the release of cholecystokinin (CCK), a trophic hormone for the pancreas (Green and Lyman, 1972).

Within 1 week of starting RSF feeding pancreatic weight, RNA and protein increase significantly, and within 2 weeks significant increases in DNA are also seen (Crass and Morgan, 1982). Pancreatic weight plateaus after 4-8 weeks and remains fairly constant for the next 35-40 weeks, but then starts to increase again in a logarithmic fashion (McGuiness et al., 1981). During the plateau phase atypical acinar cell foci and nodules (AACN, Longnecker, 1984) develop and the rapid increase in weight after about 40 weeks is associated with increased growth in these AACN. As they expand the AACN compress surrounding tissue to produce adenoma which progress to acinar cell carcinoma during the following 35-40 weeks (70-90 weeks after starting RSF, McGuiness et al., 1980). Such neoplasms are rare in animals fed diets devoid of soya bean protein (McGuiness et al., 1980, 1981, 1982). Raw soyabean flour also promotes the action of other pancreatic carcinogens, such as azaserine (Morgan et al., 1977), nitrosamines (Levison et al., 1979) and ethionine (McGuiness et al., 1983), effects which are probably linked to the ability of RSF to stimulate pancreatic growth, since other treatments which stimulate pancreatic growth have similar effects (Konishi et al., 1976, Denda et al., 1978).

In view of these pronounced effects of RSF on pancreatic growth and neoplasia in the rat and the increasing use of soyabean-derived proteins in human nutrition it is of considerable importance to define dose-response relationships for RSF (or TI) and its effect on pancreatic growth. Surprisingly few studies of this type have been published, possibly because of the length of such studies and the difficulty in accurately measuring small growth differences in such a diffuse organ. This paper will first consider 3 aspects of the problem -

(i) how much inhibitor is required?
(ii) for how long must it be fed in order to produce permanent changes in the pancreas?
(iii) the effects of repeated short term exposure to RSF.

Finally, the significance of the findings in the rat as far as human nutrition is concerned will be discussed.

HOW MUCH INHIBITOR IS REQUIRED?

Most of our studies have been performed with powdered diets of soya flour, supplemented with vitamins and minerals (Folsch and Wormsley, 1974). These diets are prepared in 5-10 kg batches, sufficient for 2-3 weeks feeding, and are stored in sealed plastic bags at 4°C. Soya flour contains approximately 40% protein, 25% carbohydrate and 20% fat, and this composition is not significantly altered by heating to destroy the trypsin inhibitor. It is possible therefore to prepare diets of the same nutrient composition by mixing RSF and HSF, with inhibitor levels linearly related to the RSF content. In studies using these mixtures pancreatic growth is proportional to the RSF content of the diet, with significant growth with a mixture of 80% HSF and 20% RSF (Fig. 1, Crass and Morgan, 1982).

In similar studies using pelleted mixtures of RSF and HSF, McGuiness, Hopwood and Wormsley found 25% RSF was sufficient to produce the full effects of RSF as far as pancreatic growth, development of spontaneous neoplasms and the potentiation of the pancreatic carcinogen azaserine was concerned (McGuiness et al., 1982). The threshold for the long term effects of RSF on the pancreas therefore appears to be below 20% RSF.

In the studies reported above, protein and fat levels, although kept constant, were necessarily high. It is quite likely that these dietary components also have an effect on the development of pancreatic neoplasms (Roebuck et al., 1981, Pour and Birt, 1983, Pour et al., 1983). Studies using defined diets of controlled protein and fat content, to which varying levels of TI are added, should determine whether this is the case and more accurately define the dose of TI needed to affect pancreatic growth. An early study of this type (Rackis et al., 1975) suggested that quite high levels of inhibitor were required for an effect on pancreatic growth, but these studies used weanling rats and were of limited (4 week) duration.

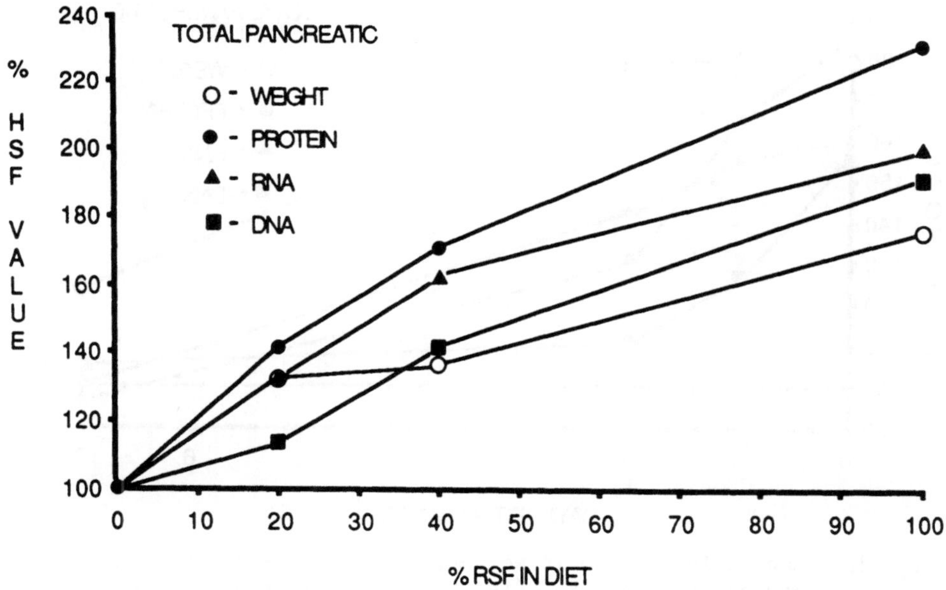

Fig. 1. Pancreatic weight, protein, RNA and DNA in rats fed 0% RSF/100% HSF, 20% RSF/80% HSF, 40% RSF/60% HSF or 100% RSF for 36 weeks.

Elsewhere in this volume Gumbmann, Rackis and Spangler report some of the findings from a more extensive long-term study which shows that there is indeed an interaction between protein and TI level and that the threshold for TI to affect the development of pancreatic neoplasms is between 215 and 337 mg TI/100 g diet, which corresponds to approximately 25% RSF.

The threshold for RSF to affect the sensitivity of the pancreas to carcinogens is not known but it is probably considerably below the threshold required to produce overt pancreatic hyperplasia. It has been pointed out that commercial rat chow in the past contained soya protein, and studies of pancreatic carcinogens in animals fed these diets may have been distorted by low levels of TI from this source. Casein, too has some anti-trypsin activity, so diets free of soya protein and casein, in which the protein component is derived from some other source, for example, yeast, should be used in such studies (McGuiness et al., 1984).

FOR HOW LONG MUST THE RSF BE FED?

If rats are fed RSF for 4 weeks and then changed to commercial rat chow or HSF there is an extremely rapid involution of the pancreas, with pancreatic DNA levels returning within 48 hours of the change to the level seen in controls fed diets containing no TI. Pancreatic weight, RNA and protein levels falls slightly slower, but all have returned to control levels within 7 days (Crass and Morgan, 1981). This tissue loss occurs by apoptosis (Oates et al., 1985), and is not accompanied by any evidence of inflammation or pancreatitis. Seven days after the involution the pancreas is histologically normal. When the animals are changed to chow after 12 weeks on RSF, involution is as rapid and complete as after 4 weeks (Fig. 2, Oates, 1983) and when RSF is fed for 12 weeks at the start of prolonged azaserine treatment, followed by chow for the rest of the animals life, there is no potentiation of this carcinogen (McGuiness et al., 1982). Other studies have also

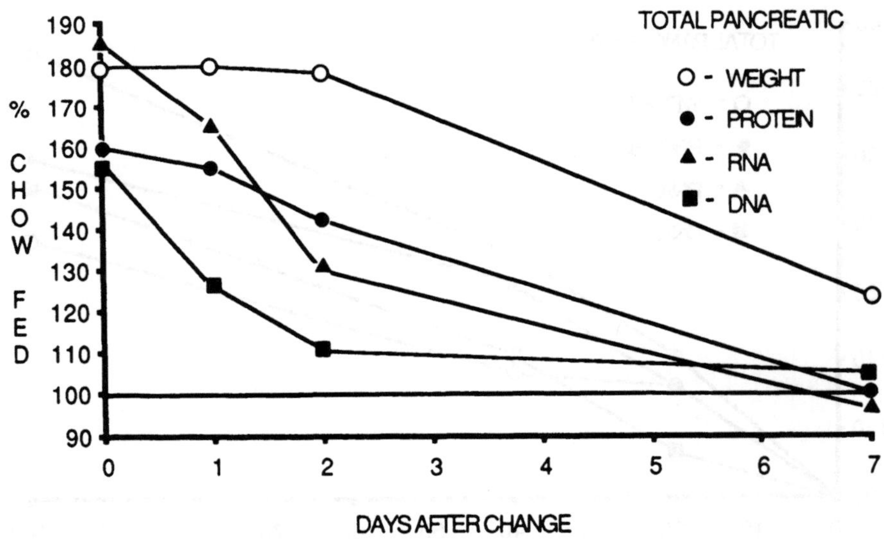

Fig. 2. Changes in pancreatic weight, protein, RNA and DNA in rats changed from RSF (fed for 12 weeks prior to the change) to chow. Data is expressed as a percentage of values in animals of the same age fed chow from weaning.

demonstrated the complete reversibility of the pancreatic hypertrophy caused by short term exposure to TI. Thus, a return of pancreatic weight to normal after 21 days (Ihse et al., 1975), 33 days (Rackis et al., 1979), or 38 days (Booth et al., 1964) of TI has been reported.

The above results show that the neoplastic response to continuous RSF feeding requires exposure to RSF for longer than 12 weeks, but just how much longer is not clear. In animals fed RSF continuously, AACN are first detected after 16 to 20 weeks on the diet. Only one study has looked at the reversibility of the changes produced by RSF feeding for longer than this. This showed that if animals are fed RSF for up to 24 weeks and then changed to chow, complete regression of the nodules occurred and when the animals were killed 1 year or more later the pancreas was histologically indistinguishable from that of rats fed chow from weaning. However if RSF was fed for longer than 24 weeks irreversible nodule formation and progression to pancreatic carcinoma occurred (McGuiness et al., 1984, K.G. Wormsley, personal communication). Studies to define more precisely the time at which the effects become irreversible and to determine what functional changes occur at this time (are oncogenes activated, for example?) are obviously required. It is apparent, however that quite prolonged continuous exposure to RSF is needed for full expression of the neoplastic process.

THE EFFECTS OF REPEATED SHORT TERM EXPOSURE TO RSF

Since all pancreatic parameters return to normal within 1 week of changing from a RSF diet to chow, and since complete reversion to an apparently normal morphology occurs after up to 24 weeks exposure to RSF (see above), it might be expected that repeated short term feeding of RSF would not produce any permanent changes in the pancreas. This does not appear to be the case. Two studies indicate that alternating the diet between RSF

Table 1. Pancreatic weights (g) in rats fed diets alternating weekly between standard rat chow and RSF, continuous RSF or continuous chow for 24, 36 or 65 weeks

Diet	24 weeks	36 weeks	65 weeks
Alternating chow-RSF			
i) Killed eating chow	1.89 ± 0.09(4)	1.94 ± 0.12(4)	2.10 ± 0.05(4)
ii) Killed eating RSF (one week later)	2.52 ± 0.11(4)	2.49 ± 0.04(4)	2.48 ± 0.10(4)
Continuous RSF	1.85 ± 0.13(9)	2.16 ± 0.17(11)	2.16 ± 0.14(4)
Continuous chow	1.31 ± 0.11(7)	1.17 ± 0.09(6)	1.14 ± 0.05(4)

Mean ± SEM
Number of rats in parenthesis

and chow leads eventually to (i) significant pancreatic growth which does not regress completely within one week, (ii) a very high incidence of spontaneous pancreatic carcinoma and (iii) an increase in the susceptibility to azaserine at least as great as that seen after continuous RSF feeding.

The first of these studies was reported as part of a more extensive study on the effects of RSF on the pancreas. Five rats were fed pelleted RSF on the weekends only, with normal rat chow for the other 5 days of the week. After 15 months of this regimen 3 of the 5 animals developed pancreatic cancer while the remaining two animals showed extensive adenomatous changes (McGuiness et al., 1982).

The second of these studies investigated some of the pancreatic growth parameters in rats fed a diet alternating between chow and powdered RSF and the effect of this diet on the action of azaserine in these animals (Crass 1983). Twelve week old rats fed chow from weaning were started on a diet alternating weekly between RSF and standard rat chow. Body weight gain on this diet was saw-toothed, with weight loss while on RSF and gain while on chow, but overall the animals grew as well as animals fed chow continuously and significantly better than rats fed RSF continuously. After 12 weeks on this regimen pancreatic weight in the animals fed the alternating diet (killed while eating RSF) was the same as in rats fed RSF continuously for 12 weeks (1.76 ± 0.09 and 1.59 ± 0.05 respectively, mean ± SEM) and significantly greater than that in control animals fed chow continuously (1.12 ± 0.11).

After 24, 36 and 65 weeks on these diets animals were killed either at the end of the week on chow, or 1 week later, at the end of the week on RSF. In animals fed the alternating diet and killed after a week eating chow, the pancreatic weight was the same as in animals fed RSF continuously for the same period and significantly greater than that in rats of similar age fed chow from weaning (Table 1).

In the animals fed the alternating diet and killed at the end of the week on RSF there was a significant increase in pancreatic weight compared with animals killed while on chow (Table 1). This may indicate that the pancreas responded, even after 65 weeks, with a vigorous growth spurt, but the increase in pancreatic weight was not associated with an increase in total pancreatic protein, RNA or DNA (Table 2) so probably much of the

Table 2. Pancreatic protein, RNA and DNA in rats fed a diet alternating weekly between RSF and rat chow (killed while eating chow or, one week later, while eating RSF), continuous RSF or continuous chow for 65 weeks.

Diet	Protein (mg)	RNA (mg)	DNA (mg)
Alternating chow-RSF			
Killed while eating chow	482 ± 16.9	50.4 ± 0.9	7.96 ± 0.92
Killed while eating RSF	397 ± 32.5	59.0 ± 3.3	7.90 ± 0.54
Continuous RSF	456 ± 33.0	55.8 ± 3.4	8.20 ± 0.38
Continuous chow	212 ± 18.2	27.8 ± 1.9	4.79 ± 0.28

Mean ± SEM, 4 rats in each group

weight gain was due to increased tissue fluid rather than tissue hypertrophy or hyperplasia.

The alternating diet also potentiated the effect of azaserine on the pancreas. From 24 weeks onwards the incidence of AACN in rats injected with azaserine (24 doses, 5 mg/kg, I.P., given twice weekly during the week on RSF) was significantly greater in animals on the alternating diet than in animals fed chow, and by 65 weeks confluent nodules and adenoma were found in all azaserine-treated animals fed the alternating diet. This potentiation of azaserine by the alternating diet was similar to that seen with continuous RSF feeding and no significant difference was found between these two diets as far as the incidence and rate of development of pancreatic neoplasms was concerned.

At 24 and 36 weeks in azaserine-treated animals fed the alternating diet, significantly more nodules/cm^2 section were seen in the pancreas from animals killed while eating RSF than in the pancreas from animals killed while eating chow. It seems unlikely that nodules resolve completely during the week on chow, but they do become smaller and, as has been pointed out elsewhere in this volume (Roebuck et al.), smaller nodules are less likely to be cut during sectioning (and therefore counted). They may also change in morphology to a more normal pattern which is more difficult to identify during the week on chow. The increased number of nodules during the week on RSF probably reflects the growth of the already present lesions rather than a true change in incidence. At later times (52 and 65 weeks) there was no change in the number of nodules with the weekly change in diet.

The fact that nodules change in size from one week to the next on the alternating diet indicates that the cells of the AACN are sensitive to the trophic stimulus for the pancreas released by RSF, probably CCK. This is supported by the finding that mitotic activity was seen in the majority of nodules from rats killed while eating RSF at 24 and 36 weeks (68% and 58%, respectively) but was rarely seen in nodules from rats killed a week later, while eating chow (0% and 1%, respectively). A similar effect was seen after 52 and 65 weeks on the diet. Although the mitotic index was not measured, increased cell division appeared to be largely confined to the nodule tissue since mitoses were rarely seen in non-nodular tissue during either the cube or the RSF phase of the diet.

When rats are started on RSF there is a short burst of cell division in the pancreas within 48 hours (as measured by the incorporation of ^{3}H thymidine into DNA and autoradiography), followed by a fall to control values and a later sustained rise beginning 1-2 weeks after starting the diet. The initial burst of cell division is though to be a response to damage as a result of the high levels of CCK released during the initial exposure to RSF (Oates and Morgan, 1984). In the alternating diet studies described above only the initial burst of cell turnover should occur, but, if this burst of increased cell division occurs with each exposure to RSF, then potentiation of azaserine might be expected, since as was pointed out earlier, any process which causes increased cell division in the pancreas seems to potentiate pancreatic carcinogenesis. However, it it difficult to explain the increased pancreatic weight, protein, RNA and DNA in animals killed after one week on chow in the long term studies (alternating-diet animals killed while eating chow compared with control animals eating chow continuously, Table 2). It seems that after repeated cycles of growth and involution regression is incomplete, unlike the situation seen after a single 4 or 12 week exposure to RSF. Possibly with each cycle a few less cells are destroyed than are formed so that after many cycles the pancreas is significantly larger. This could occur if a few cells were altered with each cycle and these cells did not completely regress with the withdrawal of the trophic stimulus. Certainly the first study reported above (McGuiness et al., 1982) indicates that dietary regimens of this type are very effective in producing initiated and transformed pancreatic acinar cells.

SIGNIFICANCE OF THESE STUDIES IN HUMAN NUTRITION

It has been estimated that the average British diet contains sufficient trypsin inhibitor to inhibit 330 mg trypsin per day (Doell et al., 1981). This is probably an overestimate, since some of the inhibitor activity may not be active against human trypsin, but it indicates that the potential exists for long term growth effects on the human pancreas. However, most of the studies on the effects of RSF on pancreatic growth and neoplasia have been performed on the rat and it is not clear whether the effects seen in the rat can be extrapolated to the human. Widespread species variation in the response to RSF occurs, as is discussed elsewhere in this volume (Schneeman et al.). The factor which determines whether RSF will produce pancreatic enlargement has not been defined, though it was suggested that the ratio of pancreatic weight to body weight might be a guide to this and, on this basis, it was predicted that the human would not respond (Liener, 1979). This is, however, purely an empirical observation and no theoretical basis for this distinction has been suggested. It seems to us that the feeding pattern of the animal might be more important, since this should determine the level of plasma CCK reached after a meal and the length of time that CCK levels would remain elevated after feeding. The rat and mouse, animals which respond to RSF, eat large amounts continuously during the night but little during the day (Cohn et al., 1970). On the other hand guinea pigs, hamsters and rabbits, which do not respond, eat several times during a 24 hour period (Horton et al., 1975). If the feeding pattern is a major factor in determining the response of the pancreas to RSF then most humans would be expected to follow the pattern of the guinea pig and not show a trophic response to soyaflour feeding, even if a negative feedback control of CCK release by trypsin is present, as seems likely (Ihse et al., 1977, Pap et al., 1981, Jacobson et al., 1982. See also Toskes, this volume). Nevertheless, as was pointed out elsewhere (McGuiness et al., 1984), one group of humans, namely infants fed soyabean derived products because of allergy to cow's milk, may be at risk. These children often receive frequent feeding because of failure to thrive and the products presently available contain significant levels of trypsin inhibitor. It may be that in these children the period of continuous exposure to TI is too short for any significant effect on the pancreas to occur, but the elimination of all TI activity in these products appears desirable.

Finally, the marked neoplastic effects of repeated short-term RSF feeding, particularly the studies of McGuiness et al. (1982), are disturbing. If repeated vigorous stimulation to secrete is equally damaging in the human then frequent radical changes in dietary composition might be undesirable. Such dietary changes occur with fad dieters, in whom rapid changes from high protein to high carbohydrate to high fat diets, interspersed with periods of semi-starvation, may occur. It is possible that this could lead to recurrent hyperstimulation of the pancreas, with an increased risk of neoplastic transformation in this organ.

REFERENCES

Booth, A.N., Robbins, D.J., Ribelin, W.E. DeEds, F., Smith, A.K. and Rakis, J.J., 1964, Prolonged pancreatic hypertrophy and reversibility in rats fed raw soybean meal, Proc. Soc. Exptl. Biol. Med., 116: 1067-1069.

Cohn, C., Webb, L. and Joseph, D., 1970, Diurnal rhythm in urinary electrolyte excretion in the rat: influence of feeding habits, Life Sci., 9:803-809.

Crass, R.A., 1983, The effects of soya flour diets and azaserine on the growth and function of the rat pancreas, PhD Thesis, University of Western Australia.

Crass, R.A. and Morgan, R.G.H., 1981, Rapid changes in pancreatic DNA, RNA and protein in the rat during pancreatic enlargement and involution, Int. J. Vit. Nutrit. Res., 51:85-91.

Crass, R.A. and Morgan, R.G.H., 1982, The effect of long-term feeding of soya-bean flour diets on pancreatic growth in the rat, Br. J. Nutr., 47:119-129.

Denda, A., Inui, S., Sunagawa, M., Takahashi, S. and Konishi, Y., 1978, Enhancing effect of partial pancreatectomy and ethionone-induced pancreatic regeneration on the tumorigenesis of azaserine in rats, Gann, 69:633-639.

Doell, B.H., Edben, C.J. and Smith, C.A., 1981, Trypsin inhibitor activity of conventional foods which are part of the British diet and some soya products, Qualitas Plantarum, 3:139-150.

Fölsch, U.R. and Wormsley, K.G., 1974, The pancreatic secretion of enzymes in rats treated with soybean diets, Scand. J. Gastroenterol., 9, 679-683.

Green, G.M. and Lyman, R.L., 1972, Feedback regulation of pancreatic enzyme secretion as a mechanism for trypsin inhibitor-induced hypersecretion in rats, Proc. Soc. Exptl. Biol. Med., 140:6-12.

Horton, B.J., West, C.E. and Turley, S.D., 1975, Diurnal variation in the feeding pattern of guinea pigs, Nutr. Metab., 18:294-301.

Ihse, I., Arnesjo, B. and Lundquist, I., 1975, Studies on the reversibility of oral trypsin inhibitor induced changes of rat pancreatic exocrine enzyme activity and insulin secretory capacity, Scand. J. Gastroenterol., 10, 321-326.

Ihse, I., Lilja, P. and Lundquist, I., 1977, Feedback regulation of pancreatic enzyme secretion by intestinal trypsin in man, Digestion, 15: 303-303.

Jacobson, D., Tillman, R., Toskes, P. and Curington, C., 1982, Acute and chronic feedback inhibition of pancreatic exocrine secretion in human subjects, Gastroenterology, 82:1091.

Konishi, Y., Denda, A., Inui, S., Takahashi, S. and Kondo, H., 1976, Pancreatic carcinoma induced by 4-hydroxyaminoquinoline-1-oxide after partial pancreatectomy and splenectomy in rats, Gann, 67:919-920.

Levison, D.A., Morgan, R.G.H., Brimacombe, J.S., Hopwood, D., Coghill, G. and Wormsley, K.G., 1979, Carcinogenic effects of di(2-hydroxypropyl)-nitrosamine (DHPN) in male wistar rats: Promotion of pancreatic cancer by a raw soya flour diet, Scand. J. Gastroenterol., 14: 217-224.

Liener, I., 1979, Significance for humans of biologically active food factors in soybeans and other food lugumes, J. Am. Oil. Chem. Soc., 56: 121-129.

Longnecker, D.S., 1984, Lesions induced in rodent pancreas by azaserine and other pancreatic carcinogens, Environ. Health Perspect., 56:245-251.

McGuiness, E.E., Hopwood, D. and Wormsley, K.G., 1982, Further studies of the effects of raw soya flour on the rat pancreas, Scand. J. Gastroenterol., 17:273-277.

McGuiness, E.E., Hopwood, D. and Wormsley, K.G., 1983, Potentiation of pancreatic carcinogenesis in the rat by DL-ethionine-induced pancreatitis, Scand. J. Gastroenterol., 18:189-192.

McGuiness, E.E., Morgan, R.G.H., Levison, D.A., Hopwood, D. and Wormsley, K.G., 1981, Interaction of azaserine and raw soya flour on the rat pancreas, Scand. J. Gastroenterol., 16:49-56.

McGuiness, E.E., Morgan, R.G.H., Levison, D.A., Frape, D.L., Hopwood, D. and Wormsley, K.G., 1980, The effect of long term feeding of soya flour on the rat pancreas, Scand. J. Gastroenterol., 15:497-502.

McGuiness, E.E., Morgan, R.G.H. and Wormsley, K.G., 1984, Effects of soybean flour on the pancreas of rats, Environ. Health Perspect., 56:205-212.

Morgan, R.G.H., Levison, D.A., Hopwood, D., Saunders, J.H.B. and Wormsley, K.G., 1977, Potentiation of the action of azaserine on the rat pancreas by raw soya bean flour, Cancer Lett., 3:87-90.

Oates, P.S., 1983, Studies on the growth and functional changes in the rat pancreas caused by raw soya flour feeding, PhD Thesis, University of Western Australia.

Oates, P.S. and Morgan, R.G.H., 1984, Short-term effects of feeding raw soya flour on pancreatic cell turnover in the rat, Am. J. Physiol., 247, G667-G673.

Oates, P.S., Morgan, R.G.H. and Light, A.M., 1985, Cell death (apoptosis) during pancreatic involution after raw soya flour feeding in the rat, In Press, Am. J. Physiol., 250.

Pap, A., Berger, Z. and Varro, V., 1981, Trophic effects of cholecystokinin-octapeptide in man - a new way in the treatment of chronic pancreatitis? Digestion, 21:163-168.

Pour, P.M. and Birt, D.F., 1983, Modifying factors in pancreatic carcinogenesis in the hamster model. IV. Effects of dietary protein, J. Nat. Cancer Inst., 71:347-353.

Pour, P.M., Birt, D.F., Salmasi, S.Z. and Götz, U., 1983, Modifying factors in pancreatic carcinogenesis in the hamster model. I. Effect of protein-free diet fed during the early stages of carcinogenesis, J. Nat. Cancer Inst., 70:141-146.

Rackis, J.J., 1965, Physiological properties of soybean trypsin inhibitors and their relationship to pancreatic hypertrophy and growth inhibition of rats, Fed. Proc., 24:1488-1493.

Rackis, J.J., 1974, Biological and physiological factors in soybeans. J. Am. Oil Chem. Soc., 51:161A-174A.

Rackis, J.J. and Anderson, R.L., 1964, Isolation of four soybean trypsin inhibitors for DEAE-cellulose chromatography, Biochem. Biophys. Res. Comm., 15, 230-235.

Rackis, J.J., McGee, J.E. and Booth, A.N., 1975, Biological threshold levels of soybean trypsin inhibitors by rat bioassay, Cereal Chem., 52:85-95.

Rackis, J.J., McGee, J.E., Gumbmann, M.R. and Booth, A.N., 1979, Effect of soy proteins containing trypsin inhibitors in long term feeding studies in rats, J. Am. Oil. Chem. Soc., 56:162-168.

Roebuck, B.D., Yager, J.D. Jr. and Longnecker, D.S., 1981, Dietary modulation of azaserine-induced pancreatic carcinogenesis in the rat, Cancer Res., 41: 888-893.

5

ENHANCEMENT OF PANCREATIC CARCINOGENESIS BY RAW SOY PROTEIN ISOLATE: QUANTITATIVE RAT MODEL AND NUTRITIONAL CONSIDERATIONS

B.D. Roebuck

Department of Pharmacology and Toxicology
Dartmouth Medical School
Hanover, New Hampshire 03756

ABSTRACT

Foods containing soybean products have been shown to modify the biochemical and physiological status of the pancreas of several species of experimental animals. Recently, these products have been implicated as a factor in the causation of pancreatic neoplasms. Extensive experimental studies into the possible mechanisms need to be undertaken. Experimental details of a rat/azaserine model for the study of pancreatic carcinogenesis are reviewed. Emphasis is given to the quantitative components of this model and the adaptation of this model to the two-stage (initiation-promotion) concept of carcinogenesis. Particular attention is devoted to considerations of the experimental diets. Application of these concepts to the study of the postinitiational effects of raw and heated soybean protein isolate with and without the addition of high levels of unsaturated fat were undertaken. The results indicate that raw soybean isolate enhanced the growth of azaserine-induced pancreatic foci; whereas, a high level of unsaturated fat had a minimal effect. The effects of the soybean isolate were abolished by heat treatments, but the effects of the unsaturated fat would not be expected to be abolished by similar treatment with heat.

INTRODUCTION

The consumption of a diet containing a high content of raw soy flour enhanced pancreatic carcinogenesis in rats when fed either concurrently with (16,20,22,25) or subsequent to (20) treatment with known pancreatic carcinogens. It is well established that the feeding of raw soy flour, particularly to the chicken (7) and the rat (6,10,24,29), leads to a rapid and dramatic enlargement of their pancreas. This is due to both hypertrophy and hyperplasia of the acinar cell component of the pancreas. Heat-treatment of the raw soy flour largely abolishes these hypertrophic, hyperplastic, and carcinogenic effects (13,21). These stimulatory effects to the pancreas generally have been attributed to the proteinaceous trypsin inhibitor (TI) content of the flour. It is hypothesized that the TI causes the release of cholecystokinin (CCK) which is known to be trophic for the pancreas (1,15,23).

The studies reported in this Proceeding (33) and a discussion of the associated methods are important for two reasons. First, we discuss in detail methodology relavent to the study of carcinogenesis and, in particular, the nutritional or dietary modulation of pancreatic carcinogenesis. Attention is focused upon models of pancreatic carcinogenesis especially the two-stage or initiation-promotion model; the quantitative analysis of pancreatic tumors using a short-term model; and consideration of the dietary and nutritional status of the animal. Second, raw soy flour has been fed in a large number of experiments investigating pancreatic carcinogenesis, but the composition of raw soy flour is not well defined. For example, not only does raw soy flour contain TI, but also it contains approximately 20% soybean oil. Soybean oil is an unsaturated fat and of similar fatty acid composition to corn oil (49) which at 20% in the diet is known to enhance pancreatic carcinogenesis (4,40,41). Thus, it is important to compare the relative contributions of TI and an unsaturated fat to the enhancement or "promotion" of pancreatic carcinogenesis.

EXPERIMENTAL MODELS OF PANCREATIC CARCINOGENESIS

Only recently have animal models been available for the study of pancreatic cancer. Of the several reasons for this situation, one is undoubtedly the relatively rare occurrence of "spontaneous" pancreatic cancers in untreated experimental animals (for references see 38). For example, an incidence of only eight pancreatic tumors (only three were malignant) was found in the routine postmortem examination of 1252 rats of the Chester Beatty strain (42). All eight rats were older than 16 months. The spontaneous occurrence of putative, preneoplastic, pancreatic acinar cell lesions (foci) in several species of rodents has been reported (34,35). The number of animals per group was too small to allow quantitative conclusion concerning species and strain susceptibility and only young rodents were evaluated thus little is known concerning the aging process on the appearance of these lesions. Recently, Chiu (8,9) observed similar foci in both male and female Sprague-Dawley rats ranging in age from 4 to 27 months. He reports an incidence of 196 out of 1124 rats. The incidence of the foci increased with age but neither the size nor the number of foci per rat increased with age. From these observations, one must conclude that the rat pancreas is relatively resistant to the occurrence of "spontaneous" tumors.

Within the last 10 years, considerable research has been devoted to the development of rodent models for the study of pancreatic cancer. Recently, the results of these efforts have been reviewed (17,27,44). The majority of attention has been devoted to two models; namely the hamster/nitrosamine and the rat/azaserine models. In the hamster/nitrosamine model the presumed cell of origin, the first histologically detectable lesions, and the bulk of the cancers are associated with cells of the ductal system. In the rat/azaserine model, the acinar cell is the presumed cell of origin and the first histologically detectable lesions possess acinar cell characteristics. A majority of the adenocarcinomas arising in this model retain at least some acinar cell characteristics. The response of these two rodent models to chemical and dietary modulation of carcinogenesis generally has been similar (for references see 19,38). The affects on the pancreas of soybean products and dietary fats have primarily been evaluated in the rat model; therefore, a more detailed discussion of the rat/azaserine model is essential.

Studies of the progressive or sequential nature of cancer development have been advanced by the somewhat artificial division of the carcinogenesis process into two stages, that is, the initiation stage and the postinitiation (promotional) stage. During the initiation stage, damage to nuclear DNA is thought to be the key event responsible for the ultimate development of cancer. Following initiation, a variety of selective processes ultimately may be responsible for cancer development (12). The stimulation or enhancement of this latter stage has been termed promotion. Historically, the phenomenon of promotion in two-stage carcinogenesis was first described for skin cancer and extensive mechanistic studies have been reported in the ensuing 40 years. Most recently the phenomenon has been demonstrated to be operative at least in part in several other organ systems (for review see 45) including the pancreas (38).

The relavence of the two-stage model to pancreatic carcinogenesis becomes quite obvious when one considers some of the known effects that soybean products have on the pancreas. For example, raw soy flour is known to stimulate pancreatic DNA synthesis and acinar cell replication (23). Stimulation in response to raw soybean flour appears to be transitory but lasts for several weeks and its intensity is several times above control. Very likely the degree of stimulation is a function of the quantity of TI, but this has not been evaluated in detail. It is almost axiomatic that replicating tissues are more responsive to carcinogens that quiescent tissues (43). Conditions which decrease the degree of cellular division during the initiation stage would be expected to prevent or reduce cancer development. Restricting the availability of diet by just 10% during azaserine initiation, significantly suppressed the development of pancreatic adenocarcinomas in the rat (41). It is not yet known if this is a specific effect on cellular division. The postinitiation stage of pancreatic carcinogenesis is also susceptible to modulation (18,39,41) and this is the focus of much of the experimental data that follow. Without the division of the process of pancreatic carcinogenesis into stages, we cannot begin to understand the phenomenon of enhancement of pancreatic cancers associated with certain soybean products.

QUANTITATIVE RAT/AZASERINE MODEL OF PANCREATIC CARCINOGENESIS

The histopathological lesions in the pancreas resulting from azaserine treatment have been reviewed (17,19). Though a fairly wide spectrum of specific tumor types result, the vast majority of neoplasms have at least some acinar cell characteristics. Considerable uncertainty and debate exists with respect to the specific criteria for the classification of individual tumors as adenomas or as adenocarcinomas in situ (5,17). Until these issues are definitively resolved careful attention must be given to fully describing the tumor classification criteria used in each pancreatic carcinogenesis study.

Quantification of pancreatic neoplasms has typically been limited to the incidence of rats with specified lesions. Fuller descriptions including the multiplicity of lesions per pancreas appear to be useful (40,41) and should be more often undertaken. This would be particularly useful in those experiments where inadequate numbers of rats per treatment group are used as it gives another index of treatment effect. Such quantitation must be undertaken with care and several considerations must be addressed: 1) One should sample for histopathological analysis as much pancreas as possible. There are few reasons not to sample the entire rat pancreas. In particular, it is difficult to justify the use of tumor-bearing pancreas for standard biochemical analyses as the organ is a mixture of apparently normal tissue (though it may have been exposed to a carcinogen) and tumors many of which are obviously at different stages of development and histological type. 2) The removal of the entire pancreas and its fixation by a standardized manner in a planar position is recommended as in Figure 1. 3) One should then section the pancreas so as to maximize the size (area) of tissue that is placed on the slide. The initial fixation of the pancreas in a flat position helps considerably. These procedures ensure that adequate (or at least all the available tissue) is sampled and eliminates the likelihood of observing and measuring a specific tumor more than once.

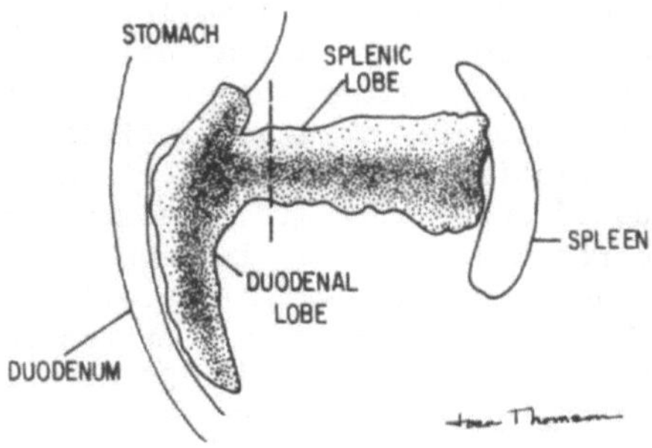

Figure 1. Orientation of pancreas in preparation for flat-fixation. For illustrative purposes only, the stomach, duodenum, and spleen are included. The line demarcates the splenic (tail) region from the remainder of the pancreas (doedenal or head region).

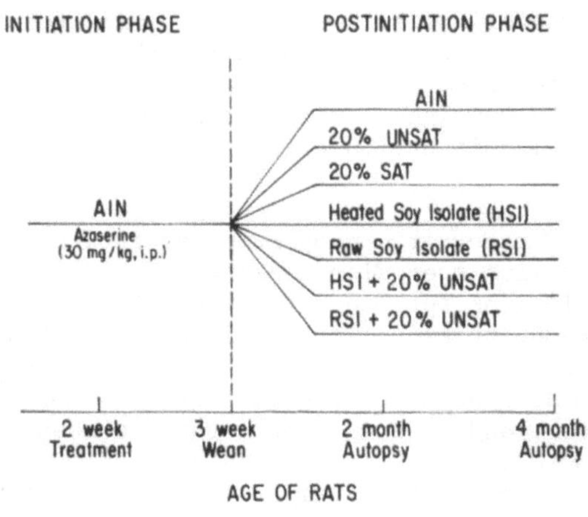

Figure 2. Schematic of experimental protocol. The abbreviations and composition of the diets are listed in Table 1. Suckling, male Wistar rats were treated only once with azaserine. The dams were fed the AIN diet.

Our laboratory (31,34,50) has developed a quantitative rat model for the study of pancreatic carcinogenesis in which azaserine is used to induce the appearance of atypical acinar cell foci. These foci represent putative, preneoplastic populations of acinar cells some of which undoubtedly give rise to adenomas and adenocarcinomas. The utility of this model is that the length of experiments can be reduced to 4 to 6 months from the 1 to 2 years required for the development of significant numbers of adenocarcinomas. The nature of the resulting data is such that fewer animals per group are required in the quantitative short-term model. Using this model, we have demonstrated inhibition of focal growth by retinoids (32) and phenolic antioxidants (39); and the enhanced development of foci by high levels of dietary unsaturated fat (31) and raw soy protein (33).

The salient features of this model are illustrated in Figure 2. The experiments are operationally divided into an initiation and a postinitiation (promotion) phase. A single, sub-necrogenic dose of azaserine is preferred though multiple doses have also been used (34,50). This protocol has been used with rats of 5 to 7 weeks of age. In the case illustrated in Figure 2 and discussed below, we treated suckling rats as they are more responsive to the carcinogenic effects of azaserine. Dietary modulation of either or both phases of carcinogenesis has been undertaken. The duration of the postinitiation phase can be varied. For a strong modulator such as raw soy isolate (see below), effects have been observed as early as 2 months, whereas the effects of high intakes of fat require longer periods to manifest themselves (33,36). At autopsy the entire pancreas is excised, weighed, and fixed flat as depicted in Figure 1. We have found that fixation in Susa's or Bouin's fixative is superior to formalin. A single pancreas usually results in 1 or 2 blocks, though a very large pancreas may require as many as 5 or 6 blocks to completely embed it. The tissue block is sectioned to attain the maximum surface area of tissue per section. Because of the number of individual slides per pancreas that sometimes results, we have on occasion systemically divided the pancreas into a splenic (tail) region and a duodenal (head) region as shown in Figure 1 and only quantified the foci from the slides of the splenic region (37).

The identification of carcinogen-induced foci and their classification by phenotypic properties has been described in detail (9,30,31,34). Typical examples of the acidophilic and basophilic foci are shown in Figures 3 and 4, respectively. Evidence has accumulated to indicate that the acidophilic population of foci are responsive to various modulators of carcinogenesis; whereas, the basophilic foci are non-responsive (9,30-32,39). The significance of these populations of foci are not completely understood.

Quantification of carcinogen-induced lesions need not require sophisticated equipment; however, as we have undertaken numerous studies, data acquisition and analysis is facilitated by our simple image analysis system (Figure 5). The size (transectional or surface area) of foci are measured with a camera lucida and an X,Y-digitizer (GTCO Corp., Rockville, MD). The size of the tissue examined is measured with the Video Van Gogh hardware and software (Tekmar, Inc., Cleveland, OH) and a video camera (Dage-MTI, Inc., Michigan City, IN). The entire system is operated by, and data processed with an IBM/XT computer (IBM Corp., Boca Raton, FL). From the observed number and size of the foci in the tissue sections, the mean number and mean size of the foci within the three-dimensional tissue are estimated by the mathematical techniques of quantitative stereology. Their application to the study of tumors within the liver (28) and pancreas (31) has been discussed in detail. Essentially, there is a greater probability for a large focus in a three-dimensional tissue to be found and identified as compared to a small focus. Finally, the volume of pancreas occupied by focal tissue as expressed as a per cent (volume %) is a calculated parameter that may be thought of as an expression of the focal or tumor burden of the pancreas.

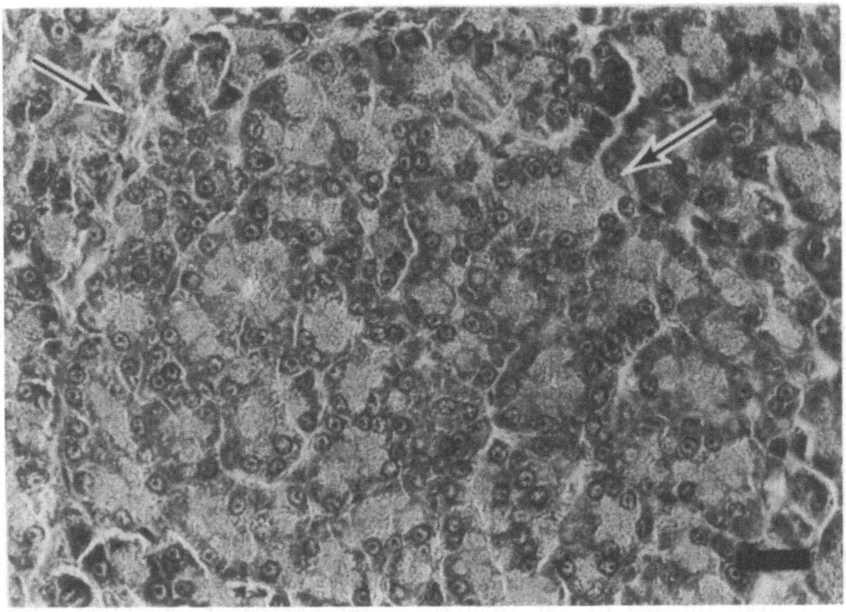

Figure 3. Acidophilic acinar cell focus (arrows). Acini are usually larger than the normal acini, the cytoplasm is as rich in zymogen as normal acinar cells, and the nuclei are densely packed and of similar or smaller size than the nuclei of normal acinar cells. Mitotic figures are common. H & E. Bar = 20 µm x 564.

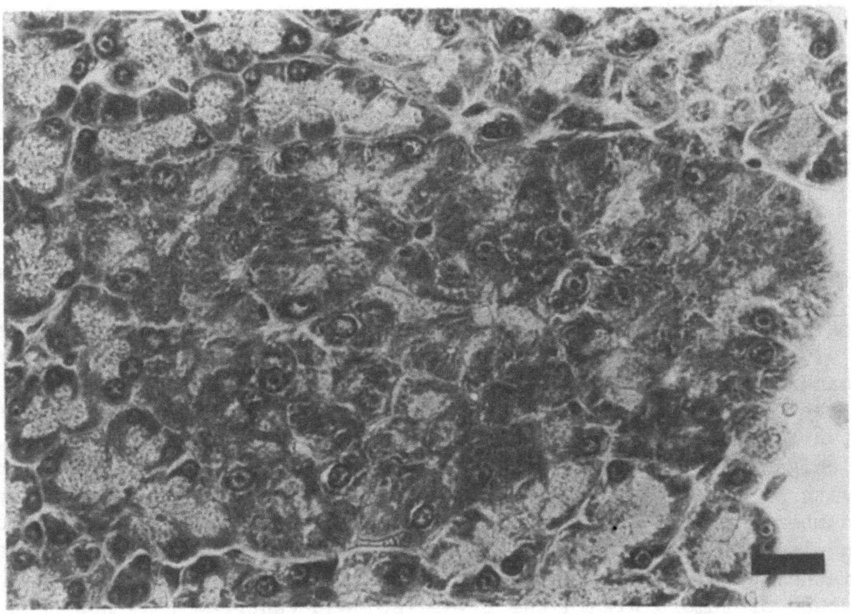

Figure 4. Basophilic acinar cell focus. The nuclei are larger and the zymogen content is reduced as compared to normal acinar cells. The cells in a basophilic focus are larger than normal acinar cells or acinar cells of acidophilic foci. H & E. Bar = 20 µm x 564.

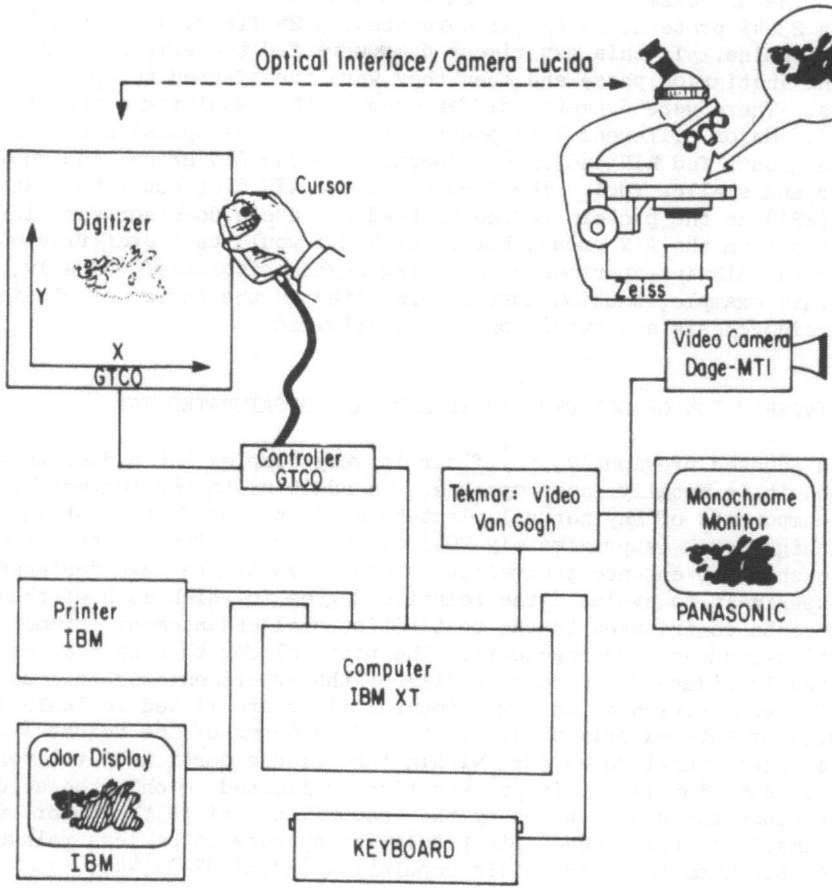

Figure 5. Image analysis system allowing semi-automatic data acquisition of tissue areas, focal counts, and the calculation of quantitative stereologic values.

NUTRITIONAL AND DIETARY CONSIDERATIONS

The consideration of this topic in relation to carcinogenesis brings to mind a quote from the French writer, Andre Gide -- "All this has been said before but since nobody listened it must be said again".

Serious concern about the nutritional and dietary status of the laboratory animals used in toxicological experiments has been expressed several years ago (14,26,51). These special concerns have been sounded again and directed specifically to nutrition and cancer experimentation (48). The need to standardize diets has brought into common usage the defined and purified diets such as the one which has been suggested by the American Institute of Nutrition (2,3). It is well known that the enzymology and physiology of the pancreas is responsive to dietary modifications (11,46,47). Demonstration that diet and nutritional status are relavent to the modulation of cancer of the pancreas has clearly been shown for effects of protein content, the quality and quantity of fat, and the availability of food (40,41). By way of example, major differences in the response of the pancreas to tumor development are shown when one compares purified diets and a commercially available laboratory chow on the postinitiational development of azaserine-induced foci. The experimental protocol was similar to the experiment illustrated in Figure 2. The composition of the purified diets are given in

Table 1. The approximate composition of the chow (Purina Laboratory Chow 5001) was 23.4% protein, 50.8% carbohydrates, 5.2% fiber, 4.5% fat, 7.3% ash and vitamins. In this experiment dams were fed the same diet (AIN) during the initiation phase and then they were transferred to the experimental diets. There were dramatic differences in the resulting pancreatic foci (Table 2). Major differences in pancreatic focal development are seen between the groups fed AIN and chow. Compared to the AIN group, the chow group had fewer and smaller foci. The feeding of the AIN diet but with heated soy isolate (HSI) as the protein source instead of casein does not show large differences from the AIN group, but clearly HSI would be a preferred control diet over the AIN in experiments involving soybean isolates. Clearly, as seen in this example, caution must be exhibited in the design of studies in which carcinogenesis and nutrition are intertwined.

DIETARY INTERACTION OF SOY PROTEIN ISOLATE AND UNSATURATED FAT

As discussed previously, soy flour is very complex and a diet largely based upon it is totally inappropriate. In addition to the innumerable undefined components of any natural plant-based diet, soy flour contains both TI and a high level (approximately 20%) of unsaturated fat. Both components have been shown to enhance pancreatic carcinogenesis. We have designed a simple experiment to evaluate the relative degree to which each of these two components contributes to the postinitiational enhancement (promotion) of azaserine-induced carcinogenesis. The protocol for this experiment is illustrated in Figure 2 and many details of the experiment were discussed above. The composition of the experimental diets are listed in Table 1. The results of this experiment are presented in terms of the volume % (tumor burden) of azaserine-induced foci within the splenic portion of the pancreas (Table 5). Additionally, this presentation is limited to only the acidophilic foci because the contribution by the basophilic foci to the tumor burden was very small and this phenotypic population appears to be less relavent to carcinogenesis than the acidophilic population (30,31,37,39,44).

A comparison of pancreatic foci between the saline-treated (non-carcinogen) and the azaserine-treated rats is important, but not the major focus of this work; therefore, the number of rats in the saline groups are quite small (Table 4). Furthermore, only selected saline controls were included. The carcinogen did not alter the growth of the rats nor did the feeding of the raw soy isolate inhibit growth of the rats (Table 3). Dietary fat at 20% of the diet supported larger weight gain than did the diets containing 5% fat. In spite of the small size of these saline groups, it is obvious that the raw soy isolate increased the pancreatic weights and that the 20% fat diets did not. The acidophilic and basophilic foci seen at 4 months postinitiation in the pancreases of the saline-treated rats were so few that statistical analyses were considered inappropriate within these groups (Table 4). A simple comparison of the volume % of foci within the pancreases of saline- and azaserine-treated groups shows the dramatic effects of this pancreatic carcinogen.

The postinitiational effects of the various diets on the azaserine initiated foci are expressed as focal volume % (tumor burden) and are presented in Table 5. The data are presented as volume % for simplicity of comparisons between the groups and over time. In specific cases the effects of the postinitiational diets were so large that the foci often coalesced preventing the measurement of individual foci. At 2 months postinitiation, we only evaluated groups fed the soy isolates as previous experience (36) had indicated that enhancement of focal growth by the 20% fat diets could not be observed that early. Raw soybean protein, whether fed in the RSI group or in the RSI +20% UNSAT group, significantly increased the tumor burden of the pancreas. There was not an effect of the unsaturated fat on the tumor burden when added

Table 1. Composition of the experimental diets expressed as percentage by weight. All diets were stored at 4°C prior to feeding. The diets were fed *ad libitum* as powders.

Ingredients	AIN-76A (AIN)	20% Saturated Fat (SAT)	20% Unsaturated Fat (UNSAT)	Raw or Heated Soy Isolate (RSI or HSI)	Raw or Heated Soy Isolate plus 20% Fat (RSI + UNSAT or HSI + UNSAT)
Casein	20.0	20.0	20.0		
Soy protein				20.0[a]	20.0[a]
Methionine	0.3	0.3	0.3	0.3	0.3
Cornstarch	15.0	11.7	11.7	15.0	11.7
Sucrose	50.0	38.3	38.3	50.0	38.3
Cellulose	5.0	5.0	5.0	5.0	5.0
Unsaturated fat[b]	5.0	2.0	20.0	5.0	20.0
Saturated fat[c]		18.0			
Micronutrients[d]	4.7	4.7	4.7	4.7	4.7

[a] The trypsin inhibition (TI) content was 600 mg TI/100 g diet for the RSI and 46 mg TI/100 g diet for the HSI.

[b] Corn oil was the unsaturated fat used in these experiments.

[c] The saturated fat was 18% hydrogenated cocoanut oil with 2.0% corn oil added to provide for the essential fatty acid requirements of the rats.

[d] The composition of the vitamin and mineral mixtures are those recommended for the AIN diet (2,3).

Table 2. Comparison of the effects on foci development of 3 different diets fed during the postinitiation phase of carcinogenesis. The experimental protocol is as illustrated in Figure 2 with the addition of one group fed a commercially available laboratory chow.

	Observed transectional data of foci		Calculated Volumetric data of foci		
Diet (No. of Rats)	No. per sq cm	Area (mm^2 x 100)	No. per cm cm	Diameter (μm)	Foci Volume Percent
AIN (7)	25 ± 3	15 ± 2	558 ± 61	448 ± 28	4.1 ± 0.8
Chow (5)	8 ± 4	6 ± 1	304 ± 96	276 ± 43	0.7 ± 0.4
HSI (10)	28 ± 3	15 ± 2	607 ± 53	460 ± 27	4.4 ± 0.8

[a] means ± SEM

to either the HSI or RSI diets. Preliminary indications are that the total number of foci observed within the pancreases of these 4 groups were the same. This is as might be expected since all the rats were initiated at the same age, while fed the same diet, and with the same dose of azaserine. The effect of the raw soy protein was to increase the size of the initiated foci.

At 4 months postinitiation, groups are included which were fed the saturated and unsaturated fat at 20%. In the UNSAT group the tumor burden was greater than in either the SAT group or the AIN group. The magnitude of this increase in the UNSAT fat group was approximately what we have observed previously (31,41), though the experiments were quite different in design and end-point. HSI did not enhance carcinogenesis in the short-term experiments nor did the addition of the unsaturated fat to the HSI diet. As we saw at 2 months, the effects of the raw soy isolate are dramatic with 20 to 30% of the pancreas occupied by foci. At 2 and 4 months postinitiation in all cases where the unsaturated fat was fed, the tumor burden was greater, but the magnitude of the effect was not large enough to achieve statistical significance. A comparison of the 2 and 4 month results show that the tumor burden increases by approximately the same fold irrespective of the diet fed. This may imply that the major effects of the raw soy isolate are on the growth of the foci within the first 2 months. More detailed experimentation must be undertaken before this issue is resolved.

The conclusions drawn from these experiments are tentative as additional analyses of the tissues are in progress. Nonetheless, it is obvious that the raw soy isolate at the level fed, had a dramatic effect upon the enhancement of growth of the carcinogen-induced pancreatic foci. The effects of unsaturated fat are comparatively quite small. The combined effects of both components may be additive, but this has to await additional experiments. The effects of the raw soy isolate are abolished, as expected, by heat treatment of the product.

RESEARCH NEEDS AND IMPLICATIONS

Concern and interest in this area of research has until recently not been strong. The USDA Trypsin Inhibitor Studies when they are fully published should stimulate considerable interest in this area. Several areas that must be addressed are: 1) The composition of the soybean products and

Table 3. Body weights and pancreatic weights at 2 and 4 months postinitiation.

At 14 days of age rats were either initiated with a single dose of azaserine (30 mg/kg b.w.) or injected with an equivalent volume of saline. They were weaned at 21 days of age and fed the test diets. At 2 and 4 months postinitiation, rats were autopsied and pancreases excised. A description of the diets and abbreviations used are in Table 1.

Diet	Saline Treatment 4 Months			Azaserine Treatment 2 Months			Azaserine Treatment 4 Months		
	No. of Rats	Body Weight (g)	Pancreas Weight (g)	No. of Rats	Body Weight (g)	Pancreas Weight (g)	No. of Rats	Body Weight (g)	Pancreas Weight (g)
AIN	4	593 ± 21[a]	1.76 ± 0.12	0			7	610 ± 27	1.80 ± 0.09
SAT	0			0			8	648 ± 15	1.58 ± 0.06
UNSAT	5	689 ± 36[b]	1.79 ± 0.03	0			9	674 ± 26[b]	1.68 ± 0.06
RSI	5	535 ± 21	2.40 ± 0.14	5	421 ± 16	2.61 ± 0.19	10	537 ± 12	2.83 ± 0.14
RSI + UNSAT	5	618 ± 22[c]	2.11 ± 0.04	5	409 ± 17	2.04 ± 0.05	10	579 ± 13[c]	3.32 ± 0.21
HSI	0			5	432 ± 14	1.68 ± 0.04	10	539 ± 16	1.73 ± 0.05
HSI + UNSAT	0			5	463 ± 21	1.50 ± 0.99	10	627 ± 18[d]	1.68 ± 0.11

[a] means ± S.E.

[b] Significantly greater than the AIN group, $p < 0.05$.

[c] Significantly greater than the RSI group, $p < 0.05$.

[d] Significantly greater than the HSI group, $p < 0.05$.

Table 4. Pancreatic foci of saline-injected, control rats at 4 months postinitiation.

No phenotypic differences were noted between acidophilic or basophilic foci of saline-injected and azaserine-initiated rats. Likewise, regional differences within the pancreases were not apparent; thus, because the total number of foci per pancreas was so small in these control rats we present data based upon both the splenic and duodenal regions.

Diet (No. of rats)	Acidophilic Foci				Basophilic Foci			
	No. of Foci Observed	No. per sq cm	Transectional Area (mm^2 x 100)	Volume as % of Pancreas	No. of Foci Observed	No. per sq cm	Transectional Area (mm^2 x 100)	Volume as % of Pancreas
AIN[b] (4)	0	0	-	0	0	0	-	0
UNSAT (5)	3	0.14 ± 0.06[a]	22.1 ± 19.1	0.033 ± 0.030	3	0.14 ± 0.06	5.05 ± 2.21	0.007 ± 0.004
RSI (5)	7	0.25 ± 0.07	13.7 ± 4.2	0.028 ± 0.008	3	0.15 ± 0.15	4.13	0.006 ± 0.006
RSI + UNSAT (5)	14	0.56 ± 0.21	25.2 ± 10.9	0.123 ± 0.066	7	0.31 ± 0.17	2.38 ± 0.53	0.009 ± 0.005

[a]means ± S.E.

[b]The abbreviations for the diet groups are defined in Table 1.

Table 5. Acidophilic focal volume as percent of pancreas volume[a]

Diet	2 Months Postinitiation		4 Months Postinitiation	
	No. of Rats	Volume as % of Pancreas	No. of Rats	Volume as % of Pancreas
AIN			7	4.1 ± 0.8
SAT			8	1.2 ± 0.3
UNSAT			9	8.8 ± 2.0
HSI	5	1.1 ± 0.3	10	3.6 ± 1.0
RSI	5	7.5 ± 2.1	10	23.0 ± 8.0
HSI + UNSAT	5	1.2 ± 0.4	10	4.9 ± 1.9
RSI + UNSAT	5	9.4 ± 1.6	10	31.7 ± 6.9

[a] means ± SEM

specific effects on pancreatic carcinogenesis must be more fully described. A wide range of soybean products must be carefully evaluated in a carcinogenesis model similar to the rat/azaserine model described. Attention should be given to not only specific inhibitor proteins but also other components of soybean seeds. 2) Effects of soybean products on the modulation of both the initiation and the postinitiation phases of carcinogenesis must be considered. These effects should be considered distinct from any direct evaluation of the carcinogenicity of a soybean product in rats not exposed to a known carcinogen. 3) The evaluation of soybean products for tumorigenicity should also consider specific non-soybean components of the diet. One example is a consideration of both the quality and quantity of dietary fat. Additionally, host-related factors as age should also be considered. 4) Finally, it is a fact that humans consume a wide variety of foods and the results from the above investigations must be put into an appropriate context. Until mechanistic studies in rodents are undertaken and directly related to effects of these and related inhibitors on the human pancreas, there will be a degree of uncertainty as to the significance of the rodent studies of pancreatic carcinogenesis.

ACKNOWLEDGEMENTS

This work was supported in part by U.S. Public Health Services Grants CA-26594 from the National Cancer Institute and ES-07104 from the National Institutes of Environmental Health Sciences. Thanks is extended to Professor Elisha R. Huggins, Department of Physics and Astronomy, Dartmouth College, and many students of his couse, Science 40--Laboratory Interfacing and Computing, for development of our Image Analysis System.

REFERENCES

1. Adrian, T.E., Pasquali, C., Pescosta, F., Bacarese-Hamilton, A.J. and Bloom, S.R., 1982, Soya induced pancreatic hypertrophy and rise in circulating cholecystokinin, Gut, 23:A889.

2. Bieri, J., 1977, Report of the American Institute of Nutrition Ad Hoc Committee on standards of nutritional studies, J. Nutr., 107:1340-1348.

3. Bieri, J., 1980, Second report of the ad hoc committee on standards for nutritional studies, J. Nutr., 110:1726.

4. Birt, D.F., Salmasi, S. and Pour, P.M., 1981, Enhancement of experimental pancreatic cancer in Syrian golden hamsters by dietary fat, J. Natl. Cancer Inst., 67:1327-1332.

5. Boorman, G.A. and Eustis, S.L., 1984, Proliferative lesions of the exocrine pancreas in male F344/N rats, Environ. Health Prospect., 56:213-217.

6. Booth, A.N., Robbins, D.J., Ribelin, W.E. and DeEds, F., 1960, Effects of raw soybean meal and amino acids on pancreatic hypertrophy in rats, Proc. Soc. Exp. Biol. Med., 104:681-683.

7. Chernick, S.S., Lepkovsky, S. and Chaikoff, I.L., 1948, A dietary factor regulating the enzyme content of the pancreas: Changes indiced in size and proteolytic activity of the chick pancreas by the ingestion of raw soy-bean meal, Am. J. Physiol., 155:33-41.

8. Chiu, T., 1983, Spontaneous hypertrophic foci of pancreatic acinar cells in CD rats, Toxicol. Pathol., 11:115-119.

9. Chiu, T., 1985, Hypertrophic foci of pancreatic acinar cells in rats, CRC Critical Rev. Toxicol., 14:133-157.

10. Crass, R.A. and Morgan, R.G.H., 1981, Rapid changes in pancreatic DNA, RNA and protein in the rat during pancreatic enlargement and involution, Int. J. Vitam. Nutr. Res., 51:85-91.

11. Dijkhof, J., Poort, S.R. and Poort, C., 1977, Effect of feeding soybean flour containing diets on the protein synthetic pattern of the rat pancreas, J. Nutr., 107:1985-1995.

12. Farber, E., 1984, The multistep nature of cancer development, Cancer Res., 44:4217-4223.

13. Folsch, U.R., Winckler, K. and Wormsley, K.G., 1974, Effect of a soybean diet on enzyme content and ultrastructure of the rat exocrine pancreas, Digestion, 11:161-171.

14. Fox, J.G., Thibert, P., Arnold, D.L., Krewski, D.R. and Grice, H.C., 1979, Toxicology studies. II. The laboratory animal, Fd. Cosmet. Toxicol., 17:661-675.

15. Green, G.M. and Lyman, R.L., 1972, Feedback regulation of pancreatic enzyme secretion as a mechanism for trypsin inhibitor-inducedhypersecretion in rats, Proc. Soc. Exp. Biol. Med., 140:6-12.

16. Levison, D.A., Morgan, R.G.H., Brimacombe, J.S., Hopwood, D., Coghill, G. and Wormsley, K.G., 1979, Carcinogenic effects of di(2-hydroxypropyl) nitrosamine (DHPH) in male Wistar rats: Promotion of pancreatic cancer by raw soya flour diet, Scand. J. Gastroent., 14:217-224.

17. Longnecker, D.S., 1984, Lesions induced in rodent pancreas by azaserine and other pancreatic carcinogens, Environ. Health Prospect., 56:245-251.

18. Longnecker, D.S., Curphey, T.J., Kuhlmann, E.T. and Roebuck, B.D., 1982, Inhibition of pancreatic carcinogenesis by retinoids in azaserine-treated rats, Cancer Res., 42:19-24.

19. Longnecker, D.S., Wiebkin, P., Shaeffer, B.K. and Roebuck, B.D., 1984, Experimental carcinogenesis in the pancreas, Int. Rev. Exp. Pathol., 26:177-229.

20. McGuinness, E.E., Hopwood, D. and Wormsley, K.G., 1982, Further studies on the effects of raw soya flour on rat pancreas, Scand. J. Gastroent., 17:273-277.

21. McGuinness, E.E., Morgan, R.G.H., Levison, D.A., Frape, D.L., Hopwood, D. and Wormsley, K.G., 1980, The effects of long-term feeding of soya flour on the rat pancreas, Scand. J. Gastroent., 15:497-502.

22. McGuinness, E.E., Morgan, R.G.H., Levison, D.A., Hopwood, D. and Wormsley, K.G., 1981, Interaction of azaserine and raw soya flour on the rat pancreas, Scand. J. Gastroent., 16:49-56.

23. McGuinness, E.E., Morgan, R.G.H. and Wormsley, K.G., 1984, Effects of soybean flour on the pancreas of rats, Environ. Health Prospect., 56: 205-212.

24. Melmed, R.N., El-Aaser, A.A.A. and Holt, S.J., 1976, Hypertrophy and hyperplasia of the neonatal rat exocrine pancreas induced by orally administered soybean trypsin inhibitor, Biophys. Acta, 421:280-288.

25. Morgan, R.G.H., Levinson, D.A., Hopwood, D., Saunders, J.H.B. and Wormsley, K.G., 1977, Potential of the action of azaserine on the rat pancreas by raw soy bean flour, Cancer Letters, 3:87-90.

26. Newberne, P.M. and Fox, J.G., 1980, Nutritional adequacy and quality control of rodent diets, Lab. Anim. Sci., 30:352-363.

27. Pour, P.M., Runge, R.G., Birt, D., Gingell, R., Lawson, T., Nagel, D., Wallcave, L. and Salmasi, S.Z., 1981, Current knowledge of pancreatic carcinogenesis in the hamster and its relevance to the human disease, Cancer, 47:1573-1587.

28. Pugh, T.D., King, J.H., Koen, H., Nychka, D., Chover, J., Wahba, G., He, Y. and Goldfarb, S., 1983, Reliable stereological method for estimating the number of microscopic hepatocellular foci from their transections, Cancer Res., 43:1261-1268.

29. Rackis, J.J., 1965, Physiological properties of soybean trypsin inhibitors and their relationship to pancreatic hypertrophy and growth inhibition of rats, Fed. Proc., 24:1488-1493.

30. Rao, M.S., Upton, M.P., Subbarao, V. and Scarpelli, D.G., 1982, Two populations of cells with differing proliferative capacities in atypical acinar cell foci induced by 4-Hydroxyaminoquinoline-1-oxide in the rat pancreas, Lab. Invest., 46:527-534.

31. Roebuck, B.D., Baumgartner, K.J. and Thron, C.D., 1984, Characterization of two populations of pancreatic atypical acinar cell foci induced by azaserine in the rat, Lab. Invest., 50:141-146.

32. Roebuck, B.D., Baumgartner, K.J., Thron, C.D. and Longnecker, D.S., 1984, Inhibition by retinoids of the growth of azaserine-induced foci in the rat pancreas, J. Natl. Cancer Inst., 73:233-236.

33. Roebuck, B.D., Kaplita, P.V. and Baumgartner, K.J., 1985, Enhancement of pancreatic carcinogenesis by raw soybean protein isolate, Fed. Proc., 44:xii.

34. Roebuck, B.D. and Longnecker, D.S., 1977, Species and rat strain variation in pancreatic nodule induction by azaserine, J. Natl. Cancer Inst., 59:1273-1277.

35. Roebuck, B.D. and Longnecker, D.S., 1979, Response of two rodents, Mastomys natalensis and Mystromys albicaudatus, to the pancreatic carcinogen azaserine, J. Natl. Cancer Inst., 62:1269-1271.

36. Roebuck, B.D. and Longnecker, D.S., 1983, Dietary lipid promotion of azaserine-induced pancreatic tumors in the rat, in: "Diet, Nutrition, and Cancer: From Basic Research to Policy Implications", D.A. Roe, ed., Alan R. Liss, Inc., New York.

37. Roebuck, B.D., Longnecker, D.S., Baumgartner, K.J. and Thron, C.D., 1985, Carcinogen-Induced lesions in the rat pancreas: Effects of varying levels of essential fatty acid, Cancer Res., in press.

38. Roebuck, B.D., Longnecker, D.S. and Yager, J.D., 1983, Initiation and promotion in pancreatic carcinogenesis, in: "Mechanisms of Tumor Promotion. Vol. 1. Tumor Promotion in Internal Organs", T.J. Slaga, ed., CRC Press, Inc., Boca Raton, FL.

39. Roebuck, B.D., MacMillan, D.L., Bush, D.M. and Kensler, T.W., 1984, Modulation of azaserine-induced pancreatic foci by phenolic antioxidants, J. Natl. Cancer Inst., 72:1405-1410.

40. Roebuck, B.D., Yager, J.D. and Longnecker, D.S., 1981, Dietary modulation of azaserine-induced pancreatic carcinogenesis in the rat, Cancer Res., 41:888-893.

41. Roebuck, B.D., Yager, J.D., Longnecker, D.S. and Wilpone, S.A., 1981, Promotion by unsaturated fat of azaserine-induced pancreatic carcinogneesis in the rat, Cancer Res., 41:3961-3966.

42. Rowlatt, U.F., 1967, Pancreatic neoplasms of rats and mice, in: "Pathology of Laboratory Rats and Mice", F.J.C. Roe, ed., Blackwell Scientific Publications, Oxford.

43. Ryser, H.J.-P., 1971, Chemical carcinogenesis, New Engl. J. Med., 285: 721-734.

44. Scarpelli, D.G., Rao, S.M. and Reddy, J.K., 1984, Studies of pancreatic carcinogenesis in different animal models, Environ. Health Prospect., 56:219-227.

45. Slaga, T.J. (ed.), 1983, "Mechanisms of Tumor Promotion. Vol. I. Tumor Promotion in Internal Organs", CRC Press, Inc., Boca Raton, FL.

46. Struthers, B.J., MacDonald, J.R., Dahlgren, R.R. and Hopkins, D.T., 1983, Effects on the moneky, pig and rat pancreas of soy products with varying levels of trypsin inhibitor and comparison with the administration of cholecystokinin, J. Nutr., 113:86-97.

47. Temler, R.S., Dormond, C.A., Simon, E., Morel, B. and Mettraux, C., 1984, Response of rat proteases to dietary proteins, their hydrolysates and soybean trypsin inhibitor, J. Nutr., 114:270-278.

48. Tove, S.B., 1981, Working group V: Hormones, fat, and cancer. Report and recommendations, Cancer Res., 41:3824.

49. Weast, R.C. (ed.), 1968, "Handbook of Chemistry and Physics, 49th Edition", The Chemical Rubber Company, Cleveland, OH.

50. Yager, J.D., Roebuck, B.D., Zurlo, J., Longnecker, D.S., Weselcouch, E.O. and Wilpone, S.A., 1981, Rationale for a single-dose azaserine protocol for initiation of pancreatic carcinogenesis in the rat, Int. J. Cancer, 28:601-606.

6

PANCREATIC CARCINOGENESIS - THE POTENTIAL OF CHOLECYSTOKININ AS A COCARCINOGEN IN THE HAMSTER - NITROSAMINE MODEL

Allan G Howatson

University Department of Pathology
Royal Infirmary, Glasgow
Scotland, U.K.

INTRODUCTION

A number of epidemiological studies have demonstrated associations which support the concept that cancer of the exocrine pancreas in man is the result of chemical carcinogenesis. Cigarette smoking doubles the risk of developing pancreatic cancer (Kahn, 1966; Doll and Peto, 1976) and is a source of many known carcinogens (Hoffmann et al., 1982). The high fat-protein diet of Western countries has also been suggested as a factor increasing the risk of developing pancreatic cancer (Wynder, 1975; Armstrong and Doll, 1976; Hirayama, 1979).

Experimental studies have shown that diet can modify carcinogenesis. In hamsters and rats pancreatic carcinogenesis is enhanced by a diet high in unsaturated fat (Birt, Salmasi and Pour, 1981; Roebuck, Yager and Longnecker, 1981). Other studies, using the rat acinar cell carcinoma model, have shown that raw soya flour containing trypsin inhibitor potentiates the action of the carcinogen, azaserine (Morgan et al., 1977).

The interpretation of these observations in terms of the multistep concept of chemical carcinogenesis (Farber, 1982) has led to a search for potential cocarcinogens in pancreatic carcinogenesis. It is well known that carcinogenic agents are more effective in tissues with increased cell metabolic activity and cell turnover both of which are secretion (Harper and Scratcherd, 1979). This paper reports a series of experiments to determine the potential of CCK to modify the pancreatic response to carcinogen in the hamster-nitrosamine model of pancreatic carcinogenesis developed by Pour and co workers (1977).

MATERIALS AND METHODS

The animals used in these experiments were 10 weeks old, male, WO strain, Syrian hamsters kept under standardised conditions in groups of four and fed Oxoid 41B diet and water ad libitum.

Effect of CCK on Pancreatic Exocrine Secretion

In order to identify a dose of CCK which had a significant pancreaticotrophic effect, the effect of step-wise increasing doses of exogenous CCK (20% natural CCK, GIH Research Unit, Karolinska Institute, Stockholm, Sweden) was studied in twelve hamsters. The animals were fasted for 24 hours with free access to water and were then anaesthetised with intraperitoneal Sagatal. Tracheostomy was performed and an intravenous cannula inserted into the left jugular vein. A saline infusion was commenced at a rate of $0.375 ml^{-1}$. A laparotomy was performed, the pylorus ligated and the common bile duct ligated in continuity distal to the entry of the cystic duct. The common bile duct, into which the two main pancreatic ducts drain, was cannulated as it entered the serosal coat of the duodenal wall and the cannula led out through the flank. A heating pad maintained body temperature at 34°C. The test was commenced after an equilibration period of one hour. The pancreatic juice was collected in preweighed tubes. A basal output was collected for 1 hour and thereafter step-wise increasing doses of CCK in saline were infused with 1 hour long collections at each dose level (0, 7.5, 15, 30, 60 $IDUkg^{-1}h^{-1}$) and 20 minutes allowed for equilibration between doses. Volume output was measured by weighing. At the end of each test a lissamine dye solution was injected up the cannula. A test was regarded as technically satisfactory if the whole pancreas was stained indicating that all of the gland had been draining into a patent duct system.

Trophic effect of CCK

Two experiments were performed to examine the trophic effect of CCK on the growth of the pancreas. The dose of CCK used was $30 IDUkg^{-1}$ (Ivy dog units, IDU) which produced a maximal response in terms of pancreatic juice volume output when administered intravenously in the experiment described above.

In the first experiment two groups of 10 animals received either CCK $30 IDUKg^{-1}$ twice daily by s.c. injection for 15 days in hydrolysed gelatin carrier or gelatin carrier alone. The CCK was made up in 10% hydrolysed gelatin in saline in order to prolong absorption (Petersen et al., 1978).

In the second experiment two groups of 10 animals received either CCK 30 $IDUkg^{-1}$ twice daily for 3 consecutive days per week for 6 weeks in gelatin carrier, by s.c. injection, or carrier alone.

At the end of each experiment the animals were killed, the pancreas removed, trimmed of fat and connective tissue and weighed. Tissue from two sites in the gastric and splenic lobes was taken for histological examination. In addition to recording the pancreatic wet weight (PWW: mg pancreas per 100g body weight), the pancreatic content of DNA (μg DNA per 100mg PWW) was also determined. The DNA was extracted by the method of Schmidt-Thanhauser as modified by Munro and Fleck (1966) and measured by the modified Burton method (Burton, 1956; Giles and Myers, 1965). The diphenylamine reaction incubation time for the DNA measurement in the first experiment was six hours and in the second experiment was fifteen hours.

Co-Carcinogenic Potential of CCK

Two groups of 100 animals received N-nitrosobis (2-oxopropyl)amine (BOP; Ash Stevens Inc., Detroit, USA) in a dose of 5mg kg^{-1} once weekly for the duration of the experiment by subcutaneous injection. One group also received CCK 30IDUkg^{-1} b.d. in hydrolysed gelatin s.c. for 3 days per week for 6 weeks on the day before, the day of, and the day after carcinogen administration; the other group received only the gelatin vehicle. CCK was administered over the 6 week period because this was the earliest stage at which premalignant lesions had been detected at microscopy in pilot studies of the model and initiation had therefore definitely occurred.

Twenty animals from each treatment group were scheduled to be killed at 5, 7.5, 10, 12.5 and 15 weeks after commencement of the experiment. A full post mortem examination was performed and all major organs examined for the presence of neoplastic lesions. The pancreas was fixed en bloc in 10% formalin and the whole organ was then blocked out, there being an average of 11 blocks for each pancreas. Three step sections were taken from each block for examination by light microscopy after staining with haematoxylin and eosin. Macroscopic tumours were excised from the main pancreatic specimen at post mortem and separately examined for histological confirmation.

Each pancreas was assessed for the presence of the following histological appearances, affecting ducts or of ductular morphology, which were used as end points for the assessment of the two groups; duct dysplasia and duct carcinoma in-situ, ductular proliferation (tubular complexes) and acinar ductular proliferation affecting an entire lobule, and pancreatic adenocarcinoma. Malignant tumours that were intralobular and those that were invasive, i.e. extending outside pancreatic lobular boundaries were classified under the single heading of pancreatic adenocarcinoma.

If any of these lesions were identified in a histological section the lesion was recorded as present for that pancreas and no formal attempt was made to record the frequency or extent of a particular lesion. Duct dysplasia and carcinoma-in-situ represent atypical duct hyperplasia of varying degress of severity with carcinoma-in-situ lesions showing all the cellular features of malignancy and the presence of mitoses. The development of ductules, or tubular complexes, as a result of ductular/centriacinar cell hyperplasia and acinar-ductular transformation was regarded as a significant premalignant lesion if virtually all of a lobule was involved (Figure 1).

All sections were examined by a single observer (AGH) and a random 20% sample selected by an individual not involved in the experimental work was examined by a trained pathologist who was not aware of the nature and duration of treatment given to any particular animal.

Statistical Analysis

All data are expressed as mean ± standard deviation. The analysis of the exocrine secretion study was by Student's t-test for paired observations. Student's t-test for unpaired value was used for the analaysis of the CCK trophism experiments. The data were normally distributed. The analysis of the histological assessment was by Fisher's exact test.

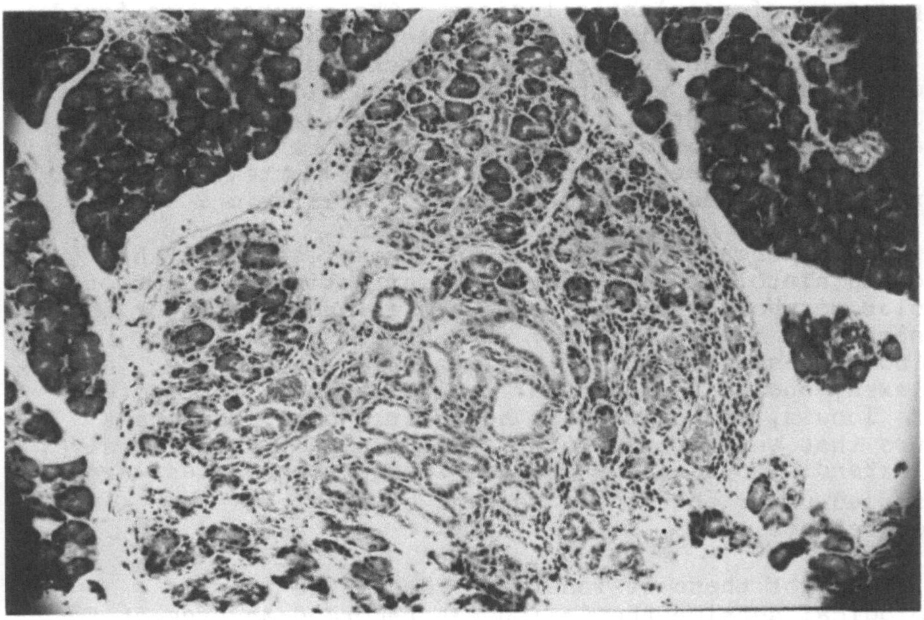

Figure I Pancreatic lobule showing ductular proliferation and morphological acinar-ductular transformation. Carcinogen + CCK treated animal after 10 weeks. H & E x 200.

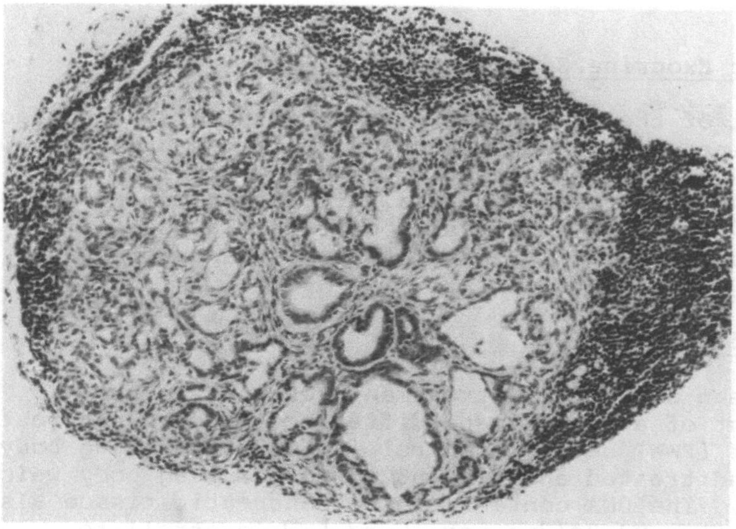

Figure 2 Metastatic adenocarcinoma in a peripancreatic lymph node draining the gastric lobe. Carcinogen + CCK treated animal after 15 weeks. H & E x 200.

Table I Effect of CCK on histological changes in hamster pancreas induced by BOP at 10 weeks

	BOP+CCK n=15	BOP n=10	P value
duct dysplasia	9	3	NS
duct carcinoma-in-situ	7	0	0.013
panlobular ductular proliferation	9	1	0.016
pancreatic adenocarcinoma	0	0	-

Table II Effect of CCK on histological changes in hamster pancreas induced by BOP at 12.5 weeks.

	BOP+CCK n=17	BOP n=13	P value
duct dysplasia	11	7	NS
duct carcinoma-in-situ	8	3	NS
panlobular ductular proliferation	9	2	0.04
pancreatic adenocarcinoma	5	0	0.043

Table III Effect of CCK on histological changes in hamster pancreas induced by BOP at 15 weeks

	BOP+CCK n=17	BOP n=10	P value
duct dysplasia	7	6	NS
duct carcinoma-in-situ	10	3	NS
panlobular ductular proliferation	8	4	NS
pancreatic adenocarcinoma	12	2	0.015

RESULTS

Pancreatic Exocrine Secretion

Eight of the twelve tests were technically satisfactory as determined by the dye injection described previously. The basal output of pancreatic juice was 215.1 ± $175\mu l kg^{-1} h^{-1}$ and rose sequentially to a maximum of 515.0± $125.5\mu l kg^{-1} h^{-1}$ with an infusion of CCK at a dose of $30 IDU kg^{-1} h^{-1}$ ($P<0.01$). This dose was subsequently used in the other experiments.

Trophic Effects of CCK

Fifteen days of CCK treatment resulted in the development of a significant difference in mean pancreatic wet weight (PWW) between control (236±27mg per 100g body weight) and treated animals (532±72mg per 100g body weight; ($P<0.001$). The DNA content of the pancreatic tissue also increased from 93.2±17μg per 100mg PWW (control) to 137.3±16μg per 100mg PWW (CCK) ($P<0.001$).

Six weeks of CCK treatment resulted in an increased PWW from 295.6±61 (control) to 466.4±77mg per 100g body weight (CCK) ($P<0.001$). The DNA content did not change significantly (Control 217.5±29.5μg per 100mg PWW; CCK 222.2± 77.7μg per 100mg PWW).

Histological examination of the tissue from the gastric and splenic lobes in animals receiving CCK for six weeks and corresponding controls showed no evidence of neoplastic lesions.

Cocarcinogenic Potential of CCK

One hundred and thirty seven animals were avilable for the histological assessment. Losses were due to infective enteritis unrelated to carcinogen treatment. All 137 animals gained weight progressively during the study, were in good health and showed no evidence of enteritic symptoms. No animals which developed enteritis survived.

Macroscopically tumours were identified only in the pancreas at post mortem.

A metastatic carcinoma deposit was present in the lymph node draining the gastric lobe in one CCK treated animal (Figure 2).

Histology of the step sections of the animals killed at 5 and 7.5 weeks showed no adenocarcinomas and no difference in the incidence of duct dysplasia or panlobular ductular proliferation between the two treatment groups. Forty percent of animals in each group showed duct dysplasia and 15% of the CCK + carcinogen group showed duct carcinoma-in-situ at 7.5 weeks. Panlobular ductular proliferation was not present in either group at 5 weeks but by 7.5 weeks this lesion was found in 15% of the carcinogen above group and 50% of the CCK + carcinogen group.

The results of the histological assessment of the groups at 10, 12.5 and 15 weeks are shown in Tables I -III. At 10 weeks there was a significant excess of duct carcinoma-in-situ and panlobular ductular proliferation in the CCK treated group. By 12.5 weeks, more animals in the carcinogen alone groups developed duct lesions and there was a significant excess of panlobular ductular proliferation and adenocarcinomas in the CCK + carcinogen group. At 15 weeks there remained a significant excess of adenocarcinomas. Premalignant and malignant lesions developed earlier and with greater frequency in the CCK + carcinogen treated animals as compared to those receiving carcinogen alone. There was full agreement between the two histological assessors on the presence and nature of the lesions in 93% of instances. The remaining cases were allocated to histological categories after discussion.

DISCUSSION

Subcutaneous injection of the carcinogenic nitrosamine N-nitrosobis(2 - oxopropyl)amine (BOP) induces the development of benign, premalignant, and malignant lesions in the pancreas of Syrian hamsters (Pour et al., 1977). The induction rate, latency period and pancreatic specificity of tumour development are dose dependent. This tumour model is widely considered to approximate most closely to the morphology and biology of human pancreatic cancer in that the tumours are of ductular morphology, the predominant histological variant in man (Cubilla and Fitzgerald, 1980), invade locally, obstruct the common bile duct causing jaundice, and metastasise to liver, lymph nodes and other sites.

Normal tissues, when treated with cocarcinogens, show increased cell metabolism, DNA synthesis and mitotic activity causing tissue hypertrophy and hyperplasia. The response of an initiated cell population is a reduction of latency period and increased tumour induction rate. These studies of the cocarcinogenic potential of CCK have examined whether these changes occur in normal hamster pancreas and in the hamster-nitrosamine pancreatic carcinoma model

The trophism experiments confirm the capacity of CCK to increase the cell metabolism and mitotic rate of the cells of the exocrine pancreas as has been previously described (Petersen et al., 1978). The short term, high dose regime of twice daily injection for fifteen days showed significant hypertrophy and hyperplasia of the pancreas with markedly increased pancreatic wet weight and DNA content. CCK treatment for three days per week for six weeks induced significant pancreatic hypertrophy (increased wet weight). The DNA content per unit mass of pancreas was not significantly increased by six weeks of CCK treatment, but the fact that the DNA content per unit mass did not decrease in the presence of significant hypertrophy proves that DNA synthesis had been increased. These results show that CCK can, in the normal pancreas, induce changes similar to the effects of cocarcinogens in other normal tissues and model systems and may therefore have cocarcinogenic potential. However, while cocarcinogens virtually always cause hyperplasia, a hyperplastic stimulant need not necessarily be an effective cocarcinogen. The determining factor is the effect on the latency period and induction rate of tumour development.

The carcinogen experiment shows that the addition of CCK to the BOP regime resulted in significant potentiation of the effect of BOP. Duct dysplasia and duct carcinoma-in-situ appeared earlier and with greater frequency in the BOP+ CCK treated animals as compared to those receiving BOP alone. The effect of carcinogen on ductular and acinar cells in causing ductular proliferation and acinar-ductular transformation with the subsequent development of pancreatic adenocarcinoma is more dramatic. At 10 weeks panlobular ductular proliferation, which subsequently leads to the development of lobular carcinoma was present in 10% of the BOP alone group and 60% of the BOP + CCK group. It was not until 15 weeks that 40% of the BOP alone group showed this lesion. The absence of any significant difference between the two treatment groups in the development of panlobular ductular proliferation at 15 weeks perhaps reflects the fact that in the CCK group many of these lesions had progressed to become adenocarcinomas. In fact, adenocarcinoma was present in several lobules in the pancreases of the 12 animals with this lesion at 15 weeks.

Pancreatic lobular carcinoma and invasive carcinoma lesions appeared earlier in the CCK + BOP group than in the BOP alone group. The reduction in latency period and increased induction rate of tumour development is consistent with cholecystokinin serving as a cocarcinogen in this model.

The ability of CCK to enhance the effectiveness of carcinogen as demonstrated above may act by at least two possible mechanisms. First, the effect may be the result of the generation of increased numbers of susceptible cells and intracellular targets for carcinogen as a result of increased cell proliferation. The experiments show increased DNA synthesis in the CCK treated animals. An alternative concept is that CCK might influence the metabolism of the parent carcinogen or a more proximate form in the target cells.

Nitrosamines, including BOP, are indirect carcinogens and must be metabolised to proximate and ultimate carcinogenic and mutagenic forms. It is now established that nitrosamines are activated by mixed function oxidase (MFO) enzyme systems which have been identified in many tissues. The organ specificity, of a carcinogen may well depend on the capability of a tissue to metabolise the carcinogen. Metabolism, in situ, of the indirect carcinogen or a proximate form transported to the target cells via the blood stream after metabolism elsewhere, liver for example, may be critical because of the very short lived nature of the resultant electrophiles, the ultimate carcinogens, which bind to macromolecular tagets, including DNA, in the cell (Miller, 1970). The original studies on the metabolism and activiation of nitrosamines focussed on hepatic microsomes and MFO systems but more recently it has been clearly shown that the pancreas also contains MFO enzymes which can metabolise carcinogenic nitrosamines and generate mutagens. The MFO enzymes are inducible, in vitro and the induction of these enzymes is associated with an increased conversion of carcinogen to mutagen as demonstrated by the Ames assay (Scarpelli et al., 1980).

Experiments using tritium labelled N-nitroso-2,6-di-methylmorpholine, a carcinogenic nitrosamine, have provided evidence that both the pancreatic acinar and ductal cells are capable of nitrosamine metabolism (Reznik-Schuller et al., 1980). The acinar cell population appears to be the major site of metabolic activation and comprises by far the largest proportion of the pancreatic cell mass. In vitro studies have shown that the electrophilic radicals, carbonium ions, cause alkylation of bases in the DNA and the evidence suggests that the alkylation of the oxygen, in the 0-6 position, of guanine is a critical step in the process of mutagenesis and carcinogenesis (Margison et al., 1976; Pegg, 1977). The presence of 0^6- methylguanine causes guanine-adenine base mispairing, instead of the normal guanine-cytosine pairing relationship, in <u>in vitro</u> studies of RNA transcription (Gerchman and Ludlum, 1973). The potential for frame shift mutations is obvious. In addition to the alkylating effects of nitrosamines, the other products of nitrosamine metabolism, aldehydes, also induce multiple lesions in the DNA including DNA-protein crosslinks, single DNA strand breaks, and inhibition of DNA repair mechanisms (Harris et al., 1982). Clearly, any increase in the capability of the pancreas to metabolise a nitrosamine carcinogen will lead to greater production of electrophilic radicals which would enhance the mutagenic effect of a given carcinogen dose by increasing the incidence of DNA lesions.

The experiments outlined above suggest that CCK, by its trophic effect on the pancreas should be a candidate for consideration as a cocarcinogen in pancreatic carcinogenesis. The trophic effect of CCK might include the stimulation of those MFO enzymes capable of nitrosamine metabolism. Support for this contention comes not only from the data presented here but also from studies of the effect of trypsin inhibitor containing soya flour diets on normal pancreas and pancreatic carcinogenesis. The pancreases of rats fed raw soya flour, containing active inhibitor, show hypertrophy, hyperplasia and increased enzyme protein (amylase, trypsin and lipase) and non enzymic protein content as compared to rats fed heat inactivated soya flour (Folsch, Winckler and Wormsley, 1974). Similar diets have been shown in the rat-azaserine acinar cell carcinoma model to enhance the effect of carcinogen (Morgan et al., 1977). The trypsin inhibitor effect is probably mediated by interference with the normal negative feedback regulation of endogenous CCK activity which induces an enhanced CCK response with subsequent and prolonged pancreatic exocrine stimulation.

There has been recent speculation on the question of soya based infant formulas which contain low levels of trypsin inhibitor and their possible cumulative effects on the pancreas. Studies in rats have shown that with long term feeding up to two years, of raw soya flour, containing more than 40,000mg/kg of trypsin inhibitor, some animals develop carcinomas. Heating the flour prevented this effect. The following points should be made. Two years is a very considerable proportion of the life expectancy of a rat and the trypsin inhibitor "dose" is very high. Neither of these parameters bear any resemblence to the human infant formula situation. Furthermore, pancreatic carcinoma is a disease of the sixth decade of life and beyond in man. This plus the epidemiological evidence suggest that the exposure to the carcinogen is the result of social/occupational habits of adult life. Initiation precedes promotion in the carcinogenesis process and therefore it is unlikely that the soya based formula represents a significant risk of the development of pancreatic cancer later in life.

The results of these studies may have important implications for our understanding of the pathogenesis of pancreatic cancer both in the hamster model and in other species including man. The physiology of the human pancreas is similar to that of the hamster and rat and many studies have shown that human tissues, including pancreas, metabolise nitrosamines to produce mutagens in an identical manner to that outlined above (Harris et al., 1982). The study presented here demonstrates that the exposure of the pancreas to a hormone normally released from the duodenum by the ingestion of fat and protein, that acts as a stimulant of pancreatic exocrine activity, potentiates the action of carcinogen and/or increases pancreatic susceptibility to carcinogen. The consumption of a fat-protein rich diet is associated with an increased incidence of pancreatic cancer (Armstrong and Doll, 1975) A mechanism is suggested by which diet may affect pancreatic carcinogenesis.

ACKNOWLEDGMENTS

This work was supported by a grant from the Cancer Research Campaign of the United Kingdom. I am grateful for the help of Professor D.C. Carter and for the assistance of technical colleagues during the performance of these studies.

REFERENCES

Armstrong, B., Doll, R., 1975, Environmental factors and cancer incidence and mortality in different countries, with special reference to dietary practices.
Internatl J Cancer., 15:617-631.

Birt, D.F., Salmasi, S., Pour, P.M., 1981, Enhancement of experimental pancreatic cancer in Syrian golden hamsters by dietary fat.
J Natl Cancer Inst., 67:1327-1332.

Burton, K., 1956, A study of the conditions and mechanisms of the diphenylamine reaction for the calorimetric estimation of deoxyribonucleic acid.
Biochem J., 62:315-323.

Cubilla, A.L., Fitzgerald, P.J., 1980, Cancer (nonendocrine) of the pancreas. A suggested classification, in: "The Pancreas," Fitzgerald, P.J., and Morrison, A.B., eds. International Academy of Pathology, Monograph No. 21, Baltimore.
Williams and Wilkins., 6:82-110.

Doll, R., Peto, R., 1976, Mortality in relation to smoking: 20 years observations on male British doctors.
Br Med J., 2:1525-1536.

Farber, E., 1982, Chemical carcinogenesis: a biological perspective.
Am J Path., 106:271-296.

Fölsch, U.R., Winckler, K., Wormsley, K.G., 1974, Effect of a soybean diet on enzyme content and ultrastructure of the rat exocrine pancreas.
Digestion., 11:161-171.

Gerchman, L.L., Ludlum, D.B., 1973, The properties of O^6-methylguanine in templates for RNA polymerase.
Biochem Biophys Acta., 308:310-316.

Giles, K.W., Myers, A., 1965, An improved diphenylamine method for the estimation of deoxyribonucleic acid.
Nature., 206:93.

Harper, A.A., Scratcherd, T., 1979, Physiology. in: "The Exocrine Pancreas," H.T. Howat, H. Sarles, eds.,
W.H. Saunders, Philadelphia.

Harris, C.C., Grafstrom, R.C., Lechner, J.F., Autrup, H., 1982, Metabolism of N-nitrosamines and repair of DNA damage in cultured human tissues and cells, in: "Nitrosamines and human cancer," P.N. Magee, ed.,
Cold Spring Harbour Laboratory., 122-134.

Hirayama, T., 1979, Diet and cancer.
Nutrit Cancer., 1:67-81.

Hoffmann, D., Brunnemann, K.D., Adams, J.D., Rivenson, A., Hecht, S.S., 1982, N-Nitrosamines in tobacco carcinogenesis, in: "Nitrosamines and human cancer," P.N. Magee, ed.
Cold Spring Harbour Laboratory., 211-220.

Kahn, H.A., 1966, The Dorn study of smoking and mortality among U.S. veterans, report on eight and one-half years of observation.
Natl Cancer Inst Monogr., 19:1-126.

Margison, G.P., Margison, J.M., Montesano, R., 1976, Methylated pruines in the deoxyribonucleic acid of various Syrian golden hamster tissues after administration of a hepatocarcinogenic dose of dimethylnitrosamine.
Biochem J., 157: 627-634.

Miller, J.A., 1970, Carcinogenesis by chemicals: An overview - G.H.A. Clowes Memorial Lecture.
Cancer Res., 30:559-576.

Morgan, R.G.H., Levinson, D.A., Hopwood, D., Saunders, J.H.B., Wormsley, K.G., 1977, Potentiation of the action of azaserine on the rat pancreas by raw soya bean flour.
Cancer Lett., 3:87-90.

Munro, H.N., Fleck, A., 1966, Recent developments in the measurement of nucleic acids in biological materials.
Analyst., 91:78-88.

Pegg, A.E., 1977, Formation and metabolism of alkylated nucleosides; possible role in carcinogenesis by nitroso compounds and alkylating agents.
Adv Cancer Res., 25: 195-269.

Petersen, H., Solomon, T., Grossman, M.I., 1978, Effect of chronic pentagastin, cholecystokinin and secretin on pancreas of rats.
Am J Physiol., 234:E286-E293.

Pour, P., Althoff, J., Kruger, F.W., Mohr, U., 1977, A potent pancreatic carcinogen in Syrian hamster: N-nitrosobis(2-oxopropyl)amine.
J Natl Cancer Inst., 58:1449-1453.

Reznik-Schuller, H.M., Lizinsky, W., Hague, B.F., 1980, Electron microscopic autoradiography of the pancreas in the hamster treated with tritiated N-nitroso-2, 6-dimethylmorpholine.
Cancer Res., 40:2245-2247.

Roebuck, B.D., Yager, J.D., Longnecker, D.S., 1981, Dietary modulation of azaserine-induced pancreatic carcinogenesis in the rat.
Cancer Res., 41:888-893.

Scarpelli, D.G., Rao, M.S., Subbarao, V., Beversluis, M., Gurka, D.P., Hollenberg, P.F., 1980, Activation of nitrosamines to mutagens by postmitochondrial fraction of hamster pancreas.
Cancer Res., 40:67-74.

Wynder, E.L., 1975, An epidemiological evaluation of the causes of cancer of the pancreas.
Cancer Res., 35:2228-2233.

7

INTERACTION OF DIETARY PROTEIN AND TRYPSIN INHIBITOR ON PLASMA CHOLECYSTOKININ AND PANCREATIC GROWTH IN RATS

Gary M. Green[1], Van H. Levan[1] and Rodger A. Liddle[2]

[1]Department of Physiology, University of Texas Health Science Center at San Antonio, San Antonio, TX 78284, and [2]Cell Biology Laboratory, Mt. Zion Hospital and Medical Center, San Francisco, CA 94120, and [2]Department of Medicine, University of California, San Francisco, CA 94143

INTRODUCTION

Pancreatic exocrine secretion in the rat is inhibited by trypsin, chymotrypsin and elastase in the proximal small intestine (Green and Lyman, 1972; Schneeman and Lyman, 1973; Green and Levan, 1985). Recent studies have supported the hypothesis that this negative feedback control of pancreatic secretion is mediated by cholecystokinin (CCK). Orogastric administration of trypsin inhibitors to rats caused a transient reduction of intestinal levels of CCK (Brand and Morgan, 1981) and greatly increased plasma levels of CCK (Liddle, Goldfine and Williams, 1984). Diversion of bile-pancreatic juice from the intestine increased plasma CCK levels, which were returned to normal by intraduodenal infusion of trypsin (Louie et al, 1985; Fölsh et al, 1984). The mechanism by which pancreatic proteases inhibit CCK release is not known. Miyasaka and Green (1983) hypothesized that trypsin inhibited pancreatic secretion by inactivating an intraluminally-secreted peptide, possibly a CCK-releasing factor.

Studies on the pancreatic secretory response to dietary protein and protein hydrolysates indicated that protein substrate was the major dietary stimulant of CCK release in the rat (Green and Miyasaka, 1983). It was proposed that protein substrate stimulated pancreatic secretion by complexing with trypsin (and other pancreatic proteases) in the intestine and thereby reversing trypsin-suppression of CCK release, analogous to the mechanism proposed for trypsin inhibitors (Green et al, 1973). Recent studies in which CCK release was measured directly have shown that intact protein is the major dietary stimulant of CCK secretion in the rat (Liddle et al, 1985b, 1985c).

We previously reported that increasing the concentration of protein (casein) in the diet caused a large, but transient, increase in plasma CCK level during feeding (Green, Levan and Liddle, 1984,1985). In that study, ad libitum-fed rats were adapted for 10 days to a control diet containing 5% casein, then were transferred to a diet containing 75% casein. They were sacrificed at 0900 hrs (without fasting) and plasma CCK level and pancreas weight were determined on the first, second, seventh and fourteenth days following introduction of the 75% casein diet, and in time-paired controls. Plasma CCK was increased approximately 3-fold in 75% casein-fed rats 24 hours after being transferred to the diet, but CCK levels fell and stabilized at values identical to controls by the seventh day. The return of plasma CCK to control values in rats fed 75% casein occurred in spite of a 15-fold increase in protein intake. Pancreas weight and protease content were greatly increased by the seventh day in 75% casein-fed rats. We proposed that equivalent plasma CCK levels in rats consuming widely different amounts of protein was a consequence of pancreatic adaptation to protein, which resulted in increased pancreatic protease synthesis and secretion and consequent increased feedback inhibition of intestinal CCK secretion. The results indicated that once pancreatic adaptation had occurred, plasma CCK concentration during meals would be the same regardless of the level of protein in the diet.

In the study described below, we determined the effect on plasma CCK and pancreatic growth of diets containing soybean trypsin inhibitor (SBTI), fed at three levels of dietary protein. The purpose of this study was to determine (1) the temporal pattern of CCK secretion in rats adapting to dietary trypsin inhibitors, and (2) to determine the influence of dietary protein on the plasma CCK response to trypsin inhibitors.

METHODS AND MATERIALS

Composition of Diets

Diet composition is shown in Table 1 below:

Table 1. Composition of Diets

Ingredient[1]	5% casein	5% Casein + SBTI	10% Casein + SBTI	20% casein + SBTI
Casein[2]	5	5	10	20
L-Methionine	0.3	0.3	0	0
L-Threonine	0.2	0.2	0	0
Dextrin	82.5	81.5	77	67
Corn Oil	5	5	5	5
Salt Mix[3]	5	5	5	5
Vitamin Mix[4]	2	2	2	2
SBTI[5]	0	1	1	1

1. Obtained from ICN Nutritional Biochemicals, Cleveland, OH, except for SBTI and corn oil.
2. Vitamin-free casein.
3. Jones and Foster Salt Mix.
4. Vitamin Diet Fortification Mixture.
5. Worthington Biochemicals, Freehold, NJ. Code SIC, 1 mg inhibited 0.8 mg bovine trypsin.

Assays

Plasma CCK was measured by a sensitive and specific bioassay as previously described by Liddle, Goldfine and Williams (1984) and Liddle et al., (1985). In brief, CCK was extracted from rat plasma by adsorption onto octadecylsilylsilica (Sep-Pak) cartridges (Waters Associates, Milford, Mass.) and eluted with 1 ml of 80% ethanol and 0.2% trifluoroacetic acid into incubation vials and dried under N_2. CCK in these extracts was quantitated by its ability to stimulate amylase release from a suspension of isolated pancreatic acini, prepared by collagenase digestion of whole rat pancreas. Values were compared to a standard curve of CCK-8 and results were expressed as CCK-8 equivalents (pM). The assay was sensitive to CCK levels as low as 1 pM and with the ability to concentrate plasma up to six fold (by adsorbing 6 ml of plasma onto a single Sep-Pak and eluting it in 1 ml) plasma CCK levels as low as 0.2 pM could be detected. There was little interference from gastrin as it was only 0.001 times as potent as CCK-8 in this system.

Trypsin activity in small intestinal washes was measured by titrimetric methods, using tosyl-L-arginine methyl ester (Green and Nasset, 1980). Activity is expressed as international units (μmole split/min).

Protocol

Seventy-two male Wistar rats, average weight 281 gm, were randomly divided into four groups of 18 and housed individually in wire-bottom cages in 24-h climate-controlled rooms with a 14-10 hr light-dark cycle (0700-2100h). Rats were fed ad libitum at all times. During an initial 10-day adaptation period, all rats were fed the control diet (5% casein without SBTI). On the morning of the 10th day, they were presented with the experimental diets. The protocol is illustrated schematically in Fig. 1. Day 0 corresponds to the end of the 10-day adaptation period. Rats were sacrificed by decapitation at 0900 hr, without fasting, on Day 1, Day 2 and Day 7, representing the number of days rats were fed the experimental diets, and corresponding time-paired controls. Trunk blood was collected in plastic heparinized centrifuge tubes. Plasma was quickly separated by centrifugation for 20 min at 10,000 x G and 4°C and CCK extracted as described above. The pancreas was removed and dissected free of visable fat, mesentery and lymph nodes, frozen and lyophilized to constant weight. The small intestine was removed and the proximal half rinsed with 6 ml of 0.15M NaCl at 0° and the rinses stored at -20°C until assayed for trypsin activity.

Statistical Analysis

Results were analyzed by Student's t test, with $P<0.05$ considered significant.

RESULTS

Plasma CCK values are shown in Fig. 2. The broken line represents

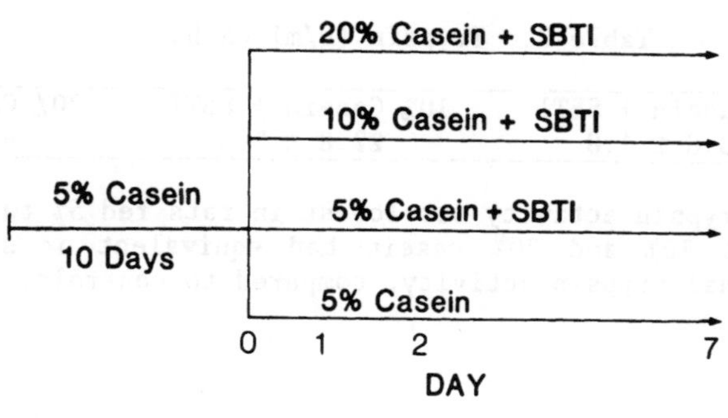

Fig. 1. Protocol for experiments. Rats were fed diets ad libitum and sacrificed at 0900 hrs in groups of six on Days 1,2 and 7.

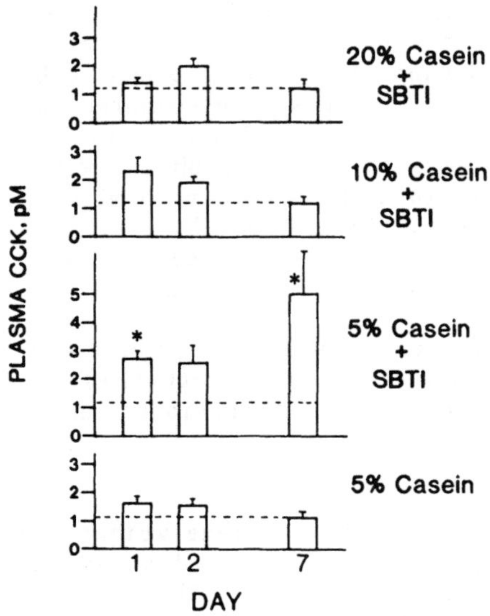

Fig. 2. Plasma cholecystokinin (CCK) expressed as pM (CCK-8 equivalents) in rats fed the diets indicated and sacrificed during ad libitum feeding. Asterisks indicate statistically significant differences (P<0.05) compared to controls (5% casein). Broken lines indicate values for controls on Day 7.

the value for plasma CCK in control animals (5% casein) on Day 7. Addition of SBTI stimulated a significant increase in plasma CCK only in rats fed the lowest level of casein. Plasma CCK levels in rats fed the higher levels of casein were significantly elevated on Day 1 (10% casein-fed) and Day 2 (10% and 20% casein-fed) when compared to their respective values on Day 7, but they were not significantly elevated compared to controls. The increase in plasma CCK in rats fed 5% casein + SBTI was greatest on Day 7. At this time, plasma CCK values in rats fed SBTI with 10% and 20% casein were identical with controls. Pancreatic size, expressed as mg dry weight/100g body weight, is illustrated in Fig. 3. The broken line here, as in Fig. 2, represents the value for controls on Day 7. By Day 7, pancreatic weight was significantly increased, compared to controls, only in rats fed SBTI at casein levels of 10% and 20%, with the greatest increase occurring in rats fed the highest level of casein. Average body weights on Day 7 were 320g, 298g, 332g and 339g for control, 5% casein + SBTI, 10% casein + SBTI, and 20% casein + SBTI, respectively.

Trypsin activities in washes of proximal small intestine in rats sacrificed on Day 7 are shown below:

Table 2. Trypsin (U/ml wash)

5% Casein	5% Casein + SBTI	10% Casein + SBTI	20% Casein + SBTI
20.8 ± 1.9	13.8 ± 1.8	22.8 ± 5.3	40.9 ± 4.0

Intestinal trypsin activity was lowest in rats fed 5% casein + SBTI. Rats fed SBTI at 10% and 20% casein had equivalent or significantly increased intestinal trypsin activity, compared to controls.

DISCUSSION

These results show that when rats adapted to 5% casein were pre-

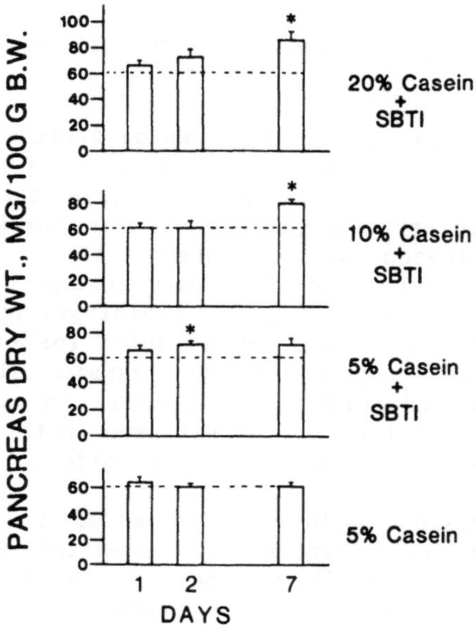

Figure 3. Pancreatic dry weights in rats in same experiment described in Fig. 2. Symbols as in Fig. 2.

sented with SBTI plus increased casein, plasma CCK was modestly increased during the initial 48 hours, and then declined to control values by the seventh day. This pattern resembles the pattern seen in our previously reported study (Green, Levan and Liddle, 1984, 1985), in which greatly increasing the protein concentration in the diet caused a transient increase in plasma CCK levels. When SBTI alone was added to the 5% casein diet, a different pattern was observed. A large and sustained increase in plasma CCK occurred, with a maximal increase on the seventh day.

We previously proposed (Green, Levan and Liddle, 1984, 1985) that the pattern of transient increase in plasma CCK in rats adapting to increased protein in the diet could be explained thusly: increased dietary protein substrate stimulates CCK secretion by binding to intraluminal pancreatic proteases and reversing protease-induced inhibition of CCK release. Increased CCK secretion then stimulates pancreatic growth and protease synthesis, and the resultant increased protease secretion suppresses CCK release, returning plasma CCK levels to normal. In this scheme, pancreatic growth and adaptation might be viewed as a homeostatic mechanism preventing persistent elevations in plasma CCK. In the present study, a pattern of transient increase in plasma CCK occurred when rats were fed SBTI along with increased casein, but a much more marked, and sustained, increase occurred when SBTI was fed without additional casein. We suggest that the sustained increased in plasma CCK in rats fed SBTI at the lowest level of casein is a consequence of inadequate pancreatic protease production secondary to impaired pancreatic growth resulting from amino acid insufficiency. According to this view, pancreatic adaptation to trypsin inhibitors is a function of increased CCK (the trophic stimulus) and amino acid availability. If amino acid availability is adequate, the pancreas responds to increased CCK secretion by increasing protease synthesis and secretion sufficient to block the increase in CCK secretion. If amino acid availability is inadequate, pancreatic protease synthesis in response to elevated circulating CCK is limited, with the result that protease secretion is inadequate to "feedback" and inhibit the increase in CCK release. This is supported by the data showing that intestinal free trypsin activity is decreased, relative to controls, only in rats fed SBTI with the lowest level of casein.

Conventional dietary protein (e.g., casein) differs importantly from trypsin inhibitors in affecting plasma CCK in that dietary protein provides both a stimulus for CCK release (protein substrate) and the building blocks for pancreatic adaptation, i.e., nitrogen and essential amino acids. For this reason, we would predict that increases in plasma CCK caused by increased consumption of dietary protein of good biological value would always be transient. Trypsin inhibitors provide a stimulus for CCK release, but do not provide amino acids for pancreatic protein synthesis. This applies to proteinaceous trypsin inhibitors such as SBTI, which are largly undegraded in the small intestine and usually constitute only a small fraction of the nitrogen in the diet, as well as to synthetic small molecular weight trypsin inhibitors.

Therefore, if trypsin inhibitors are added to a diet that is marginal or inadequate in nitrogen and essential amino acids, the protease synthetic response to CCK will be impaired, and protease secretion will be inadequate to suppress the increased CCK release. These concepts are illustrated schematically in Figures 4 and 5. Fig. 4 depicts the effect on plasma CCK of increased protein concentration in the diet. CCK in brackets denotes the relative concentration of circulating CCK, with bold letters indicating increased concentration. The initial value represents average circulating CCK concentration during feeding in rats adapted to any given level of protein in the diet. When protein intake is increased, CCK secretion increases, raising the circulating CCK concentration. This stimulates pancreatic growth and increased protease synthesis which culminates in increased intestinal protease activity and inhibition of the increased CCK secretion, returning circulating CCK concentration to normal. In Fig. 5, the effect on plasma CCK of trypsin inhibitors added to the diet is illustrated, under conditions of adequate and insufficient amino acid availability. Circulating CCK concentrations are increased by trypsin inhibitor. In the presence of adequate amino acid availability, pancreatic growth and adaptation are adequate to increase protease secretion sufficient to inhibit the increase in CCK secretion, restoring plasma CCK levels to normal. If amino acid availability is inadequate, the pancreatic growth and adaptive response to CCK is impaired and protease secretion is insufficient to normalize CCK secretion, resulting in sustained elevation of plasma CCK.

The importance of availability of nitrogen and essential amino acids as a factor in the pancreatic adaptive response to CCK releasers is not well appreciated. One reason for this is the misimpression that can arise when pancreas weights are expressed relative to body weight in rats

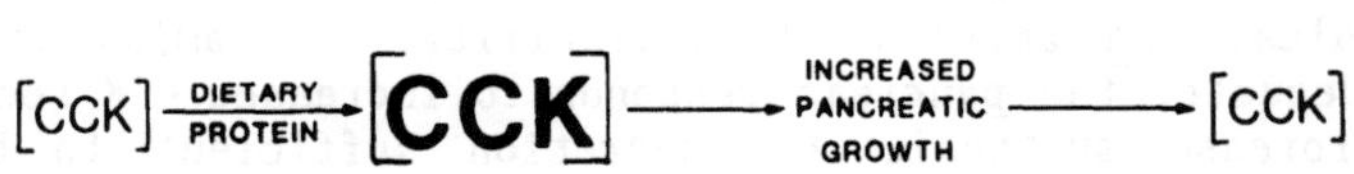

Fig. 4. Schematic representation of regulation of plasma cholecystokinin (CCK) in rats fed increased dietary protein. See text for explanation.

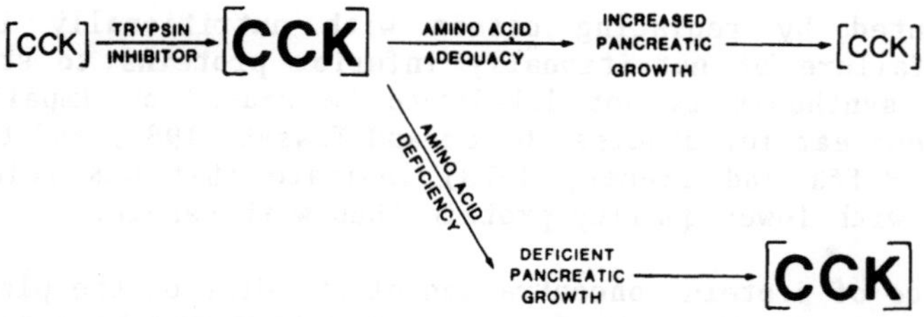

Fig. 5. Schematic representation of regulation of plasma cholecystokinin (CCK) in rats fed trypsin inhibitor and adequate or inadequate source of essential amino acids. See text for explanation.

fed nutritionally inferior proteins such as raw soybean. Booth et al. (1960) studied the effect of essential amino acid supplementation of raw soybean meal on body growth and pancreatic weight in rats over a period of 36 days. Part of their data is reproduced below in Table 3, taken from Table 1 of their report. Absolute pancreas weights were calculated by us using their data.

Table 3. Pancreatic Hypertrophy by Soybean Meal (from Booth et al., 1960)

Dietary Protein	Final body wt, g	Pancreas wt g/100g body wt	Absolute Pancreas wt., g.
(1) Casein	147	0.56	0.83
(2) Raw Soybean Meal (RSM)	89	0.85	0.76
(3) RSM + Amino Acids	142	0.93	1.32
(4) Heated Soybean Meal (HSM)	148	0.50	0.74

If only relative weight is considered, the perception is that raw soybean meal (RSM) stimulated pancreatic growth compared to heated soybean meal (HSM) and casein, and that addition of essential amino acids only modestly increased this response. When absolute values are calculated, it is clear that RSM did not increase pancreatic size compared to casein or HSM. On the other hand, addition of essential amino acids to RSM greatly increased pancreas size compared to all other diets. Expressing pancreas data as absolute weight reveals the great sensitivity of pancreatic growth to amino acid availability. Rather than stating that pancreatic growth was stimulated by RSM, it would be more accurate to state that compared to casein, RSM inhibited pancreatic growth, although to a lesser degree than it inhibited overall body growth. We would therefore interpret the results shown in Table 3 above as follows: RSM probably stimulated increased CCK release, but the pancreatic growth response to CCK was inhibited by essential amino acid insufficiency. Supplemental amino acids added to RSM permitted the full pancreatic trophic response to CCK. It is unlikely that supplemental amino acids stimulated increased pancreatic growth via enhanced CCK secretion, since amino acids do not release CCK in the rat (Liddle et al, 1985b, 1985c).

The importance of essential amino acid availability in the pancreatic adaptive response to dietary protein and carbohydrate was elegantly demonstrated by Johnson, Hurwitz and Kretchmer (1977), who showed that the large increase in pancreatic amylase synthesis caused by a high carbohydrate diet was blocked if nutritionally inferior protein (e.g., zein) was substituted for casein. Likewise, they showed that stimulation of chymotrypsinogen synthesis by increased protein in the diet was prevented by replacing casein with nutritionally inferior proteins. The failure of nutritionally inferior proteins to stimulate chymotrypsinogen synthesis is not likely to be caused by impaired CCK release because our earlier studies (Green and Nasset, 1983) and those of others (Kakade, Hoffa and Liener, 1973) indicate that CCK release is probably greater with lower quality protein than with casein.

The influence of protein concentration of the diet on the plasma CCK response to trypsin inhibitors that we reported here may help in understanding the protective effect of increased dietary protein on pancreatic carcinogenesis in rats fed diets containing soybean trypsin inhibitor. Studies conducted by the United States Department of Agriculture (USDA) examined the effect of chronic feeding of diets varying in trypsin inhibitor and protein content on incidence of the pre-cancerous lesions nodular hyperplasia and acinar adenoma. Table 4 below contains a summary of data which originally appeared in Table 19 of the USDA study (Gumbmann et al., 1985).

Table 4. Percentage Incidence of Pancreatic Acinar Adenoma in Rats Fed Diets Varying in Protein Content And Trypsin Inhibitor Content For Two Years (from Gumbmann et al., 1985)

Protein	Trypsin inhibitor (mg/100g diet)			
(%)	93	215	337	577
10	22.9	40.0	42.9	64.5
20	3.4	21.9	51.4	75.0
30	3.0	6.9	40	67.7

The data summarized in Table 4 shows the percentage incidence of pancreatic acinar adenoma after 2 years. Variable concentrations of trypsin inhibitor in the diets were accomplished by combining different ratios of raw and heated soy flour. In diets containing 10% protein, all protein came from soy flour; 20% and 30% protein levels were achieved by addition of casein. The numbers of animals in each group ranged from 29 to 37. These data are presented in greater detail in Table 7 of the report by Gumbmann and Rackis, this volume. From Table 4, it can be seen that at each protein level, the incidence of acinar adenoma varied directly with the trypsin inhibitor content of the diet. However, at the two lower levels of trypsin inhibitor, increasing the protein concentrations to 20% and 30% resulted in a progressive decrease in the incidence of acinar adenoma. This protective effect of increased dietary protein might seem to conflict with the view that increased CCK release augments pancreatic carcinogenesis (Howatson and Carter, 1985; McGuinnes et al., 1981), because we have shown that both casein and SBTI are strong CCK releasers in the rat (Liddle et al., 1985b, 1985c). However, our hypothesis illustrated in Fig. 5 and our results presented here show that increased dietary protein can lower plasma CCK in trypsin inhibitor-fed rats. Therefore, we suggest that the protective effect of increased dietary protein in the USDA Trypsin Inhibitor Study described above is mediated by reducing levels of circulating CCK. One may then ask, why is there no protective effect of increased protein at the higher levels of trypsin inhibitor? We would attribute this to chronically-elevated levels of circulating CCK, not caused by inadequate amino acids for pancreatic growth and protease synthesis, but probably due to the feeding of levels of trypsin inhibitor that exceed the ultimate protease synthetic capacity of the pancreas. Simply stated, if more protease inhibitor is fed than the pancreas can make protease, intraluminal protease activity will be chronically decreased and CCK secretion cannot be suppressed.

ACKNOWLEDGEMENTS

Supported by a Cystic Fibrosis Foundation grant and NIH grants AM01291 and AM32994.

REFERENCES

Booth, A. N., Robbins, D. J., Ribelin, W. E., and DeEds, F., 1960, Effect of raw soybean meal and amino acids on pancreatic hypertrophy in rats, Proc Soc Exptl Biol Med., 104:681-683.

Brand, S. J., and Morgan, R. G. H., 1981, The release of rat intestinal cholecystokinin after oral trypsin inhibitor measured by bio-assay, J Physiol., 319:325-353.

Fölsh, U. R., Wilms, H., Schafmayer, A., and Becker, H. D., 1984, The negative feedback mechanism of pancreatic enzyme secretion is accompanied by elevated CCK-plasma concentrations, (Abstract) Dig Dis Sci., 29:949.

Green, G. M., and Levan, V. H., 1985, Inhibition of rat pancreatic secretion by elastase, IRCS Med Sci., 13:153-154.

Green, G. M., Levan, V. H., and Liddle, R., 1984, Relationship between plasma cholecystokinin (CCK), pancreatic growth and food intake during adaptation of rats to a high protein diet, (Abstract) Dig Dis Sci., 29:31S.

Green, G. M., Levan, V. H., and Liddle, R. A., 1985, Plasma cholecystokinin and pancreatic growth during adaptation to dietary protein, (Submitted for publication).

Green, G. M., and Lyman, R. L., 1972, Feedback regulation of pancreatic enzyme secretion as a mechanism for trypsin inhibitor-induced hypersecretion in rats, Proc Soc Exptl Biol Med., 140:6-12.

Green, G. M., Olds, B. A., Matthews, G., and Lyman, R. L., 1973, Protein as a regulator of pancreatic enzyme secretion in the rat, Proc Soc Exptl Biol Med., 142:1162-1167.

Green, G. M., and Miyasaka, K., 1983, Rat pancreatic response to intestinal infusion of intact and hydrolyzed protein, Am J Physiol., 245:G394-G398.

Green, G. M., and Nasset, E. S., 1980, Importance of bile in regulation of intraluminal proteolytic enzyme activities in the rat, Gastroenterology, 79:695-702.

Green, G. M., and Nasset, E. S., 1983, Role of dietary protein in the rat pancreatic enzyme secretory response to a meal, J Nutr., 113: 2245-2252.

Gumbmann, M. R., Spangler, W. L., Dugan, G. M., Rackis, J. J., and Liener, I. E., The chronic effects of soy flour and soy protein isolate on the pancreas in rats after two years, "The USDA Trypsin Inhibitor Study, IV", Qual. Plant. Plant Foods Hum Nutr., 35:275-314, 1985.

Howatson, A. G., and Carter, D. C., 1985, Pancreatic carcinogenesis-enhancement by cholecystokinin in the hamster-nitrosamine model, Br J Cancer, 51:107-114.

Johnson, A., Hurwitz, R., and Kretchmer, N., 1977, Adaptation of rat pancreatic amylase and chymotrypsinogen to changes in diet, J Nutr., 107:87-96.

Kakade, M. L., Hoffa, D. E., and Liener, I. E., 1973, Contribution of trypsin inhibitors to the deleterious effects of unheated soybeans fed to rats, J Nutr., 193:1772-1778.

Liddle, R. A., Goldfine, I. D., and Williams, J. A., 1984, Bioassay of plasma cholecystokinin in rats: Effects of food, trypsin inhibitor and alcohol, Gastroenterology, 87:542-549.

Liddle, R. A., Goldfine, I. D., Rosen, M. S., Taplitz, R. A., and Williams, J. A., 1985, Cholecystokinin bioactivity in human plasma. Molecular forms, responses to feeding, and relationship to gallbladder contraction, J Clin Invest., 75:1144-1152.
Liddle, R. A., Green, G. M., Conrad, C. K., and Williams, J. A., 1985b, CCK response to food is species specific: fat and amino acids do not stimulate CCK release in the rat, Gastroenterology, 88:1476, (abstract).
Liddle, R. A., Green, G. M., Conrad, C. K., and Williams, J. A., 1985c, Proteins but not amino acids, carbohydrates, or fats stimulate cholecystokinin secretion in the rat. (Submitted for publication).
Louie, D., May, D., Miller, P., and Owyang, C., 1985, Cholecystokinin mediates feedback regulation of pancreatic enzyme secretion: involvement of cholinergic mechanism, (Abstract), Gastroenterology, 88:1481.
McGuinnes, E. E., Morgan, R. G. H., Levison, D. A., Hopwood, D., and Wormsley, K. G., 1981, Interaction of azaserine and raw soya flour on the rat pancreas, Scand J Gastroent., 16:49-56.
Miyasaka, K., and Green, G. M., 1983, Effect of rapid washout of proximal small intestine on pancreatic secretion in conscious rat (Abstract), Gastroenterology, 84:1251.
Schneeman, B. O., and Lyman, R. L., 1975, Factors influencing the feedback mechanism controlling pancreatic enzyme secretion in rats, Proc Soc Exptl Biol Med., 148:897-903.

8

GASTRIN AND CHOLECYSTOKININ LEVELS IN RATS FED SOYA BEAN TRYPSIN INHIBITOR

Roman S. Temler and Christine Mettraux
NESTEC LTD.
Research Department, Avenue Nestlé 55
CH-1800 Vevey, Switzerland

ABSTRACT

The effect of dietary soya bean trypsin inhibitor (SBTI) on rat pancreas, stomach, duodenum and two gastrointestinal hormones (gastrin and cholecystokinin (CCK)) levels was investigated over a 21 day period. Rats were fed either a complete diet containing 17% protein as casein (control animals), or the same diet supplemented with 1.1% SBTI (test animals). After 21 days the weight of the pancreas in the test group was 90% higher than that in the controls. The stomach antrum, and duodenum showed no change. The gastrin content in the antrum in fasted rats increased after SBTI feeding whereas plasma gastrin levels remained the same. Dietary SBTI had no effect on fasting CCK levels either in duodenal mucosa or in plasma. The results obtained on gastrin values indicate that SBTI stimulates gastrin biosynthesis in the rat antrum but does not alter its release into the circulation. Results obtained on CCK values suggested the following hypothesis : the endocrine tissue of the duodenum, where CCK is synthesized and stored, represents a relatively large reservoir of this hormone. In addition the secretory capacity of this tissue is probably much higher then is required during normal physiological conditions. Therefore, this tissue does not undergo any adaptive changes after prolonged overstimulation by SBTI. The plasma CCK peak following SBTI intake is probably much higher than after control diet, but clearance of CCK from plasma is relatively fast. Consequently, the high levels of CCK after SBTI intake produce overstimulation of pancreatic secretion and in the long term alter the pancreatic function and morphology but return to normal levels in fasting state.

INTRODUCTION

Feeding rats raw soya flour, which contains active trypsin inhibitors results in enlargement of the pancreatic gland [1,2,3,4]. The mechanism of this action is not clearly understood although it is generally believed that the presence of the inhibitor in the intestinal lumen mediates the release of cholecystokinin (CCK) which in turn acts on the pancreas [5]. To release CCK from the intestinal wall into the circulation trypsin inhibitor would act indirectly, simply by removing or blocking trypsin activity [6]. Recent studies have shown that oral administration of a single dose of trypsin inhibitor into rats was

followed by a big CCK peak in plasma [7,8]. Thus, it is conceivable that prolonged feeding a diet containing trypsin inhibitors would produce a chronic release of excessive amounts of CCK.

In accordance with this hypothesis Adrian et al. [9] reported that plasma CCK levels were increased in rats fed raw soya flour when compared to those fed heated soya flour in which trypsin inhibitors were inactivated. Our previous studies revealed that repeated injections of highly purified CCK-39 into the rat produce similar but not identical changes in the pancreas to that obtained by feeding a diet containing soya bean trypsin inhibitor (SBTI)[10]. It has been suggested therefore, that SBTI is not a specific releaser of CCK, and that probably another gastrointestinal factor is involved. Furthermore, our preliminary investigations have shown that antral gastrin content is increased in rats fed SBTI [11]. The purpose of this investigation was to study the possible adaptive changes in the gastrointestinal tract which are associated with pancreatic hypertrophy after SBTI feeding. With this aim we examined the level of CCK and gastrin in plasma and extracts of duodenum and antrum respectively. In addition, since it is known that gastrin has a trophic effect on the mucosa of stomach and duodenum, we tested the response of these two tissues to SBTI treatment.

MATERIALS AND METHODS

Soya bean trypsin inhibitor (SBTI): Lot 199 344 978 containing 12000 trypsin inhibitory units/mgN as determined by the method of Kakade et al.[12] were purchased from Fluka (Buchs, Switzerland).

Diet composition

The diet was composed as follows :

- Cellulose BW-100 (Christ, Aesch, Switzerland)	2.0 %
- Mineral mix U.S.P. XVII (1950) (Nutritional Biochemicals, Cleveland OH),	4.0 %
- Vitamin mix (Eurobio Laboratories, Paris, France)	1.25%
- Peanut oil	10.1 %
- Sucrose	25.0 %
- Casein Hammersten, protein (N x 6.25) (Merck, Darmstadt, W. Germany)	17.0 %
- DL-methionine (Kyowa Hakko Kogyo, Japan)	0.36%
- Cornstarch up to 100%	

SBTI was added to the test diet at the expense of casein, to keep the nitrogen content the same. The level of SBTI in the diet was 1.1% by weight of the total diet or 6.5% of the protein content.

Animals and treatments

Forty male rats (91.5±0.77 g) of the Sprague-Dawley strain were divided randomly into four groups of ten and housed individually in screen bottom cages in a room with a 12 hour light cycle (7.00-19.00h) and maintained at 25°C and 52% humidity. Food and water were provided ad libitum at all times. Two of the groups received casein (control) and the other two received the same diet but supplemented with SBTI.

After 18h fast (free access to water) all animals were sacrificed between 9.00-11.00h by decapitation . For determination of immunoreactive CCK, blood was collected in EDTA tubes containing 1000 U/ml of trasylol. For gastrin measurement, blood was collected in standard plastic tubes. Three organs: the antrum, pancreas and duodenum (6 cm of the small intestine starting from pylorus), were also removed from each rat. The antrum and duodenum from the first control and test group served for measurement of gastrin and CCK and from the second group of each for measurement of protein and DNA.

Antrums and duodenums in which the hormone levels were measured as well as all pancreases were weighed and frozen immediately in liquid nitrogen and then stored at -40°C. Those in which measurements of protein and DNA were made were processed immediately. This involved scraping off the full thickness of antral and duodenal mucosa by means of a surgical scalpel. The mucosa was then weighed and homogenized with a Polytron homogenizer (Kinematica, Kriens, Switzerland) at 15,000 rpm for 20 seconds at 0°C, in 150 mM NaCl solution 10 ml/500 mg and stored at -40°C.

The measurement of protein and DNA in pancreas was exactly as described previously [10].

Biochemical determinations

Extraction of CCK from plasma : CCK was extracted from plasma in 2 volumes of 96% ethanol, mixed and centrifuged. The supernatant was subsequently dried under a gently stream of N_2 at 40°C.

Extraction of CCK from duodenum : the deep-frozen duodenal tissue was boiled in deionized distilled water (10ml/g tissue) for 10 min and after homogenisation with a Polytron homogenizer at 15000 rpm for 20 sec the samples were centrifuged. The pellet was further extracted, this time in (10ml/g) boiling 0.5N acetic acid (10 ml/g) and centrifuged. Water and acetic acid extracts were then combined and lyophilized.

CCK was measured in the extracts of plasma and duodenal tissue by radioimmunoassay [13] using antiserum number T204. The anti-CCK antibodies were directed towards the sulphated tyrosyl region of CCK.

Extraction of gastrin from antrum: Frozen antrums were boiled in deionized distilled water for 20 min and cooled. They were then homogenized with a Polytron homogenizer at 15000 rpm for 20 sec before being centrifuged at low speed. The supernatant was filtered on a Millipore filter to remove debris.

The immuno-reactive gastrin was determined in plasma and antral extracts by radioimmunoassay by the method of Yalow and Berson [14] using a gastrin kit (Sorin Biomedica S.p.A., 13040 Saluggia (Vercelli), Italy).

Measurement of protein in the homogenates of pancreatic, duodenal and antral tissues was made with the method of Lowry et al.[15] with bovine albumin as a standard. DNA was measured by Burton's method[16] after extraction into 0.5N perchloric acid for 20 min. at 75°C.

Statistical Analysis

All results shown are means ± SEM from ten animals. The significance of differences between groups was assessed by Student's t-test (unpaired). A value of "p" of less than 0.05 was assumed to be significant. Values in tables different from their controls are denoted with *.

RESULTS AND DISCUSSION

All animals ate very similar amounts of food (16.7±0.21 g/day). Since there was no significant difference in growth between the control group and the group receiving SBTI (Table 1) all data obtained on the pancreas, antrum and duodenum are presented in absolute rather than relative terms. Dietary SBTI produced an increase of about 90% in pancreatic weight which was similar to that of pancreatic protein levels. On the basis that DNA content is a measure of cell number and protein /DNA is an index of cell size, the increase in the weight of pancreas appeared to result from both hypertrophy and hyperplasia of the tissue.

Table 1. Effect of feeding soya bean trypsin inhibitor (SBTI) on body weight gain and pancreas of rats.

Treatment	Body weight gain g/day	Pancreatic weight mg	Pancreatic protein mg	Pancreatic DNA mg
Control diet	7.76 ±0.22	580 ± 14.43	62.1 ± 4.97	3.27 ± 0.112
Test diet containing SBTI	8.04 ± 0.15	1089* ± 26	113.2* ± 6.72	4.07* ± 0.138

In a previous study we showed that feeding rats for ten days on a diet containing SBTI produced a 50% increase in pancreatic weight as compared to controls [10]. In the present study, which had comparable feeding conditions but lasted three weeks, the weight of the pancreas almost doubled. Results from this and the previous study, imply, therefore, that pancreatic hypertrophy develops over a period of several weeks. In addition, since pancreatic hypertrophy is a result of CCK action, they also suggest that during the feeding period the pancreas is under the continual stimulation by CCK.

The effect of dietary SBTI on antral and duodenal tissue, is shown in Tables 2 and 3. In neither organ did feeding SBTI alter tissue morphology.

Table 2. Effect of feeding soya bean trypsin inhibitor (SBTI) on the rat antral mucosa.

Treatment	Antral weight mg	Antral protein mg	Antral DNA mg
Control diet	158 ± 6.6	8.3 ± 0.47	0.513 ± 0.036
Test diet containing SBTI	178* ± 8.3	10.6 ± 0.55	0.626 ± 0.076

Table 3. Effect of feeding soya bean trypsin inhibitor (SBTI) on the rat duodenal mucosa.

Treatment	Duodenal weight mg	Duodenal protein mg	Duodenal DNA mg
Control diet	263 ± 9.0	10.17 ± 0.51	1.173 ± 0.080
Test diet containing SBTI	270 ± 11.8	8.6 ± 0.78	1.439 ± 0.082

However, SBTI did increase antral gastrin content confirming our previous observation [11] (Fig. 1A). Although the effect of diet on gastrin content has been extensively studied [17,18] there are no reports on the levels after SBTI feeding. The studies have shown that antral gastrin levels in the rat vary with protein or fat content in the diet [19]. It is known also that under appropriate conditions peptides and amino acids, as well as mechanical distention of the stomach are potent stimulants of gastrin release [20,21,22]. Under our feeding conditions both the control and test diet had exactly the same composition (Materials and Methods section). SBTI (6.5 % calculated as protein) was added to the test diet at the expense of casein so that the protein level in both diets was exactly the same and the amino acid composition was very similar.

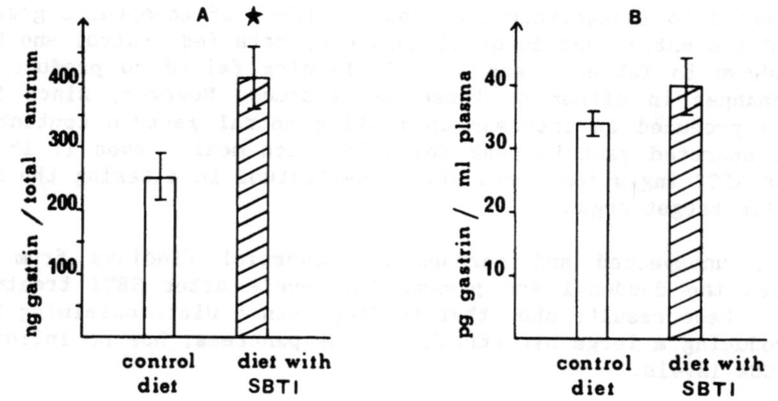

Fig. 1A,B. Effect of feeding soya bean trypsin inhibitor (SBTI) on antral (A) and plasma (B) gastrin levels. * Different ($p < 0.05$) from control.

Our data therefore clearly indicate that SBTI is a specific inducer of gastrin biosynthesis in the rat stomach. We have two possible explanations for this. The first concerns the digestibility of SBTI. Soya proteins in general, and protease inhibitors from soya beans in particular, are not easily digested by the gastrointestinal system of rats [23]. In the stomach it may only be denatured in part (if at all) by the gastric secretion, since there is still anti-tryptic activity in the intestine. Thus, the presence of SBTI in the stomach may create a larger demand for gastric secretion than other proteins would. This requirement for gastric acid would then be represented by increased gastrin levels which is the major stimulant of gastric acid release. The second possibility concerns gastric emptying. Lyman and Wilcox [24] have already shown that trypsin inhibitor from soya beans delays the rate of gastric emptying in the rat. Therefore, luminal SBTI may, in turn, alter gastrin synthesis by its ability to prolong the time in which the antral mucosa is in contact with the stimulant.

As shown in Fig. 1B gastrin levels did not change in plasma. Based on the studies of Sircar et al.[19] we concluded that changes in plasma gastrin in response to nutritional factors, are less pronounced than those produced in antral hormone levels. Consequently, rather small changes in antral gastrin produced by SBTI may have no impact on plasma gastrin levels.

The same authors also reported that biosynthesis of gastrin in the antral tissue and its release into the blood stream are two independent phenomena which are probably under the regulation of different dietary and systemic factors, hence they do not need to coincide. In such a case SBTI would be an inducer of gastrin synthesis rather than an effector of its release. It is possible also, that the postprandial plasma gastrin peak of rats fed SBTI is larger than that of control; however, the levels in the fasting state are in both cases the same.

An important property of gastrin is its ability to exert a trophic effect on antral and intestinal mucosa. This effect has been demonstrated in rats given 3 daily injections of pharmacological doses of pentagastrin over a period of 2 weeks [25]. This study raises an important question; namely whether an increase in the amount of endogenous gastrin by SBTI may also have a trophic effect on gastrointestinal tissue.

We attempted to investigate the trophic effect of endogenous gastrin and examined the antral and duodenal mucosa of rats fed control and SBTI diets. As shown in Tables 2 and 3, SBTI feeding failed to produce any important changes in either of these two tissues. However, since SBTI feeding only produced an increase in fasting antral gastrin content we assume that secreted gastrin peak following each meal - even if it was bigger after SBTI ingestion - was still ineffectual in altering the morphology of its target organ.

The most unexpected and perhaps controversial findings from our study concern the duodenal and plasma CCK levels after SBTI treatment (Fig. 2A,B). These results show that feeding rats a diet containing SBTI although producing a large hypertrophy of the pancreas, has no influence on fasting CCK levels.

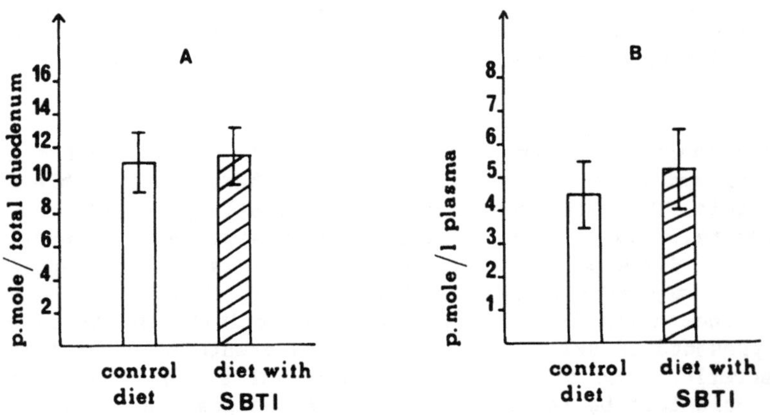

Fig. 2A,B. Effect of feeding soya bean trypsin inhibitor (SBTI) on duodenal (A) and plasma (B) CCK levels.

These data conflict with those of Adrian et al.[9], who showed that plasma CCK levels increase three-fold in rats fed a raw soya bean diet for three weeks. The more recent study of Green et al.[26] has shown, however, that CCK activity in plasma of rats fed a diet containing SBTI are higher relative to controls only during the first two days of feeding, and subsequently return to normal. Although we have no information on the dynamics of CCK, our data indirectly support the observations of Green et al.[26].

The hypertrophy of the rat pancreas in response to dietary trypsin inhibitors is well established. It has also been demonstrated in numerous studies that the hypertrophy results from overstimulation of the pancreas by high CCK levels[27,28,29]. The normal fasting CCK levels found in duodenal mucosa and plasma of rats fed SBTI in this study, however, may be explained by the fact that the endocrine cells of the duodenum where biosynthesis and storage of CCK occur, represents a relatively large reservoir of this hormone. The capacity for secretion of this tissue is probably much higher than required for normal physiological conditions, with the consequence that there is no need for it to respond to SBTI stimulation. The CCK peak in plasma which follows SBTI intake is probably much higher than after control diet[7,8], but clearance of CCK from plasma is relatively rapid. Consequently, the high levels of CCK after SBTI intake act effectively on the pancreas, stimulating its exocrine secretion and in the long term alter the pancreatic function and morphology but its levels at fasting state remain normal.

Whether gastrin activity contributes to pancreatic hypertrophy after SBTI feeding is still difficult to say from the present data. To answer this question a more detailed study would be necessary in which plasma CCK and gastrin are measured at different time points after food intake. Such a study would provide important information concerning the kinetics of gastrointestinal hormone secretion after the SBTI feeding.

ACKNOWLEDGEMENTS

We are very grateful to Dr. J.B.M.J. Jansen of the Katholieke Universiteit, Nijmegen, The Netherlands, for measurement of CCK. We would also like to thank Mr. W. Jaggi for his excellent technical assistance and Dr. E. Fern for invaluable comments and suggestions.

REFERENCES

1. J. Dijkhof, S.R. Poort, and C. Poort, Effect of feeding soybean flour containing diets on protein synthesis pattern of the rat pancreas, J. Nutr. 107:1985 (1977).
2. S. Lepkovsky, F. Furuta, and M.K. Dimick, Trypsin inhibitor and the nutritional value of soybeans, Br. J. Nutr. 25:235 (1971).
3. J.J. Rackis, Physiological properties of soybean trypsin inhibitors and their relationship to pancreatic hypertrophy and growth inhibition of rats, Fed. Proc. 24:1488 (1965).
4. R.L. Lyman, and S. Lepkovsky, The effect of raw soybean meal and trypsin inhibitor diets on pancreatic enzyme secretion in the rat, J. Nutr. 62:269 (1957).
5. H. Khayambashi, and R.L. Lyman, Secretion of rat pancreas perfused with plasma from rats fed soybean trypsin inhibitor, Am. J. Physiol. 217:646 (1969).
6. G.M. Green, and R.L. Lyman, Feedback regulation of pancreatic enzymes secretion as a mechanism of trypsin inhibitor-induced hypersecretion in rats, Proc. Soc. Exp. Biol. Med. 140:6 (1972).

7. R.A. Liddle, I.D. Goldfine, and J.A. Williams, Bioassay of plasma cholecystokinin in rats: effects of food, trypsin inhibitor, and alcohol, Gastroenterology 87:542 (1984).
8. S.J. Brand, and R.G.H. Morgan, The release of rat intestinal cholecystokinin after oral trypsin inhibitor measured by bio-assay, J. Physiol. (London) 319:325 (1981).
9. T.E. Adrian, C. Pasquali, F. Pescosta, A.J. Bacarese-Hamilton, and S.R. Bloom, Soya-induced pancreatic hypertrophy and rise of circulating cholecystokinin, (Abstract) Gut 23:A889 (1983).
10. R.S. Temler, Ch. A. Dormond, E. Simon, and B. Morel, The effect of feeding soybean trypsin inhibitor and repeated injections of cholecystokinin on rat pancreas, J. Nutr., 114:1083 (1984).
11. R.S. Temler, Ch. A. Dormond, E. Simon, and B. Morel, The effect of feeding soybean trypsin inhibitor and repeated injections of cholecystokinin on rat pancreas, Qualitas Plantarium : Plant Foods for human nutrition (in press).
12. L.M. Kakade, J.J. Rackis, J.E. McGhee, and G. Puski, Determination of trypsin inhibitor activity of soy products : a collaborative analysis of an improved procedure, Cereal Chem. 51:376 (1974).
13. J.B.M.J. Jansen, and C.B.H.W. Lamers, Radioimmunoassay of cholecystokinin in human tissue and plasma, Clin. Chim. Acta, 131:305 (1983).
14. R.S. Yalow, and S.A. Berson, Radioimmunoassay of gastrin, Gastroenterology, 58:1 (1970).
15. O.H. Lowry, N.J. Rosebrough, A.L. Farr, and R.J. Randal, Protein measurement with the Folin phenol reagent, J. Biol. Chem. 193:265 (1951).
16. L.K. Burton, Determination of DNA concentration with diphenylamine, Methods in Enzymology 12:163 (1968).
17. L.R. Johnson, L.M. Lichtenberger, E.M. Copeland, S.J. Dudrick, and G.A. Castro, Action of gastrin on gastrointestinal structure and function, Gastroenterology 68:1184 (1975).
18. L.M. Lichtenberger, J. Lechago, and L.R. Johnson, Depression of central and serum gastrin concentration by food deprivation in the rat, Gastroenterology 68:1473 (1975).
19. B. Sircar, L.R. Johnson, and L.M. Lichtenberger, Effect of chemically defined diets on antral and serum gastrin levels in rats, Am. J. Physiol. 238:G376 (1980).
20. M.G. Korman, C. Soveny, and J. Honky, Effect of food on serum gastrin evaluated by radioimmunoassay, Gut 12:614 (1971).
21. L.M. Lichtenberger, J.M. Shorey, and J.S. Trier, Organ culture studies of rat antrum: evidence for an antral inhibition of gastrin release, Am. J. Physiol. 234:E410 (1978).
22. U.T. Strunz, J.H. Walsh, and M.I. Grossman, Stimulation of gastrin release and gastric acid secretion by various amino acids, (Abstract). Clin. Res. 24:113 (1977).
23. M.L. Kakade, D.E. Hoffa, and I.E. Liener, Contribution of trypsin inhibitors to the deleterious effects of unheated soybean fed to rats, J. Nutr. 103:11772 (1973).
24. R.L. Lyman, and S.S. Wilcox, Effect of acute amino acid deficiencies on carcass composition and pancreatic function in force-fed rat, J. Nutr. 79:28 (1963).
25. M.D. Stanley, R.E. Coalson, M.I. Grossman, and L.R. Johnson, Influence of secretin and pentagastrin on acid secretion and parietal cell number in rats, Gastroenterology, 63:527 (1972).
26. G. Green, V. Levan, and R. Liddle, Homeostasis of plasma cholecystokinin (CCK) in rats, (Abstract) F.A.S.E.B. 1985.

27. L.R. Johnson, Effect of gastrointestinal hormones on pancreatic growth, Cancer 47:1640 (1981).
28. I. Ihse, B.O. Arnesjö, and I. Lundquist, Effects on exocrine and endocrine rat pancreas of long-term administration of CCK-PZ (Cholecystokinin-Pancreozymin) or synthetic octapeptide CCK-PZ. Scand. J. Gastroenterol. 11:529 (1976).
29. S.S. Rothman, and H. Wells, Enhancement of pancreatic enzyme synthesis by pancreozymin, Am. J. Phyiol. 213:215 (1967).

9

NEGATIVE FEEDBACK INHIBITION OF PANCREATIC EXOCRINE SECRETION IN HUMANS

P. P. Toskes

Department of Medicine, University of Florida
Gainesville, Florida 32610

ABSTRACT

The evidence is quite solid that a feedback inhibition system of the pancreatic exocrine secretion exists in a number of animal species. Whether or not such a regulatory system is operative in man remains controversial. Studies in normal human subjects which have been interpreted as evidence that such a regulatory system is not present in man suffer from serious flaws in experimental design, such as the inability to completely divert pancreatic enzymes (chymotrypsin, elastase) as important principles in this control mechanism. Studies in both normal human subjects and patients with chronic pancreatitis will be reviewed.

INTRODUCTION

Pancreatic trypsin and chymotrypsin, but not amylase or lipase, appear to effect feedback control of pancreatic exocrine secretion in most (Lyman, 1957; Grossman, 1958; Alumot and Nitsan, 1961; Geratz and Hurt, 1970; Green and Lyman, 1972; Lyman et al., 1974; Schneeman and Lyman, 1975; Ihse et al., 1979; Ihse and Lilia, 1970) but not all (Sale et al., 1977) animal species. These studies have shown that perfusion of the proximal small intestine with trypsin, chymotrypsin or pancreatic biliary secretions will suppress pancreatic enzyme secretion. Inactivation of the trypsin and chymotrypsin within the proximal small intestine led to an increase in enzyme secretion which returned to basal levels when trypsin or chymotrypsin were reintroduced into the lumen of the small intestine. This feedback mechanism may be mediated, at least in part, by release of cholecystokinin (CCK) from the small intestine. Rats fed soybean flour containing protease inhibitors have been shown to have elevated plasma CCK levels (Adrian et al., 1982; Folsch et al., 1984; Liddle et al., 1984) and decreased intestinal CCK content (Brand and Morgan, 1981). Finally, elevated plasma CCK levels have been observed with pancreatic exocrine insufficiency (Folsch et al., 1984). Whether or not a feedback control system for pancreatic exocrine secretion exists in humans is controversial. This review will focus on studies pertaining to such a feedback control system in normal human subjects and patients with pancreatic disease.

NORMAL HUMAN SUBJECTS

Some studies in normal human subjects have failed to demonstrate the presence of a feedback control system for pancreatic exocrine secretion but these studies all suffer from serious flaws in experimental design. Pancreatic biliary secretions infused into the jejunum did not influence pancreatic enzyme secretion (Krawisz et al., 1980). This study was not comparable to the animal studies because the techniques employed did not allow complete diversion of pancreatic enzymes from the duodenum. The conclusion from this study was that normal humans do not exhibit this feedback control mechanism in the jejunum. Animal studies have clearly demonstrated that this control mechanism exists only in the duodenum. This study in no way negated the evidence that pancreatic enzymes within the lumen of the duodenum exert a regulatory control over pancreatic exocrine secretion. Other studies (Hotz et al., 1983; Dlugosz et al., 1983) have inhibited trypsin within the duodenum with aprotinin and have been unable to influence pancreatic exocrine secretion, but these studies failed to take other proteases (e.g. chymotrypsin) into consideration. Studies in animals clearly demonstrated that chymotrypsin was effective in modulating pancreatic secretion. Thus, studies utilizing protease inhibitors to evaluate this mechanism must consider not only trypsin but also chymotrypsin and perhaps other enzymes, such as elastase. Aprotinin is not very effective in inhibiting chymotrypsin. A recent preliminary report (Goodale et al., 1985) demonstrated that this feedback control mechanism to be operative in 3 out of 5 normal subjects evaluated. In this study, a catheter was placed in the pancreatic duct endoscopically and pancreatic secretions were aspirated directly from the pancreatic duct while an inhibitor active against both trypsin and chymotrypsin (Bowman-Birk inhibitor) was infused into the duodenum. However, this study also failed to take elastase or other enzymes into consideration. Another preliminary study (Owyang et al., 1984) demonstrated that the intraduodenal perfusion of trypsin in normal subjects inhibited phenylalanine-stimulated pancreatic enzyme secretion by suppressing the release of CCK. This study also suggested that feedback regulation of pancreatic enzyme secretion was stimulus - specific; intraduodenal trypsin only inhibited pancreatic secretion mediated by CCK release, but not that mediated by cholinergic mechanisms.

PATIENTS WITH PANCREATIC DISEASE

The first study (Ihse et al., 1977) in a patient with pancreatic disease to suggest that this feedback control mechanism might be important occurred in a patient with complete obstruction of biliary and pancreatic secretions secondary to a cancer of the ampulla of Vater. Pancreatic secretions were prevented from entering the small intestine but were secreted into the bile duct via a common channel. These investigators were able to sample pancreatic biliary secretions by means of a percutaneous transhepatic cholangiography catheter, thus allowing them to study the effect of duodenal infusion of different substances on pancreatic biliary secretion. Pancreatic biliary secretions were suppressed after the infusion of trypsin of the patient's own pancreatic biliary secretion into the duodenum. The introduction of trypsin inhibitors into the duodenum reversed this suppression.

A number of anectodal reports suggested that abdominal pain in patients with chronic pancreatitis was decreased following the oral administration of pancreatic enzyme formulations, perhaps by the proteases within these enzyme preparations effecting suppression of pancreatic secretion.

Two placebo-controlled, double-blind crossover studies were reported from Sweden (Isaksson and Ishe, 1983) and one from our laboratory (Slaff et al., 1984) - both showing relief of abdominal pain in patients with chronic pancreatitis following the administration of pancreatic extract. Both studies demonstrated that the patients who experienced the most relief were those with idiopathic chronic pancreatitis who had not yet developed steatorrhea. In contrast, the patients with alcohol-induced pancreatititis and steatorrhea experienced less pain relief with pancreatic enzyme therapy.

The study from our own laboratory performed intraduodenal perfusion of pancreatic enzymes in the same patients who took part in the double-blind, crossover pain evaluation. The duodenal perfusions were performed in those patients who experienced pain relief with pancreatic enzymes and in those who did not experience relief of pain.

On day 1, the patients were perfused for 2 hours with isotonic saline and then for 2 hours with an essential amino acid solution. On day 2, the patients were perfused for 2 hours with the same amino acid solution (a known stimulus to pancreatic secretion) but with various pancreatic enzymes added to the solution. The amino acid - enzyme solution effected a significant decrease in pancreatic secretion compared to the amino acid solution alone when the enzymes added were trypsin or chymotrypsin (Figures 1 and 2), but no decrease was noted when pure amylase or lipase were added. An average decrease of 32% was noted following the infusion intraduodenally of trypsin or chymotrypsin. Two normal control subjects demonstrated a 74% decrease in pancreatic secretion in similar studies.

These studies were carried out with large doses of bovine pancreatic enzymes (e.g., 10 g/L trypsin), since the intent of the study was to show pharmacologic suppression of pancreatic secretion by pancreatic enzymes, not to evaluate this feedback phenomenon as a physiologic mechanism. Nevertheless, it is not at all clear what amount of enzymes within the duodenum is physiologic or pharmacologic.

A dose-response curve was established in several patients using various doses of trypsin at a perfusion rate 10 ml/min. The net chymotrypsin secretion was calculated by subtracting the chymotrypsin output during the isotonic baseline period (the second hour of perfusion) from the average hourly output during the particular amino acid - trypsin perfusion period that was performed on the same day (Figure 3). These experiments demonstrated 0.9 g/L to be the minimum concentration of trypsin required to suppress pancreatic secretion and 2.5 g/L to be the minimum concentration of trypsin required to maximally suppress pancreatic secretion.

The dose response studies indicated that the amount of trypsin required to suppress pancreatic secretion exceeds the amount of endogenous trypsin present in stimulated pancreatic biliary secretions. However, the ability of nonhuman proteases to modulate feedback may be different than that of endogenous human pancreatic proteases or there may be potentiation among proteases. Indeed, in this same study we demonstrated that crude amylase effected a 47% decrease in amino acid-stimulated pancreatic secretion. This amylase solution contained 0.37 g/L of trypsin and 0.53 g/L of chymotrypsin. The purified amylase solution contained no protease activity and did not suppress pancreatic secretion. The amount of trypsin in the crude amylase solution was not alone sufficient to suppress pancreatic secretion, so that either the amount of chymotrypsin contained in the solution was adequate or the trypsin and chymotrypsin were acting together, perhaps synergistically to suppress pancreatic secretion.

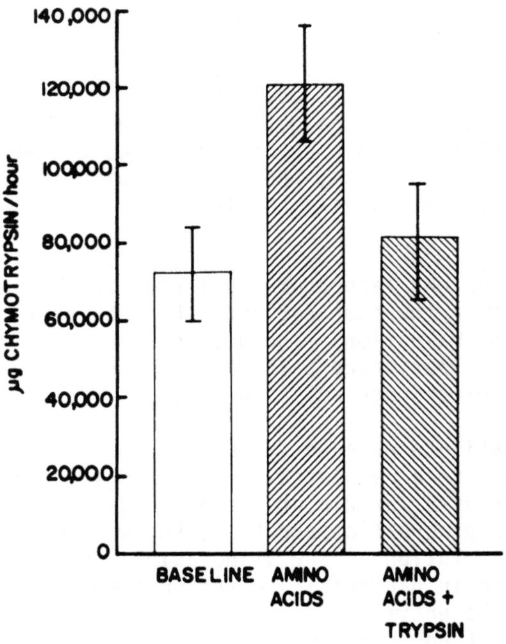

Figure 1. The effect of intraduodenally perfused trypsin on amino acid-stimulated chymotrypsin output in 3 patients with chronic pancreatitis of mild to moderate severity who experienced pain relief with pancreatic extract. Perfusion with amino acids significantly ($p<0.01$) increased chymotrypsin output over basal output whereas perfusion with amino acid-trypsin (10 g/L) mixture significantly ($p<0.01$) suppressed chymotrypsin output compared with amino acids alone. Results are expressed as $\bar{x}\pm$SEM. (Reproduced with permission, Gastroenterology, 1984;87,44.)

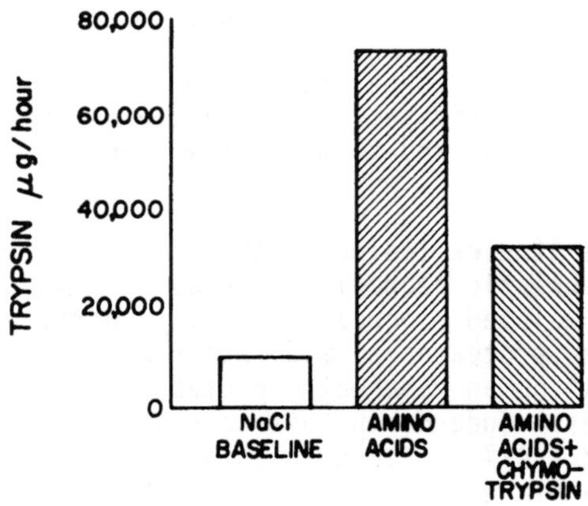

Figure 2. The effect of intraduodenally perfused chymotrypsin (10 g/L) on amino acid-stimulated trypsin output in a patient with chronic pancreatitis of mild to moderate severity who experienced pain relief with pancreatic extract. Perfusion with an amino acid-chymotrypsin mixture suppressed amino acid-stimulated trypsin output. (Reproduced with permission, Gastroenterology, 1984;87,44.)

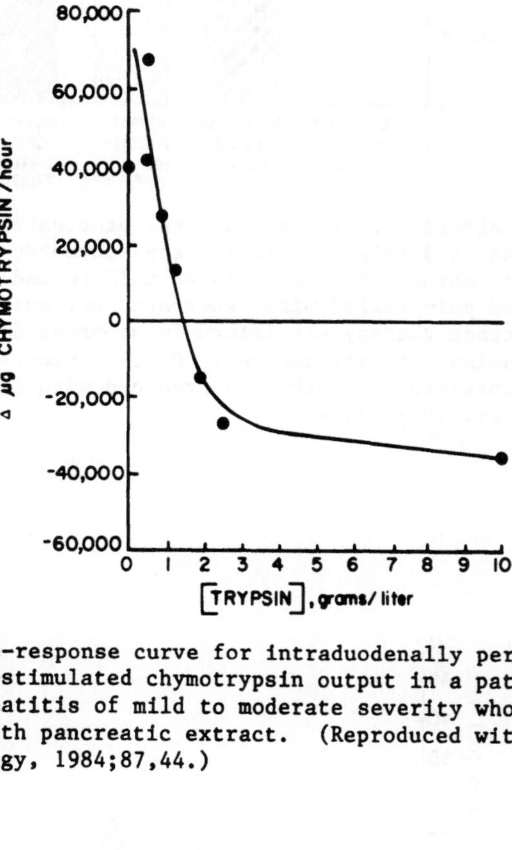

Figure 3. Dose-response curve for intraduodenally perfused trypsin on amino acid-stimulated chymotrypsin output in a patient with chronic pancreatitis of mild to moderate severity who experienced pain relief with pancreatic extract. (Reproduced with permission, Gastroenterology, 1984;87,44.)

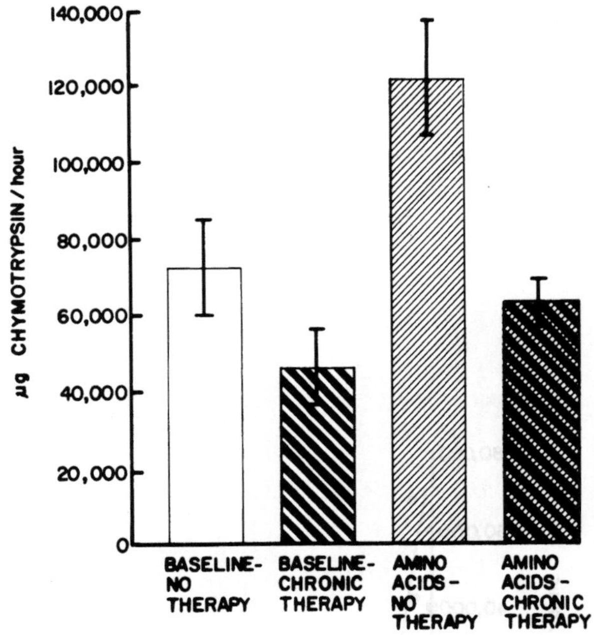

Figure 4. The effect of chronic (30 days) pancreatic extract therapy on basal and amino acid-stimulated chymotrypsin output in 3 patients with chronic pancreatitis of mild to moderate severity who experienced pain relief with pancreatic extract. Chronic pancreatic extract therapy significantly suppressed both basal ($p<0.05$) and amino acid-stimulated ($p<0.01$) chymotrypsin output. Results are expressed as $\bar{x}\pm SEM$. (Reproduced with permission, Gastroenterology, 1984;87,44.)

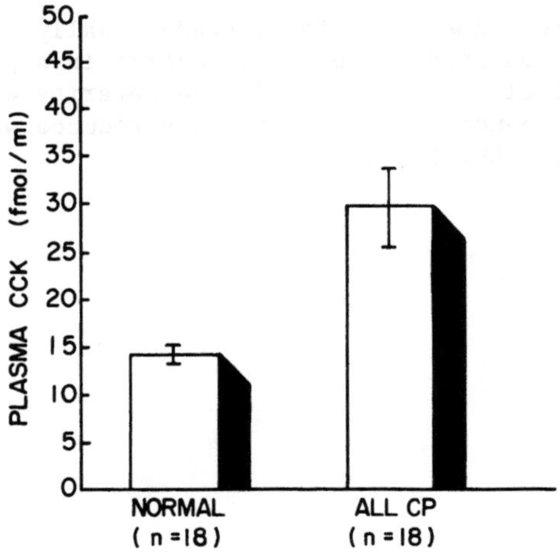

Figure 5. Basal plasma CCK levels in healthy subjects (14.3±1.3 fmol/ml) compared with patients with chronic pancreatitis (CP) (30.1±4.0 fmol/ml, $p<0.001$). Results expressed as mean±SEM. (Reproduced with permission, J Lab Clin Med, 1985;105,282.)

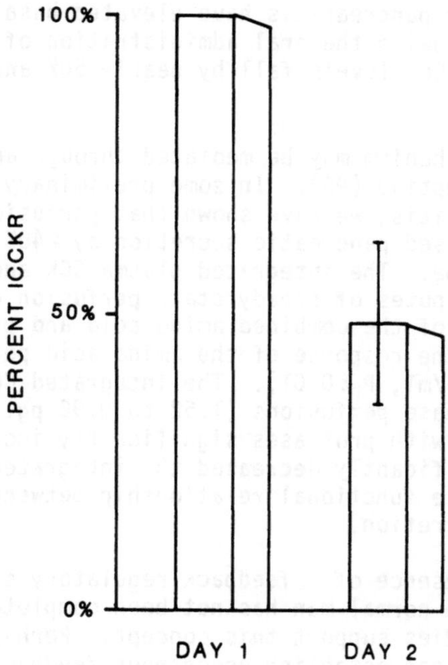

Figure 6. Basal plasma CCK levels in five patients with pancreatic extract-responsive abdominal pain without and with extract therapy. Results expressed as percent of control ± SEM. Basal value without pancreatic extract is assumed to be 100%.

The possibility that potentiation between proteases is operative in this regulatory system is an important concept since pancreatic enzyme formulations administered orally to patients with chronic pancreatitis decrease abdominal pain and suppress basal and stimulated pacreatic secretion by 30 and 40%, respectively (Figure 4), yet only a small percentage of the proteases (8 to 20%) escape destruction by gastric acid and reach the duodenum. It appears that a full complement of proteases is much more effective than individual proteases in suppressing pancreatic secretion.

The mechanism of protease - specific suppression of pancreatic secretion has yet to be defined. The observation that ovomucoid negates the suppression effect of trypsin (Slaff et al., 1984) suggests that trypsin may modulate suppression indirectly, perhaps through a digestive product, an activated proenzyme, or a released polypeptide.

Studies in our laboratory (Slaff et al., 1985) also have shown that patients with chronic pancreatitis have elevated basal plasma levels of CCK (Figure 5). Following the oral administration of pancreatic enzyme formulations, plasma CCK levels fall by nearly 50% and abdominal pain decreases (Figure 6).

The feedback mechanism may be mediated through an interplay of CCK and pancreatic polypeptide (PP). In some preliminary studies in patients with chronic pancreatitis, we have shown that perfusion of the duodenum with proteases decreased pancreatic secretion by >40% over that observed with amino acids alone. The integrated plasma CCK and PP responses were elevated during 60 minutes of steady state perfusion conditions. The integrated PP response of the combined amino acid and protease perfusion was increased above the response of the amino acid perfusion alone (3.66 to 8.06 pg. min/ml, $P < 0.01$). The integrated CCK response was reduced following protease perfusions (1.52 to 0.96 pg. min/ml, $P < 0.05$). Intestinal perfusion with proteases significantly increased the integrated PP response and significantly decreased the integrated CCK response, suggesting an intimate functional relationship between CCK and PP in modulating exocrine secretion.

Although the presence of a feedback regulatory system for pancreatic exocrine secretion in normal man has not been completely established, the preponderance of studies support this concept. Perhaps some patients with chronic pancreatitis can establish endogenous feedback control with elevated CCK levels, although possibly with increased abdominal pain. The pain relief afforded by pancreatic enzyme formulations may result from exogenous and endogenous protease inhibition of CCK release and decreased pancreatic congestion.

REFERENCES

Adrian, T. E., Pasquali, C., Pescosta, F., Bacarese-Hamilton, A. J., and Bloom, S. R. (1982). Soya-induced pancreatic hypertrophy and rise of circulating cholecystokinin. Gut, 23, A889.

Alumot, E. and Nitsan, Z. (1961). The influence of soybean antitrypsin on the intestinal proteolysis of the chick. J. Clin. Nutr., 73, 71-75.

Brand, S. J. and Morgan, R. G. H. (1981). The release of rat intestinal cholecystokinin after oral trypsin inhibitor measured by bio-assay. J. Physiol., 319, 325-343.

Dlugosz, J., Folsch, U. R., and Creutzfeldt, W. (1983). Inhibition of intraduodenal trypsin does not stimulate exocrine pancreatic secretion in man. Digestion, 26, 197-204.

Folsch, U. R., Mustroph, D., Schafmayer, A., Becker, H. D., and Creutzfeldt, W. (1984). Elevated CCK plasma concentrations during acute and chronic feeding of soybean flour. Digestion, 30, 88.

Folsch, U. R., Schafmayer, A., Ebert, R., Becker, H. D., and Creutzfeldt, W. (1984). Elevated plasma cholecystokinin concentrations in exocrine pancreatic atrophy in the rat. Digestion, 29, 60-64.

Geratz, J. D. and Hurt, J. P. (1970). Regulation of pancreatic enzyme levels by trypsin inhibitors. Am. J. Physiol., 219, 705.

Goodale, R. L., Liener, I. E., Desmukh, A., Frick, T., Satterberg, T., Ward, G., and Borner, J. W. (1985). Do trypsin inhibitors affect feedback control of human pancreatic secretion? Dig. Dis. Sci., 30, 972.

Green, G. M. and Lyman, R. L. (1971). Chymotrypsin inhibitor stimulation of pancreatic enzyme secretion in the rat. Proc. Soc. Exp. Biol. Med., 136, 649.

Grossman, M. I. (1958). Pancreatic secretion in the rat. Am J. Physiol., 194, 535-539.

Hotz, J., Ho, S. B., Go, V. L. W., and DiMagno, E. P. (1983). Short-term inhibition of duodenal tryptic activity does not affect human pancreatic, biliary, or gastric function. J. Lab. Clin. Med., 101, 488-495.

Ihse, I. and Lilja, P. (1970). Effects of intestinal amylase and trypsin on pancreatic secretion in the pig. Scand. J. Gastroenterol., 14, 1009-1013.

Ihse, I., Lilja, P., and Lundquist, I. (1977). Feedback regulation of pancreatic enzyme secretion by intestinal trypsin in man. Digestion, 15, 303-308.

Ihse, I., Lilja, P., and Lundquist, I. (1979). Trypsin as a regulator of pancreatic secretion in the rat. Scand. J. Gastroenterol., 14, 873-880.

Isaksson, G. and Ihse, I. (1983). Pain reduction by an oral pancreatic enzyme preparation in chronic pancreatitis. Dig. Dis. Sci., 28, 97-102.

Krawisz, B. R., Miller, L. J., DiMagno, E. P., and Go, V. L. W. (1980). In the absence of nutrients, pancreatic-biliary secretions in the jejunum do not exert feedback control of human pancreatic or gastric function. J. Lab. Clin. Med., 95, 13-18.

Liddle, R. A., Goldfine, I. D., and Williams, J. A. (1984). Bioassay of plasma cholecystokinin in rats: effects of food, trypsin inhibitor, and alcohol. Gastroenterology, 87, 542-549.

Lyman, R. L. (1957). The effect of raw soybean meal and trypsin inhibitor on the intestinal and pancreatic nitrogen in the rat. J. Nutr., 62, 285-294.

Lyman, R. L., Olds, B. A., and Green, G. M. (1974). Chymotrypsin in the intestine of rats fed soybean trypsin inhibitor and its inability to suppress pancreatic enzyme secretions. J. Nutr., 104, 105-110.

Owyang, C., Dismond, S., and May, D. (1984). Feedback regulation of pancreatic enzyme secretion in man is stimulus specific. Gastroneterology, 86, 1205.

Sale, J. K., Goldberg, D. M., Fawcett, A. N., and Wormsley, K. G. (1977). Chronic and acute studies indicating absence of exocrine pancreatic feedback inhibition in dogs. Digestion, 15, 540-555.

Schneeman, B. O. and Lyman, R. L. (1975). Factors involved in intestinal feedback regulation of pancreatic enzyme secretion in the rat. Proc.Soc. Exp. Med. Biol., 148, 897-903.

Slaff,J.I.,Jacobson, D., Tillman, C. R., Curington, C., and Toskes, P. P. (1984). Protease-specific suppression of pancreatic exocrine secretion. Gastroenterology, 87, 44-52.

Slaff, J.I., Wolfe, M. M., and Toskes, P. P. (1985). Elevated fasting plasma cholecystokinin levels in pancreatic exocrine impairment: evidence to support feedback regulation. J. Lab. Clin. Med., 105, 282-285.

10

PROTEASE INHIBITORS: THEIR ROLE AS MODIFIERS OF CARCINOGENIC PROCESSES

Walter Troll, Krystyna Frenkel and Rakoma Wiesner

Department of Environmental Medicine
New York University Medical Center
New York, N.Y.

INTRODUCTION

Nutrition may play a pivotal role in helping to prevent major diseases in man. For example, some constituents of our diet contribute to the prevention of cancer, whereas others have the opposite effect. Therefore, what we eat may contribute to our survival or to our demise. Evolution has not been selective in favor of the longevity of any particular species, including man. The increasing life span of man was accomplished by medical successes such as the elimination of infectious diseases through the use of antibiotics and vaccines. However, these measures have had only limited success in combating the two major causes of death in man - heart failure and cancer. Here, it is an appropriate nutrition which may offer opportunities to prevent them. The evidence for this has come from epidemiological studies which show that these diseases are directly related to the quantity of meat which is consumed. Meat is the major source of fat and protein in our diet. Thus, the Western diet, which consists of a high proportion of meat and fat, and a low proportion of vegetables, appears to contribute to a higher occurrence of breast, colon and prostate cancers (Tannenbaum and Silverstone, 1953; Wynder et al., 1971; Armstrong and Doll, 1975; Carroll, 1975; Phillips, 1975; Correa, 1981; Enstrom, 1981; Appleton and Campbell, 1983). In contrast, diets rich in certain vegetables such as rice, maize and beans, lower the incidence of these cancers (Correa, 1981). Plant seeds contain high concentrations of protease inhibitors which are particularly effective in inhibiting digestive enzymes of insects (Birk, 1974). Even in processed foods such as breads and cereals, seeds are still active in preventing cancer as has been shown by a lower incidence of oral and pharyngeal cancers in populations consuming such diets (Winn et al., 1984). These findings were corroborated by model animal experiments in which diets containing protease inhibitors inhibited the carcinogenic process(es). Among protease inhibitors, the heat-stable Bowman-Birk inhibitor isolated from soybeans was found to be one of the most potent in preventing colon cancer in mice (A. Kennedy, personal communication).

BIOLOGICAL ACTION OF PROTEASE INHIBITORS

Protease inhibitors occur in multiple forms in many tissues throughout the entire plant and animal world. Proteases and protease inhibitors are active participants in many diverse biological systems. For example, proteases are responsible for both blood clotting and for the removal of blood clots (Sherry and Troll, 1954). Blood clots are formed to prevent the excessive loss of blood caused by an injury. Thrombin induces clotting by removal of the fibrinopeptide from fibrinogen. However, clots may interfere with blood circulation and could lead to heart attacks. To prevent these dangerous side effects, plasminogen activator releases plasmin which in turn causes dissolution of circulating clots. These two processes, clotting and dissolution of clots, must be in equilibrium to assure maximum health benefits. Both thrombin and plasminogen activator hydrolyze the synthetic substrate tosyl-arginine methyl ester (TAME), which acts as an inhibitor of both clotting and lysis of clots (Sherry and Troll, 1954; Troll et al., 1954). Thus, it seems that a protease inhibitor might be involved in a fine tuning of seemingly opposing biological processes.

The role of protease inhibitors in biology and nutrition is not fully understood. Their main physiological function seems to be to prevent proteolysis. For example, the α_1-trypsin inhibitor which is present in human blood prevents emphysema by inhibiting elastase, a proteolytic enzyme. Since cigarette smoke destroys this inhibitor (Carp and Janoff, 1978; Janoff et al., 1979a; b), it potentiates the destructive activity of elastase, and thus it aggrevates emphysema. Certain components of cigarette smoke stimulate polymorphonuclear leukocytes (PMNs) and macrophages. Hypochlorous acid (HOCl), one of the active oxygen species formed by stimulated PMNs and macrophages, oxidizes the methionine residue of α_1-trypsin inhibitor, thus destroying its activity (Clark et al., 1981; Weiss et al., 1983). It is possible that inactivation of this trypsin inhibitor observed in cigarette smokers with emphysema may be an obligatory precondition in the development of lung cancer.

INHIBITION OF TUMOR PROMOTION

A variety of protease inhibitors has been shown to interfere with tumor promotion when applied to the initiated mouse skin. These include inhibitors of trypsin such as TAME and tosyl-L-lysine chloromethyl ketone (TLCK) (Troll et al., 1970), and of chymotrypsin such as tosyl-phenylalanine-chloromethyl ketone (TPCK). Other synthetic protease inhibitors block carcinogenesis when fed to animals. These include the trypsin inhibitors [N,N-dimethyl carbamoyl methyl 4(4-guanidinobenzoyloxy) phenylacetate]methane sulfate which prevents methylcholanthrene-induced skin carcinogenesis (Ohkoshi and Fujii, 1983), and N,N-dimethylamino-[p-(p'-guanidinobenzoyloxy)]benzilcarboxyglycolate which prevents 7,12-dimethyl benz[a]anthracene-induced mammary carcinogenesis in rats (Yamamura et al., 1978). The presence of ε-amino caproic acid, an inhibitor of plasminogen activator, in the drinking water blocks dimethyl hydrazine-induced colon cancer in mice (Corasanti et al., 1982).

Perhaps the most widely studied inhibitor which prevented cancer when fed to animals was leupeptin [acetyl(propyl)leucine-leucine-arginal], a naturally-occurring trypsin inhibitor elaborated by streptomycetes (Umezawa, 1972). The incorporation of leupeptin into the diet of rats and mice at a 0.1% level was found to suppress the induction of tumors in mouse skin, rat colon, esophagus and mammary gland, and leukemia. The diet appeared to be non-toxic, had no effect on weight gain or on mouse lung, rat liver or forestomach. Of some concern was the finding that the diet containing leupeptin enhanced tumor growth in experimentally-induced bladder cancer in rats (Matsushima, 1975; Kakizoe et al., 1982; 1984). However, it is likely that the enhancement of tumor growth was due to the amino acid L-leucine, the metabolic product of leupeptin excreted in the urine and not due to its trypsin inhibitory action. In a study employing acetyl-L[^{14}C]leucine-L-leucine-L-argininal, only 27.6% was excreted in the urine in the intact form plus leupeptinic acid, 16.9% was excreted as acetyl-L-leucine-L-leucine and the largest amount, 40.6%, was excreted as L-leucine. Moreover, L-leucine increased tumorigenicity in the experimentally-induced bladder cancer animals (Kakizoe et al., 1982; 1984), thus mimicking the effects of leupeptin.

Amino acid deprivation has been shown to limit tumor growth. Diets containing limited amounts of phenylalanine and tyrosine inhibited the growth of mouse hepatoma BW7756 and mouse mammary adenocarcinomas C_3HBA (Sugimura et al., 1959; Lorincz et al., 1969). A 54-year-old woman with adenocarcinoma of the ovary with abdominal metastasis showed remission of abdominal tumors when fed a diet low in these two amino acids (Lingeman, 1974). Metastases of a Lewis lung carcinoma were inhibited by a leucine-restricted diet (Chan et al., 1983). Since phenylalanine, tyrosine and leucine arise from chymotrypsin digestion of proteins, it points to the importance of this enzyme to the carcinogenic process(es). Therefore, it appears that the anticarcinogenic activity of the Bowman-Birk inhibitor is due to its ability to bind to chymotrypsin, an action leading to the inactivation of this enzyme. In fact, we have shown that the Bowman-Birk inhibitor-protease complex can be formed when mice and rats were fed diets containing this inhibitor, and it was excreted in the feces as such a complex (Yavelow et al., 1983).

It was also found that diets containing soybeans or soybean protease inhibitors reduce the appearance of tumors in mouse skin treated with 4-nitroquinoline-N-oxide and phorbol myristate acetate (PMA) (Troll et al., 1979), breast tumors in Sprague-Dawley rats subjected to ionizing radiation (Troll et al., 1980), spontaneous liver cancer in C3H mice (Becker, 1981), and colon cancer in mice (A. Kennedy, personal communication). Thus, dietary protease inhibitors may prevent a variety of cancers.

OXYGEN RADICALS IN TUMOR PROMOTION

The finding that tumor promotion can also be counteracted by agents other than protease inhibitors (e.g., anti-inflammatory hormones, retinoids and antioxidants) raised new questions of how tumor promotion can be modulated by these agents (Belman and Troll, 1972; Sporn, 1976; Boutwell, 1983). The one path that may be common to these agents could be their ability to inhibit promoter-induced stimulation of PMNs (Goldstein et al., 1979; Witz et al., 1980; Troll et al., 1982). Protease inhibitors block the influx of PMNs (Janoff et al., 1970), and also prevent the release of $O_2^{-}\cdot$ and H_2O_2 from PMNs and macrophages (Goldstein et al., 1979). Anti-inflammatory hormones block the influx of neutrophils, whereas some retinoids prevent the release of free oxygen radicals (Witz et al., 1980). The importance of oxygen radicals in tumor promotion was also shown by the finding that organic peroxides and H_2O_2 are tumor promoters (Slaga et al., 1981). H_2O_2 can also cause duodenal tumors in mice that have low levels of catalase (Ito et al., 1984). Agents such as superoxide dismutase (SOD), which interact with active oxygen species, also inhibit neoplastic transformation in tissue culture systems (Borek and Troll, 1983). Furthermore, the SOD-mimicking compound copper-diisopropyl salicylate not only blocked tumor promotion in mice but also prevented the induction of ornithine decarboxylase (Kensler et al., 1983), an indicator enzyme of tumor promotion. The carcinogenic action of oxygen radicals released by activated phagocytes also has been described (Weitzman et al., 1985).

OXYGEN RADICALS AND DNA DAMAGE

The reactive oxygen species generated by stimulated phagocytes might exert their activity by damaging or modifying cellular DNA (Emerit and Cerutti, 1981). Birnboim (1982a; b; 1983; Birnboim and Biggar, 1982) has shown unambiguously that tumor promoter-treated PMNs cause breaks in DNA through the action of oxygen radicals. Both SOD and catalase protected DNA of PMNs from that damage. However, only catalase prevented the formation of breaks in DNA of mouse erythroleukemia cells co-incubated with activated PMNs, indicating that it is H_2O_2 which is responsible for the DNA damage. This action of H_2O_2 is similar to the inactivation of excess sperm and the prevention of polyspermy in sea urchins (Coburn et al., 1981; Frenkel et al., 1985). One of the damaging agents may be the hydroxyl radical (·OH), which is formed by an interaction of $O_2^{\bar{}}\cdot$ with H_2O_2 in the presence of some transition metal ions. If tumor promotion is in certain aspects similar to the action of ionizing radiation, which is known to induce formation of ·OH and strand breaks in DNA, it also should cause the same type of modification of the DNA bases as does γ-radiation (Frenkel et al., 1981; 1985; Teebor et al., 1982; 1984). Indeed, [^{3}H]thymidine-labeled DNA incubated with PMA-stimulated PMNs, in the presence of autologous serum revealed the presence of 5-hydroxymethyl-2'-deoxyuridine (HMdU), a thymidine derivative known to be formed by γ-radiation. The amount of HMdU was determined by HPLC analysis of 2'-deoxyribonucleosides obtained by enzymatic hydrolysis of DNA. There was 0.8% HMdU formed during the 60 min incubation of DNA with PMA-stimulated PMNs (Troll et al., 1980), which is equivalent to the amount of HMdU obtained with 30 krad of γ-radiation (Teebor et al., 1984; Frenkel et al., 1985).

In addition to the radioactive fraction coinciding with the UV peak of HMdU, a larger amount of an as yet unidentified modified thymine also was observed. These results support the hypothesis that the oxidative burst of PMNs can cause a modification of DNA bases similar to that formed through the action of γ-radiation (Teebor et al., 1984; Frenkel et al., 1985).

The products of the oxidative burst of stimulated PMNs were found to be not only mutagenic (Weitzman and Stossel, 1981; Barak et al., 1983), but also carcinogenic (Weitzman et al., 1985). Therefore, it appears that at least some of the properties of tumor promoters, expressed through the PMN-mediated generation of oxygen radicals, might be similar to those of initiating carcinogens in that they cause modification of DNA constituents. Since protease inhibitors prevent the formation of oxygen radicals, we would expect that HMdU formation might also be blocked by these agents.

The active oxygen species participate in a variety of cell functions. These include glucose transport, activation of guanylate cyclase, and phagocytosis (White et al., 1976; Taylor and Halperin, 1979). In fact, a role for $O_2^{\bar{}}\cdot$ and H_2O_2 in normal cell division has been proposed (Oberley et al., 1981). However, multiple defense systems directed against the excessive reactivity of these oxygen species are evident within the cells. These include SOD and catalase which have been shown to block tumor promotion by PMA and bleomycin in hamster embryo and C3H/10T½ cells (Borek and Troll, 1983; Kennedy et al., 1984). One of the consequences of treating mouse skin with PMA seems to be the lowering of the levels of these two protective enzymes (Solanki et al., 1981).

INDUCTION OF POLY(ADP-RIBOSE) FORMATION BY OXYGEN RADICALS

Another consequence of tumor promotion by PMA, perhaps through the generation of free oxygen radicals, is the induction of poly-ADP-ribosylation (Kobayashi et al., 1984; Singh et al., 1985). Recent evidence indicates that a close relationship exists between the formation of poly-(ADP-ribose) and nuclear events involved in DNA damage and repair in mammalian cells (Sugimura and Miwa, 1983). Poly(ADP-ribose) transferase (synthetase, polymerase), the enzyme responsible for the formation of poly(ADP-ribose) from nicotinamide adenine dinucleotide (NAD), is activated in response to treatment of cells with DNA damaging agents. Treatment of cells with alkylating agents or with ionizing radiation causes a decrease in cellular NAD content with a concomitant increase in poly(ADP-ribose) synthetase activity (Shall, 1982). Cells obtained from patients with Fanconi's anemia syndrome are extremely sensitive to the damaging effects of ionizing radiation (Remsen and Cerutti, 1976). These cells were shown to contain only very low levels of NAD (Berger et al., 1982).

Protease inhibitors may play a role in controlling poly(ADP-ribose) formation. The polymer itself, poly(ADP-ribose), is a chymotrypsin inhibitor. This inhibitor was isolated from rat peritoneal macrophages and identified as a poly(ADP-ribose) of 2,000-3,500 mol. wt. [four to seven ADP-ribosyl units (Inagaki et al., 1980)]. More recently, it was shown that benzamide and 3-aminobenzamide, specific inhibitors of poly(ADP-ribose) synthetase also reduced malignant transformation in C3H/10T½ cells (Borek et al., 1984b). Therefore, it was of interest to determine whether inhibitors of poly(ADP-ribose) synthetase such as benzamide, 3-aminobenzamide and nicotinamide were also inhibitors of the proteases trypsin and chymotrypsin. We have tested the effect that these inhibitors have on the hydrolysis of [^{3}H]casein by these two enzymes. As can be seen in Table 1, all three agents blocked the activity of both chymotrypsin and trypsin. However, in every case, chymotrypsin was more sensitive than trypsin to the action of these inhibitors. We have also found that all three of these inhibitors block the production of $\cdot O_2^-$ by human PMNs stimulated by PMA (Table 2) (Wiesner and Troll, 1986).

If the induction of poly(ADP-ribose) is directly caused by the action of oxygen radicals, as has been asserted (Singh, 1985), it can be speculated that agents capable of blocking the formation of superoxide by PMNs or macrophages also may block the process of poly-ADP-ribosylation. Therefore, it is possible that it is a protease inhibitor which provides the biological signal for controlling the formation of poly(ADP-ribose). This was supported by new observations that the protease inhibitors leupeptin and antipain inhibit poly-ADP-ribosylation in C3H/10T½ cells (J.H. Cleaver, M. Bander, W. Troll and C. Borek, Some protease inhibitors are also inhibitors of poly-ADP-ribosylation, manuscript submitted; N. Singh and P. Cerutti, personal communication). Inhibitors of poly-ADP-ribosylation were also shown to block the transformation of cells in culture caused by many alkylating agents, UV light and PMA (Borek et al., 1984). These inhibitors and other drugs which modulate the action of poly(ADP-ribose) polymerase blocked transformation of carcinogen-treated human fibroblasts (Milo et al., 1985), and C3H/10T½ cells (Borek et al., 1984c). We have recently observed that even nanogram amounts of pure Bowman-Birk soybean inhibitor blocked the ionizing radiation-induced transformation of C3H/10T½ cells (Yavelow et al., 1985). The minute amount of protease inhibitor needed suggests that its action is mediated by a receptor. Only a part of the Bowman-Birk inhibitor which inhibits chymotrypsin was required for blocking transformation. The known poly-(ADP-ribose) inhibitors benzamide, 3-aminobenzamide and nicotinamide are also better inhibitors of chymotrypsin than of trypsin. Therefore, it is

Table 1. Effect of Various Inhibitors on Hydrolysis of [^{3}H]Casein by Chymotrypsin and Trypsin

		Percent Inhibition	
Inhibitor	Conc. of Inhibitor (mM)	Chymotrypsin[a]	Trypsin[a]
Nicotinamide	10	7	6
	20	22	12
	40	53	14
	60	52	40
Benzamide	10	57	24
	20	64	37
	40	76	44
	60	88	64
3-Amino-benzamide	10	34	6
	20	50	27
	40	69	31
	60	72	40

[a]Twenty ng chymotrypsin or trypsin/assay.

possible that the receptor for induction of poly-ADP-ribosylation is a chymotrypsin-type receptor.

The full biological function of poly(ADP-ribose) is not as yet known. It appears to play a role in DNA repair in mammalian cells (Sugimura and Miwa, 1983). It is possible that cancer cells are the unexpected beneficiary of this particular repair system which may have developed to assure the survival of cells at any cost.

Table 2. Inhibition of Superoxide Anion Formation in Human White Cells

	Percent Inhibition		
	Nicotinamide	Benzamide	3-Aminobenzamide
2 mM	18	23	15
5 mM	23	34	48
10 mM	35	84	73

EFFECTS OF DIETARY PROTEASE INHIBITORS

Protease inhibitors ingested by eating seeds inactivate the proteases trypsin and chymotrypsin in the duodenum as has been shown to occur in mice and rats fed the Bowman-Birk inhibitor (Yavelow et al., 1983). [^{125}I]Bowman-Birk inhibitor fed to animals was found mainly in the intestinal tract and feces. HPLC analysis of ^{125}I-labeled material isolated from the feces revealed that its molecular weight was 33,000 - precisely what is expected for a complex of trypsin or chymotrypsin with Bowman-Birk inhibitor (25,000 - 8,000). Corroboration of that composition was obtained when upon acidification of the complex free [^{125}I]Bowman-Birk inhibitor was released (Yavelow et al., 1983). The demonstration that diets containing protease inhibitors such as Bowman-Birk are capable of diminishing the occurrence of a variety of cancers in organs where the inhibitor is not present points to the important role protease inhibitors may play in the modulation of carcinogenesis.

The proteins that we eat are digested by proteases to peptides and then to amino acids which appear in the blood. If these amino acids are not oxidized or utilized for protein synthesis, they are excreted in the urine. We have already mentioned that certain amino acids present in the urine accelerate bladder cancer. A high protein (20% casein) diet started six weeks following the initiation with aflatoxin caused formation of preneoplastic lesions, whereas these lesions did not develop in animals fed a low 5% protein diet. Protein diets given during initiation with aflatoxin had no effect on the yield of preneoplastic lesions (Appelton and Campbell, 1983). These findings are similar to those of Tannenbaum and Silverstone (1953), who demonstrated that high-calorie and high-fat diets increased the yield of skin tumors caused by benzo[a]pyrene. Thus, the increased availability of amino acids in populations on a high protein diet may be a factor contributing to breast, colon, prostate, oral and pharyngeal cancers. Therefore, it is possible that the presence of an excess of amino acids in the nongrowing adult provides the needed food for the incipient or growing tumor cells, which require more nutrients than normal cells, and thus contributes to carcinogenesis.

PANCREATIC CANCER AND PROTEASE INHIBITORS

A number of questions have been raised at this Symposium regarding the deleterious effects of eating raw soybeans. For example, Rackis and coworkers (1985) have reported that the presence of trypsin inhibitors in raw soybeans was responsible for the pancreatic hypertrophy that was observed in rats fed such a diet. Moreover, Roebuck et al. (1985) showed that the diet containing both azaserine and raw soybeans resulted in an increase of pancreatic cancer in rats. Schneeman and Gallaher (1985) reported that pancreatic enlargement resulting from consumption of raw soybean meal or of purified trypsin inhibitor has been observed in rats, chicks, mice and young guinea pigs. However, it did not occur in dogs, growing swine, calf or adult guinea pigs. Furthermore, Harwood et al. (1985) found that although soy based protein diets containing trypsin inhibitor caused pancreatic hypertrophy in the rat, there was no evidence of this effect in primates even after five years of feeding such a diet. Thus, this phenomenon may be directly related to species specificity. It seems that humans are not subject to its occurrence since protease inhibitors do not contribute to pancreatic cancer in Japanese or Seventh-Day Adventists who consume large amounts of soybeans as their major source of protein (Armstrong and Doll, 1975). Rather, epidemiological studies have shown that a positive relationship exists between the consumption of meat and the incidence of pancreatic cancer (Ishii et al., 1968; Lea, 1977). Also, a direct relationship has been shown to occur between the incidence of pancreatic cancer and the consumption of coffee and decaffeinated coffee (Lin and Kessler, 1981; MacMahon et al., 1981). Thus, at this time, there does not seem to be any compelling evidence that pancreatic cancer in man is caused by protease inhibitors.

CONCLUSION

Protease inhibitors present in the diet seem to contribute to the prevention of the three major human cancers - those of breast, colon and prostate. Similarly, they also reduce the incidence of breast, liver and skin tumors in rodents. There are at least three pathways by which protease inhibitors might contribute to the prevention of cancer. They may: 1) block the formation of active oxygen species by stimulated neutrophils, 2) inhibit poly-ADP-ribosylation, and 3) prevent the digestion of proteins to amino acids, thus depriving rapidly growing cancer cells of needed nutrients. It seems that particularly effective inhibitors are those that are capable of counteracting the activity of chymotrypsin. In comparison to other chemopreventive agents, protease inhibitors show negligible toxicity. Some of those inhibitors are already present in the human diet. However, further studies are required to determine which and at what doses protease inhibitors are most effective when incorporated into the human diet.

ACKNOWLEDGMENTS

This investigation was supported by PHS grants numbers CA 16060 and CA 37858, awarded by the National Cancer Institute, DHHS.

REFERENCES

Appelton, B.S., and Campbell, T.C., 1983, Effect of high and low dietary protein on the dosing and postdosing periods of aflatoxin B_1-induced hepatic preneoplastic lesion development in the rat, Cancer Res., 43: 2150-2154.

Armstrong, B., and Doll, R., 1975, Environmental factors and cancer incidence and mortality in different countries, with special reference to dietary factors, Int. J. Cancer, 15:617-631.

Barak, M., Ulitzur, S., and Merzbach, D., 1983, Phagocytosis-induced mutagenesis in bacteria, Mutat. Res., 121:7-16.

Becker, F.F., 1981, Inhibition of spontaneous hepatocarcinogenesis in C3H/HeN mice by Edipro A, an isolated soy protein, Carcinogenesis, 2:1213-1214.

Belman, S., and Troll, W., 1972, The inhibition of croton oil-promoted mouse skin tumorigenesis by steroid hormones, Cancer Res., 32:450-454.

Berger, N.A., Berger, S.J., and Catino, D.M., 1982, Abnormal NAD^+ levels in cells from patients with Fanconi's anaemia, Nature, 299:271-273.

Birk, Y., 1974, Structure-activity relationship of several trypsin and chymotrypsin inhibitors from legume seeds, in: "Proteinase Inhibitors," N. Fritz, H. Tschesche, L.J. Green, and E. Truscheit, eds., Springer-Verlag, Berlin, pp. 355-361.

Birnboim, H.C., 1982a, DNA strand breakage in human leukocytes exposed to a tumor promoter, phorbol myristate acetate, Science, 215:1247-1249.

Birnboim, H.C., 1982b, Factors which affect DNA strand breakage in human leukocytes exposed to a tumor promoter, phorbol myristate acetate, Can. J. Physiol. Pharmacol., 60:1359-1366.

Birnboim, H.C., 1983, Importance of DNA strand-break damage in tumor promotion, in: "Radioprotectors and Anticarcinogens," O.F. Nygaard and M.G. Simic, eds., Academic Press, New York, pp. 539-556.
Birnboim, H.C., and Biggar, W.D., 1982, Failure of phorbol myristate acetate to damage DNA in leukocytes from patients with chronic granulomatous disease, Infect. Immun., 38:1299-1300.
Borek, C., Cleaver, J.E., and Fujiki, H., 1984a, Critical biochemical and regulatory events in malignant transformation in vitro, in: "Cellular Interactions by Environmental Tumor Promoters," H. Fujiki, E. Hecker, R.E. Moore, T. Sugimura, and I.B. Weinstein, eds., Japan Sci. Soc. Press, Tokyo/VNU Science Press, Utrecht, pp. 195-206.
Borek, C., Morgan, W.F., Ong, A., and Cleaver, J.E., 1984b, Inhibition of malignant transformation in vitro by inhibitors of poly(ADP-ribose) synthesis, Proc. Natl. Acad. Sci. USA, 81:243-247.
Borek, C., Ong, A., and Cleaver, J.E., 1984c, Methylating and ethylating carcinogens have different requirements for poly(ADP-ribose) synthesis during malignant transformation, Carcinogenesis, 5:1573-1576.
Borek, C., and Troll, W., 1983, Modifiers of free radicals inhibit in vitro the oncogenic actions of X-rays, bleomycin, and the tumor promoter 12-O-tetradecanoylphorbol 13-acetate, Proc. Natl. Acad. Sci. USA, 80:1304-1307.
Boutwell, R.K., 1983, Diet and anticarcinogenesis in the mouse skin two-stage model, Cancer Res., 43(suppl.):2465s-2468s.
Campbell, T.C., 1983, Mycotoxins, in: "Environmental Aspects of Cancer, the Role of Macro and Micro Components of Foods," E.L. Wynder, G.A. Leveille, J.H. Weisburger, and G.E. Livingston, eds., Food Nutr. Press, Westport, CT, pp. 187-197.
Carp, H., and Janoff, A., 1978, Possible mechanisms of emphysema in smokers. In vitro suppression of serum elastase-inhibitory capacity by fresh cigarette smoke and its prevention by antioxidants, Am. Rev. Respir. Dis., 118:617-621.
Carroll, K.K., 1975, Experimental evidence of dietary factors and hormone-dependent cancers, Cancer Res., 35:3374-3383.
Chan, W., McElhanon, S., Shu, G., and Banks, W.L., 1983, Effect of dietary leucine restriction on its utilization by tumor and host tissues, Fed. Proc., 42:670.
Clark, R.A., Stone, P.J., Hag, A.E., Calore, J.D., and Frazblau, C., 1981, Myeloperoxidase-catalyzed inactivation of α_1 protease inhibitor by human neutrophils, J. Biol. Chem., 256:3348-3353.
Coburn, M., Schuel, H., and Troll, W., 1981, A hydrogen peroxide block to polyspermy in the sea urchin Arbacia punctulata, Dev. Biol., 84:235-238.
Corasanti, J.G., Hobika, G.H., and Markus, G., 1982, Interference with dimethylhydrazine induction of colon tumors in mice by ε-aminocaproic acid, Science, 216:1020-1021.
Correa, P., 1981, Epidemiological correlations between diet and cancer frequency, Cancer Res., 41:3685-3690.
Emerit, I., and Cerutti, P.A., 1981, Tumor promoter phorbol-12-myristate-13-acetate induced chromosomal damage via indirect action, Nature, 293:144-146.
Enstrom, J.E., 1981, Reassessment of the role of dietary fat in cancer etiology, Cancer Res., 41:3722-3724.
Frenkel, K., Blum, F., and Troll, W., 1985, Copper ions and hydrogen peroxide form hypochlorite from NaCl, thereby mimicking myeloperoxidase, J. Cell. Biochem., 9C(suppl.), 1163 (abstract) (paper submitted for publication to the same Journal).
Frenkel, K., Cummings, A., Solomon, J., Cadet, J., Steinberg, J.J., and Teebor, G.W., 1985 (in press), Quantitative determination of the 5-hydroxymethyl uracil moiety in the DNA of γ-irradiated cells, Biochemistry.

Frenkel, K., Goldstein, M.S., and Teebor, G.W., 1981, Identification of the cis-thymine glycol moiety in chemically oxidized and γ-irradiated deoxyribonucleic acid by high-pressure liquid chromatography analysis, Biochemistry, 20:7566-7571.

Goldstein, B.D., Witz, G., Amoruso, M., and Troll, W., 1979, Protease inhibitors antagonize the activation of polymorphonuclear leukocyte oxygen consumption, Biochem. Biophys. Res. Commun., 88:854-860.

Harwood, J.P., Ausman, L.M., King, N.W., Sehgal, P.K., Nicolosi, R.J., Liener, I.E., Donatucci, D., and Tarcza, J., 1985, Effect of long-term feeding of soy-based diets on the pancreas of Cebus monkeys, Fed. Proc., 44(5):1496 (abstract #6349).

Inagaki, T., Miwa, K., and Murachi, T., 1980, Identification of a protease inhibitor from rat peritoneal macrophages as poly(ADP-ribose), J. Biol. Chem., 255:7746-7750.

Ishii, K., Nakamura, K., Ozaki, H., Yamada, N., and Takeuchi, T., 1968, Epidemiological problems of pancreas cancer, Jpn. J. Clin. Med., 26: 1839-1842.

Ito, A., Watanabe, H., Naito, M., Naito, Y., and Kawashima, K., 1984, Correlation between induction of duodenal tumor by hydrogen peroxide and catalase activity in mice, Gann, 75:17-21.

Janoff, A., Carp, H., Lee, D.K., and Drew, R.T., 1979a, Cigarette smoke inhalation decreases α_1-antitrypsin activity in rat lung, Science (Wash., DC), 206:1313-1314.

Janoff, A., Klassen, A., and Troll, W., 1970, Local vascular changes induced by the cocarcinogen, phorbol myristate acetate, Cancer Res., 30:2568-2571.

Janoff, A., White, R., Carp, H., Harel, S., Dearing, R., and Lee, D., 1979b, Lung injury induced by leukocytic protease, Am. J. Pathol., 97:111-136.

Kakizoe, T., Komatsu, H., Honma, Y., Niijima, T., and Sugimura, T., 1982, New promoters of bladder cancer; isoleucine, leucine and valine, Gann, 73:870-873.

Kakizoe, T., Nishio, Y., Honma, Y., Niijima, T., and Sugimura, T., 1984, L-isoleucine and L-leucine are promoters of bladder cancer in rats, in: "Cellular Interactions by Environmental Tumor Promoters," H. Fujiki, E. Hecker, R.E. Moore, T. Sugimura, and I.B. Weinstein, eds., Japan Sci. Soc. Press, Tokyo/VNU Science Press, Utrecht, pp. 373-380.

Kennedy, A.R., Troll, W., and Little, J.B., 1984, Role of free radicals in the initiation and promotion of radiation transformation in vitro, Carcinogenesis, 5(10):1213-1218.

Kensler, T.W., Bush, D.M., and Kozumbo, W.S., 1983, Inhibition of tumor promotion by a biomimetic superoxide dismutase, Science, 221:75-77.

Kobayashi, Y., Asada, M., and Osawa, T., 1984, Mechanism of phorbol myristate acetate-induced lymphotoxin production by a human T cell hybridoma, J. Biochem., 95:1775-1782.

Lea, A.J., 1977, Neoplasma and environmental factors, Ann. R. Coll. Surg. Engl., 41:432-438.

Lin, R.S., and Kessler, I.I., 1981, A multifactorial model for pancreatic cancer in man: Epidemiological evidence, JAMA, 245:147-152.

Lingeman, C.M., 1974, Etiology of cancer of the human ovary: A review, J. Natl. Cancer Inst., 53:1603-1618.

Lorincz, A.B., Kuttner, R.E., and Brandt, M.B., 1969, Tumor response to phenylalanine-tyrosine limited diets, J. Am. Dietet. Assoc., 54:198-205.

MacMahon, B., Yen, S., Trichopoulous, D., Warren, K., and Nardi, G., 1981, Coffee and cancer of the pancreas, N. Engl. J. Med., 304:630-633.

Matsushima, T., Kaziki, T., Kawachi, T., et al., 1975, Effects of protease inhibitors of microbial origin on experimental carcinogenesis, in: "Fundamentals in Cancer Prevention," P.N. Magee, S. Takayama, T. Sugimura, T. Matsushima, eds., Univ. Tokyo Press, Tokyo/Univ. Park Press, Baltimore, pp. 57-69.

Milo, G.E., Kurian, P., Kirsten, E., and Kun, E., 1985, Inhibition of carcinogen-induced cellular transformation of human fibroblasts by drugs that interact with the poly(ADP-ribose) polymerase system. Initial evidence for the development of transformation resistance, FEBS Letts., 179(2):332-336.

Oberley, L.W., Oberley, T.D., and Buettner, G.R., 1981, Cell division in normal and transformed cells: The possible role of superoxide and hydrogen peroxide, Med. Hypotheses, 7:21-24.

Ohkoshi, M., and Fujii, S., 1983, Effect of the synthetic protease inhibitor [N,N-dimethylcarbamoyl-methyl 4-(4-guanidinobenzoyloxy-phenylacetate] methanesulfate on carcinogenesis by 3-methylcholanthrene in mouse skin, J. Natl. Cancer Inst., 71:1053-1057.

Phillips, R.L., 1975, Role of life-style and dietary habits in risk of cancer among Seventh-Day Adventists, Cancer Res., 35:3513-3522.

Rackis, J.J., Baker, E.C., and Wolf, W.J., 1985, Protease inhibitors in plant foods: Content, inactivation, and removal, Fed. Proc., 44(3), AIN Symp. Abstracts.

Remsen, J.F., and Cerutti, P.A., 1976, Deficiency of gamma-ray excision repair in skin fibroblasts from patients with Fanconi's anemia, Proc. Natl. Acad. Sci. USA, 73:2419-2423.

Roebuck, B.D., Kaplita, P.V., and Baumgartner, K.J., 1985, Enhancement of pancreatic carcinogenesis by raw soybean protein isolate, Fed. Proc., 44(3), AIN Symp. Abstracts.

Schneeman, B.O., and Gallaher, D., 1985, Pancreatic response to dietary trypsin inhibitor: Variations among species, Fed. Proc., 44(5):1496 (abstract #6346).

Shall, S., 1982, ADP-ribose in DNA repair, in: "ADP-Ribosylation Reactions, Biology and Medicine," O. Hayaishi, and K. Ueda, eds., Academic Press, New York, pp. 477-520.

Sherry, S., and Troll, W., 1954, The action of thrombin on synthetic substrates, J. Biol. Chem., 208:95-105.

Singh, N., Poirier, G., and Cerutti, P., 1985, Tumor promoter phorbol-12-myristate-13-acetate induces poly ADP-ribosylation in human monocytes, Biochem. Biophys. Res. Commun., 126:1208-1214.

Slaga, T.J., Klein-Szanto, A.J., Triplett, L.L., Yotti, L.P., and Trosko, J.E., 1981, Skin tumor-promoting activity of benzoyl peroxide, a widely used free radical-generating compound, Science, 213:1023-1025.

Solanki, V., Rana, R.S., and Slaga, T.J., 1981, Diminution of mouse epidermal superoxide dismutase and catalase activities by tumor promoters, Carcinogenesis, 2:1141-1146.

Sporn, M.B., Dunlop, N.M., Newton, D.L., and Smith, J.M., 1976, Prevention of chemical carcinogenesis by vitamin A and its synthetic analogs (retinoids), Fed. Proc., 35:1332-1338.

Sugimura, T., Bernbaum, S.M., Winitz, M., and Greenstein, J.P., 1959, Quantitative nutritional studies with water-soluble chemically defined diets. 8. The forced feeding of diets each lacking one essential amino acid, Arch. Biochem., 81:448-455.

Sugimura, T., and Miwa, M., 1983, Poly(ADP-ribose) and cancer research, Carcinogenesis, 4:1503-1506.

Tannenbaum, A., and Silverstone, H., 1953, Nutrition in relation to cancer, Adv. Cancer Res., 1:451-501.

Taylor, W.M., and Halperin, M.L., 1979, Stimulation of glucose transport in rat adipocytes by insulin, adenosine, nicotinic acid and hydrogen peroxide, Biochem. J., 178:381-389.

Teebor, G.W., and Frenkel, K., 1983, The initiation of DNA excision-repair, Adv. Cancer Res., 38:23-59, 1983.

Teebor, G.W., Frenkel, K., and Goldstein, M.S., 1982, Identification of radiation-induced thymine derivative in DNA, Adv. Enzyme Regul., 20: 39-54.

Teebor, G.W., Frenkel, K., and Goldstein, M.S., 1984, Ionizing radiation and tritium transmutation both cause formation of 5-hydroxymethyl-2'-deoxyuridine in cellular DNA, Proc. Natl. Acad. Sci. USA, 81: 318-321.

Troll, W., Belman, S., Wiesner, R., and Shellabarger, C.J., 1979, Protease action in carcinogenesis, in: "Biological Function of Proteinases," H. Holzer, and H. Tschesche, eds., Springer-Verlag, Berlin, pp. 165-170.

Troll, W., Frenkel, K., and Teebor, G., 1984, Free oxygen radicals: Necessary contributors to tumor promotion and cocarcinogenesis, in: "Cellular Interactions by Environmental Tumor Promoters," H. Fujiki, E. Hecker, R.E. Moore, T. Sugimura, and I.B. Weinstein, eds., Japan Sci. Soc. Press, Tokyo/VNU Science Press, Utrecht, pp. 207-218.

Troll, W., Klassen, A., and Janoff, A., 1970, Tumorigenesis in mouse skin: Inhibition by synthetic inhibitors of proteases, Science, 169:1211-1213.

Troll, W., Sherry, S., and Wachman, J., 1954, The action of plasmin on synthetic substrates, J. Biol. Chem., 208:85-93.

Troll, W., Wiesner, R., Shellabarger, C.J., Holtzman, S., and Stone, J.P., 1980, Soybean diet lowers breast tumor incidence in irradiated rats, Carcinogenesis, 1:469-472.

Troll, W., Witz, G., Goldstein, B., Stone, D., and Sugimura, T., 1982, The role of free oxygen radicals in tumor promotion and carcinogenesis, in: "Carcinogenesis. A Comprehensive Survey. Vol. 7. Carcinogenesis and Biological Effects of Tumor Promoters," E. Hecker, N.E. Fusenig, W. Kunz, F. Marks, and H.W. Thielmann, eds., Raven Press, New York, pp. 593-597.

Umezawa, K., 1972, "Enzyme Inhibitors of Microbial Origin," Univ. Tokyo Press, Tokyo, pp. 29-32.

Weiss, S.J., Lampert, M.B., and Test, S.T., 1983, Long-lived oxidants generated by human neutrophils: Characterization and bioactivity, Science, 222:625-627.

Weitzman, S.A., and Stossel, T.P., 1981, Mutation caused by human phagocytes, Science, 212:546-547.

Weitzman, S.A., Weitberg, A.B., Clark, E.P., and Stossel, T.P., 1985, Phagocytes as carcinogens: Malignant transformation produced by human neutrophils, Science, 227:1231-1233.

White, A.A., Crawford, K.M., Pratt, C.S., and Lad, P.J., 1976, Activation of soluble guanylate cyclase from rat lung by incubation or by hydrogen peroxide, J. Biol. Chem., 251:7304-7312.

Wiesner, R., and Troll, W., 1986 (in press), Blocking superoxide anion production in polymorphonuclear leukocytes (PMNs) by inhibitors of poly(ADP-ribose): 3-Aminobenzamide, benzamide and nicotinamide, in: "Mechanism of DNA Damage and Repair," Plenum Press, New York.

Winn, D.M., Ziegler, R.G., Pickle, L.W., Gridley, G., Blot, W.J., and Hoover, R.N., 1984, Diet in the etiology of oral and pharyngeal cancer among women from the southern United States, Cancer Res., 44: 1216-1222.

Witz, G., Goldstein, B.D., Amoruso, M., Stone, D.S., and Troll, W., 1980, Retinoid inhibition of superoxide anion radical production by human polymorphonuclear leukocytes stimulated with tumor promoters, Biochem. Biophys. Res. Commun., 97:883-888.

Wynder, E., Mabuchi, K., and Whitmore, W., 1971, Epidemiology of cancer of the prostate, Cancer, 28:344-360.

Yamamura, M., Nakamura, N., Fukui, Y., et al., 1978, Inhibition of 7,12-dimethylbenz[a]anthracene-induced mammary tumorigenesis in rats by a synthetic protease inhibitor, N,N-dimethylamino-[p-(p'-guanidinobenzoyloxy)]benzilcarbonyloxyglycolate, Gann, 69:749-752.

Yavelow, J., Collins, M., Birk, Y., Troll, W., and Kennedy, A.R., 1985 (in press), Nanomolar concentrations of Bowman-Birk soybean protease inhibitor suppress X-ray induced transformation *in vitro*, *Proc. Natl. Acad. Sci. USA*.

Yavelow, J., Finlay, T.H., Kennedy, A.R., and Troll, W., 1983, Bowman-Birk soybean protease inhibitor as an anticarcinogen, Cancer Res., 43 (suppl.):2454s-2459s.

11

NUTRITIONAL AND METABOLIC RESPONSE TO PLANT INHIBITORS OF DIGESTIVE ENZYMES

Daniel Gallaher and Barbara O. Schneeman

Department of Nutrition

University of California, Davis, California 95616

INTRODUCTION

Men's food source derives predominantly from plants, from which only a relatively small proportion can be utilized nutritionally. Of these plants, many must be treated or processed in some way to improve the digestibility or remove naturally occurring toxicants. Considerable effort has been put forth in elucidating these factors in food that may constitute a hazard to man or his domesticated animals when consumed. One group of compounds that has received much attention are the digestive enzyme inhibitors. The distribution of these inhibitors among plants is widespread, and includes most agronomic crops (Liener and Kakade, 1980). The inhibitor is most often found within the edible portion. By far the most prominent of these enzyme inhibitors are the inhibitors of the proteolytic enzymes, particularly trypsin and chymotrypsin. Examples of amylase inhibitors are few, but potentially important. Although no inhibitors of pancreatic lipase have been definitely established, a compound to be discussed may fit this role. Finally, there are certain substances that inhibit all the digestive enzymes in a relatively nonspecific manner. Most notable among these are the tannins(Griffith and Mosley, 1980). The discussion to follow will consider only the more specific types of inhibitors.

Numerous studies have been conducted on the nutritional response to consumption of these digestive enzyme inhibitors. For the most part, changes in growth and in the pancreas, which synthesize the enzymes have been examined. More recently, the metabolic responses occurring secondarily to those on growth and on the pancreas have received attention. These nutritional and metabolic responses will be discussed.

GROWTH DPERESSION DUE TO PROTEOLYTIC INHIBITORS

Since the discovery by Osborne and Mendel in 1917 that raw soybeans would not support growth in rats, numerous studies using several different animal species have demonstrated that some uncooked plant products, particularly legumes, depress growth rates. Because of the economic importance of legumes as an animal feedstuff and the potential importance of soy protein in human nutrition, this effect of raw soybeans on growth has been extensively studied. Several early reports found that the slight increment in apparent digestibility caused by heating soybean meal was too small to account for the pronounced improvement in growth rate that also occurs (Melnick et al., 1946; Hayward et al., 1936; Mitchell et al., 1945). However, these studies on digestibility were based on fecal nitrogen excretion. Carroll et al. (1952) examined digestibility more directly by measuring net nitrogen absorption from a section of the samll intestine and found that rats fed raw soybean meal absorbed less than half the nitrogen of those fed heated meal. Taken together, these studies suggest that feeding raw soybean results in a considerable loss of protein from the site of absorption in the small intestine, but within the cecum bacteria metabolize the amino acids to nitrogenous compounds which are absorbed from the cecum but not utilized. Consequently, although fecal nitrogen excretion is only slightly increased in animals fed raw vs. heated soybean, less nitrogen is effectively available.

The discovery of a heat-labile trypsin inhibitor in soybeans (Bowman, 1944; Ham and Sanstedt, 1941) suggested a hypothesis for the growth depression effect. Impairment of intestinal proteolysis by the trypsin inhibitors present in soybeans could cause a ration only marginally sufficient in some essential amino acids to become deficient (Almquist and Merritt, 1951; 1953). Several studies have demonstrated decreased proteolysis in the intestinal tract of young chicks fed raw soybean meal (Alumot and Nitsan, 1961; Bielorai and Bondi, 1963), although older chicks quickly adapt and show normal levels of intestinal proteolysis (Saxena et al., 1963; Nitsan and Alumot, 1964). In the rat and the mouse, two species in which raw soybean feeding clearly causes growth depression, early studies had suggested that no difference in intestinal proteolysis occurred since intestinal proteolytic activity was generally equal or even greater in animals fed raw soybeans (Lyman, 1957; Lyman and Lepkovsky, 1957; Haines and Lyman, 1961; Nitsan and Bondi, 1965). More recently, Lyman et al. (1974) demonstrated that in the rat the presence of soybean trypsin inhibitor can interfere with protein digestion by binding trypsin and preventing the complete activation of chymotrypsinogen.

Proteolysis of endogenously secreted protein as well as dietary protein could be reduced by inhibitors. Lyman and Lepkovsky (1957) found that feeding raw soybean meal or a purified soybean trypsin inhibitor caused an exaggerated secretion of pancreatic enzymes into the small intestine. Considerable amounts of these enzymes were not degraded and reabsorbed but passed into the cecum and large intestine. This loss of essential amino acids from endogenous sources, in addition to to the incomplete intestinal proteolysis of dietary proteins, contribute to the soy-induced growth depression. The loss of sulfur amino acids would be most critical as the cystine content of pancreatic enzymes is high (Neurath, 1961). Supplementing raw soybean meal with certain essential amino acids, especially methionine, improves the growth rate associated with raw soybean meal to that of heated meal (Borchers, 1959; Booth et al., 1960; Borchers, 1961). Part of this improvement is undoubtedly due to the fact that soybean meal is limiting in methionine (Almquist et al., 1942). However, the growth response with methionine supplementation is much greater in animals fed a raw soy flour diet than the heated flour, which is consistent with an increased loss of cystine-rich endogenous protein due to trypsin inhibitors (Gertler et al., 1967; Borcher, 1961; Fisher and Johnson, 1958). In addition, rats fed purified amino acid diets supplemented with an active anti-trypsin fraction have considerably more TCA-precipitable nitrogen in their intestinal contents than those fed the same diet containing a heat-inactivated anti-trypsin fraction (de Muelanaere, 1964). Elevation of the TCA-precipitable nitrogen is most likely due to the presence of excess pancreatic enzymes in the intestine.

Several factors will mitigate the deleterious effect of trypsin inhibitors. The degree of growth depression is dependent upon the protein level of the diet. An increase in raw soybean meal from 15% to 25% of the diet was found to produce a large increase in weight gain in rats (Fisher and Johnson, 1958) and chicks (Gertler et al., 1967). In another study, mice fed a 24% casein diet containing a trypsin inhibitor fraction had a growth rate equivalent to those fed the same diet without the inhibitor (Roy and Schneeman, 1981). These results indicate that increasing the dietary supply of amino acids in terms of either protein quantity or quality supports better growth by compensating for endogenous losses. Antibiotics incorporated into the diet also have a beneficial effect on growth rates of animals fed a raw soybean diet (Carroll et al., 1953; Nitsan and Bondi, 1965; Barnes et al., 1965a). In rats, prevention of coprophagy eliminates this effect (Barnes et al., 1965b). The antibiotics appear to reduce or alter the intestinal microflora such that more endogenous protein is excreted intact in the feces. Coprophagy then makes this protein available for reabsorption. Chicks, which do not coprophagize, experience less growth rate improvement with antibiotics than rats (Nitsan and Bondi, 1965). Thus, they apparently absorb some amino acids from the lower intestine that are normally degraded by bacteria, but the amount is quantitatively less significant than that reabsorbed by the rat via coprophagy.

Most of the research involving trypsin inhibitor effects on growth have been done in rats, chicks, and mice. However, growth depression with raw soybean meal feeding has been noted in other species, such as guinea pigs (Patten et al., 1973), goslings (Nitsan and Nir, 1977), growing swine (Yen et al., 1974), and calves (Gorrill and Thomas, 1967). The effect may be age-dependent, as no difference in weight gain was found when raw soybean meal was fed to laying hens (Saxena et al., 1963) or adult dogs (Patten et al., 1971). One study found no weight difference in rats after six months of raw soybean meal feeding compared to rats fed heated soybean meal, suggesting eventual catch-up occurs (Booth et al., 1964). The effect of age may be related to the lower dietary protein requirement for adult animals. Feeding a lower protein diet may be necessary to test the effects of trypsin inhibitors on weight gain in adult animals. In general, it appears that dietary trypsin inhibitors pose less of a threat to adult animals.

Although trypsin inhibitors contribute to the growth depression experienced by growing animals fed raw soybean meal, it appears that other factors are involved. The addition of purified trypsin inhibitors to diets of heated soybean meal in amounts equivalent to that present in the raw meal did not produce growth depression to the same extent as did the raw meal (Gertler et al., 1967). Kakade et al. (1973) removed the trypsin inhibitors from the raw meal by affinity chromatograph. When this inhibitor-devoid raw meal was fed to rats, the degree of growth depression was only somewhat less than that of the raw meal containing the inhibitor. Although this experiment left no doubt that trypsin inhibitors in the diet can cause growth depression, it suggests that much if not most of the effect is due to the raw soybean protein itself. Plant proteins in their native state are known to be refractory to proteolytic attack resulting in poor amino acid availability (Grau and Carroll, 1958; Bozzini and Silano, 1978). In addition, the prolonged presence in the intestine of protein of poor digestibility may act in a similar manner to trypsin inhibitors by causing hypersecretion or reduced degradation of pancreatic enzymes (Green et al., 1973; Percival and Scheneeman, 1979). Heat treatment of soybean meal would have two effects, denaturing the trypsin inhibitors and increasing digestibility of the protein, both of which would lead to alleviation of growth depression. The degree to which each of these factors is important may be species dependent.

PANCREATIC EFFECTS OF PROTEOLYTIC INHIBITORS

Pancreatic enlargement due to raw soybean meal feeding was first reported by Chernick et al. in 1948. Since that time the observation has been repeatedly confirmed in chicks (Nitsan and Alumot, 1964; Gertler and Nitsan, 1970; Nitsan and Nir, 1977) and rats (Booth et al., 1960; Konijn and Guggenheim,1967; Rackis, 1965). Addition of purified trypsin inhibitor fractions to casein- or heated soybean-meal based diets has demonstrated that trypsin inhibitors are responsible for this phenomenon (Rackis, 1965; Khayambashi and Lyman, 1966; Gertler et al., 1967; Roy and Schneeman, 1981; Chan and de Lumen, 1982).

This enlargement has been characterized as primarily hypertrophic, with some hyperplasia occurring (Melmed et al., 1976; Crass and Morgan, 1982; Kakade et al., 1967). Concomitant with enlargement is an increase in pancreatic proteolytic activity (Chernick et al., 1948; Applegarth et al., 1964; Kakade et al., 1967; Gertler and Nitsan, 1970; Patten et al., 1973; Roy and Schneeman, 1981). This increase occurs very quickly, with only a single meal of raw soybeans needed to detect an acceleration of the trypsin synthetic rate in the rat (Dijkhof and Poort, 1978). Amylase activity in the chick pancreas may be either unchanged (Lepkovsky et al., 1966; Gerltler and Nitsan, 1970; Nitsan and Nir, 1977) or slightly reduced (Hewitt et al., 1973; Kakade, 1967) with trypsin inhibitor-containing diets, whereas in the mouse it is increased (Roy and Schneeman, 1981). Lipase activity is unaffected in both species (Nitsan and Nir, 1977; Roy and Schneeman, 1981).

How trypsin inhibitors cause pancreatic hypertrophy and increased proteolytic enzyme concentrations is best explained in terms of the mechanism of regulation of pancreatic secretion. It has been shown that levels of free intestinal trypsin and chymotrypsin act to monitor the secretion rate by a negative feedback inhibition (Green and Lyman, 1972; Lyman et al., 1974; Schneeman and Lyman, 1975; Schneeman et al., 1977). These investigators postulated that the presence of these enzymes in the upper intestine suppress release of a humoral agent that increases pancreatic secretion. The presence of protein in the tract leads to the formation of enzyme-substrate complexes allowing the release of this humoral agent, thereby increasing pancreatic juice flow. With more trypsin and chymotrypsin in the intestine, protein digestion can occur efficiently, and secretion rates will subsequently decrease. Cholecystokinin-pancreozymin (CCK-PZ) is believed to be the humoral agent released from the intestine, based on its known abilities to stimulate pancreatic secretion and cause both pancreatic hypertrophy and hyperplasia (Rothman and Wells, 1967; Mainz et al., 1973; Yanatori and Fujita, 1976).

Trypsin inhibitors in the intestine act analogous to protein, but are a much more potent stimulus (Green and Lyman, 1972; Green et al., 1973). Whereas the protein-enzyme complexes that form are reversible and fleeting, the inhibitor-enzyme complexes are essentially irreversible. Because of this, the entire complex is believed lost to the animal. In this way, a small amount of inhibitor mimics a large amount of protein and, when present in a meal, causes hypersecretion by the pancreas. The finding that purified trypsin inhibitor is a potent stimulator of the release of CCK-PZ-like activity from the upper intestine of the rat supports this hypothesis (Khayambashi and Lyman, 1969; Brand and Morgan, 1981; Schneeman and Lyman, 1975).

In addition to the rat and chick, the mouse (Schingoethe et al., 1970; Roy and Schneeman, 1981) and the young guinea-pig (Patten et al., 1973) have demonstrated pancreatic enlargement with raw soybean meal or purified trypsin inhibitor feeding. However, pancreatic enlargement failed to occur in the adult guinea-pig (Patten et al., 1973), the dog (Patten et al., 1971), the growing swine (Yen et al., 1977), or the calf (Gorrill and Thomas, 1967; Kakade et al., 1976). From this information, it was noted by Kakade et al. (1976), and reiterated by Liener (Liener, 1979a, 1979b) and Liener and Kakade (1980), that animals which have pancreata that are 0.3% of body weight or more (such as the mouse, chick, rat, and young guinea-pig) show pancreatic enlargement with trypsin inhibitor feeding. Those animals in which the pancreas weight is less than 0.3% (the adult guinea-pig, pig, dog, and calf) do not show pancreatic enlargement. These investigators suggested that, if this relationship is a valid one, then the human pancreas would not be expected to enlarge due to trypsin inhibitors, as it weighs about 0.09 to 0.12% of body weight.

A closer examination of the studies upon which this relationship is based suggests that one needs to view this relationship cautiously. To this end, several considerations should be noted. First of all, there is evidence that the negative feedback mechanism of pancreatic enzyme secretion found in the rat also exists in the pig (Corring, 1974), calf (Davicco et al., 1979), and possibly humans (Ishe et al., 1977). The existence of the same mechanism in these species implies the potential for the same effects of pancreatic enlargement by trypsin inhbitors as occurs in the rat. Another point to consider is the nutritional status of the animal during the experiment. The pancreas is a tissue that rapidly responds to starvation (Fauconneau and Michel, 1970) and to the protein content of the diet. It has been shown in both rats and chicks that increasing the protein content of the diet acts permissively to produce a greater degree of pancreatic enlargement with trypsin inhibitor feeding, compared to a heated soybean meal diet of similar protein content (Chernick et al., 1948; Fisher and Johnson, 1958; Gertler et al., 1967). The quality of the protein is also involved in the degree of the pancratic response. An animal in good nutriture responds quickly to a change in diet by altering the digestive enzyme profile (Snook, 1974; Corring, 1980). An increase in dietery protein will increase trypsinogen and chymotrypsinogen in the pancreas within a few days. Similarly, pancreatic amylase and lipase respond quickly to changes in the amount of dietary carbohydrate and fat, respectively. When only poor quality proteins are fed, however, the ability of the enzymes in the pancreas to adapt to changes in dietery composition is impaired (Ikegami et al., 1975; Johnson et al., 1977). Raw soybean meal can be considered a poor quality protein for two reasons. First of all, the protein is refractory to digestion (Kakade et al., 1973). Thus, much of it is lost to the animal by excretion in the feces or due to bacterial metabolism to compounds useless to the animal. In addition, soybean meal, and legumes in general, are relatively deficient in sulfur amino acids, a problem which is compounded by the fact that a significant proportion of the cystine in legumes is present in the trypsin inhibitors, which appear to be unavailable to the animal (Kakade et al., 1969). A final point that is important concerning the pancreatic enzyme profile in response to inhibitors is whether the animal was in an unfed or a fed state prior to removing the pancreatic tissue. The enzyme profile in the pancreas of a fed animal will reflect the secretion and synthesis that occurs in response to the diet and not necessarily the longer term adaptive changes that can be observed in an unfed animal. In many of the studies cited, it is difficult to estimate whether the animal was in a fed or unfed state when pancreatic enzyme levels were determined.

Keeping in mind the above, it is possible to re-examine the studies in which no pancreatic hypertrophy was detected from a different perspective. For example, in the case of the adult guinea-pig (Patten et al., 1973), the level of protein in the diet was adequate, but the animals fed the raw soybean meal lost considerable weight during the experiment. In the study on growing swine (Yen et al., 1977), a slight weight loss occurred in one trial and in another only a very slight weight gain occurred. Furthermore, the protein level of 16% may have been inadequate to have supported a hypertrophic response. In both studies on calves, the group fed inhibitor-containing diets lost weight. In fact, in the study by Gorrill and Thomas (1967), 3 of 9 calves on the raw soy diet died. The failure to detect pancreatic enlargement is probably related to the poor growth experienced by these animals. In the study with adult dogs (Patten et al., 1971), pancreatic enlargement, when expressed per kg body weight, was found in a group fed raw soybean meal at 15% of the diet for 12 weeks. This enlargement was accompanied by significant increases in the concentrations of trypsinogen and chymotrypsinogen. However, 8 weeks of feeding either a 15% or 30% raw soybean meal diet did not produce a pancreatic enlargement. It may be that the amount of soybean meal in the diet was too little to produce a clear-cut repsonse, as the concentration of trypsin inhibitor used in their 15% soybean meal diet was less than one-tenth that used in the unheated soy flour diet employed by Kakade et al. (1973). Two additional points are worth making. One is that the pancreatic growth response in adult animals may be species specific. Adult rats clearly showed pancreatic enlargement due to raw soybean feeding (Crass and Morgan, 1982), whereas laying hens did not (Saxena et al., 1963). The other point is that the feed back mechanism reported to control the pancreas in various species does not seem to be important in the dog, and the dog pancreas did not respond to trypsin inhibitor in the intestine with pancreatic enzyme hypersecretion (Sale et al., 1977). If these findings are correct, then dog pancreas would not be expected to be very sensitive to the presence of trypsin inhibitors in the diet.

Examining the evidence as a whole, it does not appear justified to hypothesize a relationship between the occurrence of pancreatic enlargement with raw soybean feeding and pancreas size as a percent of body weight. This relationship may, in fact, exist, but to establish its validity more definitive studies will be needed. Until that time, it is difficult to predict whether the human pancreas will enlarge with raw soybean feeding.

Another factor contributing to differences in the pancreatic response to trypsin inhibitors among species may be variations in the predominant forms of trypsin synthesized by the pancreas. One study using purified human enzymes and soybean inhibitor suggested that the predominant trypsin in human pancreatic juice is bound only weakly by the inhibitor (Figarella, 1974). In contrast, studies with crude pancreatic homogenates and crude soybean preparations suggest that substantial interference with human proteolytic activity can occur in vitro (Krogdahl and Holm, 1979). The extent of human enzyme inhibition in vivo by soybean inhibitor preparations and the physiological importance of any interactions are currently unknown. Humans are unlikely to consume a diet rich in raw soybeans, thus avoiding the extreme responses observed in experimental animal studies. Data currently available suggests that if the diet is adequate, especially with regard to protein, mechanisms exist in adult humans to adapt to small amounts of inhibitors which might be consumed in the diet (Ihse, 1977).

AMYLASE INHIBITORS

Naturally occurring inhibitors of pancreatic amylase were first discovered in aqueous extracts of wheat, rye, and kidney beans (Kneen and Sanstedt, 1943, 1946; Bowman, 1945). They were found to have the properties of a protein and to be quite thermostable. Inhibition occurred in a noncompetitive manner. The wheat-derived inhibitor is active against the amylase from a number of mammalian species, including the rat, mouse, and dog, whereas the kidney-bean derived inhibitor was found to have substantial activity only against the dog amylase (Ho et al., 1981). More recently, a new class of amylase inhibitors has been isolated from culture broths of Actinomycetes (Schmidt et al., 1977). These inhibitors are complex oligosaccharides with a molecular weight of 500-5000. Some are fully competitive inhibitors, while others are noncompetitive (Folsch et al., 1981; Suchiro et al., 1981). Although it is unlikely that this class of inhibitors would occur in a normal diet, they may find therapeutic use in the treatment of obesity or diabetes (Keup and Puls, 1974).

Due to their thermostability, proteinaceous amylase inhibitors have been detected in significant amounts in the center of baked wheat-flour goods (Bessho and Kurosawa, 1968) and in some wheat-based breakfast cereals (Marshall, 1975). However, relatively few studies have been conducted on the physiological effects of consuming these inhibitors. Using large doses of the oligosaccharide inhibitor, the growth rates of rats was reduced when fed a diet where the carbohydrate source was starch (Puls and Keup, 1974; Folsch et al., 1981). This reduction occurred in spite of an increased food intake. Evidence is conflicting as to whether the proteinaceous amylase inhibitors reduce growth rates. In one study, rats were fed a caloric-deficient, starch-free diet; addition of starch to the diet increased growth. When wheat-derived amylase inhibitor was included with the starch, the increase in growth was much less than with the starch alone (Lang et al., 1974; Saunders, 1975). Heat-inactivated inhibitor was without effect. Furthermore, if sucrose replaced starch as the carbohydrate source, then inclusion of the inhibitor did not retard growth. Conflicting with this study is one by Savaiano et al. (1977). These investigators, using a similar technique of restricted food intake, fed kidney bean-derived amylase inhibitor to rats. In two separate experiments, they found no effect on growth using an amount of inhibitor equal to that found in 11 or 82 g of kidney bean flour per day. The source of the inhibitor may explain the discrepancy between these two reports. Rat pancreatic amylase is very sensitive to wheat-derived inhibitor, but is only slightly affected by the inhibitor from kidney beans in vitro (Ho et al., 1981). This differential sensitivity is probably pH related, as the activity of the kidney bean-derived inhibitor drops off rapidly as the pH shifts from its optimum of 5.5 (Marshall and Lauda, 1975). As the intestinal lumen pH is about 6.8, one would expect the activity of the kidney bean-derived inhibitor to be low in vivo. In partial agreement with the study of Lang et al. (1974), chicks fed inhibitor-containing wheat albumin, encapsulated to prevent gastric destruction, had a reduction in initial growth rate (Macri et al., 1977). Four weeks of this regime, however, led to an apparent adaptation, with growth rates of the inhibitor-fed chicks equalling that of the control animals. Of interest is the finding that the inhibitor preparation without encapsulation had no effect on growth. It is not clear why encapsulation is necessary to produce this effect in the chick, but not in the rat.

A decreased rate of starch hydrolysis in the small intestine would logically lead to a slower absorption of glucose and possibly cause certain metabolic changes. This has been found to be the case. In rats and dogs, a wheat-derived amylase inhibitor added to a starch load attenuated both the rise in blood glucose and the fall in non-esterified fatty acids that normally occurs with such a loading (Puls and Keup, 1973). In humans, a similar effect of the inhibitor on blood glucose levels was found, as well as flattening of the serum insulin peak. The same effects on blood glucose and serum insulin are found when amylase inhibitors of microbial origin are used (Puls and Keup, 1974). Long term feeding of this type of inhibitor led to a decreased rate of incorporation of 14C-starch into lipids of epididymal adipose tissue and of aortic tissue (Puls and Keup, 1974).

Given the dramatic ability of trypsin inhibitors to increase pancreatic trypsin and chymotrypsin, one might expect amylase inhibitors to produce an analogous response. However, regulation of pancreatic amylase levels appers to be fundamentally different than that of trypsinan and chymotrypsin. As dicussed in the previous section, the presence of protein or a substrate analogue within the intestinal lumen regulates these two proteolytic enzymes. In contrast, regulation of pancreatic amylase levels is related to the quantity of digestion products absorbed (Lavau et al.,1974). Hence, increases in the quantity of absorbed glucose will increase the pancreatic amylase activity. Insulin is involved in the stimulation of pancreatic amylase synthesis, although its role may be only a permissive one (Soling and Unger, 1972). Accordingly, if starch is the only carbohydrate source, inhibition of amylase in the intestine would be expected to decrease pancreatic amylase concentrations, the opposite effect of trypsin inhibitors on trypsin and chymotrypsin. In two feeding studies with rats, using the microbial-derived amylase inhibitor, this was found to occur (Puls and Keup, 1974; Folsch et al., 1981). Pancreatic amylase levels dropped to one-half the control value in one experiment and to one-quarter in the other. However, the study by Macri et al. (1977) conflicts with these reports. These workers found that chicks receiving the encapsulated wheat albumin-inhibitor preparation showed pancreatic hypertrophy, an increased amylase activity, and no difference in pancreatic protease activity. It was claimed that the trypsin inhibitor activity in the preparation was quantitatively insignificant (Petrucci et al., 1974). Since both types of inhibitors could produce this effect, their results are difficult to explain. No reason for the apparent discrepancy is obvious to these authors. Further investigation will be required to clarify this matter.

Amylase inhibitors would seem to offer, in theory, an attractive method of decreasing glucose absorption from starch, thereby reducing the caloric content of a meal. This possibility has created interest in the scientific community for use of these inhibitors in the control of obesity (Marshall, 1975; Sullivan et al., 1980). As yet, no studies have been reported in which amylase inhibitors have been evaluated in this capacity. However, numerous commercial concerns, impatient at the imponderous pace of scientific investigation, and hearing the clear knock of opportunity, have flooded the marketplace with amylase inhibitor preparations, the so-called "starch-blockers". These preparations, derived from kidney beans, commonly claim to contain enough inhibitor activity to block the digestion of 100 g of starch, or the equivalent of 400 kcal. The enormous popularity of starch blockers is attested to by the assertion that over one million starch blocker tablets were consumed daily in the United States in the first part in 1982 (Sheff, 1982). Concern about the safety and efficacy of starch blockers has prompted the Food and Drug Administration to prohibit their sale until these issues are resolved.

Only recently has the first study on the effectivness of the starch blockers in humans been published (Bo-Linn et al., 1982). These investigators developed a one-day calorie-balance technique to allow determination of the number of calories abosrbed from a high-starch test meal with and without starch blockers. The subjects were first prepared by flushing the entire gastrointestinal tract free of residual food and fecal matter. A meal containing 97 g of starch was then consumed by each subject. Two starch blocker tablets (Carbo-Lite) were taken at the beginning of the meal and one at midmeal. Fourteen hours later, the GI tract was again flushed and the rectal effluent collected. The caloric content of the effluent, as well as any stool passed during the period, was determined by bomb calorimetry. No difference was found between subjects consuming the starch blockers or a placebo. It was concluded that the starch blocker preparation used did not inhibit the digestion and absorption of starch in vivo.

The failure of the starch blockers to decrease starch hydrolysis may be due to several factors. It is possible that much of the inhibitor, a glycoprotein (Marshall and Lauda, 1975), is destroyed in the stomach by pepsin or acid. Once in the small intestine, the pH would be sufficiently far from the optimum for enzyme-inhibitor complex formation that its activity could be reduced by two-thirds, based on in vitro studies (Marshall and Lauda, 1975). Furthermore, these same in vitro studies showed that the inhibitor must be preincubated with amylase in the absence of starch to achieve inhibition of the enzyme. Finally, one must consider that the pancreas normally secretes ten times the amount of amylase required to affect complete digestion of starch in a meal (Gray, 1971). Thus, the amount of amylase inhibitor present in the starch blocker preparations, when consumed at the prescribed dose, would at best inhibit only a fraction of the amylase present.

LIPASE INHIBITORS

In contrast to the large number of proteolytic enzyme inhibitors of plant origin, virtually no lipase inhibitors have been identified. Initial reports of a pancreatic lipase inhibitor from soybean cotyledons described a protein that decreased lipolytic activity (Mori et al., 1973; Stouchi et al., 1974). Pancreatic lipase, as all lipases, act at the oil-water interface of an emulsified oil droplet. Therefore, the substrate concentration is not equal to the amount of triacylglycerol present in the system, but to the total interfacial area. This area, for a given amount of triacylglycerol, is dependent upon the size of the emulsified oil droplets and the presence of substances that compete for the interfacial area. Inhibitors of pancreatic lipase may act as true inhibitors by forming enzyme-inhibitor complexes, or they may reduce activity by adhering to the oil-water interface and thereby effectively reducing the substrate concentration. Further study of the soybean cotyledon lipase inhibitor revealed its mode of action to be adherence to the interface (Satouchi et al., 1976) and not an enzyme inhibitor in the classical sense. Whether this protein would have any physiological effects if fed is unknown.

Research conducted in our laboratory has uncovered another inhibitor of pancreatic lipase. While examining interactions among various types of dietary fibers and digestive enzyme activities, it was found that cellulose drastically reduced lipase activity in vitro (Schneeman, 1978). In these studies, cellulose was incubated with a purified porcine lipase, then filtered and assayed for lipase activity. This method assures that inhibition is due to an effect on the enzyme itself and not to an interference with access of the enzyme to the substrate. This effect of cellulose on lipase occurs when the source of lipase is rat pancreatic juice (Schneeman and Gallaher, unpublished observations) or human pancreatic juice (Dunaif and Schneeman, 1981). Kinetic studies have not been done because of the difficulty of determining substrate concentrations. However, cellulose will reduce lipase activity in the presence of an emulsified oil substrate (Schneeman and Gallaher, unpublished observations). This suggested the possibility that cellulose may inhibit lipase activity in vivo as well. An experiment was conducted to test this hypothesis (Schneeman and Gallaher, 1980). Rats were fed either a no fiber diet or a diet containing 20% cellulose. One group of animals was killed after 12 hours without food and is designated the unfed group. Another grooup was killed 3 hours after a 3 g meal and is designated the fed group. Some results of this study are shown in Table 1. In the unfed group, both pancreatic weight and the lipase concentration were the same for both dietary treatments. Lipase activity in the fed group's intestinal contents was considerably lower in the cellulose-fed animals. However, lipase activity of the washed mucosa from the cellulose-fed group was actually slightly

Table 1. Pancreas weight and lipase activity in the intestine of rats adapted to a diet with cellulose (C) or fiber free (FF).

	Diet	
Pancreas (unfed) (mg)	FF	245.6 ± 20.2
	C	241.0 ± 19.3
Lipase in intestinal contents (fed) (units[1]/contents)	FF	43.4 ± 5.1
	C	5.7 ± 1.3[a]
Lipase in intestinal mucosa (fed) (units/mucosa)	FF	92.7 ± 9.9
	C	116.4 ± 8.8[b]

[1]Lipase unit = 1 ueq OH^- titrated/min at 30°; [a]$P < 0.001$; [b]$P < 0.10$.

higher than the control group. When the intestinal and mucosal values are combined, there is no statistically significant difference between diets. Interpretations of these last results is difficult because it is not known to what extent mucosal bound enzymes are active in the digestive process. However, the pancreatic data provide some insight. Pancreatic lipase appears to be regulated in a manner similar to amylase. That is, it is the amount of triacylglycerol digestion products absorbed, not the presence of substrate in the intestinal lumen, that influences pancreatic lipase activity (Lavau et al., 1974). The lack of an effect of the dietary cellulose on pancreatic lipase concentrations suggests that cellulose does not significantly reduce the total amount of triacylglycerol absorbed.This does not preclude the possibility that cellulose does inhibit lipase in vivo, slowing digestion rate. In humans, lipase activity must be reduced 90% to produce a significant fat malabsorption (DiMago et al., 1973). Because of this normal hypersecretion, it may be that cellulose does not inhibit lipase in vivo, but sufficient enzyme is still present to hydrolyze all the triacylglycerol present. If this is the case, absorption may simply be delayed, with greater amounts of fatty acids absorbed lower in the intestinal tract. As it has been shown that chylomicrons secreted from the distal intestine differ in composition and size from those secreted more proximally (Wu et al., 1980), such a shift in the site of absorption could affect lipid metabolism. Preliminary data indicates that cellulose may slow the disappearance of triacylglycerol from the small intestine (Schneeman et al., 1983).

CONCLUSIONS

Although enzyme inhibitors have been investigated since the early 1900's, we still have little information about their importance in the human diet. One can speculate that since humans have evolved with inhibitors of plant or animal origin in their diet, they have adapted to their presence either through physiological mechanisms or cultural food preparation practices. We are capable of estimating inhibitor levels in commercially prepared products to estimate potential intake levels. Both animal and human studies are needed to determine physiological and adaptive responses to these levels of inhibitors.

Modern science has begun to view enzyme inhibitors from perspectives other than their impact on nutrient availability. For example, naturally occurring levels of amylase inhibitors may not be nutritionally important, but researchers are examining their use as a drug in the treatment of diabetes or obesity. Likewise, similar shifts are seen in the study of proteolytic enzyme inhibitors. Experiments with diabetic rats have indicated some enhancement of endocrine function as well as exocrine function with trypsin inhibitors in the diet, suggesting a use in diabetic treatment (Ihse et al.,1976). On the other hand, concern has been raised that the pancreatic hypertrophy associated with trypsin inhibitors will enhance the incidence of pancreatic tumors (Morgan et al., 1977). These associations at the present time are unconfirmed and probably represent pharmacological responses to enzyme inhibitors. However, they do lead to new areas of investigation and emphasize the importance of designing experiments with humans to determine pancreatic response to the consumption of inhibitors.

Note: This paper is reprinted with permission from "Nutritional and Toxicological Aspects of Food Safety", M. Friedman, editor, Plenum Press, New York, Adv. Exp. Med. Biol., 177, 299, 1984. The following paper (Schneeman and Gallaher, 1986) extends and updates the discussion offered here on the comparative pancreatic response among various species to trypsin inhibitors.

REFERENCES

Almquist, H.J., Mecchi, E., Kratzer F. H. and Grau, C. R. (1942). Soybean as a source of amino acids for the chick. J. Nutr., 24, 385-392.

Almquist, H. J.and Merritt, J. B. (1953). Accentuation of dietary amino acid deficiency by raw soybean growth inhibitor. Proc. Soc. Exp. Biol. Med., 84, 333-334

Almquist, H. J. and Merritt, J. B. (1951). Effect of soybean antitrypsin on experimental amino acid deficiency in the chick. Arch. Biochem. Biophys., 31, 450-453.

Alumot, E. and Nitsan, Z. (1961). The influence of soybean antitrypsin on the intestinal proteolysis of the chick. J. Nutr., 73, 71-77.

Applegarth, A., Furuta, F. and Lepkovsky, S. (1964). Response of the chicken pancreas to raw soybeans. Poult. Sci., 43, 733-738.

Barnes, R. H., Kwong, E. and Fiala, G. (1965a). Effect of penicillin added to an unheated soybean diet on cystine excretion in feces of the rat. J. Nutr., 85, 123-126.

Barnes, R. H., Fiala, G. and Kwong, E. (1965b). Prevention of coprophagy in the rat and growth stimulating effects of methionine, cystine, and penicillin when added to diets containing unheated soybeans. J. Nutr., 85, 127-131.

Bielorai, R. and Bondi, A. (1963). Relationship between 'antitryptic factors' of some plant protein feeds and products of proteolysis precipitable by trichloroacetic acid. J. Sci. Food Agric., 14,124-132.

Bessho, H. and Kurosawa, S. (1967). Enzyme inhibitor in food. III. Effect of cooking on amylase inhibitor in floor. Eiyo To Shokuryo, 20, 317-319. Chem. Abst. 68: 113474e.

Bo-Linn, G., Santa Ana, C., Morawski, S. and Fordstran, J. (1982). Starch blockers - their effect on calorie absorption from a high-starch meal. New Engl. J. Med., 307, 1413-1416.

Booth, A. N., Robibins, D. J., Ribelin, W. E. and DeEds, F. (1960). Efffect of raw soybean meal and amino acids on pancreatic hypertrophy in rats. Proc. Soc. Exp. Med. Biol., 104, 681-683.

Booth, A. N., Robbins, D.J., Ribelin, W. E., DeEds, F., Smith, A. K. and Rackis, J. J. (1964). Prolonged pancreatic hypertrophy and reversibility in rats fed raw soybean meal. Proc. Soc. Exp. Biol. Med., 116, 1067-1068.

Borcher, R. (1961). Counteraction of the growth depression of raw soybean meal by amino acid supplements in weanling rats. J. Nutr., 75, 330-334.

Bowman, D..(1945). Amylase inhibitors of navy beans. Science, 102, 358-359.

Bowman, D. (1941). Fractions derived from soybeans and navy beans which retard tryptic digestion of casein. Proc. Soc. Exp. Biol. Med., 57, 139-140.

Bozzini, A. and Silano, V. (1978). Control through breeding methods of factors affecting nutritional quality of cereals and grain legumes. In "Nutritional Improvement of Food and Feed Proteins", M. Friedman, ed., Plenum Press, New York, pp. 249-274.

Brand, S. J. and Morgan, R. G. H. (1981). The release of rat intestinal cholecystokinin after oral trypsin inhibitor measured by bio-assay. J. Physiol., 319, 325-343.

Carroll, R. W., Hensley, G. W. and Graham, W. R., Jr. (1952). The site of nitrogen absorption in rats fed raw and heat-treated soybean meals. Science, 115, 36-39.

Carroll, R. W., Hensley, G.W., Sittler, C. L., Wilcox, E.L. and Graham, W. R., Jr. (1953). Absorption of nitrogen and amino acids from soybean meal as affected by heat treatment or supplementation with aureomycin and methionine. Arch. Biochem. Biophys., 45, 260-274.

Chan, J. and de Lumen, B. O. (1982). Biological effects of isolated trypsin inhibitor from winged bean (Psophocarpus tetragonolobus) on rats. J. Agric. Food Chem., 30, 46-50.

Chernick S. S., Lepkovsky, S. and Chaikoff, I. L. (1948). A dietary factor regulating the enzyme content of the pancreas: changes induced in size proteolytic activity by the ingestion of raw soybean meal. Am. J. Physiol., 155, 33-41.

Corring, T. (1980). The adaptation of digestive enzymes to the diet: its physiological significance. Reprod. Nutr. Develop., 20, 1217-1235.

Corring, T. (1974). Regulation de la secretion pancreatique par retroaction negative chez le porc. Ann. Biol. Anim. Biochem. Biophys., 14, 487-498.

Crass, R. A. and Morgan, R. G. H. (1982). The effect of long-term feeding of soya-bean flour on pancreatic growth in the rat. Br. J. Nutr., 47, 119-129.

Davicco, M.J., Lafaivre, J., Thiuend, P. and Barlet, J. P. (1979). Feedback regulation of pancreatic secretion in the young milk-fed calf. Ann. Biol. Anim. Bioh. Biophys., 19, 1147-1152.

de Muelenaere, H.J.H. (1964). Studies on the digestion of soybeans. J. Nutr., 82, 197-205.

Dijkhof, J. and Poort, C. (1978). Changes in rat pancreatic protein synthesis after a single feeding with diets containing raw or heated soybeans. J. Nutr., 108, 1222-1228.

DiMagno, E., Go, V. and Summerskill, W. (1973). Relations between pancreatic enzyme outputs and malabsorption in severe pancreatic insufficiency. New Engl. J. Med., 288, 813-815.

Dunaif, G. and Schneeman, B. O. (1981). The effect of dietary fiber on human pancreatic enzyme activity in vitro. Am. J. Clin. Nutr., 34, 1034-1035.

Fauconneau, G. and Michel, M. C. (1970). The role of the gastrointestinal trace in the regulation of protein metabolism. In "Mammalian Protein Metabolism", H. Munro, ed., Academic Press, New York.

Figarella, C., Negri, G. A. and Guy, O. (1974). Studies on the inhibition of the two human trypsins. In "Proteinase Inhibitors", Fritz et al., eds., Springer-Verlag, New York.

Fisher, H. and Johnson, D., Sr. (1958). The effectivness of essential amino acid supplementation in overcoming the growth depression of unheated soybean meal. Arch. Biochem. Biophys., 77, 124-128.

Folsch, U., Grieb, N., Caspary, W. and Creutzfeldt, W. (1981). Influence of short- and long-term feeding of an alpha-amylase inhibitor (BAY e 4609) on the exocrine pancreas of the rat. Digestion, 21,74-82.

Gerltler, A., Birk, Y. and Bondi, A. (1967). A comparative study of the nutritional and physiological significance of pure soybean trypsin inhibitors and of ethanol-extracted soybean meals in chicks and rats. J. Nutr., 91, 358-370.

Gertler, A. and Nitsan, Z. (1970). The effect of trypsin inhibitors on pancreotopeptidase E, trypsin, chymotrypsin, and amylase in the pancreas and intestinal tract of chicks receiving raw and heated soya-bean diets. Br. J. Nutr., 24, 803-804.

Gorrill, A. D. L. and Thomas, J. W. (1967). Body weight changes, pancreas size and enzyme activity, and proteolytic enzyme activity and protein digestion in intestinal contents from calves fed soybean and milk protein diets. J. Nutr., 92, 215-223.

Grau, C. R. and Carroll, R. W. (1958). Evaluation of protein quality of processed plant protein foodstuffs. In "Processed Plant Protein Foodstuffs," A. M. Altschul, ed., Academic Press, New York.

Gray, G. (1971). Intestinal digestion and maldigestion of dietary carbohydrates. Ann. Rev. Med., 22, 391-404.

Green, G. M., Olds, B. O., Matthews, G. and Lyman, R. L. (1973). Protein as a regulator of pancreatic enzyme secretion in the rat. Proc. Soc. Exp. Biol. Med., 142, 1162-1167.

Green, G. M. and Lyman R. L. (1972). Feedback regulation of pancreatic enzyme secretion as a mechanism for trypsin inhibitor induced hypersecretion in rats. Proc. Soc. Exp. Biol. Med., 140, 6-12.

Griffiths, D. W. and Moseley, G. (1980). The effect of diets containing field beans of high or low polyphenolic content on the activity of digestive enzymes in the intestines of rats. J. Sci. Food Agric., 31, 255-259.

Haines, P. C. and Lyman, R. L. (1961). Relationship of pancreatic enzyme secretion to growth inhibition in rats fed soybean trypsin inhibitor. J. Nutr., 74, 445-452.

Ham, W. E. and Sanstedt, R. M. (1944). A proteolytic inhibiting substance in the extract from unheated soybean meal. J. Biol. Chem., 154, 505--506.

Hayward, J.W., Steenbock, H. and Bohsted, G. (1936). The effect of heat as used in the extraction of soybean oil upon the nutritive value of the protein of soy bean meal. J. Nutr., 11, 219-234.

Hewitt, D., Coates, M. E., Kakade, M. L. and Liener, I. E. (1973). A comparison of fractions prepared from navy (haricot) beans (Phaseolus vulgaris L.) in diets from germ-free and conventional chicks. Br. J. Nutr., 29, 423-435.

Ho, R., Aranda, C. and Venico, J. (1981). Species difference in response to two naturally occurring alpha-amylase inhibitors. J. Pharm. Pharmacol., 33, 351.

Ihse, I., Lundquist, I. and Arnesijo, B. (1976). Oral trypsin inhbitor-induced improvement of the exocrine and endocrine pancreatic functions in alloxan diabetic rats. Scand. J. Gastroent., 11, 363-368

Ihse, I., Lilia, P. and Lundquist, I. (1977). Feedback regulation of pancreatic enzyme secretion by intestinal trypsin in man. Digestion, 15, 303-308.

Ikegami, S., Takai, Y. and Iwoa, H. (1975). Effect of dietary protein on proteolytic activities in the pancreatic tissue and contents of the small intestine in rats. J. Nutr. Sci. Vitaminol., 21, 287-295.

Johnson, A., Hurwitz, R. and Kretchmer, N. (1977). Adaptation of rat pancreatic amylase and chymotrypsinogen to changes in diet. J. Nutr., 107, 87-96.

Kakade, M. L., Arnold, R. L., Liener, I. E. and Waibel, P.E. (1969). Unavailability of cystine from trypsin inhibitors as a factor contributing to the poor nutritive value of navy beans. J. Nutr., 99, 34-41.

Kakade, M. L., Barton, T. L., Schaible, P. J. and Evan, R. J. (1967). Biochemical changes in the pancreas of chicks fed raw soybeans and soybean meal. Poult. Sci., 46, 1578-1585.

Kakade, M. L., Hoffa, D. E. and Liener, I. E. (1973). Contribution of trypsin inhibitors to the deleterious effects of unheated soybeans fed to rats. J. Nutr., 103, 1772-1778.

Kakade, M. L., Thompson, R. D., Engelstad, W. W., Behrens, G. C., Yoder, R. D. and Crave, F. M. (1976). Failure of soybean trypsin inhibitors to exert deleterious effects in calves. J. Dairy Sci., 59, 1484-1489.

Keup, U. and Puls, W. (1974). Influence of an amylase inhibitor,BAY e4609, on the conversion of orally applicated starch into total lipids of rat adipose tissue. In "Recent Advances in Obesity Research," A. Howard, ed., Newman Publishing Co., Ltd. London, England.

Khayambayashi, H. and Lyman, R. L. (1966). Growth depression and pancreatic and intestinal changes in rats forced-fed amino acid diets containing trypsin inhibitor. J. Nutr., 89, 455-464.

Kneen, E. and Sanstedt, R. (1943). An amylase inhibitor from certain cereals. J. Am. Chem. Soc., 65, 1247-1248.

Kneen, E. and Sanstedt, R.(1946). Distribution and general properties of an amylase inhibitor in cereals. Arch. Biochem.Biophys., 9, 235-249.

Konijn, A. M. and Guggenheim, K. (1967). Effect of raw soybean flour on the composition of rat pancreas. Proc. Soc. Expt. Biol. Med., 126, 65-67.

Krogdahl, A. and Holm, H. (1979). Inhibition of human and rat pancreatic proteinases by crude and purified soybean proteinase inhibitors. J. Nutr., 109, 551-558.

Lang, J., Chang-Hum. L., Reyes, P. and Briggs, G. (1974). Interference of starch metabolism by alpha-amylase inhibitors. Fed. Am. Soc. Exp. Biol. 33, 718 (Abstract).

Lavau, M., Bazin, R. and Herzog, J. (1974). Comparative effects of oral and parenteral feeding on pancreatic enzymes in the rat. J. Nutr., 104, 1432-1437.

Lepkovsky, S., Koike, T., Sugiura, M., Dimick, M. K. and Furuta, F. (1966). Pancreatic amylase in chickens fed on soya-bean diets. Br. J. Nutr., 20, 421-437.

Liener, I.E. (1979a). Significance for humans of biologically active factors in soybeans and other food legumes. J. Am. Oil. Chemists Soc., 56, 121-129.

Liener, I. E. (1979b). Protease inhibitors and lectins. In "International Review of Biochemistry," A. Veuberger and T. H. Jukes, eds., University Park Press, Baltimore, MD.

Liener, I. E. and Kakade, M. L. (1980). Protease inhibitors. In "Toxic Constituents of Plant Foodstuffs," I. E.Liener, ed., Academic Press, New York.

Lyman, R. L. (1957). The effect of raw soybean meal and trypsin inhibitor diets on the intestinal and pancreatic nitrogen secretion in the rat. J. Nutr., 62, 285-294.

Lyman, R. L. and Lepkovsky, S. (1957). The effect of raw soybean meal and trypsin inhibitor diets on pancreatic enzyme secretion in the rat. J. Nutr., 62, 269-284.

Lyman, R. L., Olds, B. O., and Green, G. M. (1974). Chymotrypsinogen in the intestine of rats fed soybean trypsin inhibitor and its inability to suppress pancreatic enzyme secretions. J. Nutr., 104, 105-110.

Macri, A., Parlamenti, R., Silano, V. and Valfre, F. (1977). Adaptation of the domestic chicken, Gallus Domesticus, to continuous feeding of albumin amylase inhibitors from wheat flour as gastro-intestinal microgranules. Poult. Sci., 56, 434-441.

Mainz, D. L., Black, O. and Webster, P. D. (1973). Hormonal control of pancreatic growth. J. Clin. Invest., 52, 2300-2304.

Marshall, J. (1975). Alpha-amylase inhibitors in foods. In "Physiological Effects of Food Carbohydrates", Am. Chem. Soc. Symp. Series, 115, 244-266.

Marshall, J. and Lauda, C. (1975). Purification and properties of phaseolamin, an inhibitor of alpha-amylase, from kidney bean, Phaseolus vulgaris. J. Biol. Chem., 250, 8030-8037.

Melmed, R. N., El-Aaser, A.A.A. and Holt, S. J. (1976). Hypertrophy and hyperplasia of the neonatal rat exocrine pancreas induced by orally administered soybean trypsin inhibitor. Biochem. Biophys. Acta, 421, 280-288.

Melnick, D., Oser, B.L. and Weiss, S. (1946). Rate of enzymatic digestion of proteins as a factor in nutrition. Science, 103, 326-329.

Mitchell, H. H., Hamilton, T. S. and Beadles, J. R. (1945). The importance of commercial processing for the protein value of food products. J. Nutr., 29, 13-25.

Morgan, R. G. H., Levinson, D. A., Hopwood, D., Saunders, J. H. B. and Wormsley, K. G. (1977). Potentiation of the action of azaserine on the rat pancreas by raw soybean flour. Cancer Letters, 3, 87-90.

Mori, T., Satouchi, K. and Matsushita, S. (1973). A protein inhibiting pancreatic lipase activity in soybean seeds. Agric. Biol. Chem., 37, 1225-1226.

Neurath, H. (1961). Consideration of the occurrence, structure and function of the proteolytic enzymes of the pancreas. In "The Exocrine Pancreas," A.V.S. de Reuck and M.P. Cameron, eds., Little Brown, Boston, MA.

Nitsan, Z. and Alumot, E. (1964). Overcoming the inhibition of intestinal proteolytic activity caused by raw soybean in chicks of different ages. J. Nutr., 84, 179-184.

Nitsan, Z. and Bondi, A. (1965). Comparison of nutritional effects induced in chicks, rats and mice by raw soya-bean meal. Br. J. Nutr., 19, 177-187.

Nitsan, Z. and Nir, I. (1977). A comparative study of the nutritional and physiological significance of raw and heated soya beans in chicks and goslings. Br. J. Nutr., 37, 81-91.

Osborne, T. B. and Mendel, L. S. (1917). The use of soy bean as food. J. Biol. Chem. 32, 369-387.

Patten, J. R., Patten, J. A. and Pope, H., II. (1973). Sensitivity of the guinea-pig to raw soya bean in the diet. Food Cosmet. Toxicol., 11, 577-583.

Patten, J. R., Richards, E. A. and Pope, H., II. (1971). The effect of raw soybean on the pancreas of adult dogs. Proc. Soc. Exp. Biol. Med., 137, 59-63.

Percival, S. S. and Schneeman, B. O. (1979). Long term pancreatic response to feeding heat damaged casein in rats. J. Nutr., 109, 1609-1614.

Petrucci, T., Tomasi, M., Cantagalli, P. and Silano, V. (1974). Comparison of wheat albumin inhibitors of alpha-amylase and trypsin. Phytochem., 13, 2487-2495.

Puls, W. and Keup, U. (1973). Influence of an alpha-amylase inhibitor (BAY d 7791) on blood glucose, serum insulin, and NEFA in starch loading tests in rats, dogs, and man. Diabetologia, 9, 97-101.

Puls, W. and Keup, U. (1974). Metabolic studies with an amylase inhibitor in acute starch loading tests in rats and men and its influence on the amylase content of the pancreas. In "Recent Advances in Obesity Research," A. Howard, ed., Newman Publishing Co., Lts., London.

Rackis, J. J. (1965). Physiological properties of soybean trypsin inhibitors and their relationship to pancreatic hypertrophy and growth inhibition of rats. Proc. Fed. Am. Soc. Exp. Biol., 24, 1488-1493.

Rothman, S. S. and Wells, H. (1967). Enhancement of pancreatic enzyme synthesis by pancreozymin. Am. J. Physiol., 213, 215-218.

Roy, D. M. and Schneeman, B. O. (1981). Effect of soy protein, casein and trypsin inhibitor on cholesterol, bile acids and pancreatic enzymes in mice. J. Nutr., 111, 878-885.

Sale, J. K., Goldberg, D. M., Fawcett, A. N. and Wormsley, K. G. (1977). Chronic and acute studies indicating absence of exocrine pancreatic feedback inhibition in dogs. Digestion, 15, 540-55.

Satouchi, K. and Matsushita, S. (1976). Purification and properties of a lipase inhibiting protein from soybean cotyledons. Agric. Biol. Chem., 40, 889-897.

Satouchi, K., Tomohiko, M. and Matsushita, S. (1974). Characterization of inhibitor protein for lipase in soybean seeds. Agric. Biol. Chem., 38, 97-101.

Saunders, r. M. (1975). Alpha-amylase inhibitors in wheat and other cereals. Cereal Food World, 20, 282-285.

Savaiano D., Powers, J., Costello, M., Whitaker, J.R. and Clifford, A. (1977). The effect of an alpha-amylase inhibitor on the growth rate of weanling rats. Nutr. Rept. Int., 15, 443-449.

Saxena, H. C., Jensen, L. S. and McGinnis, J. (1963). Influence of age on utilization of raw soybean meal by chickens. J. Nutr., 80, 391-396.

Schingoethe, D. J., Aust, S. D. and Thomas, J. W. (1970). Separation of a mouse growth inhibitor in soybeans from trypsin inhibitors. J. Nutr., 100, 739-748.

Schmidt, D. D., Frommer, W., Junge, B,, Miller, I., Wingender, W. and Truscheit, E. (1977). Alpha-glucosidase inhibitors. Naturwiss., 64, 535-536.

Schneeman, B. O. (1978). Effect of plant fiber on lipase, trypsin and chymotrypsin activity. J. Food Sci., 43, 634-635.

Schneeman, B. O., Forman, L. P. and Gallaher, D. (1983). Pancreatic and intestinal enzyme activity in rats fed various fibre sources. In "Fibre in Human and Animal Nutrition," G. Wallace and L. Bell, eds., Royal Society of New Zealand Bulletin 20.

Schneeman, B. O. and Gallaher, D. (1986). Pancreatic response to dietary trypsin inhibitor: variation among species. This volume.

Schneeman, B. O. and Gallaher, D. (1980). Changes in small intestinal digestive enzyme activity and bile acids with dietary cellulose in rats. J. Nutr., 110, 584-590.

Schneeman, B. O., Chang, I., Smith, L. and Lyman, L. R. (1977). Effect of dietary amino acids, casein, and soybean trypsin inhibitor on pancreatic protein secretion in rats. J. Nutr., 107, 281-288.

Schneeman, B. O. and Lyman, R. L. (1975). Factors involved in the intestinal feedback regulation on pancreatic enzyme secretion in the rat. Proc. Soc. Exp. Biol. Med., 148, 897-903.

Sheff, D. (1982). Want to have your pasta and eat it, too?. People Weekly, June,28, 30-32.

Snook, J. T. (1974). Adaptive and nonadaptive changes in digestive enzyme capacity influencing digestive function. Fed. Proc., 33, 88-89.

Soling, H. and Unger, K. (1972). The role of insulin in the regulation of alpha-amylase synthesis in the rat pancreas. Eur. J. Clin. Invest., 2, 199-212.

Suchino, I., Otsuki, M., Yamasaki, T., Ohki, A., Sakamoto, C., Yuu, H., Maeda, M. and Baba, S. (1981). Effect of alpha-glucosidase inhibitor on human pancreatic and salivary alpha-amylase. Clin. Chim. Acta, 117, 145-152.

Sullivan, I., Comai, K. and Triscari, J. (1980). Novel anti-obesity agents whose primary site of action is the gastrointestinal tract. In "Recent Advances in Obesity Research," P. Bjorntrop, M. Cairella, and A. Howard, eds., John Libby, London.

Wu, A-L., Clark, S. B. and Holt, P. (1980). Composition of lymph chylomicrons from proximal or distal rat small intestine. Am. J. Clin. Nutr., 33, 582-589.

Yanatori, Y. and Fujita, T. (1976). Hypertrophy and hyperplasia in the endocrine and exocrine pancreas of rats fed soybean trypsin inhibitor or repeatedly injected with pancreozymin. Arch. Histol. Japan, 39, 67-78.

Yen, J. T., Hymowitz, T. and Jensen, A. H. (1974). Effects of soybean of different trypsin inhibitor activities on performance of growing swine. J. Anim. Sci., 38, 304-309.

Yen, J. T., Jensen, A. H. and Simon, J. (1977). Effect of dietary raw soybean and soybean trypsin inhibitor on trypsin and chymotrypsin activities in the pancreas and in small intestinal juice of growing swine. J. Nutr., 107, 156-165.

Additional reference:

Lairon, D., Borel, P., Termine, E., Grataroli, R., Chabert, C. and Hauton, J. C. (1985). Evidence for a proteinic inhibitor of pancreatic lipase in cereals, wheat bran and wheat germ. Nutr. Reports Int., 32, 1107-1113.

12

PANCREATIC RESPONSE TO DIETARY TRYPSIN INHIBITOR:

VARIATIONS AMONG SPECIES

Barbara O. Schneeman and Daniel Gallaher

Department of Nutrition, University of California, Davis, CA 95616, and Department of Foods and Nutrition, North Dakota State University, Fargo, ND 58105

Inhibitors of proteolytic enzymes are distributed throughout various foods used for human consumption as well as throughout animal feeds. These inhibitors are predominately found in plant foodstuffs and have been associated with the lower protein digestibility of untreated or unprocessed plants used as dietary sources of protein. Because of the importance of plants for human foods as well as their importance in animal feeds, the metabolic and nutritional effects of protease inhibitors have been investigated in a variety of animal species. Most of this research has dealt with the inhibitors of trypsin and chymotrypsin that are found in soybeans. The objective of this paper is to critically examine the variability among species in the response to diets which contain trypsin inhibitors.

The two metabolic responses of nutritional importance that have been associated with trypsin inhibitor are depression of growth and enlargement of the pancreas. The effects of diets containing inhibitor on growth inhibition and pancreatic enlargement has been reviewed recently (Gallaher and Schneeman, 1984). The literature reviewed indicates that growth inhibition has been observed in rats, chicks, mice, guinea pigs, goslings, growing swine, and calves. The depression in growth appears to be dependent on several factors, which include the age of the animal, the protein quality and content of the diet, and the presence of other antinutritional factors in the diet. Adult animals appear to be less sensitive to depressions in weight gain caused by dietary trypsin inhibitors. Increasing the dietary content of protein or improving the nutritional quality of the dietary protein improves weight gain. Raw soybean meal, which is frequently used as a source of dietary trypsin inhibitors, contains other factors which are known to cause growth depression. Pancreatic enlargement due to feeding purified trypsin inhibitors or raw soybean meal has been observed in rats, chickens, mice, and the growing guinea pig (For review: Gallaher and Schneeman, 1984). The increase in pancreatic weight has been associated with elevations in the pancreatic protein content and activities of trypsin and chymotrypsin. In the rat it has been shown that trypsin inhibitors cause the release of cholecystokinin (CCK) activity from the small intestine (Review: Gallaher and Schneeman, 1984; Liddle et al., 1984). CCK is a known stimulant of pancreatic enzyme secretion and hypertrophy and appears to mediate the effects of trypsin inhibitor on the pancreas. Pancreatic enlargement failed to occur in the adult guinea pig, the dog, the pig, the calf, or the monkey (Review: Gallaher and Schneeman, 1984; Struthers et al., 1983). In monkeys some increase in pancreatic trypsin activity after consuming raw soybean flour was observed (Struthers et al., 1983). The differences among species in the pancreatic response to dietary trypsin inhibitor or raw soybean flour has lead to the suggestion that species variation in the response to inhibitors may be related to pancreatic weight in relation to body weight (Liener and Kakade, 1980). The potential value of this relationship is as a predictor of the response of the human pancreas to trypsin inhibitor. Animals in which the pancreas weight is less than 0.3% of body weight have not exhibited pancreatic enlargement with inhibitor feeding. As the adult human pancreas weighs about 0.09 to 0.12% of body weight, the human pancreas would not be expected to enlarge with trypsin inhibitor feeding if this relationship is valid.

More recent evidence on the regulation of pancreatic enzyme secretion in humans suggest that we may not yet have sufficient evidence to predict the human pancreatic response to long term feeding of dietary trypsin inhibitors. In rats, pigs, and calves pancreatic enzyme secretion is regulated by a negative feedback mechanism which is mediated by the presence of pancreatic proteolytic enzyme activity in the small intestine. Trypsin inhibitors appear to stimulate pancreatic enzyme hypersecretion and hypertrophy through this mechanism in rats. Recent studies indicate that a similar feedback mechanism may be import in regulating enzyme secretion from the human pancreas (Ihse et al., 1977, Slaff et al., 1984). The existence of such a mechanism provides a physiological means by which trypsin inhibitors can have an effect on the human pancreas; however, our knowledge of the capacity of the human pancreas to adapt to dietary factors is limited. One report suggests that a diet high in protein can lead to elevations in pancreatic trypsin and chymotrypsin activity (Bucko et al., 1982). However the more critical question concerns the ability of the human pancreas to enlarge with the chronic ingestion of a stimulant of pancreatic secretion such as trypsin inhibitor, and currently we do not know the answer to that question.

As mentioned above, inhibition of growth as well as pancreatic enlargement are responses of nutritional importance observed when raw soybean flour is incorporated into experimental diets as a source of trypsin inhibitor. In several of the studies in which pancreatic hypertrophy was not observed, the animals did respond to the presence of raw soybean flour with growth inhibition. Raw soybean flour is considered a poor quality protein source because the protein is resistant to digestion and because it is limiting in sulfur amino acids. In addition, raw soybeans are known to contain other antinutritional factors. Consequently, the poor quality of this protein source may limit the ability of the pancreas to respond to the presence of inhibitor (Gallaher and Schneeman, 1984). In this context poor weight gain, loss of body weight, and death have been reported in studies in which raw soybean flour failed to induce pancreatic hypertrophy.

In trying to compare the pancreatic response among various species to trypsin inhibitors, we are limited by the availability of information. In order to evaluate the basis for variation among species, basic information is needed on the regulation of pancreatic enzyme secretion and adaptation. Although evidence suggests that the feedback mechanism controlling pancreatic enzyme secretion, which was first observed in the rat, operates in other species, the relative importance of this mechanism under physiological conditions among species should be investigated. Likewise basic information is needed on the capacity of the pancreas to adapt to dietary factors and the impact of age and nutritional status on pancreatic adaptation.

REFERENCES

Bucko, A., Kopec, Z., Ovecka, M., and Grunt, J., 1982, Adaptability of pancreatic enzymes activity to various food nutritive values in man, Die Nahrung, 26:59.

Gallaher, D., and Schneeman, B. O., 1984, Nutritional and metabolic response to plant inhibitors of digestive enzymes, in: "Nutritional and Toxicological Aspects of Food Safety," M. Friedman, ed., Plenum Publishing Co. See also preceding paper, this volume.

Ihse, I., Lilja, P., and Lundquist, I., 1977, Feedback regulation of pancreatic enzyme secretion by intestinal trypsin in man, Digestion, 15:303.

Liddle, R. A., Goldfine, I. D., and Williams, J. A., 1984, Bioassay of plasma cholecystokinin in rats: effects of food, trypsin inhibitor, and alcohol, Gastroenterology, 87:542.

Slaff, J., Jacobson, D., Tillman, C. R., Curington, C., and Toskes, P., 1984, Protease-specific suppression of pancreatic exocrine secretion, Gastroenterology, 87:44.

Struthers, B. J., MacDonald, J. R., Dahlgren, R. R., and Hopkins, D. T., 1983, Effects on the monkey, pig and rat pancreas of soy products with varying levels of trypsin inhibitor and comparison with the administration of cholecystokinin, J. Nutr., 113:86.

13

THE EFFECT OF THE LONG-TERM FEEDING OF RAW SOYFLOUR ON THE PANCREAS OF THE MOUSE AND HAMSTER

Irvin E. Liener and Aharon Hasdai

Department of Biochemistry, College of Biological Sciences
University of Minnesota
St. Paul, MN 55108

ABSTRACT

The effects of the long-term feeding of mice and hamsters with raw (RSF) or heated (HSF) soy flours were investigated both in the presence and absence of chemical carcinogens. Mice which had been fed RSF for 18 months had enlarged pancreas but only a relatively low incidence of atypical acinar cell nodules (AACN). Mice on either RSF or HSF were also relatively resistant to the carcinogenic effects of azaserine which is known to produce a high incidence of AACN in rats. Hamsters which had been fed RSF for 15 months did not exhibit pancreatic enlargement and had a very low incidence of pancreatic tumors. Although the incidence of tumors in hamsters which had been injected with N-nitrosobis (2-oxopropyl)amine and maintained on HSF was very high (88%), those on RSF had a tumor incidence of less than 10%. Thus, there appears to be a marked difference in the response of the pancreas in different species of animals to the long-term effects of feeding RSF. This should be taken into consideration in evaluating the potential carcinogenicity of the trypsin inhibitors.

INTRODUCTION

The feeding of raw soyflour (RSF) to experimental animals such as the rat, mouse, and chick leads to growth depression and hypertrophy and/or hyperplasia of the pancreas, effects which have been attributed to the trypsin inhibitor content of soybeans (Liener and Kakade, 1980). The long-term feeding* of RSF to rats ultimately induces the formation of acinar hyperplastic nodules (McGuiness, et al., 1980, 1982) and potentiates the carcinogenicity of azaserine (McGuiness, et al., 1977; Morgan, et al., 1977) and di(2-hydroxypropyl)nitrosamine (Levison, et al., 1979). A more recent series of studies have shown that the incidence of nodular hyperplasia and acinar adenoma in the pancreas of the rat was positively

*The distinction between short- and long-term feeding as used here is an arbitrary one. In general, studies involving the feeding of RSF for 6 months or less will be referred to as "short-term", whereas feeding periods beyond 6 months will be referred to as "long-term".

associated with the level of trypsin inhibitor activity derived from soy flour (Gumbmann, et al., 1985; Liener, et al., 1985; Rackis, et al., 1985; Spangler, et al., 1985).

Although these observations leave little doubt that the long-term feeding of RSF produces abnormal pathological changes in the pancreas of the rat, the question remains as to whether other species of animals would be similarly affected. In short-term experiments RSF has been shown to have a deleterious effect on the growth and to cause pancreatic enlargement of both the mouse (Schingoethe, et al., 1970; Roy and Schneeman, 1981) and the hamster (Hasdai and Liener, 1983). The mouse and the hamster, however, are known to differ from the rat with respect to their pancreatic response to certain chemical carcinogens. The pancreas of the mouse is less sensitive than the rat to the carcinogenic effects of azaserine (Roebuck et al., 1980), and, unlike the rat, hamsters to which N-nitrosobis (2-oxopropyl)amine (BOP) has been administered develop pancreatic tumors of ductal origin (Pour, et al., 1981).

The purpose of the present study was (1) to investigate the long-term effects of feeding RSF on the pancreas of the mouse and hamster, and (2) to see whether RSF would increase the sensitivity of the pancreas of the mouse to azaserine and potentiate the carcinogenetic effect of BOP in hamsters.

EXPERIMENTAL

Studies with Mice

A total of 216 male albino mice, 6 to 7 weeks old, weighing 23 to 27 grams, were distributed into 3 groups each of which were fed diets (Table 1) containing RSF (n=84), heated soy flour (HSF) (n=80), or casein (n=52). Each of these groups was subdivided into two equal groups, one of which recovered 15 weekly intraperitoneal injections of azaserine (10 mg per kg body weight) and the other 0.9% NaCl. Body weights and food consumption were recorded weekly. Half of the animals in each group were sacrificed at the end of 10.5 months and the remaining survivors at the end of 18 months. The pancreas, liver, kidney, and spleen were removed, weighed, and fixed in 10% buffered formaldehyde. Sections were stained with haemotoxylin and eosin and examined microscopically for atypical acinar cell nodules (AACN) (Roebuck, et al., 1980).

Studies with Hamsters

A total of 54 outbred Syrian Golden male hamsters, 6 to 7 weeks old and weighing 95 to 105 grams, were divided into four groups: RSF (n=15), HSF (n=12), RSF injected with BOP (n=14), and HSF injected with BOP (13). The composition of the diets was the same as that employed in the experiment with mice (Table 1). The BOP was injected subcutaneously at a level of 10 mg per kg body weight on days 7 and 14. Animals not receiving BOP were injected at the same time with 0.9% NaCl. The experiment was terminated at the end of 15 months at which time the pancreas and other organs were excised, fixed in buffered formaldehyde, and stained with haemotoxylin and eosin. Tumors were classified as neoplastic nodules, adenomas, or adenocarcinomas according to the criteria of Mohr and Reznick (1982). Changes from the normal pathology were evaluated with respect to proliferation, hypertrophy, hyperplasia, atrophy, metaplasia, pleomorphism, fibrosis, adiposis, nuclear size and shape, and cytoplasmic vacuolation. The intensity of these changes was scored on a scale of 0 to 4, where 0 = normal (no lesions), and 1, 2, 3, and 4 represent slight, moderate, severe, and very severe lesions respectively.

Table 1. Percent Composition of diets used in experiments with mice and hamsters[a]

Dietary Components	Soyflour	Casein
Soyflour, raw[b] or heated[c]	42.1	
Casein	--	24.0
Glucose	15.0	15.0
Non-nutritive fiber	5.0	5.0
Corn oil	5.5	5.5
Lard	2.5	2.5
DL-methionine	0.1	0.1
Choline chloride	0.2	0.2
Vitamin mix[d]	2.0	2.0
Mineral mix[d]	5.0	5.0
Dextrin	22.6	40.7

[a]Prepared by Purina Test Diets, Richmond, IN.
[b]Raw defatted, dehulled soyflour, 52.5% protein (N x 6.25).
[c]Raw soyflour subjected to toasting; >90% of trypsin inhibitor activity was destroyed.
[d]Formulated according to Hasdai and Liener (1983).

Statistical Analysis

Data pertaining to body and pancreas weights and the severity of the pancreatic lesions (the latter in the case of the hamster study only) were analyzed by analysis of variance and Duncan's Multiple Range test using the Statistical Analysis System (1979). The chi-square test was used for evaluating the significance of the incidence of AACN in mice and of the various tumors observed in the hamster study.

RESULTS

Mouse Study

As shown in Figure 1 mice fed RSF grew at a significantly ($P < 0.01$) lesser rate than those fed HSF or casein. The administration of azaserine appeared to make little difference in this pattern of growth response. The mice on RSF grew so poorly that, in order to maximize their survival rate, they were allowed to consume HSF at selected intervals of two weeks (denoted by double-headed arrows in Figure 1). Each period following the change from RSF to HSF was accompanied by an immediate gain in weight and a subsequent decline in weight when these animals were returned to RSF. The RSF-fed mice consumed an average of 3.17 ± 0.03 g/day, which was significantly ($P < 0.01$) less than the 3.71 ± 0.08 and 3.91 ± 0.07 g/day consumed by the HSF- and casein-fed mice respectively. A similar pattern was observed for the azaserine-treated mice indicating that azaserine had little or no effect on food consumption.

The weights of the pancreas of mice sacrificed at the end of 10.5 and 18 months are shown in Table 2. The mice fed RSF had significantly larger pancreas than those fed HSF or casein at the end of both periods of time. The injection of azaserine had no significant effect on the pancreas weights of mice fed the different diets compared with the saline-injected control groups.

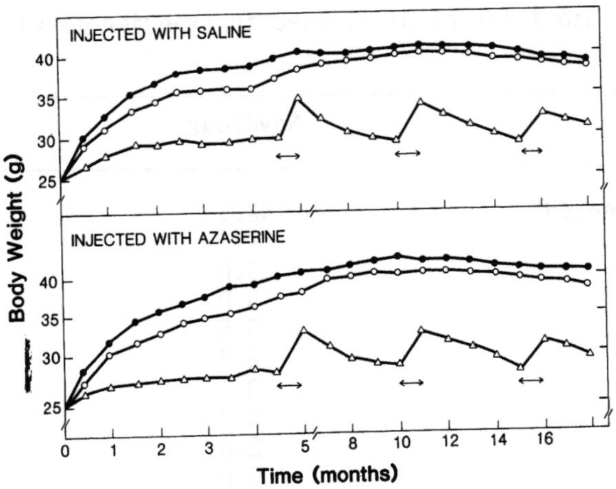

Figure 1. Growth curve of mice injected with azaserine or saline. ●: casein; O: HSF; Δ: RSF. Horizontal arrows denote period during which animals on RSF were fed HSF.

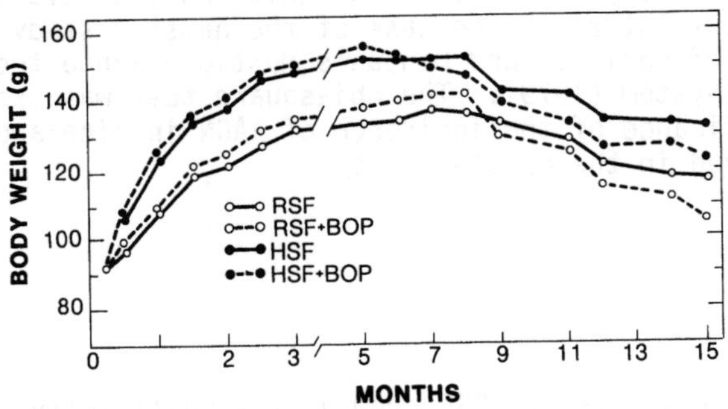

Figure 2. Growth curve of hamsters fed diets containing RSF or HSF with and without the administration of BOP.

Only one mouse out of the total number of 216 employed in this study showed any macroscopic evidence of pancreatic nodules. This one exception was a saline-injected animal which had been on RSF for 10.5 months. Nodules were found infrequently in some of the other organs; one saline-injected animal on HSF for 18 months had nodules in the stomach, one azaserine-injected mouse on RSF for 10.5 months had nodules in the spleen, and one azaserine-injected mouse on casein for 18 months had hepatic nodules. Microscopic examination of the pancreas from the various groups (Table 3) revealed a relatively low incidence of AACN, ranging from 5% to

Table 2. Weights of pancreas of mice fed RSF, HSF, or casein in the presence or absence of azaserine. Values shown are the means ± SE based on the number of mice (n) sacrificed at the end of 10.5 and 18 months.

Source of Protein	Pancreas weights (% of body weight) - azaserine	+ azaserine
	10.5 months	
RSF[a]	1.05 ± 0.05 (n=17)	0.95 ± 0.06 (n=12)
HSF	0.73 ± 0.02 (n=15)	0.78 ± 0.04 (n=16)
Casein	0.70 ± 0.03 (n=9)	0.73 ± 0.03 (n=8)
	18 months	
RSF[a]	1.13 ± 0.06 (n=7)	1.11 ± 0.04 (n=6)
HSF	0.79 ± 0.06 (n=8)	0.82 ± 0.03 (n=7)
Casein	0.81 ± 0.03 (n=8)	0.76 ± 0.06 (n=5)

[a]The difference between RSF and HSF or casein is significant by analysis of variance ($P < 0.01$). There is no significant difference due to treatment with azaserine in any of the dietary groups.

15% at the end of 10.5 months, 8% to 23% between 10.5 and 18 months, and 15% to 20% at the end of 18 months. Although the incidence of AACN was consistently higher in the azaserine-injected animals, these differences were not statistically significant ($P > 0.05$) and were unrelated to the diet.

Hamster Study

Figure 2 shows the growth curves of hamsters fed RSF or HSF with and without treatment with BOP. As might have been predicted from a previous study (Hasdai and Liener, 1983), animals on RSF grew at a significantly slower rate ($P < 0.01$) than those fed HSF. There was little difference in growth that could be attributed to the injection of BOP, except after 7 to 8 months when the BOP-injected animals lost weight at a significantly ($P < 0.05$) faster rate than their saline-injected counterparts.

The pathological data derived from the study are summarized in Table 4. No significant difference was found in the weights of the pancreas among the various groups by the end of 15 months. In the control groups which did not receive BOP, the incidence of tumors among those animals which remained alive at the end of 15 months was quite low, less than 10% in both the RSF and HSF-fed animals. On the other hand, among the BOP-treated hamsters which had been fed HSF 7 out of 8 survivors (88%) had pancreatic adenomas, and 5 (63%) had pancreatic adenocarcinomas. In sharp contrast only one of the 11 surviving BOP-injected animals on RSF (9%) had

Table 3. Incidence of atypical acinar cell nodules (AACN) in mice fed RSF, HSF, or casein in the presence or absence of azaserine.

Source of Protein	Incidence of AACN[a]	
	- azaserine	+ azaserine
	Up to 10.5 months	
RSF	1/21 (5%)	2/21 (10%)
HSF	1/22 (5%)	2/20 (10%)
Casein	1/10 (10%)	2/13 (15%)
	Between 10.5 and 18 months	
RSF	1/12 (8%)	2/17 (12%)
HSF	2/10 (20%)	3/13 (23%)
Casein	1/10 (10%)	1/16 (17%)
	At the end of 18 months	
RSF	1/7 (14%)	1/16 (17%)
HSF	1/8 (13%)	1/7 (14%)
Casein	1/8 (13%)	1/5 (20%)

[a]No. of mice with AACN/no. of mice which had died or had been sacrificed during the period indicated. Figure in parenthesis is the incidence expressed as a percentage. None of the differences are significant by chi-square analysis.

Table 4. Pancreas weights, incidence of tumors, and seventy of pancreatic lesions in BOP-injected and non-injected hamsters fed RSF and HSF for 15 months

	Non-injected		BOP-injected	
	RSF	HSF	RSF	HSF
No. of animals at start	15	12	14	13
No. of survivors at end of 15 mo.	13	7	11	8
Pancreas wgt. (% of body wgt.)	$0.32^{c}\pm0.01$	$0.30^{c}\pm0.01$	$0.32^{c}\pm0.03$	$0.36^{c}\pm0.03$
Incidence of tumors[a]				
Pancreatic adenomas	1/13 (8%)[c]	0/7 (0%)[c]	1/11 (9%)[c]	7/8 (88%)[d]
Pancreatic adenocarcinomas	0/13 (0%)[c]	0/7 (0%)[c]	0/11 (0%)[c]	5/8 (63%)[d]
Severity of pancreatic lesions[b]				
Acinar cells	$1.19^{c}\pm0.19$	$0.75^{c}\pm0.11$	$1.32^{c}\pm0.14$	$1.96^{d}\pm0.09$
Ducts	$0.44^{c}\pm0.15$	$0.36^{c}\pm0.13$	$0.79^{c}\pm0.12$	$2.43^{d}\pm0.12$
Islets	$0.42^{c}\pm0.15$	$0.27^{c}\pm0.12$	$0.70^{d}\pm0.19$	$1.58^{e}\pm0.16$
Interstitium	$0.65^{c}\pm0.13$	$0.46^{c}\pm0.17$	$0.55^{c}\pm0.17$	$1.38^{d}\pm0.16$

[a]No. of animals with tumors/no. of animals surviving at end of 15 months. Figures in parentheses denote incidence as a percentage.
[b]Based on a scale of 0 to 4 as described in text.
[c,d,e]Values in same row with different superscripts are significantly different ($P < 0.01$) by analysis of variance or the chi-squared test.

a pancreatic adenoma, and none had any adenocarcinomas. Closer microscopic inspection of the pancreas revealed that the lesions seen in the BOP-treated animals fed HSF were confined mainly to the acinar cells and the ducts. The severity of the pancreatic lesions in this group was significantly higher ($P < 0.01$) than their counterparts fed RSF. The latter group had lesions with scores which did not differ significantly from those animals which had not received BOP.

DISCUSSION

The present study has shown that, although the mouse is sensitive to RSF as manifested by retarded growth and pancreatic enlargement, there was no macroscopic evidence of nodule formation even after 18 months on a diet containing 42% RSF. Microscopic examination of the pancreatic tissue revealed a relatively low incidence of AACN which bore no relationship to whether the diet contained RSF, HSF, or casein as a source of protein. This is in marked contrast to what has been reported for the rat in which macroscopic as well as microscopic nodules are produced in the pancreas of these animals after having been fed RSF for a period as short as 30 weeks (McGuiness, et al., 1980, 1982). Although RSF likewise potentiates the action of azaserine on the rat pancreas (Morgan, et al., 1977; McGuiness, et al., 1981), no such effect was observed here with the mouse. This lack of response in the mouse may be due in part to the fact that this species is less sensitive to azaserine than the rat (Roebuck, et al., 1980). However, Levison et al. (1979) have shown that, although the pancreas of the rat is only weakly affected by di(2-hydroxypropyl)nitrosamine, the feeding of RSF potentiated the development of hyperplastic and adenomatous nodules and adenocarcinomas by this carcinogen.

The results of the experiment with hamsters show that the pancreas of this animal species also responds to the long-term feeding of RSF quite differently from the rat. Although the pancreas of the hamster undergoes enlargement when fed RSF for 32 days (Hasdai and Liener, 1983), no difference in the size of the pancreas was noted between animals fed RSF and HSF for 15 months. This would suggest an adaptation to the effects of RSF which does not occur in the rat or mouse. Unlike the rat, but similar to the mouse, long-term feeding of RSF to hamsters failed to cause any significant abnormal morphological changes in the pancreas. Thus an early increase in the size of the pancreas due to hypertrophy and/or hyperplasia may not provide a means for predicting the subsequent development of carcinogenic changes in species of animals other than the rat.

The most significant observation made in these studies was the unexpected finding that RSF not only failed to potentiate the carcinogenicity of BOP towards the pancreas of the hamster but actually caused a marked reduction in the incidence of tumors and the severity of pancreatic lesions compared to BOP-treated animals fed HSF. Whether or not this protective effect is due to the trypsin inhibitor content of RSF is not certain, although it is known that protease inhibitors have been reported to inhibit carcinogenesis in a number of instances (Troll, et al., 1970, 1979; Baugh and Schnebli, 1980). Of particular relevance in this regard are those studies which have shown that soybean diets rich in trypsin inhibitor activity protect rats from x-ray induced breast cancer (Troll, et al., 1980) and lower the incidence of spontaneous liver cancer in mice (Becker, 1981).

If indeed the trypsin inhibitor is responsible for the protective effect of RSF on BOP-induced pancreatic cancer in hamsters, the mechanism whereby this occurs is not obvious. A possible explanation lies in the fact that trypsin inhibitors are believed to exert a trophic effect on the pancreas via an inhibition of the negative feedback mechanism whereby a reduction of the level of trypsin in the small intestine causes a release of cholecystokinin (Green and Lyman, 1972; Schneeman and Lyman, 1975). Johnson, et al. (1983) have reported that the induction of pancreatic carcinomas in hamsters with di-isopropanol nitrosamine could be prevented by the administration of cholecystokinin. Assuming a negative feedback mechanism to be operative in the hamster, the action of trypsin inhibitor would be to effect the release of cholecystokinin which could then account for the inhibitory effect of RSF on BOP-induced carcinogenesis observed here. Still to be explained, however, is why the protective effect of RSF was not observed in the rat and the mouse where the feedback mechanism is also known to occur.

In conclusion, there appears to exist a marked difference in the response of the pancreas of various animal species to the long-term effects of feeding RSF. Although the rat, mouse, and hamster all exhibit pancreatic enlargement from early dietary exposure to RSF, only the pancreas of the rat undergoes transformation into a state of neoplasia with the prolonged chronic consumption of RSF. RSF may in fact exert a protective effect on the chemical induction of tumors in the hamster. This contradictory response of rats compared to mice and hamsters serves to emphasize the importance of investigating the long-term effects of feeding dietary components rich in trypsin inhibitors in several species of animals before any firm conclusions can be drawn regarding their potential risk in the human diet.

ACKNOWLEDGMENTS

The authors wish to thank Barbara Falk for her technical assistance, and Dr. Peter J. O'Brien, College of Veterinary Medicine, University of Minnesota, for the histopathological evaluation of the tissue sections. This study was supported by Cooperative Agreement No. 58-519B-9-835 between the Agricultural Experiment Station of the University of Minnesota and the Northern Regional Research Center, USDA, Peoria, Il.

REFERENCES

Baugh, R. J. and Schnebli, H. P., 1980, Role and potential therapeutic value of proteinase inhibitors on tissue destruction, in: "Proteinases and Tumor Invasion", P. Strauli, A. J. Barrett, and A. Baici, eds., Raven Press, New York, pp. 157-179.

Becker, F. F., 1981, Inhibition of spontaneous hepatogenesis in C3H/HeN mice by Edipro A, an isolated soy protein, Carcinogenesis, 2:1213-1214.

Green, G. M. and Lyman, R. L., 1972, Feedback regulation of pancreatic enzyme secretion as a mechanism for trypsin inhibitor-induced hypersecretion in rats, Proc. Soc. Exp. Biol. Med., 140:6-12.

Gumbmann, M. R., Spangler, W. L., Dugan, G. M. and Rackis, J. J., 1985, The USDA trypsin inhibitory study. IV. The chronic effects of soy flour and soy protein isolate on the pancreas in rats after two years, Qual. Plant. Plant Proteins Hum. Nutr., in press.

Hasdai, A. and Liener, I. E., 1983, Growth, digestibility, and enzymatic activities in the pancreas and intestines of hamsters fed raw and heated soy flour, J. Nutr., 113:662-668.

Johnson, F. E., LaRegina, M. C., Martin, S. A. and Bashiti, H. M., 1983, Cholecystokinin inhibits pancreatic and hepatic carcinogenesis, Cancer Detect. Prev., 6:389-402.

Levison, D. A., Morgan, R. G. H., Brimacombe, J. S., Hopwood, D., Coghill, G. and Wormsley, K. G., Carcinogenic effects of di(2-hydroxypropyl) nitrosamine (DHPN) in male Wistar rats: promotion of pancreatic cancer by a raw soya flour diet, Scand. J. Gastroenterol., 14:217-224.
Liener, I. E. and Kakade, M. L., 1980, Protease inhibitors, in: "Toxic Constituents of Plant Foodstuffs", I. E. Liener, ed., Academic Press, New York, pp. 7-71.
Liener, I. E., Nitsan, Z., Srisangnam, C., Rackis, J. J. and Gumbmann, M. R., 1985, The USDA trypsin inhibitor study. II. Time-related biochemical changes in the pancreas, Qual. Plant. Plant Foods Hum. Nutr., in press.
McGuiness, E. E., Morgan, R. G. H., Levison, D. A., Frape, D. L., Hopwood, D. and Wormsley, K. G., 1980, The effects of long-term feeding of soya flour on rat pancreas, Scand. J. Gastroenterol., 15:497-502.
McGuiness, E. E., Morgan, R. G. H., Levison, D. A., Hopwood, D. and Wormsley, K. G., 1981, Interaction of azaserine and raw soya flour on the rat pancreas, Scand. J. Gastroenterol., 16:49-56.
McGuiness, E. E., Hopwood, D. and Wormsley, K. G., 1982, Further studies on the effects of raw soya flour on the rat pancreas, Scand. J. Gastroenterol., 17:273-276.
Mohr, U. and Reznik, G., 1982, Tumors of the pancreas, in: "Pathology of Tumors in Laboratory Animals", V. S. Turusov, ed., International Agency for Research on Cancer, Lyon.
Morgan, R. G. H., Levison, D. A., Hopwood, D., Saunders, J. H. B. and Wormsley, K. G., 1977, Potentiation of the action of azaserine on the rat pancreas by raw soya bean flour, Cancer Lett., 3:87-90.
Pour, P. M., Runge, R. G., Birt, D., Gingell, R., Lawson, T., Nagel, D., Wallcave, L. and Salmasi, S. Z., 1981, Current knowledge of pancreatic carcinogenesis in the hamster and its relevance to the human disease, Cancer, 47(6, Suppl.):1573-1587.
Rackis, J. J., Gumbmann, M. R. and Liener, I. E., 1985, The USDA trypsin inhibitory study. I. Background, objectives, and procedural details, Qual. Plant. Plant Foods Hum. Nutr., in press.
Roebuck, B. D., Lilja, H. S., Curphey, T. J. and Longnecker, D. S., 1980, Pathologic and biochemical effects of azaserine in inbred Wistar/ Lewis rats and non-inbred CD-1 mice, J. Nat. Cancer Inst., 65:383-389.
SAS Users Guide, 1979, Statistical Analysis System Institute, Cary, NC.
Roy, D. M. and Schneeman, B. O., 1981, Effect of soy protein, casein, and trypsin inhibitor on cholesterol, bile acids and pancreatic enzymes in mice, J. Nutr., 111:878-885.
Schingoethe, D. J., Aust, S. D. and Thomas, J. W., 1970, Separation of mouse growth inhibitor in soybeans from trypsin inhibitors, J. Nutr., 100:739-748.
Schneeman, B. O. and Lyman, R. L., 1975, Factors involved in the intestinal regulation of enzyme secretion in the rat, Proc. Soc. Exp. Biol. Med., 148:897-903.
Spangler, W. L., Gumbmann, M. R., Liener, I. E. and Rackis, J. J., The USDA trypsin inhibitor study. III. Sequential development of pancreatic pathology in rats, Qual. Plant. Plant Foods Hum. Nutr., in press.
Troll, W., Belman, S., Wiesner, R. and Shellabarger, C. J., 1979, Protease action in carcinogenesis, in: "Biological Function of Proteases", H. Holzer and H. Tschesche, eds., Springer-Verlag, Berlin, pp. 165-170.
Troll, W., Klassen, A. and Janoff, A., 1970, Tumorigenesis in mouse skin: inhibition by synthetic inhibitors of proteases, Science, 169:1211-1213.
Troll, W., Weissner, R., Shellabarger, C. J., Holtzman, S. and Stone, J. P., 1980, Soybean diets lowers breast tumor incidence in irradiated rats, Carcinogenesis, 1:469-472.

14

ACCENTUATED RESPONSE TO SOYBEAN INHIBITORS BY MEAL-FEEDING IN VARIOUS SPECIES

Zafrira Nitsan and Israel Nir

Agricultural Research Organization, The Volcani Center, Bet Dagan; and The Hebrew University of Jerusalem, Faculty of Agriculture, Rehovot, Israel

ABSTRACT

Diets containing raw soybean products (RSD) fed <u>ad libitum</u> caused (a) a reduction in food consumption and growth, (b) pancreatic enlargement, (c) hypersecretion of digestive enzymes, and (d) enlargement of intestinal segments and their contents in rats, chicks and geese. These effects were found to be related to the concentration of trypsin inhibitor (TI) in the diet. Geese were more sensitive than rats and chicks. Calves fed RSD did not respond in pancreatic hypertrophy and hypersecretion of digestive enzymes, but growth was depressed. Meal-feeding also caused enlargement of the pancreas, digestive tract segments and their contents.The effects of feeding RSD and of meal feeding were found to be additive. In <u>ad libitum</u> feeding the animal can regulate food consumption and moderate the effects of antinutritional factors in the diet. Meal or tube-feeding interfered with this defense mechanism and the negative effects of RSD were accentuated and were lethal in non adapted rats and geese.
In long-term studies the incidence of pancreatic nodules was correlated to the level of TI in the diet. Feeding RSD potentiated the carcinogenic effect of azaserine and meal feeding enhanced the incidence and size of the pancreatic nodules in rats fed RSD.

INTRODUCTION

The effect of feeding raw soybean diets (RSD) in <u>ad libitum</u> feeding regimes is well documented, mainly in short-term experiments. Feeding raw full fat or defatted soy flour is known to inhibit growth

in many species of animals (Liener and Kakade,1980; Nitsan et al., 1971; Rackis, 1974; Struthers et al., 1983;). Pancreas enlargement and hypersecretion of pancreatic enzymes occur in young and adult chicks, turkeys, geese, rats and mice fed RSD (Alumot and Nitsan, 1961; Rackis, 1974; Nitsan and Nir, 1977;Liener and Kakade, 1980).
It was demonstrated that the secretory activity of the pancreas of the rat and the changes in RNA and DNA content, which are indicative of hypertrophy and hyperplasia, are closely related to the level of trypsin inhibitor (TI) (Liener et al., 1985). These changes were manifested within a short period of ingesting RSD (Crass and Morgan, 1981) and were maintained during a 20 month experimental period (Liener at al., 1985).

No change or reduced pancreas weight and hyposecretion of pancreatic enzymes was reported for calves (Gorrill and Thomas, 1967; Kakade et al., 1975), dogs (Patten et al., 1971), pigs (Yen et al., 1974), guinea pigs (Hasdai, 1980) and monkeys (Struthers et al., 1983).

Different results were reported when RSD were fed during prolonged periods. Adult rats fed RSD for 6 months, maintained a normal body weight in spite of prolonged pancreatic hypertrophy. Gross and microscopic examination of the pancreas revealed no abnormalities. The gross appearance of the heart, kidneys, spleen and liver was normal (Rackis et al., 1979). No adverse histopathological effects were observed in the cellular structure of the pancreas of adult dogs fed a diet containing 15% raw soybean flour (RSF) for 16 weeks (Patten et al., 1971).

In contrast to the above studies, others found that exposure of rats to diets containing RSF for 6 months induced the formation of acinar hyperplastic nodules in the pancreas and increased its sensitivity to pancreatic carcinogens such as azaserine (Morgan et al., 1977). The incidence of nodules was highly correlated with the level of TI in the diet (Liener et al., 1985).

Although toxicity and toxic factors are terms frequently used to refer to the antinutritional properties of raw soy products, such a designation is unwarranted since the growth inhibition and pancreatic hypertrophy caused by raw soybeans are reversible (Booth et al., 1964; Rackis,1974; Rackis et al., 1979; Crass and Morgan, 1982). Further confirmations of the nontoxicity of raw soybeans are: increasing the levels of (RSF) in rats' diets, or supplementing RSF-containing diets with additional calories and protein, resulted in a growth rate comparable to that achieved with heated soybean diets (HSD) (Fisher and Shapiro, 1963).

In most of the studies mentioned above, the animals were fed <u>ad libitum</u>. In meal-feeding there are periods of food deprivation and periods of intensive eating and the food remains in the gastrointestinal tract (GIT) for a prolonged period (Nir et al., 1979). Therefore, if the food contains antinutritional factors, the organism is exposed for a longer time to those factors than in _ad libitum_ feeding. Animals exposed to meal-feeding develop various adaptive mechanisms, among which some are similar to those developed by RSD feeding, such as increased pancreas weight, enhanced synthesis and secretion of pancreatic enzymes, increased GIT segments (Nir and Nitsan, 1979; Nitsan, 1985).

In the present work the combined effects of RSD and meal-feeding on various species are reported.

MEAL-FEEDING OF RATS

In a preliminary work the effect of feeding one 4-h daily meal to weanling rats, weighing 57 g, was studied. All the rats which were meal-fed diets containing 40% RSF (12.7mg TI/g diet) or over, died within 3 days. All the rats fed _ad libitum_ the same diets, survived. From the above study it was concluded that RSF, which is not considered to be toxic when fed _ad libitum_ (Fisher and Shapiro, 1963; Booth et al., 1964; Rackis, 1974; Rackis et al., 1979;), may become lethal when administered in large quantities in a meal-feeding regime.

In another study, in order to moderate the stress of meal-feeding, older rats (weighing 122 g) were brought to one 4-h meal per day gradually, their access to food being reduced by 2 h each day until a 4-h daily feeding period was reached. The rats were fed _ad libitum_ on 2 days per week. With this regime there was no mortality. Four diets containing increasing amounts of RSF were fed (Nitsan et al., 1983). The diets contained 0.9, 5.8, 12.7 and 24.1 mg TI/g diet. Increasing the RSF concentration in the diet reduced slightly food intake and body weight gain in the groups fed _ad libitum_ (_Figure 1_). At a concentration of 24.1 mg TI/g diet (total intake of about 14 g TI during the 5-week experimental period), growth was depressed by approximately 30%, as compared with counterparts fed diets containing traces of TI (0.9 mg TI/g diet). In meal-fed rats, increasing the TI dietary concentration caused a more pronounced depression in food intake and growth than in _ad libitum_-fed rats. When fed the diet containing 24.1 mg TI/g diet, almost complete cessation of growth occurred ,although the meal-fed rats consumed only 7 g TI during the experimental period.

The relative weights (g/100 g body weight) of liver and kidneys increased parallel to dietary TI concentration and were not affected by the feeding regime (_Figure 2_). The pancreas relative weight also increased parallel to the dietary TI concentration, much more in the meal-fed than in the _ad libitum_-fed rats. The relative weight of the spleen was slightly increased in _ad libitum_-fed rats, but was slightly depressed in the meal-fed ones by increasing TI concentration.

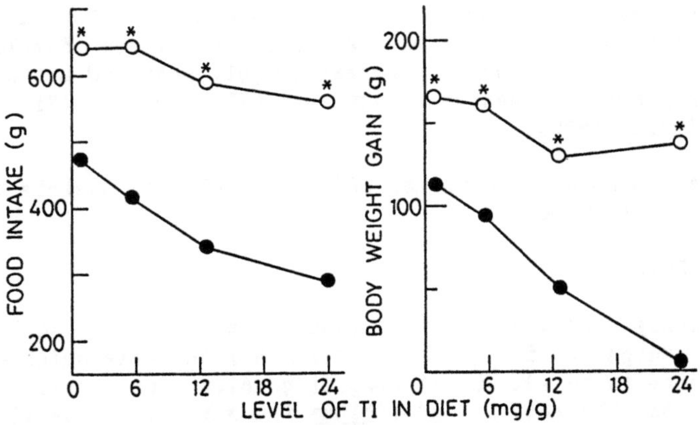

Figure 1. Food intake and body weight gain of rats fed diets containing increasing levels of Trypsin inhibitor, fed <u>ad libitum</u> (o) or in one meal per day (•). (Adapted from Nitsan et al., 1983.)
***Statistically significant differences (P<0.05) between the two feeding regimes.**

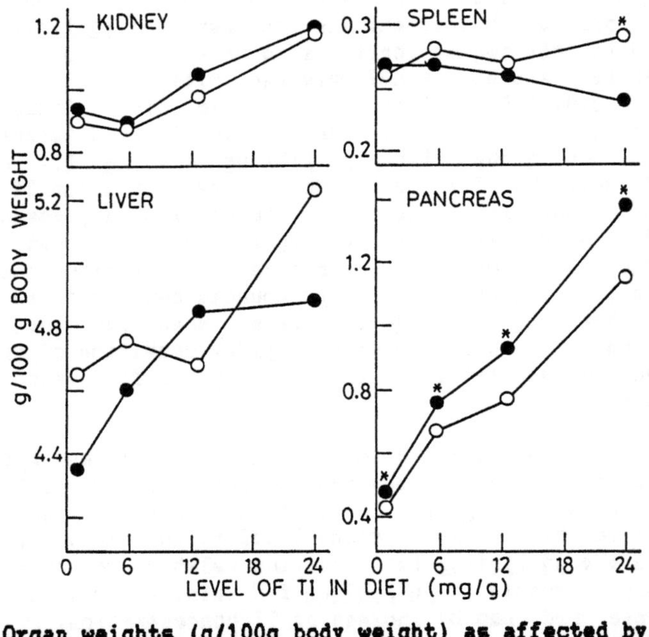

Figure 2. Organ weights (g/100g body weight) as affected by the level of dietary trypsin inhibitor in ad libitum (o) and meal-fed (•) rats. (Adapted from Nitsan et al., 1983.)
*** Statistically significant difference (P<0.05) between the two feeding regimes.**

The weight of the GIT increased along with the dietary TI concentration in both feeding regimes (Figure. 3). In the meal-fed rats the GIT segments were larger than in the *ad libitum*-fed ones, the difference increasing with the increase in dietary TI concentration (cecum excepted).

In *ad libitum-fed* rats, the length of the intestinal segments increased to a plateau at the concentration of 12.7 mg TI/g diet. In meal-fed rats, the intestine was consistently longer than in *ad libitum*-fed counterparts and the length increased parallel to the TI concentration. Since the elongation of the intestinal segments caused by meal-feeding at all TI levels exceeded the increase in weight, the thickness of the intestinal wall estimated by the weight:length ratio was reduced by meal-feeding.

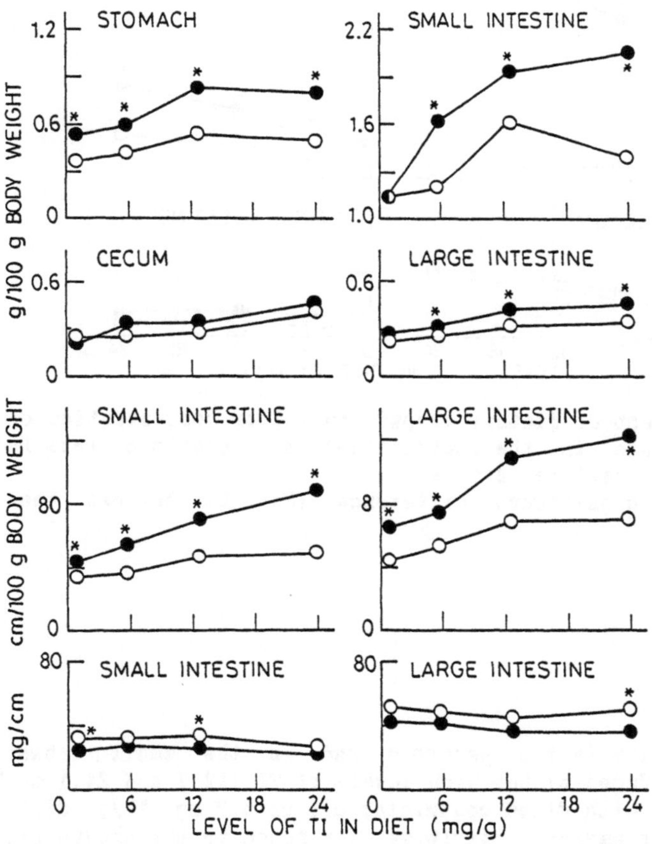

Figure 3. The effect of dietary trypsin inhibitor concentration on the relative weight of the gastrointestinal segments, the relative length and the weight:length ratio of the small and large intestines in rats fed *ad libitum* (o) or one meal per day (●).
* Statistically significant difference (P<0.05) between the two feeding regimes.

In meal-feeding the food remains in the GIT for a longer period than in *ad libitum*-fed rats, as evidenced from the markedly greater amount of contents especially in the stomach and the small intestine found in the meal-fed rats (Figure 4). This effect decreased caudally. The contents in the intestinal segments were increased also by TI dietary concentration. In the stomach and small intestine the response to TI was stronger in meal-fed than in *ad libitum*-fed rats.

Dietary TI concentration and meal-feeding were accompanied by hypersecretion of digestive enzymes from the pancreas.

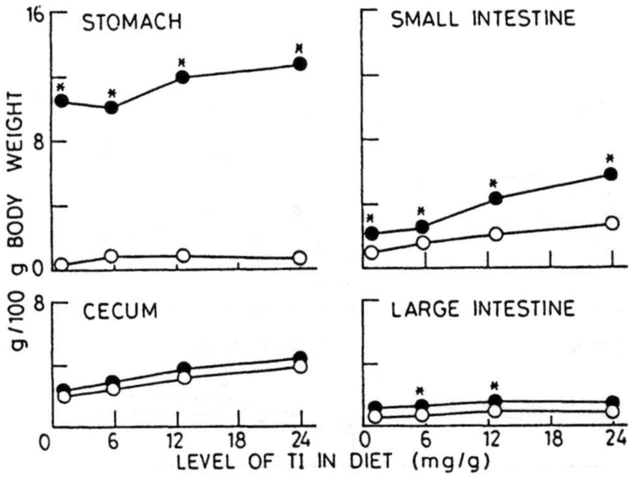

Figure 4. The effect of dietary trypsin inhibitor concentration on the amount of contents in the gastrointestinal segments of rats fed *ad libitum* (o) or one meal per day (●).
* Statistically significant difference ($P<0.05$) between the two feeding regimes.

Amylase activity in the pancreas and in the small intestinal contents was reduced by the high levels of TI (12.7 and 24.1 mg TI/g diet) as compared with diets containing 0.9 or 5.7 mg TI/g diet. In the cecum, higher amylase activities were found in the groups fed the diets containing TI (5.7-24.1 mg/g) than the diet with traces of TI (0.9 mg TI/g diet). Trypsin activity in the pancreas did not show a consistent trend when related to dietary TI concentration. In the small intestine and in the cecum the trypsin activity increased with the increase in dietary TI concentration. Chymotrypsin activity increased in the pancreas, small intestine and cecum with the increase in dietary TI concentration.

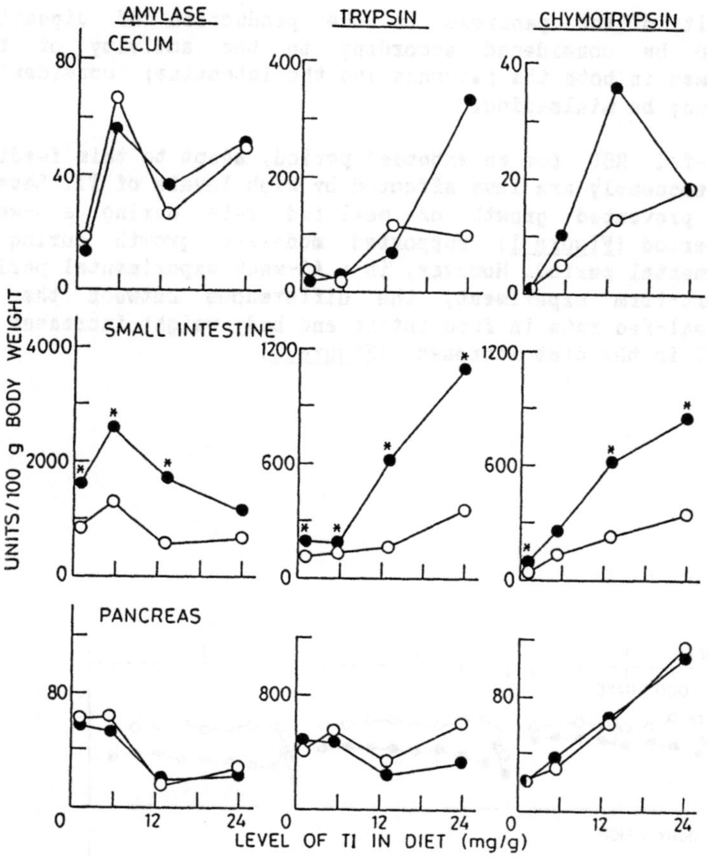

Figure 5. The effect of dietary level of trypsin inhibitor on trypsin, chymotrypsin and amylase activities in the pancreas, and in the contents of the small intestine and cecum of rats fed ad libitum (o) or one meal per day (●).
***Statistically significant difference (P<0.05) between the two feeding regimes.**

Meal feeding caused a marked increase in the synthesis and secretion of the digestive enzymes, as shown by the significantly higher levels found in the small intestine as compared with the levels in ad libitum fed rats. In spite of the excessive secretion to the intestine, no reduction occurred in the activity of the digestive enzymes in the pancreas, and the activities in the pancreas were similar in the two feeding regimes.

Hyperactivity of the pancreas in the production of digestive enzymes should be considered according to the activity of the digestive enzymes in both the pancreas and the intestine; considering only one site may be misleading.

Rats meal-fed RSD for an extended period, adapt to this feeding regime and consequently are less affected by high levels of TI. Levels of TI which prevented growth of meal-fed rats during a 5-week experimental period (Figure 1) supported moderate growth during a 40-week experimental period. However, in a 40-week experimental period as in the short-term experiment, the differences between the *ad libitum* and meal-fed rats in food intake and body weight increased as the level of TI in the diet increased (Figure 6).

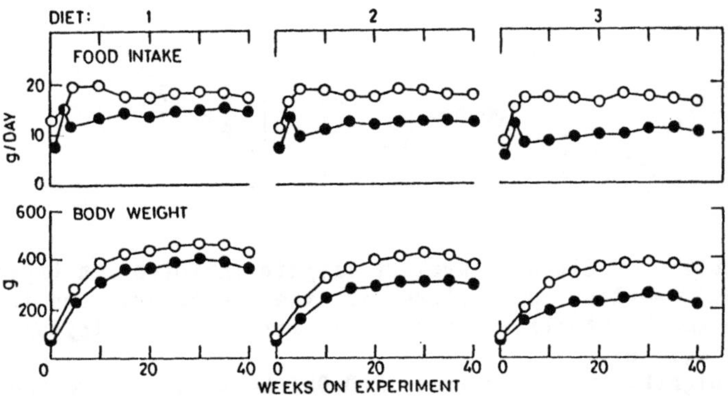

Figure 6. Food intake and body weights of rats fed diets containing different levels of trypsin inhibitor (TI), fed *ad libitum* (o) or one meal per day (•). Diets 1, 2 and 3 contained 5.7, 12.7 and 24.1 mg TI/g diet, respectively. (Adapted from Nitsan et al., 1985.)

The adaptation of rats to meal-feeding of diets containing TI was confirmed also by the pancreas weight. In a long -term experiment, as in the short-term experiment the pancreas weight was increased as the level of dietary TI increased. After a 5-week experimental period the pancreas weight was heavier in the meal-fed rats than in the *ad libitum*-fed ones (Figure 2), but no statistically significant differences were found after 40 weeks in the pancreas weight of rats on the different feeding regimes (Table 1).

Table 1. Effect of feeding regime on pancreas weight (g/100g body weight) of rats fed diets containing different levels of trypsin inhibitor (TI) during 40 weeks. (According to Nitsan et al., 1985.)

LEVEL OF TI (mg/g diet)	EFFECT OF FEEDING REGIME AD LIBITUM	ONE MEAL/DAY	SATISTICAL SIGNIFICANCE OF THE EFFECTS:p(F)	
5.7	0.71	0.56	Level of TI	0.001
12.7	1.38	0.90	Feeding regime	NS
24.1	1.73	1.53	Levelxregime	NS

Feeding RSD to rats injected with azaserine induced pancreatic nodules at an earlier age than when they were fed HSD (Morgan et al., 1977). In order to study if meal-feeding accentuates nodules formation in rats fed RSD and injected with azaserine, rats fed diets with increasing levels of TI and injected weekly with azaserine were fed ad libitum or one meal per day (Nitsan et al., 1985). The nodule score (number of nodules/pancreas x volume of nodules) at the age of 10-12 months increased with azaserine administration, dietary TI concentration and meal feeding (Figure 7). At the highest level of TI (24.1 mg TI/g diet), the nodule score in the meal-fed rats injected with azaserine was triple that found in the ad libitum-fed ones and double that in the meal-fed rats which were not injected with azaserine.

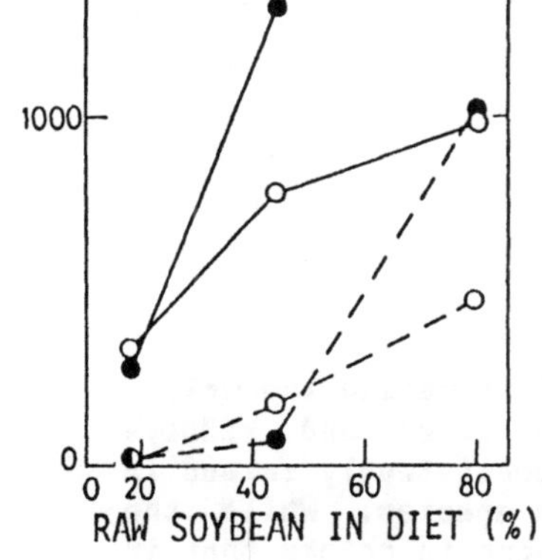

Figure 7. Effect of dietary level of raw soybean meal (RSF), azaserine and meal feeding on the nodule score in rats after 10-12 months; ad libitum+ azaserine (o——o), meal-fed + azaserine (●——●), ad libitum, no azaserine (o----o), meal-fed, no azaserine (●----●). (Adapted from Nitsan et al., 1985.)

GEESE AND CHICKS

A comparative study carried out with chicks and goslings revealed that goslings are much more sensitive to raw soybeans than chicks. A semisynthetic diet containing 80% raw or heated soybeans, supplemented with minerals, vitamins and cerelose:starch (1:1) was used (Nitsan and Nir, 1977). Food intake of the goslings given RSD was considerably more depressed during the 3 week experimental period (49% of the HSD intake) than in chicks (93% of the HSD intake). Growth depression was also greater in the goslings than in the chicks, their weight gain being 28% and 41% of the respective HSD counterparts (Table 2). The RSD-fed goslings had also inadequate feathering, ranging from bare skin on the neck and head to thin feathering on the rest of the body; the feathering of the RSD-fed chicks appeared normal.

Table 2. Food intake, body weight and relative organ and intestinal content weights (g/100 g body weight) of goslings and chicks fed diets containing raw soybeans (RSD), in relation to controls fed the same diets but containing heated soybeans (HSD) (Adapted from Nitsan and Nir, 1977.)

	RSD/HSD	
	GOSLINGS	CHICKS
Food intake	0.49*	0.93*
Body weight gain	0.28*	0.41*
Liver	1.20*	0.92
Kidneys	1.22*	1.03
Abdominal adipose tissue	1.29*	0.21
Pancreas	2.00*	2.08*
Proventriculus	1.20*	1.47*
Gizzard	1.45*	1.72*
Small intestine	1.68*	1.35
Small intestine content	3.13*	1.40
Cecum	2.32*	1.04
Cecum content	3.69*	2.15*
Large intestine	1.25*	1.05

* Statistically significant difference between RSD- and HSD- fed goslings or chicks ($P < 0.05$).

As with rats, feeding chicks and goslings RSD increased the weight of the GIT segments and contents. Although the liver and kidneys weights were not increased in chicks, they were consistently raised in goslings. The pancreas weight was doubled in both species. While the stomachs weight was increased by feeding RSD more in chicks than in goslings, the intestinal segments and the weight of their contents increased much more in goslings than in chicks (Table 2).

The over-three fold increase in the weight of the intestinal and cecal contents in geese, compared with the 1.4-2.15-fold increase in chicks, may be indicative of a slower passage rate of the digesta in the intestinal segments of the geese. It may be associated with changes in the nutritive value of the food and/or prolonged effect of the antinutritional factors in the diet.

Table 3. The activity of digestive enzymes (units/100 g body weight) in the pancreas and intestinal contents of goslings and chicks fed heated (HSD) or raw (RSD) soybean diets, and the ratio of RSD/HSD. (Adapted from Nitsan and Nir, 1977.)

	GOSLINGS			CHICKS		
	HSD	RSD	RSD/HSD	HSD	RSD	RSD/HSD
Pancreas						
Lipase	1608	1839	1.14	820	1564	1.90*
Amylase	1378	235	0.17*	1539	2436	1.58
Trypsin	92	186	2.02*	83	191	2.30*
Chymotrypsin	47	9	0.19*	21	47	2.21*
Small intestine						
Lipase	2185	10007	4.58*	1346	4422	3.28*
Amylase	84	201	2.39*	748	1391	1.86
Trypsin	29	3	0.11*	94	53	0.58
Chymotrypsin	4	3	0.81	4	8	2.00
Cecum						
Lipase	29	236	8.08*	97	302	3.10
Amylase	5	23	4.81*	703	3287	4.67*
Trypsin	0.3	0.4	1.09	21.2	16.6	0.78
Chymotrypsin	0.2	0.4	1.56	2.7	12.9	4.72*

* Statistically significant difference between RSD- and HSD- fed groups ($P<0.05$).

Amylase activity was much higher in the chick pancreas than in the goose pancreas, and this trend increased along the intestinal tract to attain a difference of about 150-fold in the cecum. The same trend, but to a lower extent, was found for the trypsin activity. No consistent differences were observed in lipase and chymotrypsin activities in the pancreas or along the intestinal tract between the two species. Activities of digestive enzymes were consistently increased in the chick pancreas by feeding RSD. In the goose, RSD depressed amylase and chymotrypsin but caused a two fold increase in trypsin activity.

Lipase and amylase activities in the small intestine and cecum contents in both species were increased substantially by feedind RSD. Trypsin activity in the small intestine was depressed by feeding RSD, more in the goslings than in the chicks, while chymotrypsin activity was depressed in the small intestine of the geese only. It seems that in chicks, the adaptive mechanism of hypersecretion of proteolytic enzymes, in order to overcome their inhibition in the small intestine when fed RSD (Nitsan and Alumot, 1964), is much greater than in geese, and may be one of the reasons for the greater growth depression in the latter.

In chicks and geese, as was found for rats, the effect of RSD was accentuated by meal-feeding.

MEAL-FEEDING GEESE

Ten 8-week-old geese weighing an average of 4.1 kg each were meal-fed by intubation a diet composed of corn, salt and vitamins. A similar group of ten geese was meal-fed the same diet supplemented with 20% raw soybeans. As in rats, the geese were very sensitive to RSD delivered in meals. When fed the corn - raw soybean diet, only one to three meals could be administered. The passage of the food through the digestive tract was blocked and 90% of the geese died within a few days. When tube-fed the corn diet, the geese developed normally. Addition of HSF to the corn diet was found to be advantageous in geese tube-fed for the production of fatty livers (Nir et al., 1972).

MEAL-FEEDING CHICKS

Fourteen-day-old chicks were fed diets containing 38% RSF or HSF (44% protein). Two groups of ten chicks each were fed *ad libitum* HSD (control group) or RSD; two other groups were tube-fed (Nir et al., 1974) twice daily the same diets, about 40% more than the amount consumed by the control group (Table 4). At the age of 23 days (9 days on treatment), the chicks were killed and the organs were dissected and weighed. In the group meal-fed RSD, three chicks died during the first week, while there was no mortality in the other groups.

Table 4. Food intake, body weight gain and organ weights (g/100 g body weight) of chicks fed heated (HSD) or raw (RSD) soybean diet *ad libitum* or in meals. The values are presented in relation to controls fed HSD *ad libitum*.

	HSD MEALS	RSD AD LIBITUM	RSD MEALS
Food intake	1.38	0.88	1.44
Body weight gain	1.15	0.59*	0.50*
Pancreas	1.37	2.00*	2.63*
Liver	1.82	0.97	1.66*
Kidneys	1.02	1.00	1.29*
Thymus	0.72*	0.75*	0.37*
Bursa of Fabricius	0.79	0.86	0.46*
Spleen	0.75	1.05	0.50*

* Statistically significant difference from *ad libitum*-fed HSD controls ($P<0.05$).

Due to higher food consumption, meal feeding HSD by intubation improved body weight gain of chicks. The depression in body weight caused by RSD was more pronounced in meal than in ad libitum feeding in spite of the higher amount of food consumed by the meal-fed chicks (Table 4).

Meal feeding HSD caused an increase in pancreas and liver weights and a slight depression in the weight of the organs associated with the immunological system: the thymus, bursa of Fabricius and spleen. Feeding RSD *ad libitum* caused a twofold increase in the weight of the pancreas and a slight depression in the weight of the thymus and bursa of Fabricius. Meal feeding RSD showed additive effects on most organs. The enlargement of the pancreas was greater by meal-feeding RSD than by feeding RSD *ad libitum* or feeding HSD in meals. Kidneys weight, which was not affected by feeding RSD *ad libitum* or by meal-feeding HSD, was increased by meal-feeding RSD. The depression of the thymus, bursa of Fabricius and spleen was most pronounced in this treatment.

Meal feeding HSD increased the weight of the crop and the duodenum significantly, and of the jejunum and ileum slightly. Feeding RSD *ad libitum* increased most of the GIT segments (statistically significant, only the ileum and the cecum). Meal-feeding RSD increased all the GIT segments (gizzard excepted) more than meal feeding HSD or feeding RSD *ad libitum*. The contents of the GIT segments was also increased more by meal-feeding RSD than by meal-feeding HSD or by feeding RSD *ad libitum* (Table 5).

Table 5. The weight of the gastrointestinal organs and their contents (g/100 g body weight), in chicks fed a heated (HSD) or raw (RSD) soybean diet, *ad libitum* or in meals. The values are presented in relation to controls fed HSD *ad libitum*.

	HSD MEALS	RSD *AD LIBITUM*	RSD MEALS
Crop	3.44*	1.00	4.74*
Proventriculus	1.01	1.05	1.33*
Gizzard	0.97	1.19	1.06
Duodenum	1.31*	1.11	1.98*
Jejunum	1.21	1.20	1.99*
Ileum	1.18	1.27*	2.22*
Cecum	1.05	1.46*	1.35*
Large intestine	0.75	0.97	1.21
CONTENT			
Duodenum	0.81	1.42	2.07*
Jejunum	2.91*	1.46	4.56*
Ileum	1.88*	1.05	2.66*
Cecum	1.36	1.38	1.92*

* Statistically significant difference from *ad libitum* fed HSD controls (P<0.05).

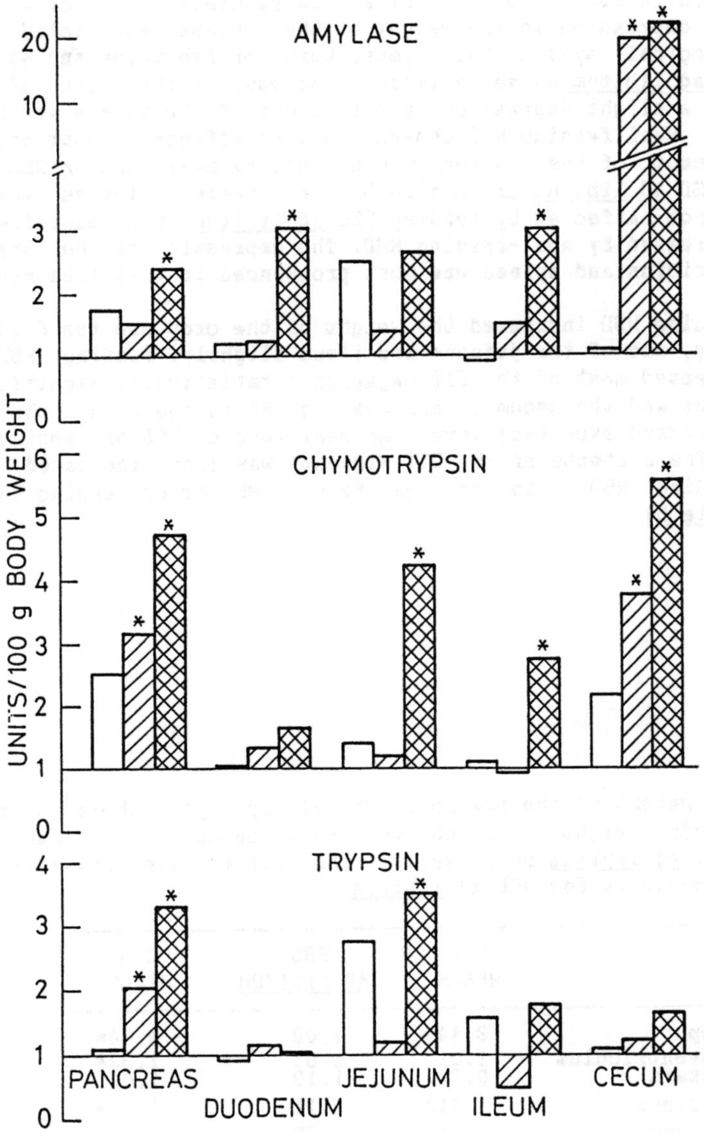

Figure 8. Trypsin, chymotrypsin and amylase activities (units/100 g body weight) in the pancreas and intestinal content of chicks fed heated (HSD) or raw (RSD) soybean diets *ad libitum* or in meals. The values are presented in relation to control chicks fed HSD *ad libitum*. (HSD-*ad libitum* = 1.00; HSD-meal-fed ☐ ; RSD-*ad libitum* ▨ ; RSD-meal-fed ⊠).

When compared with the HSD fed controls, meal-feeding HSD was accompanied by a moderate, not statistically significant increase in the activity of the digestive enzymes in the pancreas and in most parts of the intestine. Feeding RSD caused an increase in the activity of the digestive enzymes at most sites, more in meal-fed than in _ad libitum_-fed chicks (_Figure 8_). The effect of meal-feeding RSD on the activity of the digestive enzymes was additive, since at most sites the activities in this group were much higher than in either the meal-fed HSD group or the _ad libitum_-fed RSD group.

CALVES

In intensive farming, suckling calves are meal-fed one or two meals per day with milk or milk replacers. When fed the same amount of milk or milk replacer in one or two meals per day similar growth rate and food utilization were obtained (Owen et al., 1965; Ackerman et al., 1969). There are no comparative data available on the effect of feeding milk replacer containing RSF in meal or in _ad libitum_ feeding regimes.

Calves fed twice a day milk replacer containing 40% raw (RSD) or partly heated (PHSD) soybean protein concentrate (65% protein) did not grow during 30 days. When fed the same diet but with heated soybean protein concentrate (HSD), growth rate was 35% that of calves fed milk replacer only, during the first 10 days. After the adaptation period, the growth rate reached about 80% that of the milk replacer-fed calves (_Figure 9_). The effect of RSD and PHSD was reversible, as shown by the fact that when the calves were transferred to HSD or to milk replacer, compensatory growth was observed.

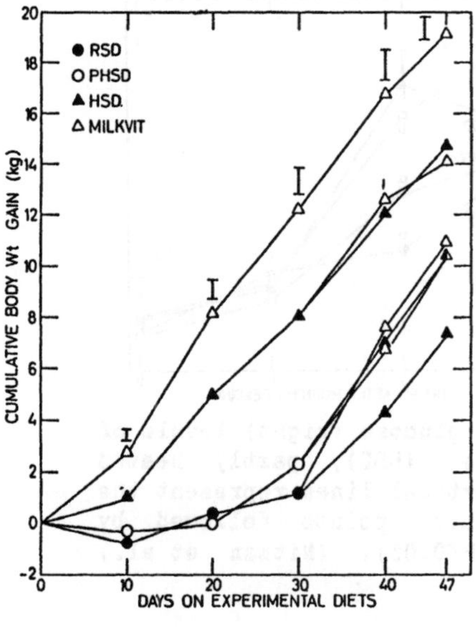

Figure 9. Body weight gains of calves fed milk replacer (Milkvit), or heated (HSD), partly heated (PHSD) or raw (RSD) soybean diets. Vertical lines represent the standard error of the mean (Nitsan et al., 1971.)

Growth retardation was associated with a decrease in protein and fat digestibility (Table 6). The improvement in digestibility of protein and fat was parallel to the degree of heat treatment of the soybean flour.

Table 6. The digestibility (%) of protein and fat in suckling calves fed milk replacer, or milk replacer containing 40% heated (HSD), partly heated (PHSD) or raw (RSD) soybean flour (Nitsan et al.,1971.)

	MILK REPLACER	HSD	PHSD	RSD
Total dietary protein	95.5a	89.8a	82.3b	76.6b
Soybean protein	----	88.5a	79.5ab	72.5b
Fat	97.0a	97.7a	96.0ab	93.7b

Within diets, figures followed by different letters differ statistically (P<0.05).

The higher blood urea level in the calves fed RSD and PHSD than in the HSD and milk replacer groups suggests a disturbed assimilation of the absorbed nitrogen (Figure 10). This could be due to a shortage of available carbohydrates-essential for optimal nitrogen utilization following consumption of RSD or PHSD, as compared with HSD or milk replacer. This assumption is supported by the blood glucose level found during the first 3 h post-feeding: high in the milk replacer-fed calves, intermediate in those fed HSD, and lowest in the RSD- and PHSD-fed calves. Interference with glucose absorption by soybean lectins or by RSD has been reported also for rats (Jaffe, 1980).

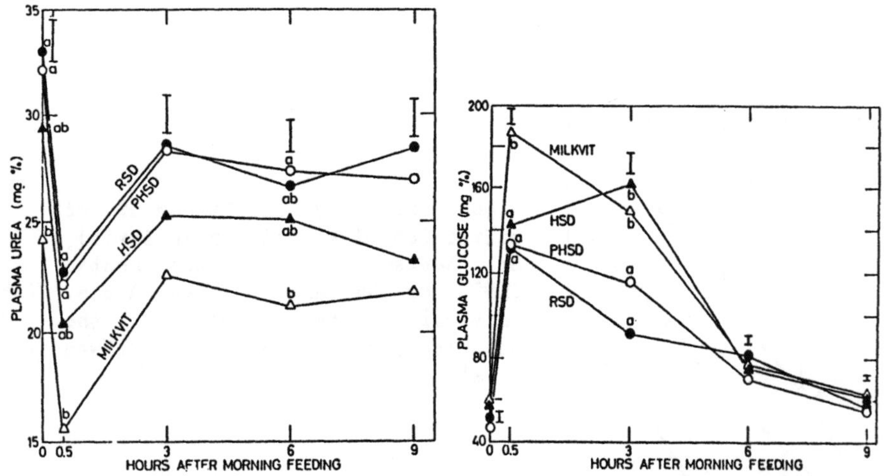

Figure 10. Blood urea (left) and blood glucose (right) levels of calves fed milk replacer (Milkvit), or heated (HSD), partly heated (PHSD) and raw (RSD) soybean diets. Vertical lines represent the standard error. In the different treatments, points followed by different letters differ statistically (P<0.05). (Nitsan et al., 1971.)

Addition of RSF or TI to milk replacer did not cause pancreatic hypertrophy in calves (Gorrill and Thomas, 1967; Kakade et al., 1975), in contrast to the marked hypertrophy observed in rats, chicks and geese (Figure 2, Table 2). In these species, the pancreatic hypertrophy is accompanied by hypersecretion of pancreatic exocrine enzymes (Figure 5, Table 3). In calves, Kakade et al.(1975) did not find any effect of dietary RSF or TI on trypsin and chymotrypsin activities in the pancreas. Gorrill and Thomas (1967) found a slight reduction in these enzymes in the pancreas but a very pronounced reduction in the activity of these enzymes in the digesta collected from the small intestine (Gorrill and Nicholson, 1971) (Table 7). The lack of adaptation of the pancreas of suckling calves could partly explain their greater sensitivity to RSD or TI ingestion, which is accompanied by a total inhibition of growth (Nitsan et al., 1971; Kakade et al., 1975).

Table 7. Pancreas weight (g/kg body weight) and trypsin and chymotrypsin activities in the pancreas and intestinal contents of calves fed raw (RSD) and heated (HSD) soybean diets, or milk supplemented with trypsin inhibitor (TI).

DIET	TI IN DIET	PANCREAS WEIGHT	UNITS OF ACTIVITY TRYPSIN	CHYMO-TRYPSIN
			PANCREAS	
RSD (1)	19 U/mg	0.67	28	15
HSD	6 U/mg	0.75	29	18
RSD (2)	6.17 U/mg	0.57	3.5	0.95
HSD	0.14 U/mg	0.47	4.9	0.72
			INTESTINAL CONTENTS	
MILK(3)	1mg/g		2.4	2.0
MILK	—		29	17

(1) According to Kakade et al. (1975).
(2) According to Gorrill and Thomas (1967).
(3) According to Gorrill and Nicholson (1971).

When adult ruminants, mainly dairy cows, are fed on full fat or extracted raw soybean meal, little or no antinutritive effect is observed (Loosli et al., 1961; Perry and Macleod, 1968; Larson and Schultz, 1970). This could be attributed to the fact that soybean TI does not suppress ruminal proteolytic activity. *In vivo* studies showed that the Kunitz inhibitor is rapidly degraded by rumen microorganisms, while the Bowman-Birk inhibitor is of relatively greater resistance (Baintner, 1981).

GENERAL REMARKS AND CONCLUSIONS

Pancreatic hypertrophy and hypersecretion of digestive enzymes in rats, mice, chicks and geese fed RSD are attributed to the presence of TI (Liener, 1979). The degree of the hypertrophy is closely related to the concentration of TI in the diet (Figure 2, and Liener et al., 1985) and can be produced by feeding animals highly purified preparations of soybean TI (Gertler et al., 1967; Nitsan and Gertler, 1972). It was suggested that the growth depression caused by TI may be a consequence of an endogenous loss of essential amino acids, deriving from a hyperactive pancreas compensating for the depletion of trypsin in the intestine caused by the inhibitor (Lyman and Lepkovsky, 1957). Since in some species, e.g. calves (Kakade et al., 1975), dogs (Patten et al., 1971), guinea pigs (Hasdai, 1980) and pigs (Yen et al., 1974), ingestion of RSD does not cause enlargement of the pancreas or hypersecretion of digestive enzymes, but growth depression remains (Figure 7), the excessive loss of endogenous proteins cannot account for all the growth depression. It seems that in spite of the loss of endogenous amino acids, hypersecretion of pancreatic enzymes may serve as a positive adaptive mechanism. Animals developing pancreas hypertrophy and hypersecretion of digestive enzymes (rats, chicks) overcome gradually the depressive effects of TI and after an adaptation period maintain almost normal proteolytic activity in the intestine (Nitsan and Alumot, 1964) and almost normal growth (Liener et al., 1985). In animals which do not develop this mechanism, such as guinea pigs (Hasdai, 1980) and calves (Kakade et al., 1975), no adaptation occurs and total growth depression persists as long as RSD is fed (Nitsan et al., 1971). The greater sensitivity of goslings than of chicks to RSD could be related to either or both of the above speculations: the slower process of overcoming the proteases inhibition in the small intestine, or the greater sensitivity of the goslings to endogenous losses of amino acids, mainly sulfur amino acids. The higher requirements for sulfur amino acids of geese than of chicks are due to differences in body composition. In goslings the relative feather weight is about twofold that of chicks (Nitsan et al., 1980). Since feather proteins are very rich in cystine, it seems that the inherent deficiency in sulfur amino acids of soybean protein, is accentuated by the endogenous loss of these amino acids, more in goslings than in chicks (although this loss may be smaller in the former). The more pronounced deficiency in sulfur amino acids was apparent in inadequate feathering of goslings, while the feathering of the chicks appeared normal (Nitsan and Nir, 1977).

Pancreatic hypertrophy and growth depression could be demonstrated also when RSF, from which the TI was removed by affinity chromatography, was fed to rats. It was suggested that the deleterious effect was due to resistance of 'native protein' to digestion by proteolytic enzymes in the intestine (Kakade et al., 1973).

That soybean TI does not account for all of the growth inhibition observed in rats has been shown already by Liener et al. (1949). This observation led to subsequent isolation from raw soybeans of phytohemagglutinins or lectins, which, like the TI, are heat-labile. This fraction inhibited growth in *ad libitum*-fed rats, due to a depressive effect on food intake; pair feeding the control group to the amounts consumed by the rats receiving the lectins, revealed no significant growth inhibition (Liener, 1953). Lectins injected to rats were lethal (Liener, 1951).

It seems that animals fed RSD *ad libitum* reduce their food intake to limit the consumption of negative factors such as TI or lectins, at levels in which their toxicity is moderated. In *ad libitum* feeding, eating is synchronized with the rate of food evacuation from the intestine via absorption or excretion. When exposed to *ad libitum* feeding, animals eat in discrete meals (Davis, 1977; Savory, 1979). Food composition and the physical nature of the food may affect eating behavior. It is suggested that the presence of toxic factors in the food may also affect eating behavior by reducing the size of the bouts and increasing their number. This in turn may reduce the time during which the toxins are present in the chyme. Therefore, feeding behavior may play an important role as a defense mechanism against dietary toxins. In *ad libitum* feeding the presence of toxic factors in the diet is generally concurrent to reduced food intake. This mechanism, in addition to others, is sufficient to prevent mortality when RSD is fed *ad libitum*. In meal-feeding the sensation of hunger interferes with this defense mechanism and excessive amounts of toxins are consumed in a short period. In this situation the animals are exposed to a higher concentration of toxins in the internal environment. Therefore, RSD, when delivered in large amounts by intubation, or when the daily food is consumed in one meal- with no adaptation period, may be lethal.

Meal feeding by intubation a diet containing HSF at levels exceeding 40% (Table 4) or 70% (Nir et al., 1974)-the *ad libitum* intake in chicks or up to 3-5-fold that in geese (Nir et al., 1973)- improved the performance. On the other hand, meal-feeding diets containing RSF was found to be highly toxic even when given in amounts lower than that consumed *ad libitum*. Rats and geese were more sensitive than chicks in this respect.

Meal feeding is associated with an increased length and reduced thickness (weight/length unit) of intestinal segments in rats (Figure 3) and in chicks (Nitsan, 1985). The reduced thickness of the intestine is in accord with the conclusion of Lepkovsky et al. (1960) that "meal feeding increases the flexibility of the intestinal wall, thus enabling accommodation of excessive amounts of chyme, and increasing the rate of absorption." Feeding RSD affects the intestinal length and wall thickness as does meal-feeding, and its effect becomes greater as the concentration of TI increases (Figure 3). On the other hand, it is well documented that ingesting RSD causes a reduction in the absorption of nutrients. Jaffe (1980) proposed that lectins, when ingested orally, can bind to specific receptor sites on the surface of intestinal epithelial cells and thus cause non-specific interference with the absorption of nutrients. This could be reflected in reduced absorption of amino acids in rats fed RSD (Nitsan and Liener, 1976). Purified lectins reduce glucose absorption through the rat intestinal wall (Jaffe, 1980) and reduce glucose absorption in calves (Figure 10).

The combined effect of meal-feeding and RSD on the absorption of nutrients and toxins is not obligatorily intermediate. The slower passage rate of the ingesta caused by feeding RSD and by meal-feeding (De Muelenaere, 1964; Nitsan, 1985) seems to be additive, since meal feeding RSD was accompanied by an excessive amount of intestinal contents (Figure 4, Table 5). If food remains in the GIT for a prolonged period, it is exposed to chemical, physical and microbiological alterations which may affect its rate of absorption and its nutritive value, and may accentuate the toxicity of negative factors such as those found in RSD.

The detrimental effects of feeding RSD are maintained in long-term experiments and are also accentuated by meal feeding. The long-term feeding of RSF to rats induced the formation of acinar hyperplastic nodules of the pancreas (McGuiness et al., 1980). The incidence of nodules was highly correlated with the level of TI in the diet (Liener et al., 1985). Feeding RSD also potentiated the carcinogenicity of azaserine toward the pancreas (Morgan et al., 1977; McGuiness et al., 1981; Nitsan et al., 1985). The number and severity of pancreatic nodules observed in azaserine-injected animals increased in relation to the level of RSF in the diet, and were enhanced at each level by meal feeding (Figure 7). The increased number and size of nodules in rats meal-fed RSD, could be associated with the depression of the immunological system as a result of the reduction in the weight of the organs associated with this system (thymus, bursa of Fabricius, and spleen) in meal-fed chicks (Table 4) and of the spleen in meal-fed rats (Figure 2).

When feeding experiments are designed with animals as models for human beings, the meal-feeding regime should be considered, since in modern society people are becoming increasingly meal eaters.

It is proposed that in studying the effects of dietary inhibitors or toxins, consideration should be given not only to what is being fed but also to how it is being fed.

REFERENCES

Ackerman, R.A., Thomas, R.O., Thayne, W.V. and Butcher, D.F. (1969). Effects of once-a-day feeding of milk replacer on body weight gain of dairy calves. J. Dairy Sci., 52, 1869-1872.

Alumot, E. and Nitsan, Z. (1961). The influence of soybean anti-trypsin on the intestinal proteolysis of the chick. J. Nutr., 73, 71-77.

Baintner, K. (1981). Determination of proteolytic activity of rumen liquor with azacasein. Vet. Med. A., 23, 796.

Booth, A.N., Robbins, D.J., Ribelin, W.E., DeEds, F., Smith, A.K. and Rackis, J.J. (1964). Prolonged pancreatic hypertrophy and reversibility in rats fed raw soybean meal. Proc. Soc. Exp. Biol. Med. 116, 1067-1069.

Crass, R.A. and Morgan, R.G.H. (1981). The effect of long-term feeding of soya-bean flour diets on pancreatic growth in the rat. Br. J. Nutr., 47, 119-129.

Davis, R.F. (1977). Long- and short-term regulation of feeding patterns in the rat. J. Comp. Physiol. Psychol. 91, 574-585.

De Muelenaere, H.J.H. (1964). Studies on the digestion of soybeans. J. Nutr., 82, 197-205.

Fisher, H. and Shapiro, R. (1963). Counteracting the growth retardation of raw soybean meal with extra protein and calories. J. Nutr., 80, 425-430.

Gertler, A., Birk, Y. and Bondi, A. (1967). A comparative study of the nutritional and physiological significance of pure soybean trypsin inhibitors and of ethanol-extracted soybean meals in chicks and rats. J. Nutr., 91, 358-370.

Gorrill, A.D.L. and Nicholson, J.W.G. (1971). Effect of soybean trypsin inhibitor, diarrhea, and diet on flow rate, pH, proteolytic enzymes and nitrogen fractions in calf intestinal digesta. Can. J. Anim. Sci., 51, 377-388.

Gorrill, A.D.L. and Thomas, J.W. (1967). Body weight changes, pancreas size and enzyme activity and proteolytic enzyme activity and protein digestion in intestinal contents from calves fed soybean and milk protein diets. J. Nutr. 92, 215-223.

Hasdai, A. (1980). Chemical, biological and mechanical modifications of soybean oil meal proteins and carbohydrates and their effect on the nutritive value for suckling calves. Ph.D Thesis. The Hebrew University of Jerusalem, Israel.

Jaffe, W.G. (1980). Hemagglutinins. In "Toxic Constituents of Plant Foodstuffs" I.E. Liener, Ed.,2nd edition, Academic Press, New York, pp. 73-98.

Kakade, M.L., Hoffa, D.E. and Liener, I.E. (1973). Contribution of trypsin inhibitors to the deleterious effects of unheated soybeans fed to rats. J. Nutr., 103, 1772-1778.

Kakade, M.l., Thompson, R.D., Engelstad, W.E., Behrens, G.C., Yoder, R.D. and Crane, F.M. (1975). Failure of soybean trypsin inhibitor to exert deleterious effects in calves. J. Dairy Sci., 59, 1484-1489.

Larson, A. and Schultz, L.H. (1970). Effects of soybeans compared to soybean oil and meal in the ration of dairy cows. J. Dairy Sci., 53, 1233-1240.

Lepkovsky, S., Chari-Briton, A., Lemmon, R.M., Ostwald, R.C. and Dimick, M.K. (1960). Metabolic and anatomic adaptation in chicks "trained" to eat their daily food in two hours. Poult. Sci. 39, 385-389.

Liener, I.E. (1951). The intraperitoneal toxicity of concentrates of the soybean trypsin inhibitor. J. Biol. Chem. 193, 183-191.

Liener, I.E. (1953). Soyin, a toxic protein from the soybean. I. Inhibition of rat growth. J. Nutr. 49, 527-539.

Liener, I.E. (1979). Protease inhibitors and lectins. In "International Review of Biochemistry, Biochemistry of Nutrition IA". A. Neuberger and T.H. Jukes, Eds. University Park Press, Baltimore, MD. 27, 97-122.

Liener, I.E., Deuel, H.J. Jr. and Fevold, H.J. (1949). The effect of supplemental methionine on the nutrituve value of diets containing concentrates of soybean trypsin inhibitors. J. Nutr., 39, 325-339.

Liener, I.E. and Kakade, M.L. (1980). Protease inhibitors. In "Toxic Constituents of Plant Foodstuffs" I.E. Liener, Ed.,2nd edition, Academic Press, New York, pp. 7-71.

Liener, I.E., Nitsan, Z., Srisangnam, C., Rackis, J.J. and Gumbman, M.R. (1985). The USDA trypsin inhibitor study. II. Timed related biochemical changes in the pancreas of rats. Qual. Pl. Foods Hum. Nutr., 35, 243-257.

Loosli, J.K., Warner, A.A. and Hintz, H.F. (1961). Value of corn distillers, dried grains, soybean oil meal, heated soybeans and soybean oil meal plus starch for milk production. J. Dairy Sci., 44, 1910-1914.

Lyman, R.H. and Lepkovsky, S. (1957). The effect of raw soybean meal and trypsin inhibitor diets on pancreatic enzyme secretion in the rat. J. Nutr., 62, 269-284.

McGuiness, E.E., Morgan, R.G.H., Levison, D.A., Frape, D.L., Hopwood, D. and Wormsley, K.G. (1980). The effects of long-term feeding of soya flour on rat pancreas. Scand. J. Gastroenterol. 15, 497-502.

McGuiness, E.E., Morgan, R.G.H., Levison, D.A., Hopwood, D. and Wormsley, K.G. (1981). Interaction of azaserine and raw soy flour on the rat pancreas. Scand. J. Gastroenterol. 16, 49-56.

Morgan, R.H., Levison, D.A., Hopwood, D., Saunders, J.H.B. and Wormsley, K.G. (1977). Potentiation of the action of azaserine on the rat pancreas by raw soyabean flour. Cancer Lett., 3, 87-90.

Nir, I. and Nitsan, Z. (1979). Metabolic and anatomical adaptations of light-bodied chicks to intermittent feeding. Br. Poult. Sci., 20, 61-71.

Nir, I., Nitsan, Z. and Vax, A. (1973). The influence of force feeding and of protein supplementation to the diet on the metabolizable energy of diets, digestibility of nutrients, nitrogen retention and digestive enzyme output in geese. Ann. Biol. Anim. Biochem. Biophys. 13, 465-479.

Nir, I., Perek, M. and Katz, Z. (1972). The influence of soybean meal supplemented to the maize diet of forced-fed geese upon their liver, organ and blood plasma components. Ann. Biol. Anim. Biochem. Biophys., 12, 77-89.

Nir, I., Petihi, I. and Nitsan, Z. (1979). Body composition, food utilization, intestinal adaptation and lipogenesis in meal-fed chicks. In "Food Intake Regulation in Poultry" K.N.Boorman and B.M. Freeman, Eds., Brit. Poultry Sci., Edinburgh. pp. 391-404.

Nir, I., Shapira, N., Nitsan, Z. and Dror, Y. (1974). Force-feeding effects on growth, carcass and blood composition in young chicks. Br. J. Nutr., 32, 229-239.

Nitsan, Z. (1985). Metabolic and anatomical adaptation to food restriction. 5th European Symposium on Poultry Nutrition. Maaleh Hachmisha, Jerusalem, Israel. pp. 53-60.

Nitsan, Z. and Alumot, E. (1964). Overcoming the inhibition of intestinal proteolytic activity caused by raw soybean in chicks of different ages. J. Nutr. 87, 179-184.

Nitsan, Z., Dvorin, A. and Nir, I. (1980). Composition and amino acid content of carcass, skin and feathers of the growing gosling (Ancer cireneus). Br. Poult. Sci., 22, 79-84.

Nitsan, Z. and Gertler, A. (1972). Enzymatic activities in the intestinal tract of chicks receiving a diet supplemented with trypsin inhibitors from soybean and egg white. Nutr. Metab., 14, 371-376.

Nitsan, Z. and Liener, I.E. (1976). Studies on the digestibility and retention of nitrogen and amino acids in rats fed raw or heated soy flour. J. Nutr., 106, 292-299.

Nitsan, Z., Hasdai, A. and Liener, I.E. (1985). Effect of raw soy flour, feeding regime, and azaserine on rat pancreas. Drug-Nutr. Interactions, 3, 223-228.

Nitsan, Z. and Nir, I. (1977). A comparative study of the nutritional and physiological significance of raw and heated soya-beans in chicks and goslings. Br. J. Nutr., 37, 86-97.

Nitsan, Z., Nir, I. and Liener, I.E. (1983). Accentuated response to raw soya-bean meal by meal feeding. Arch. Toxicol., suppl. 6, 177-181.

Nitsan, Z., Volcani, R., Gordin, S. and Hasdai, A. (1971). Growth and nutrient utilization by calves fed milk replacers containing milk or soybean protein-concentrate heated to various degrees. J. Dairy Sci., 54, 1294-1299.

Owen, F., Plum, M. and Harris, L. (1965). Once versus twice daily feeding of milk to calves weaned at 21 or 42 days of age. J. Dairy Sci., 48, 824 (Abstr.).

Patten, J.R., Richards, E.A. and Wheeler, J. (1971). The effect of dietary soybean trypsin-inhibitor on the histology of dog pancreas. Life Sci., Part II. 10, 145-150.

Perry, F.G. and Macleod, G.K. (1968). Effects of feeding raw soybeans on rumen metabolism and milk composition in dairy cows. J. Dairy Sci., 51, 1233-1238.

Rackis, J.J. (1974). Biological and physiological factors in soybeans. J. Am. Oil Chem. Soc., 51, 161A-174A.

Rackis, J.J., McGee, J.E., Gumbmann, M.R. and Booth, A.N. (1979). Effects of soy proteins containing trypsin inhibitors in long term feeding studies in rats. J. Am. Oil Chem. Soc., 56, 162-168.

Savory, C.J. (1979). Feeding behaviour. In "Food Intake Regulation in Poultry" K.N. Boorman and B.M. Freeman, Eds. Brit. Poultry Sci., Edinburgh. pp. 227-323.

Struthers, B.J., MacDonald, J.R., Dahlgren, R.R and Hopkins, D.T. (1983). Effects on the monkey, pig and rat pancreas of soy products with varying levels of trypsin inhibitor and comparison with the administration of cholecystokinin. J. Nutr., 113, 86-97.

Yen, J.T., Hymowitz, T. and Jensen, A.H. (1974). Effects of soybeans of different trypsin inhibitor activities on performance of growing swine. J. Anim. Sci., 38, 304-309.

15

EFFECT OF LONG-TERM FEEDING OF SOY-BASED DIETS ON THE PANCREAS OF CEBUS MONKEYS

J.P. Harwood[1], L.M. Ausman[2], N.W. King[2], P.K. Sehgal[2], R.J. Nicolosi[2], I.E. Liener[3], D. Donatucci[3], and J. Tarcza [3]

[1]Divsion of Nutrition, FDA, Washington, D.C.
[2]New England Regional Primate Research Center Southborough MA
[3]Dept. of Biochemistry, University of Minnesota, MN

ABSTRACT

Feeding soy-based protein containing trypsin inhibitor causes pancreatic hypertrophy in the rat, and long-term feeding (up to 2 years) has revealed a high incidence of adenoma following hypertrophy. It was therefore of interest to determine whether the ingestion of soy-based protein has any adverse effects on the primate pancreas. A resource of 27 Cebus albifrons monkeys, previously used to evaluate the protein quality of several soy and milk proteins, has been maintained on semi-synthetic diets for 3 to 4 years; the protein sources for the diets were casein, lactalbumin, soy isolate and soy concentrate. In general the monkeys were in good physical health and their weights were appropriate for age and sex. Serum biochemical and hematological profiles were normal and there were no major differences between the groups. A pancreatic biopsy from both the head and tail region of the pancreas was taken from each monkey. Visual observation of the pancreas revealed no overt pathology; two independent histological examinations indicated no diet-related differences between groups, and biochemical analyses of trypsin, chymotrypsin, protein, DNA and RNA revealed no differences. It is concluded that feeding low level trypsin inhibitor-containing diets for up to 4 years caused no adverse effects in the pancreas of the Cebus nonhuman primate.

INTRODUCTION

Use of raw unprocessed soy as a source of protein has long been associated with anti-nutritional effects including growth retardation and pancreatic hypertrophy (Liener and Kakade, 1980). These effects have been largely attributed to the content of trypsin inhibitor proteins in the soy, most of the activity of which can be readily destroyed by heat treatment. The processed soy used commercially for both human and animal nutrition contains residual quantities of trypsin inhibitor, which, until recently, was generally considered to be insignificant in terms of adverse biological effects (Rackis and Gumbmann, 1981).

During the last few years, however, as a result of long term feeding studies in rats, it has become evident that even low trypsin inhibitor levels in processed soy may have adverse nutritional consequences such as pancreatic hypertrophy, increased susceptibility to carcinogenesis and pancreatic adenoma and carcinoma (McGuinness et al., 1980; Gumbmann et al., 1984). Pancreatic hypertrophy has now been observed in a variety of species including rats, mice, hamsters, and guinea pigs, while other species, such as dog, show no such adverse effects (Rackis, 1974; Hasdai and Liener, 1983; Sale et al., 1977). Since soy-based protein products are used extensively in the United States and many other countries for human consumption, it is important to establish whether ingestion of soy has any effect on primates to aid in assessing the potential risk to the human population.

In order to ascertain this risk, the Food and Drug Administration utilized a resource of 27 Cebus monkeys which had been used previously to assess the protein quality of soy-derived proteins (Ausman et al., 1985). Since the end of the experiments, the monkeys had been maintained on soy and milk protein-based semi-synthetic diets. The monkeys were between 3 and 4 years old when the agency became aware of the carcinogenic effects of trypsin inhibitor in rats, and at that time the monkeys were viewed as an invaluable asset for investigating the effects of long-term ingestion of soy protein on the pathophysiology of the primate pancreas. In these circumstances the grouping of the animals with respect to size per group and the sex ratios, as well as the dose levels of trypsin inhibitor, were predetermined, and were not changed to avoid further compromising the interpretation of the data. The most important fact was that we had available a group of primates which had been exposed to soy protein for a long time and could therefore be used to examine whether the ingestion of soy had caused any adverse effects on the pancreas of a primate species. The monkeys were evaluated in terms of general physical health and weight, blood chemistry and hematology, and biochemical and histopathological examination of pancreatic biopsies.

METHODS

The 27 Cebus monkeys (Cebus albifrons) were used for this study following a feeding experiment to compare the protein quality of soy isolate, soy concentrate and casein with respect to lactalbumin concluded several years previously (Ausman et al., 1985). At the end of these studies, 23 of the monkeys were maintained on the same semi-synthetic diet, while four of the monkeys refused to eat their diet and were changed to a second protein source. These four animals were considered to belong to the group of their second dietary protein. This adaptation accounts for the uneven distribution of monkeys among the 4 dietary groups, which were as follows; lactalbumin (8), soy isolate (10), casein (7), soy concentrate (2). Although there were only two monkeys in the soy concentrate group, data on these two animals are included for completeness although no statistical inferences can be drawn from such a small group. The monkeys were maintained in accordance with the Guidelines of the

Committee on Animals of the Harvard Medical School and those prepared by the Committee on Care and Use of Laboratory Animals of the Institute of Laboratory Animal Resources, National Research Council (DHEW Publication # (NIH) 78-23, Revised 1978).

Diets

The four semi-synthetic diets were similar except for the source of protein, as detailed in Table I. The proteins used and the percentage composition (as percent of calories) in the final diet were as follows: lactalbumin, 12%; soy isolate, 20%; casein, 12.5%; and soy concentrate, 20%. The higher protein content of the soy diets was to compensate for the lower quality of protein and avoid addition of methionine supplements. The high fat content (21% by weight) required during infancy was maintained throughout the entire dietary period. The diets were prepared in the form of an agar cake and were prepared weekly and stored at 4° until fed. The agar cake was cut into small pieces and 250g was provided to each monkey in the morning. Monitoring of the monkeys indicated an average intake of 150g of the agar cake/day.

The trypsin inhibitor content of the proteins was determined by the Hamerstrand procedure (Hamerstrand et al., 1981), which is based on the ability of a suitable dilution of an extract to inhibit the activity of a standard amount of trypsin using a synthetic trypsin substrate. Activity was calculated as grams of trypsin inhibitor per kilogram of agar cake, as shown in Table I.

The primary pathology associated with the ingestion of trypsin inhibitor occurs in the pancreas, and we wanted to ascertain whether long-term exposure to soy diets had caused any adverse effects on the monkey pancreas without sacrificing the animals. This was achieved by surgical removal of a section of the pancreas from the head and tail part of the organ, as well as visual inspection and photography to aid in detecting any abnormal tissue.

Surgical procedure for pancreatic biopsy

For the surgical procedure, the monkeys were fasted overnight and sedated with ketamine (10 mg/kg) to permit intubation of the trachea for ventilation during surgery. A blood sample was collected for analysis, and an intravenous drip of normal saline was placed in the saphenous vein. The monkey was anesthetized with Halothane and prepared according to standard surgical procedures. A midline incision was made into the abdominal cavity to permit easy access to the pancreas. All areas of the pancreas were exposed to the extent possible for observation and photography to record color, shape, size and position. A biopsy weighing between 100 and 400 mg was surgically removed from the head and tail region of the pancreas. A silk suture was used to ligate ducts and vessels immediately proximal to the site of the biopsy. The pancreatic lesions were cauterized and a piece of sterile sponge (Gelfoam) was placed on the incision to help stop blood and exocrine seepage. This was the most difficult part of the surgery because if the pancreas is not effectively sealed, leakage of pancreatic enzymes can occur and cause major problems. By the use of cauterization and Gelfoam, most of the monkeys recovered quickly. One monkey, however, died after surgery and the pathology of disseminated intravascular coagulation may have been related to leakage of pancreatic enzymes. After removal of the biopsies, the incision was sutured in 3 planes and the animal allowed

TABLE I.

Composition of Semi-Purified Agar Cake Diets Fed to Cebus albifrons

	L - 12	SI - 20	C - 13	SC - 20
		g/kg		
Lactalbumin (12.63%N)[a]	154.9	-	-	-
Isolated soy protein[b] (13.45%N)	-	242.8	-	-
Casein (13.43%N)[c]	-	-	151.5	-
Soy concentrate[d] (10.65%N)	-	-	-	306.2
Sucrose	193.9	150.0	195.6	118.3
Dextrin	194.0	150.0	195.7	118.3
Corn oil	210.0	210.0	210.0	210.0
Ausman-Hayes salt mix	40.0	40.0	40.0	40.0
Vitamin mix I	4.0	4.0	4.0	4.0
Choline	2.4	2.4	2.4	2.4
Inositol	0.8	0.8	0.8	0.8
Non-nutritive fiber	200.0	200.0	200.0	200.0
Trypsin inhibitor (g/kg)	0.099	0.655	0.104	2.42
Protein (% kcal)	12.0	20.0	12.5	20.0
(% w/w)	12.2	20.4	12.7	20.4

[a]Spray dried lactalbumin, Teklad Inc., Madison, WI 57313, contained 0.64 mg Tl/g.

[b]Supro 710, Ralston Purina, St. Louis, MO 63188, contained 2.70 mg Tl/g.

[c]Fine-ground, Bioserve Inc., Frenchtown, NJ 08825, contained 0.69 mg Tl/g.

[d]Promosoy 100, Central Soya, Gibson City, IL 60936, contained 7.90 mg Tl/g.

to recover from the anaesthesia. The monkey was placed in its home cage, fed when fully alert and mobile, and carefully observed for the next few days to detect any problems associated with the surgery. Each monkey was treated with penicillin and streptomycin (Penstrep) at 10 mg/kg for 5 days.

Both head and tail pancreatic biopsies were weighed and divided into three pieces as follows: (1) a small, 10 mg sample was placed in glutaraldehyde for reference electron microscopy if required; (2) a section of about 80 mg was placed in 10% buffered formalin before imbedding in paraffin for sectioning and staining with hematoxylin and eosin for histological examination, and (3) the remaining tissue was immediately frozen in liquid nitrogen and stored at -70C for biochemical analysis. DNA and RNA were analyzed by the method of Hatcher and Goldstein (1969), protein by the method of Hartree (1972), trypsin by the method of Erlanger et al.(1961), and chymotrypsin by the method of Bundy (1963). Samples for analysis of trypsin and chymotrypsin were first activated with enterokinase (Nitsan and Liener, 1976). For histology, two sections were prepared from each biopsy. One set of slides was evaluated at the New England Regional Primate Research Center and a second set was sent to an independent pathologist for examination without knowledge of the dietary group to which the sample belonged.

Hematology

Blood samples were drawn before and two weeks after the surgical procedure and sent to a commercial laboratory for a standard serum chemistry profile. These analyses served to provide a comparison of the serum chemistries of the various dietary groups to determine whether the different proteins had any detectable effect on the basic parameters of the blood. In addition, it provided an index of the health status of the monkeys following recovery from the surgical procedure.

Statistical Analysis

Standard procedures were used for comparison of paired data and analysis of variance (Snedecor and Cochran, 1980). When the group main effect was significant ($p < 0.05$), differences between individual means were tested using a least significant difference procedure. An IBM 4341 computer and SAS statistical package were used for all computations.

RESULTS

Pancreatic histology and biochemistry

As indicated in the Methods, two independent histological analyses were performed on the pancreatic biopsies without knowledge of the dietary groups to which the sample belonged. In addition, as detailed in the Methods, two biopsies were taken from each animal in order to provide as large a representative sample of the total pancreatic tissue without endangering the animal. The results from both pathological examinations were similar and neither found evidence of pancreatic hypertrophy, nodules or hyperplasia in any of the animals. This is markedly different from the observations in rats where short-term feeding of trypsin inhibitor-containing diets causes hypertrophy, while long-term feeding causes nodules and adenoma (Gumbmann et al., 1984). In 26 of the 27 monkeys, the pancreatic tissue appeared normal in terms of acinar size, morphology and distribution. In 7 of the monkeys, distributed among the various diet groups, there was minimal acinar vacuolation. One monkey of two in the soy concentrate group exhibited diffuse interstitial fibrosis of the pancreas with mild to moderate atrophy of the acinar tissue.

The second index of altered pancreatic function were the biochemical parameters which can be used to detect functional change in the pancreas which might not be detectable by histological procedures. DNA and RNA measurements have been used as evidence of hyperplasia or hypertrophy and the data on the pancreatic biopsies is shown in Table II. There was essentially no biochemical evidence of either hyperplasia or hypertrophy in the Cebus pancreas as judged by the level of RNA and DNA and the RNA/DNA ratio, thus confirming the absence of pathology seen in the histological examination.

Trypsin, chymotrypsin and protein levels in pancreatic tissue have also been used as indicators of change in pancreatic function (Hasdai and Liener, 1983). The results of these measurements in the monkey biopsy samples are also shown in Table II. A one-way analysis of variance performed separately on the head and tail section and Tukey's "t" test to compare means indicated no significant differences in the biochemical measurements between the lactalbumin, soy isolate and casein groups. However the head region of the pancreas showed higher trypsin and chymotrypsin activities than the tail region in paired "t" test analysis. In the soy concentrate group, there were differences due largely to the fibrotic changes seen in one of the two animals on histological examination. These changes correlated with the morphological observations, and although adverse, they were not characteristic of the hyperplastic and nodular changes seen in rodents, as discussed earlier.

<u>Body weight parameters</u>

Table III shows the body weights of the monkeys in the various groups which were between 1.5 and 2.0 kg at the time the biopsy was performed. As stated above the animals were adopted from a previous study and since the groups were not balanced with regard to age or sex ratios direct comparisons of the weights of the monkeys in the different diet groups could not be made. It was, however, possible to compare the weight for age of each monkey with reference to data obtained on 85 monkeys over an 8 year period (Ausman et al., 1982). The body weight was compared to the expected weight for age and sex, and the deviation expressed in standard deviation units as a "z" score. The low "z" scores for the group fed lactalbumin, casein and soy isolate indicated that the weights were normal. In the soy concentrate group, one of the two animals was more than 3 standard deviations below the mean weight for age.

<u>Serum chemistry and hematological analysis</u>

Results of the serum chemical and hematological analyses for the blood sample taken before the pancreatic biopsy are presented in Table IV. Seven serum samples were recorded as hemolyzed by the laboratory; thus, values for serum K^+, LDH, and SGOT, all known to increase with hemolysis, were eliminated from these sample results before statistical analysis. The increase in circulating BUN concentration with decreasing protein quality was anticipated. In general, there were remarkably few differences between dietary groups after long-term feeding.

TABLE II. Protein, RNA, DNA, Trypsin, and Chymotrypsin Content in Both Head and Tail Pancreatic Biopsy Sections on Cebus Monkeys Fed Four Types of Diets[a]

Parameter n	Site[b]	L - 12 (8)	SI - 20 (10)	C - 13 (7)	SC - 20 (2)
Protein	H	200 ± 13#	185 ± 20	181 ± 16	139, 155
(mg/g tissue)	T	178 ± 12	177 ± 18	177 ± 14	161, 177
RNA	H	20.2 ± 2.4	18.7 ± 2.0#	17.7 ± 1.3	12.7, 15.4
(mg/g tissue)	T	20.0 ± 3.5	21.2 ± 2.0	17.8 ± 1.8	15.8, 19.2
DNA	H	8.88 ± 0.91	7.39 ± 1.48	7.99 ± 0.46	7.25, 5.75
(mg/g tissue)	T	7.18 ± 2.48	7.17 ± 0.73	7.46 ± 1.34	8.37, 5.32
Protein	H	9.99 ± 1.24	9.94 ± 1.15#	10.28 ± 0.98	10.9, 10.1
(mg/mg RNA)	T	9.11 ± 1.37	8.43 ± 1.24	10.03 ± 1.38	10.2, 9.2
Protein	H	22.70 ± 2.43	25.85 ± 5.23	22.77 ± 2.34	19.1, 27.0
(mg/mg DNA)	T	26.57 ± 6.22	24.96 ± 3.81	24.30 ± 3.77	19.2, 33.3
RNA/DNA	H	2.31 ± 0.45#	2.65 ± 0.78	2.21 ± 0.11	1.75, 2.68
	T	3.01 ± 0.82	2.98 ± 0.36	2.47 ± 0.57	1.89, 3.61
Trypsin activity	H	39.2 ± 6.5#	39.8 ± 3.4#	40.1 ± 4.3#	30.2, 41.8
(units/g tissue x 10^{-6})	T	48.3 ± 5.6	51.5 ± 5.7	50.4 ± 3.7	32.7, 52.1
Chymotrypsin activity	H	2.48 ± 0.75#	2.47 ± 0.64#	2.80 ± 0.82#	1.23, 3.77
(units/g tissue x 10^{-6})	T	2.98 ± 0.98	3.16 ± 0.67	3.60 ± 0.76	1.54, 4.64

[a]Values are means ± s.d., except the SC-20 group where the individual values are presented. No significant differences due to diet were observed using a one-way analysis of variance and Tukey's "t" test. Paired t-tests were used to compare results of the head and tail analysis and results significantly different at $p < 0.05$ were designated with a '#'.
[b]H = head of pancreas; T = tail of pancreas.

TABLE III.

Age and Weight Characteristics of Monkeys Fed Four Experimental Diets

Dietary Protein	n	Age (weeks)	Mean Body Weight (g)	Z-Score[a]
Lactalbumin (L)	5/3[b]	188 ± 37	1832/1709[c]	-0.33 ± 0.91
Isolated Soy Protein (SI)	5/5	176 ± 28	1946/1589	-0.003 ± 0.70
Casein (C)	3/4	168 ± 22	1893/1558	-0.08 ± 1.00
Soy Concentrate (SC)	1/1	159/130	1310/1506	-3.01/-0.13

[a]Mean deviation ± sd of individual body weights from a standard population as measured in standard deviation units.

[b]Numbers of males/females, respectively.

[c]Values are mean weights for males/females, respectively, except for the soy concentrate group where individual values are presented.

TABLE IV.

Serum Chemistry and Hematology Values of Cebus Monkeys (Prior to Pancreatic Biopsy) Maintained on Three Different Diets

	L - 12 (8)	SI - 20 (10)	C - 13 (7)
Iron (ug/dl)	116 ± 59	138 ± 62	90 ± 48
Na (meq/l)	143 ± 3.3	142 ± 1.9	143 ± 1.5
K (meq/l)	3.8 ± 0.2	3.6 ± 0.3	3.8 ± 0.4
Ca (mg/dl	9.0 ± 0.9	8.8 ± 1.4	8.3 ± 1.3
Inor. phosphate (mg/dl)	6.3 ± 1.2	6.8 ± 0.6	7.9 ± 2.4
Cl (meq/l)	111 ± 4.7	112 ± 4.8	106 ± 10.9
LDH (mU/ml)	205 ± 120	309 ± 92	322 ± 140
SGOT (mU/ml)	57 ± 21	50 ± 21	48 ± 15
SGPT (mU/ml)	83 ± 40	62 ± 36	61 ± 39
Alk. Phos. (mU/ml)	645 ± 126	560 ± 141	509 ± 129
Amylase (IU/l)	270 ± 73	265 ± 109	278 ± 153
Lipase (U/L)	82 ± 62	98 ± 86	190 ± 326
Cholesterol (mg/dl)	200 ± 33	161 ± 26	173 ± 47
Protein (g/dl)	6.2 ± 0.4	6.3 ± 0.3	6.5 ± 0.3
Albumin (g/dl)	3.9 ± 0.3	4.0 ± 0.3	3.9 ± 0.2
Globulin (g/dl)	2.3 ± 0.3	2.3 ± 0.3	2.6 ± 0.3
Total bilirubin (mg/dl)	0.11 ± 0.6	0.14 ± 0.07	0.16 ± 0.05
Creatinine (mg/dl)	0.65 ± 0.09	0.79 ± 0.25	0.89 ± 0.25
BUN (mg/dl)	16.8 ± 4.5	28.9 ± 7.2*	21.4 ± 5.9
Glucose (mg/dl)	104 ± 29	101 ± 30	111 ± 36
Hct (%)	45.8 ± 3.4	44.4 ± 6.9	47.0 ± 5.6
Hgb (g/dl)	14.6 ± 0.9	14.1 ± 2.2	15.0 ± 1.0
RBC (x $10^6/mm^3$)	5.7 ± 0.4	5.5 ± 0.7	5.9 ± 0.6
WBC (x $10^3/mm^3$)	4.2 ± 2.0	5.6 ± 2.4	5.8 ± 2.1
Neutrophils/mm^3	971 ± 471	1880 ± 1170	2363 ± 1736
Lymphocytes/mm^3	2934 ± 1780	3309 ± 1662	3026 ± 1263
Monocytes/mm^3	101 ± 67	126 ± 125	197 ± 91
Eosinophils/mm^3	191 ± 184	227 ± 245	90 ± 52
Basophils/mm^3	25 ± 38	34 ± 48	63 ± 90

a. Diets L-12 (n = 8), SI-20 (n = 10) and C-13 (n = 7) are detailed in Table I. Data from the two animals fed the soy concentrate diet are not presented. Values represent the mean ± SD, and the blood urea nitrogen (BUN) of the SI-20 group compared to the L-12 group was the only parameter to show a significant difference by one-way ANOVA and Tukey's "t" test at $p < 0.05$.

In most cases, serum chemistry and hematology values after pancreatic biopsy were similar to those before surgery, which suggests minimal short-term effects of the surgery. Analysis of paired data for each monkey indicated significant, although minimal, changes in 10 of the 29 indices; these are shown in Table V. The changes in the biochemical indices were small and of unknown physiological significance. An increase in total white count from a basal value of 5000 up to an average of 6600/mm³ could be expected. The mean eosinophil number increased four-fold from 175 up to 731/mm³.

DISCUSSION

A colony of 27 Cebus monkeys has been used to evaluate a key question relating to whether primates show adverse effects on ingestion of trypsin inhibitor-containing diets. The results obtained suggest that these primates might be resistant to the adverse effects of soy diets as judged by general physical health, blood chemistry and hematology, and biochemical and histopathological studies of pancreatic biopsies. These results were obtained by comparing a group of ten Cebus monkeys fed a soy isolate diet with two control groups fed casein and lactalbumin, respectively, as the protein source. A group of two monkeys, included in this report for completeness but providing no statistically significant information, were fed a higher level of trypsin inhibitor in a soy concentrate diet. While one monkey was normal, the other showed poor growth and pancreatic atrophy which was, however, not comparable to the hypertrophic and hyperplastic changes indicative of trypsin inhibitor effects in rodents. These findings of a lack of effect of soy on the primate pancreas are supported by the short-term studies of Struthers et al.(1983). In 28 day feeding studies in the rhesus monkey (Macaca mulatta), diets containing 40% of raw or heated soy flour produced no pancreatic enlargement, no growth retardation and no change in pancreatic protein, DNA or RNA. These findings are important in assessing the potential human health hazard in view of the frequently observed effects in rodents and other species where ingestion of trypsin inhibitor-containing diets frequently causes both acute and chronic changes.

Acute studies of the effects of trypsin inhibitor in soy diets have shown a number of well-recognized antinutritional effects, including growth retardation, reduced protein digestibility, increased requirements for sulfur-containing amino acids, and

TABLE V. Serum Chemistry and Hematology Values Showing a Significant Change Between the First and Second Blood Sampling[a]

Parameter	n	Post Surgery[a]	Pre Surgery[b]	Difference[c]
Albumin (g/dl)	26	4.1 ± 0.3	3.9 ± 0.3	0.19 ± 0.33
Protein (g/dl)	26	6.7 ± 0.4	6.3 ± 0.3	0.35 ± 0.53
Cholesterol (mg/dl)	26	181 ± 35	171 ± 36	10.5 ± 22
Sodium (meq/l)	26	145 ± 2.3	142.7 ± 2.2	2.9 ± 2.8
Glucose (mg/dl)	26	86 ± 25	104 ± 30	-18 ± 32
SGOT (mu/ml)	16	23 ± 14	51 ± 18	-29 ± 20
BUN (mg/dl)	26	20.3 ± 6.1	23.9 ± 8.9	-3.7 ± 5.8
WBC ($x10^3/mm^3$)	26	6.6 ± 2.3	5.0 ± 2.3	1.54 ± 1.92
Neutrophils ($x10^3/mm^3$)	26	2.88 ± 1.52	1.72 ± 1.27	1.16 ± 1.67
Eosinophils (per mm^3)	26	731 ± 847	175 ± 190	556 ± 851

a. Mean value of all monkeys, without regard to dietary group, 2 weeks postsurgery.

b. Mean value of all monkeys, without regard to dietary group, just before surgery.

c. A paired t-test showed that the mean change was significant at $p < 0.05$.

pancreatic hypertrophy and/or hyperplasia (Rackis and Gumbmann, 1981). A recent report in hamsters shows that at a level of 842 mg of trypsin inhibitor per 100 g of diet, growth was markedly reduced over a 28 day period with a substantial increase in pancreatic weight when expressed relative to body weight (Hasdai and Liener, 1983). Although 80 to 90% of the trypsin inhibitor activity is destroyed in commercial soy preparations, a question arose concerning the safety of long-term feeding of even low levels of trypsin inhibitor. Although Wormsley's studies (Levison et al., 1979) were initially focused on the potentiating effect of trypsin inhibitors on the action of pancreatic nitrosamine carcinogens, it was later shown that long-term feeding of soy protein alone caused pancreatic adenomas (McGuinness et al., 1980). Extensive studies by the USDA (Gumbmann et al., 1984) have confirmed these findings in which increasing doses of trypsin inhibitor, prepared by blending raw and heated soy flour, were fed to rats for up to two years. Hyperplastic nodules were found in the pancreas after 15 months, the incidence of which correlated with the dose of trypsin inhibitor given in the diet. In other parts of the study, acinar adenoma was also observed after long-term feeding. More recent studies by the USDA have confirmed that the adverse effects, including the adenomas, were due to the content of trypsin inhibitor rather than other factors in the soy. When the trypsin inhibitor level was reduced further in the diets, adverse pancreatic effects were still observed in the rodents, suggesting that there may not be a no-effect level for trypsin inhibitor action in the rodent pancreas (Gumbmann et al., 1984).

As indicated previously, soy preparations for human consumption are heat-treated to destroy 80 to 90% of the trypsin inhibitor activity. Further heating to destroy all the trypsin inhibitor activity results in loss of protein quality. Residual levels of trypsin inhibitor are therefore present in soy products and have been measured in a number of soy-based infant formula preparations (Churella et al., 1976, and FDA surveillance analysis). One liter of infant formula prepared for consumption contains about 61 mg of trypsin inhibitor activity (range 40 to 82) when assayed by the Hamerstrand procedure (Hamerstrand et al., 1981). Thus, assuming a daily intake of 850 ml for a 7 kg infant at 3 months of age, the daily exposure is about 9 mg/kg/day. The corresponding calculation for the Cebus monkeys on the soy isolate and soy concentrate diets, respectively, indicates a dose of 24 and 80 mg/kg/day, assuming a body weight of 2 kg and a daily intake of 150 g of the agar cake. In the USDA rat studies (Gumbmann et al., 1984) in which the dietary levels of trypsin inhibitor ranged from 93 to 577 mg/100 g of diet, assuming a daily intake of 20 g of the diet for a 333 g rat, the daily exposure ranged from 60 to 360 mg/kg/day. Hence, the studies reported here in the monkeys, in which the animals were given dose levels higher than human infant exposure together with the prolonged time of exposure from weaning up to 3 to 4 years of age, a long period relative to the estimated 20 year lifespan of this species of monkey, provide further evidence that primates may not be susceptible to the adverse effects observed in rats. The reasons for these species differences are not understood, but may relate to the gastric acidity and proteolytic action to which the trypsin inhibitor proteins are exposed in the gastrointestinal tract of the various species.

It is important to understand the mechanistic basis for the adverse effects in the rodent both as a pathological entity and to provide further information with respect to possible markers for potential changes in humans. The basic observations relating to the mechanism of action of trypsin inhibitor were provided by Green and Lyman (1972), who showed that when trypsin action in the duodenum was inhibited by the presence of trypsin inhibitor, there was an increase in the outflow of pancreatic juice. The presence of trypsin in the duodenum was postulated to inhibit the mechanism for increased pancreatic output, which is mediated by the hormone cholecystokinin (CCK). As a result, measurement of circulating CCK levels has been attempted, but since the antibodies which recognize the terminal sulfated region of the CCK's also recognize the gastrins, and since there are much higher concentrations of the latter in the circulation, it has not been possible to obtain a reliable estimate of the CCK levels. One determination has been performed by using two antibodies with differing specificities and obtaining the CCK levels by difference (Adrian et al., 1982). In their studies, rats fed for 21 days with a diet containing raw soy bean flour showed a three fold increase in circulating cholecystokinin levels. Another approach has been to use a bioassay based on the secretion of amylase from isolated cells of parotid glands. Using this technique, further evaluation of the putative mechanism needs to be undertaken to determine whether the negative feedback control is operative in monkeys. Furthermore, it should be possible to clarify the role of increased levels of cholecystokinin in promoting pancreatic hypertrophy and adenoma formation in the rat.

REFERENCES

Adrian, T.E., Pasquali, C., Pescota, F., Bacarese-Hamilton, A.J., and Bloom, S.R. (1982). Soya induced pancreatic hypertrophy and rise of circulating cholecystokinin. Gut. 23, A889.

Ausman, L.M., Gallina, D.L., Hayes, K.C. and Hegsted, D.M. (1985). Comparative assessment of soy and milk protein quality in infant cebus monkeys. Am. J. Clin. Nut., in press.

Ausman, L.M., Powell, E.M., Mercado, D.L., Samonds, K.W., el Lozy, M., and Gallina, D.L. (1982). Growth and developmental composition of the Cebus monkey (Cebus albiporis). Am. J. Primatol., 3, 211-227.

Bundy, H. (1963). Chymotrypsin-catalyzed hydrolysis of N-acetyl- and N-benzoyl-L-tyrosine p-nitroanilides. Arch. Biochem. Biophys. 102, 416-422.

Churella, H.R., Yao, B.C., and Thomson, W.A.B. (1976). Soybean trypsin inhibitor activity of soy infant formulas and its nutritional significance for the rat. J. Agric. Food Chem., 24, 393-397.

Erlanger, B.F., Kokowsky, N., and Cohen, W. (1961). The preparation and properties of two new chromogenic substrates of trypsin. Arch. Biochem. Biophys. 95, 271-278.

Green, G.M., and Lyman, R.L. (1972). Feedback regulation of pancreatic enzyme secretion as a mechanism for trypsin inhibitor-induced hypersecretion in rats. Proc. Soc. Exp. Biol. Med., 140, 6-12.

Gumbmann, M.R., Rackis, J.J., Liener, I.E., and Spangler, W.L. (1984). The USDA trypsin inhibitor study: III. Pancreatic effects of dietary trypsin inhibitor in rats for two years. Fed. Proc., 43, 792.

Hamerstrand, G.E., Black, L.T., and Glover, J.D. (1981). Trypsin inhibitors in soy products: modification of the standard analytical procedure. Cereal Chem., 58, 42-65.

Hartree, E.F. (1972). Determination of protein: a modification of the Lowry method that gives a linear photometric response. Anal. Biochem., 48, 422-427.

Hasdai, A. and Liener, I.E. (1983). Growth, digestibility and enzymatic activites in the pancreas and intestines of hamsters fed raw and heated soy flour. J. Nutrition, 113, 662-668.

Hatcher, D.W., and Goldstein, G. (1969). Improved methods for determination of RNA and DNA. Anal. Biochem., 31, 42-47.

Levison, D.A., Morgan, R.G.H., Brimacombe, J.S., Hopwood, D., Coghill, G., and Wormsley, K.G. (1979). Carcinogenic effects of di (2-hydroxypropyl) nitrosamine (DHPN) in male Wistar rats: promotion of pancreatic cancer by a raw soya flour diet. Scand. J. Gastroenterol., 14, 217-224.

Liener, I.E. and Kakade, M.L. (1980). Protease inhibitors. In "Toxic Constituents of Plant Foodstuffs", I.E. Liener, Ed., 2nd edition, Academic Press, New York, pp 7-71.

McGuinness, E.E., Morgan, R.G.H., Levison, D.A., Frape, D.L., Hopwood, D. and Wormsley, K.G. (1980). The effects of long-term feeding of soya flour on the rat pancreas. Scand. J. Gastroenterol., 15, 497-502.

Nitsan, Z. and Liener, E.D. (1976). Enzymic activities in the pancreas, digestive tract and feces of rats fed raw or heated soy flour. J. Nutr., 106, 300-305.

Rackis, J.J. (1974). Biological and physiological factors in soy beans. J. Am. Oil Chem. Soc., 51, 161A-174A.

Rackis, J.J. and Gumbmann, M.R. (1981). Protease inhibitors: physiological properties and nutritional significance. In "Antinutrients and Natural Toxicants in Foods", R.L. Ory, Ed., Food and Nutrition Press, Westport, Connecticut, pp 203-237.

Sale, J.K., Goldberg, D.M., Fawcett, A.N. and Wormsley, K.G. (1977). Chronic and acute studies indicating absence of exocrine pancreatic feedback inhibition in dogs. Digestion, 15, 540-555.

Snedecor, G.W. and Cochran, W.G., Eds. (1980). In "Statistical Methods", Seventh Edition, Iowa State University Press, Ames, Iowa, p. 41-43.

Struthers, B.J., MacDonald, J.R., Dahlgren, R.R. and Hopkins, D.T. (1983). Effects on the monkey, pig and rat pancreas of soy products with varying levels of trypsin inhibitor and comparison with the administration of cholecystokinin. J. Nutr., 113, 86-97.

16

INHIBITION OF HUMAN PROTEINASES BY GRAIN LEGUMES

Juergen K.P. Weder

Institut fuer Lebensmittelchemie
Technische Universitaet Muenchen
Lichtenbergstrasse 4, D-8046 Garching, FRG

INTRODUCTION

The inhibition of proteolytic enzymes by extracts from animal tissues was already demonstrated at the end of the last century (Fredericq, 1878). But although detailed studies have been made in this field, sixty years passed until the first indication of the inhibition of a proteinase by a plant extract (aqueous extracts from soy flour inhibited the liquefaction of gelatin by trypsin) was published (Read and Haas, 1938). This was soon followed by the first isolation of a plant proteinase inhibitor from soybeans by Kunitz (1945, 1946). A year later the first systematic study of plant proteinase inhibitors was made by Borchers and Ackerson (1947). Because of the great and increasing importance of grain legumes as food and feed, research during the next period was mainly concentrated on seeds from the legume family.

Until then, all the work on legume inhibitors was done with animal enzymes, mainly bovine, which were commercially available. In 1969, Feeney et al. were the first to systematically study the inhibition of human trypsin by various proteinase inhibitors, among them some inhibitor preparations from legumes. Later on, their results on legume inhibitors, as well as the results of others who used these inhibitors to characterize the different human pancreatic proteinases (Feinstein et al., 1974; Mallory and Travis, 1975a; Figarella et al., 1975), were often interpreted as legume proteinase inhibitors probably being of minor nutritional significance for human beings (e.g., Liener, 1979; Wilson, 1981; Flavin, 1982).

In the first part of this publication, the results on the inhibition of human trypsin, human chymotrypsin, and other human intestinal proteinases are compiled from the literature and are discussed with regards to their influence on the digestion of proteins in the human gastrointestinal tract.

The second part of this paper deals with studies on lentil proteinase inhibitors done in our laboratory. Two inhibitors from lentils, the extracts of which inhibit human chymotrypsin more effectively than the bovine enzyme and human trypsin somewhat less than its bovine counterpart, were isolated and characterized. In order to clarify the reasons for the differences in action against human and bovine proteinases, we studied the complex formation of the two purified lentil inhibitors with the various enzymes, and the influence of chemical modification of amino acid residues at the reactive sites of these inhibitors on this complex formation.

In the third part, two parameters which influence the activity of these inhibitors are discussed based on results from the literature and from our own work on lentil inhibitors. These are: a) the influence of pepsin during the passage through the stomach; and b) the effect of processing.

PART I: INHIBITION OF HUMAN TRYPSIN

The inhibition of human pancreatic trypsin by naturally occurring inhibitors was first studied by Buck et al. in 1962. They found the human trypsin to be inhibited by a bovine serum inhibitor and not to be inhibited even by great excess of chicken ovomucoid. The study, sometimes cited as the first demonstration of the inhibition of human trypsin by a soybean trypsin inhibitor (Siegelman et al., 1962), deals with the investigation of a trypsin-like activity in human blood serum. In the discussion the authors mention that the serum trypsin activity was completely inhibited by the soybean trypsin inhibitor without giving any experimental details. In reality then, it was Feeney et al. who in 1969 were the first to study the influence of legume inhibitors on the action of human pancreatic trypsin.

The available data over the past twenty years on the inhibition of human trypsin by inhibitor preparations and extracts from soybeans and lima beans are compiled in Tables 1 and 2. These two legumes have been the most extensively studied in this respect, because they are important staple foods and feeds and, also because their inhibitors were isolated and characterized very early and two of them are commercially available. These data shall be used to discuss the complexity of the inhibitor action and to show why controversies have arosen on their nutritional significance in the past.

For an evaluation of the results various aspects have to be considered. First, human trypsin is not a homogeneous compound but consists of two main components, a cationic and an anionic trypsin (Mallory and Travis, 1975a). While some authors have studied the effect on whole trypsin or tryptic activity using pancreatic juice, others have investigated the influence on the separate proteinases. The human trypsins I and II isolated by Feinstein et al. (1974) and the human trypsins 2 and 1 isolated by Figarella et al. (1975) are thought to correspond to the human anionic and the cationic trypsin, respectively, due to inhibition studies (Mallory and Travis, 1975a; Figarella et al., 1975). The preparations, though, differ in some respects; furthermore, for an overall evaluation, the relative amounts of the two components have to be considered. These vary considerably ranging from 89 % of cationic and 11 % of anionic trypsin (Mallory and Travis, 1975a) to 65 % of trypsin 1 and 35 % of trypsin 2 (Figarella et al., 1975) depending on the source they were isolated from.

Second, the type of substrate used to measure tryptic activity is of great importance. The substrates used can be roughly classified into synthetic amino acid esters, synthetic amino acid amides, and proteins. Of course, proteins are suited best to simulate the *in vivo* digestion.

Third, the experimental conditions, especially the inhibitor/enzyme ratio, will influence the results and should be taken into consideration when comparing values from different laboratories. All information needed to discuss these aspects are included in Table 1.

Soybean Inhibitors

Soybeans (*Glycine max* (L.) Merr.) contain two main proteinase inhibitors, the soybean trypsin inhibitor (Kunitz), STI, exhibiting strong inhibition of bovine trypsin and only weak inhibition of bovine chymotrypsin at the same reactive site, and the Bowman-Birk soybean trypsin inhibitor, BBI, displaying inhibition of bovine trypsin and chymotrypsin at two independent reactive sites (cf. review of Liener and Kakade, 1980, for further information).

STI. To begin with, the following discussion considers only the experiments in Table 1 in which a 1:1 inhibitor/enzyme ratio was used. Mallory and Travis (1975a) have clearly shown that STI considerably inhibits human cationic as well as anionic trypsin when casein is used as a substrate. The human trypsins were somewhat less inhibited by STI than bovine trypsin (HTI/BTI 0.82 and 0.92 for human cationic and anionic trypsin, respectively; see Table 1). A similar but somewhat lower effect had already been observed by Feeney et al. (1969) who most probably used a mixture of the two trypsins. These results, however, contrast with those of Temler et al. (1983), who compared the effect of STI and also of BBI on human cationic trypsin with that on rat trypsin. This is probably due to the substrate used, azocasein, which because of derivatization differs from the orginal casein.

As already mentioned, the substrate used plays a decisive role in the results obtained. For example, STI resulted in a HTI/BTI ratio of about 0.9 (Table 1), when an amino acid amide (BAPA; Belitz et al., 1982; Struthers and MacDonald, 1983) and a protein (casein) were used as substrates. Gertler et al. (1982) also detected a high degree of inhibition of human pancreatic juice by STI with BAPA.

In contrast, the use of amino acid esters (BAEE and TAME) resulted in a much lower HTI/BTI ratio, as can be concluded from 0.3-0.5 for the cationic and 0.9-1.0 for the anionic trypsin (Feinstein et al., 1974; Mallory and Travis, 1975a; Figarella et al., 1975). This is also the reason for the low degree of inhibition by STI observed by Temler et al. (1983) with BAPA and by Laskowski jr. et al. (1974) using methyl umbelliferyl p-guanidinobenzoate.

The results thus depend on the chemical structure of the substrate: on the one hand, the amino acid amides and casein, having similar structures, are similarly attacked; on the other, the amino acid esters, having a different type of linkage, give different results. Therefore, care should be taken when using esters as substrates in inhibitor studies. In particular, the effect on the predominant type of human trypsin, the cationic form, is inadequately detected.

Logically, the trypsin preparation used determines the substrate needed. When the purified trypsins or a mixture of the two trypsins are tested, casein can be used as shown above for STI. When human pancreatic juice is taken as enzyme source, synthetic substrates specific for trypsin must be used to discriminate between trypsin and the other proteinases present in the juice. Human pancreatic juice has been employed as a model for the digestive process in the small intestine by some authors, approximating more closely the situation *in vivo* than with purified enzymes and being more meaningful nutritionally.

Further complications may arise from the inhibitor/enzyme ratio used in the experiment. The initial slope of the inhibition curves (cf. chapter on lentil inhibitors for an example) reflects the stoichiometry of the enzyme inhibitor complex. At this stage, the enzyme is present in great excess and the inhibitor is capable of producing maximum effect.

This can be seen from the data of Figarella et al. (1975) on the inhibition of human cationic trypsin by STI, having TAME as substrate. Evaluation of the slope would lead to the conclusion that the cationic human trypsin is inhibited by STI 80 % as well as its bovine counterpart. As outlined above, 40 % will be calculated at 1:1 molar inhibitor/enzyme ratio with this substrate.

Table 1. Inhibition of Human Trypsin by Soybeans

Inhibitor sample	Substrate[a]; inhibitor/enzyme ratio	Trypsin type[b]	Inhibitor activity	HTI/BTI[c]	Ref.[d]
STI[e]		T	+[f]	ca. 0.2	(1)
	casein; 2:1 w/w	T	68 %		(1)
STI	BAEE; 1:1 mole/mole	c	30 %	0.31	(2,3)
	BAEE; 5:1 mole/mole	c	61 %		(2)
	casein; 1:1 mole/mole	c	80 %*	0.82	(2,4)
	casein; 5:1 mole/mole	c	94 %*		(2)
	BAEE; 1:1 mole/mole	a	95 %	0.97	(2,3)
	BAEE; 5:1 mole/mole	a	95 %		(2)
	casein; 1:1 mole/mole	a	89 %*	0.92	(2,4)
	casein; 5:1 mole/mole	a	95 %*		(2)
STI	TAME	I	+[g]	ca. 1.0	(5)
	TAME	II	+-[g]	ca. 0.5	(5)
STI	MUGH; burst reaction	c	weak		(6)
Lys[63]-STI[h]	MUGH; burst reaction	c	medium		(6)
STI	TAME; slope	1	0.80 mole/mole*	0.80	(7)
	TAME; 1:1 mole/mole	1	40 %	0.40	(7)
	TAME; slope	2	0.96 mole/mole*	0.96	(7)
	TAME; 1:1 mole/mole	2	88 %*	0.88	(7)
STI	BAEE; 25x10^{-2}nmole:20 μl HPJ[i]	T	91 %*		(8)
	BAEE; 125x10^{-2}nmole:20 μl HPJ	T	95 %*		(8)
STI	BAPA; 0.1 mg:0.5 ml HPJ	T	72 %*		(9)
	BAPA; 1.0 mg:0.5 ml HPJ	T	100 %*		(9)
STI	BAPA; slope	T	0.96 mg/mg	0.92	(10)
STI	BAPA; 5 μg:30 TU[j]	T	43 %*	0.88	(11)
	BAPA; 30 μg:30 TU	T	93 %*	0.95	(11)
STI	BAPA; 1.25 μg:27 μl HPJ	T	35 %*	0.60	(12)
STI	BAEE; 1:1 mole/mole	c	45 %*		(13)
	BAEE; 2:1 mole/mole	c	70 %		(13)
	azocasein; 1:1 mole/mole	c	21 %*		(13)
	azocasein; 4:1 mole/mole	c	65 %*		(13)
BBI[k]		T	+[f]	ca. 1.0	(1)
BBI	BAEE; 1:1 mole/mole	c	41 %		(2)
	BAEE; 5:1 mole/mole	c	46 %		(2)
	casein; 1:1 mole/mole	c	25 %*		(2)
	casein; 5:1 mole/mole	c	33 %*		(2)
	BAEE; 1:1 mole/mole	a	76 %		(2)
	BAEE; 5:1 mole/mole	a	79 %		(2)
	casein; 1:1 mole/mole	a	59 %*		(2)
	casein; 5:1 mole/mole	a	75 %*		(2)
BBI	MUGH; burst reaction	c	strong		(14)
BBI	BAPA; 0.1 mg:0.5 ml HPJ	T	31 %*		(9)
	BAPA; 1.0 mg:0.5 ml HPJ	T	74 %*		(9)
BBI	BAEE; 2:1 mole/mole	c	30 %*		(13)
	BAEE; 4:1 mole/mole	c	47 %*		(13)
	azocasein; 2:1 mole/mole	c	15 %*		(13)
	azocasein; 4:1 mole/mole	c	23 %*		(13)
Extract	BAEE; 170 μg soy:20 μl HPJ	T	80 %*		(8)
	BAEE; 850 μg soy:20 μl HPJ	T	88 %*		(8)
Extract	BAEE; 134 μg soy:50 μl HPJ	T	30-54 %	0.64-1.29	(15)
Extract	BAPA; slope	T	30 mg/g	1.37	(10)

Table 1 (continued).

Inhibitor sample	Substrate; inhibitor/enzyme ratio	Trypsin type	Inhibitor activity	HTI/BTI	Ref.
Extract	BAPA; 0.5 mg RSF[l]:25 TU	T	91 %*		(11)
	BAPA; 1.0 mg RSF:25 TU	T	100 %*		(11)
	BAPA; 1.0 mg USPI[m]:25 TU	T	91 %*		(11)
	BAPA; 2.0 mg USPI:25 TU	T	100 %*		(11)
	BAPA; 1.0 mg SPI[n]:25 TU	T	34 %*		(11)
	BAPA; 5.0 mg SPI:25 TU	T	99 %*		(11)
Extract	BAPA; 6.7 µg soy:27 µl HPJ	T	15 %*	0.16	(12)
Extract	BAPA; slope	T	1.80 IU[o]/mg	0.67	(16)

* Estimated by the author from the inhibition data in the original publications.

a BAEE, N^{α}-benzoyl-L-arginine ethyl ester; BAPA, N^{α}-benzoyl-DL-arginine p-nitroanilide; MUGH, methyl umbelliferyl p-guanidinobenzoate; TAME, N^{α}-tosyl-L-arginine methyl ester.

b T, non specified trypsin; c and a, human cationic and anionic trypsin according to Mallory and Travis (1975a); I and II, human trypsin I and II according to Feinstein et al. (1974); 1 and 2, human trypsin 1 and 2 according to Figarella et al. (1975).

c HTI/BTI, ratio of the inhibition of human trypsin to the inhibition of bovine trypsin (only calculated where data from the same laboratory were available).

d (1) Feeney et al. (1969); (2) Mallory and Travis (1975a); (3) Mallory and Travis (1973); (4) Coan and Travis (1971); (5) Feinstein et al. (1974); (6) Laskowski jr. et al. (1974); (7) Figarella et al. (1975); (8) Krogdahl and Holm (1979); (9) Gertler et al. (1982); (10) Belitz et al. (1982); (11) Struthers and MacDonald (1983); (12) Krogdahl and Holm (1983); (13) Temler et al. (1983); (14) Wilson (1981); (15) Holm and Krogdahl (1982); (16) Prabhu et al. (1984).

e STI, soybean trypsin inhibitor (Kunitz), commercial preparation.

f +, ++, +++, and ++++, varying degrees of inhibition from weak to strong inhibition (Feeney et al., 1969).

g +, strong inhibition; +-, weak inhibition; -, very weak or no inhibition (Feinstein et al., 1974).

h Lys^{63}-STI, "enzymatic mutant" of STI where Arg^{63} is replaced by Lys^{63} (Sealock and Laskowski jr., 1969).

i HPJ, human pancreatic juice.

j TU, trypsin unit, not defined by the authors (Struthers and MacDonald, 1983).

k BBI, Bowman-Birk soybean trypsin inhibitor, commercial or laboratory preparation.

l RSF, raw soy flour.

m USPI, unheated soy protein isolate (Edi Pro A[R], Ralston Purina Co.).

n SPI, heated soy protein isolate (Supro 620[R], Ralston Purina Co.).

o IU, inhibitor unit, generally defined as the amount of inhibitor that suppressed one unit of enzymatic activity (the amount of trypsin that produced an absorbance of 1.0; Prabhu et al., 1984).

Of course, the influence of the inhibitor/enzyme ratio is only to be observed when weak inhibitors are studied or substrates simulating weak inhibitors are used. In case of strong inhibitors, this influence is unimportant or even eliminated as can be deduced from the data of STI on

Table 2. Inhibition of Human Trypsin by Lima Beans

Inhibitor sample	Substrate; inhibitor/enzyme ratio	Trypsin type	Inhibitor activity	HTI/BTI	Ref.[a]
LBI[b]		T	++++	ca. 1.0	(1)
	casein; 2:1 w/w	T	95 %		(1)
LBI	BAEE; 1:1 mole/mole	c	95 %	1.00	(2,3)
	BAEE; 5:1 mole/mole	c	100 %	1.02	(2,3)
	casein; 1:1 mole/mole	c	80 %*	1.07	(2,3)
	casein; 5:1 mole/mole	c	98 %*	1.00	(2,3)
	BAEE; 1:1 mole/mole	a	85 %	0.89	(2,3)
	BAEE; 5:1 mole/mole	a	97 %	0.99	(2,3)
	casein; 1:1 mole/mole	a	91 %*	1.21	(2,3)
	casein; 5:1 mole/mole	a	99 %*	1.01	(2,3)
LBI	TAME	I	+	ca. 1.0	(4)
	TAME	II	+	ca. 1.0	(4)
LBI	BAEE; $25x10^{-2}$nmole:20 µl HPJ	T	82 %*		(5)
	BAEE; $125x10^{-2}$nmole:20 µl HPJ	T	88 %*		(5)
LBI	MUGH; burst reaction	c	strong		(6)
LBI	BAPA; 0.1 mg:0.5 ml HPJ	T	99 %*		(7)
	BAPA; 1.0 mg:0.5 ml HPJ	T	100 %*		(7)
LBI	BAPA; slope	T	2.44 mg/mg	1.21	(8)
LBI	BAPA; 0.54 µg:27 µl HPJ	T	20 %*	0.51	(9)
Extract	BAPA; slope	T	14 mg/g	1.56	(8)

[a] (1) Feeney et al. (1969); (2) Mallory and Travis (1975a); (3) Coan and Travis (1971); (4) Feinstein et al. (1974); (5) Krogdahl and Holm (1979); (6) Wilson (1981); (7) Gertler et al. (1982); (8) Belitz et al. (1982); (9) Krogdahl and Holm (1983).

[b] LBI, lima bean trypsin inhibitor, commercial preparation.

All other abbreviations as in Table 1.

human anionic trypsin by the same authors: HTI/BTI was 0.96 and 0.88 at both inhibitor/enzyme ratios studied (Table 1). The best way to get reliable results of practical importance is to study enzymes and inhibitors at the same molar concentrations.

For weak inhibitors, higher inhibitor/enzyme ratios will be useful to detect even small degrees of inhibition. It must be said, however, that these values can not be compared with those obtained for strong inhibitors under the same experimental conditions. Increasing inhibitor amounts will then linearly rise the effect studying a weak inhibitor, while nearly maximum inhibition will already be reached with smaller inhibitor amounts in the case of a strong inhibitor.

BBI. The double-headed legume proteinase inhibitor, member of the Bowman-Birk family, is the second most important soybean inhibitor. Although less data are available for BBI, the influence of the various features discussed above seems also to be applicable here. BBI inhibited human anionic trypsin considerably and more effectively than the cationic form (Table 1). The use of synthetic ester substrates now led to higher percentages of inhibition than did casein (Mallory and Travis, 1975a). Again, azocasein gave lower values (Temler at al., 1983) and BAPA led to results close to those with casein (Gertler et al., 1982), though the stoichiometry of their experiments remained unclear. The results of Feeney at al. (1969) with BBI and the studies of others with soybean extracts (Belitz et al., 1982; Holm and Krogdahl, 1983) indicate that BBI inhibits the human trypsins to about the same extent as the bovine enzyme.

Extracts. The experiments of Struthers and MacDonald (1983) with extracts of raw soy flour, unheated and heated soy protein isolate showed that the soybean inhibitors are able to totally suppress the proteolytic activity of the human trypsins. Krogdahl and Holm (1983) found a remarkable low inhibition of human trypsin and a very low HTI/BTI ratio in their experiments.

Lima Bean Inhibitor

LBI. The trypsin inhibitor from lima beans (*Phaseolus lunatus* L.), LBI, also belongs to the Bowman-Birk inhibitor family, like does BBI (cf. review of Liener and Kakade, 1980). Therefore, it is not astonishing that LBI resembles BBI in its action on human and bovine trypsin, though exhibiting some differences. Feeney et al. (1969) have already demonstrated that LBI severely inhibits both human and bovine trypsin. Some years later, Mallory and Travis (1975a) showed human anionic trypsin being inhibited somewhat better than the cationic and both being equally well or even better inhibited than the bovine counterpart (Table 2; casein as substrate, 1:1 molar ratio).

For LBI, the results with synthetic ester substrates, BAEE and TAME (Mallory and Travis, 1975a; Feinstein et al., 1974; Krogdahl and Holm, 1979), do not differ very much from those obtained with casein, most probably because of the strong inhibition of all enzyme components. As discussed above for STI, the usefulness of BAPA when using human pancreatic juice (Gertler et al., 1982; Belitz et al., 1982) and the effect of the inhibitor/enzyme ratio (Mallory and Travis, 1975a; Krogdahl and Holm, 1979; Gertler et al., 1982) are confirmed, although the effects are now less pronounced. As for soybeans, Krogdahl and Holm (1983) reported lower values for human trypsin inhibition and a lower HTI/BTI ratio than others for LBI.

Other Legumes

In Table 3 the results on other legumes are compiled from the literature. Besides the paper of Feeney et al. (1969), these data come mainly from two independent studies of Belitz et al. (1982) and Prabhu et al. (1984) who have comparatively investigated a great number of legume species.

Out of the 25 species (cultivars included) investigated, 14 exerted about the same degree of inhibition on human and on bovine trypsin (Table 3). Out of these fourteen species, the guar inhibitor scarcely inhibited either the human or the bovine trypsin. Therefore, besides the HTI/BTI ratios, the absolute degree of inhibition must also be taken into consideration for a discussion of the nutritional significance.

A considerable higher inhibition of the human than of the bovine trypsin was demonstrated for a crude inhibitor preparation from bush beans (PVI-Roh), though extracts from the same bean showed lower values (Belitz et al., 1982). Because other seed constituents in the extracts may interfere with the inhibitor determination, results with purer preparations are taken as more precise for bush beans and others when available. Although not important nutritionally, high relative inhibition (HTI/BTI; Table 3) was found for the seeds of *Acacia arabica*, *Inga laurina* and *Adenanthera pavonina* which were all highly potent trypsin inhibitors. In contrast, lower HTI/BTI ratios were demonstrated for broad beans, field

Table 3. Inhibition of Human Tryptic, Chymotryptic and Proteolytic Activity by Other Legumes

Inhibitor sample	Trypsin			Chymotrypsin			Proteases			Inhibitor/ enzyme ratio	Ref.[a]
	HTIA[b]	HTI/BTI	Substrate	HCTIA[c]	HCTI/BCTI[d]	Substrate[e]	HPIA[f]	HPI/BPI[g]	Substrate		
Kidney bean (*Phaseolus vulgaris* L.)											
Comp.C[h]	+	ca. 0.3									(1)
	35 %		casein							2:1 w/w	(1)
Extract	0.25 IU[i]/mg	0.68	BAPA	0.30 IU/mg	0.60	ATEE	0.30 IU/mg	0.46	casein	slope	(2)
Navy bean (*Phaseolus vulgaris* L.)											
NBI[j]	++	ca. 0.7									(1)
	72 %		casein							2:1 w/w	(1)
Garden bean (*Phaseolus vulgaris* L.)											
GBI II[k]	weak(cat. trypsin)		MUGH							burst	(3)
Bush bean (*Phaseolus vulgaris* L. var. *nanus*)											
PVI-Roh[l]	0.21 mg/mg	1.45	BAPA	0.24 mg/mg	2.39	GPPA				slope	(4)
Extract	12 mg/g	0.59	BAPA	34 mg/g	3.44	GPPA				slope	(4)
Acacia seed (*Acacia arabica* Willd.)											
Extract	1.40 IU/mg	2.00	BAPA	0.50 IU/mg	1.32	ATEE	1.00 IU/mg	0.67	casein	slope	(2)
Black gram (*Vigna mungo* (L.) Hepper)											
VMI-C[m]	0.06 mg/mg	1.16	BAPA	1 µg/mg	0.48	GPPA				slope	(4)
Broad bean (*Vicia faba* L.)											
Extract	0.8 mg/g	0.25	BAPA	0.7 mg/g	0.58	GPPA				slope	(4)
Caesalpinia bonduc (L.) Roxb. seed											
Extract	5.9 mg/g	0.57	BAPA	23 mg/g	2.14	GPPA				slope	(4)
Carob bean (*Ceratonia siliqua* L.)											
Extract	12 mg/g	1.08	BAPA	46 mg/g	4.77	GPPA				slope	(4)

Inhibitor sample	Trypsin HTIA	HTI/BTI	Substrate	Chymotrypsin HCTIA	HCTI/BCTI	Substrate	Proteases HPIA	HPI/BPI	Substrate	Inhibitor/ enzyme ratio	Ref.
Chickpea (Cicer arietinum L.)											
CPTI[n]	99 %*		BAPA	79 %*		ATPA				0.1mg:0.5ml HPJ	(5)
	100 %*		BAPA	94 %*		ATPA				1.0mg:0.5ml HPJ	(5)
Extract	0.24 IU/mg	0.40	BAPA	0.09 IU/mg	0.50	ATEE	0.30 IU/mg	0.46	casein	slope	(2)
Extract	5,8 mg/g	0.79	BAPA	27 mg/g	4.88	GPPA				slope	(4)
Common pea (Pisum sativum L.)											
PSI-Roh[o]	0.05 mg/mg	0.81	BAPA	0.18 mg/mg	2.45	GPPA				slope	(4)
Extract	1.0 mg/g	0.50	BAPA	3.7 mg/g	1.54	GPPA				slope	(4)
Cowpea (Vigna unguiculata (L.) Walp. ssp. unguiculata, formerly V. sinensis (L.) Savi ex Hassk.)											
Comp.D[p]	++	ca. 1.0									(1)
	70 %		casein							2:1 w/w	(1)
Extract	0.18 IU/mg	0.60	BAPA	0.26 IU/mg	2.36	ATEE	0.30 IU/mg	1.50	casein	slope	(2)
Elephantorrhiza burkei Benth. seed											
Extract	35 mg/g	0.95	BAPA	54 mg/g	3.19	GPPA				slope	(4)
Field bean (Lablab purpureus (L.) Sweet, formerly Dolichos lablab L.)											
Extract	0.50 IU/mg	0.62	BAPA	0.64 IU/mg	0.84	ATEE	0.70 IU/mg	0.64	casein	slope	(2)
Grasspea (Lathyrus sativus L.)											
Fract.A[q]	12 %(cat.)	0.22	casein	52 %	1.44	casein				3.2µg≙slope	(6)
	22 %(cat.)	0.23	casein	95 %	1.06	casein				16.0 µg	(6)
Extract	14 mg/g	0.74	BAPA	22 mg/g	2.47	GPPA				slope	(4)
Guar (Cyamopsis tetragonoloba (L.) Taubert)											
Extract	0.3 mg/g	0.75	BAPA	n.d.[r]	n.d./n.d.	GPPA				slope	(4)

(continued)

Table 3 (continued).

Inhibitor sample	Trypsin HTIA	HTI/BTI	Substrate	Chymotrypsin HCTIA	HCTI/BCTI	Substrate	Proteases HPIA	HPI/BPI	Substrate	Inhibitor/ enzyme ratio	Ref.
Guinea pea (Abrus precatorius L.)											
Extract	0.55 IU/mg	1.10	BAPA	0.40 IU/mg	0.80	ATEE	0.70 IU/mg	0.54	casein	slope	(2)
Inga laurina (Sw.) Willd. seed											
Extract	33 mg/g	1.75	BAPA	14 mg/g	11.8	GPPA				slope	(4)
Mungbean (Vigna radiata (L). Wilczek)											
MBI[s]	medium(cat.)		BAEE								(3)
Lentil (Lens culinaris Medik.)[t]											
Extract	1.3 mg/g	0.68	BAPA	5.2 mg/g	4.73	GPPA				slope	(4)
Red gram (Cajanus cajan (L.) Millsp.)											
Extract	0.56 IU/mg	0.43	BAPA	0.20 IU/mg	2.00	ATEE	0.30 IU/mg	0.46	casein	slope	(2)
Red sandal wood seed[u](Adenanthera pavonina L.)											
Extract	5.00 IU/mg	1.82	BAPA	5.60 IU/mg	2.00	ATEE	6.00 IU/mg	1.20	casein	slope	(2)
Scarlet runner bean (Phaseolus coccineus L.)											
PCI-Roh[v]	0.30 mg/mg	1.27	BAPA	0.36 mg/mg	2.23	GPPA				slope	(4)
Extract	19 mg/g	0.94	BAPA	41 mg/g	4.10	GPPA				slope	(4)
Sword bean (Canavalia ensiformis L.)[w]											
Extract	trace		BAPA	trace		ATEE	nil		casein	200μg protein: 0.5 enzyme units	(2)
Winged bean (Psophocarpus tetragonolobus DC.)											
Extract	16 mg/g	1.28	BAPA	35 mg/g	1.60	GPPA				slope	(4)

Table 3 (continued).

[a] (1) Feeney et al. (1969); (2) Prabhu et al. (1984); (3) Wilson (1981); (4) Belitz et al. (1982); (5) Gertler et al. (1982); (6) Bhat et al. (1976).

[b] HTIA, human trypsin inhibitor activity.

[c] HCTIA, human chymotrypsin inhibitor activity.

[d] HCTI/BCTI, ratio of the inhibition of human chymotrypsin to the inhibition of bovine chymotrypsin (see Table 1, Footnote c).

[e] ATEE, N-acetyl-L-tyrosine ethyl ester; ATPA, N-acetyl-L-tyrosine p-nitroanilide; GPPA, glutaryl-L-phenylalanine p-nitroanilide.

[f] HPIA, inhibitor activity against human proteinases.

[g] HPI/BPI, ratio of the inhibition of human proteinases to the inhibition of bovine proteinases (see Footnote d).

[h] Kidney bean component C from Dr. J.P. Riehm (Feeney et al., 1969).

[i] IU, inhibitor unit, generally defined as the amount of inhibitor that suppressed one unit of enzymatic activity (the amount of trypsin, chymotrypsin or proteinases that produced an absorbance of 1.0, 10, or 1.0, respectively; Prabhu et al., 1984).

[j] NBI, navy bean inhibitor according to Wagner and Riehm (1967).

[k] GBI II, garden bean isoinhibitor II according to Wilson and Laskowski sr. (1973).

[l] PVI-Roh, crude Phaseolus vulgaris inhibitor preparation (Belitz et al., 1972).

[m] VMI-C, crude Vigna mungo inhibitor preparation BGTI (Fraction I) according to Kamalakannan et al. (1984).

[n] CPTI, chickpeas trypsin inhibitor according to Smirnoff et al. (1976).

[o] PSI-Roh, crude Pisum sativum inhibitor preparation I 4/1 (Weder and Hory, 1972).

[p] Black-eyed pea component D according to Ventura and Xavier-Filho (1966).

[q] Fract. A, grasspea inhibitor fraction A according to Bhat et al. (1976).

[r] n.d., not detectable (activity lower than 0.05 mg chymotrypsin inhibited by 1 g of sample material).

[s] MBI, mungbean inhibitor (Wilson, 1981).

[t] For results with pure lentil inhibitors see part on lentil inhibitors.

[u] Designated Indian red wood seed (Adenanthera pavonia L.) by Prabhu et al. (1984).

[v] PCI-Roh, crude Phaseolus coccineus inhibitor preparation (Belitz et al., 1971).

[w] Designation of Prabhu et al. (1984). According to Duke (1981) Canavalia ensiformis (L.) DC. is called Jackbean, while Canavalia gladiata (Jacq.) DC. is called Swordbean.

Other abbreviations as in Table 1.

beans, sword beans, and red gram, all used as food or feed, as well as for Caesalpinia bonduc seeds.

INHIBITION OF HUMAN CHYMOTRYPSIN

Human chymotrypsin, like its trypsin counterpart, consists of several components. Travis and co-workers found two main types, which they designated chymotrypsins I and II (Coan et al., 1971). They occur in equal amounts in activated extracts from human pancreas (Mallory and Travis, 1975a). Chymotrypsin I can be separated into two further components: chymotrypsins IA and IB (Coan and Travis, 1972).

Feinstein et al. (1974) also studied the human chymotrypsins and found two components, denoted chymotrypsins 1 and 2, which differ markedly in molecular weights and amino acid composition from chymotrypsins I and II.

The study of the inhibition of human chymotrypsin by plant inhibitors parallels that of trypsin in many aspects. In addition to the different enzyme preparations used, the other parameters already discussed for the trypsins will also influence the results for chymotrypsin.

Soybean Inhibitors

The results on the inhibition of human chymotrypsin by STI and BBI, the two soybean inhibitors, are listed in Table 4. Both STI and BBI inhibited human chymotrypsin I considerably (65 and 63 %) and to a greater extent than chymotrypsin II (26 and 20 %, respectively), when casein was used as a substrate (Mallory and Travis, 1975a). But compared with bovine chymotrypsin, the common test enzyme, STI is much more effective on human chymotrypsin I and even considerably more effective on chymotrypsin II (HCTI/BCTI 4.33 and 1.73, respectively; Table 4). STI is known to be a weak inhibitor of bovine and other animal chymotrypsins (Liener and Kakade, 1980); this is obviously not true of human chymotrypsin, especially of human chymotrypsin I.

As with trypsin, differing results were obtained for the inhibition of chymotrypsin by STI and BBI with synthetic ester substrates (ATEE and BTEE, see Table 4 for abbreviations), though also indicating an equal or somewhat better inhibition of the human chymotrypsins by STI than that of the bovine counterpart (Feinstein et al., 1974; Mallory and Travis, 1975a; Krogdahl and Holm, 1979 and 1983). A clear inhibition by STI and BBI could also be found with ATPA, one of the synthetic amide substrates used (Gertler et al., 1982), while GPPA failed to detect the different action of STI on human and bovine chymotrypsin (Belitz et al., 1982).

The inhibition of human chymotrypsin by soybean extracts was between 0.6 and 8 times that of the bovine enzyme depending on the soybean preparation, and the substrate and the method used (Krogdahl and Holm, 1979 and 1983; Belitz et al., 1982; Holm and Krogdahl, 1982; Prabhu et al., 1984).

Lima Bean Inhibitor

LBI, the inhibitor from lima beans, displayed a strong inhibition of human chymotrypsin I (89 %, casein as substrate, 1:1 molar ratio; Table 5) and a considerable inhibition of chymotrypsin II (65 %, same conditions; Mallory and Travis, 1975a). Again, chymotrypsin I is more effectively inhibited than both chymotrypsin II and the bovine enzyme. The results with ATEE, BTEE, ATPA and GPPA show the same effects as discussed above for STI and BBI (for details and references see Table 5); in most cases inhibition of human chymotrypsin was found to be more than twice that of bovine enzyme. Extracts from lima beans also inhibited human chymotrypsin more effectively than its bovine counterpart (Belitz et al., 1982).

Other Legumes

The results on the inhibition of human chymotrypsin by other legumes are included in Table 3. Contrary to the results with trypsin, only three species inhibited human chymotrypsin to the same degree as the bovine counterpart. These were field beans, guinea peas and acacia seeds. Four other species (black gram, broad beans, kidney beans and sword beans) displayed a lower relative inhibition of human chymotrypsin (HCTI/BCTI in Table 3). In absolute terms, the seven species were all faintly to moderately inhibitory of chymotrypsin.

The great majority of the species studied inhibited human chymotrypsin much more effectively than the bovine enzyme. While most representatives of

Table 4. Inhibition of Human Chymotrypsin by Soybeans

Inhibitor sample	Substrate[a]; inhibitor/enzyme ratio	Chymotrypsin type[b]	Inhibitor activity	HCTI/BCTI	Ref.[c]
STI	ATEE	1	+-	ca. 1.0	(1)
	ATEE	2	+-	ca. 1.0	(1)
STI	ATEE; 1:1 mole/mole	I	92 %	1.37	(2,3)
	ATEE; 5:1 mole/mole	I	99 %	1.08	(2,3)
	casein; 1:1 mole/mole	I	65 %*	4.33	(2,3)
	casein; 5:1 mole/mole	I	79 %*	2.26	(2,3)
	ATEE; 1:1 mole/mole	II	65 %	0.97	(2,3)
	ATEE; 5:1 mole/mole	II	95 %	1.03	(2,3)
	casein; 1:1 mole/mole	II	26 %*	1.73	(2,3)
	casein; 5:1 mole/mole	II	51 %*	1.46	(2,3)
STI	BTEE; 25×10^{-2} nmole:20 µl HPJ	CT	20 %*		(4)
	BTEE; 125×10^{-2} nmole:20 µl HPJ	CT	54 %*		(4)
STI	ATPA; 0.1 mg:0.5 ml HPJ	CT	62 %*		(5)
	ATPA; 1.0 mg:0.5 ml HPJ	CT	88 %*		(5)
STI	GPPA; slope	CT	0.16 mg/mg	1.07	(6)
STI	BTEE; 1.25 µg:27 µl HPJ	CT	8 %*	1.48	(7)
BBI	ATEE; 1:1 mole/mole	I	98 %		(2)
	ATEE; 5:1 mole/mole	I	100 %		(2)
	casein; 1:1 mole/mole	I	63 %*		(2)
	casein; 5:1 mole/mole	I	70 %*		(2)
	ATEE; 1:1 mole/mole	II	69 %		(2)
	ATEE; 5:1 mole/mole	II	95 %		(2)
	casein; 1:1 mole/mole	II	20 %*		(2)
	casein; 5:1 mole/mole	II	56 %*		(2)
BBI	ATPA; 0.1 mg:0.5 ml HPJ	CT	28 %*		(5)
	ATPA; 1.0 mg:0.5 ml HPJ	CT	85 %*		(5)
Extract	BTEE; 170 µg soy:20 µl HPJ	CT	59 %*		(4)
	BTEE; 850 µg soy:20 µl HPJ	CT	86 %*		(4)
Extract	GPPA; slope	CT	6.7 mg/g	1.00	(6)
Extract	BTEE; 134 µg soy:20 µl HPJ	CT	28-56 %	2.16-8.19	(8)
Extract	BTEE; 6.7 µg soy:27 µl HPJ	CT	32 %*	0.72	(7)
Extract	ATEE; slope	CT	0.45 IU/mg	0.56	(9)

[a] ATEE, ATPA, and GPPA, see Table 3; BTEE, N-benzoyl-L-tyrosine ethyl ester.

[b] CT, non specified chymotrypsin or whole chymotryptic activity; 1 and 2 human chymotrypsin 1 and 2 according to Feinstein et al. (1974); I and II, human chymotrypsin I and II according to Mallory and Travis (1975a).

[c] (1) Feinstein et al. (1974); (2) Mallory and Travis (1975a); (3) Coan and Travis (1971); (4) Krogdahl and Holm (1979); (5) Gertler et al. (1982); (6) Belitz et al. (1982); (7) Krogdahl and Holm (1983); (8) Holm and Krogdahl (1982); (9) Prabhu el al. (1984).

All other abbreviations as in Tables 1 and 3.

this group exhibited a relative inhibition of human chymotrypsin of about 2.0 (HCTI/BCTI between 1.6 and 2.5), carob beans, chickpeas, Elephantorrhiza burkei seeds, and lentils were three to four times as effective, and

Table 5. Inhibition of Human Chymotrypsin by Lima Beans

Inhibitor sample	Substrate; inhibitor/enzyme ratio	Chymo-trypsin type	Inhibitor activity	HCTI/BCTI	Ref.[a]
LBI	ATEE	I	+	ca. 1.0	(1)
	ATEE	II	+	ca. 1.0	(1)
LBI	ATEE; 1:1 mole/mole	I	95 %	2.38	(2,3)
	ATEE; 5:1 mole/mole	I	98 %	1.03	(2,3)
	casein; 1:1 mole/mole	I	89 %*	4.45	(2,3)
	casein; 5:1 mole/mole	I	97 %*	1.14	(2,3)
	ATEE; 1:1 mole/mole	II	70 %	1.75	(2,3)
	ATEE; 5:1 mole/mole	II	95 %	1.00	(2,3)
	casein; 1:1 mole/mole	II	65 %*	3.25	(2,3)
	casein; 5:1 mole/mole	II	92 %*	1.08	(2,3)
LBI	BTEE; $25x10^{-2}$nmole:20 µl HPJ	CT	75 %*		(4)
	BTEE; $125x10^{-2}$nmole:20 µl HPJ	CT	80 %*		(4)
LBI	ATPA; 0.1 mg:0.5 ml HPJ	CT	86 %*		(5)
	ATPA; 1.0 mg:0.5 ml HPJ	CT	100 %*		(5)
LBI	GPPA; slope	CT	2.93 mg/mg	2.34	(6)
LBI	BTEE; 0.54 µg:27 µl HPJ	CT	16 %*	0.89	(7)
Extract	GPPA; slope	CT	8.2 mg/g	1.44	(6)

[a] (1) through (7), see Table 4.

All other abbreviations as in Tables 1-4.

Inga laurina seeds were almost twelve times as effective on human as on bovine chymotrypsin. It must be mentioned that Prabhu et al. (1984) have found a considerably lower inhibition of human chymotrypsin and HCTI/BCTI ratio for chickpeas than those reported by Belitz et al. (1982). This may be due to different varieties of chickpea studied or due to the different substrates used by the two groups.

Studies with isolated lentil inhibitors showed that the ratio HCTI/BCTI is somewhat lower for purified inhibitors (2.7-2.8; cf. paragraph on lentil inhibitors) than for extracts, but even then the striking feature remains that human chymotrypsin is inhibited more effectively.

In summary, the inhibition of human trypsin equals that or is somewhat lower than that of bovine trypsin, especially for some of the most important grain legumes, e.g. soybeans, lima beans, navy beans, bush beans, carob beans, scarlet runner beans, winged beans, chickpeas, cowpeas, common peas, lentils and black gram. Even more important, though largely overlooked in the past, is the fact that human chymotrypsin is often inhibited much more effectively than its bovine counterpart. This is seen for bush beans, carob beans, scarlet runner beans, winged beans, chickpeas, cowpeas, common peas, lentils, and red gram. All these legumes are capable of considerably affecting the digestion of proteins in man, if they reached the small intestine unaltered.

Table 6. Inhibition of Human Intestinal Proteolytic Activity by Soybeans and Lima Beans

Inhibitor sample	Substrate; inhibitor/enzyme ratio	Inhibitor activity	HPI/BPI	Ref.[a]
Soybean				
STI	casein; $25x10^{-2}$nmole:20 µl HPJ	35 %*		(1)
	casein; $125x10^{-2}$nmole:20 µl HPJ	43 %*		(1)
STI	casein	52 %*		(2)
STI	casein; 1.25 µg:27 µl HPJ	5 %*	0.21	(3)
Extract	casein; 170 µg soy:20 µl HPJ	31 %*		(1)
	casein; 850 µg soy:20 µl HPJ	41 %*		(1)
Extract	casein; 134 µg soy:50 µl HPJ	6-24 %		(4)
Extract	casein; 6.7 µg soy:27 µl HPJ	7 %*	0.41	(3)
Extract	casein; slope	0.90 IU/mg	0.43	(5)
Lima bean				
LBI	casein; $25x10^{-2}$nmole:20 µl HPJ	38 %*		(1)
	casein; $125x10^{-2}$nmole:20 µl HPJ	48 %*		(1)
LBI	casein; 0.54 µg:27 µl HPJ	11 %*	0.87	(3)

[a] (1) Krogdahl and Holm (1979); (2) Struthers and MacDonald (1983); (3) Krogdahl and Holm (1983); (4) Holm and Krogdahl (1982); (5) Prabhu et al. (1984).

All other abbreviations as in Tables 1-3.

INHIBITION OF OTHER MINOR INTESTINAL PROTEINASES

In addition to the trypsins and the chymotrypsins, other proteolytic enzymes act in the human small intestine. Protease E (Mallory and Travis, 1975b), elastases (Largman et al., 1976), carboxypeptidases A (Peterson et al., 1976) and B (Geokas et al., 1975), mesotrypsin (Rinderknecht et al., 1978 and 1984) and kallikrein (Amouric and Figarella, 1979) have been previously reported. More recently, two additional serine proteinases, one of them provisionally named cholesterol-binding pancreatic proteinase (Sziegoleit, 1984) and the other called ingobsin (Nexoe et al., 1984) have been found.

Mallory and Travis (1975a) have determined the relative amounts of the proteinases in crude human pancreatic extracts with synthetic inhibitors to be 99 %, 35 %, 32 % and 21 % for serine proteinases, trypsin, chymotrypsin, and protease E, respectively. Krogdahl and Holm (1979) reported the proteolytic activity of human pancreatic juice to consist of 60 % of trypsin and chymotrypsin, 25 % of carboxypeptidases and 15 % of, probably, elastase. As a consequence, all of these activities must be considered, when the influence of legume proteinase inhibitors on the total human proteolytic activity is examined. Individual activities can be tested when the specific substrates are known.

Inhibition of Proteolytic Activity

Soybean. The inhibition of the proteolytic activity of human pancreatic juice by soybeans is shown in Table 6. Depending on the inhibitor/enzyme ratio, STI inhibited between 5 and 52 % of the total proteolytic activity, while soybean extracts caused a maximum inhibition of 41 % (Krogdahl and Holm, 1979 and 1983; Holm and Krogdahl, 1982; Struthers and MacDonald, 1983). The inhibition of human proteinases by soybean extracts was 41 to 43 % that of the bovine enzyme (Krogdahl and Holm, 1983; Prabhu et al., 1984). Because the latter authors reported an inhibition of 67 % for the trypsins and 56 % for the chymotrypsins (cf. Table 1 and 4), the additional human proteinases could at most be slightly inhibited by soybean extracts.

Lima bean. Under the same experimental conditions, LBI from lima beans was only slightly more effective than STI on human proteinases (Table 6; Krogdahl and Holm, 1979 and 1983).

Other legumes. Comparison of the effect of extracts from other legumes on the total human and bovine proteolytic activities with that on the individual proteinases reported by Prabhu et al. (1984) show discrepancies in the results (Table 3), as already mentioned for soybeans. In some cases, the relative inhibition of human proteolytic activity (HPI/BPI) is lower than that of human trypsin (HTI/BTI) and human chymotrypsin (HCTI/BCTI). Besides the influence of substrates (in particular ATEE), and of other seed constituents, the possible presence of additional proteinase(s), which were much more severely inhibited in bovine than in the human preparations, is the most likely explanation for these results.

In contrast, cowpeas and *Adenanthera pavonina* seeds inhibited the human proteolytic activity more than the bovine acitivity (150 % and 120 %, respectively; Table 3).

Inhibition of the Individual Minor Proteinases

The available data on the inhibition of some additional proteinases mentioned above are summarized in Table 7.

Protease E. Protease E was only moderately inhibited by STI and BBI from soybeans and also by LBI from lima beans when tested with casein as substrate (Mallory and Travis, 1975a). In most cases, a higher inhibition was found when a synthetic substrate was used under similar conditions (Mallory and Travis, 1975a; Gertler et al., 1982). The chickpea trypsin inhibitor (CPTI) was found to be more effective than STI and BBI on human protease E by Gertler et al. (1982). No data on a corresponding bovine enzyme has been reported in the literature.

Elastase. Prabhu et al. (1984) reported bovine elastase being five times more susceptible than the human enzyme to *Adenanthera pavonina* seed extracts. Whether elastase alone was detected is doubtful, since the substrate used could also detect protease E (Mallory and Travis, 1975b).

Carboxypeptidase. Extracts from *Adenanthera pavonina* did not inhibit the carboxypeptidase activity in human or bovine pancreatic preparations (Prabhu et.al., 1984).

Enteropeptidase. Jacob et al. (1983) have published the isolation of a specific enteropeptidase inhibitor from kidney beans. Enteropeptidase (recommended name by the International Union of Biochemistry, Nomenclature Committee, 1979; also known as enterokinase) is a serine proteinase secreted by mucosa cells of the small intestine. It activates trypsinogens to trypsins. The inhibitor was most active on pig enteropeptidase, followed by bovine and human enteropeptidases.

Table 7. Inhibition of Other Human Intestinal Proteinases

Inhibitor sample	Substrate[a]; inhibitor/enzyme ratio	Enzyme	Inhibitor activity	HEI/BEI[b]	Ref.[c]
Soybean					
STI	ZAPE; 1:1 mole/mole	protease E	75 %		(1)
	ZAPE; 5:1 mole/mole	protease E	86 %		(1)
	casein; 1:1 mole/mole	protease E	17 %*		(1)
	casein; 5:1 mole/mole	protease E	37 %*		(1)
STI	AA_3PA; 0.1 mg:0.5 ml HPJ	protease E	14 %*		(2)
	AA_3PA; 1.0 mg:0.5 ml HPJ	protease E	69 %*		(2)
BBI	ZAPE; 1:1 mole/mole	protease E	1 %		(1)
	ZAPE; 5:1 mole/mole	protease E	15 %		(1)
	casein; 1:1 mole/mole	protease E	5 %*		(1)
	casein; 5:1 mole/mole	protease E	8 %*		(1)
BBI	AA_3PA; 0.1 mg:0.5 ml HPJ	protease E	34 %*		(2)
	AA_3PA; 1.0 mg:0.5 ml HPJ	protease E	83 %*		(2)
Lima bean					
LBI	ZAPE; 1:1 mole/mole	protease E	6 %		(1)
	ZAPE; 5:1 mole/mole	protease E	40 %		(1)
	casein; 1:1 mole/mole	protease E	5 %*		(1)
	casein; 5:1 mole/mole	protease E	10 %*		(1)
LBI	AA_3PA; 0.1 mg:0.5 ml HPJ	protease E	25 %*		(2)
	AA_3PA; 1.0 mg:0.5 ml HPJ	protease E	84 %*		(2)
Chickpea					
CPTI	AA_3PA; 0.1 mg:0.5 ml HPJ	protease E	47 %*		(2)
	AA_3PA; 1.0 mg:0.5 ml HPJ	protease E	97 %*		(2)
Red sandal wood seed					
Extract	SA_3PA; slope	elastase		0.20	(3)
	Hip-Arg, Hip-Phe	carboxy-peptidase	nil		(3)
Kidney bean					
KBEI[d]	trypsinogen; 20 μg:20 mU	enteropep-tidase	7.4 mU/mg	0.10	(4)

[a] AA_3PA, acetyl-L-alanyl-L-alanyl-L-alanine p-nitroanilide; Hip-Arg, hippuryl L-arginine; Hip-Phe, hippuryl L-phenylalanine; SA_3PA, succinyl-L-alanyl-L-alanyl-L-alanine p-nitroanilide; ZAPE, carboxybenzoyl-L-alanine p-nitrophenyl ester.

[b] HEI/BEI, ratio of the inhibition of human enzyme to the inhibition of bovine enzyme.

[c] (1) Mallory and Travis (1975a); (2) Gertler et al. (1982); (3) Prabhu et al. (1984); (4) Jacob et al. (1983).

[d] KBEI, kidney bean enterokinase inhibitor (Jakob et al., 1983).

All other abbreviations as in Tables 1-3.

Mesotrypsin. Although resembling the two human trypsins in many properties, human mesotrypsin is totally resistant to STI and only 5 % inhibited by twice as much LBI necessary to completely inhibit bovine trypsin (Rinderknecht et al ., 1984).

In general, the effect of the various legume inhibitors on the different minor human proteinases is clearly less pronounced than that on human trypsins and chymotrypsins.

PART II: STUDIES WITH LENTIL INHIBITORS

As already mentioned, most of the work on isolated proteinase inhibitors has been made with bovine trypsin and chymotrypsin. In addition to the studies with the human proteinases discussed in the preceeding part, the influence of legume extracts and/or inhibitor preparations on the activities of pancreatic enzymes from other mammals (e.g. pig, dog, rat, monkey, and mink) has been studied during the last years (e.g. Krogdahl and Holm, 1979; Gertler et al., 1982; Holm and Krogdahl, 1982; Struthers and MacDonald, 1983; Temler et al., 1983). This has been done to evaluate the effect of legume inhibitors in the feed of the respective animals and also to find a test animal whose response simulates best that of man.

In order to clarify the mechanism responsible for the difference in action on proteinases from different species, especially on those from man and cattle, we have isolated the proteinase inhibitors from lentils (*Lens culinaris* Medik.). After chemical characterization, the reaction of these inhibitors with human and bovine trypsin and chymotrypsin has been studied. Based on the results of Belitz et al. (1982), lentils were chosen as the source of inhibitors, because they are typical representatives of legumes displaying about the same inhibition of human and bovine trypsin and showing a much higher inhibition of human than of bovine chymotrypsin. Furthermore, detailed information on proteinase inhibitors from this species was lacking in spite of its nutritious importance and wide consumption in Asia, northern Africa, Latin America, and southern and eastern Europe (Duke, 1981).

Four proteins inhibiting bovine trypsin and chymotrypsin were found in extracts from lentils by Weder et al. (1983) and the main inhibitor was isolated from a commercial cultivar and partially characterized by the same authors.
In order to find a sample suitable for the isolation of all four inhibitors, Weder et al. (1985) have screened 38 lentil samples from different parts of the world for both inhibitor activities and electrophoretic inhibitor patterns. Disc electrophoresis and negative staining for trypsin inhibitors of the lentil seed extracts from all samples showed the presence of four inhibitors, denoted LCI-1 to LCI-4. As proven with some of the samples, these four inhibitors also inhibited chymotrypsin. The four lentil inhibitors were present in different amounts in the individual lentil samples, leading to three different inhibitor pattern types, as shown in Figure 1. Type A, with LCI-4 predominating, followed by LCI-1 and only traces of LCI-2 and -3, was found in most of the samples. Type A', with LCI-4 still predominating, differed by displaying higher amounts of LCI-1, while type B (found in 6 samples) exhibited equal amounts of LCI-1 and LCI-4 and, generally, also higher amounts of LCI-2 and -3. In contrast, type C, which was found only in three samples, contained high amounts of LCI-1, only small amounts of LCI-4 and traces of LCI-2 and -3. While type A inhibitor patterns were usually found in samples with low inhibitor contents, types B and C were mainly found in samples with high inhibitor contents.

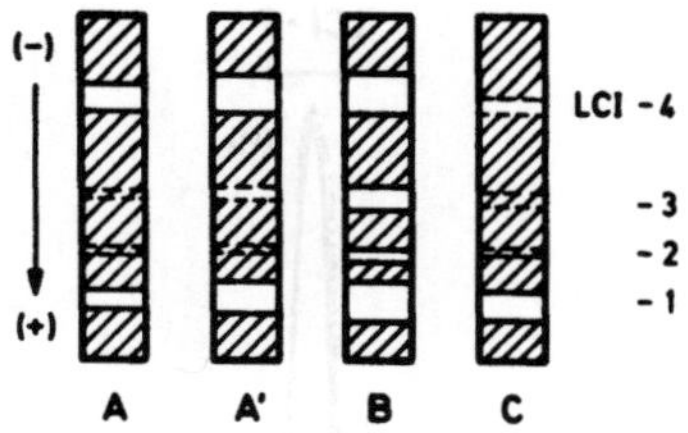

Fig. 1. Electrophoretic trypsin inhibitor pattern types of Lens culinaris. From Weder et al. (1985).

Eight lentil samples tested on human pancreatic juice all exhibited the typical behaviour against human and bovine trypsin and chymotrypsin. The screening experiments showed that highly active representatives of type C were suited best for isolating LCI-1, while type B cultivars were better for isolating all four inhibitors.

Isolation

Crude inhibitor preparation. According to the screening experiment, commercial red lentils from Italy were ground, sieved and extracted with 0.25 N sulfuric acid following the procedure of Weder et al. (1983). A crude inhibitor preparation (LCI-C) was obtained by ammonium sulfate precipitation (30-60 % saturation), dialyzation and subsequent lyophilization of the supernatant liquid.

Purified inhibitor preparation. LCI-C was then subjected to gel chromatography on Bio-Gel. Figure 2 shows the elution profile of a typical run from a commerical brown cultivar (Weder et al., 1983). Transmittance at 252 nm indicated three major protein fractions. Trypsin and chymotrypsin inhibitor activity determinations located the inhibitor at the second protein fraction (LCI-P; Fig. 2). It was eluated in one sharp peak having a shoulder on the descending slope. The active fractions (indicated by the bar in Fig. 2) were pooled and lyophilized to yield the purified inhibitor preparation (LCI-P).

Inhibitor activities, yield and enrichment of these purification steps are given in Table 8. The ammonium sulfate fractionation seems to enrich more effectively the human and bovine chymotrypsin than the two trypsin inhibitor activities; this is probably due to the inaccuracy of the inhibitor determination in the extract. The procedure achieved a sixty-six-fold purification with a recovery of 13 % of the activity.

Separation of isoinhibitors. Figure 3 shows the elution profile of LCI-P subjected to cation-exchange chromatography. Elution was performed with a pH gradient formed with volatile buffers. The proteins displaying no inhibitor activity were separated from the inhibitors. The inhibitors themselves were separated from each other and were pooled as indicated by the bars. Because of their low concentration and poor separation, LCI-2 and LCI-3 were pooled together.

The behaviour of the lentil inhibitors from Italian red lentils on a cation exchanger is shown in Figure 4. The inhibitor proteins (Figure 4, A) can be compared to the inhibitors bound to human and bovine trypsin (Figure 4, B) and chymotrypsin (Figure 4, C).

These results show that bovine trypsin is better inhibited than the human enzyme (Figure 4, B). In the case of chymotrypsin, LCI-1 and LCI-4

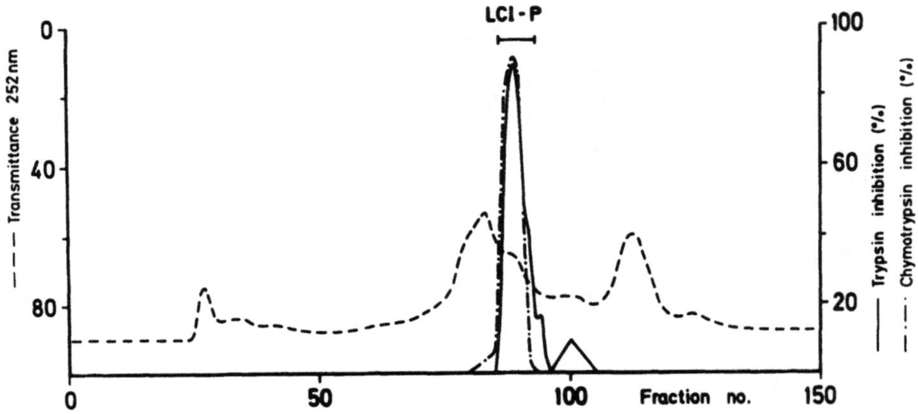

Fig. 2. Purification of the crude inhibitor preparation from lentils (LCI-C) by gel chromatography. A sample of 250 mg of LCI-C was applied to a 50x400 mm column of Bio-Gel P-100 (100-200 mesh) eluted with 0.1 N acetic acid at a flow rate of 35 ml/h (8.75 ml-fractions collected). Transmittance in the eluate was recorded at 252 nm and the inhibition of bovine trypsin (20 μg) and alpha-chymotrypsin (12 μg) by 0.1 ml of the eluate was determined. From Weder et al. (1983).

Table 8. Purification of Trypsin Chymotrypsin Inhibitors from Lentils

Inhibitor preparation	Lentil inhibitor acitivities				Yields(g)	Relative purification[e]
	BTIA[a]	HTIA[b]	BCTIA[c]	HCTIA[d]		
Extract[f]	0.005	0.005	0.002	0.004	250	1
LCI-C	0.083	0.077	0.062	0.168	1.75	26
LCI-P	0.228	0.227	0.154	0.387	0.50	66

[a] BTIA, bovine trypsin-inhibitory activity, expressed as mg of active bovine trypsin inhibited per mg of sample (BAPA as substrate).
[b] HTIA, human trypsin-inhibitory activity, activity of human trypsin in human small intestinal juice (expressed as mg of active bovine trypsin) inhibited per mg inhibitor preparation (BAPA as substrate).
[c] BCTIA, bovine chymotrypsin-inhibitory activity, expressed accordingly (GPPA as substrate).
[d] HCTIA, human chymotrypsin-inhibitory activity, expressed accordingly.
[e] Average purification, related to the activity of the seed meal extract.
[f] Extract from 250 g lentil seed meal; inhibitor activities pro mg seed meal.

cause a stronger inhibition of the human enzyme (Figure 4, C). LCI-2 and LCI-3 seem to inhibit human chymotrypsin less than the bovine enzyme; further experiments are needed in order to be able to draw final conclusions.

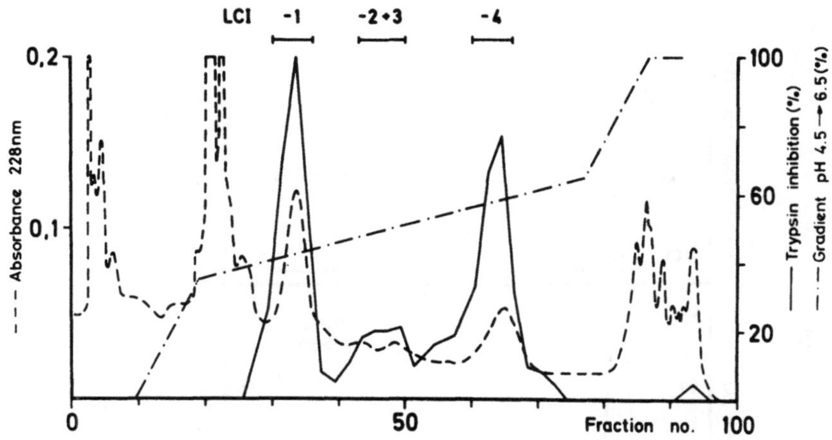

Fig. 3. Separation of the lentil proteinase inhibitors by ion-exchange chromatography. A sample of 25 mg of LCI-P was applied to a 10x250 mm column of Whatman CM-52 eluted with 0.1 M ammonium acetate buffers (gradient pH 4.5 to 6.5) with a flow rate of 24 ml/h (fraction volume: 6 ml). Proteins were detected by absorbance at 228 nm; the inhibition of bovine trypsin (15 ug) was determined on 0.1 ml aliquots (Weder and Mueller, in preparation).

Purification of LCI-1. After lyophilization, LCI-1 was finally purified by anion-exchange chromatography (Figure 5). The protein profile is highly symmetric and the trypsin inhibitor activity is very high. Only fractions showing protein purity by electrophoresis (fractions 10-14) were pooled. A total of 6.5 mg of LCI-1 were obtained from 250 g of lentil meal with an overall 500-fold purification.

Purification of LCI-4. Accordingly, LCI-4 was purified by rechromatography on the cation exchanger used before but at a constant pH of 5.5 (Figure 6). A symmetric protein peak with high inhibitor activity was obtained. The fractions pooled as indicated yielded, after lyophilization, 5.0 mg of LCI-4 from 250 g of lentils with a total 400-fold purification.

Homogeneity of LCI-1 and LCI-4 was proven by PAGE (glycine/tris pH 8.3, cf. Weder et al., 1983), SDS/PAGE according to Weber and Osborn (1969) and reversed-phase high pressure liquid chromatography (ODS Hypersil, 5 µm, 0.01 M triethylammonium formiate buffer pH 6.0 with an acetonitrile gradient) (Weder and Mueller, in preparation).

Characterization of the Purified Inhibitors

Amino acid composition. The amino acid composition of LCI-1 and LCI-4 is given in Table 9. The two inhibitors showed high cysteine contents and lack of methionine and tryptophan. LCI-4 also lacked isoleucine. No carbohydrate could be detected in the two inhibitor preparations with the phenol-sulfuric acid method (Dubois et al., 1956). Molecular weights of 9,800 and 9,400 Daltons for LCI-1 and LCI-4, respectively, could be calculated from the amino acid composition. The molecular weight and the 7 disulfide bonds found in the LCI-1 inhibitor, situates it in the Bowman-Birk inhibitor family. LCI-4 possesses one additional disulfide bond. (For a compilation of the known amino acid sequences of the Bowman-Birk inhibitors, see Weder, 1985).

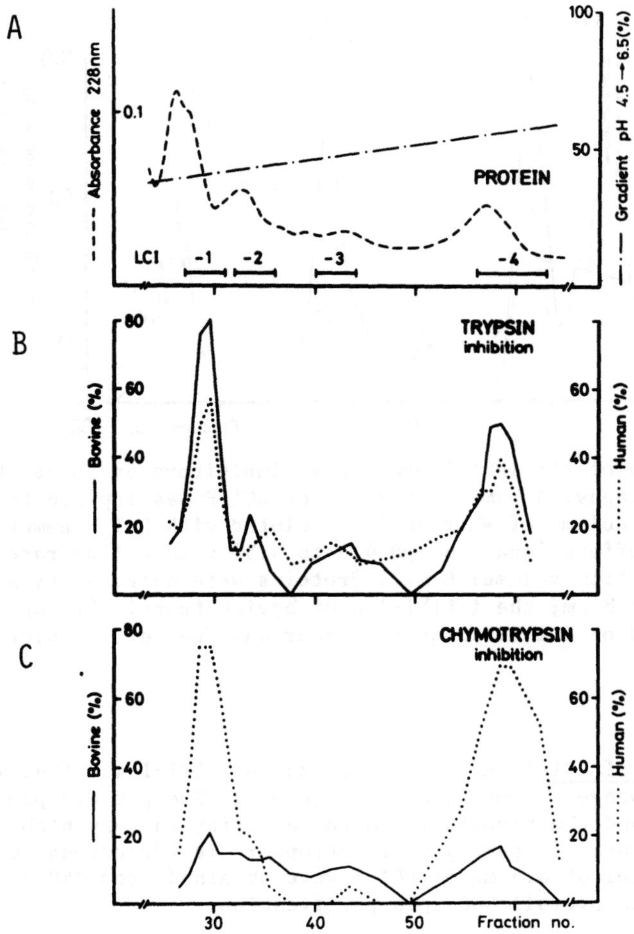

Fig. 4. Inhibition of human and bovine trypsin and chymotrypsin after ion-exchange chromatography of LCI-P. Experimental conditions as given in Figure 3. (A) Inhibitor protein; (B) inhibition of bovine trypsin (15 ug) and human trypsin (150 ul of human small intestinal juice) by 150 ul of the eluate (determined with BAPA); and (C) inhibition of bovine chymotrypsin (20 ug) and human chymotrypsin (200 ul of the human intestinal juice) by 250 ul of the eluate (determined with GPPA). (Weder and Mueller, in preparation).

Inhibitor activity. Tables 10 and 11 show the results on the inhibition of human and bovine trypsin and chymotrypsin by LCI-1 and LCI-4. These tables provide supplementary information to Table 3. Data on the experiments with BAPA and GPPA were taken from the inhibition curves (Figures 7-10).

The inhibition of human trypsin by LCI-1 and LCI-4 was about 70 % that of the bovine enzyme (Table 10). Practically the same results were obtained from the slope and from the 1:1 molar ratio value of the inhibition curve. Increasing the inhibitor/enzyme ratio resulted in a HTI/BTI ratio closer to one, especially for LCI-4. This influence was even more pronounced with the chymotrypsins (Table 11).

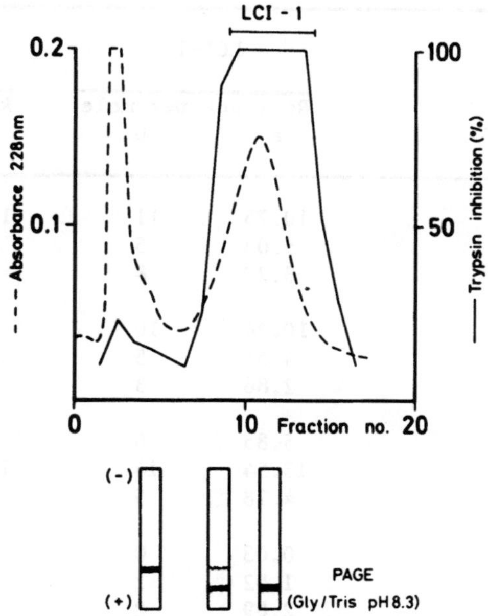

Fig. 5. Purification of LCI-1 by anion-exchange chromatography. A sample of 6 mg of LCI-1 was applied to a 11x170 mm column of Whatman DE-52 eluted with 0.1 M ammonium acetate buffer (pH 8.5) at 25 ml/h (fraction volume: 5 ml). Protein and trypsin inhibition detected as in Figure 3. PAGE (polyacrylamide gel electrophoresis; below) as described in Weder et al. (1983). From Weder and Mueller (in preparation).

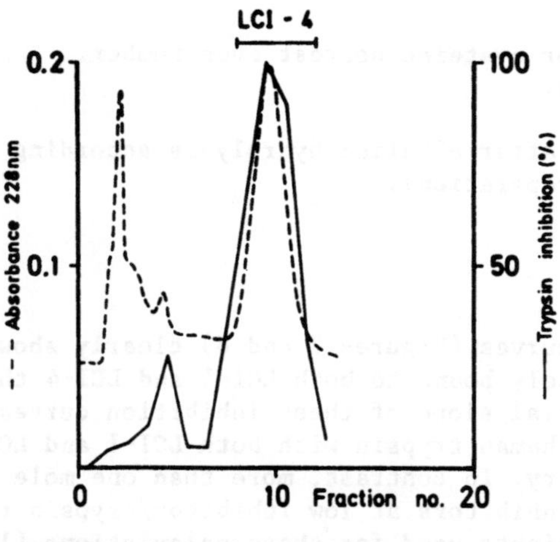

Fig. 6. Rechromatography of LCI-4. A sample of 7 mg of LCI-4 was applied to a 11x170 mm column of Whatman CM-52 eluted with 0.1 M ammonium acetate buffer (pH 5.5) at 24 ml/h (fraction volume: 6 ml). Protein and inhibition of trypsin detected as in Figure 3 (Weder and Mueller, in preparation).

Table 9. Amino Acid Composition of two Lentil Inhibitors LCI-1 and LCI-4

	LCI-1		LCI-4	
	Residues per mole		Residues per mole	
	a	b	a	b
Aspartic acid[c]	10.75	11	11.33	11
Threonine	5.03	5	5.58	6
Serine	8.20	8	7.27	7
Glutamic acid[d]	10.24	10	6.32	6
Proline	4.56	5	4.75	5
Glycine	2.86	3	2.30	2
Alanine	5.85	6	5.92	6
Cysteine	13.54	14	16.79	16
Valine	4.78	5	5.28	5
Methionine	0.03	0	0.04	0
Isoleucine	1.52	2	0.08	0
Leucine	2.09	2	1.73	2
Tyrosine	2.78	3	3.11	3
Phenylalanine	2.08	2	1.99	2
Histidine	3.60	4	4.15	4
Lysine	4.99	5	5.58	6
Arginine	3.99	4	4.06	4
Tryptophan[e]	nil	0	nil	0
Total		89		85

[a] Calculated.
[b] Nearest integer; for cysteine nearest even number.
[c] Includes asparagine.
[d] Includes glutamine.
[e] Determined by HPLC after alkaline hydrolysis according to Weder et al. (in preparation).

The inhibition curves (Figures 7 and 8) clearly show that human trypsin was more loosely bound to both LCI-1 and LCI-4 than its bovine counterpart. The initial slope of these inhibition curves indicate that the complex formation of human trypsin with both LCI-1 and LCI-4 occurs in an exact 1:1 stoichiometry. In contrast, more than one mole of bovine trypsin is bound by the two inhibitors at low inhibitor/trypsin ratios.

The molecular weights used for these calculations (10,600 and 9,900 Daltons for LCI-1 and LCI-4, respectively) are averages of values obtained by gel chromatography, size exclusion-HPLC (also of enzyme/inhibitor complexes), SDS-PAGE, SDS pore gradient-PAGE and calculated from amino acid composition (Weder and Mueller, 1985).

With respect to the chymotrypsins, the situation is entirely reverse. Both LCI-1 and LCI-4 inhibited human chymotrypsin nearly three times more than bovine chymotrypsin (slopes of the inhibition curves; Table 11). At a

Table 10. Inhibition of Human and Bovine Trypsin by the two Lentil Inhibitors LCI-1 and LCI-4

Inhibitor	Substrate; inhibitor/enzyme ratio	HTI	BTI	HTI/BTI
LCI-1	casein; slope		1.94 mg/mg	
	BAPA; slope	2.30 mg/mg	3.36 mg/mg	0.68
	BAPA; 1:1 mole/mole	68 %	96 %	0.71
	BAPA; 2:1 mole/mole	77 %	100 %	0.77
LCI-4	casein; slope		1.67 mg/mg	
	BAPA; slope	2.38 mg/mg	3.24 mg/mg	0.73
	BAPA; 1:1 mole/mole	73 %	96 %	0.76
	BAPA; 2:1 mole/mole	83 %	98 %	0.85

Abbreviations as in Table 1.

Table 11. Inhibition of Human and Bovine Chymotrypsin by the two Lentil Inhibitors LCI-1 and LCI-4

Inhibitor	Substrate; inhibitor/enzyme ratio	HCTI	BCTI	HCTI/BCTI
LCI-1	casein; slope		1.12 mg/mg	
	GPPA; slope	4.04 mg/mg	1.46 mg/mg	2.77
	GPPA; 1:1 mole/mole	86 %	57 %	1.51
	GPPA; 2:1 mole/mole	100 %	78 %	1.28
LCI-4	casein; slope		1.04 mg/mg	
	GPPA; slope	3.66 mg/mg	1.36 mg/mg	2.69
	GPPA; 1:1 mole/mole	85 %	50 %	1.70
	GPPA; 2:1 mole/mole	98 %	76 %	1.29

Abbreviations as in Table 3.

1:1 molar ratio of inhibitor/enzyme, the human enzyme is only 1.5-1.7 times more strongly inhibited and at a 2:1 molar ratio the difference is still smaller. (These data again show how inhibitor/enzyme ratios influcence the results of the inhibition on different enzymes).

The inhibition curves, Figures 9 and 10, show that bovine chymotrypsin was very loosely bound to both LCI-1 and LCI-4, and that even at very low inhibitor/enzyme ratios the 1:1 stoichiometry was not reached. Opposite to this, human chymotrypsin was more tightly bound to the two inhibitors and the inhibition observed at low inhibitor/enzyme ratios can only be explained by one mole of LCI-1 or LCI-4 binding more than one mole of human chymotrypsin.

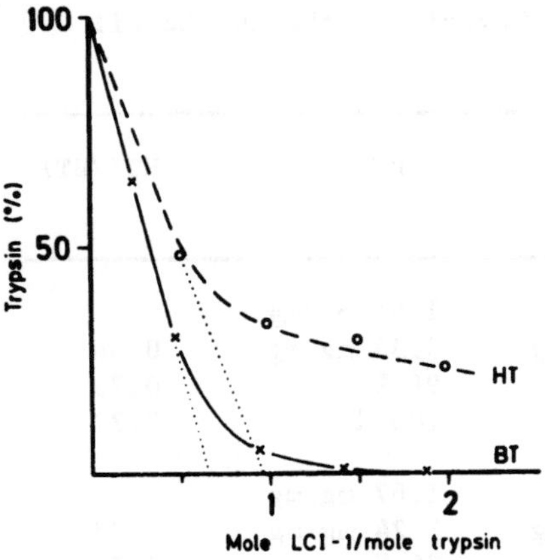

Fig. 7. Inhibition of human and bovine trypsin by LCI-1. HT, human trypsin; BT, bovine trypsin.

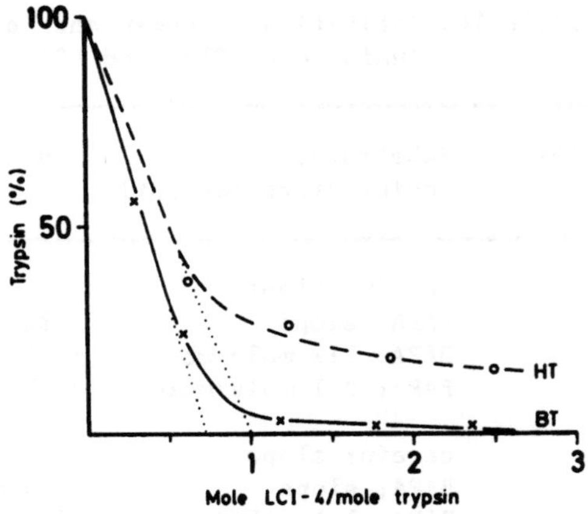

Fig. 8. Inhibition of human and bovine trypsin by LCI-4. HT and BT, see Figure 7.

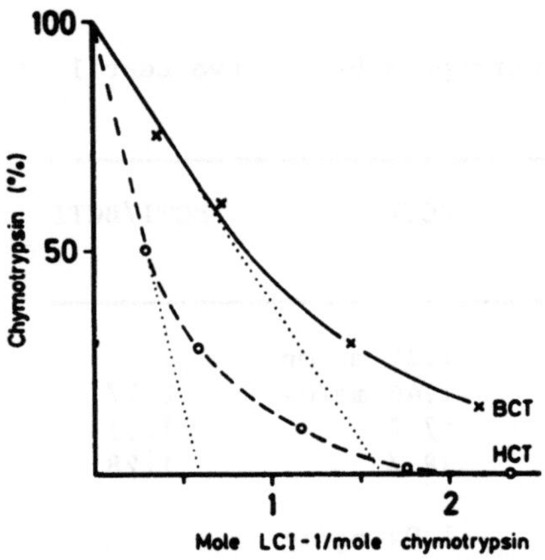

Fig. 9. Inhibition of human and bovine chymotrypsin by LCI-1. HCT, human chymotrypsin; BCT, bovine chymotrypsin.

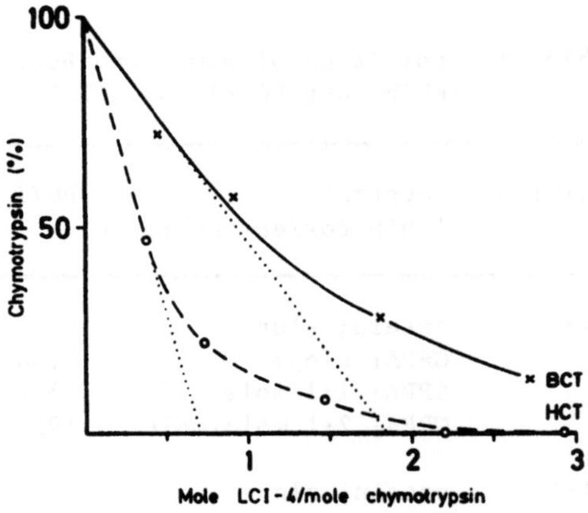

Fig. 10. Inhibition of human and bovine chymotrypsin by LCI-4. HCT and BCT, see Figure 9.

Studies on Proteinase Inhibitor Complexes

The fact that one mole of LCI-1 or LCI-4 binds more than one mole of bovine trypsin or human chymotrypsin could arise from two possible alternatives: (a) the inhibitor possesses two specific reactive sites and the enzyme reacts at its specific reactive site, but also at the second reactive site; and (b) the inhibitor possesses a third reactive site specific for bovine trypsin and human chymotrypsin. To clarify this, the capacity of LCI-1 and LCI-4 to form proteinase:inhibitor complexes with various enzymes was examined.

LCI-1. (See Table 12). Both bovine and human trypsin were inhibited by the LCI-1:BCT complex to about the same extent as by the free LCI-1 (91 and

Table 12. Activity of LCI-1 and LCI-1:Proteinase Complexes Against Bovine and Human Trypsin and Chymotrypsin

	LCI-1		LCI-1:BT[a]		LCI-1:BCT[a]		LCI-1:BT:BCT[a]	
	b	c	d	e	d	e	d	e
BT	1.53	0.96	30	0.29	91	0.87	15	0.14
HT	1.05	0.68	0	0	90	0.61	0	0
BCT	0.62	0.57	84	0.48	20	0.11	0	0
HCT	1.70	0.86	91	0.78	55	0.47	50	0.43

[a] 1:1 inhibitor:enzyme complexes; BT, bovine trypsin; BCT, bovine chymotrypsin; HT, human trypsin; HCT, human chymotrypsin.
[b] Moles enzyme that can be maximally bound per mole inhibitor (according to the inhibition curves).
[c] Moles enzyme that are bound at a molar ratio of 1:1.
[d] Percent activity of the complex related to the activity of the free inhibitor.
[e] Moles enzyme that are additionally bound by the complex.
From Weder and Mueller (1985).

90 %); human trypsin was not inhibited by either the LCI-1:BT nor by the trimeric LCI-1:BT:BCT complex. In contrast, additional bovine trypsin was inhibited by the stoichiometric LCI-1:BT (1:1) complex and, to a smaller extent, also by the trimeric complex. The inhibition of more than one mole of bovine trypsin per mole LCI-1 might be due to an unspecific binding at the chymotrypsin-reactive site of the LCI-1:BT complex, which is still unoccupied, and for which the additionally bound bovine trypsin must compete with the loosely bound bovine chymotrypsin in the trimeric complex. The total amount of enzyme bound to LCI-1 after addition of further bovine trypsin to the LCI-1:BT:BCT complex summed up to 1.67 moles (0.96 mole of bovine trypsin and 0.57 mole of bovine chymotrypsin originally bound, and 0.14 mole of newly bound bovine trypsin).

Bovine chymotrypsin was inhibited by LCI-1:BT but not by LCI-1:BT:BCT. A reaction mixture containing a 1:1 molar ratio of LCI-1:BCT could inhibit newly added bovine chymotrypsin, because only 0.57 mole of bovine chymotrypsin is originally bound pro mole LCI-1; this latter binding amounts to 0.11 mole bovine chymotrypsin.

The inhibition of human chymotrypsin by LCI-1:BT was like that by LCI-1. This was also found for bovine chymotrypsin. More human than bovine chymotrypsin (0.78 mole vs. 0.48 mole, respectively) were bound by the LCI-1:BT because of the lower inhibitor constant of the human enzyme. The LCI-1:BCT complex was able to inhibit about half the amount (55 %) of human chymotrypsin as did free LCI-1, because only 0.57 mole of bovine chymotrypsin was originally bound in this complex. As expected from our discussion above, the trimeric LCI-1:BT:BCT complex also inhibited half the amount of human chymotrypsin as free LCI-1. After addition of human chymotrypsin, a total of 1.96 moles of enzyme (0.96 mole of bovine trypsin and 0.57 mole of bovine chymotrypsin originally bound, and 0.43 mole of newly bound human chymotrypsin) were bound.

LCI-1 was never found to bind more than two moles of enzymes.

LCI-4. As shown in Table 13, similar results were obtained with LCI-4; therefore, they shall not be further discussed.

Table 13. Activity of LCI-4:Proteinase Complexes Against Various Enzymes

Complex[a]		LCI-4:BT	LCI-4:BCT	LCI-4:BT:BCT
Activity[b] against	BT	40	94	20
	HT	0	99	0
	BCT	73	10	0
	HCT	98	60	55

[a] 1:1 inhibitor:enzyme complexes; BT, HT, BCT, and HCT see Table 12.
[b] Percent activity of the complex related to the activity of the free inhibitor.
From Weder and Mueller (in preparation).

Chemical Modification

The results outlined above do not entirely exclude the possibility of an inhibitor molecule containing three reactive sites, although an inhibitor with two reactive sites is much more probable. Therefore, chemical modification of the individual amino acid residues at the reactive sites of the inhibitors were carried out and the influence of this modification on the activity against the particular enzymes was studied. The results are given in Table 14.

Lysine. Modification of lysine residues with maleic anhydride (Butler et al., 1969) did not markedly reduce the inhibitor activity of the two inhibitors against any of the enzymes. The slight alterations were probably due to conformational alterations caused by the modification.

Experiments run in parallel with LBI (lima bean inhibitor, known to contain lysine at the trypsin-reactive site; Stevens et al., 1974) and PCI-2 (trypsin chymotrypsin inhibitor from *Phaseolus coccineus* L., containing arginine at the respective reactive site; Hory et al., 1976) resulted in a total loss of trypsin inhibitor activity (against the two trypsins) of the former, while the latter was not affected by this modification.

Arginine. Modification of arginine residues with 2,3-butanedione trimer (Means and Feeney, 1971) totally suppressed the activity of both LCI-1 and LCI-4 against human trypsin, while some activity remained against the bovine enzyme; the activity against the two chymotrypsins was slightly to moderately reduced.

Similar results were observed with 1,2-cyclohexanedione (Means and Feeney, 1971), also an arginine-modifying reagent, but the chymotrypsin-inhibitory activities were less affected here.

In a parallel set of experiments, inhibitor activities of LBI were shown to be maximally 15 % altered by reaction with trimeric 2,3-butanedione.

Tyrosine. Derivatization of tyrosine residues with N-acetylimidazole (Lundblad and Noyes, 1984) did not alter the activity of LCI-1 and LCI-4 against human trypsin, but the activity against bovine trypsin as well as that against the two chymotrypsins was reduced by about half.

Reaction with tetranitromethane (Lundblad and Noyes, 1984), another tyrosine-modifying reagent, totally suppressed the activity against

Table 14. Activity of Lentil Inhibitors after Chemical Modification

Inhibitor	Modification of/with[a]	Activity[b] against	BT	HT	BCT	HCT
LCI-1	Lysine/MA		99	93	125	102
	Arginine/tBD		25	0	60	77
	Arginine/CD		20	15	104	90
	Tyrosine/AIA		50	97	43	53
	Tyrosine/TNM		92	105	14	35
LCI-4	Lysine/MA		102	91	118	104
	Arginine/tBD		13	0	85	80
	Arginine/CD		11	5	86	96
	Tyrosine/AIA		45	100	60	50
	Tyrosine/TNM		91	94	16	39

[a] MA, maleic anhydride; tBD, 2,3-butanedione trimer; CD, 1,2-cyclohexanedione; AIA, N-acetylimidazole; TNM, tetranitromethane.
[b] Percent activity related to the activity of the unmodified inhibitor. BT, HT, BCT, and HCT see Table 12.
From Weder and Mueller (in preparation).

bovine chymotrypsin in both LCI-1 and LCI-4 after longer reaction time, while still 35 and 39 % of the activity against human chymotrypsin could be determined. This modification left the activity against human trypsin almost unaltered, while it reduced the activity against ist bovine counterpart by about 10 %.

LBI contains either leucine or phenylalanine at its chymotrypsin-reactive site (due to micro-heterogeneity; Stevens et al., 1974). It was tested in parallel with the lentil inhibitors and found totally active against all the four enzymes.

Reactive sites. All the modification reactions carried out did not give any indication for the presence of a third reactive site in one of the lentil inhibitors. Thus, both LCI-1 and LCI-4 are double-headed proteinase inhibitors of the Bowman-Birk inhibitor family containing arginine at the reactive site against trypsin, and tyrosine at the reactive site against chymotrypsin.

In this respect and also with regard to amino acid composition, the lentil inhibitors seem to be closely related to the proteinase inhibitor from *Vicia angustifolia* L. (Shimokawa et al., 1983). (The latter has developed independently from the other legume species according to amino acid replacements in the primary structure (Shimokawa et al., 1984; Weder, 1985) and also at the reactive sites (Weder, 1985)).

Complex Formation

The remaining activity of both LCI-1 and LCI-4 after modification of the arginine residues with tBD and CD against bovine trypsin (Table 14) can be explained by a loose binding of this enzyme at the second unaltered chymotrypsin-reactive site of the inhibitors. This reaction would also explain the activity of the LCI-1:BT and LCI-4:BT complexes (1:1) against further bovine trypsin (Tables 12 and 13).

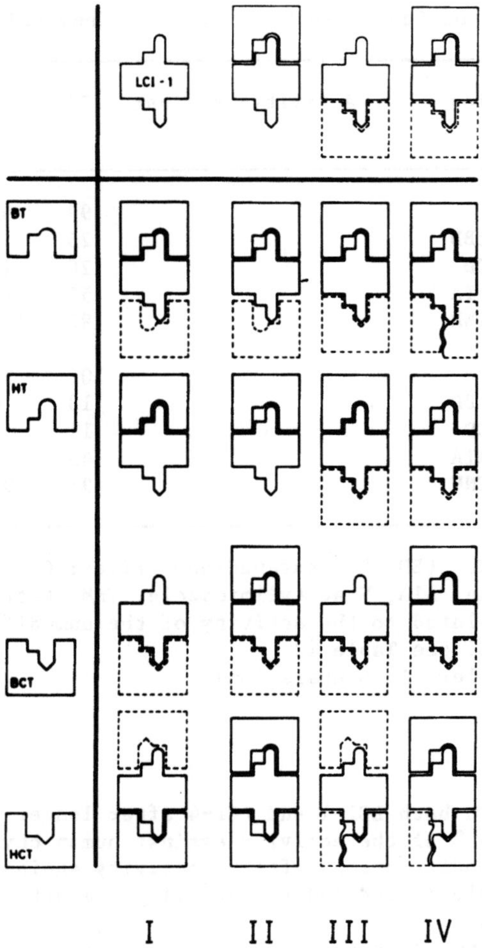

Fig. 11. Schematic representation of complex formation between lentil inhibitors and various enzymes. Schematic drawing of LCI-1 with the trypsin-reactive site at the top and the chymotrypsin-reactive site at the bottom; BT, HT, BCT, and HCT, symbols of the enzymes denoted as in Table 12 with schematic representation of the different shapes of the respective active sites; full lines, enzymes tightly bound in the complexes; broken lines, enzymes loosely bound in the complexes. Inhibitor and complexes (upper row) reacting with the individual enzymes (left-hand column) resulted in the complexes shown on columns I to IV. Two different enzymes bound at the same reactive site (fractured figures) represent simultaneous binding and/or displacement. (Weder and Mueller, in preparation).

The clearly higher activity of the two inhibitors after modification of the tyrosine residues with TNM against human than against bovine chymotrypsin is due to an additional binding of the human enzyme at the other unmodified trypsin-reactive site of the inhibitors.

The complexes that can be formed by reaction of the various enzymes with the inhibitor itself and with the different enzyme:inhibitor complexes are presented schematically in Figure 11.

As shown in column I, the lentil inhibitors primarily bind bovine trypsin at their trypsin-reactive site but, under certain circumstances, also at the reactive site against chymotrypsin. Human trypsin can only be bound by the trypsin-reactive site, as can bovine chymotrypsin by the reactive site against chymotrypsin. Human chymotrypsin primaryly reacts with the chymotrypsin-reactive but, sometimes, also with the reactive site against trypsin.

The LCI:BT complex (column II) can bind further bovine trypsin at the chymotrypsin-reactive site but cannot bind human trypsin. Bovine as well as human chymotrypsin can react with the chymotrypsin-reactive site which is free in this complex.

The LCI:BCT complex (column III) can bind bovine as well as human trypsin with its free trypsin-reactive site. Bovine chymotrypsin cannot react with this complex (the small additional inhibition observed in the experiments is due to a slight increase of the total complexed enzyme). Human chymotrypsin can be loosely bound by the free trypsin-reactive site and can be tightly bound by the reactive site against chymotrypsin using the free portion of this reactive site not complexed by bovine chymotrypsin; it can also displace the already bound bovine chymotrypsin.

The trimeric complex (column IV) can bind additional bovine trypsin with the free portion of the chymotrypsin-reactive site, in principle like the LCI:BT complex. Human trypsin as well as bovine chymotrypsin are not bound by this complex. In the experiments with human chymotrypsin, the free portion of the chymotrypsin-reactive site in the trimeric complex reacts with the added enzyme, causing some displacement of the bound bovine chymotrypsin.

Summarizing these results, the lentil proteinase inhibitors LCI-1 and LCI-4 belong to the Bowman-Birk inhibitor family. They have two different and independent reactive sites. The one, mainly reacting with trypsin, contains arginine and binds tightly to bovine trypsin, less tightly to human trypsin and probably very loosely to human chymotrypsin. The other reactive site, prefering chymotrypsin, contains tyrosine and binds tightly to human chymotrypsin, less tightly to bovine chymotrypsin and loosely to bovine trypsin. In addition to the usual enzyme specificity, the lentil inhibitors react also species-specifically. These properties, when interpreted on a molecular level, do explain the differences in action against human and bovine trypsin and chymotrypsin observed earlier with crude preparations and extracts from lentils.

PART III: INFLUENCE OF PEPSIN ON PROTEINASE INHIBITORS

The proteinase inhibitors will only inhibit the digestion of proteins by trypsin and chymotrypsin if they reach the intestine unaltered. As they are themselves proteins, they may be attacked by pepsin when passing the stomach. Therefore, the susceptibility of these inhibitors to pepsin is important for a comprehensive evaluation of their nutritional significance. In addition, the low pH of the gastric juice may also affect the inhibitors.

In order to simulate this influence, various inhibitors were exposed to the proteolytic action of pepsin and to acid conditions, and the alterations in inhibitor activity were studied. Like the inhibition studies themselves, most of these experiments were done with animal pepsin, mainly porcine, and only a few papers deal with human pepsin. Even where no information on the source of the pepsin was given, porcine pepsin was most likely used because it is commercially available.

Soybean Inhibitors

STI. The effect of bovine pepsin on STI was already studied by Kunitz (1947). Native STI was not affected by pepsin at pH 3.0, but digested to about 40 % after 150 min at pH 2.0 and 30° C. From similar experiments with heat denatured STI a gradual denaturation of STI at pH 2.0 and a subsequent digestion of the denatured protein was thougt to be responsible for this effect. The decrease in trypsin inhibitor activity paralleled the increase in digestion of the inhibitor.

These results were confirmed by Fraenkel-Conrat et al. (1952) who observed a loss of 99 % of the activity of STI caused by pepsin after 24 h in 0.01 N HCl at 40° C and about half the loss in activity (45 %) due to the acid alone, and also by Kassell and Laskowski (1956) who reported losses of 50 and 90 % of the trypsin inhibitor activity after 44 min and 2 h, respectively, at pH 1.5 and 37° C.

Using human pepsin by means of human gastric juice, Krogdahl and Holm (1981) observed a similar susceptibility of STI to human pepsin and acid as found by others with the bovine enzyme. Twenty-four hours of incubation almost eliminated the inhibitor activity of STI against human proteinases. The trypsin inhibitor activity was more slowly inactivated than the chymotrypsin inhibitor activity. In 1 and 4 h of incubation with gastric juice (pH 1.4, 37° C), the inhibitor activity against human trypsin decreased by 55 and 80 %, respectively (HCl alone caused 38 and 50 % losses under the same conditions). The effects of the treatments with pepsin plus HCl and HCl alone on the inhibitor activity of STI against human chymotrypsin did not differ significantly: about 20 and 10 % of the chymotrypsin inhibitor activity were left after 1 and 4 h of incubation with gastric juice, respectively.

BBI. In contrast to STI, BBI is not affected by pepsin. This was already found by Birk (1961) who reported no loss in inhibitor activity after incubation of purified BBI with pepsin at pH 1.5 and 37° C for 2 h. The partial loss in activity at pH 1.6 (70 % in the presence of pepsin and only 5 % without pepsin) after treatment of crude BBI, observed by the same group (Birk and Gertler, 1961), may be caused by impurities (e.g. STI) and also by the longer incubation time (48 h).

Recently, Yavelow et al. (1983) showed that BBI is resistant to stomach digestion. After gavage of BBI labeled with ^{125}I, BBI appeared to be fully active as a proteinase inhibitor in the small intestine of mice and rats, where it complexed with the proteinases occurring there, i.e., trypsin and chymotrypsin.

Extracts. As expected, the total inhibitor activity of soybean extracts is partially inactivated by pepsin because it contains both STI and BBI. Accordingly, Birk and Gertler (1961) observed a 59 % decrease of the trypsin inhibitor activity of a soy flour suspension after incubation with pepsin for 4 h at pH 1.6 and 37° C, which was partially (30 % of the total activity) caused by the acid. Incubation for 48 h led to 85 and 65 % losses for pepsin plus HCl and HCl alone, respectively.

In their study with human pepsin and human pancreatic proteinases mentioned above for STI, Krogdahl and Holm (1981) also investigated the influence of pepsin on extracts from three varieties of soybeans. After incubation with human gastric juice at pH 1.4 and 37° C, the inhibitor activities of these extracts against human trypsin and chymotrypsin were reduced to 60-70 % within 1-2 hours. Prolonged treatments (up to 24 h) did not cause any additional effect. In most cases, the influence of the acid was not increased by the addition of pepsin. From the decrease in activity caused by pepsin and acid, the authors concluded that STI contributed to about one third of the inhibitor activity in these varieties, while BBI was responsible for the remainder. The trypsin inhibitor activity was found to be lost in extracts from a soybean protein isolate, maintained in extracts from raw soybeans (Kanamori and Doi, 1981), and reduced in extracts from another soybean sample (Fukal and Sova, 1983), when subjected to similar experimental conditions.

Lima Bean Inhibitor

LBI. The action of pepsin on LBI produced similar effects like those observed with BBI. A decrease of only 7 % of the trypsin inhibitor activity after incubation with pepsin at 40° C in 0.01 N HCl for 24 h was reported by Fraenkel-Conrat et al. (1952), while acid alone caused no effect.

After incubation with human gastric juice (pH 1.4, 37° C), LBI lost about 15 and 30 % of its inhibitor activity against human trypsin and chymotrypsin, respectively, within the first hour (Krogdahl and Holm, 1981). No further decrease was observed after longer incubation times (up to 24 h), nor was there any significant difference between the effect of pH and that of pepsin at this pH.

Phaseolus vulgaris Inhibitors

Kidney bean. The protease inhibitor isolated from kidney beans is a further representative of the Bowman-Birk inhibitor family. Pusztai (1968) demonstrated that the kidney bean inhibitor was also unaffected by exposure to porcine pepsin and to low pH values (pH 2, 2 h at 37° C).

In contrast, the specific enteropeptidase inhibitor from kidney beans is digested by pepsin (Jacob et al., 1983). Incubation of the inhibitor with porcine pepsin (pH 2.0, 37° C) resulted in a gradual decrease of its anti-enteropeptidase action; 70 % of the activity was lost after pepsin treatment for 90 min. The inhibitor was stable to exposure to pH 2.0 for 34 h at 4° C.

Kintoki bean. Tsukamoto et al. (1983) isolated three isoinhibitors from Kintoki (the Japanese term for red) beans which are a subspecies of kidney beans. According to amino acid composition, molecular weight and activity against trypsin and chymotrypsin, the three isoinhibitors seem to belong to the Bowman-Birk inhibitor family, although one of them did not inhibit chymotrypsin. Treatment of the three inhibitors with pepsin at pH 2 for 1-3 h at 37° C did not affect their inhibitor activity against trypsin and chymotrypsin.

Lentil Inhibitors

LCI-1 and LCI-4. The two purified lentil inhibitors described above, LCI-1 and LCI-4, were exposed to human gastric juice for 4 h at pH 1.5 and 37° C, and the effect on their inhibitor activity was determined (Weder and Mueller, 1985). The two inhibitors displayed similar behaviour, they were scarcely affected with regard to their inhibitor action against both human and bovine trypsin and chymotrypsin (Table 15). Generally, the stability of the inhibitor activity against human proteinases paralleled that against the bovine enzymes. Taking into consideration the sites responsible for the binding of the various enzymes (cf. lentil inhibitor chapter), the reactive site preferring chymotrypsin seems to be the one altered. This would also explain the slight loss in activity against bovine trypsin. Exposure of LCI-1 and LCI-4 to pH 1.5 for the same time at the same temperature caused similar losses in the activity of both inhibitors.

Other Legumes

Chickpea. The mixture of the six isoinhibitors isolated from chickpeas (*Cicer arietinum* L.) was stable towards digestion by porcine pepsin at pH 1.8 and 37° C for 3 hours (Belew et al., 1975).

Table 15. Influence of Pepsin on Lentil Inhibitors

Inhibitor	Residual activity against (%)			
	HT	BT	HCT	BCT
LCI-1	100	89	83	85
LCI-4	100	97	89	100

HT, BT, HCT, and BCT see Table 12; experimentals are given in the text. From Weder and Mueller (1985).

Horse gram. The trypsin chymotrypsin inhibitor isolated from horse gram (*Dolichos biflorus* L.) was resistant to the action of pepsin in 0.06 N HCl at 37° C for 48 hours (Mehta and Simlot, 1982a).

Moth bean. The trypsin inhibitor purified from moth bean (*Phaseolus aconitifolius* Jacq.) was slightly affected by pepsin (Mehta and Simlot, 1982b). Incubation of the inhibitor with porcine pepsin in 0.06 N HCl at 37° C for 48 h caused 9 % inactivation.

Broad bean. The trypsin inhibitor activity of broad beans (also known as horse beans, *Vicia faba* L.) was reported to be stable against the proteolytic action of pepsin by Fukal and Sova (1983).

Pepsin Inhibitors from Legumes

Most of the proteinase inhibitors from legumes tested against pepsin failed to exhibit any inhibitory action against this enzyme. Only the trypsin inhibitor fraction I isolated from black gram (*Vigna mungo* (L.) Hepper) displayed some pepsin inhibitor activity (6 %) when tested in concentrations that caused 92 % inhibition of bovine trypsin and 10 % inhibition of bovine chymotrypsin (Kamalakannan et al., 1981).

In addition, a protein fraction containing two trypsin isoinhibitors was isolated from the vegetative portion of lucerne (*Medicago sativa* L.) which inhibited 30 % of the activity of porcine pepsin on casein when tested at the same concentration that caused 93% inactivation of the same amount of trypsin (Hazlewood et al., 1983).

In summary, most of the legume proteinase inhibitors investigated up to now in this respect are only slightly or not affected at all by pepsin and/or hydrochloric acid. All of the resistant proteinase inhibitors belong to the Bowman-Birk inhibitor family: the compact and rigid structure of the protein molecule stabilized by seven disulfide bonds is responsible for this stability. In contrast, the Kunitz soybean trypsin inhibitor (STI) is inactivated by pepsin under acidic conditions. Although most of the studies have been made with porcine pepsin, the studies with human pepsin indicate that the legume inhibitors display a similar resistance against the human enzyme. As a consequence, most of the active legume proteinase inhibitors ingested survive the passage through the human stomach and can reach the small intestine fully active.

INFLUENCE OF PROCESSING ON PROTEINASE INHIBITORS

About thirty years before the first report of a proteinase inhibitor from plants (soybeans) was published, Osborne and Mendel (1917) had already observed that rats fed raw soybean flour grew more slowly than a control group which received the same flour after it had been cooked for 3 hours. This observation has been extended since then to many other experimental animals (cf. Liener and Kakade, 1980).

There is only one corresponding study with human beings. Although both raw and autoclaved soybean protein resulted in a positive nitrogen balance in human adults, retention of nitrogen is about 20 % greater during the administration of the autoclaved material than with the raw material (Lewis and Taylor, 1947). Recent results on the effect of raw soybean extracts and soy isolate applied intraduodenally to humans are described by Holm et al. (1985).

As can be expected, then, the usefulness of heating to inactivate the proteinase inhibitors is well-known and has been studied in depth. In addition, the influence of water content and of pressure on processing has also been examined. Procedures for the inactivation of proteinase inhibitors by thermal processing and the effect on various legume inhibitors are discussed in a number of recent reviews (Askar, 1983; House and Welch, 1984; Liener and Kakade, 1980; Rackis, 1981; Sathe and Salunkhe, 1984).

Generally, the representatives of the Bowman-Birk inhibitor type need higher temperatures, higher pressures or longer exposure times for an adequate inactivation than the Kunitz soybean trypsin inhibitor. The reason for this lies on the compact and rigid structure of the former as discussed above for pepsin susceptibility. In the whole bean, though, the inactivation of the inhibitors may proceed differently as indicated by the work of Liener and Tomlinson (1981). The authors found that BBI and STI were destroyed by heat at a comparable rate.

As can be expected, limited information is available on the influence of processing on the inhibitor activity against human proteinases. The human trypsin and chymotrypsin produced daily can be inhibited by the extract from about 100 g of raw soybeans (Krogdahl and Holm, 1979), about 200 g of lentils or smaller amounts of other food legumes having a higher inhibitory activity (Belitz et al., 1982). Since the inactivation of the inhibitors is attributed to the thermal denaturation of the protein molecule, the effect on the inhibition of human proteinases should generally parallel that of bovine enzymes. This was found by Struthers and MacDonald (1983): about five times the amount of heated soy protein isolate produced the same effect on human trypsin as did unheated soy flour (Table 1).

Since severe thermal processing necessary to considerably reduce inhibitor activity may produce deleterious nutritious effects to proteins (e.g. L-amino acid racemization, lysinoalanine formation, amino acid damage etc.), attempts have been made to introduce alternative techniques.

For example, reduction of the disulfide bonds by thiols, in particular by cysteine or N-acetylcysteine, was shown to reduce inhibitor activity at moderate temperatures (Friedman et al., 1982a-c and 1984). These effects were achieved with purified inhibitors, STI and LBI, as well as with soybean, lima bean and Great Northern bean flour. In addition, the resulting products displayed better digestibility and higher cysteine content resulting in increased protein efficiency ratio.

Another interesting method for overcoming trypsin inhibition by STI was recently applied for a patent (Dennis et al., 1984). Soybean meal was treated with catalytical amounts of starfish trypsin 1 to produce modified STI (reactive site peptide bond hydrolyzed) which was then digested with a protease, e.g. carboxypeptidase B. The trypsin from starfish (*Dermasterias imbricata*) modifies STI much more rapidly than the usual trypsins (Estell and Laskowski jr., 1980). The patent also includes on oral dosage of the enzyme mixture coated in a way to traverse the stomach unscathed. This method, though, can only affect the STI portion of soybean inhibitors.

CONCLUSIONS

With respect to the nutritional significance of the legume proteinase inhibitors to humans and to the question if they mean a possible hazard for man, the following conclusions can be drawn from the facts reviewed above:

1. Legume proteinase inhibitors can inactivate human trypsins and chymotrypsins if they reached the small intestine unaltered.

2. Most of the legume proteinase inhibitors are not affected by pepsin or pH during the passage through the stomach.

3. Conventional meal preparation and modern processing techniques usually destroy the inhibitor activity of the legume proteinase inhibitors. New processes, though, should be investigated.

4. Despite of differences in action on human and bovine proteinases, the bovine enzymes can be used to evaluate the effect of a particular processing step on the inhibitor activity.

Acknowledgements. Support of the work on lentil inhibitors in the author's laboratory by the Deutsche Forschungsgemeinschaft is gratefully acknowledged.

REFERENCES

Amouric, M., and Figarella, C., 1979, Characterization and purification of a kallikrein from human pancreatic juice and immunological comparison with other kallikreins, Hoppe-Seyler's Z.Physiol.Chem., 360:457-465.

Askar, A., 1983, Schädliche Substanzen in Bohnen und ihre Eliminierung, Ernaehr.Umsch., 30:331-336.

Belew, M., Porath, J., and Sundberg, L., 1975, The trypsin and chymotrypsin inhibitors in chick peas (Cicer arietinum L.): purification and properties of the inhibitors, Eur.J.Biochem., 60:247-258.

Belitz, H.-D., Fuchs, A., and Grimm, L., 1971, Proteinasen-Inhibitoren in Lebensmitteln. III. Trypsin-Chymotrypsin-Inhibitoren von Phaseolus coccineus: Isolierung und einige Eigenschaften, Lebensm.Wiss. Technol., 4:89-92.

Belitz, H.-D., Fuchs, A., Nitsche, G., and Al-Sultan, T., 1972, Proteinasen-Inhibitoren aus Phaseolus vulgaris var. nanus: Isolierung und Vergleich mit Inhibitoren anderer Phaseolus-Arten, Z. Lebensm.Unters.Forsch., 150:215-220.

Belitz, H.-D., Lynen, F., and Weder, J.K.P., 1982, Comparative studies on the inhibitory action of some legume seeds, potato tubers, and bran against human and bovine proteinases, Z.Lebensm.Unters.Forsch., 174:442-446.

Bhat, P.G., Karnik, S.S., and Pattabiraman, T.N., 1976, Natural plant enzyme inhibitors: part III - purification of protease inhibitors from Lathyrus sativus seeds, Indian J.Biochem.Biophys., 13:339-343.

Birk, Y., 1961, Purification and some properties of a highly active inhibitor of trypsin and alpha-chymotrypsin from soybeans, Biochim. Biophys.Acta, 54:378-381.

Birk, Y., and Gertler, A., 1961, Effect of mild chemical and enzymatic treatments of soybean meal and soybean trypsin inhibitors on their nutritive and biochemical properties, J.Nutr., 75:379-387.
Borchers, R., and Ackerson, C.W., 1947, Trypsin inhibitor. IV. Occurrence in seeds of the Leguminosae and other seeds, Arch.Biochem., 13:291-293.
Buck, F.F., Bier, M., and Nord, F.F., 1962, Some properties of human trypsin, Arch.Biochem.Biophys., 98:528-530.
Butler, P.J.G., Harris, J.I., Hartley, B.S., and Leberman, R., 1969, The use of maleic anhydride for the reversible blocking of amino groups in polypeptide chains, Biochem.J., 112:679-689.
Coan, M.H., Roberts, R.C., and Travis, J., 1971, Human pancreatic enzymes. Isolation and properties of a major form of chymotrypsin, Biochemistry, 10:2711-2717.
Coan, M.H., and Travis, J., 1971, Interaction of human pancreatic proteinases with naturally occuring proteinase inhibitors, in: "Proceedings of the International Research Conference on Proteinase Inhibitors, Munich, Nov. 1970," H. Fritz and H. Tschesche, eds., Walter de Gruyter, Berlin, 294-298.
Coan, M.H., and Travis, J., 1972, Human pancreatic enzymes: properties of two minor forms of chymotrypsin, Biochim.Biophys.Acta, 268:207-211.
Dennis, M.S., Estell, D.A., and Light, D.R., 1984, A method for overcoming trypsin inhibition, Eur.Patent, 0138544 A2.
Dubois, M., Gilles, K.A., Hamilton, J.K., Rebers, P.A., and Smith, F., 1956 Colorimetric method for determination of sugars and related substances, Anal.Biochem., 28:350-356.
Duke, J.A., 1981, "Handbook of Legumes of World Economic Importance," Plenum Press, New York.
Estell, D.A., and Laskowski jr., M., 1980, Dermasterias imbricata trypsin 1: an enzyme which rapidly hydrolyzes the reactive-site peptide bonds of protein trypsin inhibitors, Biochemistry, 19:124-131.
Feeney, R.E., Means, G.E., and Bigler, J.C., 1969, Inhibition of human trypsin, plasmin, and thrombin by naturally occurring inhibitors of proteolytic enzymes, J.Biol.Chem., 244:1957-1960.
Feinstein, G., Hofstein, R., Koifmann, J., and Sokolovsky, M., 1974, Human pancreatic proteolytic enzymes and protein inhibitors. Isolation and molecular properties, Eur.J.Biochem., 43:569-581.
Figarella, C., Negri, G.A., and Guy, O., 1975, The two human trypsinogens. Inhibition spectra of the two human trypsins derived from their purified zymogens, Eur.J.Biochem., 53:457-463.
Flavin, D.F., 1982, The effects of soybean trypsin inhibitors on the pancreas of animals and man: a review, Vet.Hum.Toxicol., 24:25-28.
Fraenkel-Conrat, H., Bean, R.C., Ducay, E.D., and Olcott, H.S., 1952, Isolation and characterization of a trypsin inhibitor from lima beans, Arch.Biochem.Biophys., 37:393-407.
Fredericq, L., 1878, Sur la digestion des albuminoides chez quelques invertébrés, Bull.Acad.R.Sci.Lett.Beaux-Arts Belg., 47me année, 2. Série, 46:213-228.
Friedman, M., Grosjean, O.K., and Zahnley, J.C., 1982a, Inactivation of soya bean trypsin inhibitors by thiols, J.Sci.Food.Agric., 33:165-172.
Friedman, M., Grosjean, O.K., and Zahnley, J.C., 1982b, Effect of disulfide bond modification on the structure and activities of enzyme inhibitors, in: "Food Protein Deterioration: Mechanisms and Functionality," J.P. Cherry, ed., ACS Symp. Series, Washington, D.C., 206:359-407.
Friedman, M., Grosjean, O.K., and Zahnley, J.C., 1982c, Cooperative effects of heat and thiols in inactivating trypsin inhibitors from legumes in solution and in the solid state, Nutr.Rep.Int., 25:743-751.
Friedman, M., Gumbmann, M.R., and Grosjean, O.K., 1984, Nutritional improvement of soy flour, J.Nutr., 114:2241-2246.

Fukal, L., and Sova, Z., 1983, Effect of thermal and proteolytic effects on the trypsin-inhibiting efficiency of horse beans, soybeans and maize, Krmivarstvi Sluzby, 19:251-252; from Chem.Abstr., 1984, 100:101742g.

Geokas, M.C., Largman, C., Brodrick, J.W., Raeburn, S., and Rinderknecht, H., 1975, Human pancreatic carboxypeptidase B. I. Isolation, purification, and characterization of fraction II, Biochim.Biophys.Acta, 391:396-402.

Gertler, A., Poraz, I., Cohen, N., and Gottesfeld, F., 1982, Inhibition of human and dog serine pancreatic proteinases by naturally occurring high-molecular-weight inhibitors from plant or animal origin, Biochem.Med., 27:143-153.

Hazlewood, G.P., Horsnell, J.M., and Mangan, J.L., 1983, Trypsin isoinhibitors of lucerne: association with leaf fraction 1 protein, Phytochemistry, 22:1107-1111.

Holm, H., and Krogdahl, A., 1982, Problems in predicting the inhibition of human pancreas proteinases by soya bean proteinase inhibitors: in-vitro assays employing human, bovine and porcine proteinases, J.Sci. Food Agric., 33:1164-1171.

Holm, H., Hansen, L., and Krogdahl, A., 1985, In vivo effects of soy bean proteinase inhibitors in humans, Federation Proceedings, 44, 1946.

Hory, H.-D., Weder, J.K.P., and Belitz, H.-D., 1976, Vergleichende Untersuchungen der trypsinspezifischen reaktiven Zentren verschiedener Phaseolus coccineus und Phaseolus vulgaris-Inhibitoren, Z. Lebensm.Unters.Forsch., 162:341-347.

House, W.A., and Welch, R.M., 1984, Effects of naturally occurring antinutrients on the nutritive value of cereal grains, potato tubers, and legume seeds, ASA Spec.Publ., 48:9-35.

International Union of Biochemistry, Nomenclature Committee, 1979, "Enzyme Nomenclature, 1978," Academic Press, New York.

Jacob, R.T., Bhat, P.G., and Pattabiraman, T.N., 1983, Isolation and characterization of a specific enterokinase inhibitor from kidney bean (Phaseolus vulgaris), Biochem.J., 209:91-97.

Kamalakannan, V., Sathyamoorthy, A.V., and Motlag, D.B., 1981, Studies on black gram (Vigna mungo) trypsin inhibitor, J.Sci.Food Agric., 32:1172-1176.

Kamalakannan, V., Sathyamoorthy, V., and Motlag, D.B., 1984, Purification, characterisation and kinetics of a trypsin inhibitor from black gram (Vigna mungo), J.Sci.Food Agric., 35:199-206.

Kanamori, M., and Doi, V., 1981, Preparation and some characertistics of soybean protein from raw soy bean and soybean protein isolate, Daizu Tanpakushitsu Eiyo Kenkyukai Kaishi, 2:8-13; from Chem.Abstr., 1983, 98:52088c.

Kassell, B., and Laskowski, M., 1956, The comparative resistance to pepsin of six naturally occurring trypsin inhibitors, J.Biol.Chem., 219:203-210.

Krogdahl, A., and Holm, H., 1979, Inhibition of human and rat pancreatic proteinases by crude and purified soybean proteinase inhibitors, J.Nutr., 109:551-558.

Krogdahl, A., and Holm, H., 1981, Soybean proteinase inhibitors and human proteolytic enzymes: selective inactivation of inhibitors by treatment with human gastric juice, J.Nutr., 111:2045-2051.

Krogdahl, A., and Holm, H., 1983, Pancreatic proteinases from man, trout, rat, pig, cow, chicken, mink and fox. Enzyme activities and inhibition by soybean and lima bean proteinase inhibitors, Comp.Biochem. Physiol., 74B:403-409.

Kunitz, M., 1945, Crystallization of a trypsin inhibitor from soybean, Science, 101:668-669.

Kunitz, M., 1946, Crystalline soybean trypsin inhibitor, J.Gen.Physiol., 29:149-154.
Kunitz, M., 1947, Crystalline soybean trypsin inhibitor. II. General properties, J.Gen.Physiol., 30:291-310.
Largman, C., Brodrick, J.W., and Geokas, M.C., 1976, Purification and characterization of two human pancreatic elastases, Biochemistry, 15:2491-2500.
Laskowski jr., M., Kato, I., Leary, T.R., Schrode, J., and Sealock, R.W., 1974, Evolution of spezificity of protein proteinase inhibitors, in: "Bayer-Symposium V: Proteinase Inhibitors," Proc.Int.Res.Conf., 2nd, 1973, H. Fritz, H. Tschesche, L.J. Greene and E. Truscheit, eds., Springer-Verlag, Berlin, 597-611.
Lewis, J.H., and Taylor, F.H.L., 1947, Comparative utilization of raw and autoclaved soy bean protein by the human, Proc.Soc.Exp.Biol.Med., 64:85-87.
Liener, I., 1979, Significance for humans of biologically active factors in soybeans and other food legumes, J.Am.Oil Chem.Soc., 56:121-129.
Liener, I.E., and Kakade, M.L., 1980, Protease inhibitors, in: "Toxic constituents of plant foodstuffs, 2nd ed.," I.E. Liener, ed., Academic Press, New York, 7-71.
Liener, I.E., and Tomlinson, S., 1981, Heat inactivation of protease inhibitors in a soybean line lacking the Kunitz trypsin inhibitor, J.Food.Sci., 46:1354-1356.
Lundblad, R.L., and Noyes, C.M., 1984, "Chemical Reagents for Protein Modification, Vol, 2," CRC Press, Boca Raton, Florida.
Mallory, P.A., and Travis, J., 1973, Human pancreatic enzymes. Characterization of anionic human trypsin, Biochemistriy, 12:2847-2851.
Mallory, P.A., and Travis, J., 1975a, Inhibition spectra of the human pancreatic endopeptidases, Am.J.Clin.Nutr., 28:823-830.
Mallory, P.A., and Travis, J., 1975b, Human pancreatic enzymes: purification and characterization of a nonelastolytic enzyme, protease E, resembling elastase, Biochemistry, 14:722-730.
Means, G.E., and Feeney, R.E., 1971, "Chemical Modification of Proteins," Holden-Day, San Francisco.
Metha, S.L., and Simlot, M.M., 1982a, An acid stable trypsin-chymotrypsin inhibitor from horse gram (Dolichos biflorus), J.Biosci., 4:295-306.
Metha, S., and Simlot, M.M., 1982b, Purification and characterization of trypsin inhibitor from moth bean, Phaseolus aconitifolius, Indian J. Biochem.Biophys., 19:403-407.
Nexoe, E., Poulsen, S.S., Hansen, S.N., Kirkegaard, P., and Olsen, P.S., 1984, Characterization of a novel proteolytic enzyme localized to goblet cells in rat and man, Gut, 25:656-664; from Chem.Abstr., 1984, 101:125550q.
Osborne, T.B., and Mendel, L.B., 1917, The use of soy bean as food, J.Biol.Chem., 32:369-387.
Peterson, L.M., Sokolovsky, M., and Vallee, B.L., 1976, Purification and crystallization of human carboxypeptidase A, Biochemistry, 15:2501-2508.
Prabhu, K.S., Saldanha, K., and Pattabiraman, T.N., 1984, Natural plant enzyme inhibitors: a comparative study of the action of legume inhibitors on human and bovine pancreatic proteinases, J.Sci.Food Agric., 35:314-321.
Pusztai, A., 1968, General properties of a protease inhibitor from the seeds of kidney bean, Eur.J.Biochem., 5:252-259.
Rackis, J.J., 1981, Comparison of the food value of immature, mature, and germinated soybeans, ACS Symp.Ser., 170:183-212.
Read, J.W., and Haas, L.W., 1938, Studies on the baking quality of flour as affected by certain enzyme actions. V. Further studies concerning potassium bromate and enzyme activity, Cereal Chem., 15:59-68.

Rinderknecht, H., Renner, I.G., Carmack, C., Friedman, R., and Koyama, P., 1978, A new protease in human pancreatic juice (abstr.), Clin. Res., 26:112.
Rinderknecht, H., Renner, I.G., Abramson, S.B., and Carmack, C., 1984, Mesotrypsin: a new inhibitor-resistant protease from a zymogen in human pancreatic tissue and fluid, Gastroenterology, 86:681-692.
Sathe, S.K., and Salunkhe, D.K., 1984, Technology of removal of unwanted components of dry beans, CRC Crit.Rev. Food Sci.Nutr., 21:263-287.
Sealock, R.W., and Laskowski jr., M., 1969, Enzymatic replacement of the arginyl by a lysyl residue in the reactive site of soybean trypsin inhibitor, Biochemistry, 8:3703-3710.
Shimokawa, Y., Kuromizu, K., Araki, T., Ohata, J., and Abe, O., 1983, The complete amino acid sequence of Vicia angustifolia proteinase inhibitor, Shizen to Bunka (Fukuoka) (Nature and Culture), 10:69-73.
Shimokawa, Y., Abe, O., and Kuromizu, K., 1984, Phylogenetic relationship of legume double-headed proteinase inhibitors, Shizen to Bunka (Fukuoka) (Nature and Culture) 11:39-45.
Siegelman, A.M., Carlson, A.S., and Robertson, T., 1962, Investigation of serum trypsin and related substances. I. The quantitative demonstration of trypsinlike activity in human blood serum by a micromethod, Arch.Biochem.Biophys., 97:159-163.
Smirnoff, P., Khalef, S., Birk, Y., and Applebaum, S.W., 1976, A trypsin and chymotrypsin inhibitor from chick peas (Cicer arietinum), Biochem.J., 157:745-751.
Stevens, F.C., Wuerz, S., and Krahn, J., 1974, Structure-function relationships in lima bean protease inhibitor, in: "Bayer-Symposium V: Proteinase Inhibitors," Proc.Int.Conf., 2nd, 1973, H. Fritz, H. Tschesche, L.J., Greene and E. Truscheit, eds., Springer-Verlag, Berlin, 344-354.
Struthers, B.J., and MacDonald, J.R., 1983, Comparative inhibition of trypsins from several species by soybean trypsin inhibitors, J.Nutr., 113:800-804.
Sziegoleit, A., 1984, A novel proteinase from human pancreas, Biochem.J., 219:735-742.
Temler, R.S., Simon, E., and Amiguet, P., 1983, Comparison of the interactions of soybean protease inhibitors with rat pancreatic enzymes and human trypsin, Enzyme, 30:105-114.
Tsukamoto, I., Miyoshi, M., and Hamaguchi, Y., 1983, Purification and characterization of three trypsin inhibitors from beans, Phaseolus vulgaris 'Kintoki', Cereal Chem., 60:281-286.
Ventura, M.M., and Xavier-Filho, J., 1966, A trypsin and chymotrypsin inhibitor from black-eyed pea (Vigna sinensis). I. Purification and partial characterization, An.Acad.Brasil.Cienc., 38:553-566.
Wagner, L.P., and Riehm, J.P., 1967, Purification and partial characterization of a trypsin inhibitor isolated from the navy bean, Arch.Biochem.Biophys., 121:672-677.
Weber, K., and Osborn, M., 1969, The reliability of molecular weight determinations by dodecyl sulfate-polyacrylamide gel electrophoresis, J.Biol.Chem., 244:4406-4412.
Weder, J., and Hory, H.-D., 1972, Trypsinand Chymotrypsin-Inhibitoren in Leguminosen. I. Nachweis und Isolierung verschiedener Inhibitoren von Pisum sativum, Lebensm.Wiss.Technol., 5:54-63.
Weder, J.K.P., Hegarty, M.P., Holzner, M., and Kern-Dirndorfer, M.-L., 1983, Trypsin and chymotrypsin inhibitors in Leguminosae. VIII. Isolation and some properties of the principal inhibitor from lentils (Lens culinaris Medik.), Z.Lebensm.Unters.Forsch., 177:109-113.
Weder, J.K.P., Mueller, R., Ildefonso Mato, C., and Hegarty, M.P., 1985, Proteinase inhibitor activities and electrophoretic inhibitor patterns of lentils, LENS Newsletter, 12:in press.

Weder, J.K.P., and Mueller, R., 1985, Inhibition of human and bovine enzymes by lentil proteinase inhibitors, in: "Strategies in Food Quality Assurance: Analytical, Industrial and Legal Aspects, Vol. 2," Proc.Euro Food Chem III, March 26-29, 1985, Antwerp, Belgium, W. Baltes, P.B. Czedick-Eysenberg, H. Deelstra, P. Dirinck, W. Ooghe and W. Pfannhauser, eds., De Sikkel, Malle, Belgium, 282-287.
Weder, J.K.P., 1985, Chemistry of legume protease inhibitors and their use in taxonomy, Qual.Plant.:Plant Foods Hum.Nutr., in press.
Wilson, K.A., and Laskowski sr., M., 1973, Isolation of three isoinhibitors of trypsin from garden bean, Phaseolus vulgaris, having either lysine or arginine at the reactive site, J.Biol.Chem., 248:756-762.
Wilson, K.A., 1981, The structure, function, and evolution of legume proteinase inhibitors, in: "Antinutritients and Natural Toxicants in Foods," Pap.Symp. 1979, R.L. Ory, ed., Food Nutr. Press, Westport, Conn., 187-202.
Yavelow, J., Finlay, T.H., Kennedy, A.R., and Troll, W., 1983, Bownan-Birk soybean protease inhibitor as an anticarcinogen, Cancer Res. (Suppl.), 43:2454s-2459s.

17

REGULATION OF PROTEINASE INHIBITOR GENES IN FOOD PLANTS

Willis E. Brown[1], John S. Graham[2], Jong S. Lee and Clarence A. Ryan

Institute of Biological Chemistry
Washington State University
Pullman, WA 99164-6340

ABSTRACT

Proteinase inhibitor proteins in the families Solanaceae, Leguminosae and Graminae are stored in seeds and tubers and are also found to accumulate in leaves in response to pest attacks. The isolation of the proteinase Inhibitor I cDNA and its gene from tomato has provided information concerning proteinase inhibitor synthesis, processing and compartmentation in leaves under pest attacks. Strategies are now being developed using proteinase inhibitor genes, to genetically engineer the quantity and quality of these potentially defensive and highly nutritional proteins of important crop plants.

INTRODUCTION

Nutritional and antinutritional aspects of plant proteinase inhibitors have been a subject of considerable research over the past few decades (1-3). It is clear that the native trypsin inhibitors, when ingested by animals even as a few percent of the total protein intake, can have severe effects on their digestive physiology (4-6). On the other hand, the plant proteinase inhibitors in important crop plants, such as potatoes, are often rich in essential amino acids, such as lysine or sulfur-containing amino acids (7), and, when cooked and denatured, can be highly beneficial in animal diets. An understanding of the fundamental processes of synthesis and storage of proteinase inhibitors in plant tissues is important as a basis to employ modern DNA technology for the manipulation of inhibitor genes for potential improvement of the quantity and quality of edible plant proteins.

Proteinase inhibitors in plants are mainly found in storage organs such as seeds and tubers where they often comprise over 5% of the total proteins of the tissues (8-10). Their presence in storage organs appears to be developmentally regulated, that is, they are synthesized and stored in the seeds and tubers as the organs develop (11,12). Recently some

[1]Present address: Department of Biology, Purdue University, West Lafayette, IN 47907.
[2]Present address: CIBA-GEIGY Corp., P.O. Box 12257 Research Triangle Park, NC 27709.

proteinase inhibitors have been shown to be under environmental control in plant leaves (13-15). Potato, tomato and alfalfa leaves all respond to attacks of chewing insects or other severe mechanical damage by accumulating proteinase inhibitors throughout the plants (13-15).

A major objective of studying plant proteinase inhibitors is to isolate the individual genes and establish the basic information with which to use them and/or their promotors to transform other plants. With a successful transformation system the structural and functional characteristics of the isolated genes can be studied in a direct and useful manner. For example, with gene fusions between the wound-induced promoters and other proteinase inhibitor genes or insect toxin genes, plants may be developed with much greater defense potentials against attacking pests. On the other hand, using developmentally regulated promoters, genes might be fashioned which produce more nutritional proteins in seeds and tubers.

Another major objective of studying these proteins is to determine if the genes that are developmentally regulated are the same as those that are environmentally regulated and in each case to identify the structural components that are involved in the regulation of these genes. A complete understanding of the expression of inhibitor genes could be highly useful in improving crop plants using recombinant DNA technologies.

WOUND-INDUCTION OF PROTEINASE INHIBITORS IN LEAVES

In tomato and potato plants, two non-homologous proteinase inhibitors accumulate when the plants are attacked by insects (14,16). These inhibitors, called Inhibitor I and Inhibitor II, were previously found as two of several proteinase inhibitors that are major proteins in potato tubers (17-19). Some properties of the two inhibitors isolated from leaves of wounded tomato plants are presented in Table I. Both inhibitors are very similar in their properties to the two extensively studied proteinase inhibitors from potatoes.

Table I. Some Properties of Wound-Induced Proteinase Inhibitor from Tomato Leaves

Inhibitor	M_r		K_I	
	monomer	multimer	chymotrypsin	trypsin
I	8,100	40,000	5×10^{-9}	2×10^{-7}
II	12,300	23,000	3×10^{-8}	8×10^{-8}

Tomato Inhibitor I has a molecular weight of 40,000 and is composed of five identical subunits with molecular weights of 8100 (23). Each subunit possesses an active site specific for chymotrypsin with a K_I of about 10^{-9}M. Inhibitor I is also a weak inhibitor of trypsin. Inhibitor II has a molecular weight of about 23,000, is composed of two subunits, and strongly inhibits both trypsin and chymotrypsin with K_I values of about 10^{-8} and 10^{-7}, respectively (23). However, the two inhibitors do not share any homology. Inhibitor I, for example, possesses a single disulfide bond (20,24) while Inhibitor II has eight (25).

Inhibitors I and II in tomato plants (Lycopersicon esculentum, L. var. Bonny Best) accumulate in leaves within 6 hr after wounding. After about 60-90 min, the inhibitors commence accumulation in nearby unwounded leaves (14). Inhibitor I accumulates at about twice the rate of Inhibitor II. A signal released at the wound site is rapidly (within 1-2 hr) transported from the wounded leaves, primarily upwards, to other leaves where it activates inhibitor genes (16). The signal, called the proteinase inhibitor inducing factor, PIIF, has been tentatively identified as a fragment of the plant's own cell wall that is released during wounding (20-22).

The biochemistry of the wound-induction of proteinase inhibitors in tomato leaves has been extensively studied (14,16), although its mechanism is still not fully understood. Wounding apparently causes a mixing of cellular endopolygalacturonases (from the vacuole?) with the plant cell wall to produce small pectic fragments that are apparently transported through the phloem to other tissues where they activate the inhibitor genes (16). Pectic fragments (PIIF) supplied to excised leaves can activate the genes for Inhibitors I and II, causing the tissues to accumulate large quantities of the inhibitors (14). Within two or three days after adding PIIF the inhibitors can represent several percent of the total proteins of the leaves.

The newly synthesized inhibitors are compartmented in the central vacuole (26,27), and this process has been shown to involve synthesis of new translatable mRNA (28). In our initial studies, the inhibitors, synthesized using wound-induced mRNA in an in vitro translation system, were found to be larger than the native inhibitors. The nascent proteins contained transit sequences that were lost either during or after synthesis, presumably as part of the compartmentation process (28).

ISOLATION AND CHARACTERIZATION OF cDNAs FOR WOUND-INDUCED PROTEINASE INHIBITOR I AND II

A cDNA library was constructed in the plasmid pUC9 (24) from poly(A)+ mRNAs from leaves of tomato that were induced to synthesize Inhibitor I and II by wounding. Inhibitor I and II cDNA clones were identified among the cDNAs by colony hybridization with ^{32}P-labeled oligonucleotide probes (24,25).

A wound-induced Inhibitor I cDNA was characterized that was 571 bp in length and contained an open reading frame coding for an Inhibitor I protein of 111 amino acids (Fig. 1) (24). This is much larger than Inhibitor I found in tubers or leaves. The nascent protein contained an N-terminal 23 amino acid transit sequence followed by a 19 amino acid sequence containing 9 charged amino acids that is lost during maturation.

An Inhibitor II cDNA was 693 bp in length coded for a complete Inhibitor II protein of 148 amino acids having a transit sequence of 25 amino acids (Fig. 2) (25). This was the first determination of the complete Inhibitor II sequence. It revealed that Inhibitor II is composed of 2 domains that have apparently risen by gene duplication and elongation. Each domain possesses an active site, one specific for chymotrypsin and one for trypsin.

The acquisition of the inhibitor cDNAs has enabled us to study the temporal induction of the inhibitor mRNAs using the cDNAs as hybridization probes. Inhibitor I and II mRNAs exhibit rapid accumulation between 4-6 hrs after an initial wound, reaching a maximum level approximately 8-12 hrs after wounding. The time course of mRNA accumulation appears to be identical for the two inhibitors (29). The magnitude of the increases in mRNA levels is a function of the multiplicity of wounds. A series of 4 wounds spaced 1 hr apart resulted in a dramatic increase in accumulation of both Inhibitor I mRNA and protein as compared to that of a single wound. Thus, repeated wounding, such as inflicted by chewing insects, may

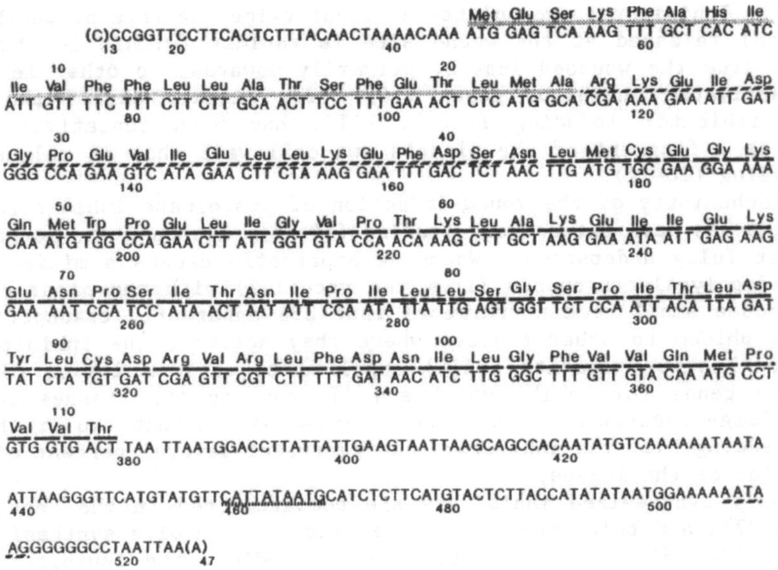

Fig. 1. Complete DNA sequence of a wound-induced tomato Inhibitor I cDNA (24). The numbering of amino acids appear on top. The numbering of nucleotides appears underneath each line of sequence data. The presumptive "transit peptide" is . The 19-amino acid segment not found in mature Inhibitor I is shown by . The amino acid sequence of the mature form of Inhibitor I is shown by . The putative polyadenylation site AATAAG is also shown by . A 10-bp palindromic sequence is shown by .

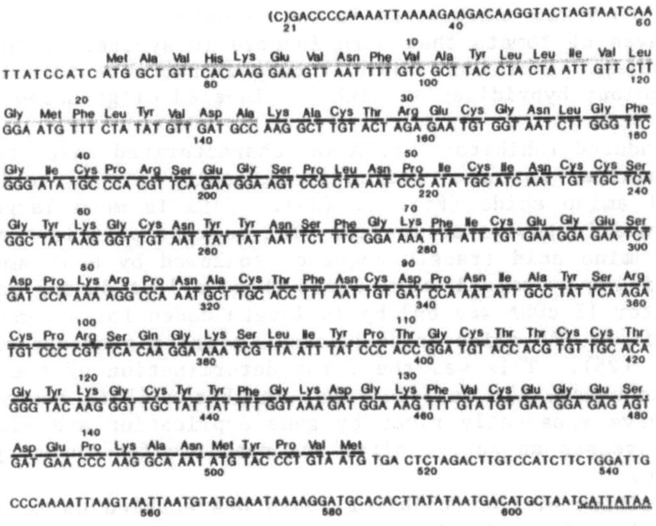

Fig. 2. Complete DNA sequence of a wound-induced tomato Inhibitor II cDNA (25). The numbering of amino acids appears on top. The numbering of nucleotides appears underneath each line of sequence data. The presumptive "signal sequence is . The amino acid sequence of the mature Inhibitor is shown by . The putative polyadenylation site is shown by . A 10-bp palindromic sequence is shown by .

result in the delivery of more PIIF to leaves and provide a reinforcement for Inhibitor I and II mRNA accumulation over long period of time.

ISOLATION OF INHIBITOR GENE FROM TOMATO

The Inhibitor I cDNA was used as a probe to select genomic clones of Inhibitor I from a tomato gene library prepared in λ phage, provided by Dr. William Briedenbach of the University of California, Davis, Calif. and Dr. Robert Goldberg of UCLA. A restriction map of a selected genomic clone was made and a fragment containing the gene was subcloned into plasmids and sequenced. The nucleotide sequence of the entire open reading frame and the 5'- and 3'-non-coding region of an Inhibitor I gene has been determined.

A full discussion of the structure and organization of the wound-induced proteinase inhibitor gene would not serve the purpose of this report, but a boxed representation of the Inhibitor I gene is presented in Fig. 3. The gene revealed the presence of several features not available from the cDNA sequence. Two intervening sequences, or introns, are

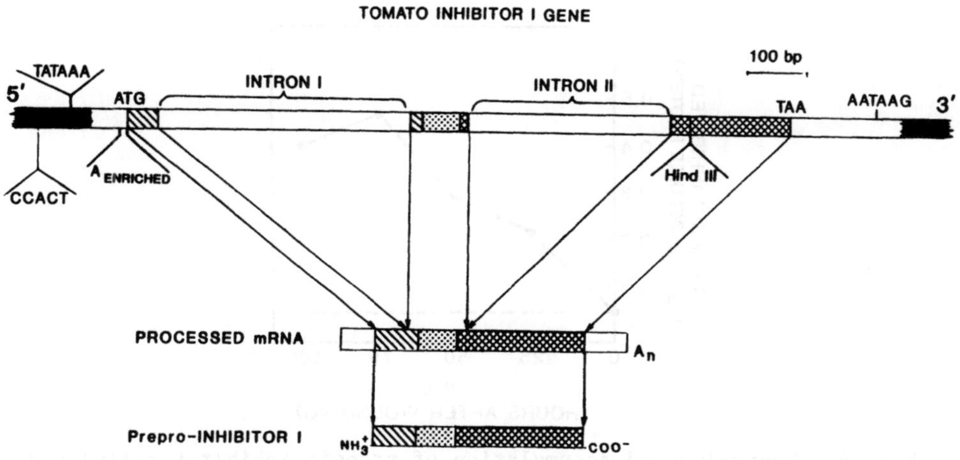

Fig. 3. The organization of a wound-induced tomato Inhibitor I gene. The region of the gene coding for the transit sequence is single hatched; for the pro-sequence, dotted, and the mature Inhibitor I, double-hatched.

present in the gene. One intron is within a sequence coding for the transit peptide while the other is within a sequence that code for the N-terminal amino acids of the inhibitor. The sequences at intron-exon junctions are typical of other eukaryotic genes. The 5'-regulatory sequences appear to include typical TATA and CAT boxes and the 3'-flanking region shows a polyadenylation signal (AATAAG) and a short palindrome sequence (CATTATAATC) that is found, either fully conserved or with minor changes, in all of the inhibitor genes. However, these assignments are only tentative since the exact sequences involved in wound induction are not yet known. In order to understand the regulation mechanism of expression of these genes, a gene fusion of the Inhibitor I gene, at the internal HindIII site with the chloramphenicol acetyltransferase gene (CAT), has been constructed by Dr. Gynheung An of our Institute of Biological Chemistry into a shuttle vector (30) to transform plants. Successful transformation will be determined by the measurement of the CAT activity of transformed tissues. The success of these experiments will be fundamental for in vitro mutagenesis of the gene to determine which regions of this and other inhibitor genes are necessary and sufficient for wound induction.

WOUND-INDUCED TRYPSIN INHIBITOR FROM ALFALFA

Recently we have found that alfalfa plants wounded in the same manner as potato and tomato plants (across the main vein of the leaf with a hemostat) accumulate trypsin inhibitor activity in upper leaves (Fig. 4) (31). The rate of induction and quantity of inhibitors accumulated in alfalfa leaves are similar to those of proteinase Inhibitors I and II in tomato and potato leaves.

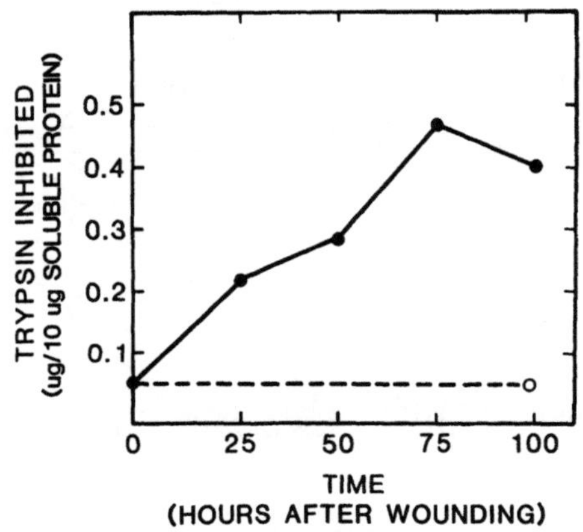

Fig. 4. Wound-induced accumulation of trypsin inhibitor activity in alfalfa leaves. The bottom two leaves of four alfalfa plants were wounded with a hemostat at zero time and again at 8 h. The three or four upper (nonwounded) leaves were collected at the time indicated, and extracts were prepared and assayed for trypsin inhibitor activity. Trypsin inhibitor activity was determined at several concentrations of each extract (31). The micrograms of trypsin inhibited by 10 µg of extract was calculated and plotted. Leaves from unwounded plants (o); unwounded leaves from wounded plants (•).

Alfalfa trypsin inhibitor, ATI, also accumulates in leaves of detached alfalfa plants supplied with tomato PIIF through the cut petioles (31). PIIF is an effective inducer of ATI in alfalfa leaves in a dose-dependent manner up to └1 mg/ml. The concentrations of PIIF that induce ATI are virtually identical to those that induce Inhibitors I and II in tomato plants. Chitosan (32), an inducer of Inhibitors I and II in tomato plants (33), is also a potent inducer of ATI (31), again suggesting possible similarities between the induction mechanism in alfalfa and tomato leaves.

The alfalfa inhibitor was purified to homogeneity (31). It contains four isoinhibitor species with Mr of 7500, a K_{diss} for trypsin of 10^{-10}M, and is rich in cysteine residues (14/mole). Immunodiffusion experiments to compare antigenic determinants of ATI with the soybean Bowman-Birk Inhibitor, BBI (34), or lima bean inhibitor, LBI (35) (a relative of BBI), did not reveal any crossreactivity with ATI (31). However, in the more sensitive immunoblot assays, in which rabbit anti-ATI IgG is visualized with goat, anti-rabbit IgG antiserum labeled with peroxidase (36), ATI was found to react weakly with LBI (37) (but still not with BBI) establishing the inhibitor as a member of the BBI family (34). The crossreactivity of ATI antibodies with LBI was confirmed using the sensitive ELISA method of detection.

```
              1       5         10        15        20        25        30        35        40
                                                                *
ATI                                       T T A C C N F C P C T R S I P P Q C R C T D I G - E T
LBI (23)      S G H H E H S T D Z P S Z S K P C(C,B)H(C,A,C)T K S I P P Q C R C T D L R L D S
BBI-III (24)    S D Q S S S Y D D D E Y S K P C C D L C M C T R S M P P Q C S C E D I R L N S
Common                                          C C     C   C T   S   P P Q C   C   D

                      45        50        55        60        65        70        75        80
                                   *
ATI           C H S A C K T C L C T K S I P P Q C H/R C A/T D I T N F C Y P K C N
LBI           C H S A C K S C I C T L S I P A Q C V(C,B,B)I B D F C Y E P C K S S H S D D D N N N
BBI-III       C H S D C K S C M C T R S Q P G Q C R C L D T N D F C Y K P C K S R D D
Common        C H S   C K   C   C T   S   P   Q C   C             F C Y     C
```

Fig. 5. The amino acid sequence of alfalfa trypsin inhibitor (ATI) (33). The sequence is compared with the Bowman-Birk inhibitor (BBI-III) and lima bean inhibitor (LBI). Residues common to all three inhibitors at similar loci are shown.

The identity of ATI as a member of the Bowman-Birk family was confirmed unambiguously from the complete amino acid sequence of one of its four isoinhibitors (37). In Fig. 5 the amino acid sequence of ATI is shown compared with those of two other members of the Bowman-Birk inhibitor family, BBI-III from soybeans and LBI from lima beans.

The weak crossreactivity of ATI antibodies with LBI, and the apparent absence of crossreactivity with BBI, were somewhat surprising considering the high percentage of amino acid sequence identity among the inhibitors. The reasons for this apparent lack of immunological homology between ATI and BBI are not yet clear but it indicates that parts of the ATI structure must be radically different from BBI.

The finding of a Bowman-Birk type inhibitor in alfalfa leaves, induced by wounding, revealed that inhibitors in legumes, previously found to be only developmentally regulated in seeds, can be induced in leaves by to mechanical wounding (pest attacks). The isolation and characterization of the ATI gene is in progress with the anticipation that it will provide valuable information concerning the regulation of synthesis of the BBI family in legumes.

SUMMARY

Synthesis and storage of proteinase inhibitors among members of the plant families Solanaceae and Leguminosae are under both developmental and environmental regulation. In seeds and tubers, synthesis is apparently regulated by signals programmed to be released into tissues by physiological events during development. In leaves, synthesis is regulated by a systemic wound signal that insect attacks or any severe damage to the tissues release into the vascular system for transport throughout the plants, presumably as a defensive response.

The proteinase inhibitors, with their unusual properties of being potential defensive chemicals as native proteins and highly nutritive proteins when denatured, provide ideal genes with an excellent potential for selection and/or manipulation to improve important food crops. DNA technology offers the possibilities of fashioning proteinase inhibitor proteins to maximize both productivity and nutritional value of plant foods. By introducing new proteinase inhibitor genes into food crops it may be possible to test whether such genes might enhance natural defense systems. It also appears possible to alter structures of the inhibitors by *in vitro* mutagenesis so that they will maintain their inhibitory properties but will be quickly and efficiently denatured during cooking. This would ensure the elimination of any potential deleterious effects to human health that might have been caused by the native inhibitors, while increasing nutritional quality. The isolation of mRNAs, cDNAs and genes that code for the proteinase inhibitors have now provided a foundation for a further understanding of the environmental and developmental regulation of the inhibitor proteins. With such information the utilization of the genes to improve the quantity and quality of human food becomes nearer to reality.

REFERENCES

1. T.B. Osborne, and L.B. Mendel, The use of soybeans as foods, J. Biol. Chem. 32 369 (1917).
2. J.R. Whitaker, and R.E. Feeney, Enzyme inhibitors in foods, in Toxicants Occurring Naturally in Foods, National Academy of Science, Washington, D.C. pp. 276-298 (1973).
3. R.R. Rackis, Protease inhibitors: physiological properties and nutritional significance, *in*: "Antinutrients and Natural Toxicants in Foods," R.L. Ory, ed. Food and Nutrition Press Inc. Westport Comm. pp. 203-238 (1981).
4. G.M. Green, W.H. Levan and R.A. Liddle, Regulation of plasma cholecystokinin and pancreatic function by substrates and inhibitors of pancreatic proteases. This volume.
5. J.J. Rackis, E.C. Baker and W.J. Wolf, Inactivation and removal of protease inhibitors from plant foods
6. D. Gallaher and B.O. Schneeman, Nutritional and metabolic response to

plant inhibitors of digestive enzymes, in Nutritional and toxicological aspects of food safety, M. Friedman ed. Plenum Press, N.Y. Vol 177, pp. 299-320 (1984).
7. G. Pearce, J. McGinnis, and C.A. Ryan, Utilization by chicks of half-cystine from native and denatured proteinase inhibitor protein from potatoes, Proc. Soc. Expt. Biol. Med. 160:180 (1979).
8. C.A. Ryan, Proteolytic enzymes and their inhibitors in plants, Ann. Rev. Plant Physiol. 24:173 (1973).
9. M. Richardson, The proteinase inhibitors of plants and microorganisms, Photochemistry 16:159 (1977).
10. J. Mikola and M. Kirsi, Differences between endospermal and embryonyl trypsin inhibitors in barley, wheat and rye, Acta. Chem. Seond. 26:787 (1972).
11. R.B. Goldberg, R.L. Fischer, J.L. Harada, D. Jafukos, and J.K. Okamuro, Organization of soybean seed protein genes and their flanking regions, in: "Structure and Function of Plant Genomes," O. Ciferri and L. Dure, eds. Plenum Press, pp. 37-45 (1983).
12. G. Pearce and C.A. Ryan, in preparation.
13. T.R. Green and C.A. Ryan, Wound-induced proteinase inhibitor in plant leaves: A possible defense mechanism against insects, Science 175:776 (1972).
14. C.A. Ryan, Proteinase inhibitors in plant leaves: A biochemical model for pest-induced natural plant protection, Trends in Biochemical Sci. 5:148 (1978).
15. W.E. Brown, K. Takio, K. Titani, and C.A. Ryan, Wound-induced trypsin inhibitor in alfalfa leaves: Identity as a member of the Bowman-Birk inhibitor family, Biochemistry 24:2105 (1985).
16. C.A. Ryan, P. Bishop, G. Pearce, and M. Walker-Simmons, Pectic fragments regulate the expression of proteinase inhibitor genes in plants, UCLA Symposium, in: "Cellular and Molecular Biology of Plant Stress," J.L. Key, and T. Kosuge, eds. Alan R. Liss Inc., (1985).
17. J.C. Melville and C.A. Ryan, Chymotrypsin inhibitor from potatoes: Large scale preparation and characterization of its subunit components, M. Biol. Chem. 247 3445 (1972).
18. J. Bryant, T. Green, T. Gurusadaaiah, and C.A. Ryan, Proteinase inhibitor II from potatoes: Isolation and characterization of the isoinhibitor subunits, Biochemistry 15:3418 (1976).22.
19. G.M. Hass and C.A. Ryan, Carboxypeptidase inhibitor from potatoes, in: Methods in Enzymology, "Proteolytic Enzymes, Part C" L. Lorand, ed. Academic Press (1981).
20. C.A. Ryan, P. Bishop, G. Pearce, A.G. Darvill, M. McNeil, and P. Albersheim, A sycamore cell wall polysaccharide and a chemically related tomato leaf polysaccharide possess similar proteinase inhibitor-inducing activities, Plant Physiol. 68:616 (1981).
21. P. Bishop, D.J. Makus, G. Pearce, and C.A. Ryan, Proteinase inhibitor-inducing factor activity in tomato leaves resides in oligosaccharides enzymically released from cell walls, Proc. Natl. Acad. Sci. 78:3536 (1981).
22. P. Bishop, G. Pearce, J.E. Bryant, and C.A. Ryan, Isolation and characterization of the proteinase inhibitor-inducing factor from tomato leaves: Identity and activity of poly- and oligogalacturonide fragments, J. Biol. Chem. 259:13172 (1984).
23. G. Plunkett, D.F. Senear, G. Zuroske, and C.A. Ryan, Proteinase Inhibitor I and II from Leaves of Wounded Tomato Plants: Purification and Properties, Arch. Biochem. Biophys. 213 463 (1982).
24. J. Graham, G. Pearce, J. Merryweather, K. Titani, and C.A. Ryan, Wound-induced proteinase inhibitor mRNA from tomato leaves: I. The cDNA-deduced primary sequence of pre-inhibitor, J. Biol. Chem. 260:6555 (1985).

25. J. Graham, G. Pearce, J. Merryweather, K. Titani, and C.A. Ryan, Wound-induced proteinase inhibitor mRNA from tomato leaves: II. The cDNA-deduced primary sequence of pre-inhibitor II, J. Biol. Chem. 260:6561 (1985)
26. M. Walker-Simmons and C.A. Ryan, Immunological identification of proteinase Inhibitors I and II in isolated tomato leaf vacuoles, Plant Physiol. 60:61 (1977).
27. H. Hollander-Czytko, J.K. Andersen and C.A. Ryan, Vacuolar localization of wound-induced carboxypeptidase inhibitor in potato leaves, Plant Physiol. 78:76 (1985).
28. C.E. Nelson and C.A. Ryan, In vitro synthesis of pre-proteins of two vacuolar compartmented proteinase inhibitors that accumulate in leaves of wounded tomato plants, Proc. Natl. Acad. Sci. 77:1975 (1980).
29. J. Graham and C.A. Ryan, in preparation.
30. G. An, B.D. Watson, S. Stachel, M. Gordon, and E.W. Nester, New cloning vehicles for transformation of higher plants, EMBO J. 4:277 (1985).
31. W. Brown and C.A. Ryan, Isolation and characterization of a wound-induced trypsin inhibitor from alfalfa leaves, Biochemistry 23:3418 (1984).
32. L. Hadwiger and J.M. Beckman, Chitosan as a component of pea-Fusarium solani interactions, Plant Physiol. 66:205 (1980).
33. M. Walker-Simmons, L. Hadwiger and C.A. Ryan, Chitosan and pectic polysaccharides both induce accumulation of the antifungal phytoalehin pisatin in pea pods and antinutrient proteinase inhibitors in tomato leaves, Biochem. Biophys. Res. Comm. 110 194 (1983).
34. S. Norioki and T. Ikenaka, Amino acid sequence of a trypsin-chymotrypsin inhibitor, B-111, of peanut (Arachis hypogea) J. Biochem. 93: 479 (1983).
35. T. Ikenaka, S. Odani and T. Koide, Chemical structure and inhibitory activities of soybean proteinase inhibitors. Bayer-Symposium V "Proteinase Inhibitors" H. Fritz, H. Tschesche, L.J. Greene and E. Truscheit eds. Springer Serlog, pp. 325-343 (1974).
36. R. Hawkes, E. Hiday and J. Gordon, A dot-immunobinding assay for monoclonal and other antibodies. Anal. Biochem. 119:142 (1982).
37. W. Brown, K. Takio, K. Titani, and C.A. Ryan, Wound-induced trypsin inhibitor in alfalfa leaves: Identity as a member of the Bowman-Birk inhibitor family, Biochemistry 24 2105 (1985).

18

GENETICS AND BREEDING OF SOYBEANS LACKING THE KUNITZ TRYPSIN INHIBITOR

Theodore Hymowitz

Department of Agronomy, University of Illinois
Urbana, Illinois 61801

ABSTRACT

The major trypsin inhibitor present in soybean [Glycine max (L.) Merrill] seed is the Kunitz trypsin inhibitor or soybean trypsin inhibitor A_2 (SBTI-A_2). Four forms of SBTI-A_2 have been identified in the U.S. soybean germplasm collection. Three of the forms designated Ti^a, Ti^b, and Ti^c are electrophoretically distinguishable from one another by their different Rf values of 0.79, 0.75, and 0.83, respectively. The three forms are inherited as codominant alleles in a multiple allelic series at a single locus. The fourth form which is the absence of SBTI-A_2 is found in P.I. 157440 and P.I. 196168. The allele lacking SBTI-A_2 is designated ti and is inherited as a simple recessive to the other three SBTI-A_2 forms. Ti^a is the most common SBTI-A_2 allele in the germplasm collection. Ti^b primarily is found in Japan and Korea. Ti^c is associated with the Tohoku District, Japan. P.I. 157440 and P.I. 196168 are from Korea. Linkage studies revealed that the Ti and Ap loci are linked by 16.2 ± 1.5 map units. Three isolines of soybeans lacking the Kunitz trypsin inhibitor, derived by backcrossing P.I. 157440 to three commercial cultivars (Ti^aTi^a), were released to the research community. Results of preliminary feeding trials, with chicks and young pigs, revealed that gain/feed was significantly higher in lines lacking the Kunitz trypsin inhibitor than cultivars containing the inhibitor. However, the gain/feed was lower in both feeds than commercially processed soybean meal.

INTRODUCTION

Almost 70 years ago, Osborne and Mendel (1917) demonstrated that unheated soybean [Glycine max (L.) Merrill] meal is inferior in nutritional quality to properly heated soybean meal. The trypsin inhibitor proteins in raw mature soybean seed have been proposed to be one of the major factors responsible for the poor nutritional value of the unheated meal (Borchers et al., 1948). Physiologically, the ingestion of raw soybeans by monogastric animals causes pancreatic hypertrophy (Chernick et al., 1948; Bray 1964; Rackis, 1972; Liener and Kakade, 1980) and is believed to inhibit growth by upsetting the balance between metheonine and cystine in the pancrease (Booth et al.,

1960). Although the trypsin inhibitors make up only ca. 6% of the total protein in soybeans (Rackis and Anderson, 1964) it has been estimated that they may be responsible for 30 to 50% of the growth inhibitory effect and much of the pancreatic hypertrophy which results when monogastric animals ingest unheated soybeans (Rackis, 1965; Kakade et al., 1973).

Several different trypsin inhibitors have been reported to be present in soybeans (Bowman, 1944; Rackis and Anderson, 1964; Yamamoto and Ikenaka, 1967). However, much of the soybean trypsin inhibitor activity (SBTI) is thought to be due to the soybean trypsin inhibitor A_2 (SBTI-A_2) (Rackis et al., 1962) which was first crystallized by Kunitz (1945) and is commonly known as the Kunitz trypsin inhibitor. The protein contains 181 amino acids and has a relative molecular mass of 21384 (Koide and Ikenada, 1973). The active center of trypsin binding in the Kunitz inhibitor is the arginine 63 - isoleucine 64 bond (Bidlingmeyer et al., 1972).

GENETIC STUDIES

Seed from the US Department of Agriculture soybean germplasm collection (2944 accessions) were screened by polyacrylamide gel electrophoresis for the presence or absence of electrophoretic forms of SBTI-A_2 (see Orf and Hymowitz, 1979a for procedures). Four electrophoretic forms of SBTI-A_2 were discovered. Three of the forms, designated $\underline{Ti}^a$, $\underline{Ti}^b$, $\underline{Ti}^c$ are electrophoretically distinguishable from one another by their different Rf values of 0.79, 0.75, and 0.83, respectively (Figure 1) (Singh et al., 1969; Hymowitz, 1973) (Rf is the mobility relative to a bromophenol blue dye front in a 10% polyacrylamide gel anodic system in which a Tris-glycine buffer, pH 8.3 is used). The three forms are controlled by a co-dominant multiple allelic system at a single locus (Hymowitz and Hadley, 1972; Orf and Hymowitz, 1977). The fourth form of SBTI-A_2 found in the soybean germplasm collection (P.I. 157440 and P.I. 196168) does not exhibit a protein band in the gels (Hymowitz et al., 1978). The lack of a protein band is inherited as a recessive allele to $\underline{Ti}^a$, $\underline{Ti}^b$, $\underline{Ti}^c$ and has been designated $\underline{ti}$ (Orf and Hymowitz, 1979b). In the homozygous recessive state ($\underline{ti}$ $\underline{ti}$) no SBTI-A_2 band is produced in the seed. A standard trypsin assay of crude protein extracts from soybeans lacking the Kunitz trypsin inhibitor had 30 to 50% less trypsin inhibitor activity per gram of protein than crude protein extracts from seeds of the cultivar Amsoy 71 which contains SBTI-A_2 ($\underline{Ti}^a$). Although other trypsin inhibitors may be present in seed of P.I. 157440 and P.I. 196168, all the experimental results indicate that the Kunitz trypsin inhibitor is lacking in those seeds which do not have the SBTI-A_2 protein.

DISTRIBUTION STUDIES

The U.S. soybean germplasm collection was screened to determine the frequency of the SBTI-A_2 forms within the collection and to determine if the SBTI-A_2 forms were associated with a particular country or geographical area (Clark et al., 1970; Hymowitz et al., 1971; Hymowitz, 1973; Orf 1976, Skorupska and Hymowitz, 1977; Hymowitz et al., 1978). Of the accessions tested 88.7% had the $\underline{Ti}^a$ allele. All of the important commercially grown soybean cultivars in the U.S. have the $\underline{Ti}^a$ allele. The $\underline{Ti}^b$ allele was found in 11% of the population and appears to be associated with the soybean accessions from Japan and Korea. The $\underline{Ti}^c$ allele was found in 0.3% of the population and is geographically associated with the Tohoku District of Japan. The $\underline{ti}$ allele occurred in 2 accessions (0.06% of the population), both of which came from Korea (Hymowitz et al., 1978).

LINKAGE STUDIES

Investigations conducted to determine whether the Ti locus was linked to several other loci revealed that the following loci are

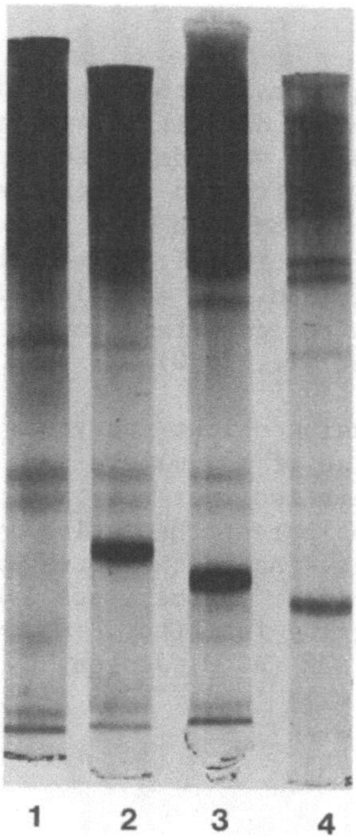

Figure 1. Polyacrylamide gels of extracts from soybean seeds showing SBTI-A_2 bands. From left to right: 1. no SBTI-A_2 band (ti); 2. Rf 0.75 (Ti^b); 3. Rf 0.79 (Ti^a); 4. Rf 0.83 (Ti^c).

inherited independently of the Ti locus: 1) the Dt_1 locus (stem termination); 2) the W_1 locus (flower color); 3) the Ep locus (seed coat peroxidase); 4) the Sp_1 locus (β-amylase) (Orf and Hymowitz, 1978); and the Le locus (seed lectin) (Orf and Hymowitz, 1979c). However, the Ti locus was found to be linked to the Ap locus (seed acid phosphatase) by 16.2 ± 1.5 map units and was assigned to Linkage Group 9 (Hildebrand et al., 1980).

ISOLINE DEVELOPMENT

In 1985 (Anon.) three soybean isolines, nulls for the Kunitz trypsin inhibitor, were released to the research community by the USDA and the University of Illiois. The allele ti from P.I. 157440 was transferred (backcrossed into the recurrent parent 5 times) into three commercial cultivars from Maturity Groups II, III, and IV having the Ti^aTi^a genotype for the Kunitz trypsin inhibitor. The parentage of line L83-4387 is Amsoy 71^6 x P.I. 157440, the parentage of line L81-4590 is Williams 82^6 x P.I. 157440, and the parentage of L81-4871 is $L6^6$ x P.I. 157440. L6 is very similar to Clark 63. The isolines are similar to their respective recurrent parents in yield, maturity date, lodging score, plant height, seed quality, seed weight, and protein and oil content in seed. Obviously, the soybean does not need the Kunitz trypsin inhibitor for its survival (Hymowitz, 1980).

FEEDING TRIALS

Two trials using 28 day-old male crossbred chicks were conducted to evaluate the nutritional value of P.I. 157440 (ti ti) versus Amsoy 71 (Ti^aTi^a). Raw mature soybeans were defatted and then compared to a commercially processed solvent extracted soybean meal (SBM) using purified diets with soybeans or SBM as a sole source of protein. In both trials, gain/feed was higher in P.I. 157440 than Amsoy 71 however, both were inferior to SBM. In addition, P.I. 157440 plus 0.3% methionine produced gain and gain/feed comparable to SBM without added methionine (Bajjalick et al., 1980).

Two preliminary trials using chicks and young pigs were conducted to evaluate the nutritive value of raw mature seed of commercial cultivars (Ti^aTi^a) versus their isolines (ti ti). The genotypes of the isolines and the recurrent parent theoretically are the same except for the SBTI-A_2 alleles. As in the above experiments, the gain/feed was significantly higher in the lines lacking the inhibitor than the cultivars containing the inhibitor. However, the gain/feed was lower in both feeds than SBM (A. H. Jensen, C. M. Parsons; pers. comm.).

CONCLUSIONS

The results reported here demonstrate the importance of collecting and maintaining extensive germplasm collections of our major crops. The availablity of soybean lines lacking the Kunitz trypsin inhibitor should be of particular interest to the scientific community for developing soybean cultivars with improved nutritional quality. Lastly, perhaps soybean seed lacking the Kunitz trypsin inhibitor does not need to be processed as extensively with moist heat as currently grown soybean seed (Liener and Tomlinson, 1981). Thus, a savings in energy and processing costs might be realized by the processors of soybean products for the feed and food industries.

United States Department of Agriculture
Agricultural Research Service
and
The Illinois Agricultural Experiment Station
University of Illinois
Urbana, Illinois

RELEASE OF FIVE SOYBEAN ISOLINES WITH KUNITZ TRYPSIN INHIBITOR VARIANTS L82-2024, L82-2051, L81-4590, L81-4871, AND L83-4387

The Agricultural Research Service, United States Department of Agriculture and the Illinois Agricultural Experiment Station announce the release of five germplasm isolines of soybeans. Each was selected for a gene controlling a Kunitz trypsin inhibitor variant or null from a fifth backcross to a commercial cultivar, Williams (Maturity Group III), Williams 82 (Group III, differs from Williams in having gene Rps^k for phytophthora rot resistance), L6 (Group IV, an experimental line apparently identical to Clark 63 but with more backcrossing to Clark), or Amsoy 71 (Group II).

Line	Parentage	Gene transferred
L82-2024	$Williams^6$ x Jefferson	Ti^b (variant)
L82-2051	$Williams^6$ x PI 196172	Ti^c (variant)
L81-4590	Williams 82^6 x PI 157440	ti (null)
L81-4871	$L6^6$ x PI 157440	ti (null)
L83-4387	Amsoy 71^6 x PI 157440	ti (null)

All of the recurrent parents are commercial cultivars having the Ti^a Ti^a genotype for the common type of Kunitz trypsin inhibitor. The donor cultivars are all introduced cultivars. Jefferson (Ti^b) was introduced in 1929 as PI 82202 from Gangweon Do, South Korea, by the Dorsett and Morse plant exploration mission and was grown for a short time in the U.S. under the cultivar name Jefferson (Maturity Group IV). PI 196172 (Ti^c) is the cultivar Huk Tae (Maturity Group IV) introduced from Korea, via Purdue AES in 1951. PI 157440 (ti) is the cultivar Kum-du introduced in 1947 from the Central Experiment Station, Suweon, South Korea.

The isolines are similar to their respective recurrent parents in field appearance and performance and in seed composition (see Table). These releases may be of interest in feeding studies to evaluate for possible economic merit and especially the nulls may be useful as parents in breeding programs.

Limited quantities of the five lines are available for research purposes upon written request to the Curator, Richard L. Bernard, Soybean Germplasm Collection, USDA-ARS, W-321 Turner Hall, Department of Agronomy, University of Illinois, 1102 South Goodwin Avenue, Urbana, IL 61801.

Table. Performance of commercial cultivars and their isolines with a variant or null for the Kunitz trypsin inhibitor in Illinois field tests with 2 replications per field.

Cultivar/Isoline	Yield	Maturity	Lodging	Height	Seed			
	bu/a	date	score	inches	Quality score	Weight cg	Protein %	Oil %
A. Average of 6 environments, 5 fields in 1983 (2 at Urbana and 1 at Girard, Belleville, and Eldorado) and 1 field at Urbana in 1984.								
Williams	36.3	9-22	1.4	38	2.8	15.8*	40.5*	21.7*
L82-2024 (Ti^b)	37.1	9-22	1.3	38	2.9	17.0	40.7	21.7
L82-2051 (Ti^c)**	36.1	9-22	1.4	38	2.9	15.1	40.3	22.0
Williams 82	34.4	9-22	1.3	37	2.8	15.5	39.7	22.1
L81-4590 (ti)**	34.5	9-21	1.5	38	2.8	16.3	41.4	21.1
* Measured at 3 environments only ** Sister line used at 3 environments only								
B. Average of 4 environments, 2 fields at Urbana in 1982 and 2 in 1983.								
L6 (Clark 63)	41.4	9-28	2.3	40*	2.4	13.6*	40.6*	21.0*
L81-4871 (ti)	40.6	9-28	2.3	40	2.2	13.1	40.7	21.0
* 1983 data only								
C. One field at Urbana in 1984.								
Amsoy 71	49.3	9-11	1.5	39	2.8	17.5	38.0	22.7
L83-4387 (ti)	51.4	9-10	1.4	40	2.5	18.3	37.6	22.6

Lodging scores range from 1 (plants erect) to 5 (plants prostrate).
Seed weight is in centigrams per seed.
Seed quality scores range from 1 (excellent) to 5 (very poor).
Protein and oil percentages are on a moisture-free basis.

REFERENCES

Anon. (1985). Release of five soybean isolines with Kunitz trypsin inhibitor variants L82-2024, L82-2051, L81-4590, L81-4871, and L83-4387. USDA-ARS and University of Illinois, Agricultural Experiment Station.

Bajjalieh, N., Orf, J.H., Hymowitz, T. and Jensen, A.H. (1980). Response of young chicks to raw, defatted, Kunitz trypsin inhibitor variant soybeans as sources of dietary protein. Poultry Sci., 59:328-332.

Bidlingmeyer, U.D., Leary, T.R. and Laskowski, M.Jr. (1972). Identity of the tryptic and α-chymotryptic reactive sites on soybean trypsin inhibitor (Kunitz). Biochemistry, 11:3303-3310.

Booth, A.W., Robbins, A.J. and Rebelin, W.E. (1960). Effect of raw soybean meal and amino acids on pancreatic hypertrophy in rats. Proc. Soc. Exp. Biol. Med., 104:681-683.

Borchers, R., Anderson, C.W., Mussehl, F.E. and Moehl, A. (1948). Trypsin inhibitors. VIII. Growth inhibiting properties of a soybean trypsin inhibitor. Arch. Biochem., 19:317-322.

Bowman, D.E. (1944). Fractions derived from soybeans and navy beans which retard the tryptic digestion of casein. Proc. Soc. Exp. Biol. Med., 57:139-140.

Bray, D.J. (1964). Pancreatic hypertrophy in layering pullets induced by unheated soybean meal. Poult. Sci., 43:382-384.

Chernick, S.S., Lepkovsky, S. and Chaikoff, I.L. (1948). A dietary factor regulating the enzyme content of the pancreas. Changes induced in size and proteolytic activity of the chick pancreas by the ingestion of raw soybean meal. Amer. J. Physiol., 155:33-41.

Clark, R.W., Mies, D.W. and Hymowitz, T. (1970). Distribution of a trypsin inhibitor variant in seed proteins of soybean varieties. Crop Sci., 10:486-487.

Hildebrand, D.F., Orf, J.H. and Hymowitz, T. (1980). Inheritance of an acid phosphatase and its linkage with the Kunitz trypsin inhibitor in seed protein of soybeans. Crop Sci., 20:83-85.

Hymowitz, T. (1973). Electrophoretic analysis of SBTI-A_2 in the USDA soybean germplasm collection. Crop Sci., 13:420-421.

Hymowitz, T. (1980). Chemical germplasm investigations in soybeans: The flotsam hypothesis. Recent Adv. Phytochemistry, 14:157-179.

Hymowitz, T. and Hadley, H.H. (1972). Inheritance of a trypsin inhibitor variant in seed protein of soybeans. Crop Sci., 12:197-198.

Hymowitz, T., Mies, D.W. and Klebek, C.J. (1971). Frequency of a trypsin inhibitor variant in seed protein of four soybean populations. East Afr. Agric. For. J., 37:63-72.

Hymowitz, T., Orf, J.H., Kaizuma, N. and Skorupska, H. (1978). Screening the USDA soybean germplasm collection for Kunitz trypsin inhibitor variants. Soybean Genet. Newsl., 5:19-22.

Kakade, M.L., Hoffa, D.E. and I.E. Liener. (1973). Contribution of trypsin inhibitors to the deleterious effects of unheated soybeans fed to rats. J. Nutr., 103:1772-1778.

Koide, T. and Ikenada, T. (1973). Studies on soybean trypsin inhibitors:3. Amino-acid sequence of the carboxyl-terminal region and the complete amino-acid seuquence of soybean trypsin inhibitor (Kunitz). Eur. J. Biochem., 32:417-431.

Kunitz, M. (1945). Crystallization of a soybean trypsin inhibitor from soybean. Science, 101:668-669.

Liener, I.E. and Kakade, M.L. (1980). Protease inhibitors. In "Toxic Constituents of Plant Foodstuffs", I.E. Liener, Ed., Academic Press, New York, pp. 7-71.

Liener, I.E. and Tomlinson, S. (1981). Heat inactivation of protease inhibitors in a soybean line lacking the Kunitz trypsin inhibitor. J. Food Sci., 46:1354-1356.
Orf, J.H. (1976). Electrophoretic studies on seed proteins of Glycine max (L.) Merrill. M.S. Thesis, Graduate College, University of Illinois at Urbana-Champaign, Urbana, Illinois.
Orf, J.H. and Hymowitz, T. (1977). Inheritance of a second trypsin inhibitor variant in seed protein of soybeans. Crop Sci., 17:811-813.
Orf, J.H. and Hymowitz, T. (1978). Soybean linkage tests between two seed proteins and other characteristics. Soybean Genet. Newsl., 5:22-24.
Orf, J.H. and Hymowitz, T. (1979a). Genetics of the Kunitz trypsin inhibitor: An antinutritional factor in soybeans. J. Amer. Oil Chemists' Soc., 56:722-726.
Orf, J.H. and Hymowitz, T. (1979b). Inheritance of the absence of the Kunitz trypsin inhibitor in seed protein of soybeans. Crop Sci., 19:107-109.
Orf, J.H. and Hymowitz, T. (1979c). Soybean linkage tests between Ti and Le seed proteins. Soybean Genet. Newsl., 6:32.
Osborne, T.B. and Mendel, L.B. (1917). The use of soybean as feed. J. Biol. Chem., 32, 369-377.
Rackis, J.J. (1965). Physiological properties of soybean trypsin inhibitors and their relationship to pancreatic phypertrophy and growth inhibition of rats. Fed. Proc., 29:1488-1493.
Rackis, J.J. (1972). Biologically active components. In "Soybeans: Chemistry and Technology. Volume 1. Proteins", A.K. Smith and S.J. Circle, Eds., The AVI Publ. Co., Westport, Connecticut, pp. 158-202.
Rackis, J.J., Sasame, H.A., Mann, R.K., Anderson, R.L. and Smith, A.K. (1962). Soybean trypsin inhibitors: isolation, purification and physical properties. Arch. Biochem. Biophys., 98:471-478.
Rackis, J.J., and Anderson, R.L. (1964). Isolation of four soybean trypsin inhibitors by DEAE-cellulose chromatography. Biochem. Biophys. Res. Commun., 15:230-235.
Singh, L.C., Wilson, C. and Hadley, H.H. (1969). Genetic differences in soybean trypsin inhibitors separated by disc electrophoresis. Crop Sci., 9:489-491.
Skorupska, H. and Hymowitz T. (1977). On the frequency distribution of alleles of two seed proteins in European soybean [Glycine max (L.) Merrill] germplasm: Implications on the origin of European soybean germplasm. Genet. Polonica, 18:217-224.
Yamamoto, M. and Ikenaka, T. (1967). Studies on soybean trypsin inhibitors. I. Purification and characterization of two soybean trypsin inhibitors. J. Biochem. (Tokyo), 62:141-149.

19

PROTEASE INHIBITORS IN PLANT FOODS: CONTENT AND INACTIVATION

J. J. Rackis, W. J. Wolf, and E. C. Baker

Northern Regional Research Center, USDA
Peoria, Illinois 61604

INTRODUCTION

Historical Perspective

To understand the present widespread interest in protease inhibitors (PI), one must look back to the vast amount of research that has been conducted for over 50 years. Intensive investigations, like wave motions on the water, followed a highly cyclical pathway depending upon the interest and support for research in various disciplines. Nutritionists were concerned with the potential adverse effects of the inhibitors of trypsin and chymotrypsin in the intestinal tract of man and animals (Liener and Kakade, 1980; Liener, 1981; Rackis and Gumbmann, 1981); biochemists recognized that the peculiar properties of PI's may yield fundamental information concerning protein-protein interactions especially the mechanism of proteolytic hydrolysis (Richardson, 1981; Ryan, 1979); medical scientists envisioned an important role for PI's in the treatment of various metabolic disorders (Katunuma et al., 1983; Vogel et al., 1968; Weyer, 1968), and biologists foresaw an important physiological role of PI's to regulate plant growth, protein synthesis and resistance to microbial and insect pests (Richardson, 1977, 1981; Ryan, 1973, 1979).

Proteolytic activity in plants was first reported in 1799 by Vauquelin, but the first recognition that plant foodstuffs contained a PI occurred when Read and Haas (1938) reported that aqueous extracts of defatted soy flour inhibited the ability of trypsin to liquefy gelatin. A crystalline trypsin inhibitor (TI) was subsequently isolated by Kunitz (1945, 1946). Kunitz (1947) also demonstrated that the soybean TI had the important property of forming a stoichiometric protein-protein complex with trypsin, resulting in competitive inhibition of its catalytic function. The existence of a natural inhibitor of proteolytic activity was first discovered by Weinland in 1903 for substances that imparted resistance of certain nematodes to digestion by gastrointestinal enzymes. He used the term "antienzymes" for such inhibitory proteins. More recently, protease and/or proteinase inhibitors are the usual terminology used to refer to proteins and polypeptides that specifically inhibit proteolytic enzymes. In addition, there is a class of low molecular weight PI's isolated from actinomycetes (Umezawa and Aoyagi, 1983), that are tri- or higher peptides.

Nutritional Aspects

From an historical perspective, the nutritional significance of PI's has commanded the greatest attention, starting with Osborne and Mendel's (1917) discovery that soybeans would not support the growth of rats unless the beans were cooked for 3 h on a steam bath. Chernick and coworkers (1948) were the first to discover that enhanced pancreatic juice secretion and enlargement of the pancreas occurred in chicks fed raw soybean meal. In the ensuing years, purified TI's from several plant sources and eggs were shown to produce the same pancreatic effects as those associated with raw soybeans (Rackis and Gumbmann, 1981; Senti, 1982).

In 1972, Green and Lyman proposed that the soybean TI's induced changes in pancreatic function and histology through the release of cholecystokinin into the blood from receptor sites in the intestional mucosa. This release mechanism presumably involves negative hormonal feedback, controlled by the amount of free trypsin existing in the small intestine. Details concerning the feedback control mechanism appear elsewhere. That the soybean TI's exert their effect at the intestinal level and not in the blood stems from the fact that the low molecular weight soybean Bowman-Birk TI is not absorbed into the blood and a major portion of the dose level of TI ends up in the feces (Madar, et al., 1979; Troll and Yavelow, 1983).

Coincident with these nutritional developments, a vast amount of information has been accumulated with regard to the development of processes to produce plant protein food products having maximum nutritive value but containing varying levels of residual TI activity (American Oil Chemists Society, 1979; Liener and Kakade, 1980; Rackis et al. 1979; Rackis, 1981; Rackis and Gumbmann, 1981).

Since TI's in plant foodstuffs are generally inactivated by heat treatment, there was a tendency to dismiss the nutritional significance of residual TI activity in processed vegetable protein foods because short-term feeding trials in various animals and humans indicated that the nutritive value of properly-processed plant proteins is nearly comparable to that of animal protein (Wilcke, et al., 1979). However, nutritional research, particularly long-term testing programs, needs to be continued based on a series of reports by European workers (McGuinness et al., 1982, 1984). It was demonstrated that in long-term feeding of up to two years in Wistar rats, raw and heated full-fat soy flour enhanced carcinogenesis of two pancreatic carcinogens, azaserine and di-(2-hydroxypropyl)nitrosamine and, even in their absence, pancreatic nodular hyperplasia and acinar adenoma occurred in rats fed the soy diets (Levison et al., 1979). The relevance to humans of such long-term effects of TI's in animals must, however, await appropriate experiments in human subjects (Liener, 1983).

Physiological Function in Plants

Following their discovery in plants (Read and Haas, 1938), importance of TI's as protection against microorganisms, insects and other pests and regulation of physiological functions was primarily in the realm of speculation with little or no research being carried out. Much attention was drawn to the fact that nearly all plant PI's are usually specific for digestive enzymes having trypsin- or chymotrypsin-like activity. Only in a few instances had it been demonstrated that these inhibitors can inhibit proteases synthesized by the plant (Richardson, 1981; Ryan, 1979). Most knowledge of regulatory functions of plant PI's is of recent origin. Foard et al. (1983) are exploring the possibility of using gene transfer of the soybean Bowman-Birk trypsin-chymotrypsin inhibitors into phylogenetically distinct plant species to confer resistance to herbivores and pathogens.

Therapeutic Aspects

There has been a long history of research and intriguing concepts for the use of PI's in the treatment of a wide range of metabolic disorders. There have been several reviews and international conferences on protease inhibitors (Fritz and Tschesche, 1971; Fritz et al., 1974; Vogel et al., 1968; Weyer, 1968). More recently, Katunuma, et al. (1983) detailed the clinical and medical aspects of several types of PI's that have been isolated and characterized from plant, microbial and animal species. The renewed interest in PI's arises from research that associates enhanced proteolytic activity with most in-vitro transformed cells, some tumors in-vivo and a wide variety of disorders, e.g., pancreatitis, emphysema, allergic and inflammatory reactions, and shock. Because the precise role and specificity of proteases in metabolic disorders have not been defined, many of the results of clinical research on PI's are conflicting. PI's possess multispecificity that may not only inhibit tumor-associated proteases, but may also adversely affect normal protease-controlled biological systems, as well as induce a cascade effect on enzymatic and hormonal changes in-vivo that cannot be explained by their pharmakinetic action on proteolytic enzymes and tumor cells in-vitro. This may account for the present dilemma that PI's appear to inhibit tumorigenesis-carcinogenesis (Troll and Yavelow, 1983), enhance carcinogenesis (Levison et al., 1979; Morgan et al., 1977) or have no effect at all.

Based on clinical results, plant PI's appear not to be very beneficial (Fritz and Tschesche, 1971; Weyer, 1968). Nevertheless, some protease inhibitors are potentially useful in chemopreventive therapy. However, the greatest problem is how to deliver them to the target organ, in their most active form and at the optimal concentration. Plant PI's are proteins or polypeptides that cannot be absorbed via the gastrointestinal tract (Madar et al., 1979; Troll and Yavelow, 1983). Over 90% of the oral dose of soybean Bowman-Birk inhibitor is recovered intact from the colon and feces of rats and mice, while the very small percent of the ^{125}I-labeled inhibitor that is absorbed and concentrated in the urine loses its original ability to complex with trypsin and chymotrypsin. Microbial PI's (tri- or tetra peptides) are also highly biodegradable regardless of the route of administration (Umezawa and Aoyagi, 1983).

Scope

This chapter will be limited primarily to the PI's in oilseeds, Leguminosae (peas and beans), Solanaceae (potatoes) and Gramineae (cereals) that are important and economical sources of food. We review the physical and chemical properties of the several types of plant PI's, many of which exist in the form of isoinhibitors and genetic variants, particularly those that inhibit trypsin and chymotrypsin. We also cover the trypsin inhibitor distribution and content of oilseeds, food legumes, potatoes and cereals and of protein products prepared from them. Also included is a discussion of the assays used in determining TIA, differentiation between proteins and non-protein constituents that inhibit trypsin, as well as the multitude of technological processes and chemical treatments that have been developed for the inactivation of TIA.

DIETARY INTAKE OF PROTEASE INHIBITORS

With the proper combination of food legumes, oilseed protein products and various cereals, protein nutritive value is raised to that of animal protein and such mixtures are eaten by a large number of people on a regular basis in developing and developed countries. This consumption represents a substantial daily intake of protein and possibly PI's. The

Table 1. Average Daily Protein and Trypsin Inhibitor Intake From the British Household Diet[a]

Food	Protein g/person/day	TIA[b] mg/person/day	TIA[b] mg/g protein
Milk and milk products	17.9	56.8	3.2
Meat	19.5	19.8	1.0
Fish	2.9	1.5	0.5
Eggs	3.7	93.6	25.3
Vegetables: Potatoes	2.8	42.5	15.2
Others	3.8	37.8	9.9
Fruit and nuts	0.7	11.8	16.9
Bread and cereals	18.7	25.3	1.4
Other foods	1.0	5.5	5.5
Total	71.3	294.6	4.1

[a]Doell et al. (1981)
[b]Trypsin inhibitor activity (mg trypsin inhibited)

1975 world production of protein from soybeans, dry beans and peanuts and of protein from the cereals, wheat, corn, and rice, amounted to 39,800 and 92,400 X 10^3 metric tons, respectively (Rockland and Radke, 1981). Excellent recent reviews of the family Leguminosae are available (Rockland and Nishi, 1979; Sgarbieri and Whitaker, 1982; Sathe et al., 1984A, B).

Data for dietary intake of PI's are not generally available, but the British household diet has been estimated to provide an average daily intake of 295 mg TIA/person/day. Such an intake is approximately equivalent to 4.1 mg TIA per g protein consumed (Doell et al., 1981). Soy-based infant formulas supply 40-60 mg TIA/day (J. Harwood, personal communication). A shift to a completely vegetarian diet (vegans) or lacto-ovo vegetarianism would greatly increase the per capita consumption of TIA based on data shown in Table 1. Milk and dairy products, eggs and vegetables (potatoes, beans, and peas) account for 78% of the total daily intake of TIA from the British diet. Except for sweetbread, other foods including meat, fresh vegetables, fruits, and nuts contain low levels of TIA. Chen and Mitchell (1973) found that potatoes, sweet corn and sweet potatoes had the highest TIA, moderate activity was present in spinach, broccoli, Brussel sprouts and cucumber, while negligible activity was present in most other vegetables and fruits.

PHYSICAL AND CHEMICAL PROPERTIES

A large number of plant protease inhibitors has been isolated and characterized. Among the inhibitors that are best characterized are those from seeds and tubers of agriculturally important crops including members of the Leguminosae, Gramineae and Solanaceae (Table 2). Although they may perform in a similar manner, (e.g., inhibit trypsin), proteinase inhibitors from various sources differ in molecular size, amino acid composition and physical structure. Nonetheless, some of the inhibitors show a high degree of amino acid sequence homology and the reactive sites in some inhibitors are identical.

Table 2. Physical and Chemical Properties of Selected Proteinase Inhibitors of Leguminosae, Gramineae and Solanaceae

Source	Inhibitor(s)	Proteinase(s) Inhibited[a]	M.W. Daltons	pI	-S-S-bonds/mole	Reactive Site(s)	References
<u>Leguminosae</u>							
Soybean	Bowman-Birk	T;C	7848	4.2	7	Lys16-Ser17; Leu43-Ser44	Birk et al. (1963); Odani and Ikenaka (1973A)
	C-II	T;C;E	8282	---	7	Ala22-Ser23; Arg49-Ser50	Odani and Ikenaka (1977)
	D-II	T	8507	---	7	Arg24-Ser25; Arg51-Ser52	Odani and Ikenaka (1978)
	Kunitz	T;C	20,083	4.5	2	Arg63-Ile64	Koide and Ikenaka (1973)
Lima bean	---	T;C	9127		7	Lys28-Ser29; Leu55-Ser56	Krahn and Stevens (1970, 1973); Tan and Stevens (1971)
Peanut	B-III	T;C	6716		7	Arg10-Arg11; Arg39-Ser40	Norioka and Ikenaka (1983)
<u>Gramineae</u>							
Corn	---	T	12,028	---	5	Arg36-Leu37	Mahoney et al. (1984)
Millet	---	T	13,300	>10	5	Arg34-Leu35; Arg91-Ala92	Campos and Richardson (1983)
Barley	---	T	13,305		5	Arg33-Leu34	Odani et al. (1983)
<u>Solanaceae</u>							
Eggplant	---	T	5552	4.7	4	Arg38-Asn39	Richardson (1979)
Potato	CI-I[b]	C	9332	---	1	Met60-Asp61	Richardson (1974); Richardson and Cossins (1974)
	PCI-I	C	5798	---	4	Leu38-Asn39	Haas et al. (1982)
	PTI	T	5593	---	4	Arg38-Asn39	" " " "
	CPI	CP	4130, 4259[c]	---	3		" " " (1975)

[a] T, C, E, and CP are trypsin, chymotrypsin, elastase and carboxypeptidase, respectively.
[b] Subunit A
[c] Sequencing revealed a mixture of two polypeptide chains differing by only one glutamyl residue.

Molecular Size

All of the inhibitors listed in Table 2 have been sequenced, hence their molecular weights are known unambiguously except for microheterogenity detected during sequencing for a few of them. All of the inhibitors in Table 2 fall into the range of 4100-21,000 daltons but most are in the narrower range of 8000-12,000 daltons. Soybeans are unusual because they contain inhibitors that fall into two widely different molecular weight classes: (a) the Bowman-Birk inhibitor type of ~ 8000 daltons and the Kunitz inhibitors of ~ 20,000 daltons. Molecular weights higher than obtained from amino acid sequence data have been reported for some inhibitors because of their tendency to associate into higher molecular weight forms. For example, the Bowman-Birk trypsin inhibitor from soybeans was originally reported to have a molecular weight of 24,000 daltons (Birk et al., 1963) but subsequent sedimentation equilibrium studies indicated that the 24,000 dalton form is a trimer of an 8000 dalton protomer (Millar et al., 1969). This estimate of the size of the Bowman-Birk protomer is in agreement with the molecular weight calculated from the amino acid sequence (Table 2). Another associating inhibitor system occurs in potatoes; they contain a chymotrypsin inhibitor (CII, Table 2) that is a tetramer of four different subunits exhibiting a high degree of amino acid sequence homology (Melville and Ryan 1972). Some 72 out of the 84-85 residues are invariant in the amino acid sequences of the four subunits (Richardson and Cossins, 1974).

Structure

The inhibitors listed in Table 2 all consist of single polypeptide chains with one to eight disulfide bonds. None of them contains carbohydrate. The Kunitz trypsin inhibitor from soybeans consisting of 181 amino acid residues contains only two disulfide bridges (Fig. 1); in contrast, the Bowman-Birk trypsin inhibitor from the same source is made up of only 71 amino acid residues but contains seven disulfide bridges (Fig. 2). The high degree of crosslinking of Bowman-Birk inhibitor is believed to be responsible for the heat stability of this molecule (Birk 1961). Several other trypsin inhibitors closely related to the Bowman-Birk inhibitor have been isolated from soybeans (C-II and D-II, Table 2). A similar family of trypsin inhibitors has been isolated from lima beans (Jones et al., 1963).

Three genetic variants of the Kunitz trypsin inhibitor of soybeans have been identified and isolated (Freed and Ryan, 1980), but the differences in the amino acid sequences responsible for the variants are still unknown. Microheterogenity was also observed in sequencing protease inhibitor component IV of lima beans; conservative substitutions of amino acids occurred in positions 26, 37 and 39 of the 84 residues that comprise the inhibitor molecule (Tan and Stevens, 1971).

As pointed out earlier, some inhibitors demonstrate a great deal of homology in their amino acid sequences. A comparison of the Bowman-Birk inhibitor of soybeans with inhibitors from lima beans, garden beans and chick peas has revealed a striking degree of homology in sequence as well as in the location of the reactive sites (Richardson, 1977). Examples of internal homology have also been noted in inhibitors that contain repetitive amino acid sequences. Internal homology was first reported for the lima bean inhibitor (Tan and Stevens, 1971) and was subsequently observed in the Bowman-Birk inhibitor of soybeans (Odani and Ikenaka, 1972). In the Bowman-Birk inhibitor, the regions of internal homology include the two reactive sites: (a) the reactive site for trypsin located in the sequence Cys 9 to Asp 26 and (b) the reactive site for chymotrypsin found in the sequence Cys 36 to Asp 53 (Fig. 2).

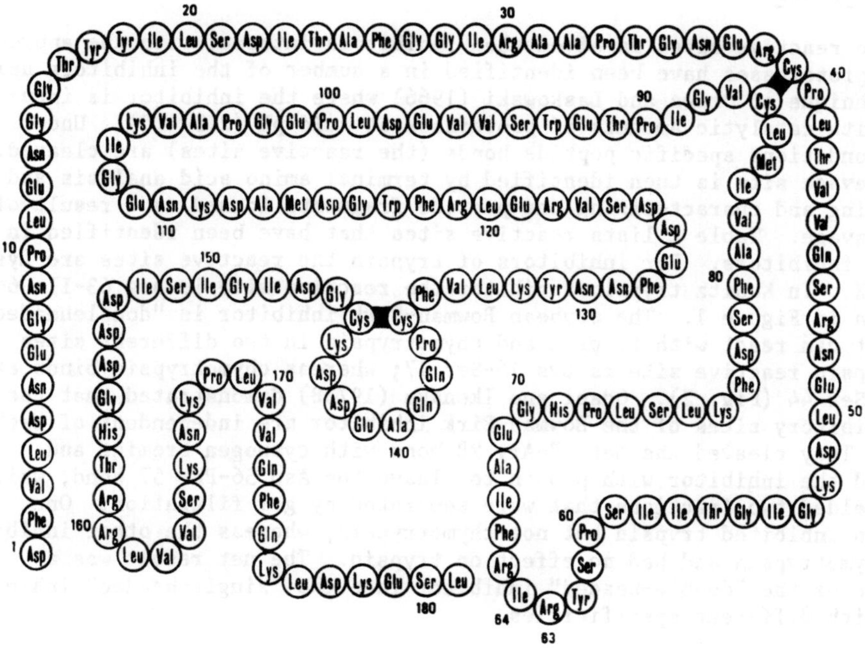

Fig. 1. Primary structure of Kunitz soybean trypsin inhibitor as determined by Koide and Ikenaka (1973). Residues are numbered beginning with N-terminal aspartic acid. Disulfide bridges are in black between half-cystine residues. The active center is Arg^{63}-Ile^{64}. Reprinted with permission from European Journal of Biochemistry.

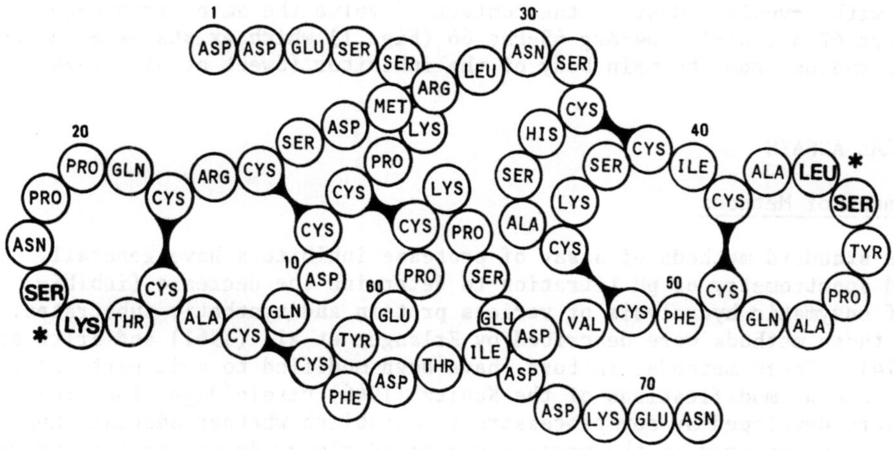

Fig. 2. Primary structure of Bowman-Birk soybean trypsin inhibitor according to Odani and Ikenaka (1973A). Residues are numbered beginning with N-terminal aspartic acid. Disulfide bridges are in black between half-cystine residues. Residues at the two reactive sites are marked in bold-faced type and have asterisks next to them. Reprinted with permission from Springer-Verlag.

Reactive Sites

The reactive sites involved when inhibitors interact with the appropriate proteinases have been identified in a number of the inhibitors using the technique of Ozawa and Laskowski (1966) where the inhibitor is incubated with catalytic amounts of proteinase at acid pH (2.5-4.0). Under these conditions specific peptide bonds (the reactive sites) are cleaved. The cleavage site is then identified by terminal amino acid analysis and by separating and characterizing the peptide fragments formed as a result of the cleavage. Table 2 lists reactive sites that have been identified in various inhibitors. For inhibitors of trypsin the reactive sites are Lys-X or Arg-X. In Kunitz trypsin inhibitor the reactive site is Arg 63-Ile 64 as shown in Figure 1. The soybean Bowman-Birk inhibitor is "doubleheaded," i.e., it can react with trypsin and chymotrypsin in two different sites. The trypsin reactive site is Lys 16-Ser 17; whereas chymotrypsin binds at Leu 43-Ser 44 (Fig. 2). Odani and Ikenaka (1973B) demonstrated that the two inhibitory sites of the Bowman-Birk inhibitor are independent of each other. They cleaved the Met 27-Arg 28 bond with cyanogen bromide and digested the inhibitor with pepsin to cleave the Asp 56-Phe 57 bond, which then yielded two fractions that were separated by gel filtration. One fraction inhibited trypsin but not chymotrypsin, whereas the other inhibited only chymotrypsin and had no effect on trypsin. The net result was a cleavage of the "double-headed" inhibitor into two "single-headed" inhibitors with different specificities.

Inhibitor-Proteinase Complexes

Although the reactive centers of interaction between proteinases and their inhibitors have been identified for a number of inhibitors, much less is known about the forces involved in the interaction between the two protein molecules during inhibition. X-ray analysis of the crystalline complex between Kunitz soybean trypsin inhibitor and trypsin revealed that the complex is a tetrahedral adduct between the active site of trypsin (serine) and the carbonyl group of Arg 63 of the inhibitor (Fig. 3). Only about 12 of the total of 181 amino acid residues in the inhibitor are in contact with trypsin. Most of the contacts involve the amino acid sequence Ser 61-Tyr 62-Arg 63-Ile 64-Arg 65-Phe 66 (Fig. 1) which exists as a curved loop protruding from the main body of the inhibitor (Sweet et al., 1974).

ANALYTICAL ASSAYS

Development of Methods

The standard methods of assay of protease inhibitors have generally involved spectrometry or pH titration to determine the decrease (inhibition) of enzymatic hydrolysis of various protein and synthetic substrates. Some of these methods were described by Erlanger et al. (1961) and Fritz et al. (1974). These methods, in turn, have been modified to suit particular needs. Several modifications of the Kunitz (1946) casein digestion procedure were developed by soy processors to establish whether adequate heat treatment was employed in the manufacture of edible-grade soy protein products of maximum protein quality. Eventually all of these methods were discarded. Many problems are associated with the use of a pH stat to measure trypsin hydrolysis and TI content of soy flours and protein products (Stinson and Snyder, 1980; Hill et al., 1982). The pH stat procedures are very seldom used. Erlanger et al. (1961) introduced the use of the synthetic substrate, benzoyl-arginine-p-nitroanilide (BAPA) to replace casein. Kakade et al. (1969) defined the inherent errors and complicating factors associated with the use of BAPA. They arbitrarily defined a trypsin unit (TU) as an increase of 0.01 absorbance units, and then determined TIA as trypsin units inhibited (TUI) or trypsin inhibitor units (TIU) based on the decrease in absorbance under the conditions of the assay. From active site titration data, Kakade et al. (1969) further established that 1 μg of pure trypsin increased absorbance 0.019 units, hence 1 μg of pure trypsin produced 1.9 TU. Assuming stoichiometric equivalence between trypsin and Kunitz soybean trypsin inhibitor, 1 μg of TI is equivalent to 1.9 TIU or TUI.

A collaborative study by a committee on Soybean Trypsin Inhibitor Analysis (Rackis et al., 1974) revealed additional problems with the BAPA procedure of Kakade et al. (1969) when analyzing raw and heat-treated soy flours. These studies resulted in a modified assay (Kakade et al., 1974) which improved both the accuracy and reproducibility. The modified assay was accepted as the official procedure for TI analysis by the American Association of Cereal Chemists (1983) and the American Oil Chemists' Society (1976). A semi-automated application of the official method was developed by Egbert et al. (1975).

As experience with the procedure of Kakade et al. (1974) increased, Smith et al. (1980) and Hamerstand et al. (1981) described further modifications and eliminated the need for extrapolation to zero inhibitor activity of four specified sample dilutions. These two groups determined TIA from a single dilution of the sample, with the specification that one

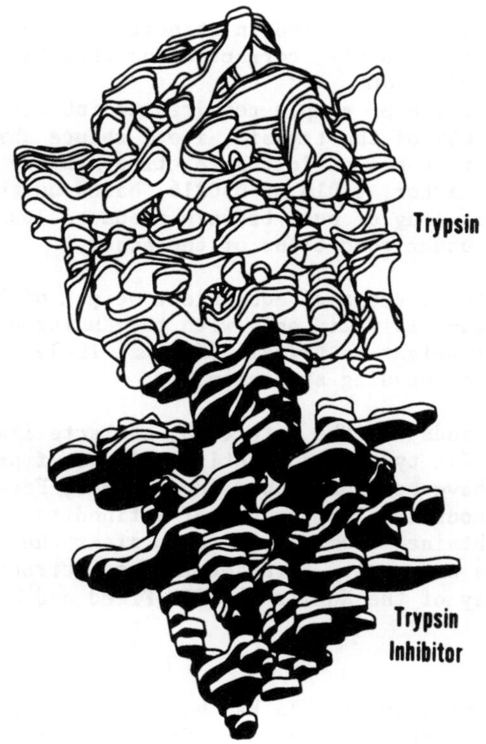

Fig. 3. Cystalline complex of porcine trypsin and Kunitz soybean inhibitor at 5 A resolution as elucidated by Sweet and coworkers (1974). Reprinted with permission; from Biochemistry 13:4212 (1974). Copyright by the American Chemical Society.

ml of diluted extract should inhibit 40 to 60% of the trypsin in the reaction mixture. Lehnhardt and Dills (1984) reported on additional changes in the procedure of Kakade et al. (1974) which allowed for analysis of a single diluted extract that inhibited 10 to 90% of the trypsin under the conditions of the assay. They have also shown that 1 μg of pure trypsin produced 2.16 TU under the conditions of the assay and this factor was substituted for the 1.9 TU factor used in the previous proce-dures (Kakade et al., 1969, 1974; Smith et al., 1980; Hamerstrand et al., 1981). An automated procedure based on the modified method of Lehnhardt and Dills (1984) was developed by Charpentier and Lemmel (1984).

It is highly recommended that new procedures be subjected to a collaborative analysis to obtain statistical data on accuracy and reproducibility of TIA in common extracts, as well as analysis of extracts of standard samples prepared by the individual collaborators in their respective laboratories (Rackis et al., 1974). Such an exercise would be valuable in selecting a new official TIA procedure.

Regardless of which modified procedure is used, the calculation of TIA (as mg trypsin inhibited) in foodstuffs is made according to the following formula:

$$\frac{\text{mg of trypsin inhibited}}{\text{per g sample}} = \frac{(A)\ (B)}{(C)\ (1000\ \mu g/mg)\ (\text{Sample wt, g})}$$

Where A = absorbance of trypsin standard - absorbance of test sample, corrected for absorbance of sample blank

B = Dilution of sample required to inhibit 40-60% of the trypsin, i.e., reduce absorbance by that amount in the reaction mixture

C = TU factor, 0.019 or 0.0216, based on the fact that 1 ug of pure trypsin produced 1.9 or 2.16 TU under conditions of the assay.

Reporting TIA data in terms of equivalent weight of "pure" native soybean trypsin inhibitor is valid only with the Kunitz soybean TI which has the same molecular weight as trypsin (Kunitz, 1947). Other PI's may exhibit widely different binding stoichiometry.

Immunological methods for the detection, characterization, and quan-titation of specific types and genetic variants of protease inhibitors in plants have been developed (Ryan, 1973; Freed and Ryan, 1978A, B). These methods involve the use of antibodies to specific protease inhibitors obtained from rabbit serum after the purified inhibitors are used as antigens. A rocket immunoelectrophoresis system was developed for assay of the Kunitz soy TI (Freed and Ryan, 1978A).

Inhibitor Units

Just as the determination of TIA has been carried out by diverse assay methodologies, so too have the units for TIA been reported in a variety of ways. As a consequence, it is frequently impossible to make direct comparisons of TI content between various foodstuffs. Some of the more common units used are summarized in Table 3.

Because several different forms of TI's are present in foodstuffs, each with different properties, binding stoichimetry and molecular weights, the equivalence of 1 μg of TI to 1 μg of trypsin inhibited may be of value for comparative purposes between foodstuffs, but, would not necessarily relate to absolute amount. A TIU obtained with BAPA or

Table 3. Trypsin Inhibitor Units

TUI/mg sample (protein)[a,b]	Kakade et al. (1969)
TIU/mg sample (protein)[a]	Kakade et al. (1974)
mgTI/g sample (protein)[c]	Kunitz (1946)
TIA/g sample (protein)[d]	Smith et al. (1980)
Trypsin Inhibitor Unit[e,f]	Chen and Mitchell (1973)
Specific activity units/mg protein	Gupta and Deodhar (1975)
HIT/mg protein[g]	Chang and Tsen (1979)

[a]One trypsin unit (TU) is arbitrarily defined as an increase of 0.01 absorbance units at 410 nm, under conditions of the assay. Activity is expressed as trypsin units inhibited (TUI) or trypsin inhibitor units (TIU), assuming a 1:1 stoichiometric ratio of trypsin:trypsin inhibitor. Since 1 µg of "pure" trypsin increases absorbance 1.9 TU, then 1.9 TIU or 1.9 TUI is equivalent to 1 µg of trypsin inhibited.
[b]Lehnhardt and Dills (1984), found that 1 µg of "pure" trypsin produced 2.16 TU, under their conditions of assay.
[c]Calculated on the basis that 1.9 TIU or TUI is equivalent to 1 µg of trypsin inhibitor, if one assumes that the inhibitor will inhibit its own weight of trypsin.
[d]TIA = trypsin inhibitor activity values in terms of "pure" trypsin inhibited rather than in arbitrary units. Since 1 µg of pure trypsin has an activity of 1.90 TU, then TIA of X mg trypsin inhibited per g is equivalent to 1.90 times TIU per mg (see footnote a).
[e]One TU was the amount of enzyme which liberated 1 µmole of p-nitroaniline per unit from BAPA. Under conditions of assay, one TU was equivalent to 0.092 absorbance at 410 nm. One inhibitor unit was defined as the amount of inhibitor which depressed the reaction by one TU, i.e., depressed absorbance by 0.092.
[f]One unit of trypsin inhibitor (TIU) is also defined in relation to µmoles of (p-toluene-sulfonyl arginine released per min by measuring absorbance at 247 nm during hydrolysis of p-toluenesulfonyl arginine methyl ester (TAME).
[g]HIT = Half inhibition of trypsin, under conditions of assay.

TAME as synthetic substances (footnotes e and f, Table 3) would not necessarily be equivalent. Trypsin obtained from commercial sources contains different levels of pure trypsin (Kakade et al., 1969; Hamerstrand et al., 1981). As noted earlier, Kakade et al. (1969) and Lehnhardt and Dills (1984) report that 1 µg of pure trypsin produced 1.9 and 2.16 TU, respectively, under the conditions of the assay. TIA activity calculated in terms of mg of pure trypsin inhibited per g sample protein is the preferable term.

Problems Associated With a General TI Assay

On the basis of previous results (Kakade et al., 1969, 1974; Rackis et al., 1974; Smith et al., 1980; Hamerstrand et al., 1981; Charpentier and Lemmel, 1984), several inherent problems associated with the determination of TI activity can be overcome with a judicious choice of experimental conditions. Details of the application and limitations of the modified methods as described in their respective reports should always be considered. Most of the methods specify that the soy products be stirred with 0.01 N sodium hydroxide at pH 8.4-10.0 for 3 h at room temperature, whereas, an overnight soak at 4°C was preferred by Smith et al. (1980).

Alkali extraction is needed for analysis of heat processed soy products because of the presence of water-insoluble forms of TI. But, the use of alkaline extraction is not universal, since TI's in some foodstuffs may be quite susceptible to inactivation in alkali. Preliminary work should always be carried out to establish conditions of extraction for maximum recovery of TIA. For example, acetate buffer at pH 3.8 was more effective than 0.01 N NaOH in extracting TIA from triticale and rye (Chang and Tsen, 1979).

For samples requiring fat extraction, both extraction of dry samples and aqueous suspensions were tried (Smith et al., 1980). Losses of TIA occurred during fat extraction of aqueous suspensions. The recommended procedure is to first freeze dry the foodstuffs and then extract them with hexane or petroleum ether before slurrying in NaOH solution. Organic solvent extraction is needed not only to remove fat, but also to remove free fatty acids which have TIA (Wang et al., 1975; Smith et al., 1980).

Alkali-treatment of casein (0.2 N NaOH, 80°C, 1 h) results in the creation of TIA (Vimont-Rispoli et al., 1980) in amounts sufficient to cause pancreatic hypertrophy in rats (Possompes et al., 1979). The chemical nature of these products having TIA has not been established, but reactions that can occur during severe treatment of proteins include cross-linking of amino acid residues, destruction of amino acid side chains, racemization of amino acids, and formation of lysinoalanine and other compounds.

Close attention must be directed towards the probability that cloudiness can occur with some samples after termination of the reaction with acetic acid (Rackis et al., 1974). Under these circumstances, one cannot properly correct for the absorbance of the sample blank. Collaborative TI analyses between our laboratory and analysts from various manufacturers of soy-based infant formulas indicated that the problems associated with cloudiness can be effectively eliminated by filtering the acidified reaction mixtures through Whatman GF/C or 934 AH glass fiber filter paper (unpublished data).

Interfering Substances

With samples containing very low TIA, the reliability of the values is uncertain because of non-specific interference of large amounts of protein in the reaction mixture. This problem and the lower limits of detection are discussed in greater detail by Smith et al. (1980).

The alkaline extraction medium as described in the standard soybean method (Kakade et al., 1974) cannot be used with *Vicia faba* because such extracts give a highly-colored reaction with BAPA (Valdebouze et al., 1980). Two major classes of tannins, termed hydrolyzable and condensed tannins, are found in many foodstuffs. In particular, the condensed tannins have a strong affinity for digestive enzymes (Price and Butler, 1980; Griffiths, 1979). Since tannins can complex with trypsin, they contribute to the total TIA under the conditions of the assay. Food legumes with colored seedcoats, some grain sorghums and other foodstuffs contain appreciable amounts of tannins (Price and Butler, 1980).

Methods to differentiate between the non-specific TI activity of tannins and specific activity of protease inhibitors are available (Valdebouze et al., 1980). The TI assay is carried out in the presence and absence of 1 to 1.5% polyvinylpyrolidone (PVP) or polyethylene glycol 4000. These two compounds complex the tannins and other phenolic compounds much more strongly than proteins. The reduction in total TIA activity in the presence of PVP is, therefore, attributed to the presence of tannins and possibly other phenolic compounds. Similar application of the PVP technique in the assay of TIA in soy flour extracts indicated that no activity was attributed to tannins (Sessa, 1984). Procedures for the removal of tannins from plant extracts have been reviewed (Price and Butler, 1980).

Green and leafy vegetables contain pigments which create special problems with the spectrophotometric assay of TIA. For example, alkali extraction may result in the conversion of leucoanthocyanins into colored products.

Free fatty acids account for a major portion of the TIA in miso (Doell et al., 1981) and in tempeh (Wang et al., 1975). Miso and tempeh, as well as soy sauce (shoyu), natto and sufu, are fermented soy foods in which appreciable amounts of free fatty acids are released during fermentation (Wang, 1983). An appropriate lipid solvent should be used to eliminate non-specific TIA associated with fatty acids (Doell et al., 1981).

Several indigestible polysaccharides (Ikeda and Kusano, 1983) and the lathrogens, p-aminopropenonitrile and 1-aminoacetonitrile (Foster et al., 1979), possess TIA under the conditions of assay. Although the contribution of these compounds to the total TIA in most foods may be minimal, it must be recognized as a potential problem.

Recommendations

It appears from the above discussion that no general TI assay can be used for all foodstuffs. Modifications in the conditions used to prepare extracts for analysis can be readily accomplished by the judicious selection of buffers and pH adjustment. With most vegetables and other foods, a simple ultrafiltration or HPLC procedure could be developed to remove pigments and other low molecular weight substances that interfere with the assay of protease inhibitors. Ultimately, an independent immunological assay for specific protease inhibitors may be required to verify the results obtained by the standard enzymatic assay.

Smith et al. (1980) discuss other sources of error and the significance of low levels of TIA in some foodstuffs. Until a more sensitive and specific method is established, we recommend that the modifications of the standard method (Kakade et al., 1974) as described by Smith et al. (1980), Hamerstrand et al. (1981), Lehnhardt and Dills (1984), and Charpentier and Lemmel (1984) be used to obtain useful data about the TI content of most of the important vegetable food proteins. Calculation of TIA according to the formula described earlier would enable direct comparison of the values obtained by various workers.

TRYPSIN INHIBITOR ACTIVITY OF PLANT FOODS

Various technologies are available for the manufacture of a wide range of food protein products based on processes developed for the production of flours, concentrates, isolates and fermented foods from soybeans (Smith and Circle, 1972; American Oil Chemists' Society, 1979, 1981). As a consequence of the multiplicity of technologies used, TIA's of these food protein products vary widely. However, because of the diverse assay procedures employed, a common TIA unit cannot be derived to compare TIA between protein products prepared from oilseeds, legumes and cereals. For these reasons, many of the data in the literature could not be used in this review. Except where noted, the TIA data reported are arbitrarily limited to those values obtained with the assay of Kakade et al. (1974) or with modifications of this procedure. In some of the tables, TIA values obtained by other assays were assembled whenever soy flour was also analyzed. The TIA values reported in the following tables represent the first comprehensive attempt to directly compare TIA of various foodstuffs.

Oilseeds

TIA's of soy protein products and of other selected oilseeds are given in Table 4. Different production lots of raw, dehulled, defatted soy flour contain 47 to 83 mg TIA/g protein, whereas the values for toasted soy flour are in the range of 9 to 16 mg TIA/g protein. Residual TIA in the toasted flours represents 13 to 16% of the original activity of raw soy flour, and can be as low as 6 % (Rackis and Gumbmann, 1981). Further reduction in residual TIA may lead to a lowering of the nutritional value (Rackis et al., 1975B). Raw soy protein isolate contains about 30 mg TIA/g protein and moist heat treatment can reduce TIA by as much as 90%. Soy protein isolates that have been heated contain 3.2 to 12.1 mg TIA/g protein. Soy protein concentrates possess low to moderate amounts of TIA, while textured soy protein products have relatively low TIA.

On a protein basis, oriental soy foods contain 0.7 to 9.2 mg TIA/g protein, approximating the residual TIA of heat-processed soy flours, concentrates and isolates. Miso, a fermented soy product, contains nearly 23 mg TIA/g protein, which can be reduced to 4.4 mg TIA after extraction with lipid solvents. This large reduction in activity is attributed to the removal of free fatty acids (Doell et al., 1981). TIA of the fermented soy product, tempeh, is higher than that of the boiled soybeans from which it was prepared (Wang et al., 1972). This increase is accounted for by the formation of free fatty acids during fermentation (Wang et al., 1975). Soy milk and soy-based infant formulas also contain relatively low levels of TIA (2.5 to 3.3 mg/g protein).

TIA in soybeans increases about twofold during maturation and was found to decrease nearly 3-fold during germination (Bates et al., 1977). On the other hand, Collins and Sanders (1976) found no significant change during germination for the 3 cultivars assayed. Soybean cultivars exhibit a wide range in TIA. Chymotrypsin inhibitor activity also varied widely in these cultivars, but the level of activity is much lower than TIA.

Cotyledons, which represent 90% of whole soybean seed, account for about 83% of the total TIA; the seed coat or hull, which represents 8.5% of the seed, accounts for 11% of the total TIA (Kapoor and Gupta, 1978). No TIA was found in the embryo (hypocotyl). The TI's apparently are synthesized at the time of seed development in the pod. Birk and Waldman (1965) did not detect any TIA in the leaves, stems or empty pods. The presence of PI's and their role in the biochemical-physiological reactions in the leaves of potatoes and tomatoes is presented elsewhere (Chapter ___).

Table 4 lists only a few other oilseeds for which reliable values are available for direct comparison with soybeans. The winged bean, Psophocarpus tetragonolobus, is a legume that is the most promising oil and protein crop in tropical countries (Kadam and Salunkhe, 1984; Sri Kantha and Erdman, 1984). The winged bean contains 27-47% protein and 15-28% oil (Khor et al., 1982). All parts of the plant, including leaves, flowers, stems, pods, seeds and tubers, are eaten raw or cooked (de Lumen and Belo, 1981). The tubers, which may contain 3-15% protein, can cause nausea when eaten raw. When cooked tubers are eaten with cocoa, severe intoxication occurs. The raw seeds are toxic (Sri Kantha and Erdman, 1984). While TIA in winged bean is comparable to that in soybeans (Table 4), the chymotrypsin inhibitor activity of winged bean is several times higher (de Lumen and Belo, 1981) than that reported in soybeans (Kakade et al., 1972). Winged bean tubers and seeds contain the same levels of trypsin and chymotrypsin inhibitor activity (de Lumen and Belo, 1981) and both activities can be reduced 82-93% by moist heat (Tan et al., 1984). Using p-tolunesulphonyl-L-arginine methyl ester as the trypsin substrate, Hildebrand et al. (1981) found that 3 winged bean genotypes contained about the same level of TIA as two soybean cultivars.

Table 4. Trypsin Inhibitor Activity (TIA) Content of Oilseeds and Their Protein Products

Name	Trypsin inhibitor activity (TIA)[a]			Reference
	mg/g sample	mg/g dry weight	mg/g Protein	
Soybeans				
Whole soybeans	18.7	22.9	49.6	Doell et al. (1981)
Defatted soy flour, raw, (1)[b]	30.2 (28-32)[b,c]	32.4	57.8	Rackis et al. (1985)
Defatted soy flour, toasted, (1)	8.5 (7.9-8.4)[c]	9.1	16.0	Rackis et al. (1985)
Defatted soy flour, over-toasted (1)	5.1 (4.4-6.3)[c]	5.4	9.4	Rackis et al. (1985)
Defatted soy flour, raw (1)[d]	23.7-28.6	---	47.2-55.4	Rackis et al. (unpublished data)
Defatted soy flour, raw (1)[d]	42.3	45.1	83.0	Rackis et al. (1979)
Defatted soy flour, toasted (1)[d]	5.6	6.0	10.5	Rackis et al. (1979)
Soy protein concentrate (2)	13.9	14.8	20.7	Rackis et al. (1979)
Soy protein concentrate (3)	6.3	---	9.2	Kakade et al. (1974)
Soy protein isolate, raw (4)	24.1 (21-30)[e]	25.5	29.4	Rackis et al. (1985)
Soy protein isolate, heated (4)	4.7 (3.3-5.7)[e]	5.1	6.1	Rackis et al. (1985)
Soy protein isolate, raw (4)	29.4	30.8	34.9	Rackis et al. (unpublished data)
Soy protein isolate, heated (4)	2.7	2.8	3.2	Rackis et al. (unpublished data)
Soy protein isolate (5)	4.4	---	5.0	Kakade et al. (1974)
Soy protein isolate (6)	10.4	---	12.1	Kakade et al. (1974)
Soy food fiber	12.3	---	---	Liener (1975)
Soy meat analogues	6.5-10.2	---	---	Liener (1975)
Textured soy concentrate	1.9	2.0	3.4	Doell et al. (1981)
Oriental soy foods[f]				
Tofu	0.6	5.5	9.2	Doell et al. (1981)
Tofu, cooked[g]	0.5	3.2	5.5	Doell et al. (1981)
Chinese bean curd (Tofu)	0.7	2.5	5.8	Doell et al. (1981)
Soy sauce	0.3	0.8	3.3	Doell et al. (1981)
Soy sauce, cooked[g]	0.1	0.1	0.7	Doell et al. (1981)

(continued)

Table 4. (continued)

Name	Trypsin inhibitor activity (TIA)[a] mg/g sample	mg/g dry weight	mg/g Protein	Reference
Miso (unextracted)	4.1	8.6	22.9[h]	Doell et al. (1981)
Miso (defatted)	---	---	4.4[i]	Doell et al. (1981)
Miso (defatted and cooked)	---	---	0.7[g]	Doell et al. (1981)
Soy-based infant formulas[j]				
Products A-E	0.06-0.33	---	2.65-3.33	Collaborative study[k]
Products F	1.4	1.5	2.5	Collaborative study[l]
Soy milk				(Rackis et al. (unpublished data)
Stage of soybean maturity				
Early milky maturation	---	---	20.9[m]	Kapoor and Gupta (1978)
Mid-milky maturation	---	---	28.6[m]	Kapoor and Gupta (1978)
Mature	---	---	45.4[m]	Kapoor and Gupta (1978)
Immature, raw[n]	---	---	49.0	Bates et al. (1977)
Mature, raw[n]	---	---	52.2	Bates et al. (1977)
Germinated, raw[n,o]	---	---	17.8	Bates et al. (1977)
Immature, cooked[n]	---	---	1.5	Bates et al. (1977)
Mature, cooked[n]	---	---	0.6	Bates et al. (1977)
Germinated, cooked[n]	---	---	1.7	Bates et al. (1977)
Soybean Cultivars				
108 strains or cultivars	---	---	35-123[p]	Kakade et al. (1972)
108 strains or cultivars	---	---	39-66[q]	Kakade et al. (1972)
32 varieties (field and vegetable type)	---	---	21-86[r]	Gupta and Deodhar (1975)
19 varieties and strains (defatted)	17-27[p]	---	---	Hafez (1983)
Other Oilseeds				
Winged bean, raw	---	30.3[p]	---	Ekpenyong and Borchers (1981)
Soybeans, raw	---	46.5[p]	---	Ekpenyong and Borchers (1981)
Winged bean, autoclaved	---	2.3[p]	---	Ekpenyong and Borchers (1981)

Table 4 continued

Name	Trypsin inhibitor activity (TIA)[a] mg/g sample	mg/g dry weight	mg/g Protein	Reference
Soybeans, autoclaved	---	6.1[p]	---	Ekpenyong and Borchers (1981)
Winged bean, defatted				
15 Thai and 15 New Guinea Cultivars	---	19.0-47.1[s]	---	Khor et al. (1982)
7 Cultivars	---	---	67-99	del Rosario et al. (1980)
Peanuts	0.4	0.4	1.4[t]	Doell et al. (1981)
Sunflower, 5 cultivars (mean values)	---	---	2.8[t]	Roy and Bhat (1974)
Soybeans, 5 cultivars (mean values)	---	---	152[t]	Roy and Bhat (1974)

[a]Except where noted, data were limited to those obtained with the procedures of Kakade et al. (1974) or with the modifications thereof as discussed in the text.
[b]Different numbers in parentheses represent products produced by different soy processors.
[c]Range in values of several composite samples taken from the bagged products representing production runs of 4000, 1500, and 1500 pounds of raw, toasted and over-toasted soy flours, respectively.
[d]Several different lots of defatted soy flours.
[e]Range in values of several composite samples taken from the bagged products representing production runs of 1500 pounds of product.
[f]See Wang (1983) for review of conditions of manufacture.
[g]Additional heat treatment (10% slurry, boiled 15 min).
[h]Total TIA of unextracted miso represents TI activity of free fatty acids plus that of the specific protease inhibitors.
[i]The difference in TIA of unextracted miso and that of miso extracted with petroleum either is attributable to the free fatty acids in the product (see also Wang et al., (1975).
[j]See Sarett (1976) for details.
[k]Results of a collaborative study between NRRC and infant formula manufacturer using the Hamerstrand et al. (1981) assay.
[l]Results of a collaborative study between NRRC and Food and Drug Administration.
[m]Activity was expressed as percent inhibition of trypsin per mg protein in a 0.05 N HCl extract.

(continued)

Table 4 (footnotes cont.)

[n]Activity given as change in absorbance at 257 mm/min/g protein, compared to control with no TI, Bragg soybeans; cooking involved autoclaving at 121°C, 15 min.

[o]Collins and Sanders (1976) reported no significant change in TIA during germination of 3 soy cultivars.

[p]The values originally expressed as trypsin units inhibited (TUI) were converted to TIA units by using the factor: one μg of pure trypsin has an activity of 1.9 trypsin units (TU). See Table 2 for details.

[q]Chymotrypsin units inhibited per mg protein (CUI)..

[r]Specific activity units per mg protein.

[s]Ranges in values were recalculated on the basis that one inhibitor unit (I.U.) is equivalent to one trypsin unit inhibited so that activity can be converted to T.I.A. units by using the factor: one μg of pure trypsin has an activity of 1.9 trypsin units (See Table 2 for details). Values are reported on a "fat-free" basis.

[t]T.I.A. as reported by these authors cannot be converted to trypsin inhibitor activity as defined in Table 2. Therefore, no direct comparison with the other oilseeds can be made.

Both peanuts and sunflower contain very low levels of TIA (Table 4); no detectable TIA was found in two of five sunflower cultivars (Roy and Bhat, 1974).

Legumes

Referred to alternatively as pulses, dry beans, food and grain legumes, only 20 of the 1300 species of the family Leguminosae are commonly consumed by humans. Only the Papilionodeae, the largest of the subfamilies, which include Mimosoideae and Caesalpinioideae, are used for human food (Rockland and Nishi, 1979; Sgarbieri and Whitaker, 1982; Sathe et al., 1984A,B). However, other Leguminosae seeds not commonly consumed are continually being evaluated as a food source because of their high protein content that can range from 18-35%.

A number of PI's have been isolated and characterized from another sub-family of Leguminosae (Joubert, 1982). Listed in Table 5 are data selected only where soybeans were also included in the analysis of TIA in food legumes. This is because direct comparisons of the values obtained by these analysts and the values reported in the other tables are not possible because of differences in conditions of assay and expression of activity units.

Examination of the data in Table 5 reveals that most legume species contain less than 50% of the TIA of soybeans. Particularly low activities are present in most cultivars of *Vicia faba*, *Pisum sativum*, *Lupinus* species and a few cultivars of *Phaseolus vulgaris*. Those legumes with at least 75% of the TIA in soybeans include: *Vigna catjung* (Soni et al., 1978); *Dolichos lablab* (Liener, 1976); *Cajanus cajans* (Valdebouze, 1977); and *Phaseolus vulgaris* (Soni et al., 1978; Liener, 1976; Jaffe and Vega Lette, 1968).

TIA's of several diverse species of legumes, determined by the Kakade procedure or by modifications thereof, are summarized in Table 6.

Of the common legumes, some cultivars of *Vigna unguiculata* and *Phaseolus lunatus* contain higher levels of TIA than soybeans (del Rosario et al., 1980). Legumes containing levels of activity comparable to soybeans include some cultivars of *Phaseolus vulgaris* (del Rosario et al., 1980), *Phaseolus acutifolius* (Thorn et al., 1983) and *Vigna sinensis* (Valdebouze et al., 1980). However, most of the legume cultivars contain low to moderate levels of TIA.

Of especial interest are the arid land legumes analyzed by Thorn et al. (1983) which may have potential importance as a food source in those parts of the world experiencing increased desertification and concern over availability of water. Of these legumes, cultivars of *Lysiloma watsonii* contain TIA comparable to or greater than that in soybeans. Protein content and presence of other antinutritional factors of these arid land legumes were also reported by these workers. Mesquite bean, a legume that also grows in desert and semi-desert areas and that is used in a number of ways as a food in Mexico, contains little TIA (del Valle et al., 1983).

Significance of Total TIA in Legumes

Protease inhibitors, defined as proteins with the peculiar property of forming reversible stoichiometric protein-protein complexes with various proteolytic enzymes to bring about competitive inhibition of their catalytic function, have been isolated and characterized from many of the plant foods discussed in this report. To what extent these protease inhibitors account for the total TIA given in Tables 4-6 cannot be quantitated by the

Table 5. Trypsin Inhibitor Activity (TIA) Content of Selected Legumes in Relation to Soybeans

Legume		TIA			
Scientific name	Common name	% Soy = 100%	TIA x 10^{-4} units/g	Units/mg extracted protein	Reference
Vigna sinensis (V. unguiculata)	cow pea	28 (11.1)[a]	---	---	Valdebouze (1977)
Vicia faba	field bean, broadbean	2-20 (0.7-36)[a]	---	---	Valdebouze (1977)
Pisum sativum	garden pea	5-13 (1.5-5.7)[a]	---	---	Valdebouze (1977)
Lupinus species	lupin beans	0[a]	---	---	Valdebouze (1977)
Phaseolus lunatus	lima bean	77[b]	---	---	Hove and King (1979)
Phaseolus vulgaris	navy, kidney bean	13-44[b]	---	---	Hove and King (1979)
Pisum sativum	pea	3-13[b]	---	---	Hove and King (1979)
Vicia faba	---	10[b]	---	---	Hove and King (1979)
Lupinus species	lupin bean	0-2[b]	---	---	Hove and King (1979)
Phaseolus vulgaris	french bean	80[c]	---	---	Soni et al. (1978)
Vigna catjung	cow pea	79[c]	---	---	Soni et al. (1978)
Cicer arietinum	bengal gram	66[c]	---	---	Soni et al. (1978)
Vigna sinensis	cow pea	64[c]	---	---	Soni et al. (1978)
Cajanus cajan	red gram	60[c]	---	---	Soni et al. (1978)
Phaseolus mungo	black gram	52[c]	---	---	Soni et al. (1978)
Phaseolus aureus	green gram	37[c]	---	---	Soni et al. (1978)
Phaseolus aconitifolus	moth bean	27[c]	---	---	Soni et al. (1978)
Lens esculenta	lentils	25[c]	---	---	Soni et al. (1978)
Pisum sativum	pea	25[c]	---	---	Soni et al. (1978)
Glycine max	soybeans	---	4.15[d]	---	Liener (1976)
Phaseolus vulgaris	kidney bean	---	4.25[d]	---	Liener (1976)
Dolichos lablab	hyacinth bean	---	4.38[d]	---	Liener (1976)
Cajanus cajan	pigeon pea	---	2.77[d]	---	Liener (1976)
Vigna sinensis	cow pea	---	1.91[d]	---	Liener (1976)
Lens esculenta	lentils	---	1.78[d]	---	Liener (1976)
Glycine max	soybeans	---	---	28.3[e]	Jaffe and Vega Lette (1968)
Phaseolus vulgaris	mottled bean	---	---	46.0[e]	Jaffe and Vega Lette (1968)
Phaseolus vulgaris	black bean	---	---	32.1[e]	Jaffe and Vega Lette (1968)
Phaseolus vulgaris	tapiaamo bean	---	---	20.2[e]	Jaffe and Vega Lette (1968)
Phaseolus vulgaris	red bean	---	---	3.8[e]	Jaffe and Vega Lette (1968)
Phaseolus vulgaris	white bean	---	---	3.3[e]	Jaffe and Vega Lette (1968)

[a]Originally reported as TUI/mg dry matter using the Kakade et al. (1974) procedure. Values in parenthesis represent additional studies (Valdebouze et al., 1980).
[b]A modified casein digestion procedure was used; values originally reported as mg inhibitor/g sample.
[c]A modified casein digestion procedure was used; values originally reported as trypsin inhibitor units (TIU)/g extracted protein.
[d]Modified casein digestion procedures were used.

Table 6. Trypsin Inhibitor Activity (TIA) Content of Legumes[a]

Scientific name[b]	Common name	TIA mg/g sample	TIA mg/g dry weight	TIA mg/g protein	Reference
Glycine max (1)	soybean, defatted flour	---	31.3	62.6[c]	Valdebouze et al. (1980)
Vigna sinensis (1)	cow pea	---	11.1	55.5[c]	Valdebouze et al. (1980)
Pisum sativum (8)	pea	---	1.5-5.7	6.0-22.8[c]	Valdebouze et al. (1980)
Vicia faba (26)	faba bean	---	1.7-3.3[d]	6.8-13.2[c]	Valdebouze et al. (1980)
Lupinus species (3)	lupin bean	---	<0.5	2.0[c]	Valdebouze et al. (1980)
-- not given	peas, fresh	0.2	0.8	2.6	Doell et al. (1981)
-- not given	peas, cooked	0.2	0.8	2.4	Doell et al. (1981)
-- not given	broad beans, fresh	0.9	2.3	7.1	Doell et al. (1981)
-- not given	broad beans, cooked	0.3	0.8	2.6	Doell et al. (1981)
Phaseolus lunatas (18)	lima	---	---	10.9-19.3	Ologhobo and Fetuga (1983)
Glycine max (1)	soybean	---	32.6	85.4	Thorn et al. (1983)
Lysiloma watsonii (1)	feather tree	---	31.8	102.0	Thorn et al. (1983)
Phaseolus acutifolius (5) Var. latifolius	Tepary	---	6.3-18.5	27.4-76.9	Thorn et al. (1983)
Phaseolus vulgaris (1)	pinto	---	8.7	39.8	Thorn et al. (1983)
Parkinsonia aculeata (3)	mexican palo verde	---	1.2-2.5	5.7-11.7	Thorn et al. (1983)
Prosopis juliflora (1)	mesquite bean	---	0.74	5.1	del Valle et al. (1983)
Phaseolus lunatus (6)	lima	---	---	84.7-122.7	del Rosario et al. (1980)
Vigna unguiculata (23)	cow pea	---	---	25.9-112.1	del Rosario et al. (1980)
Phaseolus vulgaris (6)	common bean	---	---	43.6-61.8	del Rosario et al. (1980)
Cicer arrietinum (1)	chick pea	---	---	34.2	del Rosario et al. (1980)
Vigna radiata (17)	mung bean	---	---	15.3-22.9	del Rosario et al. (1980)
Vigna umbellata (1)	rice bean	---	---	22.4	del Rosario et al. (1980)
Cicer arrietinum	chick pea	3.47	---	---	Besrat (1981)
--	haricot bean	12.81	---	---	Besrat (1981)

[a]TIA values reported here were obtained with the Kakade et al. (1974) procedure or its modification. Activity originally expressed as trypsin inhibitor units (TIU) or trypsin units inhibited (TUI) per mg was converted to TIA units per g on the basis that one μg of pure trypsin has an activity of 1.9 trypsin units (See Table 2 for details).

[b]Number in parentheses represents number of cultivars assayed.

[c]Calculated on basis that defatted soybean meal and the other legumes contain an average of 50 and 25% protein, respectively. Actual protein content of the other legumes ranged from 15 to 28%.

[d]Standard Kakade et al. (1974) assay modified to replace 0.01 N NaOH extraction with HCl extraction and addition of 1% polyvinylpyrolidone to reaction mixture to eliminate interference with pigments and tannins.

methods of assay used. Trypsin and other enzymes can interact with several food components such as phenolic compounds, particularly tannins, phytates, saponins, fatty acids, lectins and other constituents. Those compounds may contribute to the total TIA observed under the conditions of assay employed.

Tannins, for example, have been recognized as antinutritional factors in legumes and sorghum. Their inhibitory effect on activities of digestive enzymes such as trypsin, lipase, and amylase are well known. They are responsible for poor digestibility of proteins and growth inhibition when fed to animals, but they do not cause pancreatic enlargement. Tannins are usually located in the seed coat with low or negligible levels in cotyledons and embryos. Dehulling of legumes reduces tannin content and increases TIA and chymotrypsin inhibitor activity (CIA) (Table 7). The TIA attributable to tannins in these bean cultivars appears small because removing most of the tannins with the hulls increased rather than decreased the TIA. The increase in TIA is about the expected value if the hulls contain no TIA but constitute about 10% of the seed weight.

Other Plant Foods

Protease inhibitor activities of other plant foods that are consumed on a regular basis are given in Table 8. Unfortunately, several different units were used to define activity. Compared to a value of 57.8 mg TIA/g protein for raw soy flour, potatoes and potato flour may contain up to two times more TIA, while eggs contain comparable amounts of TIA. Cooked potatoes and hard-boiled eggs contain relatively higher levels of TIA than toasted soy flours.

Although protease inhibitors have been isolated and characterized from a large number of cereals, few quantitative data could be assembled here that can be directly compared with TIA in legumes and oilseeds. As given in Table 8, cereals such as corn meal, bread and rice contain low levels of TIA, while relatively high levels of TIA were found in sweet corn. Compared with corn meal, wheat and rice, rye and triticale may contain two to seven times more TIA. Distribution of inhibitor activity in wheat decreased in the following order: flour, shorts and bran (Chang and Tsen, 1979). Several protease inhibitors have been isolated and characterized from cereals which include barley, rye, wheat, triticale, corn, rice, sorghum, oats, millet and buckwheat (Boisen, 1983). Of all the cereals, rye, triticale and barley contain the highest levels of TIA. Cultivars of wheat for use in bread generally have twice the TIA of other wheats. The floury-2 gene in corn is associated with low TIA, while in high-lysine corn the opaque-2 gene is associated with high TIA. As reviewed by Boisen (1983), very high levels of chymotrypsin inhibitors were found in Hiproly barley (8 mg/g or about 5% of the total protein) as compared with much lower levels in normal barley cultivars. The increased level of PI activity is also associated with the Hiproly gene for high lysine. As a point of reference, total PI activity in barley is normally about 0.45 g/kg compared with 1.5 g/kg in chick peas and 3.6 g/kg in kidney beans (Boisen, 1983) and, in turn, these legumes normally contain much lower PI activity than soybeans and winged bean (See Tables 4 and 5).

The physiological properties and antinutritional effects of raw soybeans, potatoes and other foods have often been attributed to protease inhibitors that inhibit the pancreatic digestive enzyme trypsin (Rackis and Gumbmann, 1981). However, the antinutritional effects of protease inhibitors that specifically inhibit chymotrypsin have not been evaluated. Bananas and taro appear to contain exceptionally high levels of chymotrypsin inhibitor activity (Table 8). The TIA:CIA ratio in winged bean is about 22 (Tan and Wong, 1982; Tan et al., 1984); the ratio for Great

Table 7. Comparison of Protease Inhibitor Activity With Protein and Tannin Content of Phaseolus vulgaris Cultivars[a]

Cultivar	Protein %	Tannin(mg Catechin Equivalents)/100g		% Reduction On Dehulling	TIA, Units/gX10^{-4} [b,d]			CIA, Units/gX10^{-4} [c,d]		
		Whole Beans	Dehulled Beans		Whole Beans	Dehulled Beans	% Increase on Dehulling	Whole Beans	Dehulled Beans	% Increase On Dehulling
Small Red	---	282.8	19.0	93.3	222	245	10.4	288	360	25.0
Pinto	18.12	264.7	14.2	94.6	360	433	20.3	323	370	14.6
Light Red Kidney	20.89	152.2	22.2	85.4	216	257	19.0	318	364	14.5
Viva Pink	21.30	122.1	10.4	91.5	257	261	1.6	217	271	24.9
Dark Red Kidney	20.32	105.3	28.7	72.7	240	263	9.6	335	344	2.7
Cranberry	23.43	76.3	10.0	86.9	233	282	21.0	323	356	10.2
Black Beauty	22.87	33.7	10.8	68.0	270	289	7.0	331	371	12.1
Sanilac	18.98	ND[e]	ND	---	186	253	36.0	332	368	10.8
Great Northern	23.01	ND	ND	---	248	285	14.9	308	361	17.2
Small White	19.73	ND	ND	---	235	262	11.5	345	369	7.0

[a]Dry-basis, Deshpande et al. (1982).
[b]TIA-Trypsin inhibitor activity.
[c]CIA-chymotrypsin inhibitor activity.
[d]One unit of inhibitor activity reduced trypsin or chymotrypsin activity by one unit.
[e]ND = Not detected.

Table 8. Protease Inhibitor Activity of Potatoes, Cereals and Other Foods.

Food	Trypsin inhibitor activity (TIA) mg/g sample	mg/g dry weight	mg/g protein	HIT[a]/ mg protein	% inhibition of trypsin per mg of tissue	Chymotrypsin inhibitor activity mg/mg N (tuber)	References
Defatted soy flour[b]	30.2	32.4	57.8	---	---	---	Rackis et al. (in press)
Potatoes							
Russet Burbank cultivar[c]	---	5.19	102.6	---	---	---	Ryan and Pearce (1978)
Clones[c]	---	2.88-11.63	96.0-110.6	---	---	---	Ryan and Pearce (1978)
Potato flour	6.2-8.5	6.7-8.1	93.8-108.9	---	---	---	Doell et al. (1981)
Fresh, raw	1.3	6.4	56.5	---	---	---	Doell et al. (1981)
Fresh, cooked	0.3	1.4	14.3	---	---	---	Doell et al. (1981)
Eggs, hen's, fresh	5.7-6.7	23.6-26.8	46.0-53.2	---	---	---	Doell et al. (1981)
Eggs, hard-boiled	3.5	13.2	25.4	---	---	---	Doell et al. (1981)
Corn meal	0.74	---	---	---	---	---	Doell et al. (1981)
Bread, white	0.1	0.2	1.1	---	---	---	Doell et al. (1981)
Bread, brown	0.3	0.5	3.1	---	---	---	Doell et al. (1981)
Rice	0.1	0.1	1.4	---	---	---	Doell et al. (1981)
Wheat (1 cultivar)	---	---	---	1.0	---	---	Chang and Tsen (1979)
Triticale (11 cultivars)	---	---	---	1.8-2.4	---	---	Chang and Tsen (1979)
Rye (17 cultivars)	---	---	---	2.9-7.1	---	---	Chang and Tsen (1979)
Soybeans (2 cultivars)	---	---	---	91-97	---	---	Chang and Tsen (1979)
Potatoes	---	---	---	67	---	---	Chen and Mitchell (1973)
Sweet corn	---	---	---	13	---	---	Chen and Mitchell (1973)
Sweet potatoes	---	---	---	19	---	---	Chen and Mitchell (1973)
Banana	---	---	---	---	---	3.85	Ryan et al. (1968)
Taro	---	---	---	---	---	1.85	Ryan et al. (1968)
Potatoes	---	---	---	---	---	0.45	Ryan et al. (1968)
Sweet Potato	---	---	---	---	---	0.30	Ryan et al. (1968)

[a] Half inhibition of trypsin, BAPA substrate, modification of Kakade et al. (1974)
[b] Data taken from Table 3.
[c] Represents TIA of proteinase inhibitors I and II assayed by immunological procedures.

Northern beans is 13.1 (Sathe and Salunkhe, 1981); and that for soybeans is about 2.5 Kakade et al., 1972). According to Freed and Ryan (1978A), the soybean cultivar, Steele, contains 15 mg Kuntz TI per g of seed as measured by rocket immunoelectrophoresis, which would account for 72% of total TIA as measured by the enzymatic assay (Rackis and Gumbmann, 1981).

A specific inhibitor of the pancreatic digestive enzymes carboxypeptidase A and B had some measurable effects on protein digestibility and feed efficiency in chicks, but contributed little to the growth inhibitory and pancreatic hypertrophic effects of raw potatoes (Pearce et al., 1983).

INACTIVATION OF PROTEASE INHIBITORS

Moist Heat Treatment

General Considerations. It is now well established that most raw or underheated vegetable protein foods are of low nutritional value for use as food or feed. With few exceptions, practically all legume and oilseed protein resources require moist heat treatment to inactivate protease inhibitors, phytohemagglutinins and other heat-labile constituents, as well as to transform raw protein into more readily digestible forms (Liener, 1972; Rackis, 1972). Overheating, to completely remove protease inhibitor activity, usually results in protein degradation and other deteriorative reactions giving rise to products that cause adverse effects in animals. Therefore, precise control of the heating process is critical to the preparation of protein products with maximum nutritional value.

By live stream treatment, a process referred to as toasting in the oilseed processing industry, the inferior nutritive value of raw soy

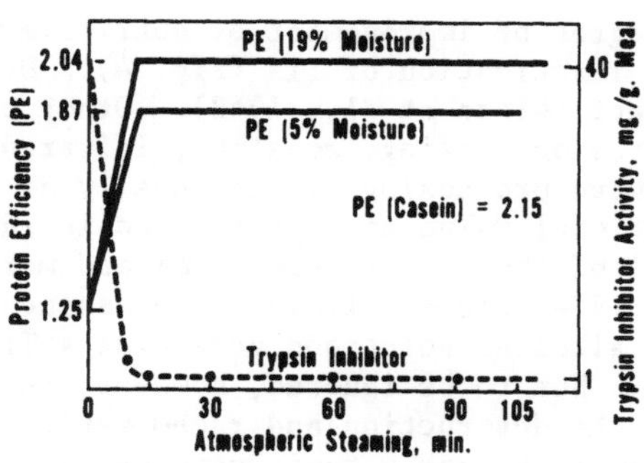

Fig. 4. Effect of autoclaving on protein efficiency (PE) and trypsin inhibitor activity of raw soybean meal. Conditions: Live steam at atmospheric pressure, 100°C: Curve A, PE of meals with 5% initial moisture before autoclaving; Curve B, PE of meals with meals tempered to 19% moisture before autoclaving; Curve C, decrease in trypsin inhibitor activity with time of autoclaving of meals at the two initial moisture levels (Rackis, 1965).

Table 9. Calculated Activation Energies (Ea) for Inactivation of TIA in Field Beans[a]

% moisture (dry basis)	T (°C)	ΔK/K Eq(1)	ΔEa/Ea (Eq(2)	Ea (Kcal/mole)
2	70	0.13		
	100	0.16	0.20	13
14	70	0.16		
	100	0.07	0.09	19
30	70	0.19		
	100	0.09	0.09	27
42	70	0.15		
	100	0.09	0.09	22
55	70	0.17		
	100			
	100	0.11	0.13	17

[a]TIA-Trypsin inhibitor activity (Buera et al. (1984).

protein products can be improved to that nearly comparable to meat and milk (Smith and Circle, 1972). Loss in TIA and inactivation of other heat-labile factors by toasting is a function of a combination of temperature, time, particle size and moisture conditions employed in the toasting process (Albrecht et al., 1966; Rackis, 1965, 1966; Baker and Mustakas, 1973; Mustakas et al., 1981; McNaughton and Reece, 1980). Excessive heat markedly impairs nutritive value of soybeans (Liener, 1972).

In general, the degree of improvement of nutritive value of soy proteins parallels the inactivation of TIA (Fig. 4). Dry heating is without beneficial effect (Klose et al., 1948). Other processing techniques such as extrusion cooking, roasting, infrared cooking, dielectric heating and microwave processing can be equally as effective as toasting in improving nutritional value of soybeans and in inactivating TIA, provided proper control of the temperature, time and moisture conditions are exercised (Rackis, 1974; Liener, 1981). Processing soybeans at elevated temperatures in mildly alkaline solutions accelerates TI inactivation (Hackler et al., 1965). Not to be ignored, however, is the fact that such conditions may lead to the destruction and racemization of amino acids and formation of lysinoalanine and other toxic components. Although the greatest amounts of research data have been accumulated with soy proteins, the principles enumerated above are readily applicable to most other plant protein foodstuffs.

Kinetic Aspects. Limited data are available concerning thermal process optimization, maximizing protease inhibitor destruction, and minimizing degradation of essential nutrients (Labuza, 1973; Cuevas and Cheryan, 1983; Buera et al., 1984; Ida et al., 1983; Phillips et al., 1983). These studies suggest that first-order reaction kinetics describe the inactivation process and establish that thermal inactivation of TIA can be quantitated in terms of the Arrhenius equation.

Based on mathematical modeling, Ida et al. (1983) predicted that blanching soy cotyledons at 96°C, at atmospheric pressure for 17.5 to 20.0 min in a 1:10 solid:$NaHCO_3$ solution ratio was sufficient to inactivate about 78% of the TIA. This would provide a defatted soy flour of maximum protein quality as measured by weight gain, nitrogen digestibility, protein efficiency ratio (PER) and pancreatic hypertrophy (Rackis et al., 1975A).

TIA loss in field beans (Phaseolus vulgaris) on heating follows first-order kinetics, with the rate of loss increasing with increasing moisture content from 2 to 55% (Buera et al., 1984). The effect of moisture content on the first order rate constant (K) for the loss in TIA as a function of temperature is shown in Figure 5. Below 14% moisture, the K value showed little change but increased by a factor of 4.5 and 6.5 at 90°C and 100°C, respectively, compared with that at 70°C. Maximum loss in TIA was about 90% at 100°C for 22-25 min. Over 100 min were required to destroy 90% of the TIA at 14% moisture and 100°C. Estimated activation energy (Ea) values for inactivating TIA in field beans at different moisture levels are given in Table 9. At 30% moisture, a maximum value of 27K cal/mole was obtained. The observed dependence of Ea on moisture content needed to inactivate TIA was similar to that found for reduction in available lysine and nitrogen solubility in heated field bean flours (Buera et al., 1983). Hackler et al. (1965) reported an Ea value of 18K cal/mole for TIA loss in heated soybean suspensions, which corresponded to a value of 17K cal at 55% moisture in field beans. Labuza (1973) estimated, from the data of Hackler et al. (1965), that doubling the rate of destruction of TIA would increase the destruction of essential nutrients in soy beverages four to fivefold.

Sanderson et al. (1982) found that thermal inactivation of Kunitz soybean TI is not a simple two-state process. At low temperature, denaturation is readily reversible with a transition enthalpy of 10-15 kcal/mole; this transition is followed by a non-reversible transition with a much larger enthalpy (approx. 100 kcal/mole). The transition temperatures for irreversible denaturation of the three genetic variants of the Kunitz soybean TI's, SBTI-Ti^a, -Ti^b, and -Ti^c were 64°C, 62°C, and 57°C, respectively.

As a general principle, most plant PI's in the purified form exhibit a much greater stability to thermal denaturation than crude forms such as

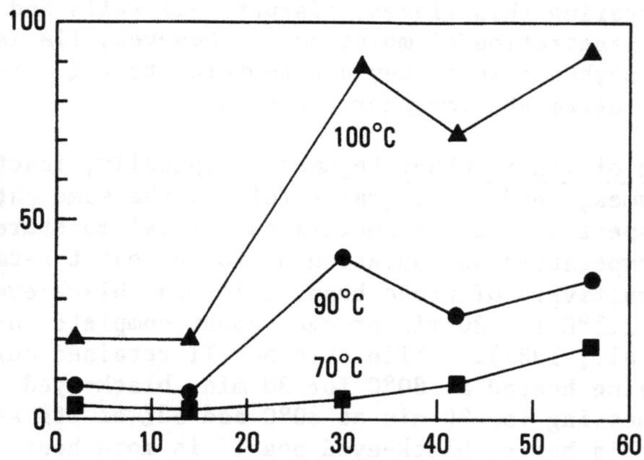

Fig. 5. Effect of moisture content on loss in trypsin inhibitor activity in soybean flours heated at three different temperatures (Buera et al, 1984).

observed in aqueous extracts of soybeans, peanuts, and kidney beans (Ellenrieder et al., 1980). Similarly, in crude extracts of Kintoki beans (Phaseolus vulgaris), 90% of the TIA was lost after heating for 100°C for 60 min, whereas, the purified TI was heat stable (Tsukamoto et al., 1983). Acceleration of thermal destruction of TIA in the crude extracts of these food legumes was attributable to the storage proteins, since addition of albumin, cytochrome c, gelatin, or the high molecular weight proteins of Kintoki beans caused inactivation of the TI's that was dependent upon protein concentration. TIA of soy protein in a meat system becomes more labile to heat inactivation than in the absence of meat (Nordal and Fossum, 1974). It has been postulated that disulfide bond interchange between TI's and several types of proteins was responsible for increased thermal inactivation. On the other hand, Kozlowska et al. (1980) reported that some meat components may exert a protective effect, since residual TIA of a mixture of meat with soy flakes, concentrate or isolate was greater than in water suspensions of the soy protein products similarly treated.

Studies on destruction of TIA in oilseeds, legumes, cereals, eggs and other foodstuffs by various types of moist heat treatment are too numerous to be cited. In practically all of these foodstuffs several molecular forms of the PI's are present. In spite of widely different degrees of heat stability of TI's in the purified state, most of the TIA present in the original food system is readily destroyed during normal cooking procedures.

Inactivation of TIA in Soybeans. Inactivation of TIA and changes in nitrogen solubility index of defatted soy flours as a function of steaming at 100°C are given in Table 10. TIA decreases rapidly with 70 to 84% of the activity being destroyed in about 10 min, but only 92% of the original activity was destroyed in 20 to 30 min. Collins and Beaty (1980) report 96% destruction of TIA in mature, green soybeans after boiling for nine min.

Results showing the inactivation of TIA in full-fat flakes and various full-fat products are given in Table 11. The rate of inactivation of TIA in full-fat flakes is comparable to that obtained with defatted flour (see Table 10). However, with an initial moisture content of 6%, only 20 to 64% of the original activity in cotyledons, chips, and whole soybeans was destroyed after 20 min of steaming. The rapid rate of inactivation of TIA in full-fat flakes, as compared with whole soybeans, cotyledons and chips, was attributed to both particle size and the flaking process. The smooth rolls used in preparing thin flakes, disrupt cell walls and other membranes, permitting rapid penetration of moist heat. However, TIA is readily destroyed in whole soybeans when they are tempered to 25% moisture, and then subjected to live steam at atmospheric pressure.

Inactivation of TIA in Other Legumes. Typically, inactivation of TIA in other legumes, seeds, and grains follows the same pattern as with soybeans with respect to such parameters as initial moisture content, particle size, temperature and duration of moist heat treatment. Heating soaked seeds of cultivars of broad bean, chickpea, black-eyed pea, lentil and mung bean at 121°C for 30 min brought about complete inactivation of TIA (Al-Bakir et al., 1982). While chickpea TI retained only 17% of its activity after being heated at 80°C for 30 min, black-eyed pea retained 94% of its TI after heating for 30 min at 80°C and 58% of its activity at 100°C and 30 min. On this basis, black-eyed pea TI is more heat stable than chick pea or soy TI. Numerous other reports indicate that inactivation of TIA in many legumes is comparable to that described for soybeans in Tables 10 and 11 while other legumes exhibit much greater stability. However, most of the TIA in all plant foodstuffs can be destroyed by autoclaving. Aqueous extracts of winged bean TI are at least as labile as the TI of extracts of soybeans (Hildebrand et al., 1981). Chymotrypsin inhibitor

Table 10. Trypsin Inhibitor (TI) Activity and Nitrogen Solubility Index (NSI) of Heat-Treated Soy Flours

Heat Treatment min[a]		TI Activity TIU/mg sample[b]		Destruction of TI Activity. %		NSI	
Trial I	Trial II	Trial I	Trial II	Trial I	Trial II		Trial II
0 (A)	0 (G)	88.2	96.6	...	...	(G)	97.2
2 (B)	1 (H)	52.1	74.9	41	23	(H)	78.2
4 (C)	3 (I)	27.8	45.0	69	54	(I)	69.6
7 (D)	6 (J)	15.5	28.0	83	71	(J)	56.5
10 (E)	9 (K)	11.8	20.5	87	79	(K)	51.3
20 (F)	20 (L)	7.2	10.1	92	90	(L)	37.9
	30 (M)	...	8.0	...	92	(M)	28.2
	0 (N)		80.4				
	3 (O)		15.3				

[a]Live steam at 100°C. Separate preparations of soy flours were used. Sample letters corresponding to the soy flours are shown in parenthesis. (Rackis et al., 1975A).
[b]TIU = trypsin inhibitor units, as-is basis. (Kakade et al., 1974).

Table 11. Effect of Steaming on Trypsin Inhibitor Activity (TIA) of Whole Soybeans and Other Full-fat Products

Product	Initial moisture, %	Steaming (min)[a]	TIA, mg/g[b]		Destruction of TIA,	
			Assay I[c]	Assay II[d]	Assay I	Assay II
Full-fat flakes	6	0	38.9[e]	53.1[e]	---	---
Full-fat flakes	6	15	2.0	7.2	94.9	86.5
Cotyledons	6	20	15.5	---	60.2	---
Cotyledons	6	30	3.5	---	91.1	---
Chips	6	20	10.0	19.8	74.3	60.3
Chips	6	30	1.5	8.5	96.2	84.6
Whole soybeans	6	20	30.9	48.1	20.6	9.5
Whole soybeans	25[f]	20	1.0	7.2	97.5	86.5

[a] Live steam treatment at atmospheric pressure at 100°C.
[b] Fat-free basis.
[c] Procedure of Kunitz (1946). See Rackis (1966) for more complete data.
[d] Procedure of Kakade et al. (1974). Rackis, unpublished data.
[e] The same average TIA value was obtained for unheated cotyledons, chips and whole soybeans.
[f] Tempered to 25% moisture prior to steaming.

activity in winged-bean extracts is heat labile as in other foods (Tan et al., 1984; de Lumen and Belo, 1981).

Complete destruction of TI in horse beans, peas, soybeans and white beans was achieved by IR heating for 40-50 sec (Kas et al., 1983). Gamma irradiation up to 400 Krad was found to have practically no effect on TIA of green gram cultivars (Haider, 1981). After evaluating quality and nutritional parameters of full-fat soy flours prepared from infrared processed soybeans, Kouzeh-Kanani et al. (1983) concluded that optimum conditions were 60 sec IR radiation at 125° and a 15 min holding period. A residual activity of 8 TUI per mg sample represented only 10% of the original activity of raw soybeans. Soni et al. (1978) found that TIA activity in 11 Indian pulses exhibited varying degrees of stability to dry heating, roasting or autoclaving (Table 12). Dry heating at 75°C for 15 min had little effect on TIA of the pulses except for field peas; heating at 100°C for 15 min destroyed considerably more TIA, except for cow peas. These studies demonstrated that TI's were the most labile in field peas and the most stable in cow peas. Roasting of seeds at 200°C for 2 min was most effective for destroying TIA in all pulses, except lentils (Soni et al., 1978). Heat treatment of arid land legumes can reduce TIA 69-97%, except for Parkinsonia aculeata (Mexican palo verde) cultivars in which 0-40% reduction in TIA was noted (Thorn et al., 1983). This stability to heat denaturation may have little nutritional significance since the TIA of these cultivars was very low.

TIA in 18 cultivars of lima beans was completely eliminated by pressure cooking at 15 lb pressure for 15 min, while a loss of 74-90% was observed by autoclaving at 105°C for 30 min (Ologhobo and Fetuga, 1983).

Inactivation of TIA in Cereals. Purified TI's from cereals are generally very heat stable. Those isolated from wheat, rye, and barley lost only 5-15% of their activity after incubation at 100°C for 30 min at pH 4.9 (Boisen, 1983). In contrast, well over 80% of the activity in the original grain was destroyed by comparable conditions.

Rye breads, made in Denmark, contain 60% rye flour and are normally baked at 170°C, wherein the internal temperature does not exceed 97°C. Of 8 different rye bread recipes examined, residual activity was 5-15% of that originally present in the rye flour (Boisen and Djurtoft, 1981). Next to wheat, rye is the most important cereal for bread making. Because rye contains 50% more TIA than barley and 5-10 times more activity than wheat or oats, it could be a more significant source of TIA in the human diet than that contributed by other cereals in a typical British diet as shown in Table 1 (Doell et al., 1981). Chang and Tsen (1981) reported that the TI in triticale exhibited more stability to heat denaturation than that of rye, since heating at 100°C for 1 h inactivated only 7% of the TIA of triticale, 36% in rye and 38-71% for wheat.

Inactivation of TIA in Potatoes. Potato tubers contain about 5% protein on a dry weight basis and 10% or more of the protein exists as protease inhibitors. There are 3 major classes of PI's referred to as Inhibitor I, a potent chymotrypsin inhibitor and a weak TI; Inhibitor II, a strong chymotrypsin inhibitor; and Carboxypeptidase Inhibitor, (CPI), a strong inhibitor of pancreatic carboxypeptidases A and B (Ryan and Pearce, 1978). Protease inhibitor content of protein juice expressed from six locations of the tissue in Russet Burbank potatoes is given in Table 13. Inhibitor II was in the highest concentration among the three protease inhibitors in raw potatoes. Variability of the three PI's in 106 clones of randomly chosen potato varieties has been reported (Ryan et al., 1976).

Table 12. Effect of Various Types of Heat Treatment on the Trypsin Inhibitor Activity (TIA) of Legumes[a]

Legume	TIA		TIA, % of original activity						
			Wet heat		Auto-claved extract, 15 lb/sq inch for 15 min	Dry heat			Roasting for two min at 200°C
	% soy=100%	units/g[b]	Heating extract in boiling water bath			Roasting for 15 min at °C			
			30 min	60 min		75	100	125	
Moth bean (Phaseolus aconitifolus)	27	1443	57.8	11.6	2.4	80.8	42.5	Nil	7.6
Cow pea (Vigna catjung)	79	4172	59.9	49.3	4.8	72.2	72.2	65.0	Nil
Cow pea (Vigna sinensis)	64	3391	96.7	36.0	3.7	94.2	91.2	71.1	8.1
Red gram (Cajanus cajan)	60	3187	71.5	64.5	12.3	80.5	54.5	17.9	11.7
French bean (Phaseolus vulgaris)	80	4239	61.6	43.2	6.2	76.1	73.5	72.0	10.6
Pea (Pisum sativum)	25	1334	20.8	12.5	Nil	42.1	27.6	4.3	Nil
Lentil (Lens esculenta)	25	1313	8.5	Nil	Nil	73.1	69.4	65.2	20.4
Green gram (Phaseolus aureus)	37	1977	11.2	5.6	Nil	97.4	69.5	24.7	Nil
Black gram (Phaseolus mungo)	52	2735	28.4	16.2	Nil	69.4	57.6	39.4	11.9
Bengal gram (Cicer arietinum)	66	3472	16.0	3.2	Nil	62.0	40.9	14.2	7.6
Soybean (Glycine max)		5296	61.9	22.1		58.9			

[a] Soni et al. (1978)
[b] One unit of trypsin activity = amount of enzyme which converts one mg casein to TCA solubles under conditions of test.

Table 13. Proteinase Inhibitor Concentrations of Juice Expressed From Six Locations Within Raw Russet Burbank Potatoes[a]

	Inhibitor concentration (mg/ml)		
Location	Inhibitor I	Inhibitor II	CPI
Apical end cortex	0.25	0.61	0.25
Apical lateral cortex	0.13	0.22	0.17
Stem lateral cortex	0.08	0.13	0.16
Stem end cortex	0.04	0.08	0.10
Apical pith	0.06	0.22	0.12
Center pith	0.03	0.03	0.09

[a]Huang et al. (1981).

The concentration of Inhibitor I and II in four selected tissue locations after boiling for 30 min, baking at 191°C for 80 min, and microwave baking at 190°C for 7 min was investigated by Huang et al. (1981). Inhibitor I was completely denatured by boiling and microwave baking, under conditions stated above. Yet, even after 80 min of oven baking, a high percentage of Inhibitor I activity remained in the cortical tissues. All three cooking methods completely denatured Inhibitor II. CPI was extremely stable during all of the cooking conditions. Boiling for 30 min or oven baking for 1 h had little effect on inhibitor activity and microwave baking for 7 min inactivated only 25% of the CPI. Among the three cooking methods, microwave baking was the most effective in denaturing Inhibitors I and II.

On a protein basis, cooked potatoes and preparations of potato flours (powders) contain appreciable levels of residual TIA (Table 8) and contribute significantly to the total intake of TIA of a typical British diet (Table 1).

Effect of Maturation and Germination on TIA

It has been suggested that immature and germinated legumes and cereals could be an even more universal and versatile protein resource than the mature seeds. Popularization of a vegetarian lifestyle and studies on health implications of vegetarian diets have focused attention on vegetable sprouts as a complement to salad bars. Green immature soybeans, known as edamame in Japan, are an important vegetable in the Far East (Rackis, 1978). Birk and Waldman (1965) did not detect TIA in roots, stems, leaves, or pods of soybeans at any stage of development. Kapoor and Gupta (1978) reported that soybean TI's are first synthesized at initiation of seed development in the pod. Activity increased during maturation (Table 4). This indicates that TI protein is synthesized exclusively in the seed, rather than as a result of translocation from other sites of the plant (Sharma et al., 1975). TIA in three winged bean cultivars increased 8-fold from maturity Stage I (40 days after flowering) to complete maturity (Kute et al., 1984). TIA was also found to increase during maturation in the pod (pericarp), cotyledon, and embryo axis of cow peas but no TIA attributable to protein could be detected in the seed coat at any state (Carasco and Xavier-Filho, 1981).

Once started, TIA synthesis in the developing soybean increases rapidly (Collins and Sanders, 1976) and reaches a maximum value coincident with maximum fresh weight, at which stage the seeds are still green and pods begin to yellow (Rackis, 1978). At this stage, soybeans are often referred to as green-mature soybeans. Mature soybeans are yellow or buff-colored.

Table 14. Effect of Maturation and Germination on Trypsin Inhibitor Activity in Soybeans[a]

Stage of Maturity	Trypsin Inhibitor Activity Reaction Rate,[b] Raw	Heated[c]
Green-Mature	49.0	1.5
Dry Mature	52.2	0.6
Sprouts	17.8	1.7

[a]Bates et al. (1977).
[b]Change in absorbance at 257 nm/min/g protein compared to control with no trypsin inhibitors, Bragg soybeans.
[c]Autoclaving 121°C for 15 min.

As shown in Table 14, immature soybeans contain levels of TIA comparable to mature seeds and activity decreases upon germination. Regardless of the stage of maturity, over 90% of the TIA can be destroyed by moist heat treatment. On the other hand, Collins and Sanders (1976) found little change in TIA in 3 soybean varieties when germinated.

In contrast to soybeans, a 40-64% loss in TIA was observed with 18 cultivars of lima beans during germination for 6 days (Ologhobo and Fetuga, 1983). The decrease in TIA during germination observed in many seeds is attributable to leaching and proteolysis (Wilson, 1980). In addition, new species of TI were formed in lima beans (Lorensen et al., 1981) and in adzuki beans (Yosikawa et al., 1979) during germination. In lima beans, the major TI species, designated TI-F, undergoes a definitive conversion in the following sequence: F→E→C→A (Lorensen et al., 1981; Wilson and Chen, 1983). An enzyme catalyzing one step in the conversion of form E to C has been identified as a carboxypeptidase (Wilson et al., 1985). The two major TI's in soybeans, Kunitz trypsin inhibitor and Bowman-Birk inhibitor, are also degraded at different rates during germination and new species are formed (Tan-Wilson et al., 1982).

An inhibitor of plant thiol protease activity in French beans decreases but TIA increases during germination. Because synthesis of TI's in legume seeds does not prevent the degradation of storage proteins, endogenous TI apparently does not participate in the control of mobilization of reserve proteins during germination (Nielsen and Liener, 1984). Little loss of TIA in mung beans was observed upon germination (Noor et al., 1980).

Other Processes for Inactivation of TIA

Reducing Agents. As described earlier, heat treatment may lead to incomplete inactivation of the inhibitors. However, the use of high temperatures to destroy TIA may also degrade amino acids, thus damaging nutritional quality. Several studies reveal that cysteine, N-acetyl-L-cysteine, and other thiols facilitate heat inactivation of TIA (Friedman et al., 1982A,B,C). The beneficial effects of thiols on the nutritional quality of soy flour may result from inactivation of enzyme inhibitors through the formation of mixed disulfides, introduction of new half-cystine residues, and reduction of methionine sulfoxide to methionine in sulfur-poor legumes (Friedman et al., 1984). The use of thiols for the denaturation of TIA and concomitant improvement in nutritional value of soy flour represents an interesting process, but large-scale commercial feasibility has yet to be determined. The effect of heat and N-acetyl-L-cysteine (NAC) on TIA in soy flour is illustrated in Figure 6. The effect of NAC was especially effective in enhancing inactivation in the temperature range 35°C to 65°C. Adding NAC increases inactivation at all pH ranges, with essentially all TIA destroyed at pH 10. The generality of the process was confirmed by the finding that thiols were also effective in the inactivation of the highly-crossed linked lima bean chymotrypsin inhibitor. Reduced glutathione, mercaptopropionylglycine and NAC were equally effective and L-cysteine was less so in destroying TIA (Friedman et al., 1982C).

Fermentation. TIA of autoclaved (121°C, 15 min) chick peas, horse beans, and soybeans, which in turn were fermented to prepare tempeh and miso, is given in Table 15. The heat treatment used in the process accounts for essentially all of the inactivation of the activity of the raw legumes (Kao and Robinson, 1978). Cultures of *Rhizopus oligosporus* are used to prepare the fermented product, tempeh, from various legumes. Wang et al. (1972) have proposed that extra-cellular proteases in *R. oligosporus* act on soybeans to release TIA and to degrade it. Similarly, Kamalakannan and Motlag (1982) found a three- to fivefold increase in TIA during 9 h of fermentation, followed by a large decrease in activity in 18 h of additional incubation. As normally carried out, most of the residual TIA in tempeh, miso, and probably all fermented products may be attributable to the presence of free fatty acids that are produced during fermentation (Doell et al., 1981; Wang et al., 1975).

Dielectric Heating. Although moist heat treatment is an effective way for inactivating TI's and producing products of high nutritional value, the process is time-consuming and requires appreciable energy. Dielectric

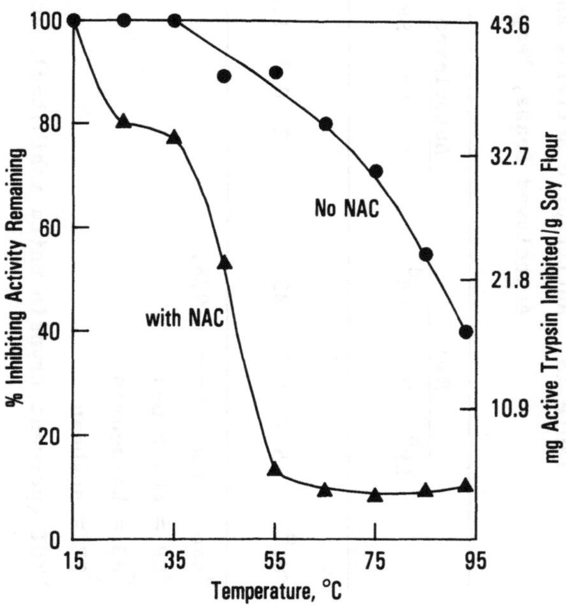

Fig. 6. Effect of heat and N-acetyl-L-cysteine (NAC) on trypsin inhibitor activity in soy flour. The 100% activity value is for raw untreated flour (Friedman et al., 1982c). Conditions: 100g soy flour in a 500 ml slurry with 0.4 mmoles of NAC in pH 8.5 Tris buffer.

Table 15. Antitryptic Activity and Aflatoxin Analyses of Raw Beans, Autoclaved Beans, Tempeh, and Miso[a]

	Raw			Autoclaved			Tempeh			Miso		
	CP[b]	HB[c]	SB[d]	CP	HB	SB	CP	HB	SB	CP	HB	SB
TUI[e]	15.7	6.7	83.3	1.2	0.0	1.2	2.4	1.2	0.0	0.0	0.0	0.0

[a]Kao and Robinson (1978)

[b]CP = chick pea

[c]HB = horsebean

[d]SB = soybean

[e]TUI (percent trypsin units inhibited), according to the method of Kakade et al. (1969)

heating may offer some advantages. The process might save energy since much less water is needed to denature TIA. The process would save time, the treatment could impart functional properties not attainable by toasting, and destruction of lipoxygenase, which has been implicated in development of off-flavors in soybeans and other food legumes, can be readily accomplished by processing the intact seed even at low moisture levels (Pour-El et al., 1981). These workers predicted that absorption of a minimum energy of 250-280 Cal/g would be sufficient to destroy over 85% of the TIA in raw soybeans (See Figure 7) which would result in maximum nutritional value (Rackis et al., 1975A).

Microwaving for 1.5 to 3.0 min was required to destroy 85 to 90% of the TIA in soaked soybeans (Hafez et al., 1983). Hamid et al. (1975) found that microwave heating of dry whole soybeans for 3 min inactivated the TI.

Extrusion-cooking. The widespread extent of serious nutritional deficiencies in less developed countries (LDC's) has been reviewed by a number of groups and needs no further discussion here. In response to this problem, a large array of cereal-legume blends was developed by many groups as well as by the Food For Peace Program. In order for most LDC's to be able to process their own blended foods from locally grown crops, a simple, inexpensive processing system referred to as a low-cost extrusion cooker (LEC) was evaluated in a series of LEC reports and publications (Harper and Jansen, 1981). The LEC's differ from typical extruders (Smith, 1976; Mustakas et al., 1970) in that they utilize little or no added water.

TIA and PER of dehulled soybeans processed by two different LEC's are summarized in Table 16. In the Brady Cooker, the maximum corrected PER of 1.94 corresponding to 57% destruction of TIA was obtained with an extrusion temperature of 143°C. Destruction of 74% TIA results in an appreciable reduction in the PER. In the Insta-Pro Cooker, maximum destruction of TIA occurred at an extrusion temperature of 149°C; over 90% of the TIA was

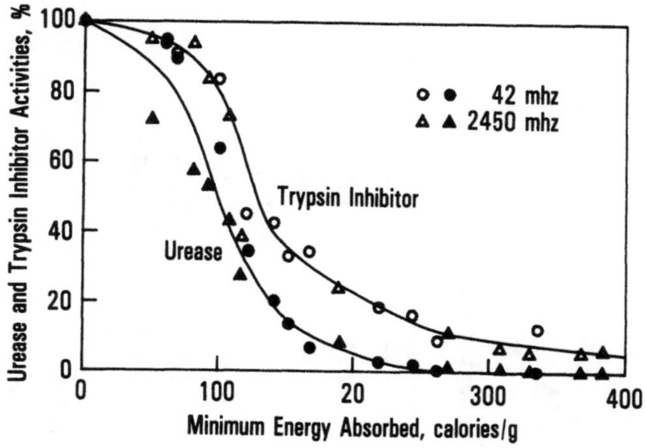

Fig. 7. Dependence of inactivation of urease and trypsin inhibitor activity (TIA) on minimum energy absorbed during dielectric heating (Pour-El et al., 1981). Arrow indicates the minimum energy absorbed required to destroy at least 85% of the TIA which is also that required for maximum protein quality for heated soy flours.

Table 16. Effect of Extrusion on Trypsin Inhibitor Activity (TIA) and Protein Quality of Soybeans

Brady Cooker				Insta-Pro Cooker					
Temperature °C	TIA TIU/mg[b]	TIA % destroyed	PER[c]	Temperature °C	$\%H_2O$, added[d]	TIA TIU/mg[b]	TIA % destroyed	PER[c]	Pancreas Weight mg/100g b.w.
Unextruded	64.5	---	1.01	Unextruded		55.3	---	1.03	416
121	64.8	0.0	1.35	139	0.0	17.6	68.2	1.94	279
127	57.2	11.1	1.42	137	3.3	15.9	71.3	2.05	270
132	45.5	29.5	1.41	143	0.0	15.6	71.8	2.09	274
138	47.2	26.7	1.55	143	2.0	14.4	73.9	2.09	263
143	28.0	56.7	1.94	149	0.0	4.9	91.1	2.13	263
149	16.8	74.0	1.78	150	3.3	9.5	82.8	2.06	273

[a]Harper and Jansen (1981)

[b]TIU = Trypsin Inhibitor Units

[c]PER = Protein Efficiency Ratio, corrected relative to casein = 2.50

[d]Original moisture content 5.9%

destroyed corresponding to a PER of 2.13. A Brady extruder can also be used to cook raw, dehulled, cracked soybeans at 137°C with a residence time of 90 sec before they are ground into full-fat soy flour. Under these conditions, nearly 70% destruction of TIA was obtained (Harper and del Valle, 1979). A number of commercial applications of the full-fat flour were tested (Harper and del Valle, 1979; del Valle et al., 1981); as with most LEC processes, residual TIA is relatively high. A more effective destruction of TIA of soybeans was obtained by the extrusion process developed by Mustakas and coworkers (1970).

Dry Roasting. This process consists of roasting beans in a heated bed of salt and then cooling and grinding (Raghavan and Harper, 1974). Instant bean powders were prepared by roasting navy beans at 190°C for 20 sec or 220°C for 10 sec (Carvalho et al., 1977). The heat treatment destroyed 70-80% of the TIA originally present. Full-fat soy flour was made by roasting whole soybeans in a heated agitated salt bed (Jansen et al., 1978). A corrected PER of 2.11-2.31 compared to 2.50 for reference casein was obtained with roasting bed temperatures of 206-234°C for 5-24 sec. The process reduced TIA 75-90%.

Alcohol Treatment. An isopropyl alcohol extraction process, coupled with desolventizing-toasting, was developed to produce defatted soy meal. Over 90% destruction of TIA was achieved (Baker and Sullivan, 1983). Raw, defatted soy flour prepared by hexane or hexane:alcohol (83:18V/V) azeotrope extraction processes contain 101-103 TIU/mg sample. Toasting reduced TIA to residual values of 12-14 TIU/mg (Rackis et al., 1975B). The flavor of soybeans, peanuts, and legumes was improved when the seeds or beans were steeped or wet-milled with aqueous ethyl alcohol at 25°C (Eldridge et al., 1977). However, only 40% of the TIA was destroyed with 20-60% aqueous ethanol. Such residual TIA values indicate that products prepared by alcohol treatment require additional toasting to ensure maximum nutritive value. Three alcohols (ethanol, methanol and isopropanol) at various aqueous concentrations were used to extract defatted soy flours and flakes at temperatures ranging from 30 to 75°C. Residual TIA ranged from 28-100% compared to the original value (Baker et al., 1979).

REFERENCES

Al-Bakir, A. Y., Sachde, A. G., and Naoum, I. E., 1982, Occurrence and stability of trypsin inhibitors in Iraqi local legumes, J. Agric. Food Chem., 30:1184.

Albrecht, W. J., Mustakas, G. C., and McGhee, J. E., 1966, Rate studies on atmospheric steaming and immersion cooking of soybeans, Cereal Chem., 43:400.

American Association of Cereal Chemists, 1983, Determination of trypsin inhibitor activity of soy products, Method 71-10, in: "Approved Methods of the American Association of Cereal Chemists", St. Paul, MN.

American Oil Chemists' Society, 1976, Trypsin inhibitor activity, Method Ba 12-75, in: "Official and Tentative Methods" Champaign, IL.

American Oil Chemists Society, 1979, Proceedings--World Conference on Vegetable Food Proteins, Amsterdam, Netherlands, November, 1978. J. Am. Oil Chem. Soc., 56:99.

American Oil Chemists' Society, 1981, Proceedings of the World Conference on Soya Processing and Utilization, Acapulco, Mexico, November, 1980, J. Am. Oil Chem. Soc., 58:121.

Baker, E. C., and Mustakas, G. C., 1973, Heat inactivation of trypsin inhibitor, lipoxygenase and urease in soybeans: Effect of acid and base additives, J. Am. Oil Chem. Soc., 50:137.

Baker, E. C., Mustakas, G. C., and Warner, K. A., 1979, Extraction of defatted soybean flours and flakes with aqueous alcohols: Evaluation

of flavor and selected properties, J. Agric. Food Chem., 27:969.
Baker E. C., and Sullivan, D. A., 1983, Development of a pilot-plant process for the extraction of soy flakes with aqueous isopropyl alcohol, J. Am. Oil Chem. Soc., 60:1271.
Bates, R. P., Knapp, F. W., and Araujo, P. E., 1977, Protein quality of green-mature, dry mature and sprouted soybeans, J. Food Sci., 42:271.
Besrat, A., 1981, Protein quality of Faffa, the Ethiopian infant formula as affected by pulse substitution, J. Food Biochem., 5:233.
Birk, Y., 1961, Purification and some properties of a highly active inhibitor of trypsin and α-chymotrypsin from soybeans, Biochem. Biophys., Acta 54:378.
Birk, Y., Gertler, A., and Khalef, S., 1963, A pure trypsin inhibitor from soya beans, Biochem. J., 87:281.
Birk, Y., and Waldman, M., 1965, Amylolytic-, trypsin-inhibiting- and urease-activity in three varieties of soybeans and in soybean plant, Qual. Plant. Mater. Veg., 12:199.
Boisen, S., 1983, Protease inhibitors in cereals. Occurrence, properties, physiological role, and nutritional significance, Acta Agric. Scand., 33:369.
Boisen S., and Djurtoft, R., 1981, Trypsin inhibitor from rye endosperm: Purification and properties, Cereal Chem., 58:194.
Buera, M. P., Pilosof, A. M. R., and Bartholomai, G. B., 1984, Kinetics of trypsin inhibitory activity loss in heated flour from bean, Phaseolus vulgaris, J. Food Sci., 49:124.
Buera, M. P., Pollio, M. L., Pilosof, A. M., and Bartholomai, G. B., 1983, Cinetica de la perdida de solubilidad y de lisina disponible en harinas de porotos trantadas termicamente, Rev. Agroquim. Tecnol. Aliment., 23:262.
Campos F. A. P., and Richardson, M., 1983, The complete amino acid sequence of the bifunctional α-amylase/trypsin inhibitor from seeds of ragi (Indian finger millet, Eleusine coracana Gaertn.) FEBS Lett., 152:300.
Carasco J. F., and Xavier-Filho, J., 1981, Sequential expression of trypsin inhibitors in developing fruit of cowpea (Vigna unguiculata (L.) Walp) Ann. Bot., 47:259.
Carvalho, C. C. C., Jansen, G. R., and Harper, J. M., 1977, Protein quality evaluation of an instant bean powder produced by dry heat processing, J. Food Sci., 42:553.
Chang, C. -R., and Tsen, C. C., 1979, Note on trypsin inhibitor activity in the acetate extract of cereal samples, Cereal Chem., 56:493.
Chang, C. -R., and Tsen, C. C., 1981, Characterization and heat stability of trypsin inhibitors from rye, triticale, and wheat samples, Cereal Chem., 58:211.
Charpentier, B. A., and Lemmel, D. E., 1984, A rapid automated procedure for the determination of trypsin inhibitor activity in soy products and common foodstuffs. J. Agric. Food Chem., 32:908.
Chen, I., and Mitchell, H. L., 1973, Trypsin inhibitors in plants, Phytochemistry, 12:327.
Chernick, S. S., Lepkovsky, S., and Chaikoff, I. L., 1948, A dietary factor regulating the enzyme content of the pancreas: Changes induced in size and proteolytic activity of the chick pancreas by the ingestion of raw soybean meal, Am. J. Physiol., 155:33.
Collins, J. L., and Beaty, B. F., 1980, Heat inactivation of trypsin inhibitor in fresh green soybeans and physiological responses of rats fed the beans, J. Food Sci., 45:542.
Collins, J. L., and Sanders, G. G., 1976, Changes in trypsin inhibitor activity in some soybean varieties during maturation and germination, J. Food Sci., 41:168.
Cuevas, R., and Cheryan, M., 1983, Inhibidores de tripsina en alimentos a base de soya: Revision critica de la cinetica de destruccion termica, y los metodos de analisis. Arch. Latinoam. Nutr., 33:902.

del Rosario, R. R., Lozano, Y., Pamorasamit, S., and Noel, M. G., 1980, The trypsin inhibitor activity of legume seeds, Philipp. Agric., 63:339.
de Lumen, B. O., and Belo, P. S., Jr., 1981, Inhibitors of trypsin and chymotrypsin in winged beans (Psophocarpus tetragonolobus) tubers, J. Agric. Food Chem., 29:884.
del Valle, F. R., Escobedo, M., Munoz, M. J., Ortega, R., and Bourges, H., 1983, Chemical and nutritional studies on mesquite beans (Prosopis juliflora), J. Food Sci., 48:914.
del Valle, F. R., Villaneuva, H., Reyes-Govea, J., Escobedo, M., Bourges, H., Ponce, J., and Munoz, M. J., 1981, Development, evaluation and industrial production of a powdered soy-oats infant formula using a low-cost extruder, J. Food Sci., 46:192.
Deshpande, S. S., Sathe, S. K., Salunkhe D. K., and Cornforth, D. P., 1982, Effects of dehulling on phytic acid, polyphenols, and enzyme inhibitors of dry beans (Phaseolus vulgaris L.), J. Food Sci., 47:1846.
Doell, B. H., Ebden, C. J., and Smith, C. A., 1981, Trypsin inhibitor activity of conventional foods which are part of the British diet and some soya products, Qual. Plant. Plant Foods Hum. Nutr., 31:139.
Egberg, D. C., Potter, R. H., and Honold, G. R., 1975, The semiautomated determination of trypsin inhibitors in textured soy protein, J. Agri. Food Chem., 23:603.
Ekpenyong, T. E., and Borchers, R. L., 1981, Some toxic factors in winged bean seeds. 1. Antitrypsin, hemagglutinating and urease activities, Nutr. Rep. Int., 23:865.
Eldridge, A. C., Warner, K., and Wolf, W. J., 1977, Alcohol treatment of soybeans and soybean protein products, Cereal Chem., 54:1229.
Ellenrieder, G., Geronazzo, H., and de Bojarski, A. B., 1980, Thermal inactivation of trypsin inhibitors in aqueous extracts of soybeans, peanuts, and kidney beans: Presence of substances that accelerate inactivation, Cereal Chem. 57:25.
Erlanger, B. F., Kokowsky, N., and Cohen, W., 1961, The preparation and properties of two new chromogenic substrates of trypsin, Arch. Biochem. Biophys. 95:271.
Foard, D. E., Murdock, L. L., and Dunn, P. E., 1983, Engineering of crop plants with resistance to herbivores and pathogens: An approach using primary gene products, in: "Plant Molecular Biology", Proceedings of the ARCO Solar-UCLA Symposium, R. B. ed., Goldberg, Keystone, Colorado, April 16-22, 1983, Alan R. Liss, Inc., New York.
Foster, J. A., Rich, C. B., Berglund, N., Huber, S., Mecham, R., and Lange, G., 1979, The anti-proteolytic behavior of lathrogens, Biochem. Biophys. Acta., 587:477.
Freed, R. C., and Ryan, D. S., 1978A, Changes in Kunitz trypsin inhibitor during germination of soybeans: An immunoelectrophoresis assay system, J. Food Sci., 43:1316.
Freed, R. C., and Ryan, D. S., 1978B, Note on modification of the Kunitz soybean trypsin inhibitor during seed germination, Cereal Chem. 55:534.
Freed, R. C., and Ryan, D. S., 1980, Isolation and characterization of genetic variants of the Kunitz soybean trypsin inhibitor, Biochim. Biophys. Acta, 624:562.
Friedman, M., Grosjean, O.-K. K., and Zahnley, J. C., 1982A, Inactivation of soya bean trypsin inhibitors by thiols, J. Sci. Food Agric., 33:165.
Friedman, M., Grosjean, O.-K. K., and Zahnley, J. C., 1982B, Cooperative effects of heat and thiols in inactivating trypsin inhibitors from legumes in solution and in the solid state, Nutr. Rep.Int., 25:743.
Friedman, M., Grosjean, O.-K. K., and Zahnley, J. C., 1982C, Effect of disulfide bond modification on the structure and activities of enzyme inhibitors. in: "Food Protein Deterioration: Mechanisms and Functionality" ACS Symp. Series, Vol. 206, American Chemical Society, J. P. Cherry, ed. Washington, D.C.

Friedman, M., Gumbmann, M. R., and Grosjean, O.-K. K., 1984, Nutritional improvement of soy flour, J. Nutr., 114:2241.

Fritz, H., Trautschold, I., and Werle, E., 1974, Protease inhibitors, in: "Methods of Enzymatic Analysis," Vol. 2, H. U. Bergmeyer, ed., Academic Press, New York.

Fritz H., and Tschesche, H., 1971, "Proceedings of the International Research Conference on Proteinase Inhibitors," Walter de Gruyter, Berlin.

Fritz, H., Tschesche, H., Greene, L. J., and Truscheit, E., 1974, "Proteinase Inhibitors," Proc 2nd Int. Res. Conf. Springer-Verlag, New York.

Green, G. M., and Lyman, R. L., 1972, Feedback regulation of pancreatic enzyme secretion as a mechanism for trypsin inhibitor-induced hypersecretion in rats, Proc. Soc. Exp. Biol. Med., 140:6.

Griffiths, D. W., 1979, The inhibition of digestive enzymes by extracts of field bean (Vicia faba), J. Sci. Food Agric., 30:458.

Gupta, A. K., and Deodhar, A. D., 1975, Variation in trypsin inhibitor activity in soybean (Glycine max), Indian J. Nutr. Diet., 12:81.

Hackler, L. R., Van Buren, J. P., Steinkraus, K. H., El Rawi, I., and Hand, D. B., 1965, Effect of heat treatment on nutritive value of soymilk protein fed to weanling rats, J. Food Sci., 30:723.

Hafez, Y. S., 1983, Nutrient composition of different varieties and strains of soybean, Nutr. Rep. Int., 28:1197.

Hafez, Y. S., Singh, G., McLellan, M. E., and Monroe-Lord, L., 1983, Effects of microwave heating on nutritional quality of soybeans, Nutr. Rep. Int., 28:413.

Haider, F. G., 1981, Inactivation studies on the trypsin inhibitor activity of green gram cultivars, Nutr. Rep. Int., 23:1167.

Hamerstrand, G. E., Black, L. T., and Glover, J. D., 1981, Trypsin inhibitors in soy products: Modification of the standard analytical procedure, Cereal Chem., 58:42.

Hamid, M. A. K., Bhartia, P., and Mostowy, N. J., 1975, Microwave bean roaster, J. Microwave Power, 10:109.

Harper, J., and del Valle, F. R., 1979, Mexican soy plant, Chilton's Food Eng. Int., 4:(6)22.

Harper, J. M., and Jansen, G. R., 1981, "Nutritious Foods Produced by Low-Cost Technology," LEC Report 10, Colorado State University, Fort Collins, CO.

Hass, G. M., Hermodson, M. A., Ryan, C. A., and Gentry, L., 1982, Primary structures of two low molecular weight proteinase inhibitors from potatoes, Biochemistry, 21:752.

Hass, G. M., Nau, H., Biemann, K., Grahn, D. T., Ericsson, L. H. and Neurath, H., 1975, The amino acid sequence of a carboxypeptidase inhibitor from potatoes, Biochemistry, 14:1334.

Hildebrand, D. F., Hettiarachchy, N. S., Hymowitz, T., and Erdman, J. W., Jr., 1981, Electrophoretic separation and properties of winged bean seed trypsin inhibitor, J. Sci. Food Agric., 32:443.

Hill, B. S., Synder, H. E., and Wiese, K. L., 1982, Use of the pH stat to evaluate trypsin inhibitor and tryptic proteolysis of soy flours, J. Food Sci., 47:2018.

Hove, E. I., and King, S., 1979, Trypsin inhibitor contents of lupin seeds and other grain legumes, N. Z. J. Agric. Res., 22:41.

Huang, D. Y., Swanson, B. G., Ryan, C. A., 1981, Stability of proteinase inhibitors in potato tubers during cooking, J. Food Sci., 46:287.

Ida, E. I., da Silva, R. S. F., Fabre-Sanches, S., Perre da Silva, M. A., and Reddy, K. V., 1983, Trypsin inhibitor inactivation during hydrothermal processing: A mathematical model, Lebensm. Wiss. Technol., 16:81.

Ikeda, K., and Kusano, T., 1983, In-vitro inhibition of digestive enzymes by indigestible polysaccharides, Cereal Chem., 60:260.

Jaffe, W. G., and Vega Lette, C. L., 1968, Heat-labile growth-inhibiting factors in beans (Phaseolus vulgaris), J. Nutr., 94:203.
Jansen, G. R., Harper, J. M. and O'Deen, L. A., 1978, Nutritional evaluation of full-fat soy flour produced by dry heat roasting, J. Food Sci., 43:1350.
Jones, G., Moore, S., and Stein, W. H., 1963, Properties of chromatographically purified trypsin inhibitors from lima beans, Biochemistry, 2:66.
Joubert, F. J., 1982, Proteinase inhibitors from Erythrina lysistemon seed, Phytochemistry, 21:1213.
Kadam, S. S., and Salunkhe, D. K., 1984, Winged bean in human nutrition, CRC Crit. Rev. Food Sci. Nutr., 21:1.
Kakade, M. L., Rackis, J. J., McGhee, J. E., and Puski, G., 1974, Determination of trypsin inhibitor activity of soy products: A collaborative analysis of an improved procedure, Cereal Chem., 51:376.
Kakade, M. L., Simons, N., and Liener, I. E., 1969, An evaluation of natural vs. synthetic substrates for measuring the antitryptic activity of soybean samples, Cereal Chem., 46:518.
Kakade, M. L., Simons, N. R., Liener, I. E., and Lambert, J. W., 1972, Biochemical and nutritional assesment of different varieties of soybeans, J. Agric. Food Chem., 20:87.
Kamalakannan, V., and Motlag, D. B., 1982, Effect of Rhizopus oligosporus (NRRL 2710) fermentation on the trypsin inhibitors of black gram (Vigna mungo), Can. J. Microbiol., 28:54.
Kao, C., and Robinson, R. J., 1978, Nutritional aspects of fermented foods from chickpea, horsebean, and soybean, Cereal Chem., 55:512.
Kapoor, A. C., and Gupta, Y. P., 1978, Trypsin inhibitor activity in soybean seed as influenced by stage of its development and different treatments, and the distribution in its anatomical parts, Indian J. Nutr. Diet., 15:429.
Kas, J., Fukal, L., Kasafirek, E., Cap, M., and Solc, J., 1983, Assay of trypsin inhibitor and its application to thermally treated horse bean (Vicia faba), Sb. Vys. Sk. Chem. Technol. Praze, Potraviny E56:177.
Katunuma, N., Umezawa, H., and Holzer, H., 1983, "Proteinase Inhibitors: Medical and Biological Aspects", Japan. Sci. Soc. Press, Tokyo.
Khor, H.-T., Tan, N.-H., and Wong, K.-C., 1982, The protein, trypsin inhibitor and lipid of the winged bean [Psophocarpus tetragonolobus (L) DC] seeds, J. Sci. Food Agric., 33:996.
Klose, A. A., Hill, B., and Fevold, H. L., 1948, Food value of soybean protein as related to processing, Food Technol., 2:201.
Koide, T., and Ikenaka, T., 1973, Studies on soybean trypsin inhibitors. 3. Amino-acid sequence of the carboxyl-terminal region and the complete amino-acid sequence of soybean trypsin inhibitor (Kunitz), Eur. J. Biochem., 32:417.
Kouzeh-Kanani, M., Van Zuilichem, D. J., Roosen, J. P., Pilnik, W., Van Delden, J. R., and Stolp, W., 1983, Infrared processing of soybeans, Qual. Plant. Plant Foods Hum. Nutr., 33:139
Kozlowska, H., Elkowicz, K., and Rutkowski, A., 1980, Thermal inactivation of trypsin inhibitors of soybean preparations added to meat, Meat Sci., 4:95.
Krahn, J., and Stevens, F. C., 1970, Lima bean trypsin inhibitor. Limited proteolysis by trypsin and chymotrypsin, Biochemistry, 9:2646.
Krahn, J., and Stevens, F. C., 1973, Peptide-bond hydrolysis equilibria in the antitrypsin site of lima bean protease inhibitor, Biochemistry, 12:1330.
Kunitz, M., 1945, Crystallization of a trypsin inhibitor from soybeans, Science, 101:668.
Kunitz, M., 1946, Crystalline soybean trypsin inhibitor, J. Gen. Physiol., 29:149.

Kunitz, M., 1947, Isolation of a crystalline protein compound of trypsin and of soybean trypsin-inhibitor, J. Gen. Physiol., 30:311.
Kute, L. S., Kadam, S. S., and Salunkhe, D. K., 1984, Changes in sugars, starch and trypsin inhibitor activity in winged bean (Psophocarpus tetragonolobus L.DC) during seed development, J. Food Sci., 49:314.
Labuza, T. P., 1973, Effects of dehydration and storage, Food Technol., 27:(1)20.
Lehnhardt, W. F., and Dills, H. G., 1984, Unpublished method, A. E. Staley Manufacturing Co., Decatur, IL.
Levison, D. A., Morgan, R. G. H., Brimacombe, J. S., Hopwood, D., Coghill, G., and Wormsley, K. G., 1979, Carcinogenic effects of di(2-hydroxypropyl)nitrosamine (DHPN) in male Wistar rats: Promotion of pancreatic cancer by a raw soya flour diet, Scand. J. Gastroenterol., 14:217.
Liener, I. E., 1972, Nutritional value of food protein products, in: "Soybeans: Chemistry and technology," Vol I, Proteins, A. K. Smith and S. J. Circle, eds. Avi Publishing, Westport CT.
Liener, I. E., 1975, Effects of anti-nutritional and toxic factors on the quality and utilization of legume proteins, in: "Protein nutritional quality of foods and feeds," Part 2, Quality Factors-Plant Breeding, Composition, Processing, and Antinutrients, M. Friedman, ed., Marcel Dekker, Inc., New York.
Liener, I. E., 1976, Legume toxins in relation to protein digestibility - A review, J. Food Sci., 41:1076.
Liener, I. E., 1981, Factors affecting the nutritional quality of soya products, J. Am. Oil Chem. Soc., 58:406.
Liener, I. E., 1983, Naturally occurring toxicants in foods and their significance in the human diet, Arch. Toxicol. Suppl., 6:153.
Liener, I. E., and Kakade, M. L., 1980, Protease inhibitors, in: "Toxic constituents of Plant Foodstuffs," 2nd ed., I. E. Liener, ed., Academic Press, New York.
Lorensen, E., Prevosto, R., and Wilson, K. A., 1981, The appearance of new active forms of trypsin inhibitor in germinating mung bean (Vigna radiata) seeds, Plant Physiol., 68:88.
Madar, Z., Gertler, A., and Birk, Y., 1979, The fate of the Bowman-Birk trypsin inhibitor from soybeans in the digestive tract of chicks, Comp. Biochem. Physiol. A, 62A:1057.
McNaughton, J. L., and Reece, F. N., 1980, Effect of moisture content and cooking time on soybean meal urease index, trypsin inhibitor content, and broiler growth, Poult. Sci., 59:2300.
McGuinness, E. E., Hopwood, D., and Wormsley, K. G., 1982, Further studies of the effects of raw soya flour on the rat pancreas, Scand. J. Gastroenterol., 17:273.
McGuinness, E. E., Morgan, R. G. H., and Wormsley, K. G., 1984, Effects of soybean flour on the pancreas of rats, EHP, Environ. Health Perspect., 56:205.
Mahoney, W. C., Hermodson, M. A., Jones, B., Powers, D. D., Corfman, R. S., and Reeck, G. R., 1984, Amino acid sequence and secondary structural analysis of the corn inhibitor of trypsin and activated Hageman factor, J. Biol. Chem., 259:8412.
Melville, J. C., and Ryan, C. A., 1972, Chymotrypsin inhibitor I from potatoes. Large scale preparation and characterization of its subunit components, J. Biol. Chem., 247:3445.
Millar, D. B. S., Willick, G. E., Steiner, R. F., and Frattali, V., 1969, Soybean inhibitors. IV. The reversible self-association of a soybean proteinase inhibitor, J. Biol. Chem., 244:281.
Morgan, R. G. H., Levison, D. A., Hopwood, D., Sanders, J. H. B., and Wormsley, K. G., 1977, Potentiation of the action of azaserine on the rat pancreas by raw soya flour, Cancer Lett., 3:87.
Mustakas, G. C., Albrecht, W. J., Bookwalter, G. M., McGhee, J. E., Kwolek, W. F., and Griffin, E. L., Jr., 1970, Extruder-processing to improve

nutritional quality, flavor, and keeping quality of full-fat soy flour, Food Technol., 24:1290.
Mustakas, G. C., Moulton, K. J., Baker, E. C., and Kwolek, W. F., 1981, Critical processing factors in desolventizing-toasting soybean meal for feed, J. Am. Oil Chem. Soc., 58:300.
Nielsen, S. S. and Liener, I. E., 1984, Degradation of the major storage protein of Phaseolus vulgaris during germination, Plant Physiol., 74:494.
Nordal, J. and Fossum, K., 1974, The heat stability of some trypsin inhibitors in meat products, with special reference to added soybean protein, Z. Lebensm. Unters.-Forsch., 154:144.
Norioka, S. and Ikenaka, T., 1983, Amino acid sequence of a trypsin-chymotrypsin inhibitor, B-III, of peanut (Arachis hypogaea), J. Biochem., 93:479.
Noor, M. I., Bressani, R., and Elias, L. G., 1980, Changes in chemical and selected biochemical components, protein quality, and digestibility of mung bean (Vigna radiata) during germination and cooking, Qual. Plant. Plant Foods Hum. Nutr., 30:135.
Odani, S., and Ikenaka, T., 1972, Studies on soybean trypsin inhibitors. IV. Complete amino acid sequence and the anti-proteinase sites of Bowman-Birk soybean proteinase inhibitor, J. Biochem., 71:839.
Odani, S., and Ikenaka, T., 1973A, Studies on soybean trypsin inhibitors. VIII. Disulfide bridges in soybean Bowman-Birk proteinase inhibitor, J. Biochem., 74:697.
Odani, S., and Ikenaka, T., 1973B, Scission of soybean Bowman-Birk proteinase inhibitor into two small fragments having either trypsin or chymotrypsin inhibitory activity, J. Biochem., 74:857.
Odani, S., and Ikenaka, T., 1977, Studies on soybean trypsin inhibitors. XI. Complete amino acid sequence of a soybean trypsin-chymotrypsin-elastase inhibitor, C-II, J. Biochem., 82:1523.
Odani, S., and Ikenaka, T., 1978, Studies on soybean trypsin inhibitors. XII. Linear sequences of two soybean double-headed trypsin inhibitors, D-II and E-I, J. Biochem., 83:737.
Odani, S., Koide, T., and Ono, T., 1983, The complete amino acid sequence of barley trypsin inhibitor, J. Biol. Chem., 258:7998.
Ologhobo, A. D., and Fetuga, B. L., 1983, Trypsin inhibitor activity in some limabean (Phaseolus lunatus) varieties as affected by different processing methods, Nutr. Rep. Int., 27:41.
Osborne, T. B., and Mendel, L. B., 1917, The use of soy bean as food, J. Biol. Chem., 32:369.
Ozawa, K., and Laskowski, M. Jr., 1966, The reactive site of trypsin inhibitors, J. Biol. Chem., 241:3955.
Pearce, G., McGinnis, J., and Ryan, C. A., 1983, Effects of feeding a carboxypeptidase inhibitor from potatoes to newly hatched chicks, Proc. Soc. Exp. Biol. Med., 173:447.
Phillips, R. D., Chhinnan, M. S., and Mendoza, L. G., 1983, Effect of temperature and moisture content on the kinetics of trypsin inhibitor activity, protein in vitro digestibility and nitrogen solubility of cowpea flour, J. Food Sci., 48:1863.
Possompes, B., Choi-Schin, I. S., and Besancon, P., 1979, Digestive transit in rats receiving an alkali-treated casein, Ann. Biol. Anim. Biochim. Biophys., 19(3B):907.
Pour-El, A., Nelson, S. O., Peck, E. E., Tjhio, B., and Stetson, L. E., 1981, Biological properties of VHF- and microwave-heated soybeans, J. Food Sci., 46:880.
Price, M. L., and Butler, L. G., 1980, Tannins and nutrition, Station Bulletin no. 272, March, 1980, Agricultural Experiment Station, Purdue University, West Lafayette, IN.
Rackis, J. J., 1965, Physiological properties of soybean trypsin inhibitors and their relationship to pancreatic hypertrophy and growth inhibition of rats, Fed. Proc., 24:1488.

Rackis, J. J., 1966, Soybean trypsin inhibitors: Their inactivation during meal processing, Food Technol., 20:1482.
Rackis, J. J., 1972, Biologically active components, in: "Soybeans: Chemistry and Technology," Vol. I, Proteins, A. K. Smith and S. J. Circle, eds., Avi Publishing, Westport, CT.
Rackis, J. J., 1974, Biological and physiological factors in soybeans, J. Am. Oil Chem. Soc., 51:161A.
Rackis, J. J., 1978, Biochemical changes in soybeans: Maturation, post-harvest storage and processing, and germination, in: "Postharvest Biology and Biotechnology," H. O. Hultin and M. Milner, eds., Food and Nutrition Press, Westport, CT.
Rackis, J. J., 1981, Comparison of the food value of immature, mature, and germinated soybeans, in: "Quality of selected fruits and vegetables of North America," R. Teranishi and H. Barrera-Benitez, eds., ACS Symposium series 170, American Chemical Society, Washington, D.C.
Rackis, J. J., and Gumbmann, M. R., 1981, Protease inhibitors: Physiological properties and nutritional significance, in: "Antinutrients and Natural Toxicants in Foods," R. L. Ory, ed., Food and Nutrition Press, Inc., Westport, CT.
Rackis, J. J., Gumbmann, M. R., and Liener, I. E., 1985, The USDA trypsin inhibitor study. I. Background, objectives, and procedural details, Qual. Plant. Plant Foods Hum. Nutr. 35: 213.
Rackis, J. J., McGhee, J. E., and Booth, A. N., 1975A, Biological threshold levels of soybean trypsin inhibitors by rat bioassay, Cereal Chem., 52:85.
Rackis, J. J., McGhee, J. E., Gumbmann, M. R., and Booth, A. N., 1979, Effects of soy proteins containing trypsin inhibitors in long term feeding studies in rats, J. Am. Oil Chem. Soc., 56:162.
Rackis, J. J., McGhee, J. E., Honig, D. H., and Booth, A. N., 1975B, Processing soybeans into foods: Selected aspects of nutrition and flavor, J. Am. Oil Chem. Soc., 52:249A.
Rackis, J. J., McGhee, J. E., Leiner, I. E., Kakade, M. L., and Puski, G., 1974, Problems encountered in measuring trypsin inhibitor activity of soy flour. Report of a collaborative analysis, Cereal Sci. Today, 19:513.
Raghavan, G. S. V., and Harper, J. M., 1974, High temperature drying using a heated bed of granular salt, Trans. ASAE, 17:108.
Read, J. W., and Haas, L. W., 1938, Studies on the baking quality of flour as affected by certain enzyme actions. V. Further studies concerning potassium bromate and enzyme activity, Cereal Chem., 15:59.
Richardson, M., 1974, Chymotryptic inhibitor I from potatoes. The amino acid sequence of subunit A, Biochem. J., 137:101.
Richardson, M., 1977, The proteinase inhibitors of plants and microorganisms, Phytochemistry, 16:159.
Richardson, M., 1979, The complete amino acid sequence and the trypsin reactive (inhibitory) site of the major proteinase inhibitor from the fruits of aubergine (Solanum melongena L.), FEBS Lett., 104:322.
Richardson, M., 1981. Protein inhibitors of enzymes, Food Chem., 6:235.
Richardson, M., and Cossins, L., 1974, Chymotryptic inhibitor I from potatoes: The amino acid sequences of subunits B, C and D, FEBS Lett., 45:11.
Rockland, L. B., and Nishi, S. K., 1979, Tropical grain legumes, in: "Tropical Foods: Chemistry and Nutrition," Vol. 2, G. E. Inglett and G. Charalambous, eds., Academic Press, New York.
Rockland, L. B., and Radke, T. M., 1981, Legume protein quality, Food Technol., 36:(3)79.
Roy, D. N., and Bhat, R. V., 1974, Trypsin inhibitor content in some varieties of soyabean (Glycine max L.) and sunflower seeds (Helianthus annus L.), J. Sci. Food Agric., 25:765.
Ryan, C. A., 1973, Proteolytic enzymes and their inhibitors in plants, Ann. Rev. Plant Physiol., 24:173.

Ryan, C. A., 1979, Proteinase inhibitors, in: "Herbivores: Their Interaction With Secondary Plant Metabolites," G. A. Rosenthal and D. H. Janzen, eds., Academic Press, New York.
Ryan, C. A., Huisman, O. C., and Van Denburgh, R. W., 1968, Transitory aspects of a single protein in tissues of Solanum tuberosum and its coincidence with the establishment of new growth, Plant Physiol., 43;589.
Ryan, C. A., Kuo, T., Pearce, G., and Kunkel, R., 1976, Variability in the concentration of three heat stable proteinase inhibitor proteins in potato tubers, Amer. Potato J., 53:443.
Ryan, C. A., and Pearce, G., 1978, Proteinase inhibitor proteins as genetic markers for identifying high protein potato clones, Am. Potato J., 55:351.
Sanderson, J. E., Freed, R. C., and Ryan, D. S., 1982, Thermal denaturation of genetic variants of the Kunitz soybean trypsin inhibitor, Biochim. Biophys. Acta, 701:237.
Sarett, H. P., 1976, Soy-based infant formulas, in: "World Soybean Research," L. D. Hill, ed., The Interstate Printers and Publishers, Danville, IL.
Sathe, S. K., Deshpande, S. S., and Salunkhe, D. K., 1984A, Dry beans of Phaseolus. A Review. Part I. Chemical composition: Proteins, CRC Crit. Rev. Food Sci. Nutr., 20:1.
Sathe, S. K., Deshpande, S. S., and Salunkhe, D. K., 1984B, Dry beans of Phaseolus. A Review. Part 2. Chemical composition: Carbohydrates, fiber, minerals, vitamins and lipids, CRC Crit. Rev. Food Sci. Nutr., 21:41.
Sathe, S. K., and Salunkhe, D. K., 1981, Studies on trypsin and chymotrypsin inhibitory activities, hemagglutinating activity, and sugars in the Great Northern beans (Phaseolus vulgaris L.), J. Food Sci., 46:626.
Senti, F. R., 1982, Annotated bibliographies on pancreatic changes in experimental animals fed soybeans, processed soybean products, soybean trypsin inhibitor, or cholecystokinin-pancreozymin, and antinutritional factors in processed soybean products, Report prepared for Bureau of Foods, Food and Drug Administration, Washington, D.C., Contract no. FDA 223-79-2295, Life Sciences Office, Federation of American Societies for Experimental Biology, Bethesda, MD.
Sessa, D. J., 1984, Are toasted soybean flour proteins responsible for trypsin inhibitor activity?, J. Am. Oil Chem. Soc., 61:Abstract no. 226, 691.
Sgarbieri, V. C., and Whitaker, J. R., 1982, Physical, chemical, and nutritional properties of common bean (Phaseolus) proteins, Adv. Food Res., 28:93.
Sharma, Y. K., Gupta, A. K., Gangrade, G. A., and Deodhar, A. D., 1975, Trypsin inhibitor activity in soybean at different stages of development, Indian J. Entomol., 36:62.
Smith, O. B., 1976, Extrusion cooking, in: "New Protein Foods," Vol. 2, Technology, Part B, A. M. Altschul, ed., Academic Press, New York.
Smith, A. K., and Circle, S. J., 1972, "Soybeans: Chemistry and Technology," Vol. 1, Proteins, Avi Publishing Co., Westport, CT.
Smith, C., Van Megen, W., Twaalhoven, L., and Hitchcock, C., 1980, The determination of trypsin inhibitor levels in foodstuffs, J. Sci. Food Agric., 31:341.
Soni, G. L., Singh, T. P., and Singh, R., 1978, Comparative studies on the effects of certain treatments on the antitryptic activity of the common Indian pulses, Ind. J. Nutr. Diet., 15:341.
Sri Kantha, S. S., and Erdman, J. W., Jr., 1984, The winged bean as an oil and protein source. A review, J. Am. Oil Chem. Soc., 61:515.
Stinson, C. T., and Snyder, H. E., 1980, Evaluation of heated soy flours by measurement of tryptic hydrolysis using a pH stat, J. Food Sci., 45:936.

Sweet, R. M., Wright, H. T., Janin, J., Chothia, C. H., and Blow, D. M., 1974, Crystal structure of the complex of porcine trypsin with soybean trypsin inhibitor (Kunitz) at 2.6-A resolution, Biochemistry, 13:4212.

Tan, C. G. L., and Stevens, F. C., 1971, Amino acid sequence of lima bean protease inhibitor component IV. 2. Isolation and sequence determination of the chymotryptic peptides and the complete amino acid sequence, Eur. J. Biochem., 18:515.

Tan, N.-H., Rahim, Z. Hj A., Khor, H.-T., and Wong, K.-C., 1984, Chymotrypsin inhibitor activity in winged beans (Psophocarpus tetragonolobus), J. Agric. Food Chem., 32:163.

Tan, N.-H., and Wong, K.-C., 1982, Thermal stability of trypsin inhibitor activity in winged bean (Psophocarpus tetragonolobus), J. Agric. Food Chem., 30:1140.

Tan-Wilson, A. L., Rightmire, B. R., and Wilson, K. A., 1982, Different rates of metabolism of soybean proteinase inhibitors during germination, Plant Physiol., 70:493.

Thorn, K. A., Tinsley, A. M., Weber, C. W., and Berry, J. W., 1983, Antinutritional factors in legumes of the Sonoran Desert, Ecol. Food Nutr., 13:251.

Troll, W., and Yavelow, J., 1983, Protease inhibitors as anticarcinogens, in: "Diet, Nutrition and Cancer: From Basic Research to Policy Implications," Proc. Workship Cornell Univ., Ithaca, N.Y., D. A. Roe, ed., Alan R. Liss, New York.

Tsukamoto, I., Miyoshi, M., and Hamaguchi, Y., 1983, Heat inactivation of trypsin inhibitor in Kintoki bean (Phaseolus vulgaris), Cereal Chem., 60:194.

Umezawa, H., and Aoyagi, T., 1983, Trends in research of low molecular weight protease inhibitors of microbial origin, in: "Proteinase Inhibitors: Medical and Biological Aspects," N. Katunuma, H. Umezawa, and H. Holzer, eds., Japan Sci. Soc. Press, Tokyo.

Valdebouze, P., 1977, Trypsin inhibitor and hemagglutinating activities in seeds of some legume species, in: "Protein Quality from Leguminous Crops," Office for Official Publication of the European Communities, Commission of the European Communities; Luxembourg.

Valdebouze, P., Bergeron, E., Gaborit, T., and Delort-Laval, J., 1980, Content and distribution of trypsin inhibitors and hemagglutinins in some legume seeds, Can. J. Plant Sci., 60:695.

Vimont-Rispoli, S., Possompes, B., and Besancon, P., 1980, Formation of trypsin inhibitors during alkaline treatments of protein, C. R. Seances Acad, Sci. Ser. D., 291:945.

Vogel, R., Trautschold, I., and Werle, E., 1968, "Natural Proteinase Inhibitors," Academic Press, New York.

Wang, H. L., 1983, Oriental soybean foods, in: "CRC Handbook of Processing and Utilization in Agriculture," Vol. II, Part 2, Plant products, I. A. Wolff, ed., CRC Press, Boca Raton, FL.

Wang, H. L., Vespa, J. B., and Hesseltine, C. W., 1972, Release of bound trypsin inhibitors in soybeans by Rhizopus oligosporus, J. Nutr., 102:1495.

Wang, H. L., Swain, E. W., Wallen, L. L., and Hesseltine, C. W., 1975, Free fatty acids identified as antitryptic factor in soybeans fermented by Rhizopus oligosporus, J. Nutr., 105:1351.

Weyer, E. M., 1968, Chemistry, pharmacology, and clinical applications of proteinase inhibitors," Ann. N.Y. Acad. Sci., 146:361.

Wilcke, H. L., Hopkins, D. T., and Waggle, D. H., 1979, "Soy Protein and Human Nutrition," Academic Press, New York.

Wilson, K. A., 1980, The release of proteinase inhibitors from legume seeds during germination, Phytochemistry, 19:2517.

Wilson, K. A., and Chen, J. C., 1983, Amino acid sequence of mung bean trypsin inhibitor and its modified forms appearing during germination, Plant Physiol., 71:341.

Wilson, K. A., Rightmire, B. R., and Tan-Wilson, A. L., 1985, Involvement of carboxypeptidase in the degradation of the mung bean (Vigna radiata) trypsin inhibitor during germination and early seedling growth, Qual. Plant. Plant Foods Hum. Nutr., In press.

Yoskikawa, M., Kiyohara, T., Iwasaki, T., and Yoshida, I., 1979, Modification of proteinase inhibitor II in adzuki beans during germination, Agric. Biol. Chem., 43:1989.

20

PREPARATION OF UNHEATED SOY PROTEIN ISOLATES WITH LOW TRYPSIN INHIBITOR CONTENT

E. C. Baker and J. J. Rackis

Northern Regional Research Center
Agricultural Research Service
U.S. Department of Agriculture
Peoria, Illinois 61604

ABSTRACT

The nutritional quality of soy protein products is affected by the processing conditions employed in their manufacture. Heat treatment during processing serves to inactivate the inhibitors of trypsin and chymotrypsin, enzymes which play a key role in the digestion of protein in animals. In the processing of defatted soy flour to prepare a soy protein isolate, generally no heat treatment is used. Instead, a high percentage of the trypsin inhibitors (TI) is physically removed in the whey fraction. However, depending on their mode of preparation, soy isolates may contain trypsin inhibitor activity as high as 40% of that found in raw soybeans. Using "salting in" techniques we found that a higher percentage (97.7%) of the TI was solubilized and removed with the whey fraction when the protein curd was precipitated from 0.1 N NaCl solution at pH 5.4. Using membrane techniques for the separation of TI from non-TI-protein, best results were obtained with 0.1 N. $MgCl_2$ where 79% of the TI was removed in the permeate.

INTRODUCTION

Edible soybean protein isolates were introduced commercially in 1957 and in the intervening years we have seen a steady increase in the use of these products as ingredients in a wide variety of fabricated or engineered foods. (Wolf, 1981) Widespread adoption of soybean proteins in foods has stimulated the need for more research on processing parameters that may be necessary to produce safe, nutritious soya foods (Wolf, 1983).

Osborne and Mendel (1917), reported that rats grew poorly when fed raw soybean meal. A number of biologically active substances have been identified in soybeans, but trypsin inhibitors (TI) have received the most attention. They not only inhibit trypsin and chymotrypsin, important enzymes involved in digestion of proteins, but indirectly, they stimulate secretion of pancreatic enzymes into the intestine which, in turn, cause pancreatic hypertrophy and loss of undigested pancreatic enzymes into the feces. Excessive loss of essential amino acids through the feces is believed responsible for inhibition of growth in young rats (Rackis, 1974).

Trypsin inhibitors are inactivated by moist-heat treatment which, at

Table 1. Composition of Commercial Defatted Soy Flour Component

Moisture %	9.5
Ash %	6.6
Protein %	51.4
Nitrogen Solubility Index	85.8
Trypsin Inhibitor mg/g Protein	55

the same time, denatures the raw proteins causing modification in protein functionality (i.e., dispersibility, viscosity, emulsification, etc.). In previous studies at this Center, (Baker and Rackis, 1985) using ultrafiltration membranes with molecular weight cutoffs (MWCO) of 100,000 daltons or less, a significant portion of the TI (MW <21,500) was rejected by the membrane. This apparent contradiction might be explained if aggregation of the TI with non-TI proteins was occurring. Consequently for this study the more porous microfilter membrane was chosen, since it presents less of a barrier to the passage of TI into the permeate.

The earlier study at this Center had also shown that the separation of TI from the isoelectric precipitated protein curd was enhanced by the addition of 0.1 N sodium chloride. Other salts which are known to have a solubilizing effect on proteins in water, such as magnesium chloride, aluminum chloride and sodium lactate, were included in this study.

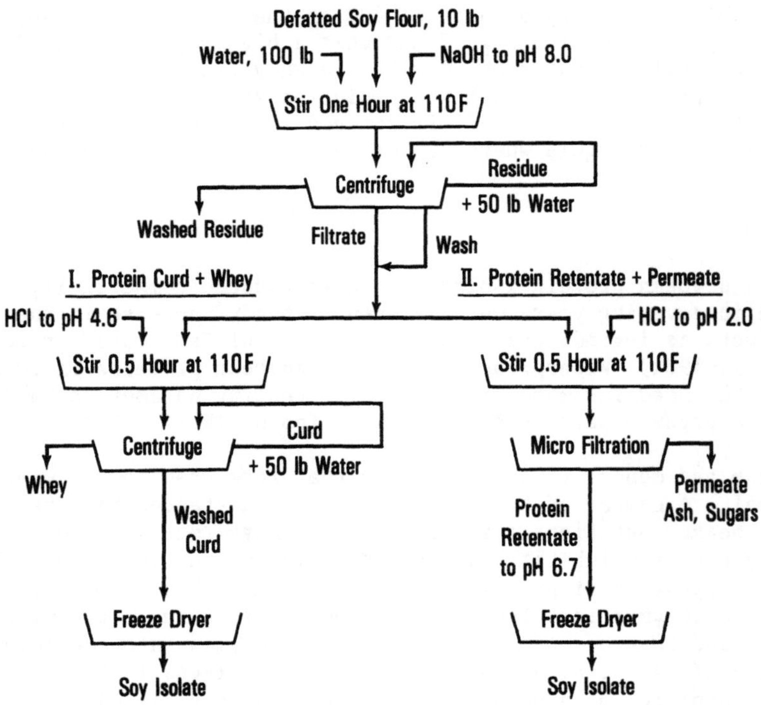

Fig. 1. Alternate procedures for preparation of soy isolates.

MATERIALS AND METHODS

Commercial defatted soy flour, Nutrisoy 7 B (Archer Daniels Midlands, Decatur, Illinois), was used as the starting material in all experiments. The commercial process for the preparation of defatted soy flour uses a mild heat treatment and results in minimal denaturation of protein as evidenced by the nitrogen solubility index of 85.8 (Table I). The TI activity of 55 mg/g protein approximates the value often observed with raw beans.

Trypsin inhibitor was determined by the method of Hammerstrand et al. (1981) which evolved primarily from the work of Kakade et al. (1974). Total solids were determined by evaporation.

Equipment and Procedure

Defatted soy flour (10 lbs) was suspended in 10 parts of water in a 30-gallon stainless steel agitated tank fitted with an internal coil for heating or cooling. Slurry pH was adjusted to 8.0 with sodium hydroxide and the temperature was regulated at 110 F. The insoluble residue, which consisted mainly of polysaccharides, was separated using a 4-inch solid bowl (17,000 rpm) centrifuge (Sharples, Pennwalt Corp., Warminister, Pennsylvania).

When the isoelectric precipitation procedure was followed (Scheme I, Figure 1), salts were added to the combined filtrate and wash and pH was adjusted with hydrochloric acid to the isoelectric region (pH 4.2 - 5.4). The precipitated curd was separated from the whey by centrifugation. The curd was resuspended with 5 parts of water and recentrifuged. The washed curd was freeze dried in a 5-shelf, 50-lb-capacity freeze dryer (Virtis Co., Gardiner, New York). Experiments were designed to test the effects on yield and TI removal of added salts, salt normality and pH.

In the second scheme (Scheme II, Figure 1) which utilizes membrane separation, salts were added to the combined filtrate and wash from the residue removal step and pH was adjusted to either 2 or 8 to achieve maximum solubility of the protein. The protein solution was then pumped to a spiral-wound microfiltration module with a molecular weight cutoff of >300,000 daltons. Feed volume was maintained at a constant level by the addition of fresh water at the same rate as permeate was being removed until permeate equal to 2 feed volumes was collected. The feed was next concentrated to <40 lb and freeze dried. In experiments run at pH 2, the retentate was adjusted to pH 6.5 before freeze drying. Experiments were designed to test the effect of added salts on yield and residual TI in the isolates.

Results and Discussion

Smith and Circle (1938) have shown that the extractability of proteins in defatted soybean meal is a function of pH (Figure 2). The region of minimum protein solubility encompasses a narrow region bounded by pH 3.8 to pH 5.2 with the isoelectric point situated near the mid point of the region at approximately pH 4.6. Precipitation of a protein curd in the isoelectric region and separation of the curd from the whey by centrifugation is one of the more common procedures used in the commercial preparation of soy protein isolates. The regions of maximum protein solubility (i.e. pH 1.5 - 2.5 and pH 7 and above) are of interest when membrane techniques are employed to separate the larger proteins from the other constituents of defatted soy flour. The membrane rejects the larger protein molecules while passing the smaller TI, sugars and ash. Wolf et. al. (1962) showed the TI (MW 8000-21,500) to be in the 2s ultracentrifuge fraction which

Table 2. Ultracentrifuge Fractions of Soybean Proteins

Fraction	Percent of Total	Component	Molecular Weight
2s	22	Trypsin inhibitors	8000-21,500
		Cytochrome C	12,000
7s	37	Hemagglutinins	110,000
		Lipoxygenases	102,000
		B-Amylase	61,700
		7s-Globulin	180,000-210,000
11s	31	11S-Globulin	350,000
15s	11	---	600,000

Source: Wolf et al. (1962)

contains the low M.W. proteins (Table 2).

Isolates Via Isoelectric Precipitation

With the control experiment (no added salts) a protein yield of 63% was obtained which was close to the expected yield in commercial processing

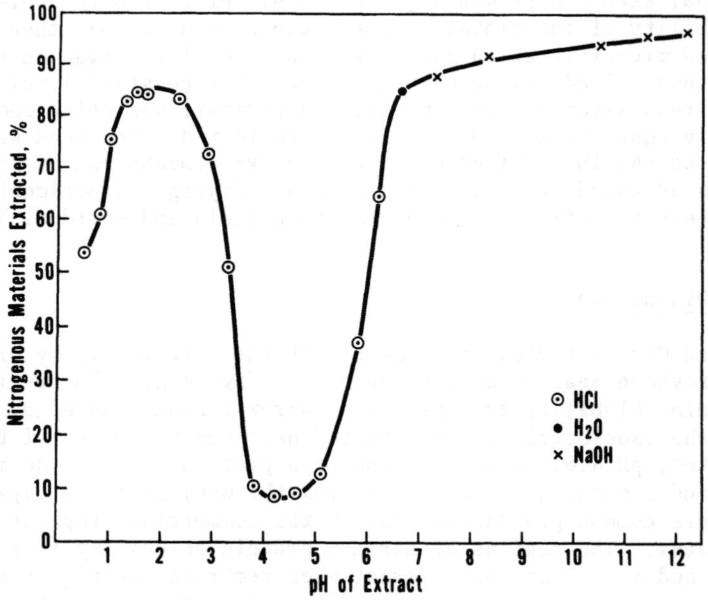

Fig. 2. Extractability of proteins in defatted soybean meal as a function of pH source, Smith and Circle (1938).

Table 3. Effect of Various Salts on Yield and Residual TI in Soy Protein Isolates Via Isoelectric Precipitation

	Curd Precipitation pH	Soy Protein Isolates			
		Protein		Trypsin Inhibitor	
		Content %	Yield %	mg Per g Protein	% of Original
Defatted Soy Flour	—	51.4	100	55	100
Soy Isolates					
Control	4.6	91.8	63	15.8	18.6
I 0.1N NaCl	5.0	91.8	52	4.6	3.9
0.1N $MgCl_2$	5.0	91.2	55	6.7	6.9
1.2% $AlCl_3$[a]	5.0	88.1	46	6.2	5.4
4.0% Na Lactate[b]	5.0	94.1	60	7.2	7.8
II 0.05N NaCl	5.0	92.5	59	13.6	14.7
0.10N NaCl	5.0	91.8	52	4.6	3.9
0.50N NaCl	5.0	81.9	14	7.8	2.3
III 0.10N NaCl	4.2	90.6	58	12.7	13.2
0.10N NaCl	4.6	91.2	60	8.2	8.9
0.10N NaCl	5.0	91.2	52	4.6	3.9
0.10N NaCl	5.4	92.5	51	2.6	2.3

[a]U.S. Patent 4,375,431, Bradford et al. 1983.
[b]Private communication, W. H. Martinez, USDA, NPS, Washington, D.C.

of soy isolates. (Meyer, 1966) The TI content of 15.8 mg/g protein (Table 3) approximated the values obtained when several commercial isolates were analyzed in our laboratory. When salts were added, residual TI content of the isolates was reduced in all cases; the best result (4.6 mg/g protein) was obtained with 0.1 N sodium chloride. When the normality of the sodium chloride was decreased to .05N, the beneficial effect of the salt was virtually eliminated. Increasing the normality to 0.50N solubilized so much of the protein that the curd yield was only 14% with most of the protein lost to the whey. Residual TI content of the isolate was shown to be a function of pH in the isoelectric region. Best results were obtained at the upper end of the region (pH 5.4) where the residual TI content was reduced to 2.6 mg/g protein with 97.7% of the TI removed in the whey. Protein yield was also a function of pH in the isoelectric region. Best yield was 60% obtained predictably at the isoelectric point, pH 4.6. At pH 5.4 the yield fell to 51% but this may be tolerable if the primary concern is low TI content in the isolate.

Isolate Via Membrane Processing

In the first series of experiments in which the feed to the microfilter was adjusted to pH 2, protein recovery was 67% of the original flour protein (no added salts) and residual TI was 39% of the original (Table 4). Both sodium and magnesium chlorides were found to be most effective at a concentration of 0.1 normal, while Bradford et. al. (1983) observed that the best results were attained with aluminum chloride at a level of 1.2 g per-100 g soy flour. Conversely, sodium lactate had little effect at concentrations below 4%. When salts were added, a greater quantity of TI was

Table 4. Effect of Various Salts on Yield and Residual TI in Soy Protein Isolates Via Membrane Processing

	Feed To Microfilter pH	Soy Protein Isolates: Protein Content %	Protein Yield %	Trypsin Inhibitor mg Per g Protein	Trypsin Inhibitor % of Original
Defatted Soy Flour	—	51.4	100	55	100
Soy Isolates					
Control	2.0	88.2	67	32	39
0.1N NaCl	2.0	86.3	69	24	29
0.1N $MgCl_2$	2.0	85.7	64	18	21
1.2% $AlCl_3$ [a]	2.0	83.1	67	22	27
4.0% Na Lactate [b]	2.0	83.8	69	28	34
Control	8.0	88.2	50	28	29
0.1N NaCl	8.0	85.6	36	27	18
0.1N $MgCl_2$	8.0	85.3	38	18	19
1.2 % $AlCl_3$ [a]	8.0	86.9	53	27	26
4.0% Na Lactate [b]	8.0	87.5	47	28	24

[a] U.S. Patent 4,375,431, Bradford et al. 1983.
[b] Unpublished data - W. H. Martinez, USDA, NPS, Washington, D.C.

removed with the permeate in all cases. If aggregation of TI with either itself or the non-TI-protein does, in fact, occur, the salts then appear to be partially responsible for dissociation of the aggregate. Best results were obtained with 0.1N $MgCl_2$ which gave a protein recovery of 64% and a residual TI content of 21% of the original.

In the second series of experiments in which the feed to the microfilter was adjusted to pH 8, protein recovery was only 50% with the control. This would indicate that mid-range MW proteins that were rejected when fed to the membrane at pH 2 were passing into the permeate at pH 8. With 0.1N NaCl or $MgCl_2$ the protein recovery was even less, (36-38%). The protein content of the isolates obtained via membrane processing ranged from 83.1 to 88.2% on a moisture-free basis (MFB). The Soybean Digest Blue Book (1966) defines isolated soy protein as containing not less than 90% protein (N X 6.25) on a moisture-free basis. The fact that the isolates produced were slightly lower in protein content indicates that the diafiltration washing procedure may have been inadequate and should probably have been extended beyond collecting permeate equal to two feed volumes, perhaps as many as three or four volumes.

SUMMARY

The isolates obtained by Scheme I and Scheme II were evaluated using two criteria as follows: first, the residual TI content of the isolate; second, the percent recovery of the original protein in the isolate.

When isolates were prepared via isoelectric precipitation (Scheme I) best results were obtained using 0.1N NaCl at pH 5.4 where 97.7% of the original TI was removed and 51% of the original protein was recovered in the isolate.

When isolates were prepared using membrane separation (Scheme II) best results were obtained using 0.1N $MgCl_2$ at pH 2.0 where 79% of the original TI was removed and 64% of the protein was recovered in the isolate.

ACKNOWLEDGMENT

R. L. Brown conducted the pilot-plant experiments; J. D. Glover, R. L. Haig, P. L. Milroy, I. M. Schulte and M. E. Riggen performed the analyses.

The mention of firm names or trade products does not imply that they are endorsed or recommended by the U.S. Department of Agriculture over other firms or similar products not mentioned.

REFERENCES

Baker, E. C., and Rackis, J. J., 1985, Development of a pilot plant process for the preparation of a soy trypsin inhibitor concentrate, JAOCS, 62:84.

Bradford, M. M., Orthoefer, F. T., and Wright, K. N., 1983, "Aluminum Treated Proteins," U.S. Patent 4,375,431.

Hamerstrand, G. E., Black, L. T., and Glover, J. D., 1981, Trypsin inhibitors in soy products: Modification of the standard analytical procedure, Cereal Chem., 58:42.

Kakade, M. L., Rackis, J. J., McGhee, J. E., and Puski, G., 1974, Determination of trypsin inhibitor activity of soy products: A collaborative analyses of an improved procedure, Cereal Chem. 51:376.

Meyer, E. W., 1966, Proc. Int. Conf. on Soy Protein Foods, Peoria, IL, ARS-71-35, pp. 142-154, USDA, Washington, D.C.

Osborne, T. B., and Mendel, L. B., 1917, The use of soybeans as food, J. Biol. Chem., 32:369.

Rackis, J. J., 1974, Biological and physiological factors in soybeans, JAOCS, 51:161 A.

Smith, A. K., and Circle, S. J., 1938, Peptization of soybean proteins. Extraction of nitrogenous constituents from oil-free meal by acids and bases with and without added salts, Ind. Eng. Chem., 30:1414.

The Soybean Digest, Blue Book Issue, 1966, 26:20.

Wolf, W. J., Babcock, G. E., and Smith, A. K., 1962, Purification and stability studies of the 11s component of soybean proteins, Arch Biochem. Biophys., 99:265.

Wolf, W. J., 1981, Progress and future needs for research in soya protein utilization and nutrition, JAOCS, 58:467.

Wolf, W. J., 1983, Edible soybean protein products, in: "CRC Handbook of Processing and Utilization in Agriculture, Vol. II, Part 2, Plant Products," Ivan A. Wolff, ed., pp. 23-55.

21

NUTRITIONAL IMPROVEMENT OF LEGUME PROTEINS THROUGH DISULFIDE INTERCHANGE

Mendel Friedman and Michael R. Gumbmann

Western Regional Research Center, Agricultural Research Service, U. S. Department of Agriculture, 800 Buchanan Street, Berkeley, California 94710

ABSTRACT

Treatment of raw soy flour with L-cysteine or N-acetyl-L-cysteine results in the introduction of new half-cystine residues into sulfur-poor legume proteins, with a corresponding improvement in nutritional quality as measured by the protein efficiency ratio (PER) in rats. The proteins are modified through formation of mixed disulfide bonds among added sulfhydryl compounds, proteolytic enzyme inhibitors, and structural legume proteins. This modification leads to loss of inhibitory activity and increased protein digestibility and nutritive value. Sodium sulfite is more effective than cysteine in facilitating inactivation of trypsin inhibitors in soy flour. The synergistic effect of sodium sulfite and heat may be due to (1) ability to induce rearrangement of protein disulfide bonds to produce new structural entities without altering the amino acid composition and to (2) the fact that the new structures lose their ability to complex with trypsin or chymotrypsin. The same treatment inactivated hemagglutinins (lectins) in lima bean flour. These considerations suggest a key role for sulfur amino acids in the nutritional quality and safety of legumes.

INTRODUCTION

Inhibitors of digestive enzymes appear in many foods (for reviews, see Bozzini and Silano, 1978; Whitaker, 1981; Liener and Kakade, 1980; Birk, 1985; Rackis et al., 1986). These foods include alfalfa (Mitchell and Parrish, 1981), barley (Mikola and Suolinna, 1969; Ogiso et al., 1975; Boisen et al., 1981), black gram (Kamalakannan and Motlag, 1982), broad pea beans (Warsy et al., 1974), chickpeas (Smirnoff et al., 1979), corn (Hochstrasser et al., 1970; Gupta et al., 1979), dry beans (Burr, 1975; Antunes and Sgarbieri, 1980), egg white (Lineweaver and Murray, 1949; Sen and Whitaker, 1973; Osuga et al., 1974; Matsuda et al., 1983), garden beans (Birk, 1976a,b), Great Northern beans (Sathe and Salunkhe, 1981; Friedman et al., 1982a), groundnuts (Tur-Sinai et al., 1972; Birk, 1976c), lima beans (Fraenkel-Conrat et al., 1952; Jones et al., 1963; Stevens et al., 1974; Friedman et al., 1982a), marama beans (Starcher et al., 1986), mustard (Menegatti et al., 1985), millet (Udupa and Pattabiraman, 1984), navy beans (Wagner and Riehm, 1967; Yadav and Liener, 1978), pineapples (Heinrikson and Kezdy, 1976), pinto beans (Wang, 1975), potatoes (Hojima et al., 1973; Iwasaki et al.,

1974), red kidney beans (Powers and Whitaker, 1977), red wood (Veerappa Gowda and Karunasagar, 1985), rye (Chang and Tsen, 1981 a,b), sorghum (Filho, 1974), soybeans (Kunitz, 1948; Liener, 1975; Birk, 1976d; Anderson et al., 1979; Rackis and Gumbmann, 1981), squash (Otlewski et al., 1984), sweet potatoes (Bradbury et al., 1985), wheat (Buonocore and Silano, 1986), and winged beans (de Lumen and Chan, 1986).

Feeding soybeans and other legumes containing active trypsin inhibitors to rats depresses growth in comparison with analogous feeding of inhibitor-free soybeans. Growth inhibition and its accompanying pancreatic hypertrophy are presumably partly or fully due to the anti-tryptic activity of the inhibitor(s) (Anderson et al., 1979; Rackis and Gumbmann, 1981).

For these reasons, extensive efforts have been made to devise processing conditions to inactivate or remove trypsin inhibitors from legumes. These approaches are based largely on heat treatments (Longnecker, 1964; Hackler et al., 1965; Borchers et al., 1972; Baker and Mustakas, 1973; Boonivust and Whitaker, 1976; Collins and Beaty, 1980; Ellenrieder et al., 1980; Naim et al., 1982; Buera et al., 1984; Chae et al., 1984; Rackis et al., 1986).

However, heat treatments do not completely inactivate all inhibitors present. The possible adverse effects of residual inhibitors in foods are largely unknown. In addition, autoclaving and other conditions used to inactivate inhibitors may destroy important amino acids such as cystine, arginine, and lysine (Rios-Iriarte and Barnes, 1966; Chae et al., 1984; Skrede and Krogdahl, 1985). In fact, the amount of heat required to destroy growth inhibitors in raw soy beans may destroy sufficient cystine to make this amino acid first-limiting (Rios-Iriarte and Barnes, 1966; Kakade et al., 1969).

Efforts to remove inhibitors by precipitation (Baker and Rackis, 1986), or by plant genetic methods (Hymowitz, 1986) also may adversely affect nutritional quality, since the inhibitors when denatured are highly nutritious, well-balanced proteins, rich in sulfur amino acids (Tan-Wilson and Wilson, 1986). Alternative strategies are, therefore, needed to completely inactivate inhibitors in foods and to avoid the cited possible adverse effects on nutritional quality and safety.

Naturally occurring protease inhibitors, such as trypsin and chymotrypsin inhibitors, require their structural integrity in order to inactivate proteolytic enzymes by complex formation (Rackis et al., this volume). Therefore, altering the inhibitor's structure will inactivate it if interaction with the enzyme is thereby prevented. A reasonable strategy is to cleave the inhibitor's disulfide bonds, which often control protein structure (Anfinsen, 1972; Friedman, 1973; Creighton, 1975; Hawkins and Freedman, 1975; Perry and Wetzel, 1984).

For several years we have been exploring some possible new approaches to facilitate inactivation of enzyme inhibitors and hemagglutinins (lectins) (Friedman et al., 1980; Friedman et al., 1982 a,b,c; 1984a; Zahnley and Friedman, 1982; Wallace and Friedman, 1985). This paper summarizes some of our relevant findings.

EXPERIMENTAL SECTION

Materials. Soy flour (soya fluff 200W) was a gift from Central Soya, Ft. Wayne, Indiana. Raw isolate was obtained from Kraft, Glenview, Illinois. Lima bean flour was prepared by milling raw seeds purchased in a local store. Soybean trypsin inhibitor (Kunitz),

L-cysteine, N-acetyl-L-cysteine, benzoyl-DL-arginine-ρ-nitroanilide (BAPNA), 5,5'-dithiobis-(2-nitro- benzoic acid) (DTNB), ethylenediamine-tetraacetic acid (EDTA), 3-(N-morpholino)propanesulfonic acid (MOPS), N, N-bis (2-hydroxyethyl) glycine (BICINE), and glutaraldehyde-stabilized human group A blood were obtained from Sigma Chemical Company, St. Louis, MO.

Treatments. In a typical experiment, soy flour (700 g) and 0.128 moles of cysteine were suspended in 2.5 liters of prewarmed 0.5 M Tris, pH 8.5 buffer. The mixture was then heated in a water bath at 45°C, 65°C, or 75°C for 1 hr with occasional stirring. The contents were then dialyzed in a cold room for 3 days with frequent changes of deionized water and lyophilized. Controls did not have cysteine or N-acetylcysteine.

Trypsin inhibitor assay. Many units are used to define the inhibitor activity of protease inhibitors (Kakade et al., 1974; Smith et al., 1980; Hammerstrand et al., 1981; Friedman et al., 1982b). The following method, based on the defined stoichiometry of enzyme, inhibitor, and substrate was adapted from Mikola and Suolinna (1969). Inhibition of trypsin activity was measured at pH 8.2 and 37°C, with BAPNA as substrate. Inhibitor was incubated with 25 µg of twice-crystallized bovine trypsin in a total volume of 0.75 ml for 5 min at 37°C. BAPNA (30 mg in 1.00 ml of DMSO) was diluted to 100 ml with 0.05 M Tris-HCl, pH 8.2, containing 0.02 M $CaCl_2$. The diluted BAPNA was kept at 37°C and used within 1 hr. To start the reaction, 3.0 ml of substrate-buffer mixture was added rapidly to the trypsin-inhibitor mixture. The reaction was stopped after 5 or 10 min by adding 0.50 ml of 30% acetic acid.

Determination of active trypsin. The concentration of active trypsin was determined by active site titration with p-nitrophenyl guanidinobenzoate (NPGB), essentially according to Chase and Shaw (1970). Absorbance at 410 nm was followed in a Cary 15 spectrophotometer, with the 0.1 absorbance slidewire. NPGB (0.01 M) was dissolved in dimethylformamide-acetonitrile (1:4). Trypsin was dissolved in 2 mM HCl/0.02 M $CaCl_2$ at 5-6 mg/ml and then diluted 1:10 in the same solvent. Each cuvette contained 0.05-0.09 M Na barbital buffer (pH 8.35 or 8.39 at 23°C), 0.02 M $CaCl_2$, 0.1 M KCl. Trypsin (3-5 nmol of active enzyme) was added to the sample cuvette as 0.20 ml of the 1:10 dilution or 25 µL of the stock solution. After the baseline was determined, 5 µL of NPGB was added to each cuvette. Total volume was 1.0 ml. In our instrument, 1 A_{410}=61.1 mol of p-nitrophenolate. The trypsin used was 61% active enzyme, based on its protein content as determined by absorbance at 280 nm and an active site titration.

Calculations. The following sample calculation was used to estimate the trypsin inhibitor activity per milligram of active enzyme, where

A_{410} = absorbance measured at 410 nm

ϵ_M = molar extinction coefficient

V_o = rate of BAPNA hydrolysis by trypsin in the absence of inhibitor.

V_i = rate of BAPNA hydrolysis by trypsin in the presence of inhibitor.

TU = trypsin unit, defined as the amount of trypsin that catalyzes the hydrolysis of 1 μmol of BAPNA per min at 37°C and pH 8.2.

TIU = trypsin inhibitor unit, defined as reduction in activity of trypsin by one trypsin unit (TU).

Using an ϵ_M for p-nitoaniline of 8800 L/mol-cm (Erlanger et al., 1961), and a total volume of 4.25 ml, 1 TU corresponds to an increase in A_{410} of 2.07 min^{-1}.

We found trypsin activity of 2.51 TU per mg of active enzyme (see sample calculation). Using molecular weights of 23,400 for trypsin and 21,500 for STI, we calculated 2.92 TIU per milligram of STI.

For trypsin only, 0.358 A_{410}/5 min; Rate, V_o = 0.0716/min; for trypsin and inhibitor, 0.389 A_{410}/10 min.

The difference in the rate of BAPNA hydrolysis by trypsin in the absence and presence of inhibitor is given by:

$$V_o - V_i = \frac{A_{410}\text{ (no inhibitor)}}{t_o} - \frac{A_{410}\text{ (inhibitor)}}{t_i} = \frac{0.358}{5} - \frac{0.389}{10}$$

$$= 0.0716 - 0.0389 = 0.0327\ A_{410}\text{min}^{-1}$$

Since 2.07 A_{410} min^{-1} is equivalent to 1 trypsin unit (TU), the amount of inhibitor present is:

$$\frac{0.0327}{2.07} = 0.0158\ \text{TIU}$$

For pure soybean trypsin inhibitor (STI), the inhibitor content of the sample used is:

$$\frac{0.0158\ \text{TIU}}{2.92\ \text{TIU/mg STI}} = 5.41 \times 10^{-3}\text{mg STI} = 5.41\ \mu\text{g STI}$$

Since 1 mg of active trypsin = 2.51 TU, 0.0158 TIU represent

$$\frac{0.0158\ \text{TU}}{2.51\ \text{TU/mg trypsin}} = 6.29 \times 10^{-3}\text{mg}$$

active trypsin inhibited in the pure soybean trypsin inhibitor sample.

If inhibitors other than STI are also present (as is apparently the case in most plant foods), a more precise estimate is:

$$\frac{\text{TIU in sample}}{\text{wt. of sample}} = \text{TIU/mg (or g)}.$$

The following average values demonstrate the reproducibility of the assay. Pure commercial STI inhibited an average of 1.23 ± 0.03 mg of active trypsin per milligram STI (eight replicates). (This value is higher than unity because the ratio of molecular weight of trypsin to STI is greater than one and because of possible minor contributions of lower molecular weight inhibitors to the total enzyme inhibitory process). Untreated soy flour (the average for four determinations), contained 109.51 ± 1.52 TIU or 43.63 ± 0.605 mg trypsin per gram. Similarly, the trypsin inhibitor assay for lima bean inhibitor (LBI) gave the following average values for four determinations: 2.76 ± 0.041 TIU/mg commercial LBI and 51.33 ± 0.994 TIU/g lima bean flour.

Chymotrypsin inhibition assay. Inhibition of chymotrypsin was determined titrimetrically, with 0.01 M ATEE as substrate. The procedure was adapted from that of Wilcox (1979). Inhibitor was incubated with chymotrypsin (25 µL of 0.44 mg/ml solution) at room temperature, then equilibrated at 25°C in 1.9 ml of 5 mM Tris/HCl, 0.1 M $CaCl_2$, pH 8.0. At least 10 min was allowed for enzyme and inhibitor to react. After the baseline uptake of titrant (0.1 N NaOH) was recorded, 100 µL of 0.2 M ATEE was injected into the solution. Initial rates were determined from the linear portion of the plot of base uptake against time. The chart speed was 2 cm/min. Measurements of the nonenzymatic breakdown of ATEE and the effect of order of adding ATEE and chymotrypsin (no inhibitor) were also made and used to correct assays of inhibition.

Hemagglutination assay. The hemagglutinating activities of lima bean and soy bean flours were determined according to the procedure of Wallace and Friedman (1985). One hundred fifty mg of flour was mixed with 1.5 ml of the appropriate buffer. The lectin was extracted for 1 hr with shaking in a water bath at the selected temperature. When tested, N-acetyl-L-cysteine was mixed with the dry flour prior to addition of the buffer. After extraction, the resulting slurry was immediately chilled and centrifuged at 9000 x g for 5 min in a Beckman Microfuge (Beckman Instruments, Palo Alto, CA). When necessary, the extracts were diluted with isotonic phosphate buffer (PBS) prior to plating so that incipient activity would fall mid-range in the plated series. One hundred µL of this extract was placed in the first U-shaped well of a microtiter plate and 50 µL was serially diluted by 10 successive transfers into 50 µL aliquots of PBS using 50 µL diluters (Flow Laboratories, McLean, VA).

To the 50 µL aliquots of serially diluted extracts and to one "blank" well containing only 50 µL PBS was added an equal volume of glutaraldehyde-stabilized human group A blood cells which had been diluted with PBS to 3.3% hematocrit immediately before use. Agglutination was observed visually after 1 hr (Lis and Sharon, 1981).

Activity is expressed as the minimum µg of flour protein required to agglutinate blood cells. Results of four separate assays conducted on each sample were averaged. Variations between replicates assays were generally within a factor of one.

Proteolytic digestion assays. The extent of peptide cleavage by trypsin was estimated on a pH-Stat (Friedman et al., 1981; 1982c). Samples were dissolved or dispersed (5 mg/ml) in 0.05 M KCl with stirring and adjusted to pH 8.5. To determine digestibility by trypsin, aliquots (5 mg of protein) were pipetted into plastic titration vessels (Radiometer V 524) and diluted to 4.0 ml with 0.05 M KCl. Samples were incubated under N_2 to minimize CO_2 absorption. In a typical run, the substrate was brought to pH 8.5 at 37°C and the baseline rate of uptake of titrant (0.02 N NaOH) was determined. Then, 20 µL of bovine trypsin (5 mg/ml in 2 mM HCl-0.02 M $CaCl_2$) was added. More was added when the uptake of alkali became very slow. The net uptake was determined when the reaction was essentially complete.

Since liberation of hydrogen ions by proteolysis is incomplete at pH values where some new α-amino groups are ionized (Rupley, 1967), calculation of the number of peptide bonds cleaved requires a value for the average pK of the newly formed peptide α-amino groups. To calculate the actual average number of peptide bonds cleaved per chain corrected for incomplete liberation of hydrogen ions, we also assumed an average molecular weight of 23,000 for the soy protein (Wolf, 1977).

The theoretical rationale for measuring the extent of peptide bond cleavages is based on potentiometric titration of hydrogen ions liberated when peptide bonds are hydrolyzed (Rupley, 1967) (Equation 1).

Equation 1.

$$\text{P-CO-NH-P'} \xrightarrow[H_2O]{\text{Trypsin}} \text{P-COO}^- + (H_3N^+\text{-P'} \overset{K}{\rightleftharpoons} H^+ + H_2N\text{-P'})$$

P, P' = Protein side chains.

Above pH 6.5, the number of hydrogen ions liberated is equal to the number of dissociated α-amino groups produced. Since the average pK of these new amino groups is near 7.5, about 90% of the splits will produce titratable hydrogen ions at the pH (8.5) used in the present study. Since pH is kept constant during the titration, the degree of ionization of lysine ε-amino groups (pK near 10) does not change. Similarly, the slight contribution of hydrogen ions from the single terminal α-amino group can also be neglected.

The relation between the number of peptide bonds cleaved and hydrogen ions produced is given by Equation 2:

Equation 2.

$$pH - pK = \log \frac{NH_2}{NH_3{}^+} = \log \frac{H^+}{NH_3{}^+}$$

The number of enzyme-catalyzed peptide bonds cleaved per unit (grams or moles) of protein is given by Equation 3:

Equation 3.

$$N = H^+ [1 + 10^{(pK - pH)}], \text{ where}$$

N = number of peptide bonds cleaved per unit of protein.

H^+= number of hydrogen ions liberated and titrated during the enzymatic hydrolysis (equation 1). This value generally ranges between 7.5 and 7.6 (Mihalyi, 1978; Rupley, 1967; Tanford, 1962).

pH = pH value at which the enzymatic digestion was carried out.

Thus, at pH 7.5,cleavage of two peptide bonds liberates one hydrogen ion since $N = H^+(1 + 1)$. At pH 8.0 and 8.5, the corresponding values are 1.32 and 1.1, respectively.

Half-cystine and methionine content. The half-cystine (cysteine plus cystine) and methionine content of treated and untreated materials was determined on an amino acid analyzer as cysteic acid and methionine sulfone, respectively, after performic acid oxidation of dialyzed and lyophilized samples (Moore, 1963; Friedman et al., 1979).

SH content. The cysteine (SH) content was determined by a modified Ellman procedure (Ellman, 1959; Sedlak and Lindsay, 1968; Friedman et al., 1984a) as follows: One gram soy flour in 9 ml 0.1M phosphate buffer (pH 8.0) and 1 ml (0.02 M) solution ethylenediamine tetraacetic acid (EDTA), were shaken together in a 37°C water bath for 10 min. The mixtures were then centrifuged for 30 min. at 8,000 rpm. Next, 0.5 ml samples of the supernatant were mixed with 1 ml 0.02M EDTA, 1 ml 0.1M phosphate buffer (pH 8), and 2.5 ml deionized H_2O. Next, 25 µL 0.01M 5,5'-dithiobis-(2-nitrobenzoic acid) dissolved in 0.1M phosphate buffer (pH 7.0) was added to each supernatant. The solution was left standing for 20 min. for maximum color development. The absorption maximum of the solution at 412 nm was then determined with a Cary Model 14 spectrophotometer against a reagent blank. A cysteine standard curve in the range 0.01 to 0.1 µmoles was prepared and the SH content calculated by means of the equation:

$$\text{SH content (mg/g soy flour)} = \frac{\text{Absorbance of sample} \times \text{Dilution factor} \times 120}{13{,}600}$$

where 120 represents the molecular weight of L-cysteine and 13,600 the molar extinction coefficient of DTNB.

Protein analysis. The percent protein (% nitrogen x 6.25) was calculated from the nitrogen values determined by an automated micro-Kjeldahl procedure (AOAC, 1980).

Protein efficiency ratios (PER). Protein efficiency ratios were determined by the official method (AOAC, 1980) using male, weanling Sprague-Dawley derived rats from Simonsen Laboratories, Inc., Gilroy, California. Each assay ran 28 days. The composition of diets was: protein, 10%; corn oil, 8%; water, 5%; cellulose (Alphacel), 3%; complete vitamin mixture, 2%; mineral mixture, 5%; glucose, 20%; and cornstarch to make 100%. Nitrogen digestibility was based on feed intake and fecal collections from the 14th to 21st day on test. The following nutritional parameters were evaluated:

PER: weight gain/protein intake.

Apparent Nitrogen Digestibility: (N intake - fecal N)/N intake x 100.

Mean body weight, feed consumption, and PER were analyzed statistically by Duncan's Multiple Range Test (Duncan, 1959).

RESULTS AND DISCUSSION

Inactivation of Enzyme Inhibitors by Thiol Amino Acids

Figure 1 shows the effect of heating STI in the presence or absence of N-acetyl-L-cysteine (NAC) in the temperature range 25°C to 93°C. The data are drawn in Figure 1 both in absolute terms (number of actual trypsin units titrated) and in percent decrease in STI activity as a function of temperature. Figure 2 illustrates an analogous study for STI in soy flour.

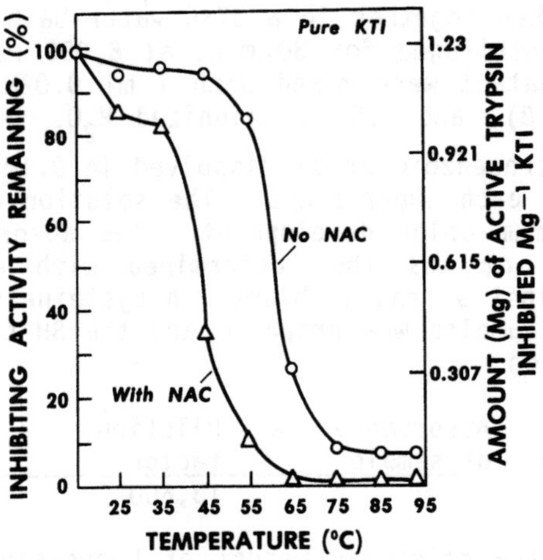

Figure 1. Effects of heat and N-acetyl-L-cysteine (NAC) on trypsin inhibiting activity of pure soybean trypsin inhibitor. The 100% value is for untreated inhibitor.

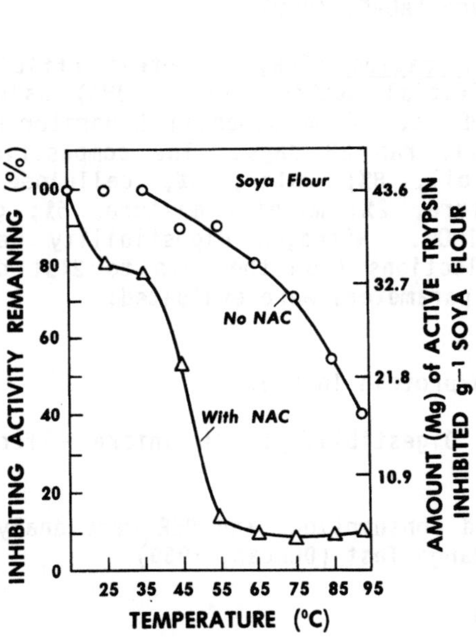

Figure 2. Effects of heat and N-acetyl-L-cysteine (NAC) on trypsin inhibiting activity of soybean trypsin inhibitor in raw soy flour. The 100% value is for untreated soy flour.

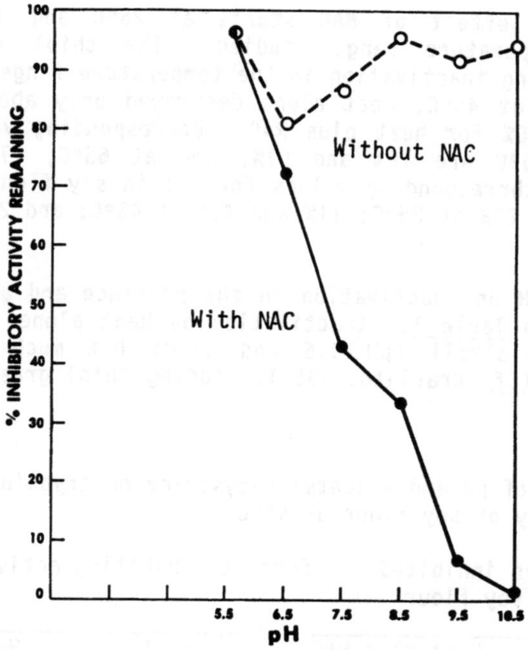

Figure 3. Effects of pH, heat (75°C), and N-acetyl-L-cysteine (NAC) on trypsin inhibiting activity of commercial lima bean inhibitor (Bowman-Birk-type).

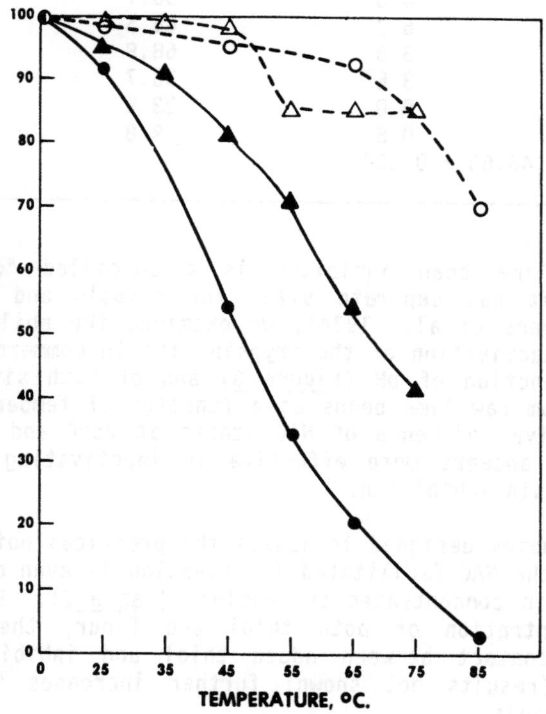

Figure 4. Effects N-acetyl-L-cysteine (NAC) on trypsin and chymotrypsin inhibiting activity in lima bean flour. Key: O, trypsin inhibitors, heat only; Δ, chymotrypsin inhibitor, heat only; ●, trypsin inhibitor, heat plus NAC; and Δ, chymotrypsin inhibitor, heat plus NAC.

The synergistic effect of NAC starts at 25°C and is maintained throughout the temperature range studied. The thiol is especially effective in enhancing inactivation in the temperature range 35°C to about 75°C. Thus, at 45°C, heat alone destroyed only about 5% of STI compared to about 60% for heat plus NAC. Corresponding values without and with NAC at 55°C are 16% and 87%, and at 65°C, 27% and 97.5%, respectively. The corresponding values for STI in soy flour without and with NAC are 0% and 23% at 35°C; 11% and 47% at 45°C; and 20% and 90% at 65°C.

The effect of pH on inactivation in the presence and absence of NAC at 75°C is shown in Table 1. Inactivation by heat alone is favored in acid (pH 3.5) and alkali (pH 9.5 and 11.0) but much less in the intermediate range (*Cf*. Craviato, 1951). Adding thiol greatly increases inactivation.

Table 1. Effect of pH and N-acetyl-L-cysteine on trypsin-inhibitory activity of soy flour at 45°C

pH	Mg active trypsin inhibited per gram soy flour		Percent inhibiting activity remaining	
	Heat	Heat plus NAC	Heat	Heat plus NAC
3.5	21.0	11.8	45.1	25.4
4.4	31.8	8.0	68.2	17.2
5.3	36.6	13.6	78.6	29.2
6.3	37.3	4.3	80.1	9.2
7.4	31.0	6.3	66.5	13.6
8.5	32.1	3.8	68.8	8.1
9.5	18.0	3.6	38.7	7.8
10.0	11.0	0.0	23.6	0
11.0	4.5	0.8	9.8	1.7
Untreated controls 46.63 ± 0.630				

Because the lima bean inhibitor is a so-called "double-headed" inhibitor, *i.e*. it has separate sites for trypsin and chymotrypsin, respectively (Stevens et al., 1974), we examined the ability of NAC to facilitate heat inactivation of the trypsin site in commercial lima bean inhibitor as a function of pH (Figure 3) and of both sites of LBI in flour prepared from raw lima beans as a function of temperature (Figure 4). The cooperative influence of NAC starts at 25°C and is maintained up to 85°C. NAC appears more effective in inactivating the trypsin than the chymotrypsin inhibition.

Additional studies designed to assess the practical potential of the method show that the NAC-facilitated inactivation is even more effective when carried out in concentrated suspensions (Table 2). Evidently, the higher the concentration of both thiol and flour, the greater the probability for contact between added thiol and inhibitor disulfide bonds. Shaking (results not shown) further increases the extent of inactivation by thiols.

Effect on disulfide content. Forming mixed disulfides with cysteine or NAC would increase the cystine content of the proteins. This is a desirable objective because methionine is a limiting amino acid in legumes and cystine has a sparing effect on methionine. This expectation was also realized, since our studies show that the half-cystine contents of treated commercial LBI and STI and of lima bean and soy flours are significantly higher than the corresponding values for the starting materials (Table 3).

Table 2. Enhancement of inactivation of trypsin inhibitors in legumes with increasing concentration of N-acetyl-L-cysteine (NAC)[a]

NAC (g)	% inhibitory activity remaining: lima bean	soybean	Great Northern bean
0	75.6	44.2	50.0
1	25.7	11.4	15.1
2	19.6	4.6	8.5
3	13.0	0.0	2.5
4	13.7	0.0	0.0
5	10.3	0.0	0.0

[a]Conditions: 100 g flour in 500 ml (slurry) Tris pH 8.5 buffer; 65°C; 1 hr.

Table 3. Half-cystine content of treated and untreated materials

Material	Treatment	Half-cystine (nmoles/mg)[a]
Lima bean inhibitor	dialyzed in H_2O	855.0
Lima bean inhibitor	36 mg LBI + 96 mg NAC; pH 8.5; 45°C; 1 hr.	1717.6
Lima bean flour	300 mg flour; pH 8.5; 65°C; 1 hr.	27.1
Lima bean flour	300 mg flour + 34 mg NAC; pH 8.5; 65°C; 1 hr.	130.5
Soy flour	dialyzed in H_2O	67.0
Soy flour	700 g + 21 g NAC; pH 8.5; 45°C; 1 hr.	139.4

[a]Determined on an amino acid analyzer as cysteic acid after performic acid oxidation and acid hydrolysis.

The action of thiols such as cysteine in inhibiting the activity of disulfide-containing enzyme inhibitors is postulated to involve formation of mixed disulfides among added thiols, enzyme inhibitors, and structural proteins. Since oxygen was not excluded, formation of new disulfide bonds probably proceeded by sulfhydryl-disulfide interchange (equations 1-5 in Scheme 1) and oxygen-catalyzed oxidation of two SH groups to an S-S bond. One of the sulfurs of the mixed disulfide originated from the protein and the other from added sulfhydryl compound. The added compound, therefore, disappears and becomes part of the protein structure. Because of structural alterations due to formation of mixed disulfide, the modified inhibitors lose their ability to complex with the active sites of trypsin and other proteolytic enzymes.

Scheme 1. Sulfhydryl-disulfide and oxidation interchange pathways. Net effects: network of new disulfide bonds and altered protein configuration.

$$\text{R-SH} + \text{In-S-S-In} \rightleftharpoons \text{R-S-S-In} + \text{HS-In} \quad (1)$$

$$\text{R-SH} + \text{Pr-S-S-Pr} \rightleftharpoons \text{R-S-S-Pr} + \text{HS-Pr} \quad (2)$$

$$\text{In-SH} + \text{PR-S-S-Pr} \rightleftharpoons \text{In-S-S-Pr} + \text{HS-Pr} \quad (3)$$

$$\text{In-SH} + \text{HS-Pr} + \tfrac{1}{2}\,O_2 \longrightarrow \text{In-S-S-Pr} + H_2O \quad (4)$$

$$\text{In-SH} + \text{HS-In} + \tfrac{1}{2}\,O_2 \longrightarrow \text{In-S-S-In} + H_2O \quad (5)$$

R-SH: Added thiol (Cysteine, N-acetyl-L-cysteine, etc.)
In-S-S-In: Inhibitor (In) disulfide bonds.
Pr-S-S-Pr: Protein (Pr) disulfide bonds.

Although equations 1-3 in Scheme 1 are, in principle, reversible, statistical considerations (Friedman, 1973) suggest a low probability for the regeneration of the native inhibitor configuration from the complex new network of mixed disulfide bonds. However, formation of mixed disulfide bonds may not occur with inhibitors with no or low amounts of disulfide bonds.

Additional evidence for the postulated reaction scheme was obtained by measuring the cysteine (SH) content of treated soybean flours by the Ellman procedure (Ellman, 1959). The results in Table 4 show that not only was there no increase in SH content, as would be expected if the added thiols did not participate in the indicated sulfhydryl-disulfide interchanges, but that the cysteine (SH) content of treated materials was also less than the value found for native soybean flour. This result, together with the observed increase in the total half-cystine content of the products determined by amino acid analysis, implies that the added cysteine or acetylcysteine (1) participated in the indicated sulfhydryl-disulfide interchange and oxidation reactions and (2) became covalently attached to soybean protein chains in the form of mixed disulfides produced by SH and S-S groups from added thiol, enzyme inhibitors, and structural protein chains, as illustrated in Scheme 1.

Significant amounts of methionine are known to be converted to the nutritionally less available oxidized forms under common food processing conditions (Marshall et al., 1982). The beneficial effects of added thiols on the nutritional quality of soy flour in which methionine is a limiting amino acid could, in principle, also partly result from thiol-initiated reduction of methionine sulfoxide to methionine residues (Snow et al., 1975; Friedman, 1977). This is illustrated in the following equation:

$$\overset{\displaystyle O}{\overset{\|}{\text{P-S-CH}_3}} \quad + \quad 2\text{R-SH} \longrightarrow \text{P-S-CH}_3 \quad + \quad \text{R-S-S-R} \quad + \text{H}_2\text{O}$$

Methionine sulfoxide residue — Thiol — Methionine residue — Disulfide

We have no direct evidence for this suggestion, however, except for a previous demonstration with model compounds that the reaction is possible (Snow et al., 1975). In the analytical procedure used here, all methionine and any methionine sulfoxide are oxidized to methionine sulfone and reported as methionine. Additional studies are, therefore, needed to differentiate methionine and methionine sulfoxide levels of native and treated soy flours.

From both chemical and nutritional viewpoints, inactivation of a protease inhibitor by an added thiol via a sulfhydryl-disulfide interchange should be distinguished from the related reductive cleavage of disulfide bonds followed by alkylation of the generated SH groups (Friedman et al., 1980; Zahnley and Friedman, 1982), illustrated in Figure 5. The former treatment produces new disulfide bonds. The latter eliminates the original ones from the protein. As mentioned earlier, possibilities for sulfhydryl-disulfide interchange include interaction between inhibitor SH or S-S bonds with soybean protein SH and S-S groups, respectively, to form a complex network that may make it difficult for the original disulfide bonds to be reformed (see below). The relative concentration of SH and S-S bonds in the thiol and protein may dictate whether the thiol is considered a catalyst or stoichiometric reagent in the inactivation. For these reasons, optimum conditions for the inactivation of disulfide-containing protease inhibitors by thiols have to be established in each case.

Moreover, the introduction of new half-cystine residues into legume proteins is an additional, desirable benefit because cystine has a sparing effect on methionine (Frost and Mann, 1966; Friedman et al., 1984 a, b).

Sulfur Amino Acids, Trypsin Inhibitors, and Nutritional Quality

Soybean trypsin inhibitor contains two disulfide bonds per mole (mol. wt. 21,500) while lima bean inhibitor (mol. wt. 9000) contains seven. The disulfide content of the Bowman-Birk-type lima bean inhibitor is nearly 10 times greater than the Kunitz-type soybean inhibitor.

The role of sulfur amino acids in soybean nutrition has been examined by numerous investigators (Hayward et al., 1936; Hayward and Hafner, 1941; Liener et al., 1949; Evans et al., 1947; McGinnis and Evans, 1947; Evans and McGinnis, 1948; Carol et al., 1953; Leveille et al., 1961; Borchers, 1962; Barnes et al., 1962, 1965a,b; Kwong and Barnes, 1963; Frost and Mann, 1966; Kakade et al., 1969; Shannon et al., 1972; Stockland et al., 1973; James and Hove, 1980; Naim et al., 1982). A general agreement seems to have evolved on the following aspects: (1) The biological or nutritional availability of both cystine and methionine in raw soybean meals is much less than in heated meals. This is ascribed to poor digestibility. (2) Fortification of raw or heated soybean meal with methionine results in significantly improved nutritional quality. The improvement is greater with raw than with heated soy products. However, there is no general agreement as to

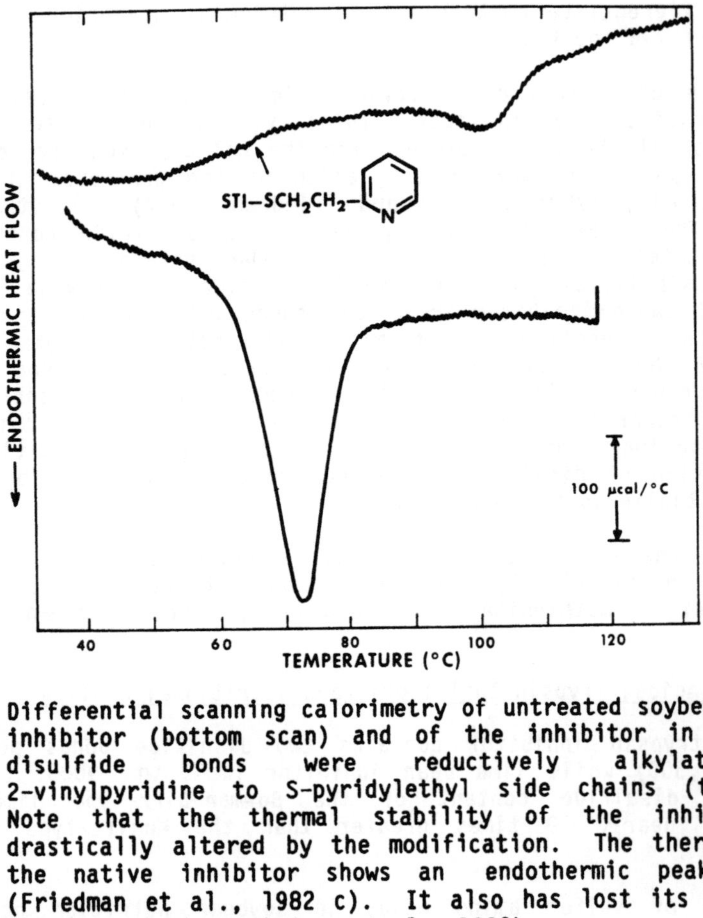

Figure 5. Differential scanning calorimetry of untreated soybean trypsin inhibitor (bottom scan) and of the inhibitor in which the disulfide bonds were reductively alkylated with 2-vinylpyridine to S-pyridylethyl side chains (top scan). Note that the thermal stability of the inhibitor was drastically altered by the modification. The thermogram for the native inhibitor shows an endothermic peak at 73°C (Friedman et al., 1982 c). It also has lost its ability to inhibit trypsin (Friedman et al., 1980).

Table 4. Trypsin inhibitor, SH, half-cystine, and methionine content of treated flours

Protein Source	Mg trypsin inhibited/g soy flour	SH content[a] (nanomoles/mg)	Half-cystine[b] (nanomoles/mg)	Methionine[c] (nanomoles/mg)
Native soy flour	37.5 ± 2.66	3.85 ± 0.12[d]	71.6	49.6
Soy flour, heated at 45°C	15.83 ± 0.77	1.29 ± 0.17	73.1	51.5
Soy flour plus cysteine heated at 45°C	6.24 ± 0.16	1.67 ± 0.57	108.9	59.7
Soy flour plus acetyl-L-cysteine, heated at 45°C	4.76 ± 0.04	1.32	147.5	51.0
Soy flour heated at 65°C	8.18 ± 0.12	1.12 ± 0.18	59.1	57.0
Soy flour plus cysteine heated at 65°C	2.87 ± 0.53	0.89 ± 0.16	108.2	57.0
Soy flour plus acetyl-cysteine, heated at 65°C	1.48 ± 0.31	0.85	86.2	47.4
Soy flour, heated at 75°C	5.84 ± 0.08	0.48 ± 0.03	63.5	50.0
Soy flour plus cysteine, heated at 75°C	2.23 ± 0.04	0.52 ± 0.16	133.9	53.8

[a]Determined by the Ellman procedure.

[b]Determined as cysteic acid.

[c]Determined as methionine sulfone.

[d]Average ± standard deviation from two separate experiments. One set of samples was fed to rats.

whether analogous fortification with cystine significantly improves nutritional quality (Barnes et al., 1962; James and Hove, 1980; Hayward and Hafner, 1941; Stockland et al., 1973).

In a preliminary study (Figures 6 and 7), we found that rats fed raw soy protein isolate supplemented with L-methionine gained much more weight than rats fed the raw isolate without methionine. This increase was observed at four protein levels (7.5, 10, 15, and 20%) supplemented with 0.3% L-methionine. In a parallel study (Figures 6 and 7) we found that analogous supplementation of raw soy protein isolate with L-cystine also resulted in significantly increased weight gain in rats. This weight gain was less than that observed with L-methionine, however.

Since high levels of L-methionine may be toxic (Benevenga, 1977; Doyle and Adams, 1980; Friedman and Gumbmann, 1984, 1985), cystine supplementation may be preferred in some cases to achieve nutritional improvement of legumes. As mentioned above, L-cystine (but not D-cystine, Friedman and Gumbmann, 1984) nutritionally spares L-methionine (Womack and Rose, 1941; Stipanuk and Benevenga, 1977; Smith el al., 1983).

Besides inactivating trypsin inhibitors, modifying disulfide bonds on inhibitors and associated structural proteins should also improve digestibility and nutritional quality since disulfide cleavage is expected to decrease protein crosslinking. The mixed-disulfide proteins should be more accessible to proteolytic enzymes, thus permitting more rapid digestion.

This expectation was realized since measurement of *in vitro* digestibilities of NAC-treated soy flour by trypsin (Table 5) show that although NAC had no effect at 25°C, its beneficial influence at higher temperatures was striking. Treatment with NAC at 45°C increases the bonds cleaved by trypsin per mole of soy protein from about 6 to 10; treatment at 75°C, from about 6 to 18. This last value is nearly equal to all the available trypsin-susceptible arginine and lysine residues in soy proteins.

Table 5. Effect of temperature and N-acetyl-L-cysteine (NAC) on trypsin inhibitor activity and trypsin digestion of dialyzed soy flour

Temperature (°C)	Percent inhibitory activity remaining		Number of peptide bonds cleaved by trypsin/mole of protein[a]	
	Heat	Heat plus NAC	Heat	Heat plus NAC
25	98.7		5.6	
25		84.4		4.9
45	82.5		3.9	
45		47.2		10.6
75	53.2		6.4	
75		0.0		18.2

[a]Based on calculated protein content (N X 6.25) and an average molecular weight per protein chain of 23,000.

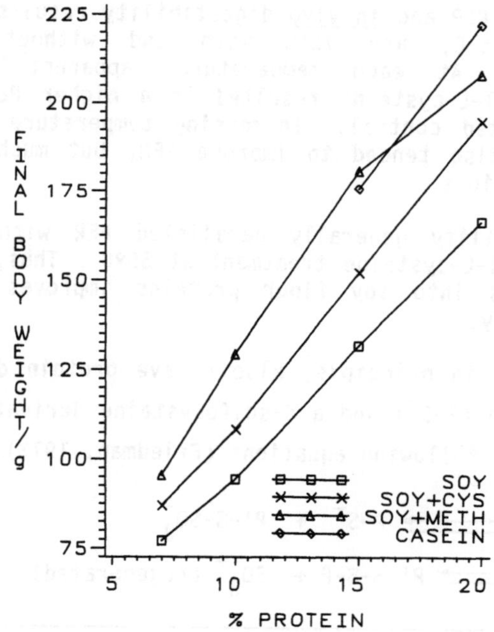

Figure 6. Effects of adding 0.3% L-cystine (Cys) or L-methionine (Meth) to raw soy protein isolate on the weight gain of rats. Weight gain with L-methionine at four protein levels (7.5, 10, 15, and 20%) was greater than corresponding gains with L-cystine.

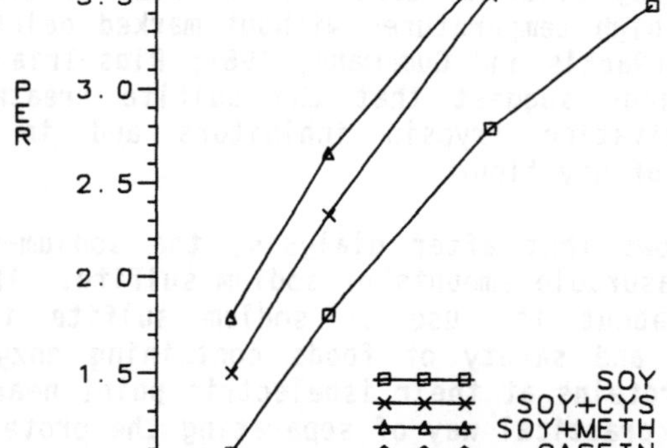

Figure 7. Effects of L-cystine (Cys) or L-methionine on the nutritional quality of raw soy protein isolate as measured by the protein efficiency ratio (PER). Note that the L-methionine-fortified soy protein was equivalent to casein in nutritional quality.

Table 6 shows the PER and *in vivo* digestibility results for soy flour heated at 45°C, 65°C, and 75°C with and without cysteine or N-acetyl-L-cysteine. At each temperature, apparent attachment of cysteine or N-acetyl-L-cysteine resulted in a higher PER than in the corresponding untreated control. Increasing temperature by itself, as would be expected, also tended to improve PER, but much less than in protein treated by thiols.

Nitrogen digestibility generally paralleled PER with the possible exception of N-acetyl-L-cysteine treatment at 65°C. Thus, incorporation of cysteine residues into soy flour proteins improves both PER and nitrogen digestibility.

Sulfite ions can, in principle, also cleave protein disulfide bonds to form a thiol anion ($P\text{-}S^-$) and a S-sulfocysteine derivative ($P\text{-}S\text{-}SO_3^-$), as illustrated by the following equations (Friedman, 1973):

$$P\text{-}S\text{-}S\text{-}P' + SO_3^{--} \rightleftharpoons P\text{-}S^- + P'\text{-}S\text{-}SO_3^-$$

$$P'\text{-}S\text{-}SO_3^- + P\text{-}S^- \rightleftharpoons P'\text{-}S\text{-}S\text{-}P + SO_3^{--} \text{ (regenerated)}$$

Net effect: rearrangement of disulfide bonds catalyzed by SO_3^{--}.

The S-sulfocysteine can interact further with the generated $P\text{-}S^-$ to form a new disulfide bond and SO_3^{--}. The net effect is a rearrangement of protein disulfide bonds catalyzed by SO_3^{--}.

Thus, exposure of disufide-containing trypsin inhibitors to sulfite ions should alter their structure and thus their inhibitory properties, in analogy with the cysteine treatments described earlier. This hypothesis was tested in a preliminary experiment by exposing soy flour to 0.03 M sodium sulfite and feeding the dialyzed and lyophilized materials to rats.

Table 7 shows that the sulfite treatment at 75°C lowers the trypsin inhibitor level in soy flour to zero. This value is usually difficult to achieve even at high temperatures without marked deleterious effects on protein quality (Rackis and Gumbmann, 1981; Rios-Iriarte and Barnes, 1966). These findings suggest that the sulfite treatment is highly effective in inactivating trypsin inhibitors and in improving the nutritional quality of soy flour.

Table 7 also shows that after dialysis, the sodium-sulfite-treated soy flour had no measurable amounts of sodium sulfite. This finding may alleviate concern about the use of sodium sulfite to improve the nutritional quality and safety of foods containing enzyme inhibitors. Precipitating soy proteins at their isoelectric point near pH 4.5 (Wolf, 1977) may be a more practical way of separating the proteins from sodium sulfite.

Table 8 shows that heat significantly improved protein quality of raw soy flour. Treatment with sodium sulfite further increased the protein efficiency ratio (PER). This improvement in protein quality is also reflected in an increase in *in vivo* nitrogen digestibilities. Table 8 shows that pancreas weights were elevated in rats fed raw soy flour, but not in those fed heated soy flour, with or without sodium sulfite.

Table 6. Effect of soy flour treated with L-cysteine and N-acetyl-L-cysteine on PER and nitrogen digestibility[1]

Diet	Soy Flour Treatment: Reaction Temperature	Soy Flour Treatment: Amino Acid	Final Body Weight (grams)	Feed Consumption (grams)	Protein Efficiency Ratio (PER)	Nitrogen Digestibility (%)
Experiment 1						
1	45°C	none	77[d]	219[d]	.95[e]	73.7[e]
2		Cysteine	113[b]	282[bc]	2.01[c]	81.7[bc]
3		Acetyl-cysteine	110[b]	242[cd]	2.20[c]	82.7[b]
4	65°C	none	97[c]	250[bcd]	1.61[d]	79.9[c]
5		Cysteine	123[b]	275[bc]	2.43[b]	82.9[b]
6		Acetyl-cysteine	115[b]	285[b]	2.02[c]	77.0[d]
7	Casein Control		156[a]	327[a]	3.05[a]	93.3[a]
Pooled Standard Error			±4	±13	±.07	±0.7[2]
Experiment 2						
1	75°C	none	118[b]	289[a]	2.14[c]	79.0[c]
2		Cysteine	123[b]	262[a]	2.53[b]	82.7[b]
3	Casein Control		154[a]	308[a]	3.19[a]	94.6[a]
Pooled Standard Error			±5	±16	±.08	±0.7[2]

[1]Means without a superscript letter in common are significantly different, $P < 0.05$, within each experimnent. Five rats per group. Mean initial body weight was 57g, exp. 1 and 56g, exp. 2.

[2]Recent average of historical standard error for digestibility assay in this laboratory based on 78 determinations on 39 diets.

Sulfur amino acids also may play an important role in the biofeedback mechanism that has been proposed to explain the adverse effects of trypsin inhibitors on the pancreas (Gallaher and Schneeman, 1984). According to this hypothesis, complexation in the intestinal tract between proteolytic enzymes (trypsin, chymotrypsin) and enzyme inhibitors (Kunitz- and Bowman-Birk-type) creates a deficiency of proteolytic enzymes. This deficiency triggers an endocrine (cholecystokinin, gastrin) sensing mechanism (Temler and Mettraux, 1986) which in turn induces increased protein synthesis in the pancreas. In this chain of events, undenatured trypsin inhibitors and proteolytic enzymes secreted by the pancreas, which are both rich in sulfur amino acids, are lost nutritionally through a tightly bound enzyme-inhibitor complex. This loss of sulfur amino acids is predicted to be an important modulating factor for the pancreas which is now under hormonal stimulation to increase synthesis of protein especially rich in sulfur amino acids.

In the preliminary study cited earlier, Gumbmann and Friedman found that feeding raw soy protein isolate fortified with L-methionine or L-cystine resulted in significantly heavier rat pancreata than when the same diets were fed without added sulfur amino acids. This result may reflect a preferential ability of the pancreas to mobilize needed sulfur amino acids for pancreatic growth (and synthesis of sulfur amino acid-containing enzymes such as trypsin and chymotrypsin) over other body organs. These observations suggest the need for additional studies to clarify the role of sulfur amino acids in pancreatic physiology and histology as related to trypsin inhibitors in the diet and their mechanism of action.

Table 7. Trypsin inhibitor, nitrogen and sulfite content of raw, heated, and sulfite-treated soy flour

Material	TIU/g	Sulfite (%)[c]
Soy flour, raw	50.43	0.0
Soy flour, heated[a]	9.72	0.0
Soy flour, sulfite-treated[b]	0.0	0.0

[a]Conditions: Soy flour (700g) was dispersed in 2.1 liters of prewarmed 0.5 M Tris buffer, pH 8.5, and heated with stirring at 75°C for 1 hr. The contents were then cooled in ice water, dialyzed for three days against water and lyophilized.

[b]Conditions: As in a above, except added 0.03 moles (3.78g) sodium sulfite before heat treatment.

[c]Analyzed according to sulfur dioxide in fruits, Industrial and Engineering Chemistry, 17 (11), 682-686.

Table 8. PER and digestibility of raw, heated, and sulfite-treated soy flour[1]

Diet[2]	PER[(2)]	Digestibility Diet (%)	Digestibility Nitrogen (%)	Pancreas Weight (% body weight)
Raw soy flour	1.55[d]	90.9[b]	78.3[c]	0.51[a]
Raw soy flour (dialyzed)	1.65[d]	90.9[b]	79.2[c]	0.56[a]
Heated soy flour	2.11[c]	91.3[b]	81.9[b]	0.43[b]
Heated soy flour (±0.03M sulfite)	2.49[b]	91.7[b]	84.0[b]	0.40[b]
ANRC Casein	3.44[a]	94.4[a]	93.0[a]	0.40[b]
S.E.	± 0.06	±0.4[3]	±0.7[3]	± 0.02

[1]Means without a superscript letter in common are significantly different. $P < 0.01$, PER; $P < 0.05$, digestibility and pancreas weight. Five rats per group.
[2]Diets contained 10% protein. Flour heated to 75°C.
[3]Recent average of historical standard error for digestibility assay in this laboratory based on 78 determinations on 39 diets.

Immunochemical Aspects

Brandon et al. (1985) prepared antibodies by inoculating rabbits with native, heat-denatured, and N-acetylcysteine-treated Kunitz soybean trypsin inhibitors. Immunochemical studies using a competitive solid-phase enzyme immunoassay and two groups of sera revealed two distinct patterns of antigenicity. Preliminary results show that antibodies elicited with the denatured inhibitor were specific for the denatured conformation of the protein. In contrast, native inhibitor elicited antibodies that selectively recognized determinants in both native and heat-treated protein, but that did not bind trypsin inhibitors treated with N-acetylcysteine. These results imply that: (1) the integrity of the disulfide bonds must be intact to maintain the native antigenic conformation and (2) the cysteine treatment may suppress possible allergic manifestations of soybean trypsin inhibitors and possibly other legume proteins.

Inactivation Of Hemagglutinins In Lima Beans

Hemagglutinins, or lectins are glycoproteins present in legumes and other plant foods (Liener, 1974, 1975). The consumption of raw or cooked kidney beans and other legumes can induce gastrointestinal disturbances, vomiting, weight loss, increases in immune response, and other adverse nutritional and toxic mansifestations in some animals and humans (Grant et al., 1985; Reaidi et al., 1981; Nachbar and Oppenheim, 1980; Noah et al., 1980; Pusztai et al., 1981).

Lectins from different sources differ in their susceptibility to heat inactivation (Reaidi et al., 1981; Zahnley, 1984). Since lectins exert their biological effects by binding to receptor sites of cells (Goldstein and Hayes, 1978; Lotan and Sharon, 1977), it was of both theoretical and practical interest to establish whether altering the structural integrity of a cyst(e)ine-containing lectin would result in a decrease in lectin-cell interactions.

To test this hypothesis, Wallace and Friedman (1985) treated lima bean flour with N-acetyl-L-cysteine in the temperature range 25°C to 85°C for 1 hr. They found that the sulfhydryl compound facilitated heat inactivation of the lectins (Table 9). Little activity was lost in flours heated for 1 hr up to 65°C. Above 65°C, activity decreased markedly. Under all conditions tested N-acetyl-L-cysteine inactivated the lectins at temperatures lower than required in its absence. The results also showed that at elevated temperatures, the lima bean lectin was most stable at pH 5 to 7, values often used in food processing.

In contrast, adding N-acetyl-L-cysteine to soybean flour within the same range of concentrations did not enhance the heat-induced inactivation of the soybean lectins. Since lima bean lectins contain cysteine or cystine residues and soybean lectins do not (Lotan and Sharon, 1977; Lotan et al., 1974), these contrasting results suggest that N-acetyl-L-cysteine facilitated the heat-inactivation of lima bean lectin by a sulfhydryl-disulfide interchange mechanism analogous to that cited earlier for the inactivation of trypsin inhibitors.

Table 9. The effect of N-acetyl-L-cysteine on lima bean lectin activity measured as the minimum amount of flour protein causing hemagglutination in 1 hr at pH 7 and 70°C

Activity (μg flour protein/50 μL)					
		N-Acetyl-L-cysteine (mM)			
Temperature (°C)	0	6.7	33	67	101
60	0.095	0.19	0.38	0.38	0.76
65	0.095	3.0	b	3.0	7.6
70	0.76	6.0	6.0	19	b
75	1.2	9.6	a	19	19
80	60	b	a	b	b

[a] Not tested.

[b] Activity values were 100 or greater, which represent less than 0.1% original activity.

Based on these findings, Wallace and Friedman (1985) suggest that lowering the temperature of inactivation of cysteine and cystine-containing lectins could enhance nutritional quality of composite flours through improved amino acid balance and decreased protein damage during processing.

CONCLUSIONS

Treatment of raw soy flour with cysteine or N-acetyl-L-cysteine results in the introduction of new half-cystine residues into native proteins, with a corresponding improvement of nutritional quality. The proteins are modified through formation of mixed disulfide bonds among added thiols, protease inhibitors, and structural proteins. This leads to decreased inhibitory activity and increased protein digestibility and nutritive value. Cysteine also facilitated heat inactivation of hemagglutinins in lima bean flour. Exposure of soy flour to sodium sulfite was also nutritionally beneficial.

Since most naturally occurring food toxicants and many antinutrients, including enzyme inhibitors and lectins, possess specific sites that are responsible for their adverse effects, modifying such sites with site-specific reagents will alter structural integrity. Such changes in the toxicophoric sites should prevent physiological interaction of the food ingredients with receptor sites of enzymes or other biomolecules *in vivo*. The results of this and previous studies (Friedman, 1984; Friedman et al., 1984 a,b; Friedman and Smith, 1984; Friedman et al., 1982 d) demonstrate that such an approach is feasible with several classes of antinutrients and food toxicants and provide experimental evidence to facilitate formulating nutritionally optimal and safe food products.

Finally, it is noteworthy that enzyme inhibitors can prevent the development of oxygen radical-induced lesions of the mucosa of the feline intestine (Parks, 1985) and of colon cancer in mice (Troll et al., 1986). They also protect fish against bacterial damage during storage (Veerappa Gowda and Karunasagar, 1985). Although the molecular basis for these beneficial effects has not been elucidated, one possibility is that inhibitors or inhibitor-protease complexes act as free radical traps, whereby the free electrons on oxygen radicals are transferred or dissipated to the sulfur atoms of the sulfur-rich inhibitors or complexes. (For a review of relevant free radical reactions of proteins, see chapter 10 in Friedman, 1973). These considerations suggest the need for further studies to enrich our knowledge about possible beneficial effects of plant protease inhibitors.

ACKNOWLEDGEMENTS

We thank MacDonald C. Calhoun and Ok-koo K. Grosjean for excellent technical assistance.

Reference to brand or firm names does not constitute endorsement by the U. S. Department of Agriculture over others not mentioned.

REFERENCES

Anderson, R. L., Rackis, J. J., and Tallent, W. H. (1979). Biologically active substances in soy proteins. In: "Soy Protein and Human Nutrition", H. L. Wilcke, D. T. Hopkins, and D. H. Waggle, eds., Academic Press, New York, p. 209-233.

Anfinsen, C. B. (1972). The formation and stabilization of protein structure. Biochem. J., 128, 737-749.

Antunes, P. L. and Sgarbieri, V. C. (1980). Effect of heat treatment on the toxicity and nutritive value of dry bean (Phaseolus vulgaris) protein. J. Agric. Food Chem., 28, 935-938.

Asquith, R. S. and Otterburn, M. S. 1977. Cystine-alkali reactions in relation to protein crosslinking. In: "Protein Crosslinking: Nutritional and Medical Consequences", M. Friedman, ed., Plenum, New York, pp. 93-122.

AOAC (1980). "Official Methods of Analysis", 13th Edition. Assoc. of Official Analytical Chemists, Washington, D.C.

Baker, G. C. and Mustakas, G. C. (1973). Heat inactivation of trypsin inhibitor lipoxygenase and urease in soybeans: effect of acid and base additives. J. Am. Oil Chem. Soc., 50, 137.

Baker, E. C. and Rackis, J. J. (1986). Preparation of unheated soy protein isolates with low trypsin inhibitor content. This volume.

Barnes, R. H., Fiala, G., and Kwong, E. (1962). Methionine supplementation of processed soybeans in the rat. J. Nutrition, 77, 278-284.

Barnes, R. H., Fiala, G., and Kwong, E. (1965a). Prevention of coprophagy in the rat and the growth-stimulating effects of methionine, cystine and penicillin when added to diets containing unheated soybeans. J. Nutrition, 86, 127-131.

Barnes, R. H., Kwong, E., and Fiala, G. (1965b). Effect of penicillin added to an unheated soybean diet on cystine excretion in feces of the rat. J. Nutrition, 85, 123-126.

Benevenga, N. J. (1977). Toxicities of methionine and other amino acids. J. Agric. Food Chem., 25, 2-9.

Birk, Y. (1985). The Bowman-Birk inhibitor. Trypsin- and chymotrypsin-inhibitor from soybeans. Int. J. Peptide Protein Res., 25, 113-121.

Birk, Y. (1976a). Proteinase inhibitors of legumes seeds. Methods Enzymol., 45B, 697-700.

Birk, Y. (1976b). Trypsin isoinhibitors from garden beans (Phaseolus vaulgaris). Methods Enzymol., 45B, 710-716.

Birk, Y. (1976c). A trypsin and chymotrypsin inhibitor from groundnuts (Arachis hypogea). Methods Enzymol., 45B, 716-722.

Birk, Y. (1976d). Trypsin and chymotrypsin inhibitors from soybeans. Methods Enzymol., 45B, 700-707.

Boisen, S., Anderson, C. Y., and Hejgaard, J. (1981). Inhibitors of chymotrypsin and microbial serine proteases in barley grains. Plant Physiol., 52, 167-176.

Boonivust, S. and Whitaker, J.R. (1976). Effect of heat, amylase and disulfide bond cleavage on the in vitro digestibility of soybean proteins. J. Agric. Food Chem., 24, 1130-1135.

Borchers, R. (1962). Supplementary methionine requirement of weanling rats fed soybean oil meal. J. Nutrition, 77, 309-311.

Borchers, R., Mange, L. D., Nelson, S. O., and Stetson, L. E. (1972). Rapid improvement in nutritional quality of soybeans by dielectric heating. J. Food Sci., 37, 333-334.

Bozzini, A. and Silano, V. (1978). Control through breeding methods of factors affecting nutritional quality of cereals and legumes. In: "Nutritional Improvement of Food and Feed Proteins", M. Friedman, ed., Plenum, New York, pp. 249-274.

Bradbury, J. H., Hammer, B., Nguyen, T., Anders, N., and Millar, J. S. (1985). Protein quantity and quality and trypsin inhibitor content of sweet-potato cultivars from the highlands of Papua New-Guinea. J. Agric. Food Chem., 33, 281-285.

Brandon, D. L., Haque, S., and Friedman, M. (1985). Antigenicity of native and modified Kunitz soybean trypsin inhibitors. Abstracts 7th European Immunology Meeting, Jerusalem, Israel, September 8-13, 1985. See also this volume.

Buera, M. P., Pilosof, A. M. R., and Bartholomai, G. B. (1984). Kinetics of trypsin inhibitor activity loss in heated flour from bean, Phaseolus vulgaris. J. Food Sci., 49, 124-126, 136.

Buonocore, V. and Silano, V. (1986). Biochemical nutritional and toxicological aspects of alpha-amylase inhibitors from plant foods. This volume.

Burr, H. K. (1975). Pulse proteins. In: "Protein Nutritional Quality of Foods and Feeds", M. Friedman, ed., Marcel Dekker, New York, Part 2, pp. 119-134.

Carol, R. W., Hensly, G. W., Sittler, C. L., Wilcox, E. L., and Graham, Jr., W. R. (1953). Absorption of nitrogen and amino acids from soybean meal as affected by heat treatment or supplementation with aureomycin and methionine. Archiv. Biochem. Biophys., 45, 260-266.

Chae, B. J., Han, I. K., and Lee, S. C. (1984). Effect of heat treatment on amino acid bioavailability of soybean meal for broiler type chicks. Korean J. Animal Sci., 26, 558-562.

Chang, C. R. and Tsen, C. C. (1981a). Isolation of trypsin from rye, triticale, and wheat samples. J. Food Sci., 58, 207-210.

Chang, C.R. and Tsen, C.C. (1981b). Characterization and heat stability of trypsin inhibitors from rye, triticale, and wheat samples. J. Food Sci., 58, 211-213.

Chase, T., Jr. and Shaw, E. 1970. Titration of trypsin, plasmin, and thrombin with p-nitrophenyl-p'-guanidinobenzoate HCl. Methods Enzymol., 19, 20-27.

Collins, J. L. and Beaty, B. F. (1980). Heat inactivation of trypsin inhibitor in fresh green soybeans and physiological responses of rats fed the beans. J. Food Sci., 45, 542-546.

Craviato, R. (1951). Majorammeinto del valor nutritive de la tortilla Ciencia (Mexico), 11, 81.

Creighton, T. E. (1975). The two-disulfide intermediates and the folding pathways of reduced pancreatic trypsin inhibitor. J. Mol. Biol., 95, 167-199.

de Lumen, B. O. and Chan, J. (1986). Antinutritional and biochemical properties of winged bean trypsin inhibitors. This volume.

Doyle, P. T. and Adams, N. R. (1980). Toxic effects of large amounts of DL-methionine infused into the rumen of sheep. Austral. Veterinary Journal, 56, 331-334.

Duncan, D. B. (1959). Multiple range and multiple tests. Biometrics, 11, 1-45.

Ellenrieder, G., Geronazzo, H., and De Bojarski, A. B. (1980). Thermal inactivation of trypsin inhibitors in aqueous extracts of soybeans, peanuts, and kidney beans: presence of substances that accelerate inactivation. Cereal Chem., 57, 25-27.

Ellman, G. L. (1959). Tissue sulfhydryl groups. Archiv. Biochem. Biophys., 82, 70-77.

Erlanger, B. F., Kokowsky, N., Cohen, W. (1961). The preparation and properties of two new chromogenic substrates of trypsin. Archiv. Biochem. Biophs., 95, 271-278.

Evans, R. J., McGinnis, J., and St. John, J. L. (1947). The influence of autoclaving soybean oil meal on the digestibility of the proteins. J. Nutrition, 33, 661-672.

Evans, R. J. and McGinnis, J. (1948). Cystine and methionine metabolism by chicks receiving raw or autoclaved soybean oil meal. J. Nutrition, 35, 477-488.

Filho, J. M. (1974). Trypsin inhibitors in sorghum grain. J. Food Sci., 39, 422-423.

Fraenkel-Conrat, H., Bean, R. C., Ducay, E. D., and Olcott, H. S. (1952). Isolation and characterization of a trypsin inhibitor from lima beans. Archiv. Biochem. Biophys., 37, 393-407.

Friedman, M. (1973). "The Chemistry and Biochemistry of the Sulfhydryl Group in Amino Acids, Peptids, and Proteins", Pergamon Press, Oxford, England and Elmsford, New York, Chapters 3, 8, and 10.

Friedman, M. (1977). Chemical basis for pharmacological and therapeutic actions of penicillamine. Proc. Royal Soc. Medicine (supplement 3), 70, 50-60.

Friedman, M. (1984). Sulfhydryl groups and food safety. Adv. Exp. Med. Biol., 177, 31-63.

Friedman, M., Grosjean, O.K., and Zahnley, J. C. (1982a). Cooperative effects of heat and thiols in inactivating trypsin inhibitors from legumes in solution and in the solid state. Nutr. Repts. Int., 25, 743-751.

Friedman, M., Grosjean, O.K., and Zahney, J. C. (1982b). Inactivation of soybean inhibitors by thiols. J. Sci. Food Agric., 33, 165-172.

Friedman, M., Grosjean, O. K., and Zahnley, J. C. (1982c). Effect of disulfide bond modification on the structure and activities of enzyme inhibitors. In: "Mechanism of Food Protein Deterioration", J. P. Cherry, ed., ACS Symposium Series, Washington, D.C. 206, 359-407.

Friedman, M. and Gumbmann, M. R. (1984). The utilization and safety of isomeric sulfur-containing amino acids in mice. J. Nutrition, 114, 2301-2310.

Friedman, M. and Gumbmann, M. R. (1985). Isomeric methionine dipeptides as a nutritional source of methionine. Abstracts, 190th American Chemical Society Meeting, Chicago, Illinois.

Friedman, M., Gumbmann, M. R., and Grosjean, O. K. (1984a). Nutritional improvement of soy flour. J. Nutrition, 114, 2241-2246.

Friedman, M., Gumbmann, M. R., and Masters, P. M. (1984b). Protein-alkali reactions: chemistry, toxicology, and nutritional consequences. Adv. Exp. Med. Biol., 177, 367-412.

Friedman, M. and Smith, G. A. (1984). Inactivation of quercetin mutagenicity. Food Chem. Toxicol., 22, 535-539.

Friedman, M., Noma, A. T., and Wagner, J. R. (1979). Ion-exchange chromatography of sulfur amino acids on a single-column amino acid analyzer. Anal. Biochem., 98, 293-304.

Friedman, M., Wehr, C. M., Schade, J. E., and MacGregor, J. T. (1982d). Inactivation of aflatoxin B_1 mutagenicity by thiols. Food Chem. Toxicol., 20, 887-892.

Friedman, M., Zahnley, J. C., and Wagner, J. R. (1980). Estimation of the disulfide content of trypsin inhibitors as S-β-(2-pyridylethyl)-L- cysteine. Anal. Biochem., 106, 27-34.

Friedman, M., Zahnley, J. C., and Masters, P. M. (1981). Relationship between in vitro digestibility of casein and its content of lysinoalanine and D-amino acids. J. Food Sci., 46, 127-131.

Frost, A. and Mann, G. V. (1966). Effect of cystine deficiency and trypsin inhibitor on the metabolism of methionine. J. Nutrition, 89, 49-54.

Gallaher, D. and Schneeman, B. O. (1984). Nutritional and metabolic response to plant inhibitors of digestive enzymes. In: "Nutritional and Toxicological Aspects of Food Safety", M. Friedman, ed., Plenum Press, New York, pp. 299-322. See also this volume.

Goldstein, I. J. and Hayes, C. E. (1978). The lectins: carbohydrate binding proteins of plants and animals. Adv. Carbohydrate Chem. and Biochem., 35, 127.

Grant, G., Greer, F., McKenzie, N. H., and Pusztai, A. (1985). Nutritional response of mature rats to kidney bean (Phaseolus vulgaris) lectins. J. Sci. Food Agric., 36, 409-414.

Gupta, H. O., Lodha, M. L., Sen. K., Joginder, S., and Mehta, S. L. (1979). Protein, lipid, mineral and trypsin inhibitor content in modified opaque-2 maize. J. Food Sci. Technol. (India), 16, 1.

Hackler, L. R., Van Buren, J. P., Steinkraus, K. H., El Rawi, I., and Hand, D. B. (1965). Effect of heat treatment on nutritive value of soy milk protein fed to weanling rats. J. Food Sci., 30, 723-728.

Hammerstrand, G. E., Black, L. T., and Glover, J. D. (1981). Trypsin inhibitors in soy products: modification of the standard analytical procedure. Cereal Chem., 58, 42-45.

Hawkins, H. C. and Freedman, R. B. (1975). Randomly reoxidized soybean trypsin inhibitor and the possiblity of conformational barriers to disulfide isomerization in proteins. FEBS Letters, 58, 7-11.

Hayward, J. W. and Hafner, F. H. (1941). The supplementary effect of cystine and methionine upon the protein of raw and cooked soybeans as determined with chicks and rats. Poultry Science, 20, 139-150.

Hayward, J. W., Steenbeck, H., and Bohsted, G. (1936). The effect of cystine and casein supplements upon the nutritive value of the protein of raw and heated soy beans. J. Nutrition, 12, 275-283.

Heinrikson, R. L. and Kezdy, F. J. (1976). Acidic cysteine protease inhibitors from pineapple stem. Methods Enzymol., 45B, 740-751.

Hochstrasser, K., Illchmann, K. K., and Werle, E. (1970). Plant protease inhibitors. Amino acid sequence of specific trypsin inhibitors from maize, characteristics of the polymer. Z. Physiol. Chem., 351, 721-728.

Hojima, Y., Moriwaki, C., and Moriya, H. (1973). Studies of kallikrein inhibitors from potatoes. Characterization of two isolated inhibitors. J. Biochem. (Tokyo), 73, 933-943.

Hymowitz, T. (1986). Genetics and breeding of soybeans lacking the Kunitz trypsin inhibitor. This volume.

Iwasaki, T., Iguchi, I., Kiyohara, T.,and Yoshikawa, M. (1974). Digestibility of the active fragment of potato proteinase inhibitor IIa by bovine chymotrypsin and a bacterial proteinase. J. Biochem. (Tokyo), 75, 1387-1390.

James, K. A. and Hove, E. L. (1980). The ineffectivness of supplementary cystine in legume-based diets. J. Nutrition, 110, 1736-1744.

Jones, G., Moore, S., and Stein, W. H. (1963). Properties of chromatographically purified trypsin inhibitors from lima beans. Biochemistry, 2, 66-71.

Kakade, M. L., Arnold, R. L., Liener, I. E., and Waibel, P. E. (1969). Unavailability of cystine from trypsin inhibitors as a factor contributing to the poor nutritive value of navy beans. J. Nutrition, 99, 34-42.

Kakade, M. L., Rackis, J. J., McGhee, J. E., and Puski, G. (1974). Determination of trypsin inhibitor activity of soy products: a collaborative analysis of an improved procedure. Cereal Chem., 51, 376-382.

Kamalakannan, V. and Motlag, D. B. (1982). Influence of dietary black gram trypsin inhibitor on the growth rate and nitrogen absorption in male albino mice. Nutr. Repts. Int., 25, 113-125.

Kunitz, M. (1948). The kinetics and thermodynamics of reversible denaturation of crystalline soybean trypsin inhibitor. J. Gen. Physiol., 32, 241-263.

Kwong, E. and Barnes, R. H. (1963). Effect of soybean trypsin inhibitor on methionine and cystine utilization. J. Nutrition, 81, 392-398.

Leveille, G. A., Sauberlich, H. E., and Shockley, J. W. (1961). Sulfur amino acid requirements for growth of mice fed two levels of nitrogen. J. Nutrition, 75, 455-458.

Liener, I. E. (1974). Phytohemagglutinins: their nutritional significance. J. Agric. Food Chem., 22, 17-22.

Liener, I. E. (1975). Effects of antinutritional and toxic factors on the quality and utilization of legume proteins. In: "Protein Nutritional Quality of Foods and Feeds", M. Friedman, ed., Marcel Dekker, New York, Part 2, pp. 523-550.

Liener, I. E., Deuel, H. H., and Fevold, H. L. (1949). The effect of supplemental methionine on the nutritive value of diets containing concentrates of the soybean trypsin inhibitor. J. Nutrition, 38, 325-339.

Liener, I. E. and Kakade, M. (1980). Protease inhibitors. In: "Toxic Constituents of Plant Foodstuffs", I. E. Liener, ed., 2nd Edition, Academic Press, New York, pp. 7-57.

Lineweaver, H. and Murray, C. W. (1949). Identification of the trypsin inhibitor of egg white with ovomucoid. J. Biol. Chem., 171, 565-581.

Lis, H. and Sharon, N. (1981). Lectins in higher plants. In: "The Biochemistry of Plants", A. Marcus, ed., Academic Press, New York, Chapter 10, p. 376.

Longnecker, M. L., Martin, W. H., and Sarrett, H. P. (1964). Improvement in the protein efficiency of soybean concentrates and isolates by heat treatment. J. Agric. Food Chem., 12, 411-412.

Lotan, R. and Sharon, N. (1977). Modification of the biological properties of plant lectins by chemical crosslinking. In: "Protein Crosslinking: Biochemical and Molecular Aspects", M. Friedman, ed., Plenum Press, New York, pp. 140-168.

Lotan, R., Siegelman, H. W., Lis, H., and Sharon, N. (1974). Subunit structure of soybean agglutinins. J. Biol. Chem., 249, 1219-1224.

Marshall, H. F., Chan, K. C., Miller, K. S., and Satterlee, L. D. (1982). Sulfur amino acid stability. Effects of processing on legume proteins. J. Food Sci., 47, 1170-1174.

Matsuda, T., Watanabe, K., and Nakamura, R. (1983). Ovomucoid and ovoinhibitor isolated from chicken egg white are immunologically cross-reactive. Biochem. Biophys. Res. Commun., 110, 75-81.

McGinnis, J. and Evans, R. J. (1947). Amino acid deficiencies of raw and heated soybean oil meal for chicks. J. Nutrition, 34, 725-732.

Menegatti, E., Palmieri, S., Walde, P., and Luisi, P. L. (1985). Isolation and characterization of a trypsin inhibitor from white mustard (Sinapis alba L.). J. Agric. Food Chem., 33, 784-789.

Mihalyi, E. (1978). "Application of Proteolytic Enzymes to Protein Structure Studies", CRC Press, Boca Baton, Florida, 2nd edition, p. 129.

Mikola, J., and Suolinna, E. M. (1969). Purification and properties of a trypsin inhibitor from barley. Europ. J. Biochem., 9, 555-560.

Mitchell, H. L. and Parrish, D. B. (1981). Effect of trypsin inhibitor from alfalfa meal on rat growth. Nutr. Repts. Int., 24, 799-801.

Moore, S. (1963). On the determination of cystine as cysteic acid. J. Biol. Chem., 238, 235-238.

Nachbar, M. S. and Oppenheim, J. D. (1980). Lectins in the United States diet: a survy of lectins in commonly consumed foods and a review of the literature. Amer. J. Clin. Nutr., 33, 2338-2345.

Naim, M., Gertler, A., and Birk, Y. (1982). The effect of dietary raw and autoclaved soya-bean protein fraction on growth, pancreatic enlargement and pancreatic enzymes in rats. British J. Nutrition, 47, 281-288.

Noah, N. D., Bender, A. E., Reaidi, G. B., and Gilbert, R. J. (1980). Food poisoning from raw red kidney beans. Brit. Medical J., p. 236.

Ogiso, T., Noda, T., Sako, Y., and Aoyama, M. (1975). Studies on trypsin inhibitor in barley. Purification and some properties. J. Biochem. (Tokyo), 78, 9-17.

Osuga, D. T., Bigler, J. C., Uy, R. L., Sjoberg, L., and Feeney, R. E. (1974). Comparative biochemistry of penguin egg-white proteins. 1. Ovomucoids: Composition and inhibitory activites for trypsin, α-chymotrypsin and subtilisin. Comp. Biochem. Physiol., 48B, 519-533.

Otlewski, J., Polanowski, A., Leluk, J., and Wilusz, T. (1984). Trypsin inhibitor in summer squash Cucurbita-Pepo seeds. Isolation, purification, and partial characterization of three inhibitors. Acta Biochim. Polonica., 31, 267-278.

Parks, D. A., Granger, D. N., Bulkley, G. B., and Shah, A. K. (1985). Soybean trypsin inhibitor attenuates ischemic injury to the feline small intestine. Gastroenterology, 89, 7-12.

Perry, L. J. and Wetzel, R. (1984). Disulfide bond engineered into T-4 lysozyme. Stabilization of the protein toward thermal inactivation. Science, 226, 555-557.

Powers, J.R. and Whitaker, J.R. (1977). Purification and some physical and chemical properties of red kidney bean (Phaseolus vulgaris) α-amylase inhibitor. J. Food Biochem., 1, 217-238.

Pusztai, A., Clarke, E. M. W., Grant, G., and King, T. P. (1981). The toxicity of Phaseolus vulgaris lectins. Nitrogen balance and immunochemical studies. J. Sci. Food Agric., 32, 1037-1046.

Rackis, J. J. and Gumbmann, M. R. (1981). Protease inhibitors: physiological properties and nutritional significance. In: "Antinutrients and Natural Toxicants in Foods", R. L. Ory, ed., Food & Nutrition Press, Westport, Connecticut. pp. 203-237.

Rackis, J. J., Wolf, W. J., and Baker, E. C. (1986). Protease inhibitors in plant foods: content and inactivation. This volume.

Reaidi, G. B., McPherson, L., and Bender, A. E. (1981). Toxicity of red kidney beans (Phaseolus vulgaris). J. Sci. Food Agric., 32, 846-847.

Rios-Iriarte, B. J. and Barnes, R. H. (1966). The effect of overheating on certain nutritional properties of the protein of soybeans. Food Technol., 20, 836-838.

Rupley, J. A. (1967). Susceptibility to attack by proteolytic enzymes. Meth. Enzymol., 11, 905-917.

Ryan, C. Y. and Hass, G. M. (1981). Structural, evolutionary and nutritional properties of proteinase inhibitors from potatoes. In: "Antinutrients and Natural Toxicants in Foods", R. L. Ory, ed., Food & Nutrition Press, Westport, Connecticut, pp. 169-185.

Sathe, S. K. and Salunkhe, D. K. (1981). Studies on trypsin and chymytropsin inhibitory activities, hemagglutinating activity, and sugars in the Great Northern bean (Phaseolus vulgaris L.). J. Food Sci., 46, 626-629.

Sedlak, J. and Lindsay, R. H. (1968). Estimation of total, protein-bound, and non-protein sulfhydryl groups in tissue with Ellman's reagent. Anal. Biochem., 25, 192-205.

Sen, L. C. and Whitaker, J.R. (1973). Some properties of a ficin-papain inhibitor from avian egg white. Arch. Biochem. Biophys., 158, 623-632.

Shannon, B. M., Howe, J. M., and Clark, H. E. (1972). Interrelationships between dietary methionine and cystine as reflected by growth, certain hepatic enzymes and liver composition of weanling rats. J. Nutrition, 102, 557-562.

Skrede, A. and Krogdahl, A. (1985). Heat affects nutritional characteristics of soybean meal and excretion of proteinases in mink and chicks. Nutr. Repts. Int., 32, 479-489.

Smirnoff, P., Khalet, S., Birk, Y., and Applebaum, S.W. (1979). Trypsin and chymotrypsin inhibitor from chick peas. Selective chemical modification of the inhibitor and isolation of two isoinhibitors. Int. J. Pept. Prot. Res., 14, 186-192.

Smith, C., Van Megen, W., Twaalfhoven, L., and Hitchcock, C. (1980). The determination of trypsin inhibitor levels in foodstuffs. J. Sci. Food Agric., 31, 341-350.

Smith, J. T., Acuff, R. V., Bittle, J. B., and Gilbert, M. L. (1983). A metabolic comparison of cysteine and methionine supplements in the diet of a rat. J. Nutrition, 113, 222-227.

Snow, J. T., Finley, J. W., and Friedman, M. (1975). Oxidation of sulfhydryl groups to disulfides by sulfoxides. Biochem. Biophys. Res. Commun., 64, 441-445.

Starcher, B., Bryant, L., and Elfant, M. (1986). Protease inhibitors of the marama bean. This volume.

Stevens, F.C., Wuerz, S., and Krahn, J. (1974). Structure-function relationships in lima bean protease inhibitor. In: "Proteinase Inhibitors", Proc. Int. Res. Conf., 2nd (Bayer Sump. V). H. Fritz, H. Tschesche, L. J. Greene, and E. Tschetsche, eds., Springer-Verlag, Berlin and New York.

Stipanuk, M. H. and Benevenga, N. J. (1977). The effect of cystine on the metabolism of methionine in rats. J. Nutrition, 107, 1455-1467.

Stockland, W. L., Meade, R. J., Wass, D. F., and Sowers, J. E. (1973). Influence of levels of methionine and cystine on the total sulfur amino acid requirement of the growing rat. J. Animal Sci., 36, 526-530.

Tan-Wilson, A. L. and Wilson, K. A. (1986). Relevance of multiple soybean trypsin inhibitor forms to nutritional quality. This volume.

Tanford, C. (1962). The interpretation of hydrogen ion titration curves of proteins. Adv. Protein Chem., 17, 69-165.

Temler, R. S. and Mettraux, C. (1986). Gastrin and cholecystokinin levels in rats fed soya bean trypsin inhibitors. This volume.

Troll, W., Frenkel, K., and Wiesner, R. (1986). Protease inhibitors: their role as modifiers of carcinogenic processes. This volume.

Tur-Sinai, A., Birk, Y., Gertler, A., and Rigbi, M. (1972). A basic trypsin and chymotrypsin-inhibitor from ground nuts (Arachis hypogaea). Biochim. Biophys. Acta, 263, 666-672.

Veerappa Gowda, H.S. and Karunasagar, I. (1985). Effect of protease inhibitors from Adenathera pavonia on the biochemical and microbiological quality of fishes. J. Sci. Food Agric., 36, 113-119.

Udupa, W. and Pattabiraman, T. N. (1984). Natural plant enzyme inhibitors. Studies on a proteinase inhibitor from Italian millet (Setaria italica). J. Sci. Food Agric., 35, 679-688. See also this volume.

Wagner, L. P. and Riehm, J. P. (1967). Purification and partial characterization of a trypsin inhibitor isolated from navy beans. Arch. Biochem. Biophys., 121, 672-677.

Wallace, J. M. and Friedman, M. (1985). Inactivation of hemagglutinins in lima bean (Phaseolus lunatus) flour by N-acetyl-L-cysteine, pH, and heat. Nutr. Repts. Int., 32, 748-753.

Wang, D. (1975). A crystalline protein-proteinase inhibitor from pinto bean seeds. Biochim. Biophys. Acta, 393, 483-496.

Warsy, A. S., Norton, G., and Stein, M. (1974). Protease inhibitors from broad bean; isolation and purification. Phytochemistry, 13, 2481-2486.

Whitaker, J. R. (1981). Naturally occuring peptide and protein inhibitors of enzymes. In: "Impact of Toxicology on Food Processing", J. C. Ayres and J. C. Kirschman, eds., AVI, Westport, Connecticut, pp. 57-104.

Wilcox, P. E. (1979). Chymotrypsinogens-chymotrypsin. Methods Enzymol., 19, 64-80.

Wolf, W. J. (1977). Legumes: seed composition and structure, processing into protein products and protein properties. In: "Food Proteins", J. R. Whitaker and S. R. Tannenbaum, eds., Avi, Westport, Connecticut, pp. 291-314.

Womack, M. and Rose, W. C. (1941). The partial replacement of dietary methionine by cystine for purposes of growth. J. Biol. Chem., 141, 375-379.

Yadav, N. R. and Liener, I. E. (1978). Nutritional evaluation of dry-roasted navy bean flour and mixtures with cereal proteins. In: "Nutritional Improvement of Food and Feed Proteins", M. Friedman, ed., Plenum Press, New York, pp. 401-413.

Zahnley, J. C. (1984). Stability of enzyme inhibitors and lectins in foods and the influence of specific binding interactions. In: "Nutritional and Toxicological Aspects of Food Safety", M. Friedman, ed., Plenum Press, New York, pp. 323-356.

Zahnley, J. C. and Friedman, M. (1982). Absorption and fluorescence spectra of S-quinolylethylated Kunitz soybean trypsin inhibitor. J. Protein Chem., 1, 225-240.

22

RELEVANCE OF MULTIPLE SOYBEAN TRYPSIN INHIBITOR FORMS TO NUTRITIONAL QUALITY

Anna L. Tan-Wilson and Karl A. Wilson

Department of Biological Sciences
State University of New York at Binghamton
Binghamton, NY 13901

ABSTRACT

Trypsin inhibitors contribute to the antinutritional component of raw soybean meal by inhibiting vertebrate pancreatic serine proteinases in the small intestine, resulting in a range of deleterious physiological effects in the animal. The variation in the nutritional quality of soybean cultivars stems partly from wide-ranging levels of trypsin inhibitor, and from varying proportions of trypsin inhibitors of two classes - the Kunitz and the Bowman-Birk inhibitor classes. The latter class is better able to survive heat processing and digestion in the stomach. Some variation in cultivars also arises from the array of isoinhibitors present in the seed. The three Kunitz isoinhibitors, Ti^a, Ti^b and Ti^c differ by as much as 1000-fold in their interaction with bovine trypsin. The Bowman-Birk isoinhibitors differ not only in their extent of interaction with trypsin, but in their spectrum of inhibition of the other pancreatic enzymes, chymotrypsin and elastase. In this chapter, we look at twenty-two Bowman -Birk inhibitors from ten soybean cultivars and find at least twelve which are different enough in amino acid composition and/or inhibitor activity to be distinct protein species. Of these, three pairs are related by proteolytic digestion. Quite ironically, the Bowman-Birk inhibitors, and to some extent the Kunitz inhibitors, contribute to the nutritional quality of soybeans by virtue of their high cystine content which supplements the low or negligible amounts of sulfur-containing amino acids in the storage proteins that comprise the bulk of the protein reserve in the seed.

INTRODUCTION: TRYPSIN INHIBITORS CONTRIBUTE TO THE ANTINUTRITIONAL COMPONENT OF RAW SOYBEAN MEAL

Utilization of the rich protein resource in soybeans is limited by antinutritional components which, unless partially destroyed, result in deleterious effects on rats, chicks and other non-ruminant animals (Rackis and Gumbmann, 1981). A significant portion of this antinutritional toxicant effect is due to proteinase inhibitors, proteins which form complexes with vertebrate pancreatic proteinases and inhibit their enzymatic activity (Rackis, 1965). These inhibitors act as pseudo-substrates, binding to the enzyme much as the normal substrate would, but not yielding products over normal time periods at physiological conditions.

The complex does dissociate, very slowly under physiological conditions but more rapidly under certain non-physiological conditions, to yield either proteinase and inhibitor, or proteinase and inhibitor in which a single specific peptide bond has been cleaved (Laskowski and Sealock, 1971; Laskowski and Kato, 1981).

There are two classes of proteinase inhibitors in the soybean - the Kunitz (Kunitz, 1947) and Bowman-Birk (Birk et al. 1963) types of trypsin inhibitors. The Kunitz soybean trypsin inhibitor has a molecular weight of approximately 21,000, consisting of a single polypeptide chain of approximately 180 amino acid residues including four residues of half-cystine and two of methionine (Koide and Ikenaka, 1973). Kunitz soybean trypsin inhibitor strongly inhibits trypsin at a single specific reactive site, while chymotrypsin is inhibited more weakly at two reactive sites, one being the same as or overlapping the trypsin reactive site (Laskowski and Sealock, 1971; Bosterling and Quast, 1981).

The Bowman-Birk type of inhibitor is considerably smaller, approximately 7,000 to 8,000 in molecular weight with a single polypeptide chain of 70 to 80 amino acid residues. Bowman-Birk inhibitors can simultaneously interact with two proteinase molecules at one time, e.g., with two trypsin molecules, a trypsin and a chymotrypsin, a trypsin and elastase, etc., depending upon the specific inhibitor (Wilson, 1981; Birk, 1985). There is no apparent homology between the Kunitz and Bowman-Birk type inhibitors as evidenced by their amino acid sequences (Laskowski and Kato, 1981).

The antinutritional effects of the soybean trypsin inhibitors are evident from a number of lines of investigation. Rat feeding experiments show that up to 30-50% of the growth inhibition and pancreatic hypertrophic effect of raw soybeans is due to the soybean trypsin inhibitors (Rackis, 1965). This was conclusively demonstrated by feeding aqueous extracts of soybeans after the selective removal of trypsin inhibitors by affinity chromatography (Kakade et al., 1973). There is an inverse relationship between the level of trypsin inhibitor activity remaining after steaming of soy flour and the protein efficiency ratio (Rackis and Gumbmann, 1981).

The deleterious physiological effects occurring in the rat upon the feeding of untreated soybean meal have been explained in terms of a model for pancreatic enzyme secretion or regulation. This model (Green and Lyman, 1972; Green et al., 1973) centers on a feedback loop between the intestinal mucosa and the exocrine pancreas. The mixing of the chyme from the stomach with the pancreatic juice in the duodenum results in the formation of proteinase-dietary protein enzyme-substrate complexes, thus lowering the concentration of free trypsin in the gut lumen. A receptor sensing the concentration of this unbound trypsin is hypothesized in the intestinal mucosa. If the free trypsin concentration falls below a minimum threshold value, the intestinal mucosa secretes cholecystokinin into the circulation. This cholecystokinin in turn stimulates the pancreas to secrete additional pancreatic enzymes, which will lead to an increase in the free trypsin level in the gut lumen.

The secretion of the pancreatic enzymes is thus in part controlled by the ability of the dietary proteins to bind the trypsin present in the intestine. The enzyme-substrate complexes formed between trypsin and the average dietary protein would be expected to be rather weak (in terms of association equilibrium constants) and transient, rapidly dissociating primarily but not exclusively to free trypsin and the hydrolyzed products. The presence of potent trypsin inhibitors in the food would obviously lead to the production of much stronger (higher K_{assoc}) trypsin-inhibitor complexes with much longer life times. The presence of such trypsin inhibitors would much more effectively lower the concentration of free trypsin in the intestinal lumen, and trigger the release of cholecystokinin. The chronic oversecretion of cholecystokinin, prompted by high levels of trypsin inhibitor in the diet, leads to the pancreatic hypertrophy in the rat. The finding of cholecystokinin release after feeding rats orally with soybean trypsin inhibitor (Brand and Morgan, 1981), increased nucleic acid and protein synthesis in rat pancreas upon feeding of soybean trypsin inhibitor (Melmed et al., 1976) and change in pancreatic blood flow with cholecystokinin treatment (Oates et al., 1974) all support this mechanism.

The growth inhibition noted under these conditions is attributable in part to the decrease in efficiency of digestion of dietary proteins. In addition, large amounts of pancreatic enzymes are lost in the feces. As the pancreatic enzymes have a higher than average proportion of sulfur-containing amino acids, their loss represents a serious loss of these essential nutrients. This is further exacerbated by the low content of sulfur-containing amino acids in the soybean storage proteins. The significance of this is demonstrated by the improvement in the weight gain/feed ratios in rats (Yen et al., 1971) and in chicks (Bajjalieh et al., 1980) fed raw soybean meal diets supplemented with D,L-methionine.

Ironically, the Bowman-Birk class of trypsin inhibitors are especially rich in sulfur-containing amino acids. These proteins have 14 half-cystine residues in a polypeptide chain of 70-80 amino acid residues. This amounts to 18-20 residue percent as compared to the half cystine contents of 0.8 and 0 residue percent and methionine contents of 0.9 and 0 residue percent in glycinin and conglycinin, the storage proteins of the soybean (Staswick et al., 1981; Thanh and Shibasaki, 1977). Efforts to develop soybean strains with low overall proteinase inhibitor activity would thus reduce a source of antitoxicant components in the bean but would possibly reduce the content of sulfur-containing amino acids in the seed protein as well. A further concern is that seed germination experiments indicate that these proteinase inhibitors might very well function as storage proteins or for defense in the seed (Ryan, 1973; Richardson, 1977; Tan-Wilson et al., 1985a) so that elimination of the inhibitors might prove deleterious to the plant.

VARIABILITY IN ANTINUTRITIONAL EFFECTS STEMS FROM DIFFERENCES IN THE ANIMALS

The pancreatic hypertrophy, increased pancreatic enzyme secretion, and growth inhibition noted in the rat do not always occur in all species (Rackis, 1974). Further, the severity does not always occur in direct relation to the extent of inhibition of the proteolytic enzymes of the animal. In the chick, intestinal proteolysis is inhibited to a greater extent than in the rat, but compensating increases in pancreatic enzyme secretion are not as great. Enlargement of the pancreas occurs in the rat and chick but not in the pig, although the latter does exhibit growth inhibition and increased pancreatic enzyme secretion (Yen et al., 1977). Some of the absence of deleterious effects no doubt stems from differences in animal physiology. However, there is no question that at least some of the deleterious effects are due to the trypsin inhibitors since feeding experiments which utilize purified trypsin inhibitors added to heated soybean flour have shown this to be so (Melmed and Bouchier, 1969; Madar, 1979; Konijn et al., 1970).

Part of the variability undoubtedly stems from the differences in the manner and extent to which particular proteolytic activities are inhibited by the soybean proteinase inhibitors. Modification of the trypsin reactive site of Bowman-Birk soybean trypsin inhibitor prevents the inhibitor-induced increase in enzyme secretion in the chick pancreas, whereas modification of the other reactive site, for chymotrypsin, does not (Madar, 1979). It is therefore trypsin inhibition, rather than chymotrypsin inhibition, that is the trigger for cholecystokinin release. Thus, responses to the inhibitors of soybean meal may differ due to differences between animal species in the relative amounts and importance of trypsin and chymotrypsin in the digestion of dietary proteins. Secondly, the response to the soybean inhibitors would be expected to vary with differences in the strength of interaction or rapidity of complex formation between the trypsin of the animal and the inhibitor complement of the soybean meal. One would expect the inhibitors to have less effect upon an animal whose trypsin interacts weakly with the soybean inhibitor than one whose trypsin rapidly forms tightly associated complexes.

In keeping with this observation, a number of investigators have made use of in vitro measurements of the interaction of different vertebrate enzymes with soybean trypsin inhibitors in an attempt to evaluate the nutritional significance of these proteins. The pancreatic enzymes of trout, fox, chick, pig, rat, cow, monkey, and man have been compared for susceptibility to inhibition by Kunitz and Bowman-Birk inhibitors (Krogdahl and Holm, 1983; Struthers and MacDonald, 1983). In one such study using pancreatic extracts, the inhibition of total caseinolytic activity by the Kunitz soybean inhibitor was ten times higher in the trout than in human. Whether the experiments employed pancreatic juice to mimic the in vivo situation or purified enzymes to allow the measurement of equilibrium and rate constants of the enzyme-inhibitor binding, the results all show significant differences in the interaction of the trypsins from these animals with the soybean trypsin inhibitors.

A complication of this type of study and its interpretation is that there are often multiple species of trypsin in the same animal. Humans, for example, have cationic (Travis and Roberts, 1969) and anionic (Mallory and Travis, 1973) trypsins, the cationic form being the major component. Soybean inhibitors exhibit different affinities for these two trypsins. Both Kunitz and Bowman-Birk inhibitors show greater inhibition for the human anionic than for the human cationic trypsin, the Kunitz class of inhibitors showing the greater discrimination. The strength of interaction between Kunitz soybean trypsin inhibitor with anionic human trypsin is approximately the same as that between the Kunitz inhibitor and bovine trypsin (Figarella et al., 1974; Mallory and Travis, 1975).

Values of equilibrium assocation constants of proteinase-trypsin inhibitor complexes are collected in Table 1.

VARIABILITY IN ANTINUTRITIONAL EFFECTS STEMS FROM DIFFERENCES IN SOYBEAN CULTIVARS

The nature and severity of deleterious physiological effects stemming from the feeding of raw soybean meal has also been observed to vary with the soybean cultivar used (Kakade et al., 1972; Holm and Krogdahl, 1982). In the latter study, the cultivars ranked differently in quality according to animal species. Some differences are accounted for by the wide range of total trypsin inhibitor activity per gram dry meal, up to seven-fold differences in one study (Kakade et al., 1972; Silva et al., 1979). In some experiments, this variable is controlled and still differences arise. These have been attributed to the varying proportions of Kunitz and Bowman-Birk inhibitors in the different strains (Kakade et al., 1972; Holm and Krogdahl, 1982). Depending upon the animal species, one or the other class of inhibitors might have the higher affinity for the mix of pancreatic serine proteinases involved.

TABLE 1

EQUILIBRIUM ASSOCIATION CONSTANTS
OF SOYBEAN INHIBITOR - PROTEINASE COMPLEXES

INHIBITOR	pH	$K_{association}$, M^{-1} BOVINE TRYPSIN	BOVINE CHYMO-TRYPSIN	BOVINE ELASTASE	HUMAN CATIONIC TRYPSIN	REFERENCE
KUNITZ						
Ti^{a}	8.35	$6.8x10^{8}$				(1)
	7.5	$3.3x10^{11}$	$1.1x10^{6}$ (1st site)			(2)
			$2.2x10^{6}$ (2nd site)			
		$1.0x10^{11}$			$1.2x10^{8}$	(3)
	8.1	$2x10^{9}$				(4)
Ti^{b}	8.1	$2x10^{7}$				(4)
Ti^{c}	8.1	$3x10^{10}$				(4)
BOWMAN-BIRK CLASSICAL						
	7.0	$2.14x10^{8}$	$4.79x10^{8}$			(5)
	8.2,T 8.0,C	$1.1x10^{8}$	$1.9x10^{7}$			(6)
		$1.0x10^{9}$			$2.5x10^{7}$	(3)
C-II	8.0	$2.7x10^{7}$	$1.2x10^{7}$	$9x10^{6}$		(7)

(1) Lebowitz and Laskowski, 1962 (2) Bosterling and Quast, 1981
(3) Wilson, 1981 (4) Freed and Ryan, 1980 (5) Harry and Steiner, 1970
(6) Turner et al., 1975 (7) Odani and Ikenaka, 1977b

The greater susceptibility of the Kunitz inhibitors to inactivation by the low pH and pepsin activity in the gastric juice is another important factor in the possibly different physiological effects of the two classes of inhibitors (Kassell and Laskowski, 1956; Krogdahl and Holm, 1981). The difference is greater still when considering that the Kunitz inhibitor is more labile in heat processing than is the Bowman-Birk inhibitor (Johnson et al., 1980) so the proportions of the two classes of inhibitor contributing to residual activity after processing of meal could be very different from the proportions in the seed.

KUNITZ ISOINHIBITOR FORMS

The main theme of this chapter deals with still another source of variation in the response of different animal species to soybean diets, i.e., multiple forms of both inhibitor classes in different soybean strains. The single gene locus coding for the Kunitz trypsin inhibitor exists in four allelic forms, Ti^a, Ti^b, and Ti^c, which are codominant, and ti, a recessive allele where no functional KSTI is synthesized (Singh et al., 1969; Clark et al., 1970; Orf and Hymowitz, 1979). The products of the Ti^a, Ti^b and Ti^c alleles are all active as trypsin inhibitors and are readily distinguishable electrophoretically. The amino acid sequence of the Ti^a form was determined by Koide and Ikenaka (1973). The Ti^b form is also a single polypeptide chain and has the same amino-terminal sequence for ten residues as does Ti^a. Ti^b is larger by sixteen amino acid residues. Although the amino acid compositions are quite similar, the data are not consistent with a structure of Ti^b like Ti^a with a 16-amino acid peptide insertion or extension at the carboxyl-terminus (Hartl et al., 1985). The amino acid compositions are listed in Table 2.

The equilibrium constants of association listed in Table 1 for the formation of the Kunitz isoinhibitor-bovine trypsin complexes at pH 8.1 differ by as much as three orders of magnitude. From the weakest to the strongest, these are $2 \times 10^7\ M^{-1}$ for Ti^b, $2 \times 10^9\ M^{-1}$ for Ti^a, and $3 \times 10^{10}\ M^{-1}$ for Ti^c (Freed and Ryan, 1980). There would also, of course, be the more pronounced effect of not having a functional trypsin inhibitor at all, as in the case of the ti allele.

A similarly wide range of affinity for the trypsins of other animal species might very well exist and could at least partly explain the better weight gain/feed ratio in rats and chicks fed with a cultivar bearing the Ti^b allele as compared to one with the Ti^a allele (Yen et al., 1971; Bajjalieh et al., 1980). The identities of the Kunitz isoinhibitor forms in a large number of soybean cultivars were determined by Orf and Hymowitz (1979).

Some feeding experiments have been done with germinated soybeans since these have lower trypsin inhibitor activities (Bates et al., 1977; Bau and Debry, 1979). Germinated soybeans have additional isoinhibitor forms - Ti^a_m, Ti^b_m and Ti^c_m in cultivars having Ti^a, Ti^b and Ti^c, respectively, in the ungerminated seed (Orf et al., 1977; Freed and Ryan, 1978; Tan-Wilson et al., 1982). We have purified Ti^a_m and Ti^b_m from cotyledons of germinated soybeans of cultivars Amsoy 71 and Fiskeby V, respectively. These inhibitors were compared to Ti^a and Ti^b from the dry seed by amino acid compositional analysis, SDS-polyacrylamide gel electrophoresis, amino and carboxyl terminal determinations and partial sequence. The results show that Ti^a_m amd Ti^b_m are still single polypeptide chains, ruling out the occurrence of an internal cleavage near the center of the sequence. Ti^a_m has the same amino acid composition as Ti^a except

TABLE 2

AMINO ACID COMPOSITIONS OF KUNITZ SOYBEAN TRYPSIN INHIBITORS

RESIDUE	RESIDUE/MOLECULE[a]			
	Ti^a	Ti^a_m	Ti^b	Ti^b_m
Asx	26	25	30	30
Thr	7	7	9	8
Ser	11	10	13	11
Glx	18	17	21	18
Pro	10	10	10	10
Gly	16	16	18	18
Ala	8	8	9	9
1/2 Cys[b]	4	4	4	4
Val	14	14	15	15
Met	2	2	1	1
Ile	14	13	15	15
Leu	15	14	14	13
Tyr	4	4	3	3
Phe	9	9	11	10
Lys	10	9	12	11
His	2	2	1	1
Arg	9	9	11	10
Trp	2	2	ND	ND
TOTAL	181	175	197+	187+

a Data for Ti^a from Koide and Ikenaka, (1973); Ti^a_m, Ti^b, Ti^b_m from Hartl et al. (1985)

b Determined as cysteic acid after performic acid oxidation by the method of Hirs (1967)

c ND - not determined

for the lack of six amino acid residues. Ti^b_m has ten amino acid residues less than Ti^b but otherwise, the two inhibitors also have the same amino acid compositions. Partial sequences at the amino-termini are the same for Ti^a_m and Ti^a as well as for Ti^b_m compared to Ti^b. The carboxyl-terminal residues, however, were different, consistent with the loss of an Asp-Lys-Glu-Ser-Leu peptide from Ti^a to yield Ti^a_m and of a decapeptide from Ti^b to yield Ti^b_m. Thus, Ti^a_m and Ti^b_m are produced during germination by the proteolytic digestion at the carboxyl-termini of the parent isoinhibitor forms. The immunological identity of Ti^b_m with Ti^b (Tan-Wilson et al., 1983) supports this conclusion.

BOWMAN-BIRK ISOINHIBITOR FORMS

The identification of the Bowman-Birk isoinhibitor forms is complicated by their multiplicity. A cultivar can have a maximum of two Kunitz inhibitor species but anywhere from five to twelve Bowman-Birk isoinhibitor forms. Odani and Ikenaka purified to homogeneity four different isoinhibitors from the Japanese soybean cultivar Sode-furi, but there might have been as many as eight inhibitors detected in the course of the chromatographic fractionation (Odani and Ikenaka, 1977a). Hwang et al. (1977) found five Bowman-Birk type inhibitors in cultivar Tracy. Kim et al. (1982) studied seven varieties of soybeans and their twenty-one F1 hybrids. They found sixteen kinds of trypsin inhibitors, with five to twelve from each soybean variety. Katagiri and Shimizu (1983) found eleven in the variety Kuromame Makino. Stahlhut and Hymowitz (1983) found five in cultivar Amsoy 71. We (Tan-Wilson et al., 1985b) have studied the isoinhibitor complements of eight soybean varieties - Altona, Amsoy 71, Black Jet, Disoy, Envy, Fiskeby V, Panther, and Wilkin. We found ten distinct Bowman-Birk isoinhibitors, with each variety having four to nine inhibitor species. There is often considerable difficulty in comparing the results of different investigators because of the different cultivars, or different chromatographic and electrophoretic separation techniques used. Ideally, the structural characterization of the purified isoinhibitor, e.g. by amino acid sequencing, would lead to the most unambiguous interpretation and correlation of these results. However, in only a few cases has this been accomplished.

In our study, the Bowman-Birk inhibitor pools were prepared by extraction of soybeans in 50 mM Tris-Cl + 0.3 mM phenylmethylsulfonyl fluoride + 0.5 mM sodium iodoacetate, pH 8. The latter two reagents were added to prevent the digestion of the Bowman-Birk inhibitors by seed proteinases and thus avoid the generation of artifactual forms. We have also found acetone extraction of lipid to generate deamidated forms and have therefore avoided the use of organic solvents. Precipitation of the inhibitors by ammonium sulfate at 85% saturation removed low molecular weight contaminants. Precipitation at pH 4 removed the bulk of the soybean storage protein. Sephadex G-75 gel filtration, monitored for cross-reactivity to Kunitz- and Bowman-Birk- specific antibodies (Tan-Wilson and Wilson, 1982) removed more storage protein and smaller components and separated the Kunitz and Bowman-Birk type inhibitors from each other.

The Bowman-Birk isoinhibitors were separated by chromatography on DEAE-cellulose in ammonium acetate buffer with a pH gradient from pH 6.5 to pH 5 as shown in Figure 1. For the purpose of this discussion, we define soybean Bowman-Birk isoinhibitors as those inhibitors which inhibit bovine trypsin and which cross-react with antibodies elicited in the rabbit by glutaraldehyde-polymerized BBSTI-E, the "classical" Bowman-Birk soybean trypsin inhibitor. The isoinhibitors were detected by their ability to inhibit bovine trypsin (Schwert and Takenaka, 1955) and by

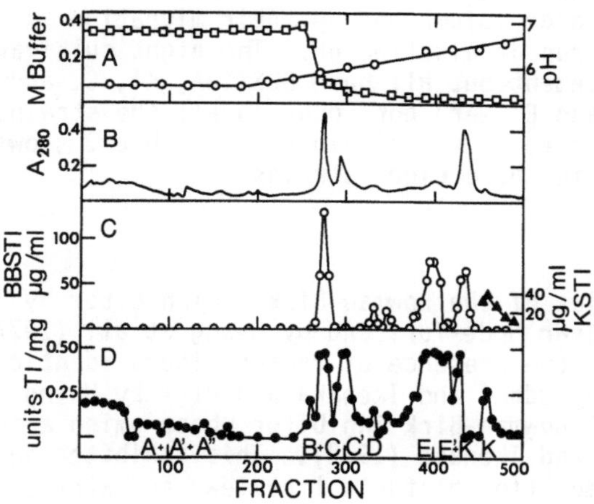

Figure 1. DEAE-cellulose (DE-52, Whatman) column chromatography of the Bowman-Birk trypsin inhibitor pool from Amsoy 71 at 25°C. The column (2.5 x 150 cm) was equilibrated in 50 mM ammonium acetate, pH 6.5. Elution was with a 5.4 liter gradient with 0.5 M ammonium acetate pH 5 as the final buffer. The pH - - and buffer concentration - - gradients are shown in panel A. The absorbance readings at 280 nm of the fractions are shown in panel B. The concentrations of Bowman-Birk - - and Kunitz - - trypsin inhibitors shown in panel C were determined by radial immunodiffusion (Mancini et al., 1965, modified as in Tan-Wilson et al., 1983). The anti-Bowman-Birk-specific and anti-Kunitz-specific antisera were prepared and characterized as described by Tan-Wilson and Wilson (1982). BBSTI-E and Ti[a] were used as standards, using extinction coefficients in Kassell (1970). Trypsin inhibitor activity - - shown in panel D was determined as inhibition of the ability of bovine trypsin to hydrolyze N-benzoyl-arginyl ethyl ester.

their cross-reaction by radial immunodiffusion (Mancini et al., 1965). Pools were prepared for electrophoresis on 10% gels in the standard Davis (1964) disc gel electrophoresis system. Ten isoinhibitors were found altogether and have been designated BBSTI-A, A', A", B, B', C, D, E and E' in order of their increasing mobility in this electrophoretic system. Since the classical Bowman-Birk inhibitor tends to show greater mobility with increasing concentration in the Davis system, another electrophoretic system, the Davis system in 15% gels with 4 M urea in the gel and buffer reservoirs, was also employed. With the addition of 4M urea to the system, the sequence was BBSTI-A, A', A", B', C', B, C, D, E and E' instead. The migration pattern in this system of the isoinhibitors which we have purified so far is shown in Figure 2. The isoinhibitors which were common across strains were identified by their similar chromatographic behavior on the ion exchange column and by their migration in the two electrophoretic systems run at alkaline pH. The eight cultivars vary in their isoinhibitor complement but all have BBSTI-A, A', C, and E. BBSTI-A", B, B', C', D and E' were not found in all the strains studied. Each strain had from four to nine isoinhibitors. Table 3 shows the isoinhibitor complement in these eight strains.

BBSTI-E

The extensive studies of the Bowman-Birk isoinhibitor by Odani and Ikenaka (1977a) on cultivar Sode-furi and by Hwang et al. (1977) on cultivar Tracy also show the presence of isoinhibitors identical to BBSTI-E. Designated A by Odani and Ikenaka and PI-V by Hwang et al., BBSTI-E is the classical Bowman-Birk inhibitor whose amino acid sequence was elucidated by Odani and Ikenaka (1972). This inhibitor has been called by many other names (for historical review, see Birk (1985)). In this chapter, we shall refer to it as BBSTI-E.

BBSTI-E has a molecular weight of approximately 8000 although its tendency toward self-association (Frattali, 1969) led earlier reports of its molecular weight to be multiples of 8000. Its amino acid sequence is homologous to that of other legume Bowman-Birk type inhibitors (Laskowski and Kato, 1981; Wilson, 1981). Like these, the first half of the molecule is homologous to the second half of the same molecule. The reactive site bond in the amino-terminal half of the molecule is Lys^{16}-Ser^{17} and accounts for the inhibition of trypsin. The reactive site bond of Leu^{43}-Ser^{44} in the carboxyl-terminal half of the molecule accounts for its inhibition of chymotrypsin (Odani and Ikenaka, 1972; Seidl and Liener, 1971; Seidl and Liner, 1972a). Its inhibition of bovine trypsin and of bovine chymotrypsin is stoichiometric and independent (Odani and Ikenaka, 1973a; Seidl and Liener, 1972b). Equilibrium and rate constants for the binding of this isoinhibitor with a number of trypsins and chymotrypsins have been determined and are summarized in Table 1. The data collected in the table shows that BBSTI-E is a better inhibitor of bovine trypsin than of bovine chymotrypsin. It is a better inhibitor of bovine trypsin than of human cationic trypsin.

The presence of BBSTI-E is not universal in all soybean cultivars. Stahlhut and Hymowitz (1983) identified their inhibitor designated III to be the classical Bowman-Birk inhibitor and found this to be absent in 3% of 470 strains studied.

Odani and Ikenaka (1977a) also describe another isoinhibitor, designated B, which appears to be derived from BBSTI-E by deamidation of asparagine and/or glutamine residues. Our (Tan-Wilson et al., 1985b) BBSTI-E', which migrates even faster than BBSTI-E on electrophoresis has not yet been purified and characterized as to amino acid composition. There is still one other isoinhibitor, called fraction V, described by

TABLE 3

BOWMAN-BIRK ISOINHIBITOR COMPOSITIONS OF SOYBEAN STRAINS[a]

SOYBEAN STRAIN	BBSTI SPECIES A	A'	A"	B	B'	C	C'	D	E	E'
Altona	x	x		x	x	x	x		x	
Amsoy 71	x	x	x	x		x	x	x	x	x
Black Jet	x	x				x			x	
Disoy	x	x	x	x		x			x	x
Envy	x	x		x		x	x	x	x	
Fiskeby V	x	x	x	x		x			x	
Panther	x	x	x	x		x			x	
Wilkin	x	x	x			x	x		x	

[a] Data from Tan-Wilson et al. (1985b).

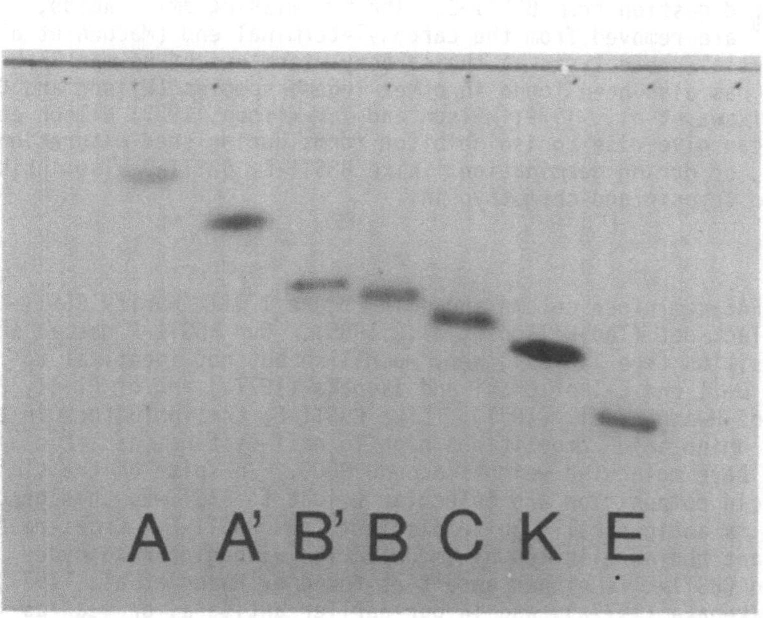

Figure 2. Polyacrylamide disc gel electrophoresis of purified Bowman-Birk trypsin inhibitors in the Davis (1964) system, except that 15% gels were used and 4 M urea was added to both gel and reservoir buffers. From left to right, BBSTI-A, BBSTI-A', BBSTI-B', BBSTI-B, BBSTI-C, Kunitz Ti[a] added for reference and BBSTI-E.

Stahlhut and Hymowitz (1983) in Amsoy 71 which migrates even faster than BBSTI-E' in their electrophoretic system. They found fraction V in 74% of 470 strains of soybean studied.

BBSTI-D

The isoinhibitor BBSTI-D is an inhibitor of bovine trypsin. It cross-reacts strongly with anti-BBSTI-E-specific antibodies. It normally arises during the process of seed germination and replaces BBSTI-E as the major component in cultivar Fiskeby V six days after the initiation of imbibition of the seed (Tan-Wilson et al., 1982). As shown in Table 3, some cultivars have trace amounts of this isoinhibitor even in the dry seed. Its nutritional significance would lie mostly in feeding experiments utilizing germinated beans.

We have purified BBSTI-D from six-day-old Fiskeby V cotyledons. The early steps were as described for the isolation of the BBSTI inhibitor pool and the chromatography in DEAE-cellulose in ammonium acetate buffer at acid pH in Figure 1. The BBSTI-D pool was identified by electrophoresis and the BBSTI-D obtained in pure form after chromatography on DEAE-cellulose in 50 mM Tris-Cl pH 7.5, eluted with a 0 to 0.5 M NaCl gradient in the same buffer. The amino acid composition, presented in Table 4, was identical to that of BBSTI-E, less one aspartyl/asparaginyl and one glutamyl/glutaminyl residue. Edman-dansyl sequence analysis demonstrated that the first five residues of the amino-terminus of BBSTI-D are the same as those of BBSTI-E. Only the tryptic peptides representing residues 38-69 of the reduced carboxy-methylated BBSTI-D and 38-71 of the corresponding derivative of BBSTI-E were different based on amino acid composition and mobility on paper and thin-layer chromatography. The difference in amino acid composition was the same difference found in the whole molecules. Thus, we concluded that BBSTI-D is derived by proteolytic digestion from BBSTI-E. The two missing amino acids, Glu^{78}-Asn^{79} are removed from the carboxyl-terminal end (Madden et al., 1985). Partial proteolysis at the carboxyl-terminus of Bowman-Birk type inhibitors has also been found in other legume species (Wilson and Chen, 1983; Yoshikawa et al., 1979; Wilson and Tan-Wilson, 1983; Wilson et al., 1985) and can give rise to isoinhibitor forms during seed maturation, quiescence, or during germination. Like BBSTI-E, BBSTI-D also inhibits both bovine trypsin and chymotrypsin.

BBSTI-C

The DEAE-cellulose column shown in Figure 1 will purify BBSTI-C from cultivar Black Jet (Tan-Wilson et al., 1985). Our BBSTI-C has an amino acid composition (see Table 4) very much like but not identical to inhibitors D-II and E-I of Odani and Ikenaka (1977a) and of PI-II, PI-III and PI-IV of Hwang et al. (1977). Like BBSTI-E, the inhibitors in this group have amino acid compositions high in half-cystine, aspartic acid and serine and have molecular weights around 8000. In spite of the similarity in amino acid composition and molecular weight to BBSTI-E, this group of inhibitors is antigenically quite distinct from BBSTI-E. Cross-reactivity to rabbit antibodies elicited by immunization with glutaraldehyde-polymerized BBSTI-E is either absent as found by Hwang et al. (1977), by Odani and Ikenaka (1977a), and in our earlier antisera, or weak as in our antisera obtained after many booster immunizations. We use only pure immunogen so the subsequent development of cross-reacting antibodies must be a result of some homology in the conformation of BBSTI-C with that of BBSTI-E. Indeed, the structures of D-II and E-I have been found to be quite homologous to that of the classical inhibitor. These isoinhibitors were also found to have internal homology, a large number of disulfide bridges and regions of sequence homology with BBSTI-E (Odani and Ikenaka, 1977a; 1978).

Amino acid sequence determination has shown PI-II to be identical to inhibitor E-I and PI-IV to be identical to inhibitor D-II (Hammond et al., 1984). E-I is a variant of D-II, lacking nine residues at the amino-terminus (Odani and Ikenaka, 1978). Our BBSTI-C is most closely similar in amino acid composition to PI-III (Tan-Wilson et al, 1985b) which in turn, is most closely similar to PI-II. The amino acid compositions are compared in Table 4.

Functionally, this group is also very different from BBSTI-E. They do not inhibit bovine chymotrypsin. Nor do these inhibit bovine elastase. They do have two reactive sites but both are reactive for bovine trypsin. The amino acid sequences of D-II and E-I show two Arg-Ser reactive site peptide bonds. A titration of trypsin with inhibitor D-II indicated a non-stoichiometric enzyme:inhibitor ratio of 1.4:1 consistent with inhibitor D-II being a double-headed inhibitor of bovine trypsin but a somewhat weak one. E-I appeared to behave much like D-II (Odani and Ikenaka, 1978). Hwang et al. (1977) compared PI-II, PI-III and PI-IV with respect to molar ratios of inhibitor to enzyme at 50% inhibition. Their numbers show PI-II to be a stronger inhibitor for bovine trypsin than PI-IV and PI-IV to be a stronger inhibitor for bovine trypsin than PI-III.

Inhibitor C-II purified from cultivar Sode-furi and characterized by Odani and Ikenaka (1977b) belongs structurally and antigenically to this group of inhibitors. There is greater homology in sequence between inhibitors C-II and D-I than between C-II and the classical Bowman-Birk soybean inhibitor. Inhibitor C-II cross-reacts with antibodies made to inhibitor D-II but not to antibodies made to BBSTI-E.

Functionally, inhibitor C-II appears to be distinct from inhibitors D-II and E-I as well as from BBSTI-E. Inhibitor C-II is also a double-headed inhibitor. It inhibits bovine elastase as well as bovine trypsin and chymotrypsin. The reactive site bond in the first domain is Ala-Ser instead of Lys-Ser as in BBSTI-E or Arg-Ser as in inhibitors D-II and E-I (Odani and Ikenaka, 1977a;b). This is the same type of reactive site found in a garden bean inhibitor also known to inhibit bovine elastase (Wilson and Laskowski, 1975). The inhibition of bovine elastase by inhibitor C-II is independent from its inhibition of bovine trypsin and chymotrypsin and is characterized by an equilibrium constant of dissociation of 1.1×10^{-7} M (Odani and Ikenaka, 1977a). The second inhibitor reactive site in the second domain is Arg-Ser and accounts for its trypsin inhibitor activity. A trypsin-C-II complex no longer inhibited bovine chymotrypsin so it is thought that this inhibitor activity is attributed to the same site as the trypsin reactive site. Although a typical trypsin reactive site, it is thought that the neighboring residues are responsible for the chymotryptic inhibitor activity. BBSTI-E has Leu-Ser, accounting for its chymotrypsin inhibitor activity, while inhibitors D-II and E-I have Arg-Ser, accounting for their second trypsin inhibitor site, at the homologous regions in their amino acid sequences. The equilibrium constant of dissociation of the inhibitor C-II-bovine trypsin complex is 3.7×10^{-8} M, indicating that it has approximately ten times weaker affinity for bovine trypsin than does BBSTI-E. The equilibrium constant of dissociation of the inhibitor C-II-bovine chymotrypsin complex was found to be 8.6×10^{-8} M, showing weaker inhibition by inhibitor C-II for bovine chymotrypsin than for bovine trypsin (Odani and Ikenaka, 1978).

Our BBSTI-C' has not yet been characterized.

TABLE 4

AMINO ACID COMPOSITIONS
OF THE BOWMAN-BIRK INHIBITORS

RESIDUE	A PI-V BB-E	BB-D	BB-C	PI-III	D-II PI-IV	E-I PI-II	C-II	PI-I	BB-B	BB-B'	BB-A	BB-A'	(a) (b) (c)
	RESIDUES/MOLECULE												
ASX	11	10	11	11	13,12	10,11	13	8	9	8	5	5	
THR	2	2	3	3	3, 2	3	4	4	3	3	3	3	
SER	9	9	8	8	12	8	12	7	9-10	6	14	12	
GLX	7	6	7	6	6	5	4	4	7	9	10	10	
PRO	6	6	4	5	5	5	5	4	4	3-4	3	2-3	
GLY	0	0	2	1	1	1	1	4	4	7	15	13-14	
ALA	4	4	1	1	0, 1	0	4	2	3	4	5	4	
1/2CYS	14	14	ND	12	14,12	14	14	10	10	ND	1	1	
VAL	1	1	0	0	0	0	0	1	1	4	3	3	
MET	1	1	2	2	3, 2	3	3	1	1	1	0-1	0-1	
ILE	2	2	1	1	1	1	1	3	2	4	2	2	
LEU	2	2	2	3	3	3	3	7	3	5	3	2	
TYR	2	2	2	2	3, 2	2	1	1	2	2	2	1	
PHE	2	2	1	1	1	1	1	1	1-2	3	2	1	
LYS	5	5	4	4	4	4	3	2	4	4	2	2	
HIS	1	1	1	1	1	1	3	1	1	1	1	1	
ARG	2	2	4	4	5	5	4	4	3	3	2	1	
TOTAL	71	69	65	65	75,70	67,66	76	64	67-69	67+-68+	73-74	63-66	

(a) Odani and Ikenaka (1977) for A, D-II, E-I, C-II
(b) Hwang et al. (1977) for PI-V, PI-III, PI-IV, PI-II, PI-I
(c) Tan-Wilson et al. (1985) for BBSTI-E and BBSTI-C, Madden et al. (1985) for BBSTI-D, recent data from Tan-Wilson and Wilson for BBSTI-B,B',A,A'

BBSTI-B and B'

These isoinhibitors were purified from cultivar Altona (Chapman and Tan-Wilson, unpublished) after chromatography as shown in Figure 1. The pools containing BBSTI-B and B' were subjected to a second DEAE-cellulose column equilibrated in 50 mM Tris-Cl, pH 8. The column was eluted with a gradient from 0 to 0.15 M NaCl. The amino acid composition of BBSTI-B presented in Table 4 shows this inhibitor to have slightly lower half-cystine content, compared to the other Bowman-Birk inhibitors. Both BBSTI-B and B'seem characteristically to have more glycine, alanine and glutamic acid/glutamine residues. BBSTI-B' has additionally more valine, leucine and isoleucine and both have fewer aspartic acid/asparagine and half-cystine residues than the other inhibitors described earlier. In many respects, such as the higher glycine, leucine and isoleucine and lower aspartic acid/asparagine content, their compositions are similar to that of PI-I in cultivar Tracy. Their extinction coefficients at 280 nm calculated from amino acid compositions differ quite significantly from that of the classical BBSTI-E and its BBSTI-D derivative. Whereas the absorbance at 280 nm for 1 mM BBSTI-E is 3.5 (calculated from the extinction coefficient given in Kassell (1970)), it is 14 for BBSTI-B and B'. These inhibitors cross-react weakly with anti-BBSTI-E antibodies.

Functionally, these inhibitors inhibit only bovine trypsin, not bovine chymotrypsin or bovine elastase. Our preliminary results show that even at ratios of inhibitor to trypsin far exceeding that needed by BBSTI-E for maximum inhibition, BBSTI-B and B' give far less than 1:1 stoichiometry, indicating very poor trypsin inhibitor activity (i.e. a low association equilibrium constant of complex formation). Hwang et al. (1977) also found PI-I to be the weakest in trypsin inhibitor activity of the isoinhibitors isolated from cultivar Tracy.

BBSTI-A and A'

These isoinhibitors were purified from the pool which elutes before the pH change in the chromatography shown in Figure 1 by a second DEAE-cellulose column in 50 mM Tris-Cl pH 8.5, eluted with a gradient of 0 to 0.5 M NaCl in the same buffer (Welsh-Duggan and Tan-Wilson, unpublished). The amino acid compositions, presented in Table 4, are very different from that of the classical Bowman-Birk inhibitor, continuing the trend toward lower half-cystine and higher content of glycine residues seen in BBSTI-B, BBSTI-B' and in PI-I. BBSTI-E, which has a composition similar to Bowman-Birk type inhibitors of other legume species, has no glycine and has fourteen half-cystine residues, the latter in disulfide bonds (Odani and Ikenaka, 1972; 1973b). In BBSTI-A and A', we found 15 glycine residues and only one half-cystine residue. This unexpected result was confirmed by hydrolysis in methanesulfonic acid + 3-(2-aminoethyl)indole (Simpson et al., 1976) in addition to the standard acid hydrolysis of performic acid-oxidized protein. If the molecular weights of BBSTI-A and A' are indeed calculated minimum molecular weights comparable to that of BBSTI-E, there would be no disulfide bond in these inhibitors. BBSTI-E has eleven aspartyl/asparginyl residues and seven glutamyl/glutaminyl residues. BBSTI-A and A' have the reverse proportion - five aspartyl/asparaginyl and ten glutamyl/glutaminyl residues. At 280 nm, the absorbance for 1 mM BBSTI-E is 3.5, that for BBSTI-A and A' is 60. These calculations were based on the minimum molecular weights calculated from the amino acid analysis data.

On the basis of these characteristics, BBSTI-A and A' would not be considered by most investigators to be Bowman-Birk inhibitors. However, they are good inhibitors of bovine trypsin. In preliminary work, we found maximum inhibition with nearly 1:1 stoichiometry at the same ratios of inhibitor to trypsin which also yield maximum inhibition by BBSTI-E. BBSTI-A and BBSTI-A' do not inhibit bovine chymotrypsin or bovine elastase. Antibodies raised to glutaraldehyde-polymerized pure BBSTI-E cross-react with BBSTI-A and A'. This cross-reaction is very clear in immunostaining with horseradish peroxidase-conjugated second antibody of Western blots after polyacrylamide gel electrophoresis (Obach and Tan-Wilson, unpublished results). Assays which require the precipitation of antigen-antibody complexes do not work well. This is why the fractions in the elution profile in Figure 1, marked BBSTI-A, A' and A" do not show BBSTI as determined by radial immunodiffusion.

The amino acid composition of BBSTI-A' is like that of BBSTI-A except for the lack of two seryl, one glycyl, one alanyl, one leucyl, one tyrosyl and one phenylalanyl residue. It is likely that BBSTI-A' is a proteolytic derivative of BBSTI-A. BBSTI-A" has not yet been purified and characterized as to composition.

Tally of Distinct Bowman-Birk Isoinhibitors

In the ten cultivars studied by Odani and Ikenaka (1977a), Hwang et al. (1977), Stahlhut and Hymowitz (1983) and Tan-Wilson et al. (1985b), there were at least twelve distinct isoinhibitors of the Bowman-Birk class. In order of increasing migration during polyacrylamide gel electrophoresis at alkaline pH, these are: (1) BBSTI-A (2) BBSTI-A' (3) BBSTI-A" (4) PI-I which may be the same as BBSTI-B (5) BBSTI-B' (6) E-I which is also PI-II (7) PI-III which may also be the same as BBSTI-C (8) D-II which is also PI-IV (9) C-II (10) A which is also PI-V, fraction III, BBSTI-E and the classical inhibitor (11) BBSTI-E' and (12) fraction V. Of these, we know of three pairs which are related by proteolytic digestion. BBSTI-D was derived from BBSTI-E by proteolytic digestion of two carboxyl-terminal residues. E-I/PI-II is derived from D-II/PI-IV by the loss of nine amino-terminal residues. BBSTI-A' is like BBSTI-A except for seven missing amino acid residues.

CONCLUSION: SIGNIFICANCE OF SOYBEAN CULTIVAR ISOINHIBITOR COMPOSITION IN NUTRITION

The significance of the Kunitz isoinhibitor composition is made clear by the thousand-fold difference in the strength of interaction of Ti^a, Ti^b and Ti^c with bovine trypsin and the better nutritional quality of strains which bear the Ti^b form as compared to those with the Ti^a form.

Although most soybean strains studied have more Kunitz than Bowman-Birk inhibitor by weight, the contribution of the Bowman-Birk inhibitors to the anti-nutritional quality of soybeans cannot be ignored. The classical Bowman-Birk inhibitor inhibits bovine chymotrypsin a hundred times more effectively than does the Kunitz inhibitor (Harry and Steiner, 1970; Turner et al., 1975; Bosterling and Quast, 1981). There is another Bowman-Birk inhibitor which inhibits bovine elastase (Odani and Ikenaka, 1977a), adding to the number of pancreatic digestive enzymes being affected. The Bowman-Birk inhibitors were more effective at inhibiting human cationic trypsin than was the Kunitz inhibitor (Mallory and Travis, 1975). The Bowman-Birk inhibitors tend to survive heat steps in soybean processing better than do the Kunitz inhibitors. After that, the Bowman-Birk inhibitors are also more resistant to the acid pH and proteolytic destruction in the gastric juice. One particular instance where the antinutritional effect due to the Kunitz trypsin inhibitor assumes greater significance is the feeding of raw soybean milk to infants, a common practice in some countries. Newborn infants might not be able to inactivate the Kunitz inhibitor as an adult would because of the lower acidity of their gastric juice and the faster rate of their gastric emptying (Flavin, 1983).

On the other hand, the contribution of the Bowman-Birk inhibitors and, to some extent, the Kunitz inhibitors, to the *nutritional* quality of soybean must also be considered. Most of the Bowman-Birk inhibitors described in this chapter have twelve or fourteen half-cystine residues out of a total of 70 or 80. Kunitz isoinhibitors have four half-cystine and either one or two methionine residues in a total of 180-200 residues. The soybean storage proteins, which contribute the bulk of the seed protein, are very low in or even lack the sulfur-containing amino acids, L-cysteine, L-cystine, or L-methionine. It is precisely the low levels of the sulfur-containing amino acids that limits the nutritional quality of soy protein for humans as compared to the protein in egg white, milk and meat. Therefore, improvement of the nutritional quality of soybean cannot be accomplished solely by eliminating the Bowman-Birk trypsin inhibitors.

It should be possible, however, to select for those Bowman-Birk inhibitors that are rich in sulfur-containing amino acids but at the same time, are relatively poor in their inhibition of the pancreatic serine proteinases. The discussion in this chapter makes clear the necessity of of making such selection specific for each animal species.

The use of germinated soybean as an alternative to meal prepared from dry seed has been considered. In such practice, it would be noted that Ti^{a}_{m}, Ti^{b}_{m}, Ti^{c}_{m} and BBSTI-D, which arise during germination, still inhibit trypsin. Eventually, these inhibitors get completely digested but the timing has to be considered in relation to the digestion of the storage proteins which form the bulk of the protein reserve in the seed.

We do not know whether the trypsin inhibitors evolved to serve in defense of the seed against predators or to serve as storage proteins to complement the low content of sulfur-containing amino acid in the storage proteins. We do know that one characteristic of these proteins - the serine proteinase inhibitory capacity - is deleterious to many domestic animal livestock species which are fed soybean meal, while the other characteristic - the high cystine content - is potentially beneficial to the same animals. Careful selection of the Kunitz and Bowman-Birk isoinhibitor composition is a good approach to the problem. This makes the characterization of all the isoinhibitors with regard to structure and function, the determination of the isoinhibitor complement in a large number of soybean cultivars, and understanding of the genetic basis for the variability of these trypsin inhibitors imperative.

REFERENCES

Bajjalieh, N., Orf, J.H., Hymowitz, T. and Jensen, A.H. (1980). Response of young chicks to raw, defatted Kunitz trypsin inhibitor variant soybeans as sources of dietary protein. *Poultry Sci.*, *59*, 328-332.

Bosterling, B. and Quast, U. (1981). Soybean trypsin inhibitor (Kunitz) is doubleheaded. Kinetics of the interaction of alpha-chymotrypsin with each side. *Biochim. Biophys. Acta*, *657*, 58-72.

Bates, R.P., Knapp, F.W. and Arajo, P.E. (1977). Protein quality of green-mature, dry mature and sprouted soybeans. *J. Food Sci.*, *42*, 271-272.

Bau, H.M. and Debry, G. (1979). Germinated soybean protein products: chemical and nutritional evaluation. *J. Am. Oil Chem. Soc.*, *56*, 160-162.

Birk, Y. (1985). The Bowman-Birk inhibitor. *Int. J. Peptide Protein Res.*, *25*, 113-131.

Birk, Y., Gertler, A. and Khalef, S. (1963). A pure trypsin inhibitor from soya beans. Biochemical J., 87, 281-284.
Brand, S.J. and Morgan, R.G.H. (1981). The release of rat intestinal cholecystokinin after oral trypsin inhibitor measured by bioassay. J. Physiol., 319, 325-343.
Clark, R.W., Mies, D.W. and Hymowitz, T. (1970). Distribution of a trypsin inhibitor variant in seed proteins of soybean varieties. Crop Sci., 10, 486-487.
Davis, B.J. 91964). Disc electrophoresis. II. Method and application to human serum proteins. Ann. N.Y. Acad. Sci., 121, 404-427.
Figarella, C., Negri, G.A. and Guy, O. (1974). Studies on inhibition of two human trypsins. In "Bayer-Symposium V-Proteinase Inhibitors," H. Fritz, H. Tschesche, L.J. Greene and E. Truscheit, Eds., pp. 213-222, Springer-Verlag, Berlin.
Flavin, D.F. (1984). Soybean trypsin inhibitors: interaction in humans. Vet. Hum. Toxicol., 26, 36-37.
Frattali, V. (1969). Soybean inhibitors. III. Properties of a low molecular weight soybean proteinase inhibitor. J. Biol. Chem., 25, 274-280.
Freed, R.C. and Ryan, D.S. (1978). Note on the modification of the Kunitz soybean trypsin inhibitor during seed germination. Cereal Chem., 55, 534-538.
Freed, R.C. and Ryan, D.S. (1980). Isolation and characterization of genetic variants of the Kunitz soybean trypsin inhibitor. Biochim. Biophys. Acta, 624, 562-572.
Green, G.M. and Lyman, R.L. (1972). Feedback regulation of pancreatic enzymes secretion as a mechanism for trypsin inhibitor-induced hypersecretion in rats. Proc. Soc. Exp. Biol. Med., 140, 6-12.
Green, G.M., Olds, B.A., Matthews, H. and Lyman, R.L. (1973). Protein as a regulator of pancreatic enzyme secretion in the rat. Proc. Soc. Exp. Biol. Med., 192, 1162-1167.
Hammond, R.W., Foard, D.E. and Larkins, B.A. (1984). Molecular cloning and analysis of a gene coding for the Bowman-Birk protease inhibitor in soybean. J. Biol. Chem., 9883-9890.
Harry, J.B. and Steiner, R.F. (1970). A soybean proeinase inhibitor. Thermodynamic and kinetic parameters of association with enzymes. Eur. J. Biochem., 16, 174-179.
Hartl, P., Tan-Wilson, A.L. and Wilson, K.A. (1985). Proteolysis of Kunitz soybean trypsin inhibitor during germination. Phytochemistry, in press.
Hirs, C.H.W. (1967). Performic acid oxidation. Methods Enzymol., 11, 199-203.
Holm, H. and Krogdahl, A. (1982). Problems in predicting the inhibition of human pancreatic proteinases by soybean proteinase inhibitors. In vitro assays employing human, bovine and porcine proteinases. J. Sci. Food Agric., 33, 1164-1171.
Hwang, D.L-R., Lin, K-T.D., Yang, W-K. and Foard, D.E. (1977). Purification, partial characterization, and immunological relationships of multiple low molecular weight protease inhibitors of soybean. Biochim. Biophys. Acta, 495, 369-382.
Johnson, L.A., Devoe, C.W., Hoover, W.J. and Schwenke, J.R. (1980). Inactivation of trypsin inhibitors in aqueous soybean extracts by direct steam infusion. Cereal Chem., 57, 376-379.
Kakade, M.L., Simons, N.R., Liener, I.E. and Lambert, J.W. (1972). Biochemical and nutritional assessment of different varieties of soybeans. J. Agric. Food Chem., 20, 87-90.
Kakade, M.L., Hoffa, D.E. and Liener, I.E. (1974). Contribution of trypsin inhibitors to the deleterious effects of unheated soybeans fed to rats. J. Nutr., 103, 1772-1778.
Kassell, B. (1970). Trypsin and chymotrypsin inhibitors from soybeans. Methods Enzymol., 9, 853-862.

Kassell, B. and Laskowski, M. (1956). The comparative resistance to pepsin of six naturally occurring trypsin inhibitors. J. Biol. Chem., 219, 203-210.
Katagiri, C. and Shimizu, M. (1983). Trypsin inhibitors of Kurodaizu (Glycine max Merrill forma Kuromame Makino) Part I. Separation and purification of trypsin inhibitors. Shokumotsu Gakkaishi, 38, 17-21.
Kim, H-G., Kim, M-C., Chang, K-Y. and Kin, J-K. (1982). Trypsin inhibitors from various soybean varieties. Korean J. of Food Sci. and Technology, 14, 106-111.
Koide, T. and Ikenaka, T. (1973). Studies on soybean trypsin inhibitors 3. Amino acid sequence of the carboxyl-terminal region and the complete amino acid sequence of the soybean trypsin inhibitor (Kunitz). Eur. J. Biochem., 32, 417-431.
Konijn, A.M., Birk, Y. and Guggenheim, K. (1970). In vitro synthesis of pancreatic enzymes. Effect of soybean trypsin inhibitor. Am. J. Physiol., 218, 1113-1117.
Krogdahl, A. and Holm, H. (1981). Soybean proteinase inhibitors and human proteolytic enzymes: selective inactivation of inhibitors by treatment with human gastric juice. J. Nutr., 111, 2045-2051.
Krogdahl, A. and Holm, H. (1983). Pancreatic proteinases from man, trout, rat, pig, cow, chicken, mink, and fox. Enzyme activities and inhibition by soybean and lima bean proteinase inhibitors. Comp. Biochem. Physiol., B, 74B, 403-409.
Kunitz, M. (1974). Isolation of a crystalline protein compound of trypsin and of soybean trypsin inhibitors. J. Gen. Physiol., 30, 311-320.
Laskowski, M. and Kato, I. (1981). Protein inhibitors of proteinases. Ann. Rev. Biochem., 49, 583-626.
Laskowski, M. and Sealock, R.W. (1971). Protein proteinase inhibitors - molecular aspects. In "The Enzymes", 3rd ed., P.D. Boyer, Ed., pp. 376-473, Academic Press, New York.
Lebowitz, J. and Laskowski, M. (1962). Potentiometric measurement of protein-protein association constants. Soybean trypsin inhibitor-trypsin association. Biochemistry, 1, 1044-1055.
Madar, Z. (1979). Kinetics of native and modified Bowman-Birk soya bean trypsin inhibitor on growth and enzyme activities of the chick pancreas. J. Nutr., 42, 121-126.
Madden, M.A., Tan-Wilson, A.L. and Wilson, K.A. (1985). Proteolysis of soybean Bowman-Birk trypsin inhibitor during germination. Phytochemistry, in press.
Mallory, P.A. and Travis, J. (1973). Human pancreatic enzymes. Characterization of anionic human trypsin. Biochemistry, 12, 2847-2851.
Mallory, P.A. and Travis, J. (1975). Inhibition spectra of the human pancreatic endopeptidases. Am. J. Clin. Nutr., 28, 823-830.
Mancini, G., Carbonara, A.O. and Heremans, J.H. (1965). Immunochemical quantitation of antigens by single radial immunodiffusion. Immunochemistry, 2, 234-254.
Melmed, R.N. and Bouchier, I.A.D. (1969). Further physiological role for naturally occurring trypsin inhibitors. Evidence for a trophic stimulant of the pancreatic acinar cell. Gut, 10, 973-979.
Melmed, R.N., El-Aaser, A.A.A. and Holt, S.J. (1976). Hypertrophy and hyperplasia of the neonatal rat exocrine pancreas induced by orally administered soybean trypsin inhibitor. Biochim. Biophys. Acta, 421, 280-288.
Oates, P.S., Bruce, N.W. and Morgan, R.G.H. (1984). Pancreatic blood flow in the rat during enlargement, involution and cholecystokinin treatment. Am. J. Physiol., 247, G457-G462.
Odani, S. and Ikenaka, T. (1972). Studies on soybean trypsin inhibitors. IV. Complete amino acid sequence and the anti-proteinase sites of Bowman-Birk soybean proteinase inhibitor. J. Biochem., 71, 839-848.

Odani, S. and Ikenaka, T. (1973a). Scission of soybean Bowman-Birk proteinase inhibitor into two small fragments having either trypsin or chymotrypsin inhibitory activity. J. Biochem., 74, 857-860.
Odani, S. and Ikenaka, T. (1973b). Studies on soybean trypsin inhibitors. VIII. Disulfide bridges in soybean Bowman-Birk proteinase inhibitor. J. Biochem., 74, 697-715.
Odani, S. and Ikenaka, T. (1977a). Studies on soybean trypsin inhibitors. X. Isolation and partial characterization of four soybean double-headed proteinase inhibitors. J. Biochem., 82, 1513-1522.
Odani, S. and Ikenaka, T. (1977b). Studies on soybean trypsin inhibitors. XI. Complete amino acid sequence of soybean trypsin-chymotrypsin-elastase inhibitor, C-II. J. Biochem., 82, 1523-1531.
Odani, S. and Ikenaka, T. (1978). Studies on soybean trypsin inhibitors. XII. Linear sequences of two soybean double-headed trypsin inhibitors, D-II and E-I. J. Biochem., 83, 737-745.
Orf, J.H. and Hymowitz, T. (1979). Genetics of the Kunitz trypsin inhibitor: an antinutritional factor in soybeans. J. Am. Oil Chem. Soc., 56, 722-726.
Orf., J.H., Mies, D.W. and Hymowitz, T. (1977). Qualitative changes of the Kunitz trypsin inhibitor in soybean seeds during germination as detected by electrophoresis. Bot. Gaz., 138, 255-260.
Rackis, J.J. (1965). Physiological properties of soybean trypsin inhibitors and their relationship to pancreatic hypertrophy and growth inhibition of rats. Fed. Proc., 24, 1488-1493.
Rackis, J.J. (1974). Biological and physiological factors in soybeans. J. Amer. Oil Chem. Soc. 51, 161A-174A.
Rackis, J.J. and Gumbmann, M.R. (1981). Protease inhibitors: physiological properties and nutritional significance. In "Antinutrients and Natural Toxicants in Foods," R.L. Ory, Ed., pp. 203-237, Food and Nutrition Press, Westport, Conn.
Richardson, M. (1977). The proteinase inhibitors in plants and microorganisms. Phytochemistry, 16, 159-169.
Ryan, C.A. (1973). Proteolytic enzymes and their inhibitors in plants. Ann. Rev. Plant Physiol., 24, 173-196.
Schwert, G.W. and Takenaka, Y. (1955). A spectrophotometric determination of trypsin and chymotrypsin. Biochim. Biophys. Acta, 16, 570-575.
Seidl, D.S. and Liener, I.E. (1971). Identification of the trypsin-reactive site of the Bowman-Birk soybean inhibitor. Biochim. Biophys. Acta, 251, 83-93.
Seidl, D.S. and Liener, I.E. (1972a). Identification of the chymotrypsin-reactive site of the Bowman-Birk soybean inhibitor. Biochim. Biophys. Acta, 258, 303-309.
Seidl, D.S. and Liener, I.E. (1972b). Isolation and properties of complexes of the Bowman-Birk soybean inhibitor with trypsin and chymotrypsin. J. Biol. Chem., 247, 3533-3538.
Silva, A.D., Barbosa, C.F. and Portela, F. (1979). Inibidores proteoliticos em variedades de soja. Cientifica, 7, 317-320.
Simpson, R.J., Neuberger, M.R. and Liu, T-Y. (1976). Complete amino acid analysis of proteins from a single hydrolysate. J. Biol. Chem., 251, 1936-1940.
Singh, L., Wilson, C.M. and Hadley, H.H. (1969). Genetic differences in soybean trypsin inhibitors separated by disc electrophoresis. Crop Sci., 9, 489-491.
Stahlhut, R.W. and Hymowitz, T. (1983). Variation in the low molecular weight proteinase inhibitors of soybeans. Crop Sci., 23, 766-769.
Struthers, B.J. and MacDonald, J.R. (1983). Comparative inhibition of trypsins from several species by soybean trypsin inhibitors. J. Nutr., 113, 800-804.

Staswick, R.E., Hermodson, M.A. and Nielsen, N.C. (1981). Identification of the acidic and basic subunit complexes of glycinin. J. Biol. Chem., 256, 8752-8755.
Tan-Wilson, A.L. and Wilson, K.A. (1982). Nature of proteinase inhibitors released from soybeans during imbibition and germination. Phytochemistry, 21, 1547-1551.
Tan-Wilson, A.L., Rightmire, B.R. and Wilson, K.A. (1982). Different rates of metabolism of soybean proteinase inhibitors during germination. Plant Physiol., 70, 493-497.
Tan-Wilson, A.L., Rightmire, B.R. and Wilson, K.A. (1983). Determination of relative antigen-antibody avidities by radial immunodiffusion. J. Immunol. Methods, 61, 99-106.
Tan-Wilson, A.L., Hartl, P., Delfel, N.E., and Wilson, K.A. (1985a). Differential expression of Kunitz and Bowman-Birk soybean proteinase inhibitors in plant and callus tissue. Plant Physiol., 78, 310-314.
Tan-Wilson, A.L., Cosgriff, S.E., Duggan, M.C., Obach, R.S. and Wilson, K.A. (1985). Bowman-Birk proteinase isoinhibitor complements of soybean strains. J. Agric. Food Chem., 133, 389-393.
Thanh, V.H. and Shibasaki, K. (1977). Beta-conglycinin from soybean proteins. Isolation and immunological and physiochemical properties of the monomeric forms. Biochim. Biophys. Acta, 490, 370-384.
Travis, J. and Roberts, R. (1969). Human trypsin isolation and physical chemical characterization. Biochemistry, 8, 2884-2889.
Turner, R., Liener, I.E. and Lovrien, R.E. (1975). Equilibria of Bowman-Birk inhibitor association with trypsin and alpha-chymotrypsin. Biochemistry, 14, 275-282.
Wilson, K.A. (1981). The structure, function and evolution of legume proteinase inhibitors. In "Antinutrients and Natural Toxicants in Foods," R. L. Ory, Ed., pp. 187-202, Food and Nutrition Press, Westport, Conn.
Wilson, K.A. and Chen, J.C. (1983). Amino acid sequence of mung bean trypsin inhibitor and its modified forms appearing during germination. Plant Physiol., 71, 341-349.
Wilson, K.A. and Laskowski, M. (1975). The partial amino acid sequence of trypsin inhibitor II from garden bean, *Phaseolus vulgaris*, with location of the trypsin and elastase reactive sites. J. Biol. Chem., 250, 4261-4267.
Wilson, K.A. and Tan-Wilson, A.L. (1983). Proteinases involved in the degradation of trypsin inhibitor in germinating mung beans. Acta Biochim. Pol., 30, 139-148.
Wilson, K.A., Rightmire, B.R. and Tan-Wilson, A.L. (1985). Involvement of carboxypeptidase in the degration of the mung bean (*Vigna radiata*) trypsin inhibitor during germination and early seedling growth. Qual. Plant. Plant Foods Hum. Nutr., in press.
Yen, J.T., Hymowitz, T., and Jensen, A.H. (1971). Utilization by rats of protein from a trypsin-inhibitor variant soybean. J. Anim. Sci., 33, 1012-1017.
Yen, J.T., Jensen, A.H., and Simon, J. (1977). Effect of dietary raw soybean and soybean trypsin inhibitor on trypsin and chymotrypsin activities in the pancreas and in small intestinal juice of growing swine. J. Nutr., 107, 156-165.
Yoshikawa, M., Kiyohara, T., Iwasaki, T. and Yoshida, I. (1979). Modification of proteinase inhibitor II in adzuki beans during germination. J. Agric. Food Chem., 43, 1989-1990.

23

ANTINUTRITIONAL AND BIOCHEMICAL PROPERTIES OF WINGED BEAN TRYPSIN INHIBITORS

Benito O. de Lumen and Jones Chan

Department of Nutritional Sciences
University of California
Berkeley, CA 94720

ABSTRACT

The trypsin inhibitors (TI) of winged bean (Psophocarpus tetragonolobus) isolated by affinity chromatography consisted of 8 protein bands by disc polyacrylamide electrophoresis. All the bands exhibited trypsin inhibitor activity (TIA) with two of the major bands also exhibiting chymotrypsin inhibitor activity (CIA). Electrofocusing separated the TI into 5 protein bands. Two bands with dual TIA/CIA activities had acidic pI while 3 bands with TIA only had alkaline pI. The TI belonged to two molecular weight groups of 20,900 and 16,600, as determined by $NaDodSO_4$ polyacrylamide electrophoresis.

Sufficient quantities of TI were isolated by affinity chromatography for rat feeding. Raw winged bean was toxic to rats, causing death after 12 days of feeding. Autoclaved winged bean was not toxic but caused growth inhibition. When fed with casein, the isolated inhibitor caused slight growth inhibition, pancreatic and spleen hypertrophy. It was concluded that TI in winged bean was not primarily responsible for the toxicity of raw winged bean.

INTRODUCTION

In 1975, after examining 400 neglected tropical plants, the National Academy of Sciences selected the winged bean, together with 35 other plants as a potential food source which warrants further research (National Academy of Sciences, 1975). The winged bean is indigenous to the tropics, chiefly in Southern Asia and the Western Pacific. While the green pods are the most commonly used part of the plant, almost all parts of the plant are considered edible. However, the greatest potential of the winged bean is in the mature seed, whose chemical composition is strikingly similar to that of soybean (Table I).

Under comparable experimental conditions, the antinutritional effects of raw winged bean such as growth inhibition, pancreatic hypertrophy, and ultimately death are more deleterious than those of raw soybean (Jaffe and Vega, 1968). In rats, raw soybeans cause low PER while raw winged beans cause death within a few days. The presence of a heat-labile proteinase inhibitor in winged bean was noted by Sohonie and Bhandakar (1954) and Cerny et al. (1971). We showed that the trypsin inhibitor activity (TIA)

Table 1. Composition of Major Nutrients in Mature Winged Bean Seeds Compared With Other Legumes (in 100 g edible portion)

	Winged Bean	Soybean	Peanut	Cowpea	Chick-pea
Moisture %	8.7-14.0	10.2	7.3	11.5	11.0
Calories	375-410	400	548	340	362
Fat g	15.0-18.3	17.7	45.3	1.6	5.6
Crude Protein g	29.8-37.4	35.1	23.4	22.7	19.4
Carbohydrates g	25.2-38.4	32.0	21.6	61.0	60.9
Fiber g	3.7-9.4	4.2	2.1	4.2	2.5
Ash g	3.3-4.3	5.0	2.4	3.2	3.1

(Cerny, 1978).

levels of two commonly studied winged bean varieties, Chimbu and Tpt-1 are higher than other beans (Table II.)

The exact nutritional significance of trypsin inhibitors to man and experimental animals has been difficult to evaluate due to the presence of other antinutritional factors in legume seeds (Liener, 1974). Isolation of inhibitors and determination of their biological activities by animal feeding will ultimately have to be done to establish their nutritional significance. Partially purified inhibitors from field beans (Vicia faba) incorporated into rat diets cause depressed growth and pancreatic hypertrophy only when fed in amounts up to 5 times the normal endogenous levels found in field beans (Abbey et al., 1979a, b). Rackis (1965) showed that 100% of the pancreatic hypertrophy and 30-60% of growth inhibition of raw soybean can be attributed to the purified trypsin inhibitor from soybean. When soybean trypsin inhibitor (TI) was removed from the water extract of raw soybean by affinity chromatography and the "TI-free extract" was fed to rats, 40% of pancreatic hypertrophy and growth depression was estimated to be due to TI (Kakade et al., 1973).

The degree to which TI is responsible for the deleterious effects of raw winged bean has not been elucidated. In this study, we isolated TI from winged bean by affinity chromatography, characterized the isolated fractions and fed them to rats to determine their antinutritional activity.

MATERIALS AND METHODS

Extraction of TI. Ground winged bean (Chimbu variety) was defatted by extracting with hexane (1:5, bean:hexane) five times. The TI were extracted by homogenizing the defatted powder in cold distilled water (5 mL water/g bean powder) for 5 min. in a Waring blender and stirring the homogenate overnight at 4°C. After centrifugation, the supernatant was collected, filtered and lyophilized.

Preparation of Affinity Column. Procedures to activate agarose and to immobilize trypsin were done as described by March et al. (1974) and Loeffler and Pierce (1973), respectively.

Isolation of TI. A 50 x 2.5 cm plastic column was packed with 200 ml of trypsin-agarose. The gel was equilibrated by washing first with 1 volume of 0.2M KCl, pH. 2.0, followed by 2 volumes of Tris buffer (0.05M Tris, 0.1M KCl, and 0.02M $CaCl_2$, pH. 8.0). A 9% bean extract was prepared by dissolving the lyophilized extract in pH 8.0 Tris buffer. The extract

Table II. Trypsin Inhibitor Activity in Winged Bean as Compared to Other Beans

bean	TIU/g of flour[a]
winged bean (Psophocarpus tetragonolobus)	
variety Chimbu	3691
variety Tpt-1	1716
large lima (Phaseolus lunatus)	1952
pink (Phaseolus vulgaris)	419
blackeye (Vigna unguiculata)	619
garbanzo (Cicer arietinum)	419
dark red kidney (Phaseolus vulgaris)	1040

[a] Average of three determinations of two levels of trypsin. One TIU (trypsin inhibitory unit) is defined as one unit of trypsin activity inhibited. One unit of trypsin activity is equal to 1 μmol of TAME hydrolyzed per minute.

(de Lumen and Salamat, 1980)

containing a total TIU less than the total trypsin unit of the column was applied to the column. The non-TI fractions were eluted with pH 8.0 Tris buffer, complete removal being monitored by low 280-nm absorption. To dissociate the trypsin-TI complex, the eluting solvent was changed to 0.2M KCl, pH. 2.0. All fractions were monitored for TIA. The TI fractions were pooled, dialyzed and lyophilized.

To prepare sufficient amounts of TI for rat feeding, the isolation procedure used in analytical affinity chromatography was scaled up 10-fold.

TI Assay. TIA was measured by the decrease in the rate by which trypsin hydrolyzed p-toluenesulfonyl-L-arginine methyl ester (TAME), as monitored at 247 nm with a Cary 14 recording spectrophotometer. One unit of trypsin activity represents the hydrolysis of 1 μmol of TAME/min. at 25^oC and pH 8.1 in the presence of 0.01 M $CaCl_2$. One unit of trypsin activity inhibited is defined as 1 TIU.

Gel Electrophoresis. Ornstein's (1964) and Davis' (1964) procedures for discontinuous polyacrylamide electrophoresis were used. The acrylamide concentration of the gel was 7.5%. The lyophilized TI was dissolved in a 1:10 dilution of the stock Tris-glycine buffer, pH 8.3, to yield a final concentration of 1 mg/ml. The system was run at pH 8.3 with a constant current of 2.7 ma/tube. The gels were stained with 0.25% Coomasie blue R-250 in methanol:acetic acid:water (5:1:5) for 30 min. and destained with 7% acetic acid and 7% methanol in water.

$NaDodSO_4$ gel electrophoresis was conducted according to the procedure described in LKB-Produkter AB (1977). Slab gels at a concentration of 10% were run in pH 7.0 phosphate buffer, at 4^oC with a constant current of 80 ma. Molecular weight standards were run alongside the samples.

Electrofocusing. Electrofocusing was carried out on ready-made ampholine polyacrylamide gel plates (LKB-Produkter AB), pH 3.5-9.5, at 10^oC under a constant current of 20 ma. The pH gradient was measured with a Corning microelectrode. The staining and destaining were performed as described in

Table III. Composition of Diets

ingredients	g/100 g diet			
	casein	RWB	AWB	casein + TI (WBTI)[a]
protein	12.0	28.0	30.0	12.0
corn oil	5.0	1.0	1.0	5.0
vitamin premixture (2 x NRC)	1.0	1.0	1.0	1.0
mineral mixture	4.0	4.0	4.0	4.0
choline bitartrate mixture	1.0	1.0	1.0	1.0
DL-methionine	0.3	0.3	0.3	0.3
powdered sucrose	75.7	65.3	63.3	75.7

[a] Freeze-dried isolated winged bean trypsin inhibitor was added in an amount (0.67 g/100 g of diet) equivalent to the total inhibitor activities of RWB without replacing other diet components.

(Chan and de Lumen, 1982)

LKB Produkter AB (1977).

Chromatic Protease Staining. TIA and CIA were detected in polyacrylamide gels using acetyl-DL-lysyl-B-naphthyl ester and acetyl-DL-phenylalanine-β-naphthyl ester with a trypsin concentration of 0.01 mg/ml or a chymotrypsin concentration of 0.02 mg/ml (Uriel and Berges, 1968: Rosenberg et al., 1976). The β-naphthol formed was detected by reaction with o-dianisidine. The gels were destained with 2% acetic acid for 30 min.

Amino Acid Analysis. Four hundred mg of lyophilized WBTI were hydrolyzed, in vacuo, in 6N HCl for 22 hrs. at 110°C. Amino acid analysis was done with a Beckman Amino Acid Analyzer, Model 121, according to the method of Moore and Stein (1963).

pH and Thermal Stability of WBTI. A 20 mg/ml WBTI solution was prepared by dissolving lyophilized WBTI in Tris-glycine, pH 8.3. Three buffer systems were employed to establish a pH range of 1-12 (Gomori, 1955). One-tenth ml of WBTI solution was added to 9.9 ml of buffer solution at certain pH, incubated for 20 min. at 25°C, and diluted 100X with the assay buffer. This dilution was necessary to eliminate the effect of pH on the TAME assay.

To determine thermal stability, 2 ml aliquots of 0.002 mg/ml WBTI solution were immersed in a 100°C waterbath for 0-60 min. and at 60°C for 0-2 hrs., cooled to 25°C and assayed for TIA.

Feeding Study. The rats were fed for 28 days on either casein, raw winged bean (RWB), autoclaved winged bean (AWB) or casein plus isolated trypsin inhibitor (WBTI) as the sole protein source (Table III). The WBTI diet contained 0.67 g isolated WBTI/100 g diet to provide a level of TIU equivalent to that of RWB. Since the WBTI was added at a low level, there was no need to replace other ingredients in the diet. For each diet, a 10% level of protein was used. AWB, which was prepared by autoclaving ground raw winged bean at 121°C for 25 min., contained negligible TIA. The protein and fat were adjusted to the desired level at the expense of sucrose. Six 21-day old, male Sprague-Dawley rats were randomly assigned to each treatment group. They were fed ad libitum and weighed every other day.

At the end of 28 days, the rats were sacrificed and the pancreas, liver and spleen were excised and weighed immediately. The data were analyzed with one-way analysis of variance and Tukey's Studentized Range Test to determine differences among groups.

RESULTS

Extraction and Isolation of WBTI. The water extraction yielded and average of 8800 TIU/g of winged bean. The results were similar for both small and large scale extractions.

The affinity chromatographic elution profile is shown in Fig. 1. The major protein peak eluted by the first solvent contained no TIA. The minor protein peak eluted by the second solvent had TIA, and the 280 nm-absorbance coincided exactly with the TIA. Measured by its specific activity, the isolated WBTI was purified 8-fold over the extract before chromatography. Approximately, a total of 8.5×10^{6} TIU in 18.5 g of lyophilized samples were isolated for the feeding study. This represented an average yield of 19.2 g of lyophilized WBTI/kg winged bean.

Electrophoretic Properties. Eight protein bands, each exhibiting TIA, were separated on non-denaturing disc gel electrophoresis (Fig. 2). Two of the major protein bands showed both TIA and CIA.

With electrofocusing, the WBTI fraction was separated into five components, each exhibiting TIA (Fig. 3). The two protein bands that contained both TIA and CIA had isoelectric points of pH 5.0 and 6.0, whereas the three protein bands that contained only TIA had isoelectric points of pH 8.8, 9.6 and 10.2. Thus, the dual activity components with lower isoelectric points are more acidic than the components containing only TIA.

Once denatured by $NaDodSO_4$, the heterogeneity of WBTI was reduced to two bands with molecular weights of 20,900 and 16,000, regardless of whether mercaptoethanol was added or not (Fig. 4 and 5).

Effects of pH and Temperature on WBTI. WBTI was stable at acidic conditions, but became labile in alkaline conditions beginning at pH greater than 8.0 (Fig. 6).

Thermal stability of WBTI. The WBTI was stable at 60°C, losing no activity after 120 min. of heating (Fig. 7). However, at 100°C the TIA decreased with time in a triphasic manner. During the first 7 min. of heating, TIA decreased with time, leveled off between 10 and 20 min., and decreased gradually thereafter.

Amino Acid Composition. Taking into considerations that the fraction contained 8 protein components, the amino acid composition was determined. WBTI had high levels of acidic and basic amino acids, proline and serine and a low level of methionine (Table IV). Compared with the whole bean, the methionine and lysine content of the WBTI was higher.

Antinutritional Effects on Rats. The weekly weight gains of the rats are shown in Fig. 8. At any week, the weight gains were highest in the casein control, followed by WBTI and then AWB. The rats fed RWB died between days 7 and 12. These differences are also reflected on the PER values in Table V.

The WBTI group had significantly higher relative pancreatic weight and lower PER than the control group. The PER of the heated bean group

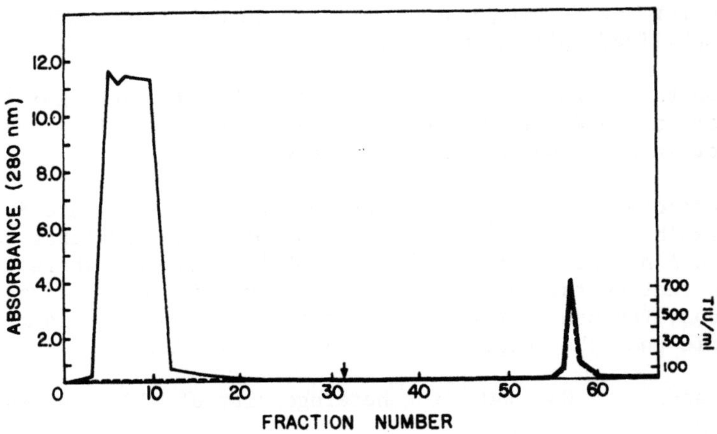

Figure 1. Affinity column separation of winged bean trypsin inhibitor. Two hundred milligrams of the lyophilized winged bean extract in 20 mL of Tris buffer, pH 8.0, was added to a column (50 x 2.5 cm) of agarose-bound trypsin. The arrow indicates a change of eluent from Tris buffer, pH 8.0, to 0.2 M KCl, pH 2.0. Four-milliliter fractions were collected. (—) indicates 280-nm absorption and (---) indicates TIA.

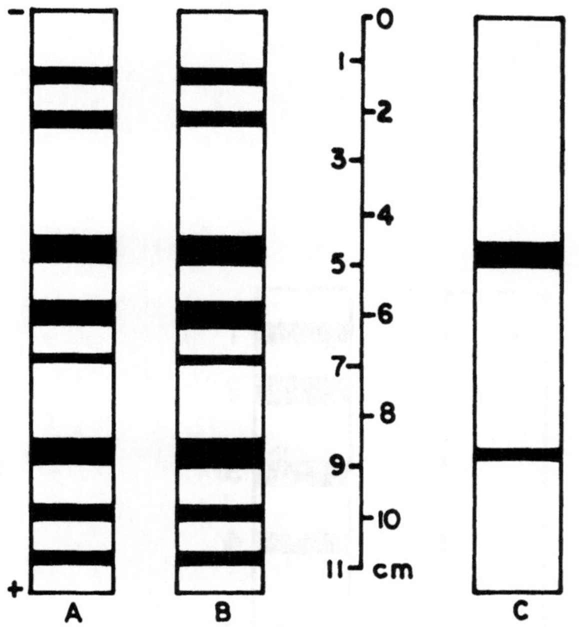

Figure 2.
Electrophoresis of WBTI isolated by affinity column. The acrylamide concentration of the gels was 7.5%. By use of a Tris-glycine buffer, the gel was run at a pH of 8.3 for 4.5h. A constant current of 2.7 mA was applied per tube. (A) Trypsin inhibitory activity. (B) Protein bands. (C) Chymotrypsin inhibitory activity. The bands in gels A and C actually represent the absence of staining.

Figure 3.
Electrofocusing of WBTI. The WBTI was run on a ready-made ampholine plate with a pH range of 3.5-9.5. Twenty milliamperes was applied per tube for 3.5h. (A) Trypsin inhibitory activity. (B) Protein bands. (C) Chymotrypsin inhibitory activity. The bands in gels A and C actually represent the absence of staining.

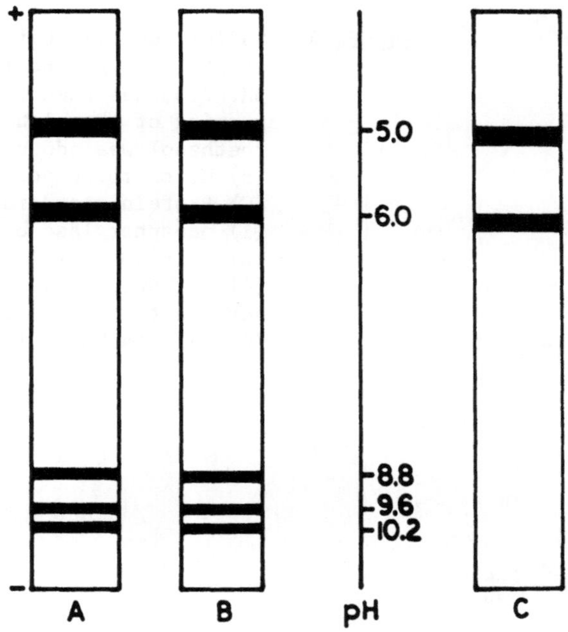

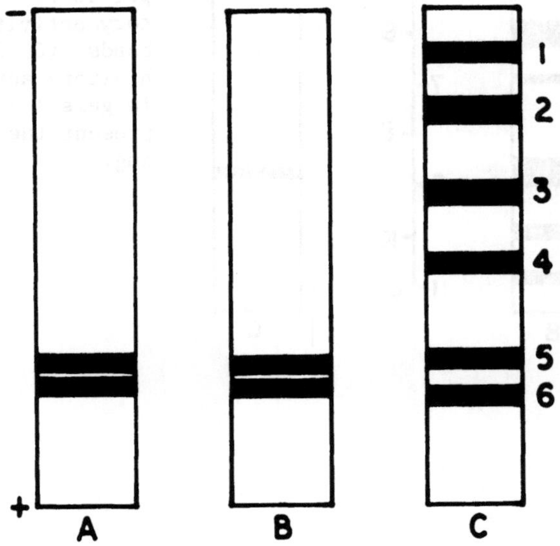

Figure 4. Sodium dodecyl sulfate electrophoresis of WBTI. A phosphate buffer system, pH. 7.0, was used for electrophoretic running of the slab gel. (A) Mercaptoethanol was added to the system. (B) Mercaptoethanol was not added. (C) Protein standards were applied: (1) phosphorylase b (94000); (2) albumin (67000); (3) ovalbumin (43000); (4) carbonic anhydrase (30000); (5) soybean trypsin inhibitor (20100); (6) α-lactalbumin (14400).

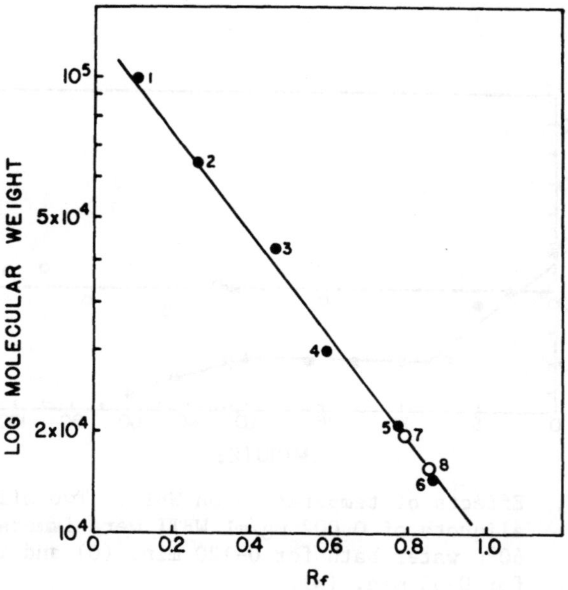

Figure 5. R_fs (relative mobility) of protein standards plotted against the log of the molecular weight. Numbers 7 and 8 correspond to the two protein bands displayed on the gel when $NaDodSO_4$ electrophoresis was performed on the WBTI.

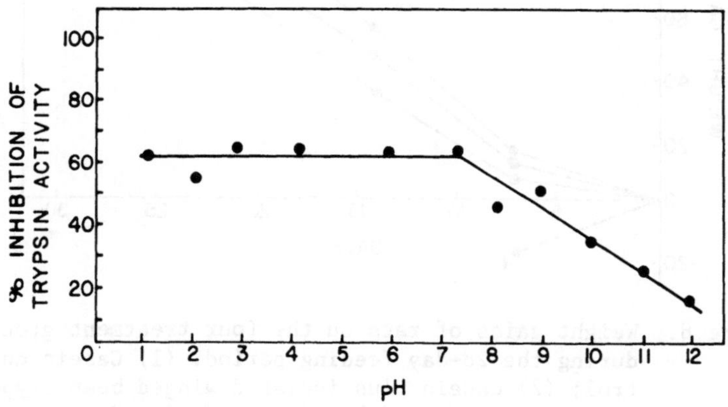

Figure 6. Effects of pH on trypsin inhibitory activity of the WBTI. The WBTI solution was incubated in different pHs from 1 to 12 for 20 min. at 25°C, after which the TIA of the solution was assayed.

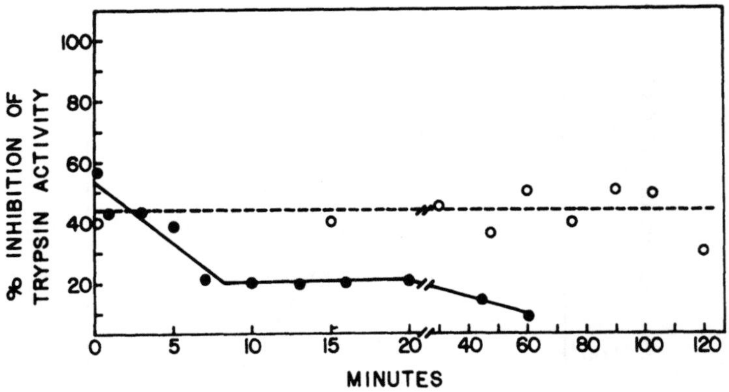

Figure 7. Effects of temperature on WBTI. Two-milliliter aliquots of 0.002 mg/mL WBTI were immersed in a 60^oC water bath for 0-120 min. (o) and in 100^oC for 0-60 min. (•).

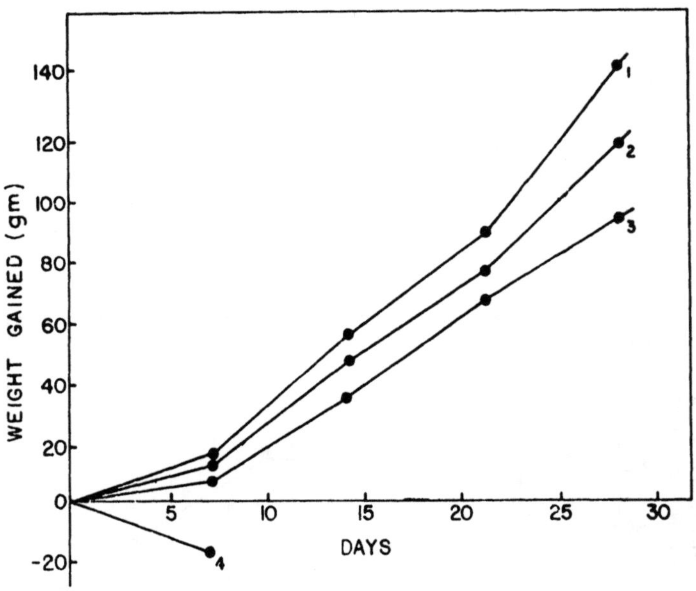

Figure 8. Weight gains of rats on the four treatment groups during the 28-day feeding period. (1) Casein control; (2) casein plus isolated winged bean trypsin inhibitor (WBTI); (3) autoclaved winged bean (AWB); (4) raw winged bean (RWB).

Table IV. Amino Acid Composition of WBTI and Winged Bean Seed Protein

amino acid	winged bean trypsin inhibitor, mg/g of N[c]	winged bean[a] seeds, mg/g of N
isoleucine	171	306-350
leucine	723	462-564
lysine	699	413-600
methionine	118	75-87
cystine	b	73
phenylalanine	34	214-362
tyrosine	506	195-200
threonine	231	269-287
tryptophan	b	47-63
valine	190	242-306
arginine	434	400-440
histidine	89	169-183
alanine	325	230-269
aspartic acid	882	719-781
glutamic acid	1142	856-1154
glycine	445	259-269
proline	776	425-431
serine	602	306-327

[a] Cerny, 1978

[b] Not determined

[c] Chan and de Lumen, 1982

Table V. Relative Organ Weights, PER, and First Week Feed Consumption of Rats After Consuming the Four Treatment Diets for a Period of 28 Days[a]

	g/100 g of body wt					
	liver	spleen	pancreas	wt. gained, g	PER	1st week feed consumed, g
casein control	5.35 ± 0.28[a]	0.277 ± 0.0083[a]	0.493 ± 0.49[a]	133.0 ± 5.84[a]	3.97 ± 0.072[a]	50.2
raw bean (RWB)	2.73 ± 0.06[b]	0.100 ± 0.0081[b]	0.518 ± 0.05[a]	b	b	22.2
heated bean (AWB)	4.65 ± 0.18[a]	0.230 ± 0.0093[a]	0.533 ± 0.04[a]	92.75 ± 8.15[b]	3.05 ± 0.050[b]	43.0
casein + TI (WBTI)	4.78 ± 0.26[a]	0.338 ± 0.0270[c]	0.740 ± 0.02[b]	115.2 ± 2.40[a]	3.72 ± 0.048[c]	45.7

[a] All the values are expressed as mean ± SEM. All groups not designated with the same letter within the same column are statistically different at $p < 0.05$. [b] Rats died between day 7 and day 12.

(Chan and de Lumen, 1982)

(AWB) was lower than the WBTI group, yet no pancreatic hypertrophy was observed. The relative weight of the spleen was significantly highest in the WBTI group and lowest in the RWB group. The liver weight and the first week feed consumption in the RWB group were lower than the rest.

DISCUSSION

Through affinity chromatography, we were able to prepare sufficient amounts of WBTI for rat feeding and characterize them. The WBTI was heterogeneous, showing 8 protein bands on non-denaturing polyacrylamide electrophoresis. The trypsin inhibitors of legumes are generally made up of several protein components. For example, the TI fractions from lima beans (Frankel-Conrat et al., 1952; Jones et al., 1963; Haynes and Feeney, 1967), navy bean (Bowman, 1971), soybean (liener and Kakade, 1969) and Lathyrus savitus seed (Roy, 1980) contain 6, 5, 8 and 5 components respectively.

Since the ultimate objective of this experiment was to determine the antinutritional effects of the isolated WBTI, it was important to establish that all the protein components contained TIA. Indeed, all 8 bands possessed TIA, while two bands had chymotrypsin inhibitor activity (CIA) as well. With electrofocusing, the dual activity proteins were shown to be acidic, while the bands with TIA only were highly alkaline.

Once denatured by $NaDodSO_4$, the number of bands was reduced to 2, with molecular weights of 20,900 and 16,600 regardless of whether mercaptoethanol was added or not, suggesting that all the 8 protein components belong to either one of these two molecular weight classes. These are large compared to the majority of plant proteinase inhibitors which have molecular weights between 8000 and 10000 (Richardson, 1981). The 8 soybean TI are also large with two molecular weight classes of 20000-24000 and 14000-18000. The WBTI fraction isolated by Kortt (1979) also yielded large molecular weight TI. The WBTI was, however, not isolated from the whole bean but from a protein fraction recovered by various isoelectric precipitations of the winged bean extract (Gillespie and Blagrove, 1978). From this fraction, Kortt isolated 3 TI: WBTI-1, WBTI-2 and WBTI-3. WBTI-2 and WBTI-3 are basic proteins with molecular weights of 20500, and account for 60% of the protein in the TI fraction. WBTI-1 is an acidic protein with a molecular weight of 10500 and accounts for 10% of the WBTI fraction. All 3 WBTI exhibit both TIA and CIA.

Since the number of bands in $NaDodSO_4$ electrophoresis was the same whether mercaptoethanol was added or not, we concluded that none of the subunits of WBTI are linked by disulfide linkages. This is contrary to the findings of Kortt (1979), who found additional minor bands when mercaptoethanol was added during $NaDodSO_4$ electrophoresis.

WBTI was stable under acidic conditions but became labile beginning at pH greater than 8.0. Many inhibitors are stable at low pH but a few are quickly destroyed (Richardson, 1977). The ability of inhibitors to remain active during passage through the stomach determines to a large extent whether they would have adverse physiological effects in the intestine. Our data indicate that WBTI can survive the acid environment in the stomach.

The WBTI remained active even after 2 hrs. of heating at 60°C. At 100°C, inactivation proceeded in a triphasic manner, suggesting that the 8 protein components possess varying heat stabilities. Overall, the WBTI is less heat stable than other plant inhibitors which remain active after heating at 80-100°C for 10 min. (Richardson, 1981).

In designing these experiments, we had the choice of adding the isolated WBTI to the heated winged bean or to casein. We added the WBTI to casein to separate out the effects of any residual antinutritional factor in the heated winged bean and to allow direct comparison with the casein group without the complicating differences in protein quality and digestibility.

Our study shows that isolated WBTI not only caused pancreatic hypertrophy but also depressed growth. This agrees with the observation of Rackis (1965) that TI was responsible for 100% of pancreatic hypertrophy and 30-60% of the growth inhibition observed in raw soybean and that of Kakade et al. (1973), who concluded that 40% of pancreatic hypertrophy and 40% of growth inhibition was due to TI. As far as we know, this is the only study of its kind where sufficient quantities of TI were isolated by affinity chromatography, their in vitro activity confirmed, and fed to rats under controlled conditions.

It is interesting that the PER of the heated bean group (AWB) was lower than the WBTI group, yet no pancreatic hypertrophy was observed in AWB. One explanation is the presence of a heat stable factor that inhibited growth without pancreatic hypertrophy, such as found in soybean by Liener (1951). It may also be due to the lower quality and availability of the protein in AWB than in WBTI, since casein provided the protein in the WBTI group.

WBTI also appeared to cause spleen hypertrophy. No other study has shown this effect, and before any conclusions can be made, it needs to be confirmed.

Along with liver and spleen atrophy in the RWB group, pancreatic hypertrophy was not observed, contrary to the findings of Jaffe and Korte (1976). We were unable to excise the organs immediately since the rats died unexpectedly, often at night and dehydration could have taken place reducing the organ weights and masking the hypertrophy of the pancreas. In addition, since the majority of the rats died within 7 days of the study, the time needed for significant hypertrophy of the pancreas to occur may be insufficient.

Since the lethal effects observed in the RWB group were not found in the WBTI group, the isolated WBTI was apparently not responsible for the toxicity of raw winged bean. Strong evidence suggests that lectins play a primary role in this deleterious effect. Severe damage to small intestines were found in rats fed raw winged bean and hemagglutinating activity was detected in the intestinal mucosa and faeces (Higuchi et al., 1983; Kimura et al., 1982).

The main objective of this study is to demonstrate the antinutritional activity of isolated WBTI and determine its relative contribution to the toxicity of raw winged bean. The limitation of this approach is that one cannot observe the combined and possibly synergistic effects of all the antinutritional factors in the raw bean. Such effects can put sufficient stress on the rat, leading to death as opposed to the individual and separate effects of the isolated factors. Within this limitation, we conclude that WBTI was not primarily responsible for the toxic effects of raw winged bean but caused pancreatic hypertrophy, growth depression and possibly spleen hypertrophy.

REFERENCES

Abbey, B., Neale, R.J., Norton, G., Br. J. Nutr., 1979a, 41:31.
Abbey, B., Norton, G., Neale, R.J., Br. J. Nutr., 1979b, 41:39.
Bowman, D.E., Arch Biochem. Biophys., 1971, 144:541.
Cerny, K., Kordylas, M., Pospisil, F., Svabensky, O., Zajic, D., Br. J. Nutr., 1971, 26:293.
Cerny, K., Proc. Int. Symp. Dev. Potentials Winged Bean, 1st., 1978, 281.
Davis, B.J., Ann. N.Y. Acad. Sci., 1964, 121:404.
de Lumen, B.O., Salamat, L.A., Agric. Food Chem., 1980, 28:533.
Fraenkel-Conrat, H., Bean, R.C., Ducay, E.D., Olcott, H.S., Arch. Biochem. Biophys., 1952, 37:393.
Gillespie, J.M., Blagrove, R.J., Aust. J. Plant Physiol., 1978, 5:357.
Gomori, G., Methods Enzymol., 1955, 1:138.
Haynes, R., Feeney, R.E., J. Biol. Chem., 1967, 242:5378.
Higuchi, M., Suga, M., Iwai, K., Agr. Biol. Chem., 1983, 47:1879-1886.
Jaffe, W.G., Vega, C.L, J. Nutr., 1968, 94:203.
Jaffe, W.G., Korte, R., Nutr. Rep. Int., 1976, 14:449.
Jones, G., Moore, S., Stein, W.H., Biochemistry, 1963, 2:66.
Kakade, M.L., Hoffa, D.E., Liener, I.E., J. Nutr., 1973, 103:1772.
Kimura, T., Satanachote, C., Yoshida, A., J. Nutr. Sci. Vitaminol., 1982, 28:27.
Kortt, A.A., Biochim. Biophys. Acta, 1979, 577:371.
Liener, I.E., J. Biol. Chem., 1951, 193:183.
Liener, I.E., J. Agric. Food. Chem., 1974, 22:17.
Liener, I.E., Kakade, M.L., "Protease Inhibitors in Toxic Constituents in Plant Foodstuffs", Liener, I.E., Ed., Academic Press: New York, 1969, p. 16.
LKB-ProdukterAB, Application Notes, No. 250, Bromma, Sweden, 1977.
Loeffler, L.J., Pierce, J.V., Biochim. Biophys. Acta, 1973, 317:20.
March, S.C., Parikh, I. Cuatrecasas, P., Anal. Biochem., 1974, 60:149.
Moore, S., Stein, W.H., Methods Enzymol., 1963, 6:819.
National Academy of Science "The Winged Bean: A high-protein crop for the tropics", National Academy of Sciences: Washington, DC, 1975.
Ornstein, L., Ann. N.Y. Acad. Sci., 1964, 121:321.
Rackis, J., J.Fed. Proc., Fed. Am. Soc. Exp. Biol., 1965, 24:1488.
Richardson, M., Phytochemistry, 1977, 16:159.
Richardson, M., Food Chem., 1981, 6:235.
Rosenberg, M., Roegner, V., Becker, F.F., Am. Rev. Respir. Dis., 1976, 113:779.
Roy, D.N., J. Agric. Food Chem., 1980, 28:48.
Sohonie, K., Bhandakar, A.P., J. Sci. Ind. Res., 1954, 138:500.
Uriel, J., Berges, J., Nature (London), 1968, 218:578.

24

PROTEASE INHIBITORS OF THE MARAMA BEAN

Barry Starcher and Leah Bryant

University of Texas Health Center
Tyler, Texas

Michale Elfant

University of California at Davis
Davis, California

The marama bean, which thrives in semi-desert climates, has the potential to provide a high protein supplement in areas of the world, where many crops will not grow. Nutritional studies are presented that suggest the marama bean contains an antinutritional factor, in addition to very high levels of antiproteinases. Five trypsin inhibitors on two elastase inhibitors have been identified to date.

The marama bean, Tylosema esculentum, is classified in the Leguminosae family and the Caesalpinioideae sub family (Watt and Bryer - Brandwijk, 1962). The bean grows on thin runner vines up to 18 feet in length with soft, reddish-brown leaves. In April, the beans ripen and are contained in woody pods with up to six beans per pod. The beans, weighing two to three grams each, are covered with a hard brown shell which encases a firm, cream colored and rather oily kernel. In addition to the bean, the marama plant contains a larger edible underground tuber that stores water for the plant (Vietmeyer, 1978).

The nourishing vines are indigenous to the semi-desert regions of the Kalahari desert and other sandy areas of South Africa. As a result of its uneven distribution, its importance in the diets of these hunter-gatherer tribes varies extensively. To some tribes in the Kalahari, the marama bean is a major food source, contributing 75% of the total vegetable content of the diet (Wehmeyer et al, 1969). The bean is praised highly by the natives who use the bean as a main staple, considering it as satisfying as well as nourishing (Marshall, 1976). Although marama beans are used primarily as a seasonal food, they can be stored with no difficulty for long periods of time (Tanaka, 1976).

Rodents have been observed eating the raw marama beans and apparently suffer no ill effects. When consumed raw, they have a slimy or oily texture with a rather bitter taste. The natives roast the beans in hot sand or coals and they are either eaten at this stage or pounded and boiled with water to make a porridge (Marshall, 1976; Storey, 1958; Wehmyer et al, 1969).

Flavors have been reported that range from almond or cashew-like to coffee-like, with a slightly bitter after taste.

The natives have made no attempts to cultivate the marama bean with the possible exception that, it has been reported to be sown directly onto the sand in Transvaal, South Africa. The only reported use as an animal supplement is by some native farmers who use them to fatten pigs.

There have been several partial analyses of marama beans, starting with the Imperial Institute in 1921 (Bray, 1921). The most recent analysis is from a masters thesis from The University of Texas (Ripperger-Suhler, 1983), and these values are shown in tables 1 and 2. As we see from the analysis, the marama is quite high in protein (34%). However, the sulfur amino acids are fairly low, resulting in a chemical score of only 64. Supplementation with methioneine should result in a good quality protein and very similar to soybean. Only one animal study has been reported which tests the nutritional value of the marama bean (Ripperger-Suhler, 1983). Young growing rats were fed raw or cooked bean meal, contributing 10% protein to a purified diet. Food consumption was very poor, resulting in an overall weight loss over the four week test period. Mortality was not significant and pancreatic hypertrophy was minimal. Additional studies were conducted searching for antinutritional factors that might cause the rats to refuse diets containing marama beans. No hemagglutinins were found using blood samples from rats, rabbits, sheep and humans. On the other hand, trypsin inhibitors were found in extremely high levels, 239 TUI/mg protein, which is close to twice that reported for soybeans (Kakade et al, 1973). Whether these high levels of trypsin inhibitor contributed to the rats refusal to consume the diet could not be determined. However, diets containing roasted beans that no longer contained antitrypsin activity were also not consumed.

For whatever reasons, humans that consume marama beans have done so with delight over many years with no apparent adverse effects.

Table I. Analysis of Raw Marama Bean

Component	Marama Bean
Protein	34.30 %
Fat	34.80 %
CHO	21.10 %
Moisture	5.64 %
Ash	3.15 %
Energy	6.43 kcal/gm

Table II. Essential Amino Acid Composition of Marama Bean Compared with Soybean and Their Chemical Scores

	g/100g		Amino Acid Score	
	Marama Bean	Soybean	Marama Bean	Soybean
Isoleucine	4.5	5.9	111	109
Leucine	6.3	8.4	95	98
Lysine	6.8	6.9	99	99
Methionine & Cysteine	3.4	3.4	60	60
Phenylalamine & Tyrosine	18.8	8.9	232	96
Threonine	3.6	4.3	120	89
Tryptophan	1.3	1.5	77	92
Valine	4.7	5.8	98	87

The antiprotease present in marama beans has recently been isolated and partially characterized. Beans were obtained from Southern Africa and were part of the same lot used in the nutritional study previously mentioned. The hard shells were cracked, the oily kernel removed and the inhibitor extracted essentially as described by Wagner and Rheim, 1967, using an ethanol wash and solubilization in dilute HCl. An ammonium sulfate cut was taken between 40 and 80 percent saturation, dialyzed and freeze dried. This fraction represented approximately 10% of the total protein in the bean. Analysis of this fraction on SDS acrylamide gels revealed a single but rather diffuse band with a molecular weight around 18,000 (Fig 1). When compared directly on the same gel with soybean inhibitor (mw 21,500), the marama bean migrated with an apparent molecular weight of 23,000. A similar discrepancy in molecular weight determination on SDS gels was reported for soybean inhibitor by Gomes et al, 1979. It is probable that the marama bean inhibitor shows the same anomalous behavior.

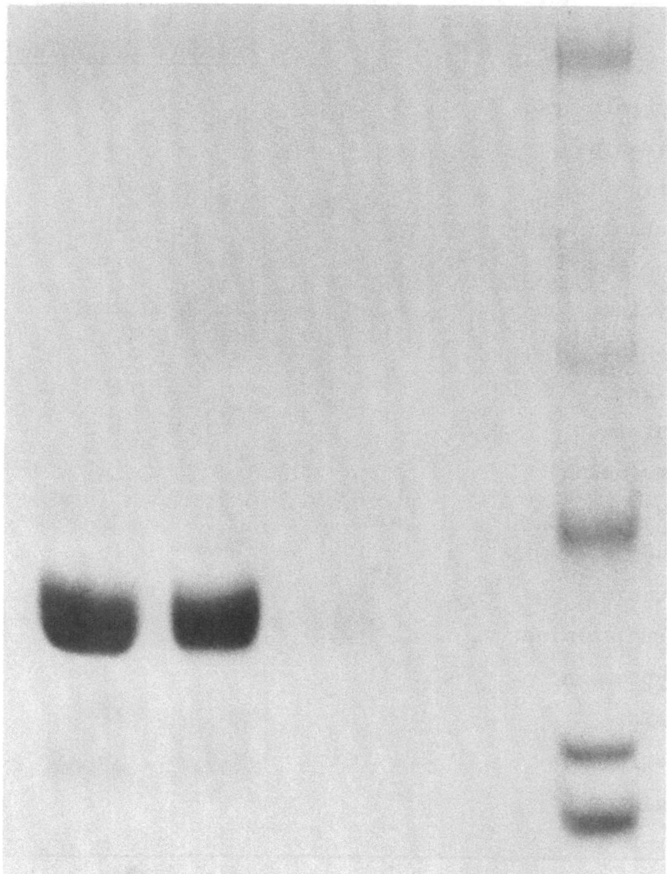

Figure 1

SDS Acrylamide gel electrophoresis of the marama bean extract (left). Standards (right) were BSA ovalbumen, pepsin, trypsin, lysozyme and cytochrome C.

Amino acid analysis of the ammonium sulfate fraction is shown in table 3. Perhaps the most notable feature of the analysis is the low cystein content representing only one or two residues/mole.

Table 3

Amino Acid Composition of Marama Bean Inhibitor

Amino Acid	Residues per Molecule
Aspartic Acid	25
Threonine	8
Serine	19
Glutamic Acid	18
Proline	15
Glycine	25
Alanine	12
Valine	14
Cysteine	1-2
Methionine	3
Isoleucine	13
Leucine	16
Tyrosine	13
Phenylalanine	10
Tryptophan	1
Lysine	17
Histidine	6
Arginine	8
Molecular Weight	24,500

Inhibition curves against trypsin, chymotrypsin and elastase are shown in figures 1-3 and are compared with soybean inhibitor.

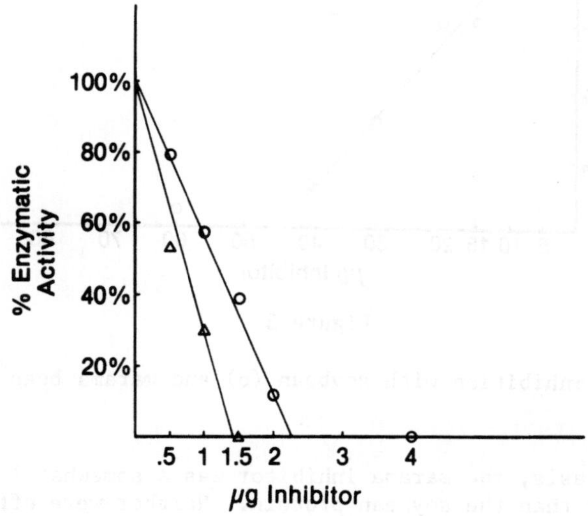

Figure 1

Trypsin (2 μg) inhibition with soybean (△) and marama bean (○) inhibitors.

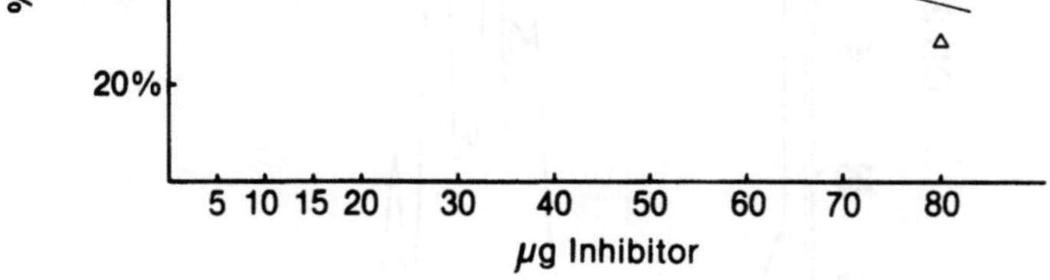

Figure 2

Chymotrypsin (20 μg) inhibition with soybean (△) and marama bean (○) inhibitors.

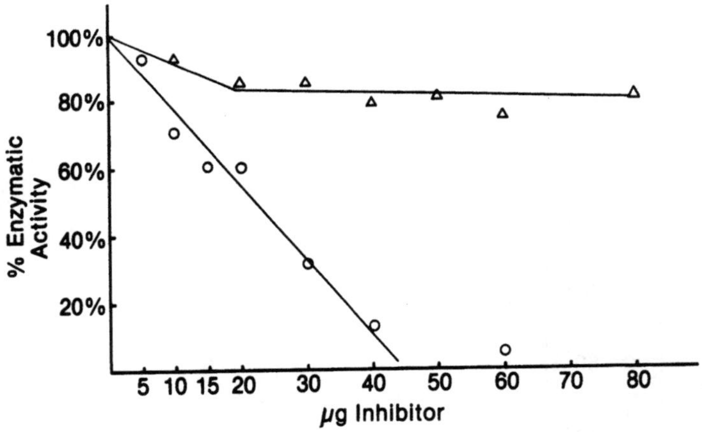

Figure 3

Elastase (20 μg) inhibition with soybean (o) and marama bean (Δ) inhibitors.

On a molar basis, the marama inhibitor was a somewhat less effective trypsin inhibitor than the soybean protein. Neither were effective chymotrypsin inhibitors. Elastase activity, however, was inhibited rather effectively by the marama inhibitor, requiring about 2 moles of inhibitor/mole of enzyme for complete inhibition.

Attempts to further purify the marama inhibitor were unsuccessful, using classical methods such as gel filtration, ion exchange and affinity chromatography as assessed by acrylamide gels. Recently, however, striking results were obtained using a Mono S column and FPLC.

Figure 4 shows the elution pattern obtained following chromatography of the ammonium sulfate fraction on the Mono S column using a linear salt gradient. At least seven distinct protein peaks were observed, all with very similar molecular weights as determined with SDS gels. Trypsin and elastase inhibitory capacity was measured for each fraction from the FPLC column and is illustrated in figure 5.

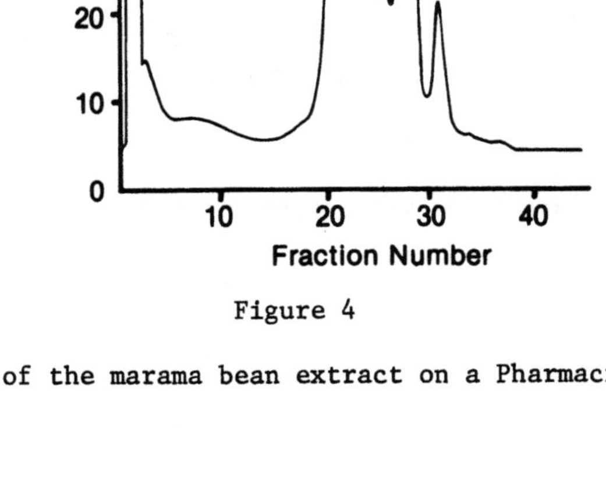

Figure 4

Gradient elution of the marama bean extract on a Pharmacia Mono S column.

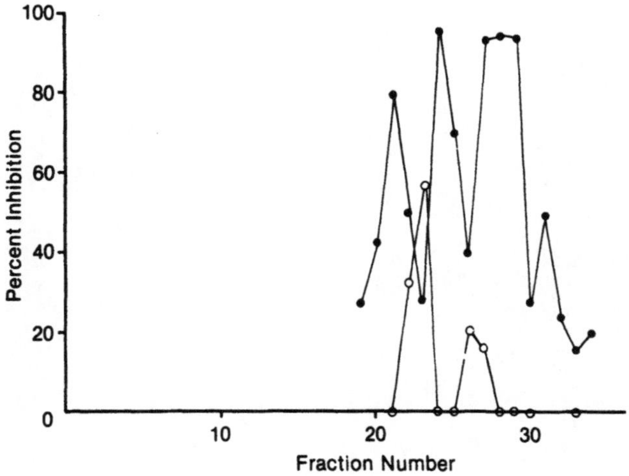

Figure 5

Inhibitory capacity of fractions from the Mono S column for trypsin (●) and elastase (o) activity. Trypsin inhibitor capacity was present in five of the protein peaks, while elastase inhibition was present in two peaks and quite distinct from the trypsin inhibitor.

To further separate the inhibitors, the ammonium sulfate fraction was applied to an agarose-trypsin affinity column. Absorbed protein was then eluted with glycine buffer pH 2.3, dialyzed against water and freeze dried. This protein (2.5 mg) was then dissolved in phosphate buffer and applied to the Mono S column under identical conditions as described previously. Only five of the proteins were now evident and all show trypsin inhibitory capacity (fig 6). No elastase inhibition was present. Electrophoresis on SDS gels indicates that the trypsin inhibitors differ only slightly in molecular weight (fig. 7).

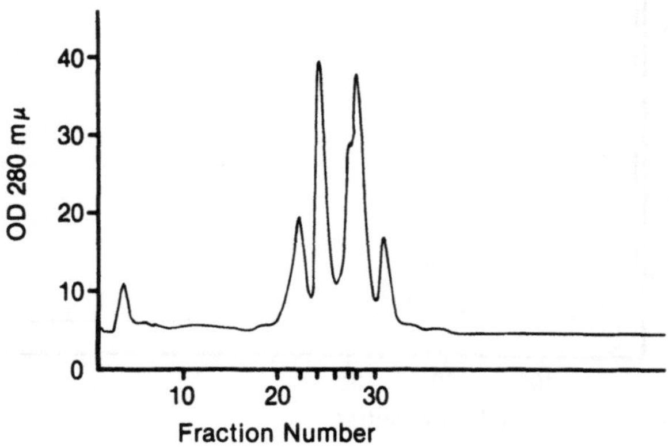

Figure 6

Elution profile of trypsin-agarose affinity purified marama bean inhibitor from the Mono S column.

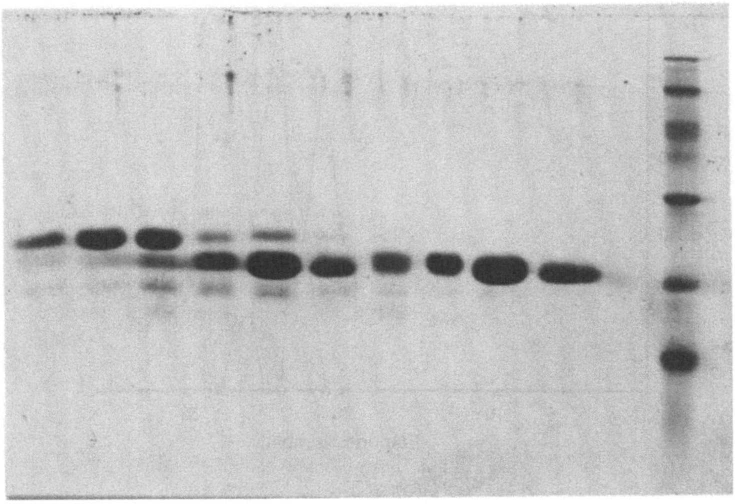

Figure 7

SDS gel of fractions from the Mono S column in figure 6. From the left, fractions number 20 to 30. Standard contains: phosphorylase B, BSA, ovalbumen, carbonic anhydrase, SBTI and α-lactalbumin.

The same approach was taken using an elastase-agarose affinity column. The results illustrated in fig 8 show two proteins eluting from the Mono S column and each showing one major band on SDS gels (fig 9). No trypsin inhibition was present in either of these two proteins. Molecular weights for all of the proteinase inhibitors ranges from 20,000 to 27,000.

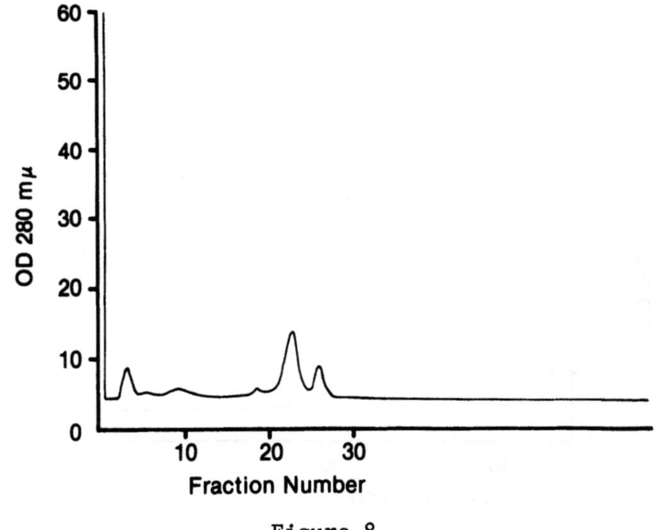

Figure 8

Elution profile of elastase-agarose affinity pruified marama bean inhibitor.

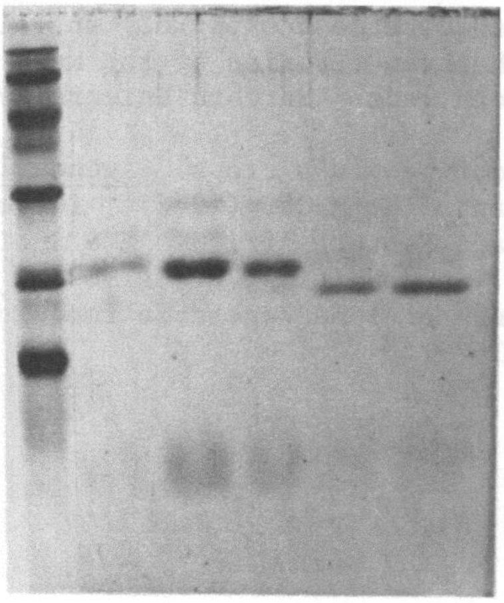

Figure 9

SDS gel of fractions from the Mono S column in figure 8. Fractions starting 2nd lane from the left: 21,22,23,26 and 27. Standard contains: phosphorylase B, BSA, ovalbumin, carbonic anhydrase, SBTI and α-lactalbumin.

Amino acid analysis and inhibitor constants of each of the different species of the trypsin and elastase inhibitors have not yet been determined.

To summarize, we can say that the marama bean has great potential as a high protein food source in arid, desert areas where other crops will not survive. Man has eaten these beans for centuries with no apparent ill effects. Conversely, rats will not eat diets containing marama beans, leaving nutritional studies hard to assess. At least five distinct trypsin inhibitors are present and two elastase inhibitors. They are present in large amounts but their nutritional or toxicological significance are unknown.

REFERENCES

Bray, G.T., (1921). Gemsbok beans. The Analyst 46:401.

Gomes, J.C., Koch, U., Brunner, J.R., (1979). Isolation of a trypsin inhibitor from navy beans by affinity chromatography. Cereal Chem., 56: 525-529.

Kakade, M.L., Hoffa, D.E. and Liener, I.E., (1973). Contribution of trypsin inhibitors to the deleterious effects of unheated soybeans fed to rats. J. Nutr., 103:1772.

Marshall, L., (1976). The kung of Nyae Nyae. Harvard University Press, Cambridge, Massachusetts.

Ripperger-Suhler, J.A., (1983). Protein quality of the African marama bean. Masters Thesis, University of Texas at Austin.

Storey, R., (1958). Some plants used by the bushmen in obtaining food and water. Bot. Surv. Memoir 30:25.

Tanaka, Jiro, (1976). Subsistence ecology of central Kalahari San. In Kalahari Hunter-Gatherers: Studies of the Kung San and their Neighbors, R.B. Lee and I. DeVore, eds. Harvard University Press, Cambridge, Massachusetts.

Vietmeyer, N, (1978). Poor people's crops. _Agenda_ _1_:12.

Watt, J.M. and Breyer-Brandwijk, M.G., (1962). The medicinal and poisonous plants of Southern and Eastern Africa. E & S Linvingston, London.

Wehmeyer, A.S., Lee, R.B., and Whiting, M., (1969). The nutrient composition and dietary importance of some vegetable foods eaten by the Kung bushmen. _S. Afr. Med. J._ _43_:1529.

25

TRYPSIN/CHYMOTRYPSIN INHIBITORS FROM MILLETS

Thillaisthanam N. Pattabiraman

Department of Biochemistry
Katsurba Medical College
Manipal - 576 119, INDIA

ABSTRACT

Properties of purified proteinase inhibitors from the small-seeded grains, finger millet, sorghum, pearl millet, Italian millet, and barnyard millet are compared. The changes in the levels of inhibitory activities during germination and plant growth of finger millet and barnyard millet are discussed. The action of millet proteinase inhibitors on human and other animal pancreatic enzymes is detailed. The role of substrates in determining the magnitude of the inhibition of proteolytic enzymes by the millet inhibitors is emphasized.

INTRODUCTION

Millets, the small-seeded cereal grains are well adapted to a wide range of ecological conditions and are able to produce good yields of grain under conditions unfavourable to most other cereals. Millets are the most important food crops of the semi-arid tropics. While in economically developed countries, the millets are mostly used as animal feed, in developing countries, they are predominantly a source of food for humans in terms of caloric and protein requirements. Because of th importance of sorghum and other millets in the diets of some of the poorest people, the nutritional quality of these grains is a subject deserving serious study. The polyphenols in millets which are considered to impair nutritional quality, have received considerable attention in recent years (Price and Butler, 1980). The proteinase inhibitors, which are also viewed as antinutritional factors, have been extensively investigated for their structure and function, especially from legume seeds (Liener, 1979; Richardson, 1981). The millet proteinase inhibitors have attracted attention relatively recently. The available information on the millet inhibitors is reviewed in this paper.

Table 1 lists some of the common millet varieties used in animal feeds and in human diets.

MILLET INHIBITORS

Although an early survey failed to detect proteinase inhibitory activity in sorghum (Laskowski and Laskowski, 1955), later studies established the preesence of antichymtryptic and antitryptic activities in the same

Table 1. Common millets

Scientific name	Common name
Sorghum bicolor L.	Sorghuhm, Jowar (Hindi)
Eleusine coracana, Gaertn.	Finger millet, Ragi (Hindi)
Pennisetum typhoides, Stapf.	Pearl millet, Bajra (Hindi)
Setaria italica, Beauv.	Italian millet, Thenai (Tamil)
Echinocloa fruneutacea, (Roxb) Link.	Barnyard millet, Sanva (Hindi)
Paspalum scorbiculatum, Linn.	Kodo, Ditch millet, Varagu (Tamil)
Panicium miliaceum, Linn.	Proso, Common millet, Panivaragu (Tamil)
Panicium miliare, Lam.	Little millet, Samai (Tamil)

millet (Harish Kumar et al., 1978; Chandrasekher et al., 1982). Chandrasekher et al. (1982) observed that two varieties out of thirty one cultivars of sorghum tested did not possess inhibitory activity. While they could not detect trypsin or chymotrypsin inhibitors in Proso, Kodo, and little millet, Veerabhadrappa and coworkers (1978) and Udupa and associates (1984) identified antitryptic activity in the latter two millets, although at concentrations much lower than those found in sorghum, finger millet, and pearl millet. The apparent discrepancies could be attributed to varietal differences, extraction procedures, sensitivity of assay procedures, age of seeds, and stability of inhibitors in neutral buffer extracts. In this context, it is interesting to note while Chandrasekher et al. (1982) failed to detect antichymtryptic activity in barnyard millet, Udupa and Pattabiraman (1985) demonstrated the presence of both antitryptic and antichymotryptic activities in this millet variety. The latter workers also showed that the antitryptic activity in crude extracts of barnyard millet is stable during storage at ambient temperature, whereas the antichymotryptic activity diminished rapidly as a function of time. Among the several millets surveyed, finger millet was found to possess the highest inhibitory activity (Chandrasekher et al., 1982; Veerabhadrappa et el., 1978).

FINGER MILLET INHIBITORS

Antitryptic activity is more pronounced than antichymotryptic activity in all the twenty eight varieties of finger millet tested (Veerabhadrappa et al., 1978). Shivaraj and Pattabiraman (1981) isolated a homogenous form of a protein that had the bifunctional ability of inhibiting bovine trypsin and porcine pancreatic alpha amylase. It had no action on bovine chymotrypsin. The inhibitor (I-1) was found to be stable to heat treatment at 100°C for one hour and to exposure to a wide range of pH's (1 to 10). Demonstration of the formation of the trimer, trypsin: I-1: amylase complex, and retention of both antitryptic and antiamylase activities on a column of trypsin: Sepharose, when the purified inhibitor was subjected to affinity chromatography, conclusively proved that a single protein is reponsible for both inhibitory activities. Modification of arginyl residues by cyclohexanedione resulted in loss of trypsin inhibition without any effect on amylase inhibition of I-1. This result indicated that the binding sites on the inhibitor for the two enzymes are different. The conclusion was supported by the additional observation that pepsin preferentially destroyed the amylase inhibitory activity, whereas pronase was more potent in abolishing the antitryptic activity of I-1. Treatment of the inhibitor with starch masked the amylase inhibitory activity leaving the trypsin inhibitory activity unaffected.

Shivaraj et al. (1982) separated I-1 from a second inhibitor (I-2) by successive chromatography of finger millet extracts on chymotrypsin: Sepharose and trypsin: Sepharose columns. Inhibitor-2 had comparable activity with bovine trypsin and chymotrypsin and did not affect amylase activity. The molecular weight (M_r) of I-1 (14.3 K), determined by SDS-polyacrylamide gel electrophoresis,was not very different from that of I-2 (12.0 K). Both inhibitors, which are basic proteins, were retained on cross-linked Dextran matrices during gel chromatography. Inhibitor-2, like I-1, was highly heat stable ($100^{o}C$ for one hour), and also stable to acidic and alkaline conditions (pH 1.0 - 12.6). Pepsin treatment diminished the antichymotryptic activity of I-2 faster than its antitryptic activity. On the other hand, modification of arginyl residues with cyclohexane dione caused a more rapid loss of antitryptic than antichymotryptic activities.

Campos and Richardson (1983) determined the amino acid sequence of I-1 from finger millet. The inhibitor was shown to consist of a single polypeptide chain of 122 amino acids, corresponding to a molecular weight of 13.3 K. Further, I-1 was shown to be a basic protein with a pI value above pH 10. Alignment of the sequence of I-1 with barley trypsin inhibitor (Odani et al., 1982) revealed approximately 60% homology. Comparison of the sequence between I-1 and a wheat amylase inhibitor (Kashlan and Richardson, 1981) indicated a homology of about 30%. Furthermore, the wheat alpha-amylase inhibitor, which did not inactivate trypsin, does not contain a recognizable reactive sequence in the region corresponding to the trypsin reactive sites of barley trypsin inhibitor and finger millet I-1. In spite of the high amino acid sequence homology between I-I and the barley inhibitor, the latter did not inactivate alpha-amylase.

SORGHUM INHIBITORS

Unlike finger millet, sorghum contains inhibitory factors which are more active against alpha-chymotrypsin than trypsin (Harish Kumar et al., 1978; Chandrasekher et al., 1982). Harish Kumar and associates (1978) separated three inhibitors (I, II, III) starting with neutral aqueous extracts followed by salt fractionation and chromatography on DEAE-cellulose, Sepahdex G-100, and QAE-Sephadex A-25. Inhibitor III was purified to homogeneity. Inhibitors I, II, and III were, respectively, 2.0, 1.4, and 10.7 times more active against alpha-chymotrypsin than against trypsin. From amino acid analysis it was concluded that III had a minimal molecular weight of 12.5 K (114 amino acid residues plus one half cystine). Based on gel chromatographic studies, it was be concluded that the native inhibitor is a dimer consisting of two, possibly identical, polypeptide chains linked by a disulphide bridge. A preparation containing all three inhibitors was found to inhibit chymotrypsin, elastase, trypsin, and pronase to different degrees; but not thiol proteinases or carboxypeptidase A.

PEARL MILLET INHIBITORS

Chandrasekher and Pattabiraman (1982) isolated two homogenous trypsin inhibitors, TI-1 and TI-2, from pearl millet. They were inactive against chymotrypsin. Both are basic proteins with a molecular weight of approximately of 11.0 K. The inhibitors resisted inactivation by pepsin and alpha-chymotrypsin and were stable to highly acid or alkaline conditions. However, heat treatment at $100^{o}C$ for one hour inactivated both inhibitors. Arginyl residues were found to be essential for the activity of both inhibitors.

ITALIAN MILLET INHIBITOR

Udupa and Pattabiraman (1984) isolated a homogenous inhibitor protein from Italian millet. The inhibitor was resolved into two charge isomers with equal activity during electrophoresis on polyacrylamide gel. Like the pearl millet inhibitor, the Italian millet factor was found to be highly specific in its action towards trypsin. It did not inactivate chymotrypsin, elastase, and subtilisin BPN. The inhibitor has a molecular weight of 14 K, as determined by gel chromatography and SDS-polyacrylamide gel electrophoresis. Amino groups are essential for its action against trypsin. The inhibitor resisted inactivation by pepsin and alpha-chymotrypsin. However, treatment with these two enzymes caused a limited proteolysis of the inhibitor protein and resulted in an increase in trypsin inhibitory activity when assayed with the amide substrate. The inhibitor protein was fully stable when exposed to 90°C for one hour. It was also stable to pH in the range 1 to 10. The inhibitor formed a stable, stoichiometric complex with bovine trypsin.

BARNYARD MILLET INHIBITORS

Udupa and Pattabiraman (1985) isolated and characterized a trypsin/chymotrypsin inhibitor from barnyard millet with some unusual properties. The protein has a molecular weight of 13.5 K. It was twice as active against trypsin than against chymotrypsin. The antichymotryptic activity was more labile to heat treatment and to alkaline pH values. Similarly, modification of arginyl residues of the inhibitor resulted in a more rapid loss of antichymotryptic acitivity. On the othe hand, treatment of the inhibitor with trinitrobenzene sulphonate, ethyl acetamidate, or O-methyl isourea resulted in a selective loss of antitryptic activity. The data indicate that the trypsin and chymotrypsin binding sites on the inhibitor are different, and that the latter is more exposed and hence more easily susceptible to alterations. The inhibition profile with respect to inhibitor concentration was sigmoidal when the purified inhibitor was directly added to the assay medium containing trypsin or chymotrypsin. However, preincubation of the inhibitor with the target enzyme for ten minutes gave linear inhibitory patterns. The stoichiometric ratios between the inhibitor and trypsin or chymotrypsin were 1: 2 and 1: 1, respectively. The trypsin: inhibitor complex had not antichymotryptic activity. The chymotrypsin: inhibitor complex, on the other hand, displayed tryptic inhibition.

COMPARISON

A comparison of the properties of the different millet inhibitors indicates many similarities (Table 2). All the well characterized inhibitors have molecular weight values around 12-13 K. In this respect, they resemble the proteinase inhibitors from other cereals like wheat (Boisen and Djurtoft, 1981), rice bran (Maki et al., 1980), barley (Odani et al., 1983), rye (Boisen and Djurtoft, 1981 a), and corn (Johnson et al., 1980). Many of the cereal inhibitors are also basic proteins. The millet inhibitors are generally heat and acid stable. Most of them are resistant to the action of proteolytic enzymes. The pearl millet and Italian millet inhibitors are highly specific in their action twoard trypsin. Finger millet inhibitor-2 and barnyard millet inhibitor are active against both trypsin and chymotrypsin. Finger millet inhibitor is unique in that it acts on trypsin and alpha-amylase. The sorghum inhibitors, on the other hand, are more active against chymotrypsin.

Table 2. An over-view of millet proteinase inhibitors

Inhibitor	Molecular size and nature	Target enzymes	Reactive site(s)	Stability
Finger millet I-1	13.3 K, basic protein	Trypsin; alpha-amylase	Arg^{34}-Leu^{35}, Arg^{91}-Ala^{92} for trypsin	Stable at 100°C, 1 hr and at pH 1-10; inactivated by pepsin
Finger millet I-2	12.0 K	Trypsin > chymotrypsin	--	Stable at 100°C, 1 hr and at pH 1-12.6; resistant to pepsin
Sorghum I	--	Chymotrypsin > trypsin	--	--
Sorghum II	--	Chymotrypsin > trypsin	--	--
Sorghum III	12.5K, dimer 25 K, acidic protein	Chymotrypsin	--	--
Pearl millet I	11.0 K, basic protein	Trypsin	Arg - X, Lys - X	Inactivated at 100oC, 1 hr; stable at pH 1-9; resistant to chymotrypsin and pepsin
Pearl millet II	11.0 K, basic protein	Trypsin	Arg - X	Inactivated at 100oC, 1 hr; stable at pH 1-9; resistant to chymotrypsin and pepsin
Italian millet	14.0 K, basic protein	Trypsin	Lys - X	Stable at 100°C, 30 min and at pH 2-10; resistant to chymotrypsin and pepsin
Barnyard millet	13.5 K	Trypsin chymotrypsin	Arg - X, Lys - X	40% loss of anti-tryptic and 100% loss of antichy-motryptic activi-ties at 115°C, 10^5 pa, 15 min; resistant to pep-sin.

GERMINATION STUDIES

Veerabhadrappa et al.(1978) showed that during germination of finger millet seeds, both antitryptic and antichymotryptic activities declined gradually in the endosperm. However, considerable antitryptic activity remained in the axis, even on the sixth day of germination, even though antichymotryptic activity completely disappeared. The authors suggested that the inhibitor in the endosperm is degraded during germination and is resynthesized in the growing part of the plant. At the time of this investigation, the presence of two inhibitors in finger millet with differing specificities was not known. Vidyavathi and associates (1983) used finger millet inhibitor-1 as the substrate for the proteinases formed during germination of the seeds. The proteolytic activity was maximal on the third day of germination. Both the antitryptic activity and antiamylase activity of I-1 was destroyed by a proteinase that required sulfhydryl groups for maximal activity. The proteinase that displayed an optimum pH of 2.5 was inhibited by diazoacetyl norleucine methyl ester and peptstatin, indicating that it is a carboxyl proteinase. The data indicate that both I-1 and I-2 in finger millet were mobilized during germination and that the residual activity observed in the axis could be due to a different inhibitor synthesized during plant growth.

Udupa and Pattabiraman (unpublished observations) studied the changes in the levels of antitryptic activity and antichymotryptic activity in barnyard millet during germination. The results are shown in Table 3.

Table 3. Proteinase inhibitory activities in barnyard millet during germination

Days of germination	Antitryptic activity, units*/mg protein	Antichymotryptic activity, units*/mg protein	Protein in plant, mg/g dry weight
0	1.04	0.44	18.0
1	1.12	0.47	16.0
2	1.04	0.51	18.0
3	0.80	0.40	19.5
4	0.52	0.33	24.0
5	0.21	0.28	28.0
6	0.13	0.27	26.2
7	0.00	0.28	27.3

* Values are in optical density units.

It was found that antitryptic activity disappeared completely on the seventh day, whereas two thirds of the antichymotryptic activity still persisted. Since both activities in barnyard millet are known to reside in the same protein, the purified barnyard inhibitor was tested as the substrate for the proteinases arising during germination (seventh day). Maximal inactivation of the antitryptic activity was observed at pH 5.0.

Under the same assay conditions, antichymotryptic activity was not abolished. Pepstatin prevented the action of the proteinase. The proteinase activity was localized in the shoot portion of the seedlings. The root portion contained antichymotryptic activity. In addition, it also contained lesser amounts of antitryptic acitivity. On the mixing the shoot and root extracts, the antitryptic activity, as expected, disappeared. These data suggest that during germination the trypsin/chymotrypsin inhibitor in the dormant seed is mobilized, and most probably a new inhibitor with a greater activity towards chymotrypsin appeared in the shoot portion of the plant.

ACTION OF MILLET INHIBITORS ON DIFFERENT PROTEINASES

Harish Kumar et al. (1978) observed that the sorghum inhibitors inactivated bovine alpha-chymotrypsin, porcine elastase, pronase, bovine trypsin, and porcine trypsin in the decreasing order of effectivness. In terms of micrograms of enzyme inhibited per one microgram of inhibitor, the values are, respectively, 0.50, 0.38, 0.37, 0.06, and 0.025. Chandrasekher et al. (1982) showed that crude extracts of pearl millet, sorghum, Italian millet, and barnyard millet inhibited the caseinolytic activity of 'activated' human and bovine pancreatic extracts to comparable degrees. Finger millet inhibitors were two times more active against the bovine system. These results are in agreement with observations by Shivaraj et al. (1982) that purified I-1 and I-2 of finger millet suppressed the caseinolytic activity of the bovine pancreatic preparation more effectively. Table 4 shows the action of aqueous extracts of millets on trypsin, chymotrypsin and elastase. Many of the millets possess powerful antielastase activities. It appears that the Kodo and sorghum are basically elastase inhibitors.

In addition, Udupa et al. (1984) showed that Kodo extracts inhibited the caseinolytic activity of a human pancreatic preparation two to three times more effectively than that of a bovine pancreatic extract. The action of purified barnyard millet inhibitor on tryptic and chymotryptic activities of bovine, human, rabbit, and porcine pancreatic activities was studied by Udupa and Pattabiraman (1985). While no significant difference in tryptic inhibition was observed, rabbit and porcine chymotrypsins were poorly affected by this inhibitor. The relative ratio of inhibition of bovine, human, porcine, and rabbit chymotrypsin was 237: 115: 13.8: 1, respectively.

Table 4. Relative inhibition of proteinases by millet extracts

Millet	Inhibition of bovine trypsin	Inhibition of bovine chymotrypsin	Inhibition of porcine elastase
	micrograms enzyme inhibited/mg protein of seed extract*		
Kodo	0.83	1.30	3.47
Sorghum	2.59	6.48	7.41
Finger millet	33.30	12.50	8.33
Barnyard millet	10.00	4.86	6.94
Italian millet	7.18	0.00	1.28
Pearl millet	6.60	0.00	3.30

* Values obtained when casein was used as substrate.

EFFECT OF SUBSTRATE INHIBITION

The magnitudes of inhibition of proteolytic enzymes by some millet inhibitors depend very much on the nature of the substrates used in the assay systems. Harish Kumar et al. (1978) provide some interesting data on inhibition by a mixture of sorghum inhibitors I, II, and III. While the inhibitors did not affect the caseinolytic activity of subtilisyn A and subtilisin BPN', with acetyltyrosine ethyl ester as substrate, these two enzymes were as powerfully inhibited as bovine alpha-chymotrypsin. The inhibition of bovine chymotrypsin itself was two times greater with the ester as a substrate than with casein. Bovine trypsin was five times more powerfully inhibited when benzoylarginine ethyl ester was used as a substrate than with casein. Inhibition of pronase, however, did not depend on the nature of the substrate used. Udapa and Pattabiraman (1984) showed that the Italian millet inhibitor, which did not act against chymotrypsin with casein as the substrate, showed weak inhibition of the enzyme when acetyltyrosine ethyl ester was used as a substrate. The relative raio of inhibition of trypsin and chymotrypsin by Italian millet inhibitor was found to be 36: 1 when the synthetic substrate was used. The inhibitory activity against trypsin was also found to be two-fold greater when benzoylarginine p-nitroanilide was used as the substrate rather than casein. Shivaraj et al. (1982), on the other hand, observed that the inhibitory action of I-2 from finger millet was weak against trypsin when benzoylarginine ethyl ester was used as the substrate. The relative ratios of inhibition of trypsin with casein, benzoylarginine p-nitroanilide, and benzoylarginine ethyl ester as substrates were found to be 23: 33 : 1, respectively. However, no difference in magnitude of inhibition of alpha-chymotrypsin by I-2 was observed with either casein or acetyltyrosine ethyl ester as substrates. Similarly, the action of purified barnyard millet inhibitor on chymotrypsin and trypsin did not depend on the nature of the substrate used (Udupa and Pattabiraman, 1985). Udupa et al. (1984) showed that inhibition by Kodo extracts of bovine trypsin was eight times greater when benzoylarginine was used as the substrate in place of casein. No such difference was observed with crude inhibitor preparations of sorghum, finger millet, Italian millet, barnyard millet, and pearl millet. Inhibition of chymotrypsin by these six millets was not significantly different, whether casein or acetyl tyrosine ethyl ester was used as the substrate. With all these millet inhibitors, inhibition of porcine elastase was greater (2.2 to 13.0 times) with casein rather than succinyl trialanyl p-nitroanilide as the substrate.

It is difficult to explain the widely varying degrees of inhibition of proteinases by millet inhibitors in terms of a unified hypothesis. However, the available data suggest that extrapolation of information obtained on the inhibition of purified enzymes with ester and amide substrates to evaluate antinutritional effects of trypsin/chymotrypsin inhibitors is fraught with uncertainties.

REFERENCES

Boisen, S. and Djurtoft, R. (1981). Trypsin inhibitor from wheat endosperm: Purification and characterization. Cereal Chem., 58, 460-463.

Boisen, S. and Djurtoft, R. (1981 a). Trypsin inhbitor from rye endosperm: Purification and properties. Cereal Chem., 58, 194-198.

Campos, F. A. P. and Richardson, M. (1983). The complete amino acid sequence of the bifunctional alpha-amylase/trypsin inhibitor from seeds of ragi (Indian finger millet, Eleusine coracana, Gaertn). FEBS Letters, 167, 221-225.

Chandrasekher, G. and Pattabiraman, T. N. (1982). Natural plant enzyme inhibitors: Isolation and characterization of two trypsin inhibitors from bajra (Pennisetum typhoideum). Indian J. Biochem. Biophys., 19, 1-7

Chandrasekher, G., Raju, D. S. and Pattabiraman, T. N. (1982). Natural plant enzyme inhibitors. Protease inhibitors in millets. J. Sci. Food Agric., 33, 447-450.

Harish Kumar, P. M., Virupaksha, T. K. and Vithayathil, P. J. (1978). Sorghum protease inhbitors. Purification and some biochemical properties. Intl. J. Peptide Protein Res., 12, 185-196.

Johnson, R., Mitchell, H. L., Wassom, C. E. and Reeck, R. (1980). Comparison of trypsin inhibitors from several types of corn. Phytochemistry, 19, 2749-2750.

Kashlan, N. and Richardson, M. (1981). The complete amino acid sequence of a major wheat protein inhibitor of amylase. Phytochemistry, 20, 1781-1784.

Laskowski, M. and Laskowski, M., Jr. (1954). Naturally occurring trypsin inhibitors. In "Advances in Protein Chemistry", M. L. Anson, K. Bailey and J. T. Edsall, eds., Academic Press, New York, Vol. 9, pp. 203-242.

Liener, I. E. (1979). Protease inhibitors and lectins. Intl. Rev. Biochem., 27, 97-122.

Maki, Z., Tashiro, M., Sugihara, N. and Kanamori, M. (1980). Double-headed nature of a trypsin inhibitor from rice bran. Agric. Biol. Chem. (Japan), 44, 953-955.

Mundy, J., Hejgaard, J. and Svendsen, I. (1984). Characterization of a bifunctional wheat inhibitor of endogenous alpha-amylase and subtilisin. FEBS Letters, 167, 210-214.

Odani, S., Koide, T. and Ono, T. (1982). Sequence homology between barley trypsin inhibitor and wheat alpha-amylase inhibitors. FEBS Letters, 141, 279-282.

Odani, S., Koide, T., Ono, T. and Ohnishi, K. (1983). Structural relationship between barley (Hordeum vulgare) trypsin inhibitor and castor bean (Ricinus communis) storage protein. Biochem. J., 213, 543-545.

Price, M. L. and Butler, L. G. (1980). Treatment of sorghum grains that reduce the assayable tannin content and their effect on the nutritional value of the grain. In "Polyphenols in Cereals and Legumes", J. H. Hulse, ed., International Development Research Centre, Ottawa, Canada, pp. 39-42.

Richardson, M. (1981). Protein inhibitors of enzymes. Food Chemistry, 6, 235-253.

Shivaraj, B. and Pattabiraman, T. N. (1981). Natural plant enzyme inhibitors. Characterization of an unusual alpha-amylase/trypsin inhibitor from ragi (Ekeusine coracana, Geartn). Biochem. J., 193, 29-36.

Shivaraj, B., Rao, H. N. and Pattabiraman, T. N. (1982). Natural plant enzyme inhibitors. Isolation of a trypsin/alpha-amylase inhibitor and a chymotrypsin/trypsin inhibitor from ragi (Eleusine coracana) grain by affinity chromatography and study of their properties. J. Sci. Food Agric., 33, 1080-1091.

Udupa, S. L., Khanum, F. and Pattabiraman, T. N. (1984). Comparative study of the action of cereal protease inhibitors on human, bovine and porcine pancreatic enzymes. J. Indian Inst. Sci. 65(C), 1-9.

Udupa, S. L. and Pattabiraman, T. N. (1985). Isolation and characterization of a trypsin/chymotrypsin inhibitor from the millet Echinocloa fruneutacea. J. Agric. Food Chem., 33, 642-646.

Veerabhadrappa, P. S., Manjunath, N. H. and Virupaksh, T. K. (1978). Proteinase inhibitor of finger millets (Eluesine corcana Gaertn). J. Sci. Food Agric., 29, 353-358.

Vidyavathi, U., Shivaraj, B. and Pattabiraman, T. N. (1983). Proteases in germinating finger millet (Eleusine coracana) seeds. J. Bioscience, 5, 219-224.

26

ANTIGENICITY OF NATIVE AND MODIFIED KUNITZ SOYBEAN TRYPSIN INHIBITORS

David L. Brandon, Sakhina Haque, and Mendel Friedman

Western Regional Research Center, Agricultural Research Service, U.S. Department of Agriculture, 800 Buchanan Street, Berkeley, California 94710

ABSTRACT

Food provides a continuous antigenic stimulus to the immune system and the antigenicity of processed food proteins should be considered in toxicological evaluations. The antigenicity of the Kunitz trypsin inhibitor was studied using antibodies prepared by inoculating rabbits with native, heat-denatured, and N-acetylcysteine-treated Kunitz soybean trypsin inhibitors. Immunochemical studies using a competitive solid-phase enzyme immunoassay and two groups of sera revealed two patterns of antigenicity. Antibodies elicited with the denatured inhibitor were specific for the denatured conformation of the protein. In contrast, native inhibitor elicited antibodies that selectively recognized determinants in both native and heat-treated protein, but that did not bind trypsin inhibitors treated with N-acetylcysteine. These results imply that: (1) the disulfide bonds must be intact to maintain the native antigenic conformation and (2) the cysteine treatment may suppress allergic manifestations of soybean trypsin inhibitors and possibly other food proteins. These studies were extended by analyzing a panel of monoclonal antibodies prepared against native Kunitz trypsin inhibitor. The inhibitor has at least two distinct antigenic sites (epitopes), one of which is retained under denaturing conditions. The measurement of native Kunitz trypsin inhibitor in food samples by immunoassay appears practical. The relevance of these findings to food processing, food safety, and health is also discussed.

INTRODUCTION

General Considerations

Food constitutes one of the major features of our environment and provides a continuous antigenic stimulus to the immune system. In addition to strictly antigenic components, toxic factors within the food supply can interact immediately with cells of the mucosal immune system and, later, with the systemic immune system. Without a more complete understanding of the mechanisms governing the immunopharmacological activity of food ingredients, we cannot gauge the potential health consequences of various dietary choices, new food sources, and alternative food processing techniques.

Two categories of immune responses to food proteins are of interest in connection with the toxicological properties of soy proteins: (1) the immune response to food proteins may cause adverse reactions, such as intestinal pathology associated with severe soy intolerance; and (2) antibodies elicited against soy proteins can be used to develop analytical methods for measuring individual constituents.

Because of the problems associated with the consumption of soy protein, particularly due to the presence of protease inhibitors, we are interested in immunochemical approaches to (1) assessment of toxicity; (2) quantitation of individual protease inhibitors; and (3) monitoring the inactivation of these inhibitors during processing. Although this paper focuses on the Kunitz trypsin inhibitor, the approaches could also be applied to studying the antigenicity of other food proteins.

Immunity, Immunotoxicology, and Diet

The link between immunity and diet is generally considered to be indirect, and the immunosuppression associated with severe malnutrition is often cited as an example of the bridge between nutrition and immunology (Brandon, 1984). More cogent is the observation that nearly 25% of intestinal tissue is lymphoid in origin (Kagnoff, 1981); this includes the lymph node-like Peyer's patches as well as diffuse tissue like the intraepithelial lymphocytes. The intraepithelial lymphocytes can account for as much as 17% of the surface of the intestine (Kagnoff, 1981). Clearly, the intestinal tract is part of the immune system, generally referred to as the gut-associated lymphoid tissue (GALT). Another aspect of the immune system which overlaps with the GALT is the secretory immune system (McGhee and Mestecky, 1983). Secretory immunity refers to the production of antibodies in secretions such as tears and saliva. The intestinal tract is protected by such antibodies, predominantly secretory IgA and IgM, and is also a key element in the development of secretory immunity in other organs.

Food constitutes the most concentrated antigenic stimulus that an organism normally encounters, but the basis for the immune response to food proteins is poorly understood. Normal adult plasma contains antibodies directed against common food proteins (e.g., milk, eggs, and soy proteins). Since the smallest protein antigen is approximately eight amino acids in length, the intestinal barrier must permit sizeable peptides to transit the mucosal surface or immunization must occur at the mucosal surface. Transit of peptides across the mucosa or a pinocytotic mechanism for delivering macromolecules into the lamina propria may be necessary to achieve sensitization (Walker and Bloch, 1983). Since macromolecular and particulate antigens have been localized within the GALT following enteric exposure (Bockman and Winborn, 1966), it seems reasonable to assume that all components of the diet can interact with the immune system. Hence, the lack of immune response by a normal individual to a food protein can be attributed to intrinsic antigenicity of the protein or operation of a control system which effectively suppresses the immune response.

Some immune responses to food proteins may be deleterious particularly because immune complexes can be deposited in tissues (Van der Woude et al., 1983; Gallagher and Gibney, 1983; Allison, 1982). Immune complexes resulting from enterically encountered antigen have been implicated in renal disease (Emancipator and Gallo, 1983). Bowel diseases such as celiac disease involve an immunological component, though the initial events in pathogenesis are not clear (Scott et al., 1983). The correlation between the immune response to foods (especially secretory immunity) and pathogenesis is limited and circumstantial. For example, patients with selective IgA deficiency have a general increase in autoimmunity and an abnormally high mucosal response to enteric antigens (Cunningham-Rundles et al., 1978). Immune complexes containing food proteins, food protein-specific antibodies, and anti-antibodies are found in the circulation of these patients, and the quantity and composition of the complexes depends on diet (Cunningham-Rundles, 1983). There are few data implicating specific food antigens in pathogenicity, but food proteins encountered in high concentration, and which are poorly digestible and acid-stable, are thought most likely to evoke a response measured as circulating antibodies (Moroz and Yang, 1980). In addition, direct interaction of proteins with components of the intestinal mucosa has been implicated in bowel disease and subsequent autoimmunity. Soy proteins fit into these categories. Soy antigens appear to be unusually antigenic via the intestinal mucosa and soy intolerance is among the best-documented food intolerances (Barratt et al., 1979).

The immune response to dietary soy protein has been studied, particularly in cattle, and the presence of certain protein components correlates with both antigenicity and digestive disorders (Sissons et al., 1982; Seegraber and Morrill, 1982). In humans, a strong immune response to soy proteins has been implicated in the failure of some celiac patients to respond favorably to a wheat gluten-free diet (Haeney et al., 1982). However, there is little information on the influence of soy proteins in the diet on immune complex formation or immune response to other dietary proteins. Problems associated with quantitation of immune complexes have been amply documented in the literature (Peeters, 1979), but the advent of more specific immunoassays (Doi et al., 1984), and application of high-resolution physicochemical techniques to protein fractionation (Regnier, 1983) offer the possibility of unequivocal results.

The possible long-term hazards resulting from the immune response to food antigens would not affect all segments of the population equally. For example, persons suffering from immune deficiencies or experiencing immunosuppression due to disease or drug intervention may have altered responsiveness to enteric antigens. In addition, it appears that bowel disease, even commonly occuring diarrhea, can alter the permeability of the intestinal mucosa (Aioti and Paganelli, 1983). Presumably, the entire human population experiences periods of enhanced susceptibility to sensitization by enteric antigens. The health risk associated with immunotoxicity of food components could thus apply to large segments of the population. It is likely that those affected adversely by exposure to potential immunotoxins comprise a subpopulation whose genetically determined immune response results in pathology. For example, celiac disease may arise in a subset of the population whose intestinal epithelial cells express surface antigens which function as receptors for gliadin proteins and whose immune system recognizes as alien a particular combination of gliadin and cell surface antigens (Cole and Kagnoff, 1985). It is important to note the presence of lectins in processed soy as well as in other foodstuffs (Toms, 1981; Concon et al., 1983). These could have a direct immunopharmacological activity or alter mucosal permeability to food proteins as a consequence of binding to cell surface components.

A better understanding of antigenic and immunomodulatory aspects of food proteins will permit a better estimate of health risks associated with dietary proteins. There may be health risks attendant upon introduction of new food proteins or proteins modified by processing. These new or modified proteins could differ from conventional proteins in their interaction with the cell surface of the intestinal mucosa, in their digestibility, in the transit of immunoreactive peptides across the intestinal epithelium, or in their ability to evoke a cellular or humoral immune response. Changes in antigenicity of proteins upon processing have been documented (Morton and Deutsch, 1961; Kato et al., 1983), but the consequences of enteric encounter with modified proteins represent a gap in scientific knowledge.

Relationship Between Antigenicity and Toxicity of Soy Proteins

Although Kunitz trypsin inhibitor (KTI) has been established as a human allergen (Moroz and Yang, 1980), no definite evidence links its chronic or acute toxicity to its antigenicity. Other soy proteins that have been implicated as toxicants may affect the immune system differently. For example, adverse intestinal responses to soy protein limit the use of soybean products as economical milk replacers in young calves (Barratt et al., 1978; Seegraber and Morrill, 1982). The digestive disturbances appear to correlate with immunological sensitization of the calf to soy protein, resulting in the production of IgG_1 antibodies. Though the immune mechanisms have not been delineated, the intestinal disturbances correlate with the level of glycinin and β-conglycinin antigens (Sissons et al., 1982). Extraction of defatted soybean flakes with hot ethanol reduces the antigenicity of the diet _in vitro_ and eliminates the toxicity, although the process does not apparently remove the protein components.

Since processing conditions also influence the digestibility of soy proteins (Barratt et al., 1978), the potential toxicity of food proteins cannot be assessed solely by measuring the antigen remaining in the processed product. The concentration of soy antigens surviving proteolytic digestion depends on the prior processing conditions (Srihara, 1984). Extrapolating from the results in animal systems, it is clear that the antigenicity and potential allergenicity of soy-based infant formulas may depend on the processing techniques (Heppell et al, 1985).

The mechanism of toxicity will determine the selection of approaches to evaluate and possibly eliminate toxic food proteins. For example, some food intolerance, such as gluten intolerance or celiac disease, involves T cell responses (Cole and Kagnoff, 1985). In such cases, application of antibody-based analysis of antigen denaturation may fail, since T cells distinguish poorly between native and denatured protein (Streicher et al., 1984). A more systematic approach such as that perfected by the Atassi laboratory may be necessary (Atassi, 1985). This technique involves screening antibodies or T cell clones with a comprehensive set of overlapping peptides derived from the sequence of the antigenic protein. However, this peptide mapping procedure is very expensive and requires knowledge of the amino acid sequence of the protein of interest. In addition, their approach seems likely to miss conformation dependent antigens formed by amino acid residues not adjacent in the unfolded protein. The method should probably be reserved for selected proteins of known sequence and toxicity. Though less elaborate, the monoclonal antibody-based epitope mapping technique (see below) may be effective for evaluating the antigenicity of food proteins. The structure of the Kunitz inhibitor has been elucidated (Rackis et al., 1986).

Previous Studies of Kunitz Trypsin Inhibitor

Immunochemical analysis of the Kunitz trypsin inhibitor has thus far employed antisera of various specificities. The data are very sketchy.

Catsimpoolas and Leuthner (1969) developed an immunoassay for the Kunitz inhibitor based on radial immunodiffusion. In this procedure, antigen is allowed to diffuse into an agarose gel containing antibody specific for the component of interest. The interaction of multivalent antigens and specific antibodies results in the formation of insoluble complexes which precipitate. Increasing antigen concentration leads to increased penetration of antigen into the gel before a precipitating lattice forms. The radius of the zone of precipitation thus can be related to antigen concentration. Catsimpoolas and Leuthner (1969) were able to measure samples containing 0.3 µg of inhibitor by this procedure. However, the slope of the standard curve was low, and it is likely that analyses of complex samples would have been subject to large errors.

The same laboratory described a microcomplement fixation assay (Catsimpoolas et al., 1969). This assay format is highly subject to interference, due to the wide variety of naturally occurring substances that influence the activity and stability of the multiple proteins comprising complement. For this reason, complement fixation assays have been largely supplanted by radioimmunoassays and enzyme immunoassays. Rossebo and Nordal (1971) demonstrated that antibodies can inhibit the activity of Kunitz trypsin inhibitor and that native and heat-denatured inhibitor differ antigenically. Our own findings generally support and extend their conclusion.

Antigenicity Through Enteric Exposure

Various soybean proteins have been implicated in adverse immunological reactions. Moroz and Yang (1980) described a patient who had soybean-specific food allergy. They demonstrated the presence of immunoglobulin E antibodies which bound Kunitz trypsin inhibitor specifically. Thus, Kunitz trypsin inhibitor can stimulate the immune system through enteric exposure, as well as by parenteral administration.

Studies Focused on the Function of Kunitz Trypsin Inhibitor

While the Kunitz trypsin inhibitor may represent a problem to nutritionists and toxicologists, to plant biochemists it is one of a number of protease inhibitors still in search of a function. Freed and Ryan (1978 a, b) sought to establish the function of the Kunitz trypsin inhibitor by analyzing the protease inhibitors of soybeans during germination. They found no change in antigenicity or quantity of the Kunitz inhibitor during germination. There was, however, some indication that the electrophoretic mobility of the inhibitor may change during germination.

Horisberger and Tacchini-Vonlanthen (1983 a, b) studied the localization of protease inhibitors within plant cells, employing immunocytochemical methods. They found the Kunitz trypsin inhibitor primarily within protein bodies. However, even lines lacking this inhibitor exhibited some staining.

Rationale for Immunochemical Analysis of Kunitz Trypsin Inhibitor

The importance of soybeans as a human foodstuff, as a component of animal feeds, and as a major export commodity of the United States requires that we understand the mechanism of the toxic and anti-nutritional effects and devise methods for overcoming them. The potential importance of the problem, combined with our knowledge of the role of disulfides in maintaining the activity of the Kunitz trypsin inhibitor (Friedman and Gumbmann, 1986), impelled us to explore the antigenic structure of this protein and the feasibility of immunoassays to quantitate the inhibitor and monitor its inactivation. The epitope mapping technique, recently applied to the study of α_1- antitrypsin (Herion et al., 1984), seemed an appropriate tool to use for this study.

MATERIALS AND METHODS

Kunitz soybean trypsin inhibitor (Sigma Chemical Co., St. Louis, MO) was characterized by SDS-polyacrylamide gel electrophoresis and by inhibition of trypsin (Friedman et al., 1982). It was used without further purification. Heat-treated and N-acetyl-L-cysteine (NAC)-modified forms of KTI were prepared as described by Friedman and collaborators (Friedman et al., 1982; Friedman and Gumbmann, 1986). Soy flour was a gift from Central Soya, Ft. Wayne, Indiana. Other chemicals were obtained from commercial sources.

Treatment with heat and N-acetylcysteine. A sample containing 70 mg of commercial Kunitz trypsin inhibitor and 70 mg of NAC and one containing KTI alone were each dissolved in 10 ml of 0.5 M Tris buffer, pH 8.5. The solutions were heated in a water bath at 65°C for 1 hr. The contents were then dialyzed in the coldroom for 3 days with frequent changes of deionized water and lyophilized. The trypsin inhibitor content expressed in trypsin inhibitor units (TIU) per milligram (Friedman et al., 1982) was 2.29 for the untreated material, 0.09 for the heat + NAC-treated sample, and 0.84 for the heat-treated sample.

Preparation of biotinylated derivatives. The N-hydroxysuccinimide ester of biotin (Calbiochem, La Jolla, CA) was used to label the partially purified monoclonal antibodies essentially according to Wofsy (1983). The antibodies were purified by dialysis against phosphate-buffered saline (PBS, 5 mM sodium phosphate and 150 mM sodium chloride, pH 7.0) containing 0.01% sodium azide and characterized by titration in a solid-phase enzyme immunoassay (EIA) using horseradish peroxidase conjugates of rabbit anti-mouse Ig (Zymed Laboratories, South San Francisco, CA) and streptavidin (Amersham Corp., Arlington Heights, IL).

Antibody preparation. Rabbit antisera were elicited in New Zealand white rabbits by a course of subcutaneous injections of KTI (native or treated) emulsified in complete Freund's adjuvant. Sera were characterized initially by immunodiffusion, and later, by solid-phase EIA. Mouse antibodies were elicited in Balb/c mice by two intraperitoneal (i.p.) inoculations at 3-week intervals with native KTI emulsified in complete Freund's adjuvant. Antibody production was assessed by solid-phase EIA. Monoclonal antibodies were produced essentially according to Oi and Herzenberg (1980). In brief, an antibody-producing mouse was inoculated i.v. with 50 µg of KTI 3 days prior to cell fusion. Equal numbers of immune spleen cells and NS-1 myeloma cells were fused by treatment with 50% polyethyleneglycol 1450 (Bethesda Research Laboratories, Bethesda, MD), and the fusion suspension was dispersed into 96-well tissue culture plates, 0.2 ml/well, at a cell density of 1.5×10^6 cells/ml. Hybridomas secreting KTI-specific antibody were identified by solid-phase EIA. Following transfer and expansion of the cultures, hybridomas were cloned by limiting dilution. Clones were expanded in culture, permitting harvest of supernatant, freezing of cell lines, and preparation of ascites fluid. Antibodies are denoted by the number assigned to the cloned hybridomas. Monoclonal antibodies were assayed for isotype and light chain composition by solid-phase assay similar to the direct binding assay, but using isotype- and light chain-specific antibodies, conjugated to β-galactosidase (Southern Biotechnology, Birmingham, AL). Myeloma proteins (Litton Bionetics, Kensington, MD), were used to assess the specificity of these antibodies.

Ascites production. Balb/c mice were injected with Pristane^R (Aldrich Chemical Co., St. Louis, MO) one week prior to i.p. inoculation with 10^6 viable hybridomas. High-titer ascitic fluid and sera were obtained from four of six mice within 3 weeks of inoculation. For use in assays, the IgG fraction was purified by ammonium sulfate precipitation and chromatography on diethylaminoethyl-cellulose (DE-52, Whatman, Ltd., Maidstone, U.K.).

Immunoassays

Plate preparation. Polyvinylchloride assay plates (Costar, Cambridge, MA) were coated with Kunitz trypsin inhibitor (10 µg/ml; 50 µl/well). The plates were incubated for four hours and then washed with PBS containing 0.05% Tween-20 (Sigma). Remaining sites were blocked by incubating for 1 hr with 100 µl/well of 1% bovine serum albumin (BSA, Miles Laboratories, Elkhart, IN) in PBS-Tween solution. Plates were then rewashed and used immediately or stored for up to one week in PBS with azide. Assay plates stored longer were less efficient in binding assays.

Direct assays. Antibody samples to be assayed were diluted in PBS. The resulting samples (50-µl/well) were applied to KTI-coated plates, generally in triplicate, and incubated with shaking for 2 hr. Plates were emptied, washed, rinsed with distilled water, and drained. Labeled reagent (50 µl/well of peroxidase-labeled rabbit-anti-mouse IgG or streptavidin) at appropriate dilution in PBS-Tween-BSA was added and incubated for 1 hr with shaking. The wells were again emptied, washed, rinsed, and drained. Next, 50 µl of substrate (1 mM 2,2'-azinobis-3-ethylbenzthiazoline sulfonic acid) and 6.7 mM H_2O_2 in 60 mM sodium citrate buffer, pH 4.2) were added, and the absorbance at 415 nm was determined using an EIA reader (Model 308, Bio-Tek, Burlington, VT). Assays were generally read after 15 min or terminated by addition of 50 µl/well of 10% sodium dodecylsulfate.

Inhibition assays. Antibody and antigen (Kunitz trypsin inhibitor) were mixed and preincubated in microtitration dishes for 2 hr at room temperature prior to assay. To maximize sensitivity of the assay, antibodies were used at a concentration within the linear range of the direct binding titration curve. Aliquots of the preincubation mixture were assayed on KTI-coated plates as described above, except that the incubation with antibody was for 1 hr only.

Epitope mapping assays. Monoclonal antibodies were tested for their ability to inhibit the binding of other antibodies as outlined in Table 1. Unlabeled antibodies were incubated in KTI-coated wells, at a standard concentration based on their titers (reciprocal of dilution giving 0.8 absorbance units after 15 min under standard assay conditions). Following a 3 hr incubation at room temperature, the assay plates were washed, rinsed, and drained, and the binding of various biotinylated antibodies was determined. Mutual inhibition of antibodies suggested that the two antibodies bound to the same or nearby sites.

Table 1. Epitope Mapping Procedure

1. Coat assay plate with KTI.
2. Incubate assay wells with excess unlabeled monoclonal anti-KTI.
3. Wash and drain plate.
4. Add biotinylated monoclonal antibodies to assay wells (in triplicate) and incubate.
5. Wash and drain plate.
6. Add peroxidase-labeled streptavidin to all assay wells and incubate.
7. Wash and drain plate.
8. Add substrate solution to all wells and incubate.
9. Read absorbance at 415 nm. Binding of biotinylated antibody will be revealed by green color development.
10. Calculate binding in presence of unlabeled monoclonal antibody as a percent of control performed without unlabeled antibody.

RESULTS AND DISCUSSION

Rabbit antibodies. Figure 1 illustrates the results obtained using rabbit antisera with (a) modified (heat-denatured or N-acetylcysteine-treated) or (b) native KTI. There were two distinct patterns of reactivity. Antisera elicited with native KTI recognized determinants in both native and heat-treated proteins, as indicated by the similar degree of inhibition at low concentrations of KTI. In contrast, NAC-modified KTI inhibited binding only at higher concentrations, indicating that the treatment destroys or greatly alters the antigenic determinants. Antibodies elicited with modified inhibitor selectively recognized determinants in denatured KTI.

Studies with monoclonal antibodies. The results with rabbit antisera suggested that at least some antigenic determinants in native and denatured KTI differ. These differences were explored using monoclonal antibodies derived from mice inoculated with native KTI. Fusion cultures were screened for antibody production using assay plates coated with native KTI. Positive cultures were expanded and cloned by limiting dilution. Groups of cloned hybridomas were defined on the basis of specificity.

The antibodies were tested by direct binding to assay plates coated with native KTI or samples of KTI treated with heat or NAC + heat, and having low residual trypsin-inhibitor activity (Friedman et al., 1982). Representative results are illustrated in Figure 2. These assays demonstrated that both of the monoclonals had similar binding patterns, with poor discrimination between native and modified KTI. Antibody 136 was exceptional due to its low binding to denatured KTI. The most likely explanation for the relatively poor discrimination is the nature of the assay itself. The plates are coated with a saturating concentration of KTI, so that even the residual active molecules of KTI (approximatley 5%) in the denatured samples would be sufficient to permit substantial binding by antibodies specific for native KTI as well as antibodies that recognize determinants on the modified inhibitor.

Selected antibodies were then compared by inhibition assays. Representative results are shown in Figure 3. Native KTI inhibits the binding of all the antibodies to the solid-phase coating. Heat-treated KTI is similarly inhibitory, but KTI treated with NAC at elevated temperature is less effective in this assay, although the relative inhibition varies among the different antibodies. Thus, these results

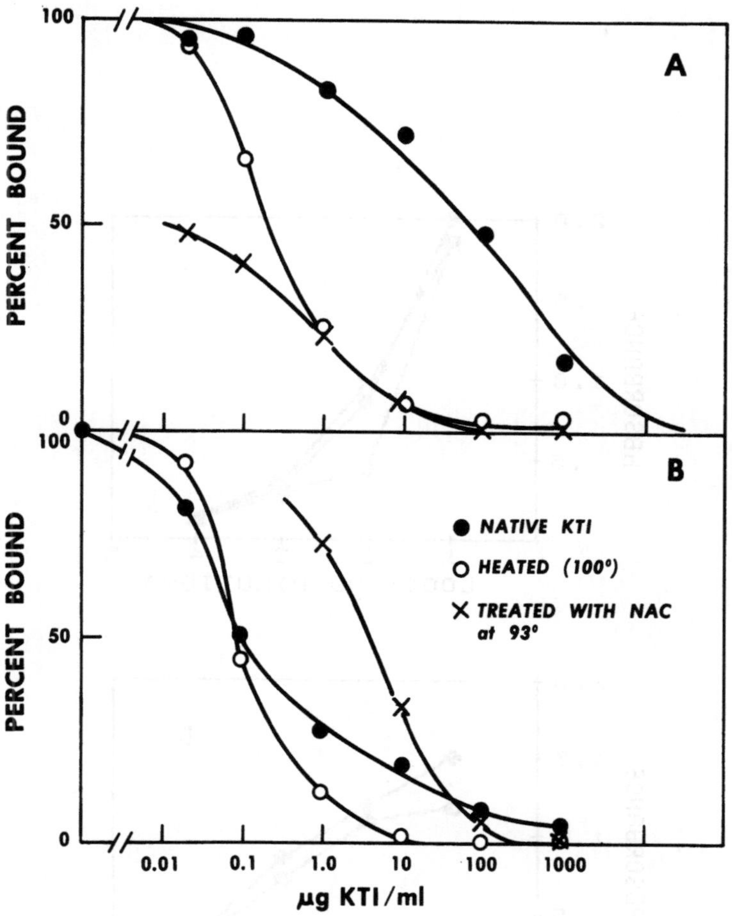

Figure 1. Analysis of native and modified Kunitz trypsin inhibitor with rabbit antibodies.

Competitive binding assays were performed on KTI-coated assay plates. Trypsin inhibitor was preincubated with antibodies before they were applied to the assay plate. Peroxidase-conjugated goat anti-rabbit IgG was used to detect rabbit anti-KTI bound to the plate.

(A) Antibodies elicited with heat-inactivated KTI (10% residual activity) recognize determinants shared by heat-treated KTI (in the presence or absence of NAC). These antibodies bind native KTI more weakly by 2- to 3- orders of magnitude.

(B) Antibodies elicited with native KTI bind NAC-treated KTI about 50-fold more weakly than native or heat-inactivated KTI.

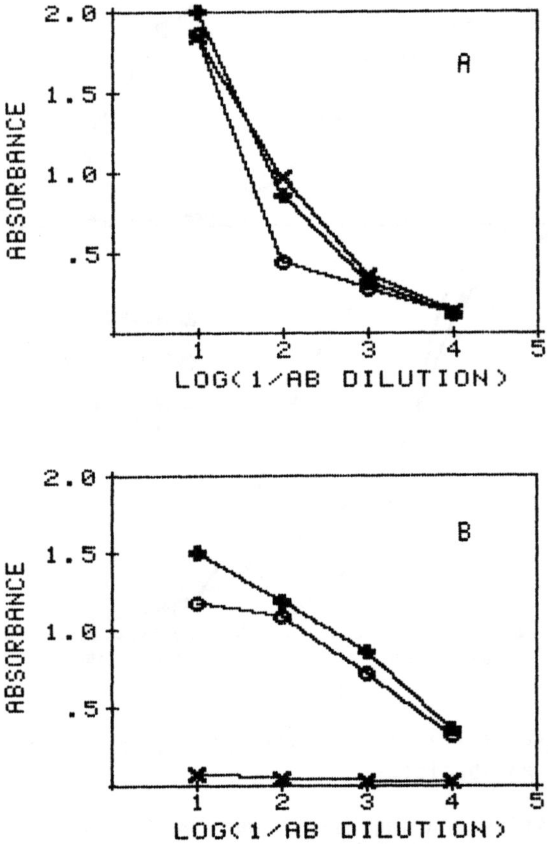

Figure 2. Titration of KTI-specific antibodies. The graphs illustrate binding of antibodies to assay plates coated with native KTI (O), heat-treated KTI (+), or heat + NAC-treated KTI (X). Antibody dilutions are expressed in arbitrary units. (A) clone 129; (B) clone 136.

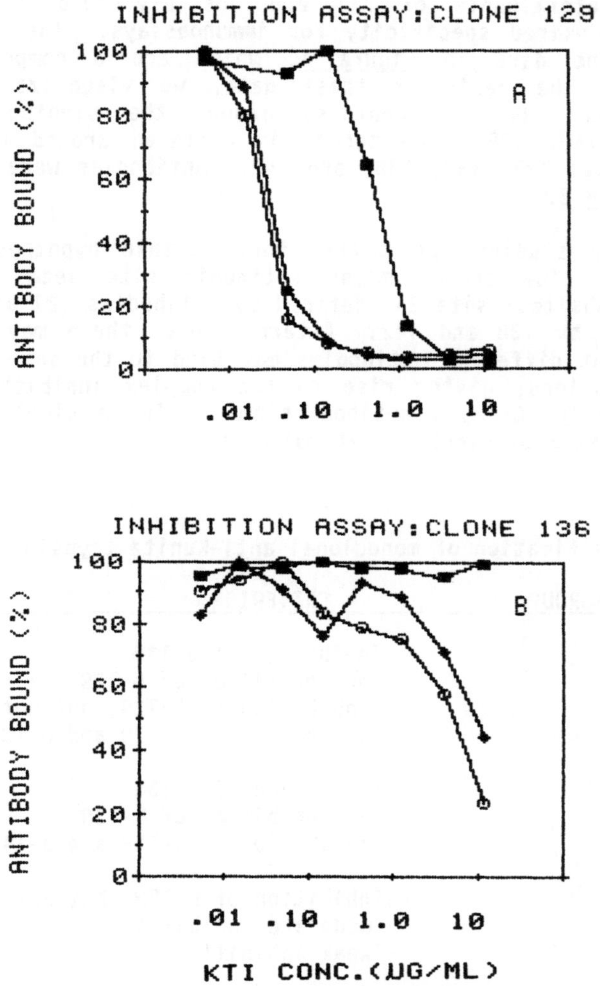

Figure 3. Competitive inhibition assay. The graphs illustrate that the binding of antibodies to assay plates depends on the concentration of KTI and processing condition: native KTI (O), heat-treated KTI (◇), heat + NAC-treated KTI (□). (A) clone 129; (B) clone 136.

lead to conclusions similar to those drawn from the experiments with polyclonal antibodies. Native and heat-denatured KTI are similar antigenically, whereas disruption of the disulfide bonds to form mixed disulfide forms between protein and added cysteine sulfhydryl groups causes a major change in antigenic determinants.

Epitope mapping. The results obtained with the direct binding assays and the inhibition assays motivated a higher-resolution analysis of antibody specificity. This would enable a more detailed definition of the antigenic determinants of KTI and possibly help in selecting antibodies with desired specificity for immunoassays. The results are summarized by the data in Figure 4, which are a composite of 3 experiments. On the basis of these data, we visualize two major antigenic regions. One is centered around the binding sites of antibodies 126, 140, 129. The second is centered around the binding sites of antibodies 127, 130, 138, and 142. Antibodies were grouped as tabulated in Table 2.

While further studies are called for, certain hypotheses can be entertained now. The second major antigenic site seems likely to consist of two subsites: site 2A, defined by antibodies 127 and 130; and site 2B, defined by 138 and 142. Alternatively, there may be only a single site 2, but different antibodies may bind to the same epitope in different orientations, giving rise to the complex inhibition pattern observed. Similarly, Group 1 antibodies may define a single site or a cluster of at least 3 overlapping determinants.

Table 2. Classification of monoclonal anti-Kunitz trypsin inhibitors

GROUP OR SUBGROUP	CRITERION
GROUP 1	Inhibition of b-140
A	no inhibition of b-129
B	inhibition of b-129, but not b-138
C	inhibition of b-129 and b-138
GROUP 2	Inhibition of b-130
A	no inhibition of b-138
B	inhibition of b-138 and b-142
GROUP 3	Inhibition of b-138, but not b-130
A	moderate inhibition
B	weak inhibition

As a test of our hypothesis that the three groups of antibodies bind to different antigenic regions of the molecule, we compared the binding of three representative antibodies--129, 127, and 136--from groups 1, 2, and 3, respectively. These antibodies were titered on solid-phase assay plates coated either with native KTI or KTI reduced and denatured by treatment with sodium dodecylsulfate and mercaptoethanol. The results are summarized in Table 3. It is apparent that 136 has the greatest specificity for native KTI, and that 127 has the least. We hypothesize that Site 2 is similar in native KTI and denatured KTI, and probably consists of a linear sequence of amino acid residues. Site 3 appears to differ most in native and denatured KTI, both in this assay and the inhibition assays. We hypothesize that Site 3 is unique to the native protein and depends on amino acid residues adjacent in the tertiary

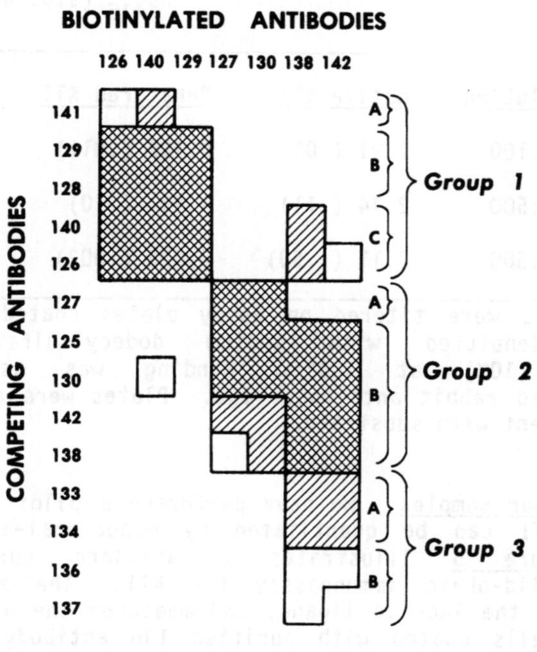

Figure 4. Epitope mapping using mouse monoclonal antibodies.

Monoclonal antibodies were purified and derivatized with the N-hydroxysuccinimide ester of biotin. In the mapping experiment (see Table 1), labeled antibodies (listed horizontally) were tested for binding to KTI-coated wells which had been pretreated with unlabeled antibodies (listed vertically). The results are derived from 3 separate experiments, and are scored as strong inhibition (cross-hatching), moderate inhibition (shading), or weak inhibition (blank). Unmarked areas indicate lack of inhibition. The criteria for grouping antibodies are described in Table 2.

structure of native KTI. These results suggest that Group 2 antibodies should be most useful in assaying mixtures of native and denatured KTI for total inhibitor content, or for detecting KTI in polyacrylamide gels under denaturing conditions. On the other hand, Group 3 antibodies may be most useful in detecting native KTI. The relationship among the subgroups of antibodies requires further study now in progress.

Table 3. Binding of monoclonal antibodies to native and SDS-denatured Kunitz trypsin inhibitors (KTI)

			OD_{415} (std. dev.)		
Group	Clone	Dilution	Native KTI	Denatured KTI	Binding ratio
1	129	1:100	2.21 (.01)	0.129 (.011)	17.1
2	127	1:500	2.74 (.11)	1.49 (.10)	1.84
3	136	1:500	2.11 (.50)	0.023 (.002)	91.7

Culture supernatants were titered on assay plates coated with native KTI, or KTI denatured with sodium dodecylsulfate (1%) + 2-mercaptoethanol (10%) at 100°.C. Binding was assessed with peroxidase-conjugated rabbit anti-mouse IgG. Plates were read after 15 minutes of development with substrate.

Analysis of soy flour samples. We have performed a pilot experiment to examine whether KTI can be quantitated by monoclonal-antibody-based immunoassay. Figure 5 illustrates a standard curve for an inhibition-type solid-phase immunoassay for KTI. The assay employs biotinylated KTI as the labeled ligand, and measures the binding of the ligand to assay wells coated with purified 136 antibody. The bound biotinylated KTI is detected with peroxidase-conjugated streptavidin, which binds specifically to biotin and does not adhere to assay wells coated with 136 or with unmodified KTI. Preliminary data on the assay of various soy proteins is illustrated in Figure 5 (inset). It appears that this assay format may be practical for complex mixtures of soy proteins. The presence of heat-denatured soy proteins does not interfere with the binding of biotinylated KTI to 136. In addition, these proteins do not appear to affect substantially the competition between native and biotinylated KTI for 136 binding sites. We are encouraged by these results, which suggest that active KTI can be assayed in the presence of inactivated protein and other protease inhibitors.

The experiments described above point the way to understanding the antigenic structure of KTI. Extension of the epitope mapping procedure with more antibodies could give a higher-resolution picture of the protein. In addition, more antibodies, derived from independent cell fusions, must be analyzed to establish whether all the antigenic sites have been defined. (The probability that all antibody specificities have been sampled can be tested statistically.) This work is in progress. In characterizing antibody specificities, it will be essential to determine the affinity of the antibodies. It seems likely that 136, and perhaps other Group 3 antibodies, are low-affinity antibodies. This conclusion is drawn from the inhibition assay data (Figure 3), which showed that about ten-fold higher concentration of KTI

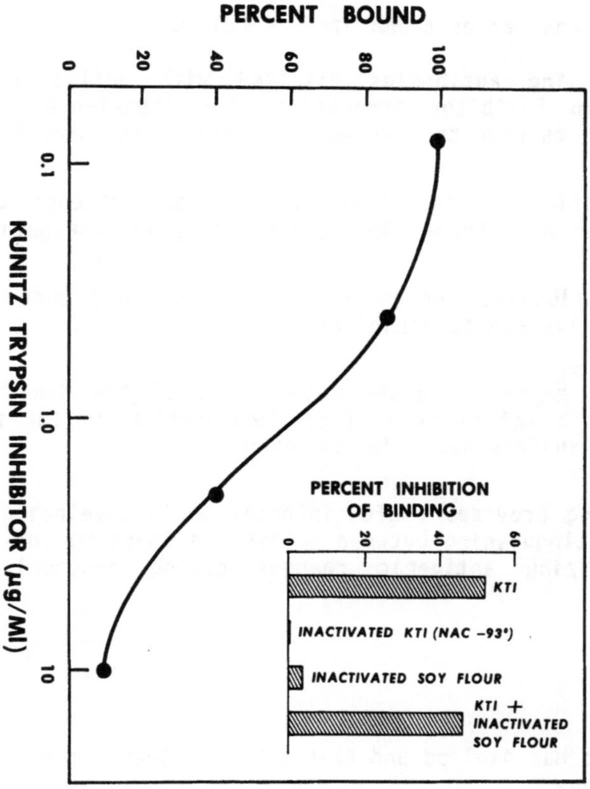

Figure 5. Assay of Kunitz trypsin inhibitor.

A solid-phase assay plate was coated with 136 antibody. This coated plate specifically bound biotinylated KTI, which was detected with peroxidase-streptavidin conjugate. In the presence of unlabelled KTI, the binding of biotinylated KTI was reduced, resulting in the standard curve.

Inset: Denatured KTI and heat-treated soy flour lacking KTI do not appear to interfere with the assay of KTI.

is needed to inhibit the binding of 136 than of 129 for example. Some of the differences between Group 2 and Group 3 antibodies, or between subgroups A and B of Group 3, may be due to affinity differences rather than differences in localization of binding. A higher affinity antibody with Group 3 specificity should enable a more sensitive assay of KTI. However, if Group 3 is actually a low-affinity subset of Group 2, it may be difficult to find the desired combination of affinity and specificity.

CONCLUSIONS

The major conclusions can be summarized as follows:

- Experiments using antibodies elicited with native and modified Kunitz trypsin inhibitor demonstrate the importance of disulfide bonding in maintaining the native structure of the protein.

- Heat denaturation of the inhibitor, in the absence of reducing agent, produces a conformation closely resembling the native form.

- Monoclonal antibodies recognize two, or possibly three, antigenic regions of native Kunitz inhibitor.

- At least one major antigenic determinant of the Kunitz inhibitor appears to be retained even after denaturation in the presence of sodium dodecylsulfate and reducing agent.

- Epitope mapping provides useful information in developing analytical tools for distinguishing between active and inactive inhibitors, and in characterizing antigenic changes during processing of food proteins.

ACKNOWLEDGEMENTS

We are grateful to Mac Axelrod and Ok-koo K. Grosjean for excellent technical assistance.

REFERENCES

Aioti, F. and Paganelli, R. (1983). Food allergy and gastrointestinal diseases. Ann. Allergy, 51, 275-280.

Allison, R. G. (ed.) (1982). "An Evaluation of the Potential for Dietary Protein to Contribute to Systemic Diseases", FASEB, Rockville, MD, 1-53.

Atassi, M.Z. (1985). Surface-simulation synthesis of the substrate-binding site of an enzyme. Demonstration with trypsin. Biochem. J., 226, 477-485.

Barratt, M.E.J., Strachan, P. J., and Porter, P. (1978). Antibody mechanisms implicated in digestive disturbances following ingestion of soya protein in calves and piglets. Clin. Exp. Immunol., 31, 305-312.

Barratt, M.E.J., Strachan, P.J., and Porter, P. (1979). Immunologically mediated nutritional disturbances associated with soya-protein antigens. Proc. Nutr. Soc., 38, 143-150.

Bockman, D. E. and Winborn, W. B. (1966). Light and electron microscopy of intestinal ferritin absorption. Observations in sensitized and non-sensitized hamsters. Anat. Rec., 155, 603-622.

Brandon, D. L. (1984). Interactions of diet and immunity. In: "Nutritional and Toxicological Aspects of Food Safety", M. Friedman, ed., Plenum Press, New York, pp. 65-90.

Catsimpoolas, N. and Leuthner, E. (1969). Immunochemical methods for detection and quantitation of Kunitz soybean trypsin inhibitor. Anal. Biochem., 31, 437-447.

Catsimpoolas, N., Rogers, D. A., and Meyer, E. W. (1969). Immunochemical and disc electrophoresis study of soybean trypsin inhibitor SBTIA-2. Cereal Chem., 46, 136-144.

Cole, S. G. and Kagnoff, M. F. (1985). Celiac disease. Ann. Rev. Nutr., 5, 241-266.

Concon, J. M., Newburg, D. S., and Eades, S. N. (1983). Lectins in wheat gluten proteins. J. Agric. Food Chem., 31, 939-941.

Cunningham-Rundles, C. (1983). Isolation and analysis of anti-idiotypic antibodies from IgA-deficient sera. Ann. N.Y. Acad. Sci., 409, 469-477.

Cunningham-Rundles, C., Brandeis, W. E., Good, R. A., and Day, N. K. (1978). Milk precipitins, circulating immune complexes and IgA deficiency. Adv. Exp. Med. Biol., 107, 523-530.

Doi, T., Kanatsu, K., Sekita, K., Yoshida, H., Nagai, H., and Hamashima, Y. (1984). Detection of IgA class circulating immune complexes bound to anti-C3d antibody in patients with IgA nephropathy. J. Immunol. Methods, 69, 95-104.

Emancipator, S. N. and Gallo, G. R. (1983). IgA-immune complex renal disease induced by mucosal immunization. Ann. N.Y. Acad. Sci., 409, 171-175.

Freed, R. C. and Ryan, D. S. (1978 a). Changes in Kunitz trypsin inhibitor during germination of soybeans: an immunoelectrophoresis assay system. J. Food Sci., 43, 1316-1319.

Freed, R. C. and Ryan, D. S. (1978 b). Note on modification of the Kunitz soybean trypsin inhibitor during seed germination. Cereal Chem., 55, 534-538.

Friedman, M. and Gumbmann, M. R. (1986). Nutritional improvement of legumes through disulfide interchange. This volume.

Friedman, M., Grosjean, O. K., and Zahnley, J. C. (1982). Inactivation of soya bean trypsin inhibitor by thiols. J. Sci. Food Agric., 33, 165-168.

Gallagher, P. J. and Gibney, M. J. (1983). Immunological aspects of atherosclerosis: Role of dietary protein. In: "Current Topics in Nutrition and Disease", M. J. Gibney and D. Kritschevsky, eds., Alan R. Liss, New York, Vol 8, pp. 149-168.

Haeney, M.R., Goodwin, B.J., Barratt, M.E.J., Mike, N., and Asquith, P. (1982). Soy protein antibodies in man: their occurrence and possible relevance in celiac disease. J. Clin. Pathol., 35, 319-322.

Heppell, L.M.J., Pederson, H.E., and Sissons, J.W. (1985). Potential allergenicity of soya-based infant formulas. Proceedings 13th Int. Congr. Nutr., Brighton, U.K., 190.

Herion, P., Siberdt, D., Francotte, M., Urbain, J., and Bollen, A. (1984). Monoclonal antibodies against plasma protease inhibitors: II. Production and characterization of 25 monoclonal antibodies against human α_1-antitrypsin. Correlation between antigenic structure and functional sites. Biosci. Rep., 4, 139-148.

Horisberger, M. and Tacchini-Vonlanthen, M. (1983 a). Ultrastructural localization of Kunitz inhibitors on thin sections of Glycine max (soybean) cv. Maple Arrow by the gold method. Histochemistry, 77, 37-50.

Horisberger, M. and Tacchini-Vonlanthen, M. (1983 b). Ultrastructural localization of Bowman-Birk inhibitor on thin sections of Glycine max (soybean) cv. Maple Arrow by the gold method. Histochemistry, 77, 313-321.

Kagnoff, M. F. (1981). Immunology of the digestive system. In: "Physiology of the Gastrointestinal Tract", L.R. Johnson, ed., Raven Press, New York, 1337-1359.

Kato, Y., Matsuda, T., Watanabe, K., and Nakamura, R. (1983). Immunochemical studies on the denaturation of ovalbumin stored with glucose. J. Food Sci., 48, 769-772.

McGhee, J. R. and Mestecky, J. (eds.) (1983). "The Secretory Immune System". Ann. N.Y. Acad. Sci., 409, 1-869.

Moroz, L. A. and Yang, W. H. (1980). A specific allergen in food anaphylaxis. New Engl. J. Med., 302, 1126-1128.

Morton, J. I. and Deutsch, H. F. (1961). Immunochemical studies of modified ovomucoids. Arch. Biochem. Biophys., 93, 661-665.

Oi, V. T. and Herzenberg, L. A., (1980). Immunoglobulin-producing hybrid cell lines. In: "Selected Methods in Cellular Immunology", B. Mishell and S. Shiigi, eds., W.H. Freeman, San Francisco, CA., 351-372.

Peeters, H. (ed.) (1979). "Peptides of Biological Fluids", Pergamon Press, Oxford, England, Vol. 26, 1-720.

Rackis, J. J., Wolf, W. J., and Baker, E. C. (1986). Protease inhibitors in plant foods: content and inactivation. This volume.

Regnier, F. (1983). High-performance liquid chromatography of biopolymers. Science, 222, 245-252.

Rossebo, L. and Nordal, J. (1971). Soybean trypsin inhibitor: neutralization of its inhibitory effect upon the casein precipitating reaction of trypsin by rabbit antiserum. Z. Lebensmitt. Untersuch. Forsch., 147, 335-338.

Scott, H., Brandtzaeg, P., Thorsby, E., Baklien, K., Fausa, O., and Ek, J. (1983). Mucosal and systemic immune response patterns in celiac disease. Ann. Allergy, 51, 233-239.

Seegraber, F. J. and Morrill, J. L. (1982). Effect of soy protein on calves' intestinal absorptive ability and morphology determined by scanning electron microscopy. J. Dairy Sci., 65, 1962-1970.

Sissons, J. W., Smith, R. H., and Hewitt, D. (1982). Prediction of the suitability of soya-bean products for feeding to preruminant calves by an in vitro immunochemical method. Brit. J. Nutr., 47, 311-318.

Srihara, P. (1984). Processing to reduce the antigenicity of soybean products for preruminant calf diets. Dissertation, University of Guelph, Guelph, Ontario, 1-141.

Streicher, H. Z., Berkower, I. J., Busch, M., Gurd, F.R.N., and Berzofsky, J. A. (1984). Antigen conformation determines processing requirements for T cell activation. Proc. Natl. Acad. Sci. USA, 81, 6831-6835.

Toms, G. C. (1981). Lectins in Leguminosae. In: "Advances in Legume Systematics," R. M. Polhill and P. H. Raven, eds., Royal Botanic Gardens, Kew, 561-577.

Van der Woude, F. J., Hoedemaeker, P. J., van der Giessen, M., de Graeff, P. A., de Monchy, J, The, T. H., and van der Hem, G. K. (1983). Do food antigens play a role in the pathogenesis of some cases of human glomerulonephritis? Clin. Exp. Immunol., 51, 587-594.

Walker, W. A. and Bloch, K. J. (1983). Intestinal uptake of macromolecules: In vitro and in vivo studies. Ann. N.Y. Acad. Sci., 409, 593-601.

Wofsy, L. (1983). Methods and applications of hapten-sandwich labeling. Meth. Enzymol., 92, 471-488.

27

PHOTOREACTIVE, ACTIVE DERIVATIVES OF TRYPSIN AND CHYMOTRYPSIN INHIBITORS FROM SOYBEANS AND CHICKPEAS

Yehudith Birk*, Patricia Smirnoff* and J. Ramachandran**

*Faculty of Agriculture, The Hebrew University of Jerusalem Rehovot, Israel and ** Department of Biochemistry and Biophysics, University of California, San Francisco, CA 94143

ABSTRACT

The photoreactive arylsufenyl chloride 2-nitro-4-azidophenylsulfenyl chloride (2,4-NAPS-Cl) has been used for the selective modification of tryptophan in Kunitz's soybean trypsin inhibitor (STI). The ultraviolet absorption spectrum and amino acid analysis of 2,4-NAPS-STI indicated that only one of the two tryptophans, 93 or 117, present in STI was modified. Amino acid anlysis of the two separated CNBr-cleavage products of 2,4-NAPS-STI showed that only tryptophan 93 underwent modification. 2,4-NAPS-STI fully retained its inhibitory activity against trypsin. The covalent attachment of 2,4-NAPS-STI to tritiated trypsin after photolysis was demonstrated by exclusion chromatography on Sephadex G-50 in the presence of guanidine hydrochloride. Photoreactive derivatives of the Bowman-Birk trypsin-chymotrypsin inhibitor (BBI) from soybeans and of CI, the trypsin-chymotrypsin inhibitor from chick peas were prepared by selective modification of the ε-amino groups of lysine residues with 2,4(5)-NAPS-Cl. The ultraviolet absorption spectra of the photolabeled inhibitors indicated that three out of the five lysines of BBI and one of the seven lysines of CI were modified. The inhibitory activity of the modified inhibitors towards trypsin and chymotrypsin was not reduced even after photolysis. The specific lysine residues that constitute the trypsin-inhibitory sites of BBI and CI did not react with the photoreactive reagents. Further modification of the photoreactive derivatives of BBI and CI with maleic anhydride, directed towards the trypsin-reactive sites, resulted in almost complete loss of trypsin-inhibiting activity without reducing the ability to inhibit chymotrypsin. A pronounced potentiation effect (∿2x) of the chymotrypsin inhibiting activity was noted for 2,5-NAPS-CI and it was retained even after maleylation followed by photolysis, raising the possibility of exposure of an additional chymotrypsin inhibitory site in CI.

INTRODUCTION

Proteinlike proteinase inhibitors are widely distributed in the plant kingdom. The current knowledge of their distribution among plants and within the plant reveals hundreds of inhibitors dispersed among different botanical families (1). The most frequent source of the inhibitors is the seed, but their location is not necessarily restricted to this part of the plant. The physiological significance of plant proteinase inhibitors has been questioned for a long time. The fact that certain seeds, such as soybeans and wheat grains contain inhibitors of growth and of larval gut proteinases of the insects Tribolium and Tenebrio, suggest the possibility that these inhibitors may have evolved as a defense mechanism against predatory insects (2). The inhibitors have been the object of research in many disciplines. They have attracted the attention of nutritionists due to their presence in valuable plant foods and their subsequent, possible, involvement in nutritive properties. Inhibitor proteins have been studied as model systems for elucidation of the mechanism of inhibition of proteinases, as well as for studies of protein-protein association. The inhibitors are also being used as valuable tools in medical research because of unique pharmacological properties that suggest clinical application. Since the inhibitory capacities are usually evaluated on bovine pancreatic proteinases, the relevance of clinical and nutritional uses of the inhibitors in other species is being questioned.

The presence of protein-like proteinase inhibitors in all legume seeds is a well established fact. Of these, the Kunitz soybean trypsin inhibitor (STI) and the Bowman-Birk trypsin and chymotrypsin inhibitor (BBI) from soybeans have been most thoroughly studied. They have shown pronounced physiological effects, such as pancreatic enlargement and increased production and secretion of pancreatic proteinases, when ingested by mammals and birds (3-6). Recently, attention has been focused on trypsin- and chymotrypsin-inhibitors from legume seeds as possible dietary anti-carcinogens (7,8), stimulating a new look at proteinases and proteinase-inhibitors in relation to carcinogenesis.

The mechanisms by which proteinase inhibitors exert their different physiological effects in vivo have not been elucidated so far. Specific, covalent attachment of the inhibitors to target proteinases, and perhaps also to zymogens, by affinity labeling under physiological conditions, may shed some light on their mechanisms of action.

THE TRYPSIN AND CHYMOTRYPSIN INHIBITORS FROM SOYBEANS

Molecular properties of STI and BBI

The first plant proteinase inhibitor to be isolated and well-characterized was soybean trypsin inhibitor (Kunitz), henceforth STI. Surprisingly, only a few homologous inhibitors to STI have been found in the common legume seeds. The complete amino acid sequence of STI which consists of 181 amino acid residues and two disulfide bridges is shown in Fig. 1 (9). STI served as the model for the establishment of the Standard Mechanism as well as for the development of reactive site replacement techniques (10-13). The primary structure of the reactive site of the inhibitor modified with catalytic amounts of trypsin at acid pH is:

60	61	62	63		64	65	66	67
-Pro-	Ser-	Tyr-	Arg-OH		H-Ile-	Arg-	Phe-	Ile-

Dietary STI induced an increase in rat pancreatic weight which led to both cellular hypertrophy and hyperplasia. Trypsin and chymotrypsin activity increased with the increase in pancreatic weight whereas amylase and lipase activities were not affected. Insulin content showed a slight tendency to increase and glucagon content was the same as the control (14).

A soybean proteinase inhibitor that differs markedly from STI was first noted by Bowman (15) and subsequently purified and characterized by Birk and associates (16,17), and later on by Frattali and Steiner (18). BBI has served as the phototype for a family of inhibitors that are present in the seeds of all leguminous plants. They consist ot two tandem homology

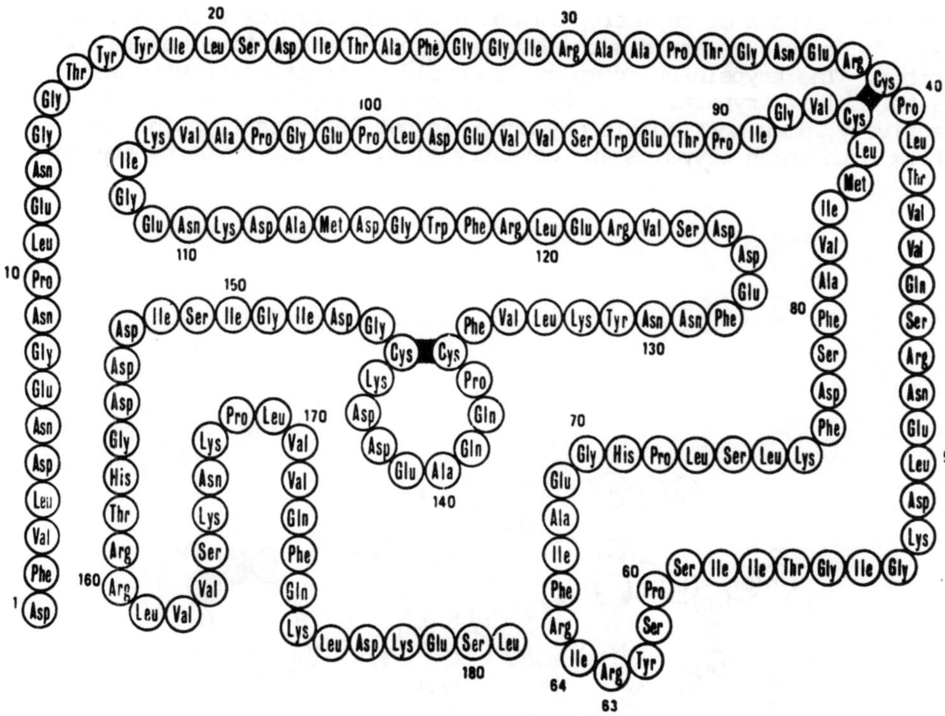

Fig. 1 Covalent structure of Kunitz soybean trypsin inhibitor (STI) (9).

regions on the same polypeptide chain, each with a reactive, inhibitory site, as exemplified by BBI from soybeans.

BBI is a protein molecule consisting of a chain of 71 amino acids crosslinked by 7 disulfide bonds. Its complete amino acid sequence and covalent structure, shown in Fig. 2, were revealed by Ikenaka and associates (19). It contains no tryptophan. It possesses two independent sites of inhibition, one at Lys 16-Ser 17 against trypsin and the other at Leu 43-Ser 44 against chymotrypsin. BBI has been found to be a highly specific limited proteolysis substrate for both of its target enzymes, trypsin and chymotrypsin (20), and therefore obeys the Standard Mechanism (21). Indeed, conversion of virgin BBI to trypsin-modified BBI did not affect the inhibitory activity against trypsin or chymotrypsin. Further treatment of the reaction product with carboxypeptidase B resulted in removal of the newly-formed COOH-terminal lysine and consequent loss of trypsin-inhibitory activity. However, parallel modification of virgin BBI with chymotrypsin abolished most of the inhibitory activity against chymotrypsin while the activity against trypsin was not affected at all (20). Thus, limited proteolysis may be utilized to modify the double headed BBI into a "single headed" inhibitor that inhibits solely either trypsin or chymotrypsin. The minimum molecular weight of BBI, based on amino acid composition, is 7975. In aqueous solutions, BBI (isoelectric point pH 4.2) undergoes self-association, which is concentration dependent (22,23). The inhibitory capacities of BBI, usually evaluated on bovine trypsin and chymotrypsin, reveal the formation of a 1:1 complex with either trypsin or chymotrypsin and a ternary complex with both enzymes. The complexes are dissociable below pH 5 into active enzyme and inhibitor. The trypsin-BBI complex is devoid of tryptic activity and fails to inhibit trypsin. The chymotrypsin-BBI complex lacks chymotryptic activity, does not inhibit chymotrypsin but inhibits trypsin. BBI inhibits both human trypsins and chymotrypsins, but the trypsins are more

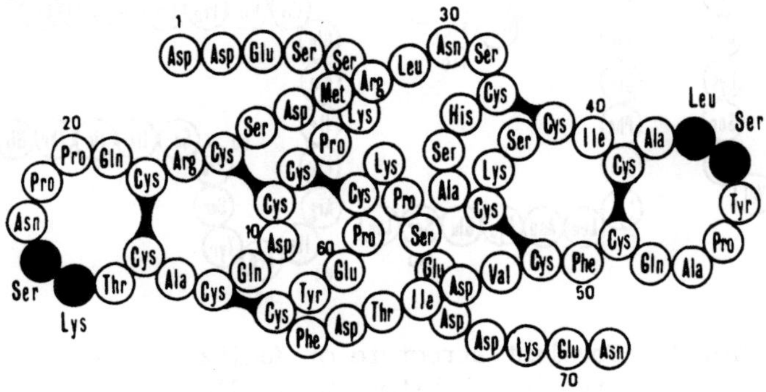

Fig. 2 Complete covalent structure of the soybean Bowman-Birk trypsin-chymotrypsin inhibitor (BBI). Residues at the two reactive sites are shown as solid black circles (19)

susceptible to the inhibitor than the chymotrypsins. BBI is also a potent inhibitor of trypsins and chymotrypsins from the digestive tracts of insects, such as Tenebrio molitor. BBI has an unusual resistance to various proteolytic enzymes including pepsin and pronase. Although BBI does not inhibit pancreatic elastase, the latter hydrolyses the Leu 43-Ser 44 bond of BBI, abolishing most of the chymotrypsin-inhibitory activity without affecting the trypsin-inhibiting activity (reviewed (24)).

Numerous modification studies were undertaken to relate structure and composition of BBI to function and activity, and to investigate the role of the reactive site amino acids (24). The existence, in BBI, of two homologous inhibitory regions, the almost symmetrical structure of the inhibitor and the presence of a single methionine in its center triggered the attempt to cleave the inhibitor with the aid of cyanogen bromide. The scission of BBI, with cyanogen bromide followed by pepsin, resulted in two active fragments, one with trypsin inhibitory activity and the other with chymotrypsin inhibitory activity (Fig. 3) (25,26). In contrast to the unusual stability of the native inhibitor, the separated trypsin- and chymotrypsin-inhibitory domains were less stable, suggesting that the legume double headed inhibitors acquired their unusual stability by duplicating an ancestral single-headed structure.

Nutritional and physiological significance

The availability in pure form of STI and BBI ultimately enabled the resolution of the effect of ingested inhibitors on growth rate, pancreatic hypertrophy and intestinal proteolysis of mammals and birds. Feeding experiments of chicks and rats on diets supplemented with STI, BBI or with both, resulted in an insignificant depression of animal growth rate but the inhibitors were fully responsible for causing hypertrophy of the pancreas. The pancreases of rats and chicks adapted to raw soybean meals were synthesizing more trypsinogen and chymotrypsinogen and less amylase than pancreases of rats adapted to heated soybean meal (reviewed, (24)). Ingestion of diets supplemented with BBI modified by maleylation or succinilation, or with the complex of BBI plus trypsin (in which the trypsin-inhibitory site is masked) did not cause pancreatic hypertrophy and had no significant effect on the amount of pancreatic proteinases. This indicates that the trypsin inhibitory site, rather than the chymotrypsin inhibitory site of BBI is involved in the enlargement of the pancreas and in the increase of pancreatic proteolytic activity (5). The fate of BBI in the digestive tract of chicks was studied by direct follow-up of radioactive inhibitor and by radioimmunological estimation of the unlabeled inhibitor. The *in vitro* experiments using the inverted sac technique suggested that both the

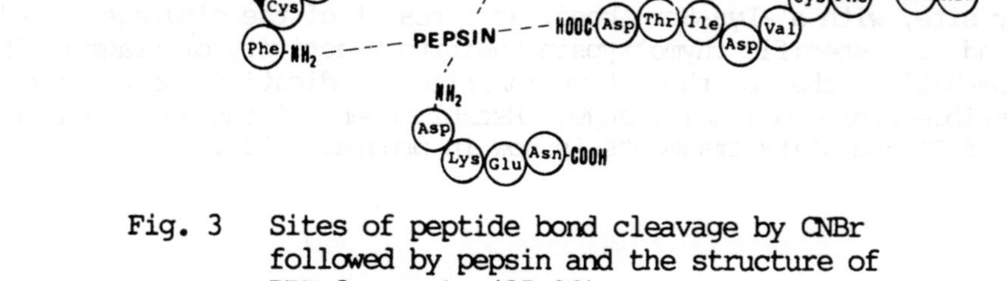

Fig. 3 Sites of peptide bond cleavage by CNBr followed by pepsin and the structure of BBI fragments (25,26)

inhibitor and its degradation products were absorbed, whereas the results of in situ and in vivo experiments indicated that only a small amount of the native inhibitor was absorbed. Most BBI is degraded during its passage through the intestine, and degradation products are excreted in the feces (6). Troll and associates, who have been studying dietary proteinase inhibitors as possible anticarcinogens, investigated the metabolic fate of BBI fed to mice and rats. They showed that the inhibitor survives inactivation by stomach digestion and it appears to be fully active as a proteinase inhibitor in the small intestine, where it complexes with the trypsin and chymotrypsin occurring there. The majority of the inhibitor was excreted in combined form in the feces (27). They have also reported that BBI inhibits X-ray transformation of cells specifically at the initiation stage. A recent report revealed that nanomolar concentrations of BBI suppress X-ray induced transformation in vitro and that the effect of BBI on the transformation process is irreversible (28). The significance of possible, direct and indirect, anticarcinogenic action of BBI and other proteinase inhibitors still remains to be established. As to other possible therapeutic applications of the inhibitor, such as inactivation of granulocytic proteinases which are able to degrade connective tissue in pathological states, BBI showed a strong interaction with elastases from human and canine granulocytes (29). BBI was also highly effective in preventing human antithrombin III inactivation by degradation, due to interaction with elastases from human granulocytes (30).

THE TRYPSIN AND CHYMOTRYPSIN INHIBITOR FROM CHICK PEAS

CI, the doubleheaded trypsin and chymotrypsin inhibitor from chick peas resembles BBI in chemical and physical properties (31,32). Based on amino acid analysis, CI has a molecular weight of 8188 and appears as two isoinhibitors that differ from one another in a glutamyl, aspartyl, glycyl and alanyl residue. The trypsin- and chymotrypsin-reactive sites of CI are -Lys-Ser- and -Tyr-Ser-, respectively. However, limited proteolysis of CI with trypsin at the specific -Lys-Ser- bond abolishes most of the trypsin inhibitory activity and affects also, to a considerable extent, the chymotrypsin inhibitory activity. In a similar manner, limited proteolysis of CI with chymotrypsin, directed towards the chymotrypsin-reactive site -Tyr-Ser- bond, completely suppresses the chymotrypsin inhibitory activity and also considerably affects the ability of CI to inhibit trypsin. These findings suggest that the two inhibitory sites of CI are sterically close to each other and that opening of a single reactive bond at either of the two inhibitory sites allowed a conformational change, such as "masking" the other, intact, inhibitory site. This hypothesis was supported by modifications of CI with maleic or succinic anhydride, which react selectively with Lys residues, including the one at the trypsin-inhibitory site, and resulted in an almost full loss of the trypsin inhibiting activity of CI but had no effect on the chymotrypsin inhibitory activity. Furthermore, scission of CI with cyanogen bromide followed by pepsin yielded two active fragments, A and B. Fragment A inhibited trypsin but not chymotrypsin, and fragment B inhibited chymotrypsin but not trypsin. The specific trypsin inhibitory activity, on a molar ratio, of fragment A was twice that of the native inhibitor. This suggests the unveiling of another trypsin inhibitory site, with a -Lys-Ser- bond, as a result of the cleavage. On the other hand, the specific chymotrypsin inhibitory activity of fragment B was about one-half of that of the native inhibitor, indicating the occurrence of a possible conformational change. Establishment of the amino acid sequence of CI and of its fragments is now in progress (33).

PHOTOREACTIVE REAGENTS AND PHOTOAFFINITY LABELING

Photoaffinity labeling is being used increasingly in studies of polypeptide hormone receptors (34,35). Such an approach has been practiced successfully for the identification of specific receptors of ACTH by their covalent interaction with a photoaffinity probe, prepared by selective modification of the single Trp residue in ACTH (36,37). Arylsulfenyl chlorides have been found to be specific mild reagents for the modification of tryptophan and cysteine residues of polypeptides in acidic media (38-40). Tryptophan is converted into a derivative with thioether function in the 2 position of the indole nucleus, and cysteine is converted to an unsymmetrical disulfide. In aqueous alkaline solution, sulfenyl chlorides react with α-amino acids and peptides as well as ε-amino groups of lysine residues (41). Hence, a variety of photoreactive derivatives of polypeptides and proteins may be obtained by selective modification with arylsulfenyl chlorides containing azido groups. It became apparent that a more stable photoreactive arylsulfenyl chloride would be highly useful for the introduction of photoreactive groups into proteins without significantly altering the biological activities. This led to the synthesis of 2-nitro-4-azidophenylsulfenyl chloride (2,4-NAPS-Cl) and the isomer 2-nitro-5-azidophenylsulfenyl chloride (2,5-NAPS-Cl) (36). The location of the azido group in O-nitrophenylsulfenyl chloride is limited to the 4,5 or 6 positions on the phenyl rings. Since the 6 position is ortho to the SCl group and may lead to steric hindrance as well as reaction with the indole during photolysis of the NAPS derivative of tryptophan compounds, only the nitrophenylsulfenyl chlorides containing the azido group in the 4 or 5 positions were synthesized. Both 2,4-NAPS-Cl and the isomer 2,5-NAPS-Cl reacted rapidly with the indole group of tryptophan residues. The 2,4-NAPS-Cl showed a faster rate of reaction than its isomer, suggesting that this reagent may be more useful when short reaction times are necessary. The nucleophilic reactivity of the sulfenyl chloride is increased by the presence of the azido-group in the para position in 2,4-NAPS-Cl. The absorption spectra of the 2,4-NAPS-Trp derivatives are also considerably different from the spectra of the 2,5-NAPS-Trp compounds due to the differences in the electronic interaction of the azido and nitro groups in the two isomeric compounds (36). The 2,4-NAPS chlorides exhibit a broad absorption band near 400nm and an absorption band at 300 - 350 nm is observed in the spectrum of 2,5-NAPS-Cl.

A similar study performed for the introduction of a photo reactive group at thiol groups of peptides shows that these reagents cause rapid and quantitative modification of thiol groups of cysteine residues (42). 2,4-NAPS-Cl appears to react faster than the isomeric reagent - 2,5-NAPS-Cl, as was observed in the modification of the indole moiety of tryptophan (36). Although the reaction with the indole group is slower than that with the thiol group, the difference in the rates may not be sufficient to permit selective modification under acidic conditions.

The use of these photoreactive reagents for selective modification of amino groups has also been investigated (43). The two isomers appear to react at similar rates in aqueous alkaline solutions. The spectral characteristics of the NAPS groups bound to ε-amino groups showed that the 2,4-NAPS derivatives exhibit one broad absorption band at 375-460 nm ($\varepsilon \simeq 3{,}000$) and the 2,5-NAPS derivatives show one absorption band at 320-330 nm ($\varepsilon \simeq 4{,}000$).

Covalent attachment of the photolabeled protein proteinase inhibitor to the target proteinase is performed by photolysis, the process of electron activation of an azide group of the NAPS derivative by ultraviolet irradiation at ∿ 366 nm (44).

PHOTOREACTIVE DERIVATIVE OF KUNITZ'S SOYBEAN TRYPSIN INHIBITOR (STI)

STI contains two tryptophans, in positions 93 and 117, as shown by the complete amino acid sequence (Fig. 1). Since STI contains no cysteine, the modification with 2,4-NAPS-Cl at pH 3.5 would be limited to one of the two tryptophan residues or both. The preparation of 2,4-NAPS-STI, the photoreactive derivatives of STI has been described by Ramachandran et al. (45). The spectrum of 2,4-NAPS-STI exhibits one broad absorption band at 360-420 nm, which disappears after photolysis. From the absorption spectral maxima of 2,4-NAPS-STI and the molar extinction coefficient of 2,4-NAPS-tryptophan (36), it has been estimated that only one tryptophan of the two present in STI was modified. This has been supported by the amino acid analysis of the photoreactive derivative. In order to find out which of the two tryptophans was modified, CNBr was used to split both STI, as control, and 2,4-NAPS-STI at Met 114 into two fragments: fragment 1-114, which contains Trp 93, and fragment 115-181 with Trp 117. The two distinct fractions which were separated on a Sephadex G-50 column and then analysed showed that only the tryptophan of fragment 1-114 underwent modification. The modification with 2,4-NAPS-Cl should not interfere with the inhibitory activity of STI since Trp 93 is not among the residues which are in contact with trypsin in the enzyme-inhibitor complex, as shown by the crystal structure of the complex of STI with porcine trypsin (46). Indeed, examination of the esterolytic inhibitory activity of 2,4-NAPS-STI showed that the photoreactive compound and STI had similar inhibitory activities against trypsin prior to photolysis. However, 2,4-NAPS-STI lost 70% of the inhibitory activity upon photolysis.

Covalent attachment of 2,4-NAPS-STI to trypsin

The photoaffinity labeling of trypsin with 2,4-NAPS-STI was performed on tritiated trypsin prepared by reacting bovine trypsin with (^{3}H)-succinimidyl propionate. The covalent attachment of 2,4-NAPS-STI to tritiated trypsin after photolysis was demonstrated by gel filtration on Sephadex G-75 in the presence of 5 M guanidine hydrochloride, showing the presence of a radioactive compound of molecular weight of $\sim$ 50,000. The covalent attachment was also demonstrated by SDS polyacrylamide gel electrophoresis which showed the presence of a single peak of radioactivity at a position corresponding to an apparent molecular weight of nearly 50,000. These results clearly show that 2,4-NAPS-STI has been covalently linked to the tritiated trypsin (45).

PHOTOREACTIVE DERIVATIVES OF TRYPSIN-CHYMOTRYPSIN INHIBITORS FROM SOYBEANS (BBI) AND CHICK PEAS (CI)

As already mentioned BBI and CI lack tryptophan and contain 5 and 7 lysine residues, respectively, including the lysine residue that constitutes the trypsin inhibitory site (33,47). Since arylsulfenyl chlorides can react, in aqueous alkaline solutions, with ε-amino groups of lysine residues, BBI and CI have been modified under these conditions (48).

Photoreactive derivative of BBI

The preparation of 2,5-NAPS-BBI, the photoreactive derivative of BBI, has been described by Smirnoff et al. (48). The ultraviolet absorption spectrum of 2,5-NAPS-BBI exhibits one absorption band at 320-330 nm, which disappears after photolysis. From the absorption spectral maximum of 2,5-NAPS-BBI and the molar extinction coefficient of 2,5-NAPS lysine it has been estimated that three of the five lysines of BBI were modified. The selective introduction of photoreactive groups into some, but not all of the lysine residues, did not affect the inhibitory effect of the photolabeled inhibitor on the esterolytic and proteolytic activity of trypsin or chymotrypsin (Table 1). Inhibitory activity was not lost even after submission of the photoreactive inhibitor to photolysis, differing in this respect, from 2,4-NAPS-STI. These results strongly suggest that the specific lysine residue that constitutes the trypsin-inhibitory site, did not react with the photoreactive reagent. More direct evidence supporting this conclusion was provided by further modification with maleic anhydride. It has been shown previously (5) that maleylation of BBI is directed towards the lysine residue in the trypsin inhibitory site, and results in almost complete loss of trypsin-inhibitory activity, without affecting the ability to inhibit chymotrypsin. A similar maleylation experiment on 2,5-NAPS-BBI has shown that the maleyl group reacted, as expected, with the lysine residue in the trypsin-inhibitory site and affected the trypsin inhibiting activity accordingly. No change was observed in the ultraviolet absorption spectrum of the photolabeled and maleylated inhibitor. For hitherto unknown reasons BBI did not react with 2,4-NAPS-Cl.

Table 1. Inhibitory activities of NAPS derivatives of BBI and CI on trypsin and chymotrypsin when assayed on tosyl-L-arginine methyl ester and acetyl-L-tyrosine ethyl ester, respectively (48).

Derivative assayed	Residual inhibitory activity against	
	Trypsin	Chymotrypsin
	(expressed as % of the control)	
BBI (control)	100	100
2,5-NAPS-BBI	100	100
2,5-NAPS-BBI, photolysed	100	100
2,5-NAPS-BBI, maleylated	7	100
2,5-NAPS-BBI, maleylated, photolysed	14	100
CI (control)	100	100
2,5-NAPS-CI	100	200
2,5-NAPS-CI, photolysed	100	100
2,5-NAPS-CI, maleylated	2	250
2,5-NAPS-CI, maleylated, photolysed	2	250
2,4-NAPS-CI	100	100
2,4-NAPS-CI, photolysed	100	100
2,4-NAPS-CI, maleylated	10	250
2,4-NAPS-CI, maleylated, photolysed	10	250

Photoreactive derivatives of CI

The trypsin-chymotrypsin inhibitor, CI, was submitted to photolabeling modifications with 2,4-NAPS-Cl and 2,5-NAPS-Cl. The ultraviolet absorption spectrum of the photoreactive derivative of 2,4-NAPS-CI showed one broad absorption band at 375-460 nm, which disappears after photolysis. Based on the absorption spectral maxima of the modified inhibitor and on the molar extinction coefficient for 2,4-NAPS-lysine, it has been estimated that only one lysine of the seven present in CI was modified. A similar analysis of the 2,5-NAPS-CI, showed a spectrum with an absorption band at 320-330 nm, which also disappeared after photolysis. The modified derivatives of CI did not lose their inhibitory activity on trypsin or chymotrypsin, even after submission of the photoreactive inhibitors to photolysis (Table 1). Presumably, the hydrophobic nature of the reagent and the conditions under which the modifications took place favoured lysine residues other than those in the hydrophylic inhibitory sites of CI and BBI. Maleylation experiments of the photoreactive derivatives of CI have shown that the maleyl group reacted, as expected, with the lysine residue in the trypsin-inhibitory site and decreased the trypsin inhibiting activity of both 2,4-NAPS-CI and 2,5-NAPS-CI, accordingly. A striking finding has been the potentiation (x2) of the chymotrypsin-inhibiting activity of CI as a result of the modification with 2,5-NAPS-Cl. This potentiation disappears after photolysis. Moreover, maleylation of both 2,5-NAPS-CI and 2,4-NAPS-CI, before and after photolysis, results in a similar, more pronounced effect. So far, no explanation for this phenomenon has been proposed. A possible occurrence of a conformational change and/or the introduction of an additional chymotrypsin-inhibiting site into CI as a result of these modifications is being examined.

In summary, 2,4-NAPS Cl has been found to be a useful reagent for the selective introduction of a photoreactive group in a tryptophan residue in the Kunitz STI-type proteinase inhibitors. The covalent attachment of the modified trypsin inhibitor to its target proteinase can perhaps be used for probing the three dimensional structure of the inhibitor-enzyme complex. 2,4(5)-NAPS-Cl are also useful reagents for the selective introduction of photoreactive groups in lysine residues, other than those present in the trypsin-inhibitory sites of BBI and CI. They may be applied for the study of trypsins and and chymotrypsins from various sources via photoaffinity labeling and may help to identify the amino acid residues in the active sites of the enzymes. They may perhaps shed some light on the species specificity of these naturally occurring protein proteinase inhibitors.

ACKNOWLEDGEMENT

The support of grant 2113 from the National Council for Research and Development, Israel is appreciated.

References

1. I.E Liener, and M. Kakade, Protease inhibitors, in: "Toxic Constituents of Plant Foodstuffs," I.E. Liener, ed., Academic Press, New York (1980).
2. S.W. Applebaum, and Y. Birk, Natural mechanisms of resistance to insects in legume seeds, in: "Insect and Mite Nutrition," J.G. Rodriguez, ed., North-Holland, Amsterdam (1972).
3. A. Gertler, Y. Birk, and A. Bondi, A comparative study of the nutritional and physiological significance of pure soybean trypsin inhibitors and ethanol -extracted soybean meals in chicks and rats, J. Nutr. 91:358 (1967).
4. A.M. Konijin, Y. Birk and K. Guggenheim, In vitro synthesis of pancreatic enzymes: effect of soybean trypsin inhibitor, Am. J. Physiol. 218:1113 (1970).
5. Z. Madar, Y. Birk and A. Gertler, Native and modified Bowman-Birk trypsin inhibition.Comparative effect on pancreatic enzymes upon ingestion by quails (Conturnix coturnix japonica), Comp. Biochem. Physiol. 48B:251 (1974).

6. Z. Madar, A. Gertler, and Y. Birk, The fate of the Bowman-Birk trypsin inhibitor from soybeans in the digestive tract of chicks, Comp. Biochem. Physiol. 62A, 1057 (1979).
7. W. Troll, R. Wiesner, C.J. Shellabarger, S. Holtzman, and J.P. Stone, Soybean diet lowers breast tumor incidence in irradiated rats, Carcinogenesis. 1:469 (1980).
8. J. Yavelow, M. Gidlund and W. Troll, Protease inhibitors from processed legumes effectively inhibit superoxide generation in response to TPA, Carcinogenesis. 3:135 (1982).
9. T. Koide, and T. Ikenaka, Studies on soybean trypsin inhibitors. 3. Amino acid sequence of the carboxyl-terminal region and the complete amino acid sequence of soybean trypsin inhibitor (Kunitz), Eur. J. Biochem. 32:417 (1973).
10. R.W. Sealock, and M. Laskowski, Jr., Enzymatic replacement of the arginyl by a lysyl residue in the reactive site of soybean trypsin inhibitor, Biochemistry 8:3703 (1969).
11. D. Kowalski and M. Laskowski, Jr., Chemical-enzymatic replacement of Ile^{64} in the reactive site of soybean trypsin inhibitor (Kunitz), Biochemistry 15:1300 (1976).
12. V. Quast, J. Engel, E, Steffen, H. Tschesche and S. Kupfer, Stopped-flow kinetics of the resynthesis of the reactive site peptide bond in kallikrein inhibitor (Kunitz) by β-trypsin, Biochemistry 17:1675 (1978).
13. M.W. Baillargeon, M. Laskowski, Jr., D.E. Neves, A. Porubcan, and R.E. Santini Soybean trypsin inhibitor (Kunitz) and its complex with trypsin. Carbon-13 nuclear magnetic resonance studies of the reactive site arginine. Biochemistry 19:5703 (1980).
14. R.S. Temler, C.A. Dormond, E. Simon, and B. Morel, The effect of feeding soybean trypsin inhibitor and repeated injections of cholecystokinin on rat pancreas, J. Nutr. 114:1083 (1984).
15. D.E. Bowman, Fractions derived from soybeans and navy beans which retard the tryptic digestion of casein, Proc. Soc. Exp. Biol. Med. 57:139 (1944).
16. Y. Birk, Purification and some properties of a highly active inhibitor of trypsin and α-chymotrypsin from soybeans, Biochim. Biophys. Acta 54:378 (1961).
17. Y. Birk, A.Gertler and S. Khalef, A pure trypsin inhibitor from soya beans, Biochem. J. 87:281 (1963).
18. V. Frattali, and R.F. Steiner, Soybean inhibitors. II. Preparative electrophoretic purification of soybean proteinase inhibitors on polyacrylamide gel. Anal. Biochem. 27:285 (1969).
19. S. Odani and T. Ikenaka, Studies on soybean trypsin inhibitors. IV. Complete amino acid sequence and the anti-proteinase sites of Bowman-Birk soybean proteinase inhibitor, J. Biochem (Tokyo) 71:839 (1972).
20. Y. Birk, A. Gertler and S. Khalef, Further evidence for a dual, independent, activity against trypsin and α-chymotrypsin of inhibitor AA from soybeans. Biochim. Biophys. Acta 147:402 (1967).
21. M. Laskowski, Jr., and I. Kato, Protein inhibitors of proteinases. Ann. Rev. Biochem. 49:593 (1980).
22. D.B.S. Millar, G.E. Willick, R.F. Steiner and V. Frattali, Soybean inhibitors. IV. The reversible self-association of a soybean proteinase inhibitor. J. Biol. Chem. 244: 281 (1969).
23. M.L. Kakade, N.R. Simons, and I.E. Liener, The molecular weight of the Bowman-Birk soybean protease inhibitor, Biochim. Biophys. Acta 200:166 (1970).
24. Y. Birk, The Bowman-Birk inhibitor. Trypsin- and chymotrypsin-inhibitor from soybeans. Int. J. Peptide Protein Res. 25:113 (1985).

25. S. Odani and T. Ikenaka, Scission of soybean Bowman-Birk proteinase inhibitor into two small fragments having either trypsin or chymotrypsin inhibitory activity, J. Biochem. (Tokyo) 74:857 (1973).
26. S. Odani and T. Ikenaka, Studies on soybean trypsin inhibitors. XIII. Preparation and characterization of active fragments from Bowman-Birk proteinase inhibitor, J. Biochem. (Tokyo) 83: 747 (1978).
27. J. Yavelow, T.H. Finlay, A.R. Kennedy, and W. Troll, Bowman-Birk soybean protease inhibitor as an anticarcinogen, Cancer Res. 43: 2454 (1983).
28. J. Yavelow, M. Collins, Y. Birk, W, Troll and A.R. Kennedy, Nanomolar concentrations of Bowman-Birk soybean protease inhibitor suppress X-ray induced transformation in vitro, Proc. Natl, Acad. Sci. USA, 82:5395 (1985).
29. H. Schiessler, K. Ohlsson, I. Olsson, M. Arnhold, Y. Birk and H. Fritz, Elastases from human and canine granulocytes. II. Interaction with protease inhibitors of animal, plant and microbial origin, Hoppe-Seylers Z. Physiol. Chem. 358:53 (1977).
30. M. Jochum, S. Lander, N. Heimburger and H. Fritz, Effect of human granulocytic elastase on isolated human antithrombin III. Hoppe-Seylers Z. Physiol. Chem. 362:103 (1981).
31. P. Smirnoff, S. Khalef, Y. Birk and S.W. Applebaum, A trypsin and chymotrypsin inhibitor from chick peas (Cicer arietinum), Biochem. J. 157:745 (1976).
32. P. Smirnoff, S. Khalef, Y. Birk and S.W. Applebaum, Trypsin and chymotrypsin inhibitor from chick peas. Selective chemical modifications of the inhibitor and isolation of two isoinhibitors. Int. J. Peptide Protein Res. 14:186 (1979).
33. Y. Birk, W.J. Kohr, J. Ramachandran and P. Smirnoff, unpublished data (1986).
34. H. Bayley, and J.R. Knowles, Photoaffinity labeling, Methods Enzymol. 46:69 (1977).
35. V, Chaudry, and F.H. Westheimer, Photoaffinity labeling of biological systems, Annu. Rev. Biochem. 48:293 (1979).
36. K. Muramoto, and J. Ramachandran, Photoreactive derivatives of corticotropin. 2. Preparation and characterization of 2-nitro-4(5)-azidophenylsulfenyl derivatives of corticotropin, Biochemistry 19:3280 (1980).
37. J. Ramachandran, K. Muramoto, M. Kénez-Kéri, G. Kéri and D.I. Buckley, Photoaffinity labeling of corticotropin receptors. Proc. Natl. Acad. Sci. USA, 77:3967 (1980).

38. E. Scoffone, A. Fontana, and R. Rocchi, Sulfenyl halides as modifying reagents for polypeptides and proteins. I. Modification of tryptophan residues, Biochemistry 7:971 (1968).
39. A. Fontana, E. Scoffone and C.A. Benassi, Sulfenyl halides as modifying reagents for polypeptides and proteins. II. Modification of cysteinyl residues, Biochemistry 7:980 (1968).
40. E. Canova-Davis and J. Ramachandran, Chemical modification of the tryptophan residue in adrenocorticotropin, Biochemistry 15:921 (1976).
41. L. Zervas, D. Borovas and E. Gazis, New methods in peptide synthesis. I. Tritylsulfenyl and O-nitrophenylsulfenyl groups as N-protecting groups, J. Am. Chem. Soc. 85:3660 (1963).
42. K. Muramoto and J. Ramachandran, Preparation of photoreactive derivatives of glutathione and (9-(2-mercaptotryptophan)) corticotropin by selective modification of the sulfhydryl group, Biochemistry 20:3376 (1981).
43. K. Muramoto, D. Buckley and J. Ramachandran, Preparation of photoreactive derivatives of α-melanotropin by selective modification of the lysine or tryptophan residue, Int. J. Peptide Protein Res. 20:366 (1982).
44. E. Canova-Davis and J. Ramachandran, Photoreactive derivatives of corticotropin. 1. Preparation and characterization of 2,4-dinitro-5-azidophenylsulfenyl derivative of corticotropin, Biochemistry 19:3275 (1980).
45. J. Ramachandran, P. Smirnoff and Y. Birk, Preparation of a photoreactive derivative of Kunitz's soybean trypsin inhibitor by selective modification of a tryptophan residue and formation of a covalent complex of the modified inhibitor with trypsin, Int. J. Peptide Protein Res. 23:72 (1984).
46. R.M. Sweet, H.T Wright, J. Janin, C.H. Chothia and D.M. Blow, Crystal structure of the complex of porcine trypsin with soybean trypsin inhibitor (Kunitz) at 2.6 - Å resolution, Biochemistry 13: 4212 (1974).
47. Y. Birk and A. Gertler, Chemistry and biology of proteinase inhibitors from soybeans and groundnuts, in "Proteinase Inhibitors," Proc. Int. Res. Conf. 1st. H. Fritz and H. Tschesche, eds., de Gruyter, Berlin (1970).
48. P. Smirnoff, J. Ramachandran and Y. Birk, Preparation of photoreactive derivatives of trypsin-chymotrypsin inhibitors from soybeans and chickpeas by selective modification of lysine residues, Int. J. Peptide Protein Res. 26:274 (1985)

28

BIOCHEMICAL, NUTRITIONAL AND TOXICOLOGICAL ASPECTS OF ALPHA-AMYLASE INHIBITORS FROM PLANT FOODS

Vincenzo Buonocore and Vittorio Silano*

Department of Organic and Biological Chemistry, Naples University, Naples and *Department of Comparative Toxicology and Ecotoxicology, Istituto Superiore di Sanità, Rome (Italy)

ABSTRACT

This paper is a critical review of the available data on plant protein inhibitors active either on animal or endogenous plant alpha-amylases. The First Section is a review of available data on molecular properties of the purified inhibitors from cereals, legumes, colocasia and yam. The Second Section deals with properties of amylase-inhibitor complexes and parameters controlling the interaction between amylases and inhibitors. The Third Section discusses possible roles of these inhibitors in the plant, whereas Section Four focusses on nutritional and toxicological significance of amylase inhibitors for human beings and other mammals. Lastly Section Five examines some applications in medicine of alpha-amylase inhibitors from plants.

INTRODUCTION

Protein inhibitors of alpha-amylase (1,4-alpha-D-glucan glucanohydrolase, EC 3.2.1.1) are widely distributed throughout the plant kingdom and some of them have been studied in detail with respect to their structure, chemico-physical properties and mechanism of action (Marshall, 1975; Saunders, 1975; Buonocore et al., 1977; Silano, 1978; Richardson, 1981; Warchalewski, 1983; Whitaker, 1983).

Reports on natural alpha-amylase inhibitors of protein nature date back to the early forties of this century. In 1943 Kneen and Sandstedt described a water-soluble amylase inhibitor with protein character occurring in wheat, rye and some sorghum cultivars; since then, alpha-amylase inhibiting activity has been detected in a number of cereals, legumes and other plants including wheat, barley, rye, sorghum, maize, oat, pearl millet (*Pennisetum typhoideum*), setaria (*Setaria italica*), ragi (*Eleusine coracana*), *Triticale*, *Aegilops*, kidney bean (*Phaseolus vulgaris*), chickpea (*Cicer arietinum*), colocasia (*Colocasia esculenta*), yam (*Dioscorea alata*), potato (*Solanum tuberosum*), peanut (*Arachis hypogoea*), unripe mango (*Mangifera indica*) and banana (*Musa paradisiaca*).

Alpha-amylase inhibitors from wheat, barley, rye, maize, millet, kidney bean, peanut, colocasia and yam have been purified and characterized; most of them are glycoproteins and occur as multiple forms in the same plant source. Among all those studied, the alpha-amylase inhibitors from wheat have received more attention than any other.

This chapter will concentrate mainly on questions related to the molecular properties, inhibition specificity and mechanism of action of the most thoroughly investigated alpha-amylase inhibitors, as well as on the significance of these inhibitors in diets and on their potential in diagnostics and in the treatment of some disorders.

MOLECULAR PROPERTIES

1. Alpha-amylase inhibitors from wheat

From the appearance of the first report on the detection of alpha-amylase inhibitors of albumin nature in wheat kernel (Kneen and Sandstedt, 1943), several authors have worked on the purification and characterization of such proteins from the same source. The use of different purification approaches based, according to the choice, on either size or charge of the protein molecules led to the isolation of a large number of inhibitors with more or less similar characteristics. In the early 1970's, the situation regarding the number and the relationships among so many functionally-related proteins isolated from wheat was rather confused, but following the rational purification approach described by Deponte et al. (1976) it was possible to put some order in the matter.

It is now generally recognized that wheat kernel contains several related forms of inhibitors active towards mammalian, insect and avian alpha-amylases but inactive on plant, bacterial and fungal enzymes. These inhibitor molecules may be grouped into three albumin fractions (families) with apparent molecular weight (M_r) 50000, 26000 and 13000. Each albumin family is highly heterogeneous consisting of several components (isoinhibitors) which may be separated by electrophoretic or ion-exchange chromatographic techniques. The isoinhibitors from the same family are strictly related: they have similar molecular properties (size, subunit composition, association-dissociation ability) and inhibition specificities and close homology among the primary structures. The minor differences in the sequence are reflected, of course, in differences in the amino acid composition,electric charge and related properties. Conversely, inhibitors from different families show distinct molecular properties and specificity of amylase inhibition, even though some homology is still apparent among their primary structures, thus suggesting a common genetic origin of their subunits.

In the 1970's, other types of alpha-amylase protein inhibitors, differing from the well-known inhibitors of exogenous alpha-amylases for both molecular and inhibition properties, have been detected in wheat kernel. Warchalewski (1976) provided first evidence that wheat and malted wheat contain multiple forms of protein inhibitors highly active on bacterial, fungal and endogenous alpha-amylases. Other authors (Peruanskii et al., 1980; Konarev, 1982) described similar inhibitors consisting of one polypeptide chain of M_r close to 20000. More recently, bifunctional alpha-amylase inhibitors have been detected in wheat kernel: Mundy at al. (1984) reported the presence of a double-headed inhibitor very active on both wheat alpha-amylase and bacterial subtilisin. This protein consists of a single chain of M_r close to 20000, thus resembling the inhibitors isolated by Warchalewski, but inactive on bacterial and fungal alpha-amylase. The relationships existing among all the wheat inhibitors of the endogenous alpha-amylase need to be clarified.

1.1. Inhibitors of exogenous alpha-amylases

The fist reports on purification and partial characterization of some components of the isoinhibitor family with M_r close to 13000 were by Shainkin and Birk (1970) and Cantagalli et al. (1971). As observed by polyacrylamide-gel electrophoresis at alkaline pH, this family contains five closely-related anodic components (designated 0.28, 0.32, 0.35, 0.39 and 0.48 according to their gel-electrophoretic mobilities relative to bromophenol blue) which have been individually isolated and characterized (Silano et al., 1973; Deponte et al., 1976; Petrucci et al.,1978). Features common to the five isoinhibitors are, in addition to the M_r, amino acid composition, which is characterized by the lack of phenylalanine and histidine residues (Table I), the presence of a single subunit, CD spectra in the far and near ultraviolet. The isoinhibitors are highly active against alpha-amylases

Table I

Amino acid compositions (mol %) of alpha-amylase inhibitors from plants

Source	Amino Acid																
	Lys	His	Arg	Asp	Thr	Ser	Glu	Pro	Gly	Ala	Cys	Val	Met	Ile	Leu	Tyr	Phe
Wheat (0.28)	5.0	trace	6.4	7.5	2.7	7.3	11.4	7.6	7.8	7.8	8.2	10.8	2.6	1.8	8.2	3.3	trace
Wheat (0.30)	4.9	trace	5.7	7.4	2.5	8.2	10.8	8.2	8.2	8.2	8.2	9.8	2.5	1.6	7.4	3.3	trace
Wheat (0.19)	2.6	1.6	5.6	5.9	2.9	6.3	12.8	7.3	8.4	13.4	8.1	7.4	2.0	2.4	8.3	3.3	1.8
Wheat (0.53)	3.1	0.9	5.3	6.8	2.5	6.5	10.3	7.7	10.0	12.9	8.1	8.4	2.0	2.1	8.5	4.0	1.7
Wheat (tetrameric)	2.3	1.6	5.7	6.9	5.1	6.0	13.6	9.6	7.4	6.4	7.4	6.9	2.1	3.2	8.5	3.7	1.8
Wheat (bifunctional)	3.3	4.1	7.4	10.2	3.7	5.1	8.2	8.5	11.0	9.1	2.2	7.8	8.5	7.8	3.8	3.2	3.2
Barley (bifunctional)	3.6	5.0	6.3	9.2	4.6	5.3	8.4	8.1	11.0	8.9	2.2	7.0	1.2	3.7	6.5	3.5	3.8
Ragi (bifunctional)	1.6	2.5	8.2	4.1	5.7	4.9	10.6	8.2	8.2	8.2	7.4	7.4	2.5	4.1	10.0	1.6	1.6
Ragi	2.1	trace	9.4	7.4	2.1	16.8	2.1	4.2	9.4	22.0	7.4	5.3	trace	3.3	5.5	2.1	trace
Rye	3.5	1.0	6.2	6.6	4.2	5.9	12.3	7.7	8.6	9.6	7.7	8.7	2.4	2.4	7.7	4.3	1.8
Kidney bean	4.0	1.2	2.6	15.3	8.7	17.6	8.0	trace	4.4	4.9	0.8	8.7	0.6	4.2	4.9	5.5	5.2

from Tenebrio molitor (yellow mealworm) and other insects acting as pests of stored grains and from a number of marine species (octopus, squid, cuttlefish, etc.); they are weaker inhibitors of human saliva and some avian enzymes and are inactive against bacterial (Bacillus subtilis) and fungal (Aspergillus oryzae) alpha-amylases (Silano et al., 1975; Buonocore et al., 1985). The isoinhibitors are not deamidated forms of the component with the highest pI value. Redman (1975,1976) cleaved three components of the M_r 13000 family with cyanogen bromide at the three methionine residues and isolated the resulting four peptides. Evidence from amino acid analyses, N-terminal amino acids and electrophoretic mobilities of the peptides and from sequencing studies of the N-terminal peptides confirmed the high degree of homology among the three isoinhibitors.

Investigations on specific properties have been carried out on some isolated components of this family (Shainkin and Birk, 1970; Granum and Whitaker, 1977; Petrucci et al., 1978; Silano and Zahnley, 1978; Buonocore et al., 1980). It has been shown that these isoinhibitors have five disulfide bridges and one residue of reducing sugar covalently bound per molecule. They have a good stability to heat, pH and dissociating agent treatments, but are irreversibly inactivated by the presence of disulfide reducing agents as dithiothreitol or 2-mercaptoethanol. When investigated by differential scanning calorimetry, the inhibitors show a high thermal stability (denaturation temperature close to 93°C) which agrees well with the thermostability of the inhibitory activity and may reflect the high number of disulfide crosslinking and the large fraction of ordered structure (about 50%).

The complete amino acid sequence of the major component of this family indicates that the protein consists of 123 residues with only one position (number 65) exhibiting significant polymorphism with equal amount of serine and alanine being found in this position; the primary structure studies confirmed the absence of phenylalanine and histidine residues in the molecule of the M_r 13000 isoinhibitors (Kashlan and Richardson, 1981).

Extensive investigations have also been carried out on the isoinhibitor family with M_r 26000. When purified by molecular sieving and analysed by polyacrylamide-gel electrophoresis at alkaline pH (Deponte et al., 1976), this family appears highly heterogeneous: at least ten anodic components are evident, all of which active towards T. molitor and human salivary alpha-amylases. The inhibitors are not simply deamidated forms of the component with the highest pI value. In the presence of guanidine hydrochloride or sodium dodecylsulphate, the isoinhibitors dissociate reversibly into two subunits with M_r close to 13000. Several components of this family have been purified with different approaches and characterized (Inhibitor AmI_2, Shainkin and Birk, 1970; Inhibitors I and II, Saunders and Lang, 1973; Inhibitor 0.55, Granum and Whitaker, 1977; In2 and In3, O'Connor and McGeeney, 1981a; Inhibitor 0.53, Maeda et al., 1982,1983a,b; Takagi et al., 1982).All these inhibitors show very close similarity of their chemicophysical properties, amino acid (Table 1) and subunit composition and alpha-amylase inhibition specificities.

The main component of this family, the inhibitor 0.19, has been purified independently by several authors and completely characterized(Sodini , 1970; Silano et al., 1973; Petrucci et al., 1976,1978; O'Donnell and McGeeney, 1976; Granum and Whitaker, 1977; Silano and Zahnley, 1978; Buonocore et al., 1980; O'Connor and McGeeney, 1981a;Maeda et al., 1985). The protein molecule has M_r 26000 and consists of two identical polypeptide chains; as the inhibitor 0.28,it has five disulfide bridges and one molecule of reducing sugar bound per protomer. Its stability with respect to extreme pH values, high temperature, dissociating and reducing agents and proteases resembles that of inhibitor 0.28, with the exception that the dimeric inhibitor appears more resistant to tryptic and chymotryptic digestion as compared to the monomeric inhibitor. Inhibitor 0.19 resembles inhibitor 0.28 also in its amylase inhibition specificity but is more active than 0.28 against avian and mammalian (salivary and pancreatic) alpha-amylases. Denaturation enthalpies and thermograms and CD spectra of inhibitor 0.19 are also strictly related to those of inhibitor 0.28. The complete sequence of inhibitor 0.19 has been recently reported (Maeda et al., 1985) and compared to that of another component (inhibitor 0.53) of the same family

(Fig. 1); this study confirmed the identity of the two 0.19 subunits, at least for the peptide moiety, and the high homology (about 94%) between the primary structures of components of the same family. The amino acid substitutions observed between inhibitor 0.19 and 0.53 justify the differences in the electrophoretic mobility and are likely responsible of the slightly different inhibiting specificity of the two isoinhibitors. The location of disulfide bridges in inhibitors 0.19 and 0.53 has been partially elucidated (Maeda et al., 1983b, 1985).

Among the inhibitor families active against exogenous amylases, that with M_r about 50000 received little attention until recently. Its occurrence in wheat was recognized by Petrucci et al. (1974); later, Deponte et al. (1976) showed that most of this fraction dissociated reversibly into subunits of M_r 13000. More recently, Buonocore et al. (1985) have shown that this family consists at least of four components with very close pI values (from 4.55 to 4.90); the isoinhibitors have M_r close to 50000 and reversibly dissociate, in the presence of guanidine hydrochloride, into subunits of M_r 13000. Amino acid composition (Table I) is similar to that of monomeric and dimeric inhibitors; no free thiol groups are present and the half-cystine content is consistent with an average of four cystine residues per monomer. CD spectra resemble those of inhibitor 0.19 but with a much smaller intensity. The tetrameric inhibitors are highly active against insect and mammalian alpha-amylases but inactive towards microbial, fungal and plant enzymes. O'Connor and McGeeney (1981a) isolated a protein inhibitor which shares several properties with those of the M_r 50000 family.

The data reported so far for a number of wheat inhibitors of exogenous alpha-amylases suggest the existence of several structural and functional relationships among the components of the three isoinhibitor families (Fig. 2). A strong support to this suggestion comes from the sequence comparison of inhibitor 0.30 (M_r 13000 family, Kashlan and Richardson, 1981) from one hand and inhibitors 0.19 and 0.53 (M_r 26000 family, Maeda et al., 1983a,b, 1985) from the other (Fig. 1). A significant homology (about 60%) between the inhibitor sequences is evident; furthermore, most of the substituted amino acids are coded by codons differing by only one nucleotide. All the available data are consistent with the hypothesis that the albumin alpha-amylase inhibitors are coded by a small number of structurally related genes, possibly derived from a single ancestor. Replication of this ancestral gene and divergence by various mechanisms have likely given rise to the three subfamilies of genes coding for the many subunits of the different inhibitors. Whereas gene sequences coding for inhibitor domains responsible of the functional properties of the proteins were highly conservative, major events in the gene evolution were the mutations yielding polypeptide chains able to associate, forming dimeric and tetrameric structures. These events can be regarded as an important advantage acquired by wheat through its evolution as the spectrum of inhibited alpha-amylases was considerably broadened: monomeric inhibitors are mainly active against insect alpha-amylases, whereas dimeric and tetrameric inhibitors are also very active on avian and mammalian enzymes (Table II).

It has been recently observed that limited sequence homology (below 50% in some cases) is not only occurring among functionally-related proteins from the same source, but also among proteins with unknown functional relationships and present in quite different species as monocotyledonous and dicotyledonous plants (Shewry et al., 1984). It is the case of a "superfamily" of homologous proteins which, at the moment, includes wheat alpha-amylase inhibitors 0.28, 0.19 and 0.53, barley trypsin inhibitor, a group of low-M_r salt-soluble proteins of unknown functions from wheat and barley, the bifunctional inhibitor from ragi, three albumin proteins from diploid wheats (Triticum monococcum and Aegilops squarrosa) and two storage proteins of castor bean (Ricinus communis) and oilseed rape (Brassica napus). Although the structural genes of all the above mentioned proteins have diverged to give rise to molecules with different sequences and functions, they share the property of being expressed only in storage tissues of seeds (Shewry et al., 1984).

1.2 Inhibitors of endogenous amylases

Warchalewski (1977a,b) partially purified from wheat and malted wheat a number of

Source	Inhibitor	Sequence
Wheat	(0.19)	S G P W M - C Y P G Q A F Q V P A L P A C R P L L R L Q - C N - G S Q V P E A V L - R D C
Wheat	(0.53)	S G P W M - C Y P G Q A F Q V P A L P G C R P L L K L Q - C N - G S Q V P E A V L - R D C
Wheat	(0.30)	S G P W S W C N P A T G Y K V S A L T G C R A M V K L Q - C V - G S Q V P E A V L - R D C
Barley	(tryp. inhib.)	F G - - D S C A P G D A L P H N P L R A C R T Y V V S Q I C H Q G P R L L T S D M K R R C
Ragi	(bifunct.)	S V G - - T S C I P G M A I P H N P L D S C R W Y V S/A T/R R T/A C G V G P R L A T Q E M K A R C
Wheat	(bifunct.)	D P P P V H D T D G N E L R A D A N Y Y V L P A N R A H G G G L T M A P G H G R R ? P L F
Barley	(bifunct.)	A D P P P V H D T D G H E L R A D A N Y Y V L S D N R A H G G G L T M A P G H G R H ? P L F
Soya	(tryp. inhib.)	D F V L D N E G N P L E N G G T Y Y I L S D I T A F G G - I R A A P T G D E R C P L T
Ragi	(am. inhib.)	A I S C G Q V S S A I G P C L A Y A R G A G A A P S A S C Q S G V R S L N A A A R T T A D R
Wheat	(0.19)	C Q Q L A H I S - E W C R C G A L Y S M L D S M Y K E H G A Q E G Q A G T G A F P R C R R E
Wheat	(0.53)	C Q Q L A D I S - E W P R C G A L Y S M L D S M Y K E H G V S E G Q A G T G A F P S C R R E
Wheat	(0.30)	C Q Q L A D I N N E W C R C G D L S S M L R A V Y Q E L G V R E G - - - K E V L P G C R K E
Barley	(tryp. inhib.)	C D E L S A I P A - Y C R C E A L R I I M Q G V V T W Q G A F E G A Y F K D S - P N C P R E
Ragi	(bifunct.)	C R Q L E A I P A - Y C R C E A V R I L M D G V V T P/S S G Q H E G R L L Q D - L P G C P K Q
Ragi	(am. inhib.)	R A A C N C S L K S A A S R V S G L N A G K A S S I P G R C G V R L P Y A I S A S I D C S R
Wheat	(0.19)	V V K L T A A S - I T A V - C R L P I V V D A S G D G A Y V C K - D V A A Y P D A
Wheat	(0.53)	V V K L T A A S - I T A V - C R L P I V V D A S G D G A Y V C K - D V A A Y P D A
Wheat	(0.30)	V M K L T A A S - V P E V - C K V P I P - N P S G D R A G V C Y G D W A A Y P D V
Barley	(tryp. inhib.)	R Q T S Y A A N L V T P Q E C N L G T I - - - - - H G S A Y C P E L Q P G Y G
Ragi	(bifunct.)	V Q R A F A P K L V T E V E C N L A T I - - - - - H G G P F C L S L L - G A G E
Ragi	(am. inhib.)	V N N

Fig. 1 Comparison of amino acid sequences of alpha-amylase and/or protease inhibitors from plants. Sequences have been taken from: Maeda et al., 1985 (inhibitor 0.19) and 1983a(inhibitor 0.53); Kashlan and Richardson, 1981 (inhibitor 0.30), Odani et al., 1982, 1983 (barley trypsin inhibitor); Campos and Richardson 1983 (ragi bifunctional inhibitor) and 1984 (ragi amylase inhibitor); Mundy et al., 1984 (wheat bifunctional inhibitor); Hejgaard et al., 1983 and Weselake et al., 1983b (barley bifunctional inhibitor); Koide and Ikenaka, 1973 (soya trypsin inhibitor).
Areas of high homology in sequences are boxed. Single - letter codes for amino acids are A = Ala, C = Cys, D = Asp, E = Glu, F = Phe, G = Gly, H = His, I = Ile, K = Lys, L = Leu, M = Met, N = Asn, P = Pro, Q = Glu, R = Arg, S = Ser, T = Thr, V = Val, W = Trp, Y = Tyr.

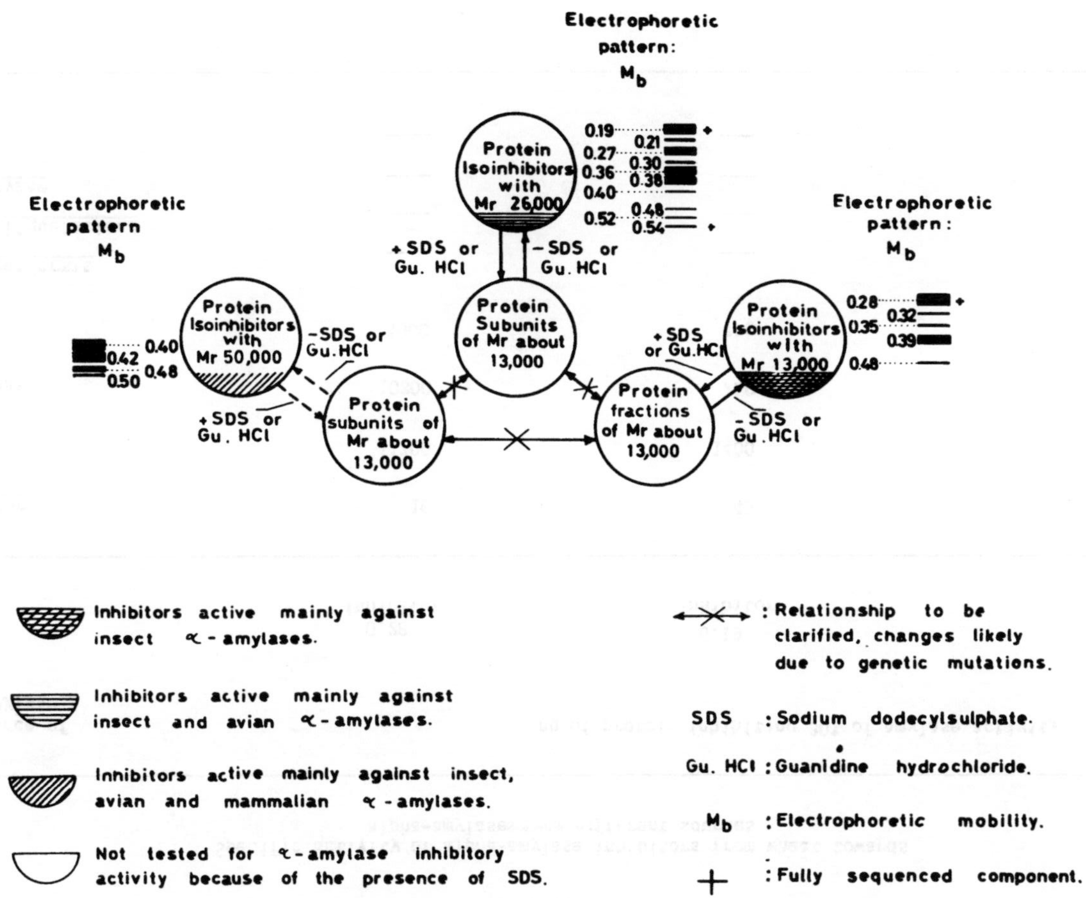

Fig. 2. Relationships among α-amylase isoinhibitor families from wheat kernel.

Table II

Specific activity of alpha-amylase inhibitors from wheat towards alpha-amylases from different sources

Source of amylase	ng of protein inhibiting 30% of amylase activity		
	0.28 inhibitor	0.19 inhibitor	tetrameric inhibitor
Tenebrio molitor	18	49	130
Pig pancreas	11700	1400	65
Chicken pancreas	10300	750	48
Human saliva	4000	770	46
Bacillus licheniformis	—	—	—
Bacillus amyloliquefaciens	—	—	—
Aspergillus oryzae	—	—	—
Barley malt	—	—	—

inhibitor forms active against fungal, bacterial and wheat alpha-amylases. The fractionation procedure includes steps as heat treatment and acetone fractionation at acidic pH which are probably requested to dissociate endogenous amylase-inhibitor complexes occurring in the wheat flour extracts. According to Peruanskii et al. (1980), the family of endogenous alpha-amylase inhibitors consists of five components with similar M_r, amino acid composition and inhibition specificity; the relative amounts of the five isoinhibitors, but not their properties, are affected by growth conditions of wheat. The M_r of the inhibitors is close to 20000 and is not affected by dissociating agents (Konarev, 1982).

1.3 Double-headed inhibitors

A bifunctional alpha-amylase/subtilisin inhibitor has been recently purified from wheat kernel by Mundy et al. (1984); it is active specifically against alpha-amylases from wheat and closely related cereals (barley, rye, oat) and against bacterial subtilisin. It is inactive towards sorghum, rice, mammalian, insect, fungal and bacterial alpha-amylases. As no competition was observed between amylase and subtilisin for the inhibitor binding sites, it has been suggested that the inhibitor is "double-headed", containing separate sites for the amylase and the protease. The bifunctional inhibitor has a single polypeptide chain of M_r close to 20000; its amino acid composition (Table I) and N-terminal sequence (45 residues, Fig. 1) are strictly related to those of the bifunctional inhibitor from barley and to the soybean trypsin inhibitor (Kunitz) family, and are distinctly different from those of inhibitors 0.28 and 0.19. It is not clear whether this inhibitor is relat d to any other of those discussed in the previous paragraph.

2. Inhibitors from cereals other than wheat

An inhibitor of plant alpha-amylases has been isolated from barley by two independent research groups (Weselake et al., 1983a,b; Mundy et al., 1983,1984; Hejgaard et al., 1983; Mundy, 1984). The inhibitor consists of a single polypeptide chain of M_r close to 20000 and has an isoelectric point of 7.3. It inhibits specifically only one of the two endogenous alpha-amylases from both germinated barley and wheat, but has no effect on the other isozyme from the same sources, on sorghum malt, hog pancreas, *Aspergillus oryzae* and *Bacillus subtilis* alpha-amylases and on barley beta-amylase. Mundy et al. (1983) have shown that the same inhibitor protein is active towards bacterial subtilisin; the inhibitor is "double-headed" in the sense that different sites interact with the susceptible amylase and protease molecules. The barley inhibitor is strictly related to the bifunctional inhibitor isolated from wheat (Table I) (Mundy et al., 1984). The homology of the N-terminal sequence (40 residues, Fig. 1) of the barley inhibitor with the soybean Kunitz-type protease inhibitors suggests its belonging to this family of enzyme inhibitors (Hejgaard et al., 1983).

A "double-headed" inhibitor has also been detected in ragi (*Eleusine coracana*) kernel (Shivaraj and Pattabiraman, 1980; Shivaraj et al., 1982). It is active towards trypsin and alpha-amylases of animal origin and does not inhibit chymotrypsin; it is a glycoprotein with 2% carbohydrates and consists of a single polypeptide chain of M_r 14000 (Table I). The molecule is remarkably resistant to heat treatments and exposure to pH values ranging from 1 to 10; it is slowly degraded by pepsin and pronase, but is resistant to tryptic attack. Modification of specific amino acid residues affects differently the inhibiting capacity of the inhibitor and suggests that arginine residues are essential for the trypsin-inhibiting activity. The existence of two distinct binding sites on the inhibitor molecule was confirmed by isolation by means of gel-permeation chromatography of a trimer complex alpha-amylase-inhibitor-trypsin. The ragi bifunctional inhibitor has been completely sequenced by Campos and Richardson (1983): the protein molecule consists of 122 amino acid residues and exhibits high homology with a trypsin inhibitor from barley (60%) and, to a lesser extent (30%), with a wheat alpha-amylase inhibitor of the M_r 13000 family (Fig. 1). Among the 26 amino acid residues which can be regarded as highly conserved (invariant) are the ten cysteine residues which are involved in the formation of intrachain disulfide bridges. It is interesting to note that the wheat alpha-amylase inhibitor, which has no activity against trypsin, lacks a specific sequence in the region that has been tentatively identified in the ragi bifunctional inhibitor as the trypsin binding site.

A second alpha-amylase inhibitor has been purified from ragi by Shivaraj and Pattabiraman (1980); even though it has an alpha-amylase inhibiting specificity similar to that of the bifunctional inhibitor, it does not inhibit trypsin and is susceptible to tryptic degradation. Structural properties (Table I) of this alpha-amylase inhibitor are also distinct from those of the double-headed inhibitor: the M_r is 10000 and the molecule, with 93 amino acid residues (Fig. 1), shows no sequence homology with any of the alpha-amylase inhibitors from other plants (Campos and Richardson, 1984).

Although early reports (Kneen and Sandstedt, 1946) indicated that maize kernel was lacking alpha-amylase inhibiting activity, more recently it has been established the presence of amylase inhibitors in this cereal (Blanco-Labra and Iturbe-Chinas, 1981). Actually, a family of at least three isoinhibitors, distinguishable easily by their electrophoretic mobilities but with similar structural properties, was extracted and purified from maize flour. The inhibitors have M_r close to 30000 and show an inhibition specificity somewhat unusual, being active towards alpha-amylases from maize itself, B.subtilis and a number of insects which act as pests of stored grains, but inactive on alpha-amylases from other cereals, mammalian and A. oryzae.

Granum (1978) purified to homogeneity from rye an alpha-amylase inhibitor with chemico-physical properties (M_r, amino acid and subunit composition, electrophoretic mobility, stability to heat and pH) and inhibition specificity strictly resembling those of inhibitor 0.19 from wheat.

3. Inhibitors from kidney bean

A protein inhibitor of exogenous alpha-amylases has been purified from both white and red kidney beans by several authors (Marshall and Lauda, 1975; Powers and Whitaker, 1977a,b; Pick and Wöber, 1978,1979). It is a glycoprotein with a high content (about 10%) of carbohydrates, probably linked to an asparagine residue via an amide link; it has M_r close to 49000 and consists of four subunits. The kidney bean inhibitor shows activity against mammalian, insect and Helix pomatia alpha-amylases, but is inactive towards fungal (Aspergillus), bacterial (Bacillus) and plant alpha-amylases, including that from kidney bean. It has also no activity towards beta-amylases and other hydrolytic enzymes as acid and neutral alpha-glucosidases and trypsin. The inhibitor is resistant to pepsin and trypsin attack, but is rapidly inactivated by chymotrypsin or by acidic pH (below 3) (Andriolo et al., 1984). The amino acid composition is shown in Table 1.

4. Inhibitor from peanut

A thermostable protein inhibitor of exogenous alpha-amylases has been purified from an aqueous extract of peanut (Arachis hypogaea) cotyledons by Irshad and Sharma (1981). The peanut inhibitor has M_r 25000 and shows an inhibition specificity similar to that of the kidney bean inhibitor.

5. Inhibitor from Colocasia

Two basic proteins, inhibiting mammalian alpha-amylases, have been purified by Krishna Sharma and Pattabiraman (1980) from Colocasia antiquorum tubers. In the presence of dissociating agents they show M_r of 12500 and 14300, respectively, and the latter is a glycoprotein. The two inhibitors are relatively stable to heat and to pH variations, and highly resistant to proteolytic attack by excess pepsin, trypsin, chymotrypsin and pronase. An acidic inhibitor active against human salivary alpha-amylase has been purified from Colocasia esculenta by Narayana Rao et al. (1967,1970); this protein appears rich in aspartic and glutamic acids and particularly resistant to heat treatments.

6. Inhibitors from yam

From mature tubers of yam (Dioscorea alata) an exogenous alpha-amylase inhibitor has been partially purified which appears to have a high content (about 60%) of carbohydrates (Krishna Sharma and Pattabiraman, 1982b). The inhibitor has an apparent M_r close to 70000 and an inhibition specificity resembling that of the kidney bean inhibitor.

INTERACTION WITH AMYLASES

As mentioned previously, most amylase inhibitors detected in plants show activity against animal amylases and are inactive towards bacterial, fungal and plant enzymes. Few exceptions are the wheat inhibitors of endogenous alpha-amylase first isolated by Warchalewski (1977a,b),the double-headed inhibitors from wheat, barley and ragi and an inhibitor from maize with an unusual spectrum of activity, being able to inhibit insect, bacterial and endogenous alpha-amylases, but inactive towards mammalian and fungal enzymes. It may be worth to recall that one has to be cautious in assessing the inhibition specificity of a protein inhibitor. Since ionic bonds are essential in protein-protein interactions, assays of amylase inhibition by protein inhibitors must be performed at different pH values. For instance, the wheat inhibitor 0.28 was reported to be inactive at pH 7.0 on chicken pancreas alpha-amylase, but it was found active against this enzyme when tested at pH 5.8 (Silano et al., 1973; Buonocore et al., 1984). Moreover, since most amylase inhibitors display their action at concentrations as low as 0.01 mg/l, the possible presence of active impurities, hardly detectable with the usual homogeneity tests, should be considered when assessing the inhibition specificity of an inhibitor.

1. Formation of amylase-inhibitor complexes

The extent of amylase inhibition by protein inhibitors depends on a number of external variables, as pH, temperature, ionic strenth, time of enzyme-inhibitor preincubation, order of addition of reactants and, of course, inhibitor concentration. With the exception of inhibition of human pancreatic amylase by two wheat inhibitors belonging to the M_r 26000 family (O'Connor and McGeeney, 1981b), in all other cases tested preincubation of inhibitor and enzyme was required to achieve maximum inhibition. These findings indicate that the rate of complex formation between the protein inhibitors and the susceptible alpha-amylases is somewhat slow. Inhibitors from wheat (Buonocore et al., 1980,1984; O'Connor and McGeeney, 1981b; Belisario et al., 1981), rye (Granum, 1978), maize (Blanco-Labra and Iturbe-Chinas, 1981), yam (Krishna Sharma and Pattabiraman, 1982b) and peanut (Irshad and Sharma, 1981) active against exogenous alpha-amylases as well as the endogenous alpha-amylase inhibitor from barley (Weselake et al., 1983b) require preincubation times ranging from 5 to 30 minutes to achieve maximum inhibition (Table III). A larger variability (10-120 min) has been reported by Warchalewski (1976,1978) for wheat inhibitors of endogenous alpha-amylase.

Several reports have shown that the kidney bean inhibitor can take hours to inactivate susceptible alpha-amylases (Marshall and Lauda, 1975; Powers and Whitaker, 1977b; Powers and Culbertson, 1982; Wilcox and Whitaker, 1984). The reaction of complex formation between red kidney bean inhibitor and porcine pancreatic alpha-amylase follows second order rate kinetics with a constant of $1.1\ 10^4\ M^{-1}\ sec^{-1}$ at pH 6.9 and 37°C. (Wilcox and Whitaker, 1984). The interaction proceeds through two steps: the first step (second order) has a calculated Keq of $3.09\ 10^{-5}$ M, the second (first order, unimolecular) has a rate constant of 3.05 min^{-1}. These rate constants are too small for a diffusion-controlled reaction involving two protein molecules and suggest that a conformational change may be the rate-controlling step in the complexation.

The rate and extent of interaction between protein inhibitors and amylases are largely dependent on pH. Optimal pH values for inhibition generally range from 5.5 to 7.5 and,in many cases, correspond to intermediate values between inhibitor and enzyme isoelectric points, where the two molecules present an opposite net electric charge (Table III).

The effect of temperature and ionic strenght on amylase inhibition by protein inhibitors has been studied in a few cases. In general, amylase inhibition increases slowly with temperature, with a maximum in the range 30-40°C (Table III); in the case of the kidney

Table III

Some properties of inhibitor : alpha-amylase complexes

source of inhibitor : alpha-amylase	inhibitor pI	amylase pI	pH of maximal inhibition	temperature of maximal inhibition(°C)	preinc:time for maximal inhibition (min.)	molar ratio in the I:E complex	dissociation constant (Ki) (nM)
Wheat (0.28): Tenebrio molitor	6.2	4.0	6.2	37	5	2 : 1	0.13
Wheat (0.19): Tenebrio molitor	7.3	4.0	5.0	37	10	1 : 1	0.85
Wheat (0.28): Chicken pancreas	6.2	5.0/5.3[a]	5.8	30	40	2 : 1	570.
Wheat (0.19): Chicken pancreas	7.3	5.0/5.3[a]	5.8	30	30	1 : 1	3.70
Wheat (0.28): Octopus vulgaris	6.2	4.8	6.0	30	30	—	—
Wheat (0.19): Octopus vulgaris	7.3	4.8	6.5	30	30	—	—
Kidney bean: Tenebrio molitor	4.7	4.0	4.4	—	60	1 : 1	0.03
Barley (bifunct): barley	7.2	6.2	—	—	5	—	40

[a]two isozymes present

bean inhibitor, a temperature increase of 11°C, from 26 to 37°C, caused an about 11-fold increase in the rate of complex formation with porcine pancreatic alpha-amylase (Marshall and Lauda, 1975; Powers and Whitaker, 1977b). Different results have been reported about effects of ionic strenght on activity of amylase inhibitors. Increasing phosphate buffer concentration from 0.1 to 50 mM in the absence of chloride ions did not affect the extent of salivary amylase inhibition by wheat inhibitors of the M_r 26000 family (O'Connor and McGeeney, 1976). Chloride ions increased inhibition activity of the kidney bean inhibitor (Powers and Whitaker, 1977b), whereas these ions did not exert any effect on salivary amylase inhibition by the rye inhibitor (Granum, 1978).

Several authors have stressed the importance of the addition order of reactants (inhibitor, enzyme and starch) on the extent of inhibition (Shainkin and Birk, 1970; O'Donnell and McGeeney, 1976; O'Connor and McGeeney, 1981b). The decrease of inhibition observed by preincubating inhibitor and starch before the enzyme addition has been attributed to the tendency of protein inhibitors to bind polysaccharides. However, it is more likely that the observed effects are a consequence of the slower rate of inhibitor-amylase complexation as compared to starch hydrolysis (Buonocore et al., 1977; Blanco-Labra and Iturbe-Chinas, 1981).

Lastly, the extent of amylase inhibition depends on inhibitor concentration. In general, the relationship is linear up to 60-80% of enzyme inhibition; in many cases, a complete inactivation of the amylase was not observed even at very high inhibitor levels. These findings may suggest that in the amylase-inhibitor complex the enzyme active site is still accessible to starch which, when present at high concentration, may induce a slow dissociation of the amylase from the complex.

2. Properties of amylase-inhibitor complexes

Amylases combine with protein inhibitors in stoichiometric ratios forming either binary or ternary complexes characterized by dissociation constants ranging from 10^{-8} to 10^{-10} M (Table III). Complex formation may be obtained by incubating for a few minutes, at the optimal conditions for inhibition, mixtures of the amylase and the inhibitor in the appropriate molar ratio; the complex formed can be freed from the component in excess by gel filtration or equivalent techniques.

Complex formation between wheat inhibitors and a number of exogenous alpha-amylases has been studied (Table III). Wheat inhibitor 0.28 forms stable complexes with either *T. molitor* or chicken pancreas alpha-amylases (Buonocore et al., 1980,1984), each complex showing a molar stoichiometric ratio of 2:1 (inhibitor:amylase). This combination ratio has been determined by means of several techniques as gel filtration, difference spectroscopy, kinetic studies (Buonocore et al., 1980,1984) and differential scanning calorimetry (Silano and Zahnley, 1978). With the same techniques it has been observed that inhibitor 0.19 forms with either amylase two stable complexes, each one containing one mole of inhibitor per mole of enzyme, which are likely conformational isomers. The formation of a ternary complex among *T. molitor* amylase and inhibitors 0.28 and 0.19 has been ruled out.

O'Connor and McGeeney (1981b) demonstrated by gel-filtration studies that human salivary but not human pancreatic amylase forms stable complexes with each of three components of the M_r 26000 family from wheat. Lastly, both the human alpha-amylases were able to bind in 1:1 ratios the tetrameric wheat inhibitor.

Complex formation with a 1:1 molar ratio was observed between porcine pancreatic amylase and two inhibitors from *Colocasia* by means of gel-filtration experiments (Krishna Sharma and Pattabiraman, 1980). A 1:1 stoichiometry has also been suggested for porcine pancreatic amylase and kidney bean inhibitor (Table III) by measuring the extent of inhibition in correspondence of increasing amounts of inhibitor (Marshall and Lauda, 1975: Powers and Whitaker, 1977b). Other amylase-inhibitor complexes have been only visualized by means of electrophoretic or chromatographic procedures without determining combination ratios. This is the case of barley and wheat inhibitors with the corresponding endogenous alpha-amylases (Mundy et al., 1983,1984; Weselake et al., 1983b) and of ragi bifunctional inhibitor with either porcine pancreatic amylase or bovine trypsin and with both enzymes at the same time (Shivaraj and Pattabiraman, 1981).

The dissociation constants (K_i) of a number of amylase-inhibitor complexes have been calculated by means of graphical methods (Lineweaver-Burk or Dixon plots) based on Michaelis-Menten kinetics; typical uncompetitive or non-competitive inhibition patterns were observed with dissociation constants ranging from 10^{-7} to 10^{-9} M. However these data give little informations in terms of inhibition mechanism because the inhibitors are tightly-bound and their concentrations, in the experimental conditions used in assaying amylase inhibition, are not much higher than enzyme concentrations, as it is assumed in the classical steady-state kinetics. More meaningful data can be obtained by applying kinetic treatments for slow-acting, tight-binding inhibitors in reversible and mutual-depletion systems. The treatments have been applied to few inhibitor-amylase systems (Table III), such as those involving wheat inhibitors of exogenous alpha-amylases (Buonocore et al., 1980,1984), the inhibitor from barley (Mundy et al., 1983) and that from red kidney bean (Powers and Whitaker, 1977b; Wilcox and Whitaker, 1984).

3. Dissociation of amylase-inhibitor complexes

Even though the interaction of protein inhibitors with alpha-amylases is characterized by a tight binding, the inhibition reaction can be reversed by several factors such as pH, amylase substrates and reaction products of starch hydrolysis. Direct evidence of that was given from the affinity chromatography studies by Buonocore et al. (1975) who purified several susceptible alpha-amylases (from human saliva, chicken pancreas, octopus digestive gland and T. molitor) by absorption of crude enzymatic extracts on a Sepharose-coupled wheat inhibitor matrix. Desorption of the enzyme could be obtained, even though at a slow rate, by eluting with high concentrations of maltose or gelatinised starch. Since then, similar procedures have been used to purify either alpha-amylases or plant protein inhibitors (Pick and Wöber, 1979; Shivaraj and Pattabiraman, 1981; Krishna Sharma and Pattabiraman, 1982a).

Maltose, a competitive inhibitor of alpha-amylases, not only causes dissociation of preformed amylase-inhibitor complexes but, when added to enzyme solutions simultaneously to the protein inhibitor, prevents its inhibition. For instance, maltose partly prevents amylase inhibition by wheat inhibitors 0.28 and 0.19 at concentrations that only slightly inhibit the enzymes (Petrucci et al., 1976; Buonocore et al., 1976,1980; O'Connor and McGeeney, 1981b).

For several amylase-wheat inhibitor systems, time-dependence of starch hydrolysis upon addition of starch to the amylase-inhibitor complex was compared with that observed upon addition of starch to the amylase alone (Buonocore et al., 1980,1984). When the inhibitor is present, production of reducing sugars after starch addition became linear after a log phase whereas, in the absence of the inhibitor, production of reducing sugars was linear throughout the entire time-range tested. These findings are consistent with a partial dissociation of the amylase-inhibitor complex induced by starch addition. Binding of starch to the kidney bean inhibitor-porcine pancreatic amylase complex has been demonstrated by absorption of the complex to starch columns (Wilcox and Whitaker, 1984); the complex can be desorbed by maltose elution.

The ability of maltose and starch of inducing partial dissociation of inhibitor-amylase complexes may suggest that the protein inhibitors are not bound to the enzyme active site which is then still available for binding substrates or inhibitors. Steric effects due to the hindrance of the inhibitor molecule being close to the enzyme active site might explain a lower affinity of the complex for starch as compared to smaller molecules as maltose or synthetic substrates.

4. The model

The 1:1 stoichiometry between wheat isoinhibitors of the 0.19 family on one hand and insect, avian or mammalian alpha-amylases on the other (Buonocore et al., 1980,1984; O'Connor and McGeeney, 1981b), and the 2:1 stoichiometry between the wheat inhibitor 0.28 and insect or avian alpha-amylases (Buonocore et al., 1980,1984) indicate that each 0.19 subunit and 0.28 molecule have one binding site for the amylases and that each amylase has two binding sites for the inhibitors, one for each protomer of M_r 13000. The occurrence of two slightly different inhibitor 0.19-amylase complexes, each with a 1:1 stoichiometry, suggests that two non-equivalent 0.19-to-amylase binding positions are involved. This model might explain the higher inhibiting activity and heat stabilization effects of the T. molitor enzyme displayed by inhibitor 0.28 as compared to inhibitor 0.19.

Two molecules of inhibitor 0.28 should in fact have more freedom of adjusting themselves to complementary parts of the amylase molecule than two subunits linked in a 0.19 molecule, thus giving stronger and possibly faster interactions. The model does not imply significant structural differences between the binding sites of each 0.19 subunit and 0.28 molecule, being in line with the close structural similarity of these polypeptide chains.

The ability of maltose to reduce the affinity of amylases for the inhibitors might indicate that the reducing sugars covalently bound to the inhibitor polypeptide chains play a role in the inhibition mechanism, specifically recognizing on the amylase molecules a site where maltose also binds.

On the other hand, the possibility that maltose acts by stabilizing an amylase conformation less favourable to the binding of the protein inhibitors should not be overlooked. Some observations indicate that the complex between red kidney inhibitor and porcine pancreatic amylase is still able to bind both substrates (starch, p-nitrophenyl-alpha-D-maltoside) and competitive inhibitors (maltose), thus suggesting that the protein inhibitor is not bound to the active site of the enzyme (Powers and Whitaker, 1977b; Wilcox and Whitaker, 1984). These findings led the authors to suggest that the slow steps involved in getting the protein inhibitor on and off alpha-amylases are probably due to conformational changes in the enzyme molecule.

No similar studies have been performed on amylase inhibitors other than those from wheat and kidney bean; however, considering the many similarities existing among the plant inhibitors of exogenous alpha-amylases, it is likely that the models proposed are applicable to other components of this group of amylase inhibitors as well.

ROLE IN THE PLANT

The function of most alpha-amylase inhibitors in plants is not known yet. Since most natural inhibitors are only active against animal alpha-amylases, it appears that they do not act as antimicrobial or antifungal agents, nor do they serve to regulate the carbohydrate metabolism of the plant. Nevertheless, alternative roles have been postulated: these inhibitors might function as storage proteins and to protect the seeds against insects and animals. On the other hand, different functions may be attributed to plant inhibitors active against the endogenous alpha-amylases. For such inhibitors an involvment in the regulation of digestion and mobilization of the reserve stores of starch may be easily inferred.

1. Inhibitors of exogenous alpha-amylases

Amylase inhibitors from cereals and legumes have been detected in the kernel only. Albumin inhibitors from wheat kernel are located in the endosperm and appear to be closely associated to starch (Saunders, 1975; Pace et al., 1978). It is known that albumin molecules strongly adhere to starch granules and make a significant contribution to endosperm hardness playing the role of cementing agents between starch granules and storage proteins (Barlow et al., 1973; Simmonds et al., 1973; Minetti et al., 1973).

The production of wheat inhibitors starts about eight days after fertilization and rapidly increases with maturation up to a maximum at full maturity; inhibitor content rapidly decreases after germination. No detectable amount of inhibitors has been found in roots or coleoptiles from germinated seeds (Pace et al., 1978). Similar patterns have been observed for other amylase inhibitors. The ragi inhibitor decreases upon seed germination up to undetectable levels 48 hours after germination (Shivaraj and Pattabiraman 1980). Colocasia inhibitors, which represent a significant percent of the total extractable proteins of tubers, slowly disappear as the plant grows. The production pattern, the high level found in seeds and tubers and the high heterogeneity of amylase inhibitors are features typical of storage proteins whose physiological role is essentially that of being degraded and utilized as a source of amino acids during germination and the early stages of seedling growth. A tendency towards multiplicity is usually considered as the expression of a high tolerance for mutations deriving from a loose relationships between structure and function.

As mentioned previously, most plant inhibitors of exogenous alpha-amylases are strongly active in vitro against amylases from a number of insect species (Silano et al., 1975; Blanco-Labra and Iturbe-Chinas, 1981; Powers and Culbertson, 1982). It has been postulated by Applebaum and coworkers that the wheat inhibitor activity may be expressed also in vivo towards the intestinal alpha-amylases of some stored wheat and wheat product pests, including T. molitor, Tribolium castaneum and Prodernia litura larvae (Applebaum, 1964; Applebaum et al., 1964; Applebaum and Konijn, 1965,1967). In fact, the addition of wheat flour albumins to a synthetic diet adversely affected development and greatly increased mortality of pest larvae. Thus, the speculation can be made that amylase inhibitors exist in plant seeds as naturally-occurring insect resistance factors. As a support to this hypothesis, Silano et al. (1975) have reported that insect species normally attacking wheat grains and wheat products have higher amylase activities and are more susceptible to inhibition by wheat inhibitors than are the amylases of those insects that do not attack wheat. If amylase-inhibiting activity in wheat could be shown to be directly related to resistance to insect pests, it could serve as an important index to breeders for selecting improved wheat varieties and for a rapid prediction of variety resistance to insects. The development of more insect-resistant wheat varieties could have an important impact on the world food supply by reducing the volume of stored wheat and wheat products lost each year to insect infestation (Yetter et al., 1979). Preliminary data, obtained by correlating progeny development of two predatory insects, T. molitor and Sitophilus oryzae (rice weevil), with alpha-amylase-inhibiting activities of wheat varieties grown at ten different locations over two crop years, are encouraging in the direction that inhibitor levels may represent a valid measure of wheat resistance to postharvest insects (Yetter et al., 1979).

Even though more comprehensive studies are needed to clarify the role of amylase inhibitors in insect resistance, the data available at the moment suggest that plants have evolved these toxic factors as part of their overall defense strategy.

2. Inhibitors of endogenous alpha-amylases

It has been suggested that the physiological role of protein inhibitors of endogenous alpha-amylases consists in regulating starch metabolism in the kernel by controlling the endogenous alpha-amylase activity. It can be recalled that, to date, endogenous alpha-amylase inhibitors have been detected only in quiescent and germinated wheat, barley and maize kernels.

The amylase/subtilisin inhibitor is an endosperm protein present in relatively large amounts in barley grain; the inhibitor is actively synthesized in the developing endosperm, but not during germination, at which time it apparently has only a passive function; its synthesis can be induced in germinated seeds by abscisic acid (Mundy et al., 1983; Mundy, 1984). It has been postulated that the inhibitor: i) may serve a regulatory role during seed development by inhibiting endogenous alpha-amylase during starch granule synthesis; ii) may delay starch degradation caused by premature sprouting; iii) may have a protective function by inhibiting microbial proteases.

NUTRITIONAL AND TOXICOLOGICAL SIGNIFICANCE

The possibly deleterious nutritional effects on human and animals of amylase inhibitors present in plant foods are difficult to assess. Many plant foodstuffs containing these inhibitors undergo heat treatments before they are eaten but, due to their thermostability, protein inhibitors might survive at least in part in an active form in the cooked food. This is particularly the case of wheat and rye-based products as bread, biscuits, cakes, breakfasts, instant baby foods, and animal feeds (Marshall, 1975). About 20% of the amylase-inhibiting activity present in the corresponding flour has been detected in white and rye breads, with the peripheral part of the bread showing lower activity than the central part due to a higher baking temperature (Granum, 1979). Conversely, spaghetti and kidney beans showed very low activity after boiling.

Furthermore, amylase inhibitors exhibit varying degrees of susceptibility to inactivation by animal proteases and acidic pH values. However, it has been shown that, when supplied in large quantities, wheat inhibitors can survive gastric digestion in man and other mammals (Puls and Keup, 1973; Lang et al., 1974; Saunders, 1975). These effects may be amplified when the amylase inhibitors are ingested as integral components of particulate foods or in the presence of other proteins which could be preferentially hydrolysed. Bean inhibitors also have been reported as susceptible both to acidic pH values (below 3) and to chymotryptic attack at pH 8 (Andriolo et al., 1984), but Singh et al. (1982) found significant negative correlations between amylase-inhibiting activity of bean-based feeds and in vitro meal digestibility.

In vivo experimentations with either raw materials or purified inhibitors have contributed to clarify the marginal nutritional significance of the alpha-amylase inhibitors. Red kidney bean inhibitor did not affect the rate of growth of weanling rats, thus suggesting that it is not toxic, nor did it alter the availability of energy from dietary starch (Savaiano et al., 1977). Addition of high levels of wheat alpha-amylase inhibitors to starch-enriched diets decreased starch availability to rats as compared to inhibitors-free control diets; these effects on rat growth were not observed when inhibitors had been inactivated by autoclaving or when glucose replaced starch in the diet (Lang et al., 1974). Continuous intake of gastro-resistant wheat inhibitors significantly depressed chicken growth rate, whereas native inhibitors did not. After four weeks of treatment, treated chicken showed a growth rate identical to that of control animals, indicating adaptation to inhibitor intake. Chickens treated with gastro-resistant inhibitors showed pancreas hypertrophy and histological changes of pancreatic tissue indicating degenerative processes in progress (Macrì et al., 1977). Dose-dependent pancreas hyperplasia, but not reduction of growth rate, were reported in rats and chickens fed for three weeks with moderate amounts of wheat inhibitors (Granum and Eskeland, 1981). Retardation of the digestion of a raw starch load by alpha-amylase inhibitors from wheat was observed in rats, dogs and healthy volunteers (Puls and Keup, 1973).

In conclusion, available data do not appear to indicate cereal and legume amylase inhibitors, eventually occurring in diets in an active form, as particularly deleterious for nutrition of healthy humans and animals. However, in vivo experiments have been carried out for short periods so that it is not known whether chronic ingestions of active alpha-amylase inhibitors may result in an excessive stimulation of the pancreatic function to an extent of being toxicologically significant. In this light, when the maximal nutritional value of a foodstuff is required, the presence of amylase inhibitors may be undesirable and special attention has to be devoted to ensure their inactivation during food processing. This might be of special significance in diets for patients with impaired pepsin or gastric acid secretion or for infants who have a much lower production of pancreatic alpha-amylase than adults.

ALPHA-AMYLASE INHIBITORS IN MEDICINE

1. Alpha-amylase inhibitors as therapeutic agents

The early results obtained in animal experimentation with plant inhibitors prompted many researchers to consider these naturally-occurring substances as possible agents in the treatment of obesity and diabetes.

The first report on effects of alpha-amylase inhibitor administration to humans was by Puls and Keup (1973). An alpha-amylase inhibitor isolated from wheat was used in experiments involving healthy volunteers, obese and diabetic patients, as well as rats and dogs. The hyperglycaemia and hyperinsulinaemia resulting from raw, but not cooked, starch loading could be reduced dose-dependently by adding the wheat inhibitors to the diet.

A large number of alpha-amylase inhibitor formulations (called "starch blockers") derived from kidney bean began to appear in 1981 on the market as dietary supplements. Advertisements for these products claimed that they, taken either slightly before or during a starch-containing meal, would decrease starch metabolism, thus resulting in weight loss. In the U.S.A., more than 200 different starch blocker formulations were on the market by mid 1982, being consumed at a rate of about one million tablets per day (Rosenberg, 1982); government officials estimated the annual sales to be approximately $ 350 million (Granum et al., 1983). However, a convincing scientific evidence that these formulations are active in vivo has not been produced yet. Actually, several recent reports did not confirm the efficacy of starch blockers in reducing the digestion of starch in man. Bo-Linn et al.(1982) demonstrated that fecal caloric content was not increased in normal subjects given starch-blocker tablets with a high starch meal. Carlson et al. (1983) tested some commercially available starch blockers in a randomized double-blind, cross-over clinical trial in six nonobese, healthy adult males and found that the amylase inhibitor had no effect on the response of blood glucose, insulin and breath hydrogen to a standardised starch meal. Similar results were obtained by Garrow et al. (1983a,b) on obese subjects.

The inefficacy in vivo of starch blockers may be attributed to several causes: i) secretion by pancreas of a large excess of alpha-amylase with regard to the amount needed for hydrolysis of ingested starch; ii) inadequacy of lumen conditions for the formation of inactive inhibitor-amylase complexes; iii) digestion of ingested starch by mucosal glycoamylases or intestinal microflora; iv) inactivation of amylase inhibitors by gastric and pancreatic proteases. As already mentioned, the last possibility has been tested in vitro by Andriolo et al. (1984) who found that the kidney bean inhibitor was readily inactivated at pH below 3.0 and by chymotrypsin at pH 8.

Another aspect that deserves attention is the fact that alpha-amylase inhibitors represent often only a minor constituent of many commercial starch blockers, which also contain significant levels of toxic substances commonly present in raw beans (Granum et al., 1983; Kilpatrick et al., 1983; Taylor and Barker, 1983; Liener et al., 1984).

2. Alpha-amylase inhibitors in diagnostics

The alpha-amylase activity of human serum is due to several isozymic forms which have been coded as S-type (salivary type) and P-type (pancreatic type) amylases. The percent of P-type amylase in healthy individuals varies in a wide interval, with reported mean values ranging from 41 to 55%, depending on the methodology used for the isozyme differentiation. Although several different tissues (e.g. salivary glands, lungs, leukocytes, thrombocytes, genital tissue) synthesize the S-type amylase, the P-type is essentially tissue-specific. For this reason, hyperamylasemia not only may occur in pancreatic disorders, but also in mumps, intestinal disorders, tumors of lungs and ovaries; it also has been observed post-operatively as a concomitant of urinary retention, in diabetic ketoacidosis and in macroamylasemia (Huang and Tietz, 1982). Therefore, determination of the total amylase activity in serum is of utility as a screening test but of limited usefulness in hyperamylasemia for clinic diagnosis; in these cases, separation of amylase isozymes is mandatory as aid for a differential diagnosis.

Amylase isozymes could be separated by ion-exchange chromatography, electrophoresis or electrofocusing, all these techniques being relatively time consuming and unsuited for emergency applications. O'Donnell et al. (1977) developed a method based on the use of a wheat inhibitor of the M_r 26000 family (coded inhibitor 0.20) which inhibits human salivary amylase much more effectively than pancreatic amylase. The method allows the rapid and easy differential determination of serum amylase isozymes, thereby fulfilling some important diagnostic needs. Originally, the wheat inhibitor was incorporated into an amylase-testing procedure in which insoluble dye-coupled starch was the substrate (O'Donnell et al., 1977). Later, the inhibitor method has been favourably adapted to other manual and automated assay procedures involving maltotetraose (Huang and Tietz, 1982), maltopentaose (Takagi et al., 1982) or 4-nitrophenylmaltoheptaoside (Okabe et al., 1984) as substrates. At the moment, not only a preparation of the amylase inhibitor from wheat germ, but also several isoamylase kits based on the use of such inhibitor for the differential diagnosis of acute pancreatitis are commercially available.

The wheat inhibitor method for distinguishing S-type from P-type amylase has been compared with chromatographic, electrophoretic and electrofocusing techniques. Good correlations have been reported with the three methods (Berk et al. 1981; Benoit et al., 1981; Ellis et al., 1982; Kameya et al., 1982); however, the inhibitor method appears somewhat less reliable for samples with very high or very low percentages of P-type amylase (Berk et al., 1981). Moreover, the animal alpha-amylases (e.g. bovine, porcine) usually incorporated in many commercial control sera show a relative insensitivity to the wheat inhibitor as compared to human alpha-amylases. Therefore, for a better accuracy of the wheat inhibitor-based amylase isozyme assays, serum and amylases of human origin have to be used for quality control and standardization (O'Donnell and McGeeney, 1983).

CONCLUSIVE REMARKS

After their discovery in the early 1940s and a long phase of low interest lasting up to the 1960s, alpha-amylase inhibitors naturally occurring in plants have stimulated an exponentially-growing interest in the last two decades. Although the first reports on alpha-amylase inhibitors were restricted to those from wheat active on exogenous alpha-amylases, more recent studies have shown that alpha-amylase inhibitors are relatively widespread in nature and that they act on endogenous alpha-amylases and that bifunctional inhibitors act on both amylases and proteases.It is easy to predict that the future interest of researchers will shift to these new-discovered classes of inhibitors and to the relationships existing among inhibitors of the different classes, expecially when they occur in the same plant tissue.

The nutritional and toxicological significance of alpha-amylase inhibitors are no doubt areas deserving additional investigations. In particular, precautions should be taken on foods deriving from plants containing inhibitors particularly resistant to denaturation agents as, for instance, the Colocasia inhibitors which are resistant to high temperatures, extreme pH values and to a number of proteases including pepsin. In view of the relatively high content of lysine in alpha-amylase inhibitors, it would be desirable to select wheat varieties with higher inhibitor contents to improve wheat nutritional value. A higher content of amylase inhibitors might also provide to new varieties a higher resistance to insects, as the bulk of wheat kernel albumins consists of exogenous alpha-amylase inhibitors; on the other hand, higher levels of inhibitors active on endogenous amylases might provide a better resistance to preharvest sprouting. However, due to actual or potential antinutritional effects of alpha-amylase inhibitors, foods deriving from varieties containing high albumin levels should be checked for adequate inhibitor inactivation during processing.

Finally, in view of the possible utilization of alpha-amylase inhibitors in medicine, a chronic toxicity study on rodents would be of great value as it might be able to show whether prolonged ingestion of active inhibitors could stimulate the pancreatic function to an extent that they become toxicologically significant.

ACKNOWLEDGEMENTS

This work was supported in part by C.N.R., Italy, Special grant I.P.R.A. -Subproject 1, Paper N. 520.

REFERENCES

Andriolo, S., Rouanet, J.M., Lafont, J. and Besançon, P. (1984). Inactivation of Phaseolamin, an alpha-amylase inhibitor from Phaseolus vulgaris by gastric acid and digestive proteases. Nutr. Rep. Int., 29,149-156

Applebaum, S.W. (1964). The action pattern and physiological rôle of Tenebrio molitor larval amylase. J. Ins. Physiol., 10,897-905

Applebaum, S.W. and Konijn, A.M. (1965). The utilization of starch by larvae of the flour beetle, Tribolium castaneum. J. Nutrition., 85,275-282

Applebaum, S.W. and Konijn, A.M. (1967). Factors affecting the development of Tribolium castaneum (Herbst) on wheat: J. Stored Prod. Res., 2,323-329

Applebaum, S.W., Harpaz, I. and Bondi, A. (1964). Amylase secretion in the larvae of Prodernia litura F. (insecta). Comp. Biochem. Physiol., 13,107-111

Barlow, K.K., Buttrose, M.S., Simmonds, D.H. and Vesk, M. (1973). The nature of the starch-protein interface in wheat endosperm. Cereal Chem., 50,443-454.

Belisario, M.A., Buonocore, V., Cantarella, M., Scardi, V. and Silano, V. (1981). Inhibition of alpha-amylase purified from Octopus vulgaris Lam. by albumin inhibitors from wheat flour. Qual. Plant. Plant foods Hum. Nutr., 31,21-30.

Benoit, M.O., Joffray, P., Caillens, H., Souciet, C., Ekindjian, O.G. and Douris, R.G. (1981). Evaluation des isoamylases seriques. Etude analytique et critique d'une methode nouvelle utilisant une proteine inhibitrice. Path. Biol., 29,223-228.

Berk, J.E., Simon, D. and Fridhandler, L. (1981). Inhibitor test for amylase isoenzymes. Am. J. Gastroenterol., 75,128-131.

Blanco-Labra, A. and Iturbe-Chinas, F.A. (1981). Purification and characterization of an alpha-amylase inhibitor from maize (Zea maize). J. Food Biochem., 5,1-17.

Bo-Linn, G.W., Santa Ana, C.A., Morawski, S.G. and Fordtran, J.S. (1982). Starch Blocker - their effect on calorie absorption from a high-starch meal. N. Engl. J. Med., 307,1413-1416.

Buonocore, V., Giardina, P., Parlamenti, R.,Poerio, E. and Silano, V. (1984). Characterization of chicken pancreas alpha-amylase isozymes and interaction with protein inhibitors from wheat kernel. J. Sci. Food Agric., 35,225-232.

Buonocore, V., Gramenzi, F., Pace, W., Petrucci, T., Poerio, E. and Silano, V. (1980). Interaction of wheat monomeric and dimeric protein inhibitors with alpha-amylase from yellow mealworm (Tenebrio molitor L. Larva). Biochem. J., 187,637-645.

Buonocore, V., Petrucci, T. and Silano, V. (1977). Wheat protein inhibitors of alpha-amylase. Phytochemistry, 16,811-820.

Buonocore, V., Poerio, E., Gramenzi, F. and Silano, V. (1975). Affinity column purification of amylases on protein inhibitors from wheat kernel. J. Chromatog., 114,109-114.

Buonocore, V., Poerio, E., Pace, W., Petrucci, T., Silano, V. and Tomasi, M. (1976). Interaction of Tenebrio molitor L. alpha-amylase with a wheat flour protein inhibitor. FEBS Lett., 67,202-206.

Buonocore, V., De Biasi, M.G., Giardina, P.,Poerio, E. and Silano, V. (1985). Purification and properties of an alpha-amylase tetrameric inhibitor from wheat kernel. Biochim. Biophys. Acta, in press.

Campos, F.A.P. and Richardson, M. (1983). The complete amino acid sequence of the bifunctional alpha-amylase/trypsin inhibitor from seeds of ragi (Indian finger millet, Eleusine coracana Gaertn.). FEBS Lett., 152,300-304.

Campos, F.A.P. and Richardson, M. (1984). The complete amino acid sequence of the alpha-amylase inhibitor 1-2 from seeds of ragi (Indian finger millet, Eleusine coracana Gaertn.). FEBS Lett.,167,221-225.

Cantagalli, P., Di Giorgio, G.,Morisi, G., Pocchiari, F. and Silano, V. (1971). Purification and properties of three albumins from Triticum aestivum seeds. J. Sci. Food Agric. 22,256-259.

Carlson, G.L., Li, B.U.K., Bass, P. and Olsen, W.A. (1983). A bean alpha-amylase inhibitor formulation (starch blocker) is ineffective in man. Science, 219,393-395.

Deponte, R., Parlamenti, R., Petrucci, T.,Silano, V. and Tomasi, M. (1976). Albumin alpha-amylase inhibitor families from wheat flour. Cereal Chem., 53,805-820.

Ellis, C., Koehler, D.F., Eckfeldt, J.H., Levitt, M.D. (1982). Evaluation of an inhibitor assay to determine serum isoamylase distribution. Dig. Dis. Sci., 27, 897-901.

Garrow, J.S., Scott, P.F., Heels, S., Nair, K.S. and Halliday, D. (1983a). "Starch blockers" are ineffective in man. Lancet. 1,60-61.

Garrow, J.S., Scott, P.F., Heels, S., Nair, K.S. and Halliday, D. (1983b). A study of "starch blockers" in man using ^{13}C-enriched starch as a tracer. Hum. Nutr. Clin. Nutr., 37,301-305.

Granum, P.E. (1978). Purification and characterization of an alpha-amylase inhibitor from rye (Secale cereale) flour. J. Food Biochem., 2,103-120.

Granum, P.E. (1979). Studies on alpha-amylase inhibitors in foods. Food Chem., 4,173-178.

Granum, P.E. and Eskeland, B. (1981). Nutritional significance of alpha-amylase inhibitors from wheat. Nutr. Rep. Int., 23,155-162.

Granum, P.E. and Whitaker, J.R. (1977). Purification and characterization of alpha-amylase inhibitors in wheat (Triticum aestivum var. Anza). J. Food Biochem., 1,385-401.

Granum, P.E., Holm, H., Wilcox, E. and Whitaker, J.R. (1983). Inhibitory properties of two commercially available starch blocker preparations. Nutr. Rep. Int., 28,1233-1244.

Hejgaard,J., Svedsen, I. and Mundy, J. (1983). Barley alpha-amylase/subtilisin inhibitor. II. N-terminal amino acid sequence and homology with inhibitors of the soybean trypsin inhibitor (Kunitz) family. Carlsberg Res. Comm., 48,91-94.

Huang, W.Y. and Tietz, N.W. (1982). Determinations of amylase enzymes in serum by use of a selective inhibitor. Clin. Chem., 28,1525-1527.

Irshad, M. and Sharma, C.B. (1981). Purification and properties of an alpha-amylase protein-inhibitor from Arachis hypogaea seeds. Biochim. Biophys. Acta, 659,326-333.

Kameya, A., Hayakawa, T., Noda, A. and Kondo, T. (1982). Isoamylase analysis using an amylase inhibitor-comparison with an electrophoretic method. Clin. Chim. Acta, 125,77-80.

Kashlan, N. and Richardson, M. (1981). The complete amino acid sequence of a major wheat protein inhibitor of alpha-amylase. Phytochemistry, 20,1781-1784.

Kilpatrick, D.C., Green, C. and Yap, P.L. (1983). Lectin content of slimming pills. Letter. Brit. Med. J., 286,305.

Kneen, E. and Sandstedt, R.M. (1943). An amylase inhibitor from certain cereals. J. Am. Chem. Soc., 65,1247.

Kneen, E. and Sandstedt, R.M. (1946). Distribution and general properties of an amylase inhibitor in cereals. Arch. Biochem., 9,235-249.

Koide, T. and Ikenaka, T. (1973). Studies on soybean trypsin inhibitors 3. Amino-acid sequence of the carboxyl-terminal region and the complete amino-acid sequence of soybean trypsin inhibitor (Kunits). Eur. J. Biochem., 32,417-431.

Konarev, Al.V. (1982). Identification of inhibitors of own and alien alpha-amylase in proteins of wheat. In "Pecularities of biology and utilization of species and varieties of cultivated plants in breeding". Bulletin of the N.I. Vavilov Institute of Plant Industry. Fasc. 118, ISSN0202 - 5361,11-12 (In Russian).

Krishna Sharma, K. and Pattabiraman, T.N. (1980). Natural plant enzyme Inhibitors. Isolation and characterization of two alpha-amylase inhibitors from Colocasia antiquorum tubers. J. Sci. Food Agric., 31,981-991.

Krishna Sharma, K. and Pattabiraman, T.N. (1982a). Purification of porcine pancreatic amylase by affinity chromatography with amylase inhibitor of Dioscorea alata immobilized on DEAE-cellulose. Indian J. Biochem. Biophys., 19,54-56.

Krishna Sharma, K. and Pattabiraman, T.N. (1982b).Natural plant enzyme inhibitors. Purification and properties of an amylase inhibitor from yam (Dioscorea alata). J. Sci. Food Agric., 33,255-262.

Lang, J.A., Chang-Hum, L.E., Reyes, P.S. and Briggs, G.M. (1974). Interference of starch metabolism by alpha-amylase inhibitors. Fed. Proc., 33,718.

Liener, I.E., Donatucci, D.A. and Tarcza, J.C. (1984). Starch blockers: a potential source of trypsin inhibitors and lectins. Am. J. Clin. Nutr., 39,196-200.

Macrì, A., Parlamenti, R., Silano, V. and Valfrè, F. (1977). Adaptation of the domestic chicken, Gallus domesticus, to continuous feeding of albumin amylase inhibitors from wheat flour as gastro-resistant microgranules. Poultry Sci., 56,434-441.

Maeda, K., Hase, T. and Matsubara, H. (1983a). Complete amino acid sequence of an alpha-amylase inhibitor in wheat kernel. Biochim. Biophys. Acta, 743,52-57.

Maeda, K., Takamori, Y. and Oka, O. (1982). Isolation and properties of an alpha-amylase inhibitor (0.53) from wheat (Triticum aestivum). Agric. Biol. Chem., 46,2873-2875.

Maeda, K., Wakabayashi, S. and Matsubara, H. (1983b). Disulfide bridges in an alpha-amylase inhibitor from wheat kernel. J. Biochem., 94,865-870.

Maeda, K., Wakabayashi, S. and Matsubara, H. (1985). Complete amino acid sequence of an alpha-amylase inhibitor in wheat kernel (0.19 inhibitor). Biochim. Biophys. Acta, 828, 213-221.

Marshall, J.J. (1975). Alpha-amylase inhibitors from plants. ACS Symp. Ser., 15,244-266.

Marshall, J.J. and Lauda, C.M. (1975). Purification and properties of phaseolamin, an inhibitor of alpha-amylase, from the kidney bean, Phaseolus vulgaris. J. Biol. Chem., 250,8030-8037.

Minetti, M., Petrucci, T., Cattaneo, S., Pocchiari, F. and Silano, V. (1973). Studies of the differential staining of wheat albumins, globulins and gliadins in polyacrylamide gel by aniline blueblack. Cereal Chem., 50,198-209.

Mundy, J. (1984). Hormonal regulation of alpha-amylase inhibitor synthesis in germinated barley. Carlsberg Res. Comm., 49,439-444.

Mundy, J.,Hejgaard, J. and Svendsen, I. (1984). Characterization of a bifunctional wheat inhibitor of endogenous alpha-amylase and subtilisin. FEBS Lett., 167, 210-214.

Mundy, J., Svendsen, I. and Hejgaard, J. (1983). Barley alpha-amylase/subtilisin inhibitor. I. Isolation and characterization. Carlsberg Res. Comm., 48,81-90.

Narayana Rao, M., Shurpalekar, K.S. and Sundaravalli, O.E. (1967). An amylase inhibitor in Colocasia esculenta. Indian J. Biochem., 4,185.

Narayana Rao, M., Shurpalekar, K.S. and Sundaravalli, O.E. (1970). Purification and properties of an amylase inhibitor from colocasia (Colocasia esculenta) tubers. Indian J. Biochem., 7,241-243.

O'Connor, C.M. and McGeeney, K.F. (1981a). Isolation and characterization of four inhibitors from wheat flour which display differential inhibition specificities for human salivary and human pancreatic alpha-amylases. Biochim. Biophys. Acta, 658,387-396.

O'Connor, C.M. and McGeeney, K.F. (1981b). Interaction of human alpha-amylases with inhibitors from wheat flour. Biochim. Biophys. Acta, 658, 397-405.

Odani, S., Koide, T. and Ono, T. (1982). Sequence homology between barley trypsin inhibitor and wheat alpha-amylase inhibitors. FEBS Lett., 141, 279-282.

Odani, S., Koide, T. and Ono, T. (1983). The complete amino acid sequence of barley trypsin inhibitor. J. Biol. Chem., 258,7998-8003.

O'Donnell, M.D., Fitzgerald, O. and McGeeney, K.F. (1977). Differential serum amylase determination by use of an inhibitor, and design of a routine procedure. Clin. Chem., 23,560-566.

O'Donnell, M.D. and McGeeney, K.F. (1976). Purification and properties of an alpha-amylase inhibitor from wheat. Biochim. Biophys. Acta, 422,159-169.

O'Donnell,M.D. and McGeeney, K.F. (1983). Suitability of control materials in the differential inhibition assay for human pancreatic and salivary amylase. Clin. Chem., 29,510-512.

Okabe, H., Uji, Y., Netsu, K. and Noma, A. (1984). Automated measurement of amylase isoenzymes with 4-nitrophenylmaltoheptaoside as substrate and use of a selective amylase inhibitor. Clin. Chem., 30,1219-1222.

Pace, W., Parlamenti, R., Ur Rab, A., Silano, V. and Vittozzi, L. (1978). Protein alpha-amylase inhibitors from wheat flour. Cereal Chem., 55,244-254.

Peruanskii, Yu. V., Gabsattarova, B.S. and Peruanskaya, O.N. (1980). Studies of the composition and activity of wheat alpha-amylase inhibitors in relation to growth conditions. Fiziol. Biokhim. Kul't. Rast., 12,358-363 (in Russian).

Petrucci, T., Sannia, G., Parlamenti, R. and Silano, V. (1978). Structural studies of wheat monomeric and dimeric protein inhibitors of alpha-amylase. Biochem. J., 173,229-235.

Petrucci, T., Tomasi, M., Cantagalli, P. and Silano, V. (1974). Comparison of wheat albumin inhibitors of alpha-amylase and trypsin. Phytochemistry, 13,2487-2495.

Petrucci, T. Rab, A., Tomasi, M. and Silano, V. (1976). Further characterization studies of the alpha-amylase protein inhibitor of gel electrophoretic mobility 0.19 from the wheat kernel. Biochim. Biophys. Acta, 420,288-297.

Pick, K.H. and Wöber, G. (1978). Proteinaceous alpha-amylase inhibitor from beans (Phaseolus vulgaris). Purification and partial characterization. Hoppe-Seyler's Z. Physiol. Chem., 359,1371-1378.

Pick, K.H. and Wöber, G. (1979). Purification of the proteinaceous alpha-amylase inhibitor from Phaseolus vulgaris by affinity chromatography with immobilized enzyme or antibody. Preparative Biochem., 9,293-302.

Powers, J.R. and Culbertson, J.D. (1982). In vitro effect of bean amylase inhibitor on insect amylases. J. Food Protec., 45,655-657.

Powers, J.R. and Whitaker, J.R. (1977a). Purification and some physical and chemical properties of red kidney bean (Phaseolus vulgaris) alpha-amylase inhibitor. J. Food Biochem., 1,217-238.

Powers, J.R. and Whitaker, J.R. (1977b) Effect of several experimental parameters on combination of red kidney bean (Phaseolus vulgaris) alpha-amylase inhibitor with porcine pancreatic alpha-amylase. J. Food Biochem., 1,239-260.

Puls, W. and Keup, V. (1973). Influence of an alpha-amylase inhibitor (Bay d 7791) on blood glucose, serum insulin, and NEFA in starch loading tests in rats, dogs and man. Diabetologia, 9,97-101.

Redman, D.G. (1975). Structural studies on wheat (Triticum aestivum) proteins lacking phenylalanine and histidine residues. Biochem. J., 149,725-732.

Redman, D.G. (1976). N-terminal amino acid sequence of wheat proteins that lack phenylalanine and histidine residues. Biochem. J., 155,193-195.

Richardson, M. (1981). Plant protein inhibitors of enzymes: their role in animal nutrition and plant defence. J. Biol. Educ., 15,178-182.

Rosenberg, I.H. (1982). Starch blockers - Still no calorie-free lunch. N. Engl. J. Med., 307,1444-1445.

Saunders, R.M. (1975). Alpha-amylase inhibitors in wheat and other cereals. Cereal Fd. World, 20,282-285.

Saunders, R.M. and Lang, J.A. (1973). Alpha-amylase inhibitors in Triticum aestivum: purification and physical-chemical properties. Phytochemistry, 12,1237-1241.

Savaiano, D.A., Powers, J.R., Costello, M.J., Whitaker, J.R. and Clifford, A.J. (1977). The effect of an alpha-amylase inhibitor on the growth rate of weanling rats. Nutr. Rep. Int., 15,443-449.

Shainkin, R. and Birk, Y. (1970). Alpha-amylase inhibitors from wheat: Isolation and characterization. Biochim. Biophys. Acta, 221,502-513.

Shewry, P.R., Lafiandra, D., Salcedo, G., Aragoncillo, C., Garcia-Olmedo, F., Lew, E.J.L., Dietler, M.D. and Kasarda, D.D. (1984). N-terminal amino acid sequences of chloroform/methanol-soluble proteins and albumins from endosperm of wheat, barley and related species.FEBS Lett., 175,359-363.

Shivaraj, B., Nayana Rao, H. and Pattabiraman,T.N. (1982). Natural plant enzyme inhibitors. Isolation of a trypsin/alpha-amylase inhibitor and a chymotrypsin/trypsin inhibitor from ragi (Eleusine coracana) grains by affinity chromatography and study of their properties. J. Sci. Food Agric., 33,1080-1091.

Shivaraj, B. and Pattabiraman,T.N. (1980). Natural plant enzyme inhibitors. Part VIII.

Purification and properties of two alpha-amylase inhibitors from ragi (Eleusine caracana) grains. Indian J. Biochem. Biophys., 17,181-185.

Shivaraj, B. and Pattabiraman,T.N. (1981). Natural plant enzyme inhibitors. Biochem. J., 193,29-36.

Silano, V. (1978). Biochemical and nutritional significance of wheat albumin inhibitors of alpha-amylase. Cereal Chem., 55,722-731.

Silano, V., Furia, M., Gianfreda, L., Macrì, A., Palescandolo, R., Rab, A., Scardi, V., Stella, V. and Valfrè, F. (1975). Inhibition of amylases from different origins by albumins from the wheat kernel. Biochim. Biophys. Acta, 391,170-178.

Silano, V., Pocchiari, F. and Kasarda, D.D. (1973). Physical characterization of alpha-amylase inhibitors from wheat. Biochim. Biophys. Acta, 317,139-148.

Silano, V. and Zahnley, J.C. (1978). Association of Tenebrio molitor L. alpha-amylase with two protein inhibitors - one monomeric, one dimeric - from wheat flour. Differential scanning calorimetric comparison of heat stabilities. Biochim. Biophys. Acta, 533,181-185.

Simmonds, D.H., Barlow, K.K. and Wrigley, C.W. (1973). The biochemical basis of grain hardness in wheat. Cereal Chem., 50,553-562.

Singh, C., Kherdekar, M.S. and Jambunathan, R. (1982). Studies on desi and kabuli chickpea (Cicer arietinum L.) cultivars. The levels of amylase inhibitors, levels of oligosaccharides and in vitro starch digestibility. J. Food Sci., 47, 510-512.

Sodini, G., Silano, V., De Agazio, M., Pocchiari, F.,Tentori, L. and Vivaldi, G. (1970). Purification and properties of a Triticum aestivum specific albumin. Phytochemistry, 9,1167-1172.

Takagi, Y., Gomi, K., Senju, O., Uzawa, R., Ishii, T., Motonaga, H., Noraka, N. and Noguchi, Y. (1982). Studies on the amylase inhibitor from wheat. I - Fundamental studies on its biochemical properties and application for amylase isoenzyme analysis. Rinsho Byori, 30,204-208.

Taylor, R.H. and Barker, H.M. (1983). Alpha-amylase inhibitors. Letter. Lancet May, 28, 1228-1229.

Warchalewski, J.R. (1976). Preliminary investigation on purification of native alpha-amylase inhibitors from durum wheat. Bull. Acad. Polon. Sci. Ser. Sci. Biol., 24,559-563.

Warchalewski, J.R. (1977a). Isolation and purification of native alpha-amylase inhibitors from winter wheat. Bull. Acad. Polon. Sci, Ser. Sci. Biol., 25,725-729.

Warchalewski, J.R. (1977b). Isolation and purification of native alpha-amylase inhibitors from malted winter wheat. Bull. Acad. Polon. Sci. Ser. Sci. Biol., 25,731-735.

Warchalewski, J.R. (1978). Formy isoenzymatyczne alpha-amylaz oraz ich rodzime biakowe inhibitory wystepujace wpszenicy i sozie pszennym. Roczniki AR Poznan, 82,1-6.

Warchalewski, J.R. (1983). Present day studies on cereals protein nature alpha-amylase inhibitors. Nahrung, 27,103-117.

Weselake, R.J., Macgregor, A.W. and Hill, R.D. (1983a). An endogenous alpha-amylase inhibitor in barley kernels.Plant Physiol., 72,809 812.

Weselake, R.J., Macgregor, A. W., Hill, R.D. and Duckworth, H.W. (1983b). Purification and characteristics of an endogenous alpha-amylase inhibitor from barley kernels. Plant Physiol., 73, 1008-1012.

Whitaker, J.R. (1983). Protease and amylase inhibitors in biological materials. In "Xenobiotics in foods and feeds". ACS Symposium Series No., 234,15-46.

Wilcox, E.R. and Whitaker, J.R. (1984). Some aspects of the mechanism of complexation of red kidney bean alpha-amylase inhibitor and alpha-amylase. Biochemistry, 23,1783-1791.

Yetter, M.A., Saunders, R.M. and Boles, H.P. (1979). Alpha-amylase inhibitors from wheat kernels as factors in resistance to postharvest insects. Cereal Chem., 56,243-244.

NOTE ADDED IN PROOF

After this review had been completed several papers concerning alpha-amylase inhibitors from kidney beans were published. Three of them deal with purification and characterization of inhibitors from black bean:

Frels, J.M. and Rupnow, J.H. (1984) Purification and partial characterization of two alpha-amylase inhibitors from black bean (Phaseolus vulgaris). J. Food Biochem. 8,281-287. Frels, J.M. and Rupnow, J.H. (1985) Characterization of two alpha-amylase inhibitors from black bean (Phaseolus vulgaris) J. Food Science, 50,72-77.

Lajolo, F.M. and Filho, F.F. (1985) Partial characterization of the amylase inhibitor of black bean (Phaseolus vulgaris), variety Rico 23. J.Agric. Food Chem.33,132-136.

A fourth contribution on amylase inhibitors from white and red kidney bean describes their thermal stability:

Cinco, F.J., Frels, J.M., Douglas, L.H. and Rupnow, J.H. (1985) Determination of the number and heat stability of alpha-amylase inhibitors in white and red kidney bean (Phaseolus vulgaris). J. Food Science,50,533-534.

The last two papers report nutritional studies carried out with rats fed with protein preparations from white bean containing high levels of amylase inhibitors:

Maranesi M., Carenini, G. and Gentili, P. (1984) Nutritional studies on anti-alpha-amylase. I) Influence on the growth rate, blood picture and biochemistry and histological parameters in rats. Acta Vitaminol. Enzymol. 6,259-270.

Maranesi, M., Barzanti, V., Biagi, P.L., Carenini, G. and Gentili, P. (1984) Nutritional studies on anti-alpha-amylase. II) Lipid Metabolism investigation: fatty acid composition of organs and tissue. Acta Vitaminol. Enzymol. 6,347-354.

Arai, M., Oouchi, N., and Muro, S. (1985). Inhibitory properties of an alpha-amylase inhibitor form Steptomyces-Corchorushii. Agric. Biol. Chem., 49, 987-992.

29

THE INHIBITION OF DIGESTIVE ENZYMES BY POLYPHENOLIC COMPOUNDS

David Wynne Griffiths

Scottish Crop Research Institute
Pentlandfield
Roslin
Midlothian EH25 9RF
Scotland

ABSTRACT

The ability of polyphenolic compounds to form insoluble complexes with other macro-molecules such as proteins has long been associated with the observed reduction in nutritive value resulting from their inclusion in animal diets. Naturally occurring polyphenols, and in particular condensed tannins isolated from various plant sources, have been shown to inhibit in vitro a number of digestive enzymes including trypsin, α-amylase and lipase. In addition, the results of various feeding trials suggest that similar reductions in intestinal digestive enzyme activity may result from the feeding of high polyphenolic diets. The evidence for this is summarised and discussed in relation to the possible effect of enzyme inhibition on reduced nutritive value and it is concluded that the observed reduction in protein availability found in vivo on consuming high tannin diets cannot simply be explained by the formation of dietary protein tannin complexes and that the ability of polyphenolic compounds to inhibit digestive enzymes may be of greater significance than realised previously.

INTRODUCTION

Naturally occurring polyphenolic compounds represent a diverse group of secondary plant products, which have been isolated and identified from a wide range of plant species. Their chemical structures may vary from the comparatively simple C_{15} flavanoid type compounds to the highly complex polymeric structures associated with the lignins and tannins.

Many polyphenolic compounds have been linked (Friend, 1979) with plant defence mechanisms against both fungal and microbial infection and have been postulated as both retardants and stimulants to insect feeding (Harborne, 1979). Minor modifications in their structure may completely reverse their biological activity as is demonstrated by the effect of quercetin (Harborne, 1979); in the form of the 3-glucoside it may act as a stimulant to the feeding of silk worm larvae but when the harbohydrate side chain is modified to either the 3-rhamnoside or 3-rutinoside larval feeding may be inhibited. A number of flavanoids and simple phenolic compounds have been shown (Van Sumere, et. al., 1975) to be activators of enzyme activity in vitro but with increasing complexity of structure enzyme inhibition appears to dominate and it has been suggested (Williams,1963) that in vivo more efficient inhibition results from the formation of tannin type compounds produced by oxidative coupling or natural polymerisation of the simpler flavanoid compounds.

The presence of tannins in the vegetative parts of plants have also been related to a reduction in palatability (Burns, 1966), resulting from the precipitation of salivary proteins and associated bitter taste. This ability of the tannins to interact with proteins of various kinds was initially responsible for the developing interest in their chemistry and the remainder of this paper is primarily concerned with their interaction with enzymic proteins and in particular those which may adversely affect nutritional value.

The Vegetable Tannins

Historically the term tannin has been applied to any compound with the ability to render the collagen fibres present in animal skins impervious to microbial attack and resistant to degradation by extremes of temperature and moisture. The majority of such compounds derived from vegetable sources were identified as being of a phenolic nature and consequently produced the characteristic reaction associated with phenols, including the formation of coloured ferric ion salts and the reduction of oxidising agents. The development of various screening methods based on these and other phenolic reactions led to considerable confusion since no distinction was made between simple phenolic compounds and flavanols and the true vegetable tannins. Consequently more rigorous definitions (Swain, 1965) have been proposed for the term vegetable tannin and this term is now generally accepted as referring to any naturally occurring compound which is of a high molecular weight ($>$ 500) and contains a sufficient number of hydroxyl groups (1-2 per 100 units of molecular weight) to enable it to form effective cross linkages between proteins and other macro-molecules.

When the chemical structures of the true vegetable tannins were examined (Haslam, 1966) two distinct classes were identified, namely the hydrolysable or gallo-tannins and the non-hydrolysable or condensed tannins. The former are compounds which, as their name suggests, are readily hydrolysed to yield a mixture of carbohydrates and gallic acid and other closely related phenolic acids such as ellagic acid. Their structures consist of a carbohydrate core to which gallic acid molecules are linked through ester bonds, formed between the hydroxyl group of the carbohydrate and the carbonyl group of the phenolic acid. The phenolic acid molecules can also polymerise via a depside bond to produce long chains of gallic acid spreading out from the carbohydrate centre. Although the hydrolysable tannins are comparatively widespread in nature their relative importance in animal nutrition is limited.

The second class of tannins, namely the non-hydrolysable or condensed tannins consist of dimers and higher oligomers of variously substituted flavan-3-ols, linked via a carbon-carbon bond normally at the C_4 and C_8 positions of the adjacent flavanoid constituents. The stability of the intra-flavanoid carbon-carbon bond is considerably greater than that of either the depside or ester linkages found in the hydrolysable tannins and consequently their stability in vivo and effect on nutritive value is considerably greater. The stereo-chemistry of many of the naturally occurring dimers has been extensively studied (Thompson et al., 1972) but little information appears available relating structural and substitutional differences to their affinity and interaction with proteins and other macro-molecules.

In contrast to the hydrolysable tannins, the condensed tannins have been identified (Price and Butler, 1980) in a wide range of plant species commonly consumed by both animals and man (Table 1).

Table 1. Selection of edible⁺ plant species containing condensed tannins

Seed Grains	Fruits	Forages
Barley	Apple	Crownvetch
Sorghum	Banana	Lespedeza
Field beans	Pear	Lotus
Pea	Blackberry	Sainfoin
Oilseed Rape	Strawberry	Trefoil

⁺ ruminant and non-ruminant animals

It should, however, be noted that for some of the species listed varieties have been found that contain little or no tannin, typical examples being field beans (Rowlands and Corner, 1962) and peas (Griffiths, 1981) where condensed tannin content is restricted to those cultivars having coloured flowers.

The nutritional significance of condensed tannins, particularly with regard to the feeding of non-ruminants has largely been associated with their ability to form insoluble tannin-protein complexes, which reduce the availability of dietary protein to the animal *in vivo*. High tannin diets have been shown to decrease egg production in laying hens (Martin-Tanguy et al., 1977) and live weight gain in rats (Jambunathan and Mertz, 1973) and growing chicks (Armstrong et al., 1973) and in a comparison of various grain legumes (Lindgren, 1975) a statistically significant negative correlation was found between tannin content (when expressed as a percentage of total protein) and the digestibility coefficients for crude protein. In addition the presence of high tannin seed coats from coloured flowered cultivars of field beans (*Vicia faba*) were seen (Moseley and Griffiths, 1979) to reduce not only protein availability but also to reduce significantly the apparent availability of both the soluble carbohydrate and lipid fractions of the diet when consumed by rats. Thus the role of the condensed tannins *in vivo* may not be solely restricted to the formation of dietary protein-tannin complexes but inhibition of endogenous digestive enzymes may also be of importance.

Enzyme inhibition by condensed tannins *in vitro*

In a study of the variation in the dry matter and organic matter digestibilities of various field bean (*Vicia faba*) varieties and selections, consistently lower values, as determined by the pepsin-rumen liquor *in vitro* method of Tilley and Terry, were found for the tannin-containing coloured flowered varieties and selections (Bond, 1976). Further investigations (Griffiths and Jones, 1977) revealed that the condensed tannins present in the seed coats significantly reduced the digestibility of cellulose by the rumen liquor, whilst only minor reduction in the solubilisation of dietary protein by the pepsin solution was observed. A highly significant negative correlation (r - 0·93) was also found between total phenolic content and seed coat solubility in cellulase (*Trichoderma viride*) solution strongly suggesting that the reduced *in vitro* dry matter digestibilities seen in the tannin-containing varieties was due to the inhibition of the cell wall digesting enzymes present in the rumen liquor rather than an effect of protein solubility *per se*.

Earlier investigations (Mandels and Reese,1962) into natural inhibitors of cellulases, particularly those present in the bayberry (*Myrica* pennsylvanica) had also revealed the presence of phenolic substances capable of significantly reducing the activity of Trichoderma viride cellulase in vitro. These compounds also gave reactions with various metal ions and acids consistent with the properties expected of condensed tannins and in a study of the kinetics of the inhibition, the results on the basis of a Linweaver-Burke plot suggested the mode of action of the inhibitor was non-competitive. The inhibitor present in the bayberry appeared to have a certain degree of specificity with regard to the source of the enzyme preparation with that from Trichoderma viride being the most susceptible to inhibition, whilst a cellulase preparation from Penicillium pusillum was unaffected by the inhibitor.

The presence of condensed tannins in some bird resistant cultivars of sorghum has been shown to be responsible for their poor brewing qualities for the production of sorghum based beers (Daiber, 1975). At the malting stage, when the grain remains intact and there is rigid segregation of the polyphenolic compounds in the nucellus and testa of the seed, no effect was found in either enzyme production or activity. However, during the mashing stage at which the malted grain is finely ground and steeped a significant loss in endogenous α- amylase activity resulted and this reduction was believed to account tor the poor brewing value of high tannin sorghum beers. This ability to inhibit amylase has been developed as a possible screening method for the determination of biologically active polyphenols in sorghum grain and found to correlate well with other chemical methods of determination (Daiber, 1975).

The condensed tannins present in the seed coats of coloured flowered cultivars of field beans and peas have also been examined for possible enzyme inhibitory activities in vitro (Griffiths, 1981). Hot water extracts prepared from both high-tannin and zero-tannin, white flowered cultivars were prepared and aliquots added to three different digestive enzyme systems. The results, some of which are shown in the table below (Table 2), indicated that the white flowered testa extracts had little if any effect on trypsin, α-amylase and lipase, whilst the extracts from the foloured flowered tannin-containing cultivars significantly inhibited all three enzymes.

To confirm that the observed reduction in trypsin activity was due to the presence of condensed tannins rather than to any contamination from the coyledons, which are known to contain a proteinic protease inhibitor (Wilson et al., 1972), basic kinetic studies were also undertaken. The results based on a Linweaver-Burke plot revealed that the inhibitor present in the seed coat displayed non-competitive characteristics, whilst the proteinic

Table 2. The inhibition of various digestive enzymes by hot water extracts prepared from the seed coats of white and coloured flowered cultivars of field beans (Vicia faba) and peas (Pisum spp.)

Species	Flower Colour	(%) Inhibition Trypsin PVP(-)	PVP(+)	α-Amylase PVP(-)	PVP(+)	Lipase PVP(-)	PVP(+)
Field beans	White	5	6	4	3	11	N.A.
	Coloured	51	0	43	2	69	33
Peas	White	7	8	5	5	N.A.	N.A.
	Coloured	66	9	55	1	N.A.	N.A.

N.A. = data not available

protease inhibitor showed competitive inhibition kinetics. Non-competitive inhibition was also found in an earlier study of the condensed tannins of carobs (Tamir and Alumot, 1969) and as was also found in this study the addition of polyvinyl pyrrolidone (PVP) significantly reduced the inhibitory properties of the condensed tannins in both trypsin and lipase. In contrast, the addition of PVP to the α- amylase-carob condensed tannin system did not fully restore amylase activity but for both pea and field bean tannins almost full activity was restored on the addition of this tannin-binding agent.

It would appear that condensed tannins extracted from widely differing plant sources all appear to have the ability _in vitro_ to inhibit a number of nutritionally important enzyme systems ranging from the cell wall digesting enzymes present in the rumen of herbivores to the endogenous intestinal enzymes, common to the majority of the domesticated animals and man.

Enzyme inhibition by condensed tannins _in vivo_

In contrast to the situation _in vitro_ condensed tannins, when consumed as a part of the normal diet of either animals or man, would be accompanied by a wide range of alternative binding sites. These include not only proteins and possibly other macro-molecules present in the feed but also salivary proteins released in the mouth during the processes of ingestion. The procedures employed during the extraction of condensed tannins for _in vitro_ assays, such as fine grinding and subsequent exposure to solvents and heat, may also significantly alter the degree of polymerisation and consequently the activity of plant phenolics to an extent that might not be reproduced _in vitro_. Therefore, in order to determine if naturally occurring condensed tannins reduced digestive enzymes _in vivo_, a rat feeding trial was undertaken utilising seed coats from high tannin varieties of field beans (Griffiths and Moseley, 1980). A tannin-free control diet was also fed using the seed coats of white flowered varieties which were previously known to be almost completely devoid of tannins (Griffiths and Jones, 1977).

The rats were fed their respective diets for a total of 15 days and at the termination of the trial, sacrificed and the intestinal contents removed and analysed individually for trypsin, α- amylase and lipase activities. The results, part of which are reproduced in Table 3, indicated significant differences in activities between the samples taken from the two groups, with those consuming the high tannin diets having statistically significant lower activities for both trypsin and α-amylase.

Statistically significant negative correlations were also obtained between both trypsin and α- amylase intestinal activities and polyphenolic intake per rat and, to confirm that the observed trypsin inhibition was due to the condensed tannins, the assays were repeated in the presence of added

Table 3. The effect of high tannin diets on enzyme activities in the intestines of rats

	Diet 1	Diet 2
Tannin content (g/kg)	Trace	4
Enzyme activity+		
Trypsin	20	12
α- amylase	12	4
Lipase	5	19

\+ A unit of activity was defined as that producing an increase of 10^{-2} in absorbance under the assay conditions used; all results are means for six rats.

polyvinyl pyrrolidone (PVP). The addition of PVP restored enzyme activity in the samples of intestinal fluid taken from the high tannin diet fed to rats to levels comparable to that found in the tannin-free diet fed rats, suggesting that intestinal contents of high-tannin fed rats contained a reversible trypsin-tannin complex.

In complete contrast to the results obtained for both trypsin and α-amylase, lipase activity was significantly increased in the high-tannin diet fed rats and although these results are difficult to explain satisfactorily, particularly in view of the results obtained in vitro, it may be that the situation is similar to that reported for enzyme activity in the intestines of chicks fed raw soya beans (Gertler and Nitsan, 1970). The presence of soya bean protease inhibitors reduced the activities of both trypsin and chymotrypsin but increased pancreatopeptidase E activity. This was concluded as resulting from an increased secretion of all the pancreatic enzymes in response to the presence of the protease inhibitors but since the interaction between the inhibitor and pancreatopeptidase E was less than that with either trypsin or chymotrypsin, the net effect was an increase in pancreatopeptidase E activity. An analogous situation may have arisen during the feeding of high tannin diets to rats resulting in an increased secretion of all the pancreatic enzymes but under intestinal conditions the affinity of tannin for lipase may be less than that for either trypsin or α-amylase and consequenstly results in a net increase in lipase activity. However, no evidence of pancreatic hypertrophy was observed in the tannin fed rats and further studies are clearly required to explain this anomaly.

Further evidence for the ability of tannins to interact with digestive enzymes in vivo was found in a study of protein absorbtion and utilisation in rats fed with different phenolic and polyphenolic compounds (Glick and Joslyn, 1970). Significant increases in nitrogen excretion were observed when condensed tannins or tannic acid of unspecified origin were included in the diets but no effects were seen with either catachin or gallic acid. Increased intestinal proteolytic enzyme activities were also found with both condensed tannins and tannic acid, and when C^{14} labelled casein was fed no trend towards an increased recovery of C^{14} in faecal material of the tannic acid fed rats was seen. The amino acid patterns of the faeces from both the two groups of rats were similar and quite different from that of casein. From this the authors concluded that the increased nitrogen excreted by the tannin fed rats was due mainly to proteins from enzymatic or other endogenous sources rather than to undigested dietary proteins.

The existence of uncomplexed tannin in the intestinal tract would require that the originally formed complexes between condensed tannins and dietary or salivary proteins produced during the initial chewing and ingestion processes be rendered unstable by the acidic conditions of the stomach. In an investigation of sainfoin tannins (Jones and Mangan, 1977), the complex produced with submaxillary mucoproteins collected from Jersey cows was found to produce insoluble complexes only at temperatures below $25^{o}C$, whilst the complex formed with fraction 1 leaf protein in vitro was stable between pH3·5 and pH7·0 but between pH1·0 and pH3·0, 95% of the fraction 1 protein was released from the complex. In vitro samples taken from a sheep fed sainfoin indicated that at normal rumen pH the protein-tannin complex was stable but at the abomasal end of the duodenum (pH2·5) the tannins were readily extractable suggesting that they were not associated with protein. This would appear to confirm that free tannins are released in the abomasum and as the pH increases along the intestinal tract conditions suitable for reforming tannin-protein complexes are likely to develop, although more detailed animal investigations with both ruminants and non-ruminants are required to confirm this.

In conclusion it would appear that the reduction in nutritive value associated with the presence of plant polyphenolic compounds in animal diets may not be simply explained in terms of the formation of insoluble, and consequently, unavailable dietary protein complexes and their ability to inhibit digestive enzymes in vivo may be of greater consequence than hitherto realised.

REFERENCES

Armstrong, W.D., Featherstone, W.R. and Rogler, J.C., 1973, Influence of methionine and other dietary additions on the performance of chicks fed bird resistant sorghum grain diets, Poultry Sci., 52:1592-1599.

Bond, D.A., 1976, In vitro digestibility of the testa in tannin-free field beans (Vicia faba L.), J. Agric. Sci., 86:561-566.

Burns, R.E., 1966, Tannin in Sericea lespedeza, Georgia Agricultural Research Bulletin NS 164.

Daiber, K.H., 1975, Enzyme inhibition by polyphenols of sorghum grain and malt, J. Sci. Fd. Agric., 26:1399-1411.

Gertler, A. and Nitsan, Z., 1970, The effect of trypsin inhibitors on pancreatopeptidase E, trypsin, chymotrypsin and amylase in the pancreas and intestinal tract of chicks receiving raw and heated soyabean diets, Br. J. Nutr. 24:893-904.

Glick, Z. and Joslyn, M.A., 1970, Effect of tannic acid and related compounds on the absorbtion and utilisation of proteins in the rat, J. Nutr., 100:516-520.

Griffiths, D.W., 1981, The polyphenolic content and enzyme inhibiting activity of testas from bean (Vicia faba) and pea (Pisum spp.) varieties, J. Sci.Fd. Agric., 32:797-804.

Griffiths, D.W. and Jones, D.I.H., 1977, Cellulase inhibition by tannins in the testa of field beans (Vicia faba), J. Sci. Fd. Agric., 28:983-989.

Griffiths, D.W. and Moseley, G. 1980, The effect of diets containing field beans of high or low polyphenolic content on the activity of digestive enzymes in the intestines of rats, J. Sci. Fd. Agric., 31:225-259.

Harborne, J.B., 1979, Flavanoid Pigments, In Herbivores: Their interaction with secondary plant metabolites (Rosenthal, G.A. and Janzen, D.H. Eds.) Academic Press, London and New York, pp.619-656.

Haslam, E., 1966, The scope of vegetable tannin chemistry In The Vegetable Tannins (Haslam, E., Ed.) Academic Press, London and New York, pp. 1-13.

Jambunathan, R. and Mertz, E.T., 1973, Relationship between tannin levels, rat growth and diets, of proteins in sorghum, J. Agric. Food Chem., 28:692-696.

Jones, W.T. and Mangan, J.L., 1977, Complexes of the condensed tannins of sainfoin (Onobrychis viciifolia Scop.) with Fraction 1 leaf protein with submaxillary mucoprotein, and their reversal by polyethylene glycol and pH, J. Sci., Fd. Agric., 28:126-136.

Lindgren, E., 1975, The nutritive value of peas and field beans for hens, Swed. J. Agric. Res., 5:159:161.

Martin-Tanguy, J., Guillaune, J and Kossa, A., 1977, Condensed tannins in horse bean seeds: chemical composition and apparent effect on poultry, J. Sci. Fd. Agric., 28:757-765.

Moseley, G. and Griffiths, D.W., 1979, Varietal variation in the anti-nutritive effects of field beans (Vicia faba) when fed to rats. J. Sci. Fd. Agric., 30:772-778.

Price, M.L. and Butler, L.G., 1980, Tannins and nutrition, Purdue University Agricultural Experiment Stations Bulletin No. 272.

Rowlands, D.G. and Corner, J.J., 1962, Genetics of pigmentation in field beans (Vicia faba), Eucarpia, 3:229-234.

Swain, T., 1965, The tannins. In Plant Biochemistry (Bonner, J. and Varner, J.E., Eds.), Acadamic Press, London and New York, pp.552-580.

Tamir, M. and Alumot, E., 1969, Inhibition of digestive enzymes by condensed tannins from green and ripe carobs, J. Sci. Fd. Agric., 20:199-202.

Thompson, R.S., Jacques, D., Haslam, E. and Tanner, R.J.N., 1972, Part 1, The isolation, structure and distribution in nature of plant procyanidins, J. Chem. Soc. Perkin Transactions I, pp.1387-1399.

Tilley, J.M.A. and Terry, R.A., 1963, A two stage technique for the in vitro digestion of forage crops, J. Br.Grassland Soc. 18:104-111.

Van Sumere, C.F., Albrect, J., Dedonder, H., De Pooter, H, and Pe, I., 1975, Plant proteins and pehnolics, In The chemistry and biochemistry of plant proteins (Harborne, J.B. and Van Sumere, C.F., Eds.) Academic Press, London and New York, pp. 211-264.

Williams, A.H., 1963, Enzyme inhibition by phenolic compounds, In Chemistry of phenolic compounds (Pridham, J.B., Ed.) Pergammon Press, Oxford and London, pp.87-95.

Wilson, B.J., McNab, J.M. and Bentley, H., 1972, Trypsin inhibitor activity in the field bean (Vicia faba L.), J. Sci. Fd. Agric., 23:679-684.

30

EFFECT OF SEVERELY ALKALI-TREATED CASEIN ON GASTROINTESTINAL TRANSIT AND SELECTED INTESTINAL ENZYME ACTIVITIES

Bernard Possompes and Jacques Berger

Laboratoire de Physiologie de la Nutrition
Universite des Sciences et Techniques du Languedoc
34060 Montpellier Cedex, France

Alkaline treatment of proteins leads to chemical changes which alter the proteins' digestibility. Severely alkali-treated casein (0.2N NaOH, 80°C, 1 hour) in the diet reduces food intake and growth of young but not of adult Sprague Dawley rats. Gastrointestinal transit time is not reduced significantly in either young or adult rats.

Food intake and growth rate are improved by amino acid supplementation. In this case, protein content and total leucine aminopeptidase activity are increased in the distal part of the small intestine whereas gamma-glutamyltranspeptidase and maltase activities are increased in both the proximal and distal parts. Alkaline phosphatase activity remains unchanged.

These intestinal adaptations differ from those observed in rats receiving a diet containing untreated casein and graded levels of a synthetic trypsin inhibitor. In the latter, protein digestibility remains high, gamma-glutamyltranspeptidase and also maltase activities are increased in the proximal and medial parts of the small intestine only.

Intestinal adaptation in rats receiving alkali-treated casein does not result from a deficiency of pancreatic proteases activity. Ileal accumulation of undigested peptides from insufficient hydrolysis of alkali-treated casein may account for these muscosal adaptations.

INTRODUCTION

Severe alkaline treatment alters the chemical composition and nutritional value of proteins (DeGroot et al, 1969). Toxicological studies (Gould et al, 1977) reveal that intake of such proteins induces renal lesions (Woodard et al, 1973) which are reversible (Struthers et al, 1978). Although they can induce diarrhea (Karayiannis et al, 1976), their effects on the pancreas and the gastrointestinal tract remain unclear.

The chemical modifications induced by severe alkaline treatment alter protein digestibility <u>in vitro</u> (Hayashi et al, 1980 a and b; Friedman et al, 1981; Vachon et al, 1982; Possompes et al,1982) and <u>in vivo</u> (De Groot et al, 1969). That can be explained by the destruction of substrate amino acids such as arginine, their isomerization, the formation of intra and intermolecular covalent bonds that decrease the accessibility of the enzymes (Hurrel et al, 1977) and also by the synthesis of lysinoalanine that could inhibit metalloenzymes by complexing, for instance, with the zinc required for carboxypeptidase function (Hayashi, 1982; Friedman et al, 1985). In addition, previous reports show that hydrolysis of severely treated casein induces the formation of peptides having a high affinity for trypsin and chymotrypsin and capable of inhibiting these enzymes (Vimont-Rispoli et al, 1980). The consumption of such treated casein by adult Sprague Dawley rats, in specific experimental conditions, induces hypertrophy of the pancreas (Possompes et al, 1979).

Insufficient hydrolysis of proteins causes an accumulation of nitrogen compounds in the distal parts of the small intestine (Diaz, 1978) and in particular an accumulation of peptides (Abe et al, 1984). This phenomenon can be intensified because the absorption of D amino acids by the enterocyte is low, either in free form (Gibson et al, 1951; Berg, 1959; Stegink, 1977; Baker, 1979; Winne et al, 1979) or in peptide form (Das et al, 1976). The level of isomerization of amino acids is different for each one and depends on the nature of the protein (Masters et al, 1979) and on the severity of the treatment (Liardon et al, 1983). It must be stressed that some amino acids are toxic, thus D-serine given by subcutaneous injection leads very rapidly to necrosis of the cells of the proximal tubule of the nephron (Ganote et al, 1974). Products derived from lysine by an addition reaction to the εNH_2 are poorly absorbed by the enterocyte (Yasumoto et al, 1977); however lysinoalanine is absorbed bound to other amino acids (Abe et al, 1981, 1984).

The small intestine can adapt to the physiological and nutritional condition of the individual (Williamson, 1978; Karasov et al, 1983). These adaptations consist of a modulation of the digestive and absorptive area caused by a change in the size of the villi and microvilli, and possibly involve an elongation of the small intestine. Systems of transport could be affected more or less directly (Karasov et al, 1983).

One of the means used to verify an adaptation is to measure enzyme activity in intestinal mucosa homogenates or in isolated intestinal brush border preparations.

The enzymatic equipment of the intestinal brush border depends on the nutritional condition (Raul et al, 1980), on bile and pancreatic enzyme secretions (Alpers et al, 1975; Batt et al, 1979; Weser et al, 1977, 1982), on hormonal secretions of the pancreas (Nakabou et al, 1980; Madara et al, 1982) and of the small intestine (Weser et al, 1981). These factors do not affect all brush border enzymes in the same way but they seem to stimulate certain among them more or less specifically. Experiments using parenteral nutrition show that infusion of glucose and/or amino acids avoids atrophy of the intestinal mucosa (Spector et al, 1977). To achieve this, nutrients do not need to be absorbed (Weser et al, 1982). We can speculate that D amino acid, synthetized during the treatment, could stimulate mucosal development.

In the first experiment, by establishing the kinetics of excretion of a marker ingested with their diet by rats fed ad libitum, we determined to what degree the ingestion of severely alkali-treated casein could modify gastrointestinal transit. Then, in the second experiment we examined, by analyzing enzyme activities, whether the enzymatic equipment of the enterocyte is modified. We also hypothesized that the modifications induced by intake of such casein could be due to peptides inhibiting trypsin, i.e. they would be the consequence of a pancreatic insufficiency. To verify this, we gave rats a diet containing untreated casein as a protein source and graded levels of a synthetic trypsin inhibitor, benzamidine (Mares-Guia et al, 1965).

MATERIAL AND METHODS

Alkaline Treatment

Casein (according to Hammarsten, Merck, N.J.) is suspended, at 10% concentration, in 0.2N sodium hydroxyde, and heated at 80°C for 1 hour. The mixture is then cooled, neutralized to pH 7.0 with 6N HCl, and lyophilized. This casein served as the protein source for rat diets. Chemical modifications induced by this treatment have been described (Possompes et al, 1983).

Diets

Diets contained 10 per cent protein, provided by alkali-treated casein or by untreated casein (control group). When supplementation was used, 90% of the protein was provided by treated casein and 10% by an amino acid mixture, i.e. 2% methionine, 1% cysteine, 3% threonine, 3% tyrosine and 1% valine. The remainder of the diet was a protein free flour containing glucose (5%), wheat starch (70%), vegetabel and animal fat (8%), cellulose (5%), minerals (11%) and vitamins (1%).

Depending upon the experiment, we added to the diet either chromic oxide (Cr_2O_3) at a concentration of 0.2% of dry matter, or benzamidine (Merck).

Animals and Experiment Design

Study of transit. Male Sprague Dawley rats were housed in individual metabolic cages in a 26±1°C thermostat controlled room, with a 12 hour reverse light cycle (light period 0800-2000). These cages had a mechanism capable of changing food cups every two hours. That mechanism also allowed us to collect feces for each two hour period. Rats were adapted to that mechanism for 4 days. They were fed the diets and water ad libitum. On day five, they then received the diet containing chromic oxide between 2000 to 2200 and after that the plain diet again. Feces were collected for 72 hours. Values are means of 8 measurements.

Chromic oxide was chosen because it fulfills the necessary criteria of a transit marker (Francois et al, 1968), and its use does not require any special caution or any rat handling.

The transit time was determined by the time needed for the appearance of the maximum concentration of the marker in the feces. Overall transit times were calculated according to the equations from Castle (1956) and from Zierler (1958). Chromic oxide was analyzed according to the method of Bolin (1952).

Intestinal adaptation. Rats were housed in individual metabolic cages, in a 26±1°C thermostat controlled room. They were given experimental diets for 21 days. Then they were killed at 1500 after a 7 hour fast. Ten centimeter segments were collected at the beginning, middle and end of the small intestine. The contents were flushed out with 9% sodium chloride. Mucosa were collected by gentle scraping, then frozen in liquid nitrogen until analysis.

Enzyme activities were assayed according to Dahlquist (1968) modified for maltase, to Bessey et al (1946) for alkaline phosphatase (AP), to Naftalin (1969) for gamma-glutamyltranspeptidase (γ-GTP), to Goldbard et al. (1958) for leucine aminopeptidase (LAP). Proteins were assayed according to Lowry et al (1951). Activities are expressed in μ-moles of produce produced during the hydrolysis, per minute, per mg of protein, at 37°C. Activities are also expressed per segment.

Values are expressed as means ± SEM and comparisons between the values are done either by analysis of variance or by t-test (Schwartz, 1963).

RESULTS

Influence of Alkali-treated Casein Intake on the Gastrointestinal Transit

Prepubescent rats, receiving a diet containing 10% protein supplied by alkali-treated casein, reduce food intake to 45% of that of rats receiving the control diet, whereas adult rats do not change their feeding behavior (table 1).

Food intake and fecal excretion present a marked nychtemeral rhythm; however with young rats receiving the alkali-treated casein, 45% of the feces (dry weight) are collected during the light period in contrast to 21% with control rats.

The most representative value of gastrointestinal transit time is the period required to give the maximal concentration of marker in the feces. This period equals 9.5 ± 2.2 hours in young rats receiving the alkali-treated casein vs 10.5 ± 1.3 hours for control (no statistically significant difference). In adult rats values are 11.5 ± 1.2 and 12.0 ± 2.0 hours respectively (table 1).

The spontaneous food intake reduction seen in young rats receiving treated casein does not affect transit time. When imposed on adult rats, that limitation induces a statistically significant decrease ($p<0.05$) in the time required for the appearance of the maximal concentration of the marker which then averages 7.8 ± 1.1 hours.

In all experiments, overall retention times of the marker expressed according to the Castle and Zierler equation are similar. These values take the possibility of coprophagy into account.

Since transit time is less than 12 hours, the percentage of chromic oxide recovered in a period of 12 or 24 hours must be proportional to transit time. Coprophagy affects this relationship. In the first 12 and 24 hour periods following the beninning of the ingestion of the diet containing the marked, the percentage of marker recovered with the alkali-teated casein is 34 and 64% of ingested amount in young rats and 21 and 54% in adult rats. With untreated casein these values are 44 and 63% in young rats, 32 and 64% in adult rats and 35 and 45% in restricted fed rats (table 1). The presence of alkali-treated casein in the diet seems to enhance coprophagy.

Addition of 0.2 or 0.4% benzamidine to the control diet leads to a statistically significant reduction of the transit time to 8.5 ± 2.5 and 8.4 ± 1.7 respectively in adult rats. Amounts of marker recovered at 12 and 24 hours were 24 and 56% respectively with 0.2% trypsin inhibitor, and 32 and 64% with 0.4% of trypsin inhibitor. These values are low when compared to control.

In adult control rats, excretion of the marker shows two peaks, one at the end of the first dark period, the second at the beginning of the following dark period. In adult rats receiving benzamidine or alkali-treated casein, a third peak appears between the other two during the light period.

In adult rats receiving treated casein or 0.2% benzamidine, the time needed for the marker to go through the digestive tract is shorter than in control rats, however the percentage of chromic oxide recovered after 12 and 24 hours remains lower than that observed in controls. That means coprophagy is more important in these experimental rats (when the concentration of the benzamidine is 0.4 per cent, coprophagy decreases). This increase of coprophagy shows that, despite transmit acceleration, overall retention time, expressed according to Castle (1956) and Zierler (1958), is similar in all groups.

Table 1. Influence of alkali-treated casein and trypsin inhibitor intake on gastrointestinal transit time of chromic oxide (means ± SEM)

Rat Final body weight (g)	Food Intake (g dry matter per day)	Time needed for the maximal concentration of the marker (hours)	Overall retention time calculated according to Zierler (hours)	Overall retention time calculated according to Castle (hours)
Control diet: untreated casein. Rats fed ad libitum.				
85	21.5 ± 0.6	10.5 ± 1.3	17.6 ± 1.3	21.3 ± 4.0
270	27.0 ± 0.8	11.5 ± 1.2	17.7 ± 2.1	19.9 ± 1.6
Control diet: untreated casein. Restricted rats.				
220	12.0 ± 0.2	7.8 ± 1.1	17.2 ± 2.7	21.2 ± 4.6
Alkali-treated casein.				
70	9.3 ± 0.4	9.5 ± 2.2	16.6 ± 3.1	19.8 ± 4.8
270	25.9 ± 0.8	12.0 ± 2.0	18.7 ± 1.7	23.5 ± 3.9
Untreated casein + 0.2 per cent benzamidine.				
280	22.0 ± 0.7	8.5 ± 2.5	14.7 ± 1.2	20.3 ± 1.3
Untreated casein + 0.4 per cent benzamidine.				
300	21.1 ± 1.2	8.4 ± 1.7	16.8 ± 0.7	20.7 ± 2.6

Intestinal Enzyme Adaptations

Effect of alkali-treated casein. Food intake and growth of young rats fed alkali-treated casein are improved by amino acid supplementation (table 2).

Overall digestibility of dry matter is not significantly affected; thus we can conclude that absorption of nutrients other than protein is not affected. The apparent digestibility of the protein decreases which is manifested by a significant increase in fecal nitrogen (table 2).

The increase in relative pancreas weight in young rats receiving alkali-treated casein non-supplemented is not due essentially to the poor digestibility of the protein but rather to the fact that relative pancreas weight is higher in young rats (a condition that persists until puberty, i.e., body weight around 110gm).

Rats receiving supplemented alkali-treated casein show food intake and growth rates similar to control rats. Ingestion of treated casein induces a considerable increase in the amount of protein in the mucosa of the terminal segment of the small intestine (table 3), and a very slight increase in the proximal segment. Leucine aminopeptidase specific activity is stimulated in the terminal part of the ileum. γ-GTP and maltase specific activity are likewise increased in that region and also in the duodenum. For both these enzymes, the distribution gradient of the activities along the small intestine is modified compared to control.

Table 2. Influence of intake of alkali-treated casein on food intake, growth rate, notrogen digestibility and pancreas weight.

	Control	Alkali-treated casein	Alkali-treated casein + amino acids
Initial body weight (g)	69 ± 1	71 ± 1	71 ± 2
Growth rate (g/day)	3.30 ± 0.24^{a}	0.10 ± 0.005^{ab}	3.08 ± 0.14^{b}
Food intake (g dry matter per day)	10.4 ± 0.6^{a}	5.5 ± 0.6^{ab}	9.5 ± 0.2^{b}
Dry matter (%) digestibility	91 ± 0.5	93 ± 1.0	90 ± 0.6
Apparent nitrogen digestibility (%)	88 ± 1.0^{ab}	75 ± 3.8^{a}	72 ± 2.1^{b}
Fecal nitrogen (% of dry matter)	2.32 ± 0.2^{ab}	5.96 ± 0.24^{a}	4.68 ± 0.11^{b}
NPU	3.15 ± 0.12^{a}	0.20 ± 0.14^{ab}	3.22 ± 0.02^{b}
Pancreas weight (g)	0.48 ± 0.03^{a}	0.35 ± 0.02^{ab}	0.51 ± 0.03^{b}
Relative pancreas weight (% body weight)	0.34 ± 0.03^{a}	0.48 ± 0.03^{ab}	0.37 ± 0.02^{a}

NPU = Net Protein Utilization

Apparent nitrogen digestibility $= \frac{I-F}{I} \times 100$

I = Nitrogen intake
F = Nitrogen in feces

Mean ± SEM
Numbers with same letters are significantly different ($p<0.05$)

		Control	Alkali-treated casein	Alkali-treated casein + amino acids
Total protein (mg)	1	73.1 ± 4.5^{a}	52.8 ± 4.7^{ab}	82.8 ± 4.9^{b}
	2	78.0 ± 10.1^{a}	53.8 ± 2.6^{a}	71.7 ± 5.5
	3	39.6 ± 3.5^{a}	31.5 ± 4.9^{b}	55.8 ± 4.1^{ab}
LAP U/mg protein	1	0.14 ± 0.04^{a}	0.22 ± 0.01^{ab}	0.16 ± 0.01^{b}
	2	0.63 ± 0.17^{a}	0.87 ± 0.05^{ab}	0.71 ± 0.05^{b}
	3	0.33 ± 0.07^{ab}	0.46 ± 0.05^{a}	0.42 ± 0.05^{b}
γ-GTP U/mg protein (.10-3)	1	11.4 ± 2.1^{a}	24.0 ± 3.5^{a}	17.8 ± 2.8^{a}
	2	13.9 ± 1.1^{a}	17.2 ± 1.2^{ab}	11.9 ± 1.7^{b}
	3	9.6 ± 2.2^{ab}	20.2 ± 1.6^{a}	18.0 ± 3.7^{b}
Enterokinase U/mg protein (.10-3)	1	27.9 ± 4.1^{a}	43.5 ± 5.7^{ab}	31.2 ± 1.9^{b}
Alkaline Phophatase U/mg protein	1	3.26 ± 0.50^{a}	2.21 ± 0.19^{a}	2.89 ± 0.30
	2	0.34 ± 0.08	0.28 ± 0.04	0.30 ± 0.03
	3	0.14 ± 0.11	0.15 ± 0.04	0.38 ± 0.01
Maltase U/mg protein	1	1.08 ± 0.10^{a}	2.04 ± 0.08^{a}	1.68 ± 0.09^{a}
	2	1.06 ± 0.05^{ab}	1.46 ± 0.05^{a}	1.39 ± 0.08^{b}
	3	---	1.95 ± 0.19	1.69 ± 0.41

Table 3. Influence of intake of alkali-treated casein on specific activities of selected intestinal enzymes.

Values (means ± SEM) with the same letter are significantly different ($p < 0.05$).

Number 1 represents segment collected at beginning, number 2 at the middle and number 3 at the end of the small intestine.

Examination of total activity per segment (table 4) corroborates this observation. Not all enzyme activities are stimulated since alkaline phosphatase activity remains unchanged.

		Control	Alkali-treated casein	Alkali-treated casein + amino acids
	1	9.9 ± 1.3	11.5 ± 1.1	12.3 ± 1.4
LAP	2	48.3 ± 6.5	46.5 ± 4.4	47.1 ± 4.2
U/segment	3	13.6 ± 1.7[a]	14.8 ± 3.2[b]	25.3 ± 3.2[ab]
	1	0.85 ± 0.02[ab]	1.28 ± 0.24[a]	1.48 ± 0.29[b]
γ-GTP	2	1.06 ± 0.10	0.93 ± 0.09	0.94 ± 0.18
U/segment	3	0.37 ± 0.06[a]	0.64 ± 0.13[a]	1.00 ± 0.21[a]
Alkaline	1	233 ± 22	118 ± 14	242 ± 35
Phosphatase	2	24 ± 3	15 ± 3	21 ± 2
U/segment	3	4 ± 3	6 ± 1	20 ± 6
	1	78 ± 1[a]	108 ± 13[a]	141 ± 14[a]
Maltase	2	84 ± 13	78 ± 4	96 ± 10
U/segment	3	--	62 ± 13[a]	93 ± 15[a]

Table 4. Influence of intake of alkali-treated casein on total activity per segment of enzymes in the different segments of the small intestine.

Values (means ± SEM) with the same letter are significantly different.

The malnourished condition of rats receiving treated casein without supplementation induces a marked decrease in the amount of protein in the intestinal mucosa. Specific activity of the enzymes studied, except for alkaline phosphatase, increases. γ-GTP and maltase activity are especially stimulated and the total activity per segment is higher for proximal and distal segments than in control rats.

Influence of benzamidine. Enzymatic adaptations in rats receiving the trypsin inhibitor are more marked in the medial part and to a lesser extent at the beginning of the small intestine. Leucine aminopeptidase specific activity is stimulated in the medial part of the small intestine in rats receiving 0.3 and 0.4% benzamidine (table 5). γ-GTP and maltase specific activities are higher than in control rats in proximal and medial part of the small intestine. The adaptation of maltase appears clearly in animals receiving the diet containing 0.4% benzamidine. The distribution along the small intestine of γ-GTP and multase activity is not modified by the ingestion of benzamidine. Total activity of alkaline phosphatase shows a tendency to decrease in experimental rats.

Table 5. Influence of benzamidine intake on specific activities of intestinal enzymes.

Values (mean ± SEM) with the same letter are significantly different ($p < 0.05$).

		Control	Benzamidine 0.15%	Benzamidine 0.30%	Benzamidine 0.40%
Protein mg/segment	1	93.7 ± 10.0	109.6 ± 5.3^{a}	92.1 ± 5.4	83.2 ± 2.1^{a}
	2	75.5 ± 4.7^{a}	68.4 ± 9.3	59.1 ± 1.0^{a}	64.4 ± 4.4
	3	28.0 ± 3.6	47.9 ± 12.0^{ab}	17.6 ± 2.8^{a}	24.5 ± 3.3^{b}
LAP U/mg protein	1	0.13 ± 0.01^{a}	0.11 ± 0.01	0.11 ± 0.01	0.08 ± 0.01
	2	0.54 ± 0.04^{ab}	0.56 ± 0.06	0.74 ± 0.05^{a}	0.74 ± 0.03^{b}
	3	0.34 ± 0.03	0.35 ± 0.04	0.41 ± 0.04	0.36 ± 0.08
γ- GTP U/mg protein (.10-3)	1	4.5 ± 1.0^{abc}	10.8 ± 2.6^{a}	12.1 ± 3.7^{b}	7.4 ± 1.2^{c}
	2	12.4 ± 1.0^{abc}	19.2 ± 0.7^{a}	21.9 ± 1.1^{a}	17.2 ± 1.4^{c}
	3	11.2 ± 0.5	12.4 ± 0.9	15.0 ± 1.1^{a}	10.2 ± 0.8^{a}
Alkaline Phosphatase U/mg protein	1	2.51 ± 0.33^{a}	1.17 ± 0.20^{abc}	2.13 ± 0.29^{b}	2.28 ± 0.20^{c}
	2	0.38 ± 0.06	0.17 ± 0.04	0.32 ± 0.03	0.34 ± 0.07
	3	0.16 ± 0.04^{a}	0.03 ± 0.01^{abc}	0.11 ± 0.01^{b}	0.06 ± 0.02^{c}
Maltase U/mg protein	1	0.77 ± 0.06^{a}	0.62 ± 0.07^{b}	0.81 ± 0.08^{c}	1.21 ± 0.11^{abc}
	2	0.91 ± 0.09^{ab}	1.12 ± 0.08^{c}	1.38 ± 0.09^{ad}	1.61 ± 0.05^{abd}
	3	0.58 ± 0.19	0.65 ± 0.21	0.54 ± 0.07	0.57 ± 0.06

Table 6. Influence of intake of benzamidine on total activity per segment of enzymes in the different parts of the small intestine.

Values (mean ± SEM) with the same letter are significantly different ($p<0.05$).

		Control	Benzamidine 0.15%	Benzamidine 0.30%	Benzamidine 0.40%
LAP U/segment	1	12.1 ± 1.2	12.0 ± 1.7	9.4 ± 0.9	6.3 ± 0.4
	2	40.3 ± 1.8	39.2 ± 4.1	43.4 ± 6.0	47.6 ± 3.8
	3	8.1 ± 1.3	11.8 ± 3.9	7.1 ± 1.1	8.4 ± 2.2
γ- GTP U/segment	1	0.32 ± 0.06^{ab}	--	0.99 ± 0.30^{a}	0.70 ± 0.17^{b}
	2	0.91 ± 0.07^{ab}	1.50 ± 0.16^{bd}	1.34 ± 0.19^{c}	1.11 ± 0.05^{d}
	3	0.45 ± 0.03^{ab}	0.52 ± 0.10^{cd}	0.29 ± 0.04^{ac}	0.24 ± 0.04^{bd}
Phosphatase Alkaline U/segment	1	244 ± 7^{a}	230 ± 15^{b}	201 ± 24	190 ± 21^{ab}
	2	30 ± 3	15 ± 4	18 ± 2	21 ± 3
	3	4 ± 1	1 ± 1	2 ± 1	2 ± 0
Maltase U/segment	1	71 ± 6^{a}	61 ± 8^{b}	68 ± 6^{c}	90 ± 7^{abc}
	2	63 ± 2^{a}	81 ± 6	86 ± 13	109 ± 8^{a}
	3	15 ± 3	13 ± 4	10 ± 2	13 ± 1

DISCUSSION

Consumption of severely alkali-treated casein does not modify significantly the overall retention of chromic oxide in the digestive tract of Sprague Dawley rats. The same alkali-treated casein induces severe and sometimes fatal diarrhea in young Wistar rats (Berger, 1983). Study of intestinal transit using the method described above is not practical with these rats. Sensitivity to severely treated proteins appears to be dependent on the strain of rats.

Analysis of gastrointestinal transit reveals several changes from the normal. In young rats, the amount of feces recovered during the light period is higher. On the other hand, in control rats two peaks of excretion of the marker appear, one at the end of the first dark period, the other one at the beginning of the following dark period. In adult rats receiving alkali-treated casein or benzamidine, an other peak appears between these two, during the light period.

Rats bred in normal conditions, fed ad libitum, ate mainly during the dark period. Food eaten at the beginning of the dark period (2000 hrs) goes through the digestive tract rapidly since the marker appears in feces 6 hours later. During the second part of the dark period, 70 percent of feces excreted are reingested. At the beginning of light period (0800 hrs), fecal excretion decreases and dry matter content in caecum and in colon rises throughout the dark period. Fecal excretion begins quickly again around 2000 hours (Possompes, unpublished).

Coprophagy is a natural phenomenon in rats. It allows improvement of protein digestion but involves also the reabsorption of potentially toxic compounds, i.e. in our experiments, benzamidine and peptides and amino acids resulting from hydrolysis of alkali-treated casein, and not absorbed by the intestinal mucosa.

The percentage of recovered marker at 12 and 24 hours gives an idea of the extent of coprophagy. The time needed for the appearance of the maximal concentration of the marker in the feces is less than 12 hours. The lack of correlation between this time and the percentage of chromic oxide recovered in the first 12 hours is due to coprophagy. Our experiment shows that rats are able to modify their coprophagy. In adult rats receiving treated casein or benzamidine, the pattern of chromic oxide excretion are similar, however, the marker goes through the digestive tract faster with benzamidine than with treated casein (8.4 vs 12 hours). It may be concluded that the trypsin inhibitor and alkali-treated casein do not act upon the digestive tract in the same way. The overall retention time alone, however, does not define this difference. For this reason we preferred to express transit time as the delay needed for the appearance of the maximal concentration of the marker in the feces.

Destruction and isomerization of amino acids by alkaline treatment decrease the nutritional value of the protein. This is partially restored by supplementation, but, in our experiment, the analysis of plasma amino acids shows a lysine deficiency (Possompes et al, 1985). This deficiency coul explain the lack of pancreatic hypertrophy which is observed in rats receivin a diet with 20% protein supplied by treated casein (Possompes, unpublished). Pancreatic hypertrophy occurs in rats receiving a pancreatic proteases inhibitor (Liener, 1981), a finding that is confirmed in rats receiving benzamidine with untreated casein.

As a rule intestinal adaptations affect the length of villi and microvilli (Williamson et al, 1978) which is indicated by the change in mucosal proteins. When treated casein is consumed, mucosal protein content is significantly increased in the terminal ileum and to a lesser degree in the duodenum, but only in animals in an adequate nutritional condition (i.e., with amino acid supplementation). In rats fed parenterally, infusion of nutrients into the ileum induces a hypertrophy of the mucosa in the ileum and also to a lesser extent in the duodenum (Spector et al, 1977); the nutrients do not need to be absorbed (Weser et al, 1982). Due to its low digestibility, alkali-treated casein leads to a significant accumulation of non-absorbed nitrogen compounds in the ileum, hence the high nitrogen concentration in the feces. This accumulation could be the cause of the hypertrophy of the mucosa. Benzamidine does not alter protein digestibility and nitrogen concentration in feces remains low, so ileal hypertrophy is not observed in rat receiving this compound.

Specific activities of enzymes are located along the small intestine according to a gradient that changes with enzymes and species (Laganiere et al, 1984). Maltase and γ-GTP gradients are modified in rats receiving the alkali-treated casein when compared to control rats.

Maltase activity is increased in the jejunum of diabetic rats consuming glucose (Nakabou et al, 1980; Madara et al, 1982). In this case, alkaline phosphatase activity does not change when animals are fed ad libitum, but decreases when their food intake is restricted (Nakabou et al, 1974). Hyperinsulinemia (Berger, 1983) and increased maltase activity in the medial part of the small intestine are observed in rats receiving 0.3 and 0.4% benzamidine in the diet. Adaptation in rats receiving alkali-treated casein is quite different and has certainly other causes.

Leucine aminopeptidase whose activity is evaluated by its hydrolysis of leucine β- naphtylamine, plays an essential role in the hydrolysis of peptides in the brush border membrane (Adibi et al, 1981). It cannot hydrolyze dipeptides containing D amino acids (Kim et al, 1976); however, these dipeptides may be absorbed in the ileum. This enzyme is inhibited by products of the Maillard reaction (Oste et al, 1982). The formation of such compounds cannot be excluded during alkaline treatments of protein where reducing sugars are present. The activity of this enzyme is especially stimulated in the distal ileum of rats receiving alkali-treated casein and in the medial part of the small intestine of rats receiving only benzamidine.

It may be concluded that the effects of severely alkali-treated casein in the gastrointestinal tract are different from those due to benzamidine. These effects do not derive from the inhibitory pancreatic protease properties of peptides isolated from alkali-heated casein hydrolyzates by affinity chromatography (Vimont-Rispoli et al, 1980).

The adaptive mechanisms of the intestine are not well understood and depend on a number of regulatory factors (Williamson et al, 1978). The consumption of treated protein induces an increase in plasma urea (Karayiannis et al, 1979) and modifications of the enzymatic equipment of enterocyte have been described in uremic rats (Sterner et al, 1983). It seems that regulatory systems for the growth rate of intestinal mucosa are different in the proximal and distal parts of the small intestine (Weser et al, 1982). Bile and pancreatic secretions as well as accumulation of undigested compounds have an important part in the induction of such mechanism.

Alterations in gastrointestinal transit of markers and adaptations of brush border enzymes could be explained by the accumulation of undigested nitrogen compounds. Physiological effects of an unknown compound are not necessarily involved here.

References

Abe, K., Arai, H., Homma, S., Fujimaki, M., Arai, S. (1981) Peptide bound lysinoalanine absorbed and transported to the kidney. Observations in a feeding test with rats. Agric. Biol. Chem. 45 (8), 1921-1923.

Abe, K., Homma, S., Fujimaki, M., Arai, S. (1984) Administration of lysinoalanine-containing proteins to rats and the characterization of their small intestinal contents. Agric. Biol. Chem. 48 (3), 573-578.

Abe, K., Suehara, S., Homma, S., Fujimaki, J., Arai, S. (1984) Metabolic transit of lysinoalanine in the rat body: observation of feeding tests with alkali-treated casein and gluten. Agric. Biol. Chem. 48 (3), 579-584.

Abidi, S.A., Kim, Y.S. (1981) Peptide absorption and hydrolysis. In: "Physiology of the gastrointestinal tract", p. 1073-1095, ed. by Leonard R. Johnson, Raven Press, New York.

Alpers, D.H., Tedusco, R.J. (1975) The possible role of pancreatic proteases on the turnover of intestinal brush border proteins. Biochem. Biophys. Acta. 401, 28-40.

Baker, D.H. (1979) Efficacity of D and L-isomers of N-acetylmethionine for chicks fed diets containing either crystalline amino acids or intact protein. J. Nutr. 109, 970-974.

Batt, R.M., Bush, B.M., Peters, T.J. (1979) Biochemical changes in the jejunal mucosa of dogs with naturally occuring exocrine pancreatic insufficiency. GUT 20, 709-715.

Bert, C.P. (1959) Utilization of D amino acids. In: "Proteins and amino acid nutrition", p. 57-96, ed. by Albanese, A.A., Academic Press, New York.

Berger, J. (1983) Effets physiopathologiques dus à l'ingestion de protéines modifiées en milieu alcalin. Etude sur le pancréas endocrine et l'intestin chez le rat. Thèse de 3eme cycle, U.S.T.L., Montpellier.

Bessey, O.A., Lowry, O.H., Brock, M. (1946) A method for the rapid determination of alkaline phosphatase with five cubic millimeters of serum. J. Biol. Chem. 164, 321-329.

Bolin, D.W., King, R.P., Klosterman, E.W. (1952) A simplified method for determination of chromic oxide (Cr_2O_3) when used as an index substance. Science 116, 634-635.

Castle, E.J. (1956) The rate of passage of foodstuffs through the alimentary tract of goat. I. Studies on the adult animal fed on hay and concentrates. Brit. J. Nutr. 10, 15-23.

Dahlquist, A. (1968) Assay of intestinal dissacharidases. Anal. Biochem. 22, 97-107.

Das, M., Radhakrisnan, A.N. (1976) Role of peptidases and peptide transport in the intestinal absorption of protein. Wl. Rev. Nutr. Diet. 24, 58-87.

DeGroot, A.P., Slump, P. (1969) Effects of severe alkali-treatments of proteins on amino acid composition and nutritive value. J. Nutr. 98, 45-56.

Diaz, B.M. (1978) Etude des facteurs toxiques et de la valeur nutritionnelle d'isolats protéiques du haricot (Phaseolus vulgaris L): effets physiologiques chez le rat. Thèse de doctorat de spécialité, U.S.T.L., Montpellier.

Francois, E., Compere, R., Rondia, G. (1968) Etude comparée de la vitesse de passage des aliments et des résidus alimentaires non digérés dans le tractus digestif du rat et du mouton. Bul. Rech. Agron. Gembloux 3 (4), 655-688.

Friedman, M., Zahnley, T.C., Masters, P.M. (1981) Relationship between in vitro digestibility of casein and its content of lysinoalanine and D amino acids. J. Food Sci. 46, 127-131.

Friedman, M., Gros Jean, O-K, Zahnley, J.C. (1985) Metalloenzyme inhibition by lysinoalanine, phenylethylaminoalanine and alkali-treated food proteins. Fed. Proc., 44, 1497. See also, this volume.

Ganote, E.E., Peterson, D.R., Carone, F.A. (1974) The nature of D-serine induced nephrotoxicity. Am. J. Pathol. 77, 269-282.

Gibson, Q.H., Wiseman, G. (1951) Selective absorption of stereoisomers of amino acids from loops of the small intestine of the rat. Biochem. J. 48, 426-429.

Goldbarg, J.A., Rutenburg, A.M. (1958) The colorimetric determination of leucine aminopeptidase in urine and serum of normal subjects and in patients with cancer and other diseases. Cancer 11, 283-291.

Gould, D.H., McGregor, J.T. (1977) Biological effects of alkali-treated protein and lysinoalanine: an overview. In: "Protein Crosslinking Nutritional and Medical Consequences", p. 29-49, ed. by Friedman, M., Plenum Press, New York. London.

Hayashi, R., Kameda, I. (1980a) Racemization of amino acid residues during alkalitreatment of protein and its adverse effect on peptide digestibility. Agric. Biol. Chem. 44 (4), 891-895.

Hayashi, R., Kameda, I. (1980b) Decreased proteolysis of alkalitreated protein: consequences of racemization in food processing. J. Food Sci. 45, 1430-1431.

Hayashi, R. (1982) Lysinoalanine as a metal chelator. J. Biol. Chem. 257 (23), 13896-13898.

Hurrel, F.R., Carpenter, K.J. (1977) Nutritional significance of cross-link formation during food processing. In: "Protein crosslinking, nutritional and medical consequences", p. 225-238, ed. by Friedman, M., Plenum Press, New York.

Karasov, W.H., Diamond, J.M. (1983) Adaptive regulation of sugar and amino acid transport by vertebrate intestine. Am. J. Physiol. 245, G443-G462.

Karayiannis, N., McGregor, J.T. (1976) Lysinoalanine content and biological effects of alkali-treated proteins. 172eme congres of American Chemical Society, San Francisco.

Karayiannis, N., McGregor, J.T., Bjeldanes, L.F. (1979) Lysinoalanine formation in alkali-treated proteins and model peptides. Fd. Cosmet. Toxicol. 17, 585-590.

Kim, Y.S., Brophy, E.J., Nicholson, J.A. (1976) Rat intestinal brush border membrane peptidases. 2. Enzymatic properties, immunochemistry and interaction with lectins of two different forms of the enzyme. J. Biol. Chem. 251, 3206-3212.

Laganiere, S., Berteldot, A., Maestracci, D. (1984) Digestive and absorptive functions along dog small intestine: comparative distributions in relation to biochemical and morphological parameters. Comp. Biochem. Physiol. 79A, 463-472.

Liardon, R., Hurrel, R.F. (1983) Amino acid racemization in heated and alkali-treated proteins. J. Agric. Food Chem. 31, 432-437.

Liener, I.E. (1981) Factors affecting the nutritional quality of soja projects. J. AOCS, 406-415.

Lowry, O.H., Rosebrough, N.J., Farr, A.L., Randall, R.J. (1951) Protein measurements with the folin-phenolreagent. J. Biol. Chem. 193, 265-275.

Madara, J.L., Wolf, J.L., Trier, J.S. (1982) Structural features of the rat small intestinal microvillus membrane in acute experimental diabetes. Digest Dis. Sci. 27 (9), 801-806.

Mares-Guia, M., Shaw, E. (1965) Studies on the active center of trypsin. The binding of amidines and guanidines as models of the substrates side chain. J. Biol. Chem. 240, 1579-1585.

Masters, P.M., Friedman, M. (1979) Racemization of amino acids in alkali-treated food proteins. Agric. Food Chem. 27 (3), 507-511.

Naftalin, P.M., Sexton, M., Whitaker, J.F., Tracey, D. (1969) A routine procedure for estimating γ-glutamyltranspeptidase activity. Clin. Chim. Acta. 26, 293-296.

Nakabou, Y., Okita, C., Takano, Y. (1974) Hyperplastic and hypertrophic changes of the small intestine in alloxan diabetic rats. J. Nutr. Sci. Vitaminol. 20, 227-234.
Nakabou, Y., Ishikawa, Y., Misake, A., Hagihira, H. (1980) Effect of food intake on intestinal absorption and mucosal hydrolazes in alloxan diabetic rats. Metabolism 29 (2), 181-185.
Oste, R., Bjork, I. Dahlquist, A., Jagerstad, M., Sjodin, P., Sjostrom, H. (1983) Effect of glucose-lysine reaction mixture on protein and carbohydrate digestion and absorption in the Maillard reaction. In: "Food and Nutrition", p. 405-419, ed. by Waller, G.R., Reather, M.S., ACS Symposium Series. American Chemical Society, Washington, D.C.
Possompes, B., Vimont-Rispoli, S., Cabon, M., Besancon, P. (1979) Effet des traitements alcalins sévères sur la digestibilité in vivo et l'hydrolyse trypsique in vitro de la caséine. Ann. Nutr. Alim. 33, 539-543.
Possompes, B., Diaz, B., Mahe, Y., Vimont-Rispoli, S., Besançon, P. (1982) Hydrolyse enzymatique de divers isolats protéiques de haricot: comportement des inhibiteurs. Sci. Aliments 2, 169-172.
Possompes, B., Cuq, J.L, Guenoun, P., Besançon, P. (1983) Use of cysteine as a reducing agent during alkali treatment of proteins. Biochemical and nutritional effects. Food Chemistry 11, 15-26.
Possompes, B., Berger, J., Cuz, J.L. (1985) Traitement alcalin et inhibiteur de la trypsine: effet des supplémentations en acides aminés sur la croissance des rats. Journées Sciences des Aliments. Association Française de Nutrition, Paris.
Raul, F., Simon, P.M., Kedinger, M., Grenier, J.F,. Haffen, K. (1980) Effect of sucrose refeeding on dissacharidases and aminopeptidase activities of intestinal villus and crypt cells in adult rats. Evidence for a sucrose-dependent induction of sucrase in the crypt cells. Biochem. Biophys. Acta. 630, 1-9.
Schwartz, D. (1963) Methodes statistiques a l'usage des medecins et des biologistes. Ed. Medicale Flammarion (Paris).
Spector, M.H., Levine, G.M., Deren, J.J. (1977) Direct and indirect effects of dextrose and amino acids on gut man. Gastroenterology 72, 706-710.
Stegink, L.D. (1977) D amino acids. In: "Clinical Nutrition Update: amino acids", p. 198-205, American Medical Assoc., ed. by Greene, M.L., Holliday, M.A., Munro, H.N., Chicago.
Struthers, B.J., Hopkins, D.T., Dahlgreen, R.P. (1978) Reversibility of nephrocytomegaly caused in rats by lysinoalanine (N^{ϵ}-(D-L-2-amino-2-carboxcyethyl)-L-Lysine. J. Food Sci. 43, 616-618.
Vachon, C., Gauthier, S.F., Jones, J.D., Savoie, L. (1982) Enzymatic digestion method with dialysis to assess protein damage: application to alkali-treated protein containing lysinoalanine. Nutr. Res. 2, 675-688.
Vimont-Rispoli, S., Possompes, B., Besançon, P. (1980) Mise en évidence de la formation d'inactivateurs de la trypsine au cours des traitements alcalins des protéines. C.R. Acad. Sci. Paris 291, 945-948.
Weser, E., Heller, R., Tawil, T. (1977) Stimulation of mucosa growth in the rat ileum by bile and pancreatic secretions after jejunal secretion. Gastroenterology 73, 524-529.
Weser, E., Bell, D., Tawil, T. (1981) Effects of octapeptide-cholecystokinin, secretion and glucagon on intestinal mucosa growth in parenterally nourished rats. Digest. Dis. Sci. 26 (5), 409-416.
Weser, E., Vandeventer, A., Tawil, T. (1982) Non-hormonal regulation of intestinal adaptation. Scand. J. Gastroenterol. 17 (suppl. 74) 105-113.
Williamson, R.C.N., Chir, M. (1978) Intestinal adaptation: structural, functional and cytokinetic changes. New England J. Med. 298 (25), 1393-1402.
Williamson, R.C.N., Chir, M. (1978) Intestinal adaptation: mechanisms of control. New England J. Med. 298 (25), 1444-1450.
Winne, D., Koff, S., Ulmer, M.L. (1979) Role of unstirred layer in intestinal absorption of phenylalanine in vivo. Biochem. Biophys. Acta. 550, 120-130.

Woodard, J.C., Short, D.D. (1973) Toxicity of alkali-treated soy protein in rats. J. Nutr. 103, 569-574.
Yasumoto, K., Sugiyama, K., Mitsuda, H. (1977) Inhibition of amino acid transport in rat small intestine by N ε- substituted-L-lysine derivatives. Agric. Biol. Chem. 41 (1), 195-200.
Zierler, K.L. (1958) A simplified explanation of the theory of indicator-dilution for measurement of fluid flow and volume and other distributive phenomena. Johns Hopkins Medical Journal. 103, 199-217.

31

INACTIVATION OF METALLOENZYMES BY LYSINOALANINE, PHENYLETHYLAMINOALANINE, ALKALI-TREATED FOOD PROTEINS, AND SULFUR AMINO ACIDS

Mendel Friedman, Ok-Koo Grosjean, and James C. Zahnley

Western Regional Research Center, Agricultural Research Service, U. S. Department of Agriculture
800 Buchanan Street, Berkeley, California 94710

ABSTRACT

Synthetic lysinoalanine (LAL) may be a more effective inhibitor of the zinc-containing enzyme carboxypeptidase A than is ethylenediamine tetraacetic acid (EDTA). The enzyme is also inactivated by alkali-treated, lysinoalanine-containing food proteins such as casein, high-lysine corn protein, lactalbumin, soy protein isolate, and wheat gluten, and by alkali-treated zein, which contains no lysinoalanine. Zinc sulfate regenerates only part of the enzymatic activity after exposure to the treated proteins. The extent of inhibition increases with protein concentration and time of treatment. Any inhibition due to phytate is distinct from that due to the treatment. Phenylethylaminoalanine (PEAA), derived from biogenic phenylethylamine, inhibited enzymatic activity of the metalloenzyme carboxypeptidase A (CPA). The inhibition was maximal at pH 7.0 in the pH range 7 to 8.5. The extent of inhibition increased with time of treatment and PEAA concentration. N-acetyl-PEAA did not inhibit the enzyme, suggesting that the free α-NH_2 group is required for inhibition. PEAA, LAL, sodium phytate, and cysteine also inactivated the copper enzyme, polyphenol, oxidase (tyrosinase) which plays a major role in enzymatic (oxidative) browning of foods. Analogous comparative studies with LAL, EDTA, and sodium phytate suggest that the potency of PEAA as an inhibitor of CPA is similar to that of sodium phytate, and that of the four compounds tested, PEAA is least effective against tyrosinase. Related studies of the iron and copper containing enzyme cytochrome C oxidase showed that EDTA was not inhibitory, PEAA was slightly inhibitory, and LAL and sodium phytate were stronger inhibitors. Mechanistic explanations are offered to account for some of these observations. The possible relevance of these findings to <u>in vivo</u> protein digestion, enzymatic (oxidative) browning of foods, and the mechanism of the lysinoalanine effect on kidney cells are also discussed.

INTRODUCTION

Food processing conditions that use heat and alkali may result in compositional changes including formation of lysinoalanine (LAL) and phenylethylaminoalanine (PEAA) crosslinks and racemization of L-amino acid residues to D-isomers (Masters and Friedman, 1979; Friedman et al.,

1981; Jones et al., 1981; Liardon and Hurrell, 1983; Tucker et al., 1983; Friedman et al., 1984 a,b; Friedman and Noma, 1986).

Feeding proteins containing LAL or D-serine to rats induces changes in kidney cells. These changes are characterized by enlargement of the nucleus and cytoplasm, increased nucleoprotein content, and disturbances of DNA synthesis and mitosis (Woodard et al., 1975; Gould and MacGregor, 1977; Friedman et al., 1984 a,b; Pfaff, 1984, 1985; Ganote et al., 1974). The molecular mechanism of the observed cellular action is still not well understood. One possibility, however, is that since LAL contains three amino and two carboxyl groups and structurally resembles ethylenediamine tetraacetic acid (EDTA), a well known metal chelator, LAL could chelate essential trace elements in vivo (Friedman, 1977). This prediction was partly confirmed by Hayashi (1982), who found that synthetic LAL has a strong affinity for metal ions and inactivates metalloenzymes such as carboxypeptidase A and B and alcohol dehydrogenase in vitro. These observations raise the question whether LAL-containing food proteins, in contrast to free LAL, also inhibit metalloenzymes.

The main objectives of our studies were to find out (a) whether LAL and PEAA can inhibit zinc-, copper-, and iron- containing metalloenzymes such as carboxypeptidase A (CPA), alkaline phosphatase, polyphenol oxidase, and cytochrome C oxidase; (b) the relative effectiveness of PEAA, LAL, EDTA, and sodium phytate in inhibiting the three enzymes; (c) whether alkali-treated food proteins inhibit the activity of the zinc-containing metalloenzyme, carboxypeptidase A (CPA); (d) whether the inhibition is related to the LAL content of the proteins; and (e) whether the inhibitions can be reversed by metal ions.

EXPERIMENTAL SECTION

Reagents. Carboxypeptidase A (bovine, crystalline), alkaline phosphatase, polyphenol oxidase (from mushrooms), cytochrome C oxidase, β-alanine, hippuryl-L-phenylalanine, p-nitrophenyl phosphate, L-tyrosine, and Tris base were from Sigma, St. Louis, MO. Phenylethylamine and N-acetyl-dehydroalanine were from Aldrich Chemical Co, Milwaukee, Wisconsin. Copper and zinc sulfate were from Fisher, Fair Lawn, New Jersey. Casein, lactalbumin, soy protein isolate, and wheat gluten were from U.S. Biochemical Corp., Cleveland, Ohio. High lysine corn was from Crow's Hybrid Corn Co., Milford, Illinois. High lysine corn protein isolate was prepared from the corn seeds as previously described (Friedman et al., 1985). Synthetic PEAA was prepared by an adaptation of the procedure described by Jones et al. (1981) as follows:

3-(N-β-Phenylethylamino)-N-acetyl-DL-alanine (acetyl-PEAA). With stirring, 85.4 ml of phenylethylamine was slowly added to 17.7 g of N-acetyl-dehydroalanine dissolved in 136 ml of H_2O in a 500 ml Erlenmeyer flask. The resulting clear solution was then placed in a 50°C water bath for 72 hr. Next, the excess amine was removed by rotary evaporation, repeated washing with water, and further evaporation until the amine odor was imperceptible. The remaining yellow oil was triturated with 200 ml chilled acetone to precipitate the amino acid derivative. The flask was kept in the freezer for 16 hr, the precipitate filtered, and the filtrate recrystallized from a small amount of hot methanol. The yield was 13.0 g. The NMR spectrum, determined in trifluoroacetic acid (TFA) with tetramethylsilane (TMS) as a reference, was consistent with the assigned structure.

3-(N-β-Phenylethylamino)-DL-alanine monohydrochloride (PEAA). Three grams of acetyl-PEAA was dissolved in 105 ml 2 N HCl in a 500 ml round-bottom flask. The solution was refluxed for 3 hr; then excess HCl was removed by rotary evaporation. The product was kept overnight in a freezer and recrystallized from 95% ethanol. The yield of PEAA was 1.2 g. The NMR spectrum in TFA was consistent with the assigned structure. Elemental analysis showed that the compound crystallized as a monohydrochloride.

Alkali treatment. Five grams of soy protein was suspended in 500 ml 0.1 N NaOH in a 1 L Erlenmeyer flask. Initial pH was checked. The flask was then stoppered with a Nalgene cap, covered with aluminum foil, and kept in a water bath at the appropriate temperature for various time periods listed in the Tables. At the end of the treatment, the flask was cooled under running water and the final pH was measured. The pH values of the protein suspensions ranged from 12.5 to 12.7 at 25°C, which corresponds to an effective pH near 11.5 at 75°C (Masters and Friedman, 1980; Friedman and Liardon, 1985). The material was then placed in Spectra/Por 4 dialysis tubing, molecular weight cut off 12,000 (Spectrum Medical Industries, Los Angeles, California) and dialyzed against dilute acetic acid (0.25%) for the first day, and thereafter against distilled water for another two days. The samples were then lyophilized.

Autoclaved soy protein. One hundred milligrams of soy protein in a 50-ml Erlenmeyer flask was heated in an autoclave at 121°C for 10 min. Assay of inhibitory activity against CPA was carried out in the flask.

Denatured soy protein. One gram of soy protein was suspended in 10 ml of water and boiled for 2 min. Inhibitory activity against CPA was assayed with the cooled suspension.

Amino acid analysis. Amino acid analyses were carried out on a single-column Durrum 500 amino acid analyzer, as previously described (Friedman et al., 1981). In this system, LAL elutes as a single peak just before histidine.

Phytate analysis. Low-phytate soy protein and soy flour were prepared as described by de Rham and Jost (1979). Phytic acid content of native and alkali-treated soy proteins and soy flours was measured by an HPLC procedure described by Knuckles et al. (1985).

Carboxypeptidase assay. CPA activity was assayed by the method of Folk and Schirmer (1963) using hippuryl-L-phenylalanine as substrate. The substrate (3.26 mg) was first dissolved in 100 μL of 95% ethanol and diluted to 10 ml in pH 7.5 Tris-HCl (0.025 M containing 0.5 M NaCl). Next, 2.9 ml of the substrate solution was pipetted into a cuvette and incubated at 25°C for 3 min in a Cary 14 spectrophotometer with temperature control. To the substrate solution was then added 0.1 ml of the enzyme solution, prepared by dissolving 10 μM CPA in 2 ml of 10% LiCl. The resulting increase in absorbance at 254 nm was followed for 1 min. Enzyme activity was calculated as follows:

$$\text{Units (U) of activity/mg} = \frac{\text{A254/min}}{0.36 \times \text{mg enzyme/ml reaction mixture}}$$

The undiluted stock solution contained 0.88 U enzyme/μL, or 47.37 U/mg protein.

Inhibitors. LAL solutions (5mM) were prepared by dissolving 3.1 mg samples in 2 ml each of 0.025 M Tris-HCl buffers, pH 6.5, 7.0, 7.5, 8.0, and 8.5, containing 0.5 M NaCl. To each of these solutions was then added 100 μL of CPA stock solution, and the mixtures were left standing at room temperature for 3 minutes. Then, 100 μL aliquots were added to 2.9 ml of 0.5 or 1 mM substrate in pH 7.5 Tris buffer for assay of enzymatic activity. Identical experiments were carried out with EDTA (3.7 mg/2 ml) except that the CPA plus EDTA solutions were left standing for 60 minutes.

Effect of proteins. One hundred milligrams of ground protein was dissolved or suspended in 2 ml Tris buffer at pH 7.5. The solution or suspension was then added to 10 μL of undiluted enzyme solution and the mixture shaken for various periods, usually 35 min, in a 25°C water bath. The reaction was terminated by immersing the flasks in crushed ice for 5 min and then centrifuging them at 0°C for 5 min in a Beckman micro B centrifuge. Enzymatic activity was determined with the supernatant solution.

The influence of $ZnSO_4$ was studied by replacing 1 ml of the buffer with 1 ml of 1 mM $ZnSO_4$. Additional details are given in the footnotes to the Tables.

Alkaline phosphatase assay. Either 21 or 42 μmoles PEAA was dissolved in 1 ml of 0.75 M β-alanine with 50 mM NaCl, pH 10.2. Next, 50 μL of the enzyme stock solution (1 mg/ml) was added and the mixture was incubated at 37°C for 1 hr. Alkaline phosphatase activity was determined as described in Worthington (1982).

Polyphenol oxidase assay. The assay was carried out as described in Worthington (1982) with some modifications as follows:
One milligram of tyrosinase containing 2000 U was dissolved in 1 ml of deionized, distilled water and diluted to a concentration of 400 U/ml H_2O. One milliliter of 0.5 M phosphate buffer, pH 6.5, 1 ml of 1 mM L-tyrosine, and 0.9 ml of distilled, deionized H_2O were mixed and oxygenated by bubbling oxygen into the solution through a capillary tube for 5 min. The oxygen-saturated substrate solution was then transferred into a cuvette, the cuvette was placed into the spectrophotometer, and the A_{280} values were recorded for 5 min at 25°C until equilibrium was reached. To the substrate solution was then added 100μL of the enzyme solution and the A_{280} values were recorded for 12 min. Enzyme activity was then calculated by:

$$\text{Activity (U/mg)} = \frac{A_{280}/\text{min} \times 1000}{\text{mg enzyme in reaction mixture}}$$

Effect of inhibitors. An appropriate amount of the inhibitor was mixed with 400 U of the enzyme in 1 ml H_2O and the mixture incubated for 15 or 60 min at 37°C. At the end of the reaction, 100 μL of the solution was pippetted into 2.9 ml of oxygen-saturated substrate solution and the remaining activity of the tyrosinase was determined as described above.

Cytochrome C oxidase assay. The assay was adapted from Hodges and Leonard (1974). Cytochrome C (0.91 mM or 11.3 mg/ml in 0.05 M phosphate buffer, pH 7.5) was reduced with a few crystals of $NaBH_4$. The solution was then neutralized with dilute HCl and saturated with nitrogen gas for 1 min. The extent of reduction was estimated by measuring the A_{550}/A_{565} ratio from a spectrum of 50 μL reduced cytochrome C diluted in 2.95 ml phosphate buffer. The rate of oxidation of reduced cytochrome C was then determined spectrophotometrically as a decrease in A_{550}. Specifically, 2.7 ml of phosphate buffer and 0.2 ml of reduced cytochrome C (final concentration, 60 μM) were mixed in a 3 ml cuvette and the absorbance at 550 nm was measured against a buffer blank. Then, 0.1 ml of enzyme was added, the solution mixed, and the absorbance recorded for 2 min.

Effect of inhibitors. One milligram of enzyme and an appropriate amount of one of the four potential inhibitors (1.53 or 7.65 mg LAL; 16.5 mg sodium phytate; 7.05 mg phenylethylaminoalanine; 9.30 mg EDTA) were dissolved in 1 ml of phosphate buffer and the mixture was incubated at 37°C for 1 hr. At the end of this period, 0.1 ml of enzyme-inhibitor solution was added to the cuvette and the activity of the enzyme determined as described earlier. Calculations (based on Smith, 1957) were:

$$\text{Cytochrome C oxidized (in } \mu\text{mol)} = \Delta A_{550} (\text{Volume}/\Delta \in) = \Delta A_{550} \times 0.162, \text{ where volume} = 3 \text{ ml}; \Delta \in = \in_{\text{Cyt C (reduced)}} - \in_{\text{Cyt C (oxidized)}} = 18.5 \text{ mM}^{-1} \text{ cm}^{-1}.$$

Trypsin assay. Trypsin activity in the presence of alkali-treated (0.1 N NaOH; 75°C; 3 hr) soy protein and casein was assayed as follows: 300 mg of the protein was dissolved in 15 ml 0.5 M Tris buffer, pH 8.5. The solution was then diluted with 0.05 M Tris, pH 8.2, containing 0.02 M $CaCl_2$ to a concentration of 1 mg protein/0.5 ml. Duplicate test tubes containing 0.5 ml of the protein solution, 25 μg of trypsin and 3 ml of the substrate (BAPNA) were then incubated in a 37°C water bath for 10 min. Trypsin activity was measured as described previously (Friedman et al., 1982). Under these conditions, the treated soy protein did not affect trypsin activity and the treated casein lowered it by about 5%.

RESULTS AND DISCUSSION

Inactivation of Carboxypeptidase A

Figure 1 illustrates the formation of LAL and PEAA in a food or other protein. The amino groups of lysine or phenylethylamine participate with the double bond of dehydroalanine in nucleophilic addition reactions to form the corresponding protein adducts. Hydrolysis with HCl produces free LAL and PEAA. Since a new asymmetric carbon is created during the addition, LAL consists of a mixture of two isomers (L-L and L-D) and PEAA consists of an equimolar mixture of D- and L-enantiomers.

Effect of free lysinoalanine. Table 1 shows that (a) 5 mM LAL completely inhibited CPA activity in the pH range 6.5 to 7.0; (b) inhibition at pH 8.0 requires about twice the amount of LAL and depends on substrate concentration; (c) LAL does not inhibit CPA at pH 8.5; and (d) LAL may be a more potent inhibitor than EDTA in the pH range 6.5-7.5 since additional rate studies revealed that it takes about 60 minutes for EDTA to completely inactivate the enzyme compared to only about 3 minutes for LAL.

These results and additional studies on the concentration-dependence of the inactivation suggest that inhibitory effectiveness of LAL against CPA decreases with increasing pH and that inhibition requires a threshold concentration of LAL of about 5 mM at 2.6 μM CPA. They also support the suggestion that LAL-induced cytomegaly in the rat kidney

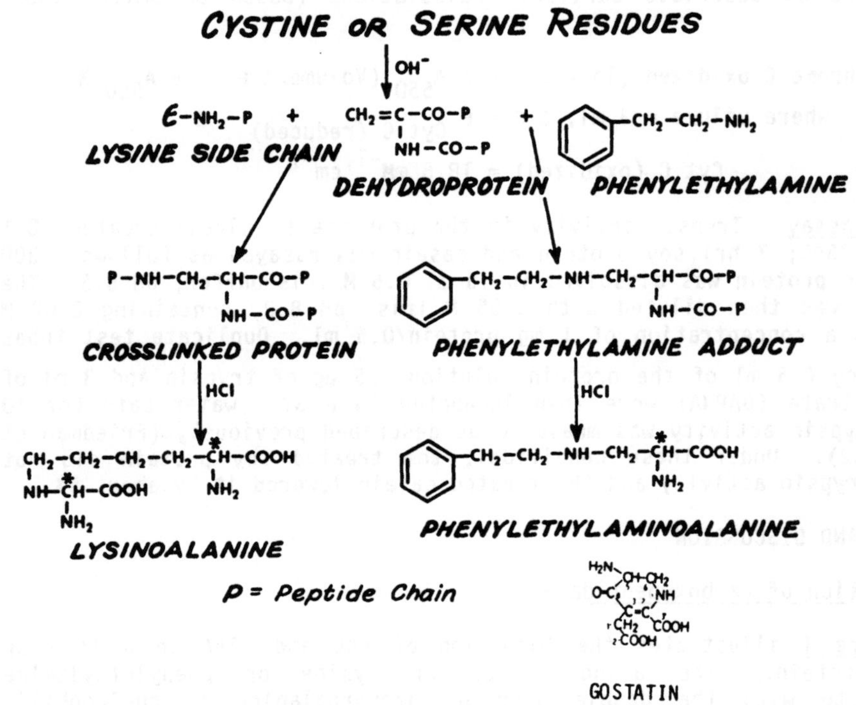

Figure 1. Transformation of cystine or serine residues in a protein to lysinoalanine and phenylethylaminoalanine side chains. Newly created asymmetric centers are indicated by asterisks. The structure of a naturally occuring enzyme inhibitor, gostatin (Nishino and Murao, 1983), is included for comparison.

could be due to binding of LAL or LAL-containing peptides to structural and functional metalloproteins in the kidneys cells (Friedman, 1977).

Table 1. Effect of lysinoalanine and EDTA on carboxypeptidase activity[a]

Substrate conc, mM	% Carboxypeptidase inhibited at pH 6.5	7.0	7.5	8.0	8.5
		LAL			
0.5	100	100	100	0.0(11.5)	0.0(0.0)
1.0	100	100	100	0.0(98.6)	0.0(0.0)
		EDTA			
0.5	100	100	100	68.6(100)	16.1(61.5)
1.0	100	100	100	71.3(100)	36.3(68.7)

[a] Listed values were obtained with 5 mM LAL or EDTA except those in parentheses, which were obtained with 10 mM inhibitor concentration.

Effect of protein-bound lysinoalanine. Two sets of experiments were carried out to assess the influence of protein-bound lysinoalanine on inhibition of CPA by alkali-treated soy protein. In the first, alkali treatment was carried out at 10°C intervals in the temperature range 25-95°C. The LAL content of these samples ranged from about 2 to 10 mmoles/100 g (Table 2). In the second, the time of treatment varied from 10 to 480 min. The LAL content of these samples ranged from about 3 to 12 mmoles/100 g.

Table 2 shows that the inhibition of CPA by both sets of samples increased with LAL content, reaching a maximum at about 10 mmoles LAL/100 g protein. Significant inhibition occured even with relatively low levels (2 to 3 mmoles/100 g) of LAL.

The results of the second set of experiments, summarized in Table 3, show that different amounts of LAL are produced when structurally different proteins are exposed to the same alkaline conditions. For casein, the alkaline treatment (0.1 N NaOH; 65°C; 3 hr) generated 17.4 mmoles of LAL per 100 g of protein. The corresponding value for lactalbumin was 18.0; for soybean protein, 7.9; for wheat gluten, 4.4; for high-lysine corn protein, 4.25; and for zein, 0. The data in this table also show no direct relationship between the LAL contents of these proteins and their extent of inhibition of CPA. For example, alkali-treated wheat gluten, with a low LAL content, completely inhibited CPA, whereas alkali-treated lactalbumin, with a high LAL content, inhibited only 57.5% of the enzyme activity.

Effect of heat denaturation. Using autoclaved or heated soy protein had no or only a minor influence on inhibitory activity against CPA (results not shown). This result implies that the inhibition is not caused by heat-induced conformational changes in the macrostructure of the treated protein.

Effect of protein concentration. To assess the influence of protein concentration on the extent of inhibition of CPA by alkali-treated soy protein, four experiments were carried out with concentrations varying from 25 to 200 mg. The results in Table 4 show that the extent of inhibition was 48.3% with 25 mg of protein, 75.6% with 50 mg, 88.9% with 100 mg, and 92.4% with 200 mg. All of the values were higher than those observed with the untreated (21.5%) and dialyzed (24.2%) controls. Since 100 mg of alkali-treated soy protein produced nearly maximal inhibition, this amount was used in all further studies.

Table 2. Relationship between lysinoalanine content and carboxypeptidase-inhibitory activities of alkali-treated soybean proteins

Treatment	Carboxypeptidase A inhibited, % Without $ZnSO_4$[a]	With $ZnSO_4$[b]	Lysinoalanine content (mmoles/100) g
Untreated protein	21.5 ± 0.20[c]	13.7 ± 3.66	0
Dialyzed protein	24.2 ± 0.80	17.6 ± 1.71	0
1% protein; 0.1 N NaOH; 3 hrs			
25°C	60.3 ± 1.02	7.25 ± 2.05	2.05
35°C	75.4 ± 2.04	8.71 ± 2.04	3.24
45°C	75.4 ± 0.00	27.2 ± 2.60	3.95
55°C	89.9 ± 0.00	34.8 ± 2.68	6.09
65°C	89.9 ± 0.00	48.1 ± 0.00	7.90
75°C	88.6 ± 1.79	58.2 ± 1.80	9.33
85°C	100.0 ± 0.00	76.8 ± 0.98	10.73
95°C	100.0 ± 0.00	89.0 ± 0.00	9.69
1% protein; 0.1 NaOH; 75°C			
10 min	34.4 ± 0.97	12.4 ± 0.97	3.22
30 min	61.4 ± 0.00	15.9 ± 1.95	6.78
60 min	79.6 ± 0.99	33.8 ± 2.00	6.65
180 min	88.9 ± 0.93	49.0 ± 1.85	9.63
300 min	93.1 ± 1.72	57.3 ± 0.00	10.52
480 min	100.1 ± 0.00	66.2 ± 2.00	12.09

[a] The protein and enzyme were incubated for 35 minutes.

[b] The protein and enzyme were incubated for 5 minutes, zinc sulfate was added, and incubation continued for another 30 minutes.

[c] Average from two separate determinations ± standard deviation.

Effect of time. To assess the influence of time of exposure on the extent of inhibition of CPA by alkali-treated soy protein, a series of experiments were carried out with reaction times of 10 to 120 min. The results in Table 5 suggest that inactivation takes place relatively rapidly since time of exposure had a relatively minor influence on inhibition (which ranged from 85 to 94%). A 35-minute time period was adopted as standard for most of the experiment.

Effect of phytate. Since soy products contain phytic acid, which is known to chelate zinc and other transition metal salts (Clydesdale and Camire, 1983; Graf and Eaton, 1984), we tested the influence of phytate removal on the inhibitory activities of both soy protein isolate and soy flour. Table 6 shows that (a) subjecting commerical soy protein isolate to the phytate removal procedure of de Rham and Jost (1979) lowered the phytate content from 0.81 to 0.68%; (b) the corresponding values for soy flour were 2.62 and 0.48%; (c) phytate removal lessened the inhibition of CPA from 23.8 to 1.5% for the protein and from 28.7 to 14.7% for the soy flour; (d) alkali treatment of either product increased inhibition to 88.9% for the soy protein and 100% for the soy flour; (e) the presence of zinc sulfate prevented the inhibitory action of native soy products; and (f) the presence of zinc sulfate only partly prevented the inhibition by alkali-treated soy protein but did not prevent that by treated soy flour.

Table 3. Inhibition of carboxypeptidase A by structurally different native and alkali-treated food proteins[a]

Protein	Carboxypeptidase inhibited, % Without zinc sulfate[b]	With zinc sulfate[c]	Lysinoalanine content (mmoles/100g)
Casein, native	28.3 ± 0.0[d]	31.8 ± 0.0	0.0
Casein, alkali-treated	50.3 ± 1.63	12.5 ± 1.85	17.4
Lactalbumin, native	16.7 ± 0.76	4.58 ± 0.78	0.0
Lactalbumin, alkali-treated	57.5 ± 0.76	36.0 ± 2.99	18.0
Soy protein, native	21.5 ± 0.20	13.7 ± 3.66	0.0
Soy protein, alkali-treated	89.9 ± 0.00	48.1 ± 0.00	7.90
Wheat gluten, native	23.7 ± 1.65	16.8 ± 1.63	0.0
Wheat gluten, alkali-treated	100.0 ± 0.0	85.0 ± 1.63	4.40
Corn protein, high lysine, native	17.6 ± 2.61	14.5 ± 0.60	0.0
Corn protein, high lysine, alkali-treated	62.4 ± 7.45	56.7 ± 1.69	4.85
Zein, native	16.0 ± 1.33	15.5 ± 1.44	0.0
Zein, alkali-treated, 65°	21.4 ± 0.97	24.9 ± 1.44	0.0
Zein, alkali-treated, 75°	50.3 ± 1.44	34.5 ± 0.72	0.0
Zein, alkali-treated, 85°	71.0 ± 0.81	57.4 ± 0.81	0.0

[a] Alkali-treatment: 1% protein; 65°C; 3 hrs except for zein which was also treated at 75 and 85°C.

[b] The protein and enzyme were incubated for 35 min.

[c] The protein and enzyme were incubated for 5 min, zinc sulfate was added, and incubation was continued for another 30 minutes.

[d] Averages from two separate determinations ± standard deviation.

Table 4. Effect of concentration of alkali-treated and untreated soybean protein on the inhibition of carboxypeptidase A[a]

Protein concentration (mg)	Carboxypeptidase A inhibited, % Without $ZnSO_4$[b]	With $ZnSO_4$[c]
25	48.3 ± 2.07[d]	21.5 ± 0.82
50	75.6 ± 1.65	23.2 ± 1.64
100	88.9 ± 0.93	49.0 ± 1.85
200	92.4 ± 0.83	52.9 ± 2.47
Untreated control, 100 mg	21.5 ± 0.20	13.7 ± 3.66
Dialyzed control, 100 mg	24.2 ± 0.80	17.6 ± 1.71

[a]Conditions of alkali-treatment: 1% protein; 75°C; 3 hours.

[b]The protein and enzyme were incubated for 35 minutes.

[c]The protein and enzyme were incubated for 5 minutes, zinc sulfate was added, and incubation was continued for another 30 minutes.

[d]Average from two separate determinations ± standard deviation.

Table 5. Effect of time on the inhibition of carboxypeptidase A activity by alkali-treated soybean protein[a]

Time (min)	Carboxypeptidase inhibited, %	
	Without $ZnSO_4$	With $ZnSO_4$[b]
10	85.0 ± 0.92[c]	41.2 ± 1.85
20	86.9 ± 1.85	42.5 ± 1.85
30	88.9 ± 0.93	49.0 ± 1.85
60	92.2 ± 0.00	47.0 ± 1.02
120	94.2 ± 0.93	58.8 ± 2.77

[a]Conditions of alkali-treatment: 1% protein; 75°C; 3 hours.

[b]Incubation with $ZnSO_4$ was for an additional 30 minutes after the indicated exposures without $ZnSO_4$

[c]Average from two separate determinations ± standard deviation.

Table 6. Effect of phytate and alkali-treatment on carboxypeptidase A activities of soy products[a]

Soy product	Carboxypeptidase A activity inhibited, %		
	Without $ZnSO_4$[b]	With $ZnSO_4$[c]	Phytate content, %
Native-soy protein	23.8	0.90	0.81
Low-phytate protein	1.5	0.90	0.68
Alkali-treated soy protein	88.9	49.0	0.83
Native soy flour	28.7	2.9	2.62
Low-phytate soy flour	14.7	0.0	0.48
Alkali-treated soy flour	100.0	100.0	3.94
Sodium phytate, 4 mg[d]	0.0		
Sodium phytate, 40 mg[e]	19.1		

[a]Conditions of alkali-treatment: 1% protein or flour; 0.1 N NaOH; 75°C; 3 hours.

[b]The soy protein or flour was incubated with the enzyme for 35 minutes.

[c]The soy protein or flour was incubated with the enzyme for 5 minutes, zinc sulfate was added, and incubation was continued for another 30 minutes.

[d]4 mg of sodium phytate and 10 μL CPA stock enzyme solution were incubated at 25°C for 1 hr.

[e]As above, except used 40 mg sodium phytate.

Additional evidence that protein-bound LAL, in contrast to the free amino acid, may not be directly involved in the mechanism of inhibition comes from observations with the corn protein, zein. This protein has no lysine and therefore does not form LAL during exposure to alkali. Nevertheless, the data in Table 3 show that exposure of zein to 0.1 N NaOH for 3 hr at 65, 75, or 85°C increased the inhibitory capacity from 16.0% for the untreated protein to 21.4, 50.3, and 71.0%, respectively. These results strikingly demonstrate the complex nature of the inhibitory process.

Table 7. Effect of zinc sulfate on inactivation of carboxypeptidase A (CPA) activity by phenylethylaminoalanine (PEAA)

PEAA: zinc sulfate molar ratio	Carboxypeptidase A inhibited, %			
	CPA + PEAA (no Zn^{+2})[a]	CPA + PEAA + Zn^{+2} (simultaneous)[b]	CPA + PEAA + Zn^{+2} (sequential)[c]	PEAA + Zn^{+2} + CPA (preincubation)[d]
1: 0	60.2 ± 1.44[e]			
1: 0.05		23.1 ± 0.0	90.4 ± 0.0	26.2 ± 1.41
1: 0.10		33.9 ± 1.22	82.4 ± 0.62	46.8 ± 1.21
1: 0.2		84.5 ± 0.67	90.0 ± 0.66	70.0 ± 0.65
1: 1		100 ± 0.0	100 ± 0.0	100 ± 0.0

[a] 1 ml of the PEAA solution (21 mM) was incubated with 5 μL of undiluted CPA for 1 hr; pH 7.0; 37°C.

[b] CPA, PEAA, and zinc sulfate incubated simultaneously for 1 hr; pH 7.0; 37°C.

[c] CPA and PEAA were incubated for 1 hr followed by addition of zinc sulfate and incubation for another hr; pH 7.0; 37°C.

[d] PEAA and zinc sulfate were incubated together for 10 min followed by addition of 5 μL CPA and incubation for another hr; pH 7.0; 37°C.

[e] Average ± standard deviation from duplicate determinations.

Effect of zinc sulfate. The effect of adding of $ZnSO_4$ was investigated to ascertain whether the inhibition is the result of chelation or absorption of the zinc atom that is part of the active site of CPA (Chlebowski and Coleman, 1976). Since concurrent or sequential addition of zinc sulfate prevented or reversed the inhibition of metalloenzymes by LAL (Hayashi, 1982), it was of interest to establish whether this process also operates with alkali-treated proteins. Tables 2-7 and Figure 2 show that zinc sulfate minimized but did not completely prevent inhibition of CPA by all of the alkali-treated food proteins tested.

Zinc sulfate alone seems to have an inhibitory effect on the enzyme, since exposure of 5 µL CPA of stock to 0.5 ml of 1 mM zinc sulfate lowered CPA activity by about 6%. The listed values were not corrected for this effect. Since the reversible inhibition by free LAL could involve abstraction or removal of the zinc moiety from the active site of the metalloenzymes, followed by restoration of activity after readdition of zinc to the active sites, our studies imply that free LAL and the alkali-treated proteins may operate by different mechanisms.

Inhibition by phenylethylaminoalanine, sodium phytate, and EDTA. The influence of several parameters expected to govern the inhibition of CPA by PEAA was explored to define the scope of the reaction.

Table 8 summarizes the extent of inhibition of CPA by PEAA in Tris buffers of pH 7.0, 7.5, 8.0, and 8.5 at 25 and 37°C, respectively. Since inhibition within the range studied was highest at pH 7, this pH at 37°C was adopted as standard for most of the other studies.

Table 8. Inactivation of carboxypeptidase A with phenylethylaminoalanine[a]

Temperature (°C)	pH	Carboxypeptidase A inhibited, %
25	7.0	83.7
25	7.5	43.5
25	8.0	38.0
25	8.5	28.3
37	7.0	100.0
37	7.5	100.0
37	8.0	79.6
37	8.5	61.2

[a]Conditions: 84 mM PEAA ad 5 µL (5 IU) CPA/ml; 1 hr. No inhibition was observed in enzyme controls without PEAA.

A series of experiments with reaction times of 30 to 480 min was carried out to assess the influence of time on the extent of inhibition of CPA by PEAA. Table 9 shows that under the cited conditions, inhibition increased continuously with time, ranging from 54.3% after 30 min exposure to 97.2% after 480 min. For convenience, a 60 min exposure was adopted as standard for most of the experiment.

Five experiments were carried out to assess the influence of PEAA concentration on the extent of inhibition of CPA, with concentrations of PEAA ranging from 0.84 to 84 mM. Table 10 shows that the extent of inhibition markedly increases with increased PEAA concentration, reaching a maximum at about 42 mM PEAA.

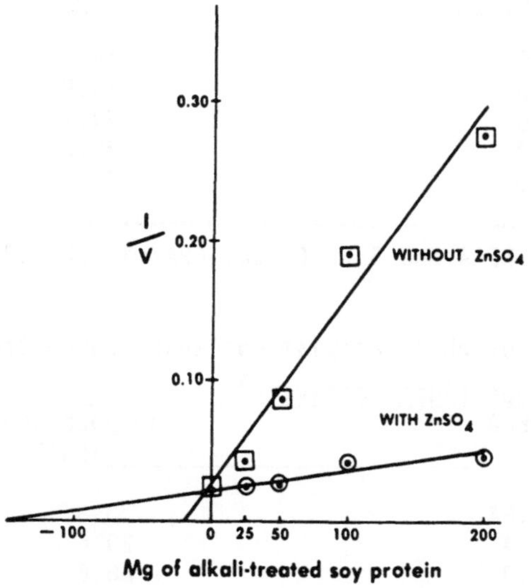

Figure 2. Plot of the reciprocal velocity (1/v) against protein concentration for the inhibition of carboxypeptidase A by alkali-treated soy protein in the presence and absence of zinc sulfate.

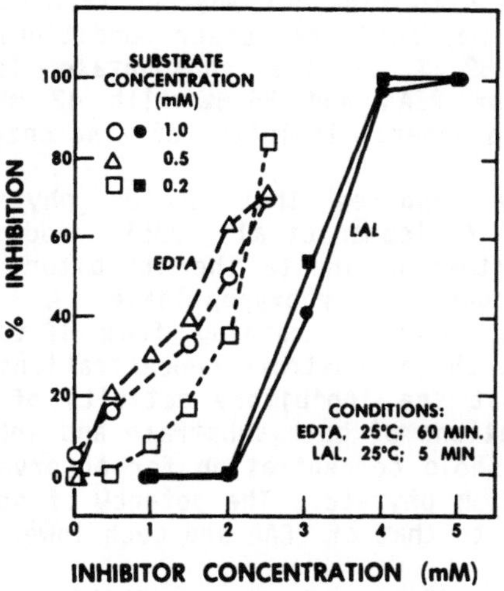

Figure 3. Inhibition of carboxypeptidase A activity by ethylenediamine tetraacetic acid (EDTA) and lysinoalanine (LAL) as a function of inhibitor and substrate concentrations.

Table 9. Effect of time of incubation on carboxypeptidase inhibition by phenylethylaminoalanine[a]

Time (min)	Carboxypeptidase A inhibited, %
30	54.3
60	64.9
120	77.1
240	89.8
480	97.2

[a]Conditions: 21 mM of PEAA and 5 μL (5 IU) CPA/ml; pH 7.0; 37°C.

Table 10. Effect of phenylethylaminoalanine concentration on carboxypeptidase activity[a]

PEAA (mM)	Carboxypeptidase A inhibited, %
0.84	3.8
8.4	29.6
21.0	66.8
42.0	100.0
84.0	100.0 (1.05)[b]

[a]1 ml of PEAA solution was incubated with 5 μL of undiluted CPA; pH 7.0; 1 hr; 37°C.

[b]Value in parenthesis is for N-acetyl-PEAA.

Since CPA is a member of a class of zinc enzymes that includes several other enzymes, such as alkaline phosphatase (Vallee and Galdes, 1984), it was of interest to find out whether PEAA would inhibit other members of this class. We found that under conditions described in the experimental section, 82.2% of the phosphatase is inactivated or inhibited with 21 mM of PEAA and 88.8% with 42 mM. These results suggest the PEAA may be a general inhibitor of zinc enzymes.

We have previously reported that sodium phytate has a small inhibitory effect on CPA (Friedman et al. 1985). Additional experiments were carried out to further delineate the inhibitory action of sodium phytate on this enzyme. Carboxypeptidase A was exposed to concentrations of sodium phytate ranging from 1 to 50 mM. Enzyme activity was measured at three substrate concentrations: 0.2, 0.5, and 1 mM. Table 11 shows that the inhibitory activity of the phytate salt depends on the concentrations of both substrate and inhibitor. The data also show that the threshold concentration for observable inhibition is between 1 and 10 mM sodium phytate. The potency of sodium phytate as a CPA inhibitor is similar to that of PEAA and much lower than that of LAL.

Studies on relative effects of three inhibitors (PEAA, LAL, and EDTA) on CPA activity at three substrate levels are plotted in Figures 3 and 4. Because of its low inhibitory potency at 25°C, the PEAA experiments were carried out at 37°C. In contrast, because of rapid inactivation of CPA by LAL at 37°C, these LAL experiments were carried out at 25°C. The data for the known metalloenzyme inhibitor, EDTA, obtained at 25°C are included for comparison. Additional studies showed that inhibition by EDTA at 37°C was greater than at 25°C, reaching 84% at 0.5 mM and 100% at 1.5 mM EDTA.

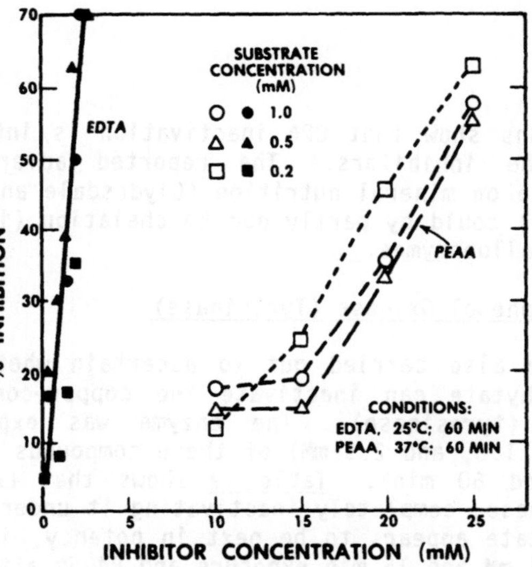

Figure 4. Inhibition of carboxypeptidase A activity by ethylenediamine tetraacetic acid (EDTA) and phenylethylaminoalanine (PEAA) as a function of inhibitor and substrate concentrations.

Table 11. Effect of sodium phytate concentration on carboxypeptidase activity at three substrate concentrations [a,b,c]

Sodium phytate concentration (mM)	% CPA inhibited at the following substrate concentrations (mM): 0.2	0.5	1.0
1.0	0.0	0.0	0.0
10	31.7 ± 3.16	30.1 ± 1.25	42.6 ± 2.08
15	38.3 ± 2.37	44.3 ± 1.25	55.4 ± 0.69
25	55.0 ± 2.36	56.6 ± 1.30	70.6 ± 0
50	66.2 ± 0.0	59.1 ± 3.04	80.3 ± 0.80

[c]Conditions: 37°C; 60 min.

[b]Average from two separate determinations ± standard deviation.

[c]Corrected for any loss of activity with enzyme controls.

These observations show that CPA inactivation is influenced by the concentration of the inhibitors. The reported adverse effects of protein-bound phytate on mineral nutrition (Clydesdale and Camire, 1983; Graf and Eaton, 1984) could by partly due to chelation (inactivation) of bound phytate to metalloenzymes.

Inactivation of Polyphenol Oxidase (Tyrosinase)

Experiments were also carried out to ascertain whether LAL, PEAA, EDTA, and sodium phytate can inactivate the copper-containing enzyme polyphenol oxidase (tyrosinase). The enzyme was exposed to three concentrations (0.4, 1.0, and 2.5 mM) of these compounds at 37°C for two time periods (15 and 60 min). Table 12 shows that LAL is a potent inhibitor of tyrosinase, completely inactivating it under all conditions tested. Sodium phytate appears to be next in potency, inactivating 62% of the enzyme at 0.4 mM and 15 min exposure and 84.5% after 60 min. The corresponding values for 1 mM phytate are 75.3% and 98.7%, and for 2.5 mM, 93.7% and 100%, respectively. These data show that inactivation is both concentration- and time-dependent. The next most potent inhibitor was EDTA followed by PEAA, the least potent one of the four tested. The extent of inactivation was also a direct function of inhibitor concentration and time of exposure.

Additional experiments summarized in Tables 13 and 14 show that the extent of inactivation of tyrosinase is strongly dependent on the concentration of sodium phytate and to some extent on the concentration of the substrate. Generally, in classical enzyme-inhibitor interactions (Segal, 1975), the higher the substrate concentration, the lower the inhibition. Since our data show that this is apparently not the case for the inhibition of tyrosinase by sodium phytate, the apparent inactivation appears to be the result of a combined effect of inhibitor and substrate.

The data also show that sodium phytate more effectively inhibits the copper enzyme (tyrosinase) than the zinc-containing enzyme (CPA). The results listed in the tables and Figures 3 and 4 suggest that LAL, EDTA, and PEAA are also more effective as inhibitors against tyrosinase than against CPA.

Table 12. Effects of time and concentration on the extent of inactivation of polyphenol oxidase (tyrosinase) by four inhibitors at 37°C

	Inhibitor concentration					
	0.4 mM		1.0 mM		2.5 mM	
Inhibitor	15 min	60 min	15 min	60 min	15 min	60 min
LAL	100 ± 0	100 ± 0	100 ± 0	100 ± 0	100 ± 0	100 ± 0
Sodium phytate	62.1 ± 4.88	84.5 ± 2.44	75.3 ± 1.83	98.7 ± 1.84	93.8 ± 0	100 ± 0
EDTA	39.4 ± 0	74.2 ± 2.14	51.9 ± 1.84	74.0 ± 0	70.1 ± 0	72.4 ± 3.54
PEAA	25.0 ± 2.36	28.3 ± 7.07	37.6 ± 0	42.9 ± 7.35	57.8 ± 2.21	100 ± 0

[a]Listed values are % inhibition from two separate experiments ± standard deviation.

[b]Enzyme controls run simultaneously showed no loss in activity after 15 or 60 min.

Table 13. Effect of temperature and concentration of sodium phytate on the inhibition of tyrosinase

Conc. sodium phytate (mM)	% Tyrosinase inhibited[a]		
	25°C	37°C	47°C
0.1 M	0.0	35.9	53.1
0.25 M	33.8	81.3	100.0
0.50 M	72.1	100.0	100.0

[a] % Tyrosinase Inhibited = Activity of enzyme + sodium phytate/ activity of enzyme alone carried through the same procedure.

Table 14. Effect of inhibitor and substrate concentration on the inactivation of tyrosinase by sodium phytate

Conc. sodium phytate (mM)	% Inhibition at the following substrate concentrations[a,b]	
	0.25mM	1.0 mM
0.1	0.0	0.0
0.25	0.0	26.6 ± 9.33
0.5	65.0 ± 7.1	66.0 ± 8.51
1.0	90.0 ± 0	78.0 ± 2.8

[a]Conditions: 200 U tyrosinase; 25°C; 1 hr.

[b]Average from two experiments ± standard deviation.

A series of experiments were also carried out to find out whether $CuSO_4$ would alter or reverse the inhibition of tyrosinase by LAL, PEAA, sodium phytate, and EDTA. The results in Table 15 show that (a) for LAL, inhibition was not affected by adding the copper salt after the enzyme was inactivated (sequential addition), exposing the enzyme to both LAL and copper sulfate (simultaneous addition), or exposing the enzyme to the inhibitor plus copper sulfate (reverse addition); (b) for PEAA and EDTA, the presence of copper sulfate enhanced the inhibition under all three conditions; (c) for sodium phytate, the presence of the copper salt lowered the extent of inhibition in the simultaneous and reverse additions. These results imply that copper ions bind or chelate to the inhibitory sites of sodium phytate, lowering its inhibitory potency. Evidently, the copper of the active site of the enzyme and added copper compete for metal-chelating sites on sodium phytate. This is apparently not the case for the other three inhibitors under the conditions shown in Table 15. Finally, Table 16 shows that thiols such as cysteine and N-acetyl-L-cysteine are potent inhibitors of tyrosinase (cf. also, Golan-Goldhirsch and Whitaker, 1984). In contrast, lysine up to 10 mM concentration had no measurable effect on the activity of tyrosinase.

Inactivation of Cytochrome C Oxidase

Cytochrome C oxidase is a terminal enzyme in the respiratory chain whose main function is to reduce oxygen to water. The respiratory chain transports electrons from the substrates of the citric acid cycle through a series of electron carriers which are mainly cytochromes (Powers and Chance, 1985).

Table 15. Effect of copper sulfate on the inhibition of tyrosinase by lysinoalanine, (LAL), phenylethylaminoalanine (PEAA), sodium phytate, and EDTA

Conditions[a]	% Tyrosinase inhibited[c]			
	LAL	PEAA	Sodium phytate	EDTA
Tyrosinase + inhibitor, 1 hr	100 ± 0	50 ± 0	100 ± 0	91.2 ± 0
Tyrosinase + inhibitor, 1 hr;followed by $CuSO_4$, 5 min (sequential addition)	100 ± 0	100 ± 0	100 ± 0	100 ± 0
Tyrosinase + inhibitor + $CuSO_4$, 1 hr (simultaneous addition)	100 ± 0	100 ± 0	57.6 ± 0.5	100 ± 0
Inhibitor + $CuSO_4$, 1 hr; followed by tyrosinase, 1 hr (reverse addition)	100 ± 0	100 ± 0	42.5 ± 0.5	100 ± 0
Tyrosinase + inhibitor[b], 1 hr	4.3 ± 6.1	54.3 ± 3.1	38.8 ± 5.7	17.4 ± 6.1

[a] Conditions: 200 units of tyrosinase; 0.5 mM inhibitor; 0.5 mM $CuSO_4$; 37°C. In control experiments it was found that in the absence of inhibitors, tyrosinase lost 16.2% of its activity after 1 hr at 37°C.

[b] Inhibitor concentration was reduced from 0.5 to 0.1 mM.

[c] Average from two determinations ± standard deviation.

Table 16. Inhibition of tyrosinase by L-cysteine, N-acetyl-L-cysteine, and reduced glutathione[a,b]

Inhibitor	% Tyrosinase inhibited at the following inhibitor concentrations (mM)		
	0.05	0.1	0.5
L-Cysteine	0.0	37.5; 39.1	100.0; 100.0
N-acetyl-L-cysteine	0.0	25.0; 25.7	100.0; 100.0
Reduced glutathione	0.0	12.5; 13.0	100.0; 100.0

[a]Conditions: 200 units of tyrosinase; 25°C at the indicated inhibitor concentrations.

[b]Results from two separate experiments.

According to Thompson et al. (1985), oxygen reduction by cytochrome C oxidase takes place at a binuclear metal site containing the iron of cytochrome a_3 and copper ion. Thompson et al. also report that certain small molecules, such as CN^- and NO, can form bridged complexes with cytochrome C which can be represented by Fe^{III}-CN-Cu^{II}, respectively. Similar bridged species are formed with azide (N_3) and sulfide (S^-) ions. The formation of such species suppresses the oxygen-reducing activity of the enzyme.

Results of the present study (Table 17) show that EDTA did not inhibit cytochrome C oxidase activity, PEAA produced slight but discernable inhibition, while lysinoalanine and sodium phytate produced strong inhibition under the described conditions. These latter findings contrast with the statement by Wharton and Tzagoloff (1967) that most metal chelating agents cause little or no inhibition of cytochrome oxidase.

The cited findings with cytochrome C oxidase imply that lysinoalanine and sodium phytate should also inhibit the copper-, zinc- and manganese-containing superoxide dismutase (Fee and Bull, 1985; Bull and Fee, 1985).

Table 17. Effect of lysinoalanine, sodium phytate, phenylethylaminoalanine, and EDTA on the enzymatic activity of cytochrome C oxidase [a]

Compound	% Inhibition [b]
LAL (5 mM)	8.8; 21.5
LAL (25 mM)	100.0;100.0
Sodium phytate (25 mM)	58.0; 61.3
PEAA (25 mM)	8.8; 10.1
EDTA (25 mM)	0.0; 0.0

[a] Conditions: 37°C 1 hr. [b] Results from two separate experiments.

Mechanistic Possibilities

Carboxypeptidase A hydrolyzes amino acids one at a time from the C-terminal end of the polypeptide chain. The enzyme differs from most other proteases in two ways: it has zinc as part of its active site and it undergoes a large conformational change during the binding of the substrate in order to bring together the components of the active site for efficient activity (Zubay, 1983).

Inhibition of CPA activity could occur by any of several mechanisms: (a) competitive inhibition by a substrate-like molecule, including the carboxyl end of the polypeptide chain; (b) blockage (masking) of the active site of CPA by noncompetitive inhibition; or (c) removal of the co-factor essential for catalytic activity, namely, Zn^{2+}. Free LAL appears to act, at least in part, by mechanism (c) through chelation of Zn^{2+}.

Our results, suggest, however, that LAL bound in the polypeptide chain acts predominately by other mechanism(s). Non-terminal LAL differs from free LAL in that four of the five groups that potentially can participate in metal binding (2 COOH, 2-NH_2) participate in amide (peptide) linkages. The inability of Zn^{2+} to completely reverse inhibition of CPA activity by alkali-treated soy protein suggests that the above-named groups are, indeed, important for Zn^{2+} binding.

Denaturation of protein by alkali or heat, or both, exposes a number of side chain groups that might react with CPA. The COOH-terminal region may be rendered accessible to the active site of CPA. If this sequence binds with much greater affinity to the active site than to the substrate, it could hinder substrate access and inhibit activity competitively (mechanism a). This mechanism appears to resemble the action of the naturally occuring carboxypeptidase inhibitors from potatoes (Pearce et al., 1983) and kidney beans (Hojima et al., 1979).

If the alkali treatment partially hydrolyzes the test protein, the new number of COOH termini will be increased in proportion. This will increase the probability that one or more termini can bind to CPA but be hydrolyzed slowly. Groups exposed by unfolding may interact with side chains on CPA external to the active site. If such interactions between denatured protein and CPA result in masking of the active site of CPA, noncompetitive inhibition may be observed (mechanism b).

Zinc binding involving amino acid side chains exposed or generated by alkali treatment is theoretically possible because of three hydrolytic events: (a) disruption of hydrogen bonds or salt bridges ($-COO^{-}NH_3^{+}-$) between -COOH groups of aspartic and glutamic acid and ϵ-NH_2 groups of lysine residues; (b) deamidation of asparagine or glutamine to produce new -COOH groups, or (c) hydrolysis of peptide bonds to -COOH and -NH_2 residues. On the basis of reported model studies with peptides, we have previously suggested that the extent of hydrolysis of primary amide groups of glutamine residues might not proceed to any significant extent at pH 12 (Masters and Friedman, 1979). In a related study, Friedman and Liardon (1985) also addressed the problem of peptide bond hydrolysis during exposure of food proteins to alkaline conditions. These considerations suggest the need for additional studies to define the possible importance of the cited hydrolytic events in both racemization and metalloenzyme inhibition.

Extensive studies of the mechanism of CPA action and structure of the active site (see Lipscomb, 1982; Vallee and Galdes, 1984, for reviews) indicate that zinc is coordinated by Glu, 2 His, and H_2O, and that Glu, Arg, and Tyr residues are involved in substrate binding or catalysis. If these or other side chains in the treated protein can achieve proper orientation to furnish multiple ligands for Zn^{2+}, the protein could compete with CPA for the metal ion.

Plots of reciprocal velocities (1/v) against inhibitor (treated protein) concentration depicted in Figure 2 permit some conclusions about the general aspects of the CPA inhibition. Figure 2 shows a greater inhibitory effect of treated soy protein on CPA in the absence of zinc ions. Limits on K_i (the inhibition constant) can be determined from the plots in Figure 1. These are K< 10 mg in the absence of Zn^{2+} ions and K_i> 100 mg in the presence of Zn^{2+} ions. These results suggest that zinc ions eliminate one mode of inhibition or greatly weaken the binding of the protein to the enzyme.

Another possibility is that the D-amino acid residues formed by the alkali treatment bind nonspecifically to the active site of CPA, thus inhibiting the enzyme. This could occur since (a) CPA has its primary specificity for aromatic amino acid side chains such as PHe and Tyr; and (b) these residues rapidly racemize to D-isomers under the influence of alkali and heat (Masters and Friedman, 1979, 1980; Liardon and Hurrell, 1983; Friedman and Liardon, 1985). The extent of this type of inhibition should depend on the number and location of such residues formed during the alkaline treatment.

This suggestion is reinforced by our observation that the extent of inhibition may not depend on protein-bound LAL, since inhibition by alkali-treated zein, which has no lysine and thus cannot produce LAL, progressively increased with increasing temperature of the alkali treatment from 65 to 85°C (Table 2). The temperature dependence of the inactivation may be partly due to increased racemization of amino acid residues at the higher temperatures (Friedman and Liardon, 1985).

It is noteworthy that the natural amino acid gostatin, which was isolated from cultures of Streptomyces sumanensis and which structurally resembles lysinoalanine (Figure 1), inhibits the enzymes asparate aminotransferase and alanine aminotransferase (Nishino and Murao, 1983; Jung, 1985).

Figures 3 and 4 suggest that rates of enzyme inhibitory reaction and equilibrium position show different patterns. Inhibition by LAL reaches equilibrium in ~3 min, but complete inhibition by PEAA requires about an hour (see Experimental Section). Concentration effects do not show simple behaviour. LAL displays sharp 'titration' of CPA activity between about 2 mM (threshold) and 4 mM inhibitor. EDTA inhibits at even lower concentration that LAL. PEAA inhibition of CPA is more concentration-dependent above 15 mM, suggesting two classes of inhibiting sites. Effects of substrate concentration over a five fold range for both LAL and PEAA are complex and not clear-cut. Differences in shapes of the inhibition curves suggest that no single mechanism is operating (Segel, 1975). Differences in mode of action of the three inhibitors seem reasonable, because structural features differ. Thus, (a) EDTA may act primarily by chelating to zinc or copper of the enzymes; (b) action of LAL may be complex and depend on pH (Friedman et al., 1985); and (c) PEAA, unlike the others, has a Phe-CH_2-CH_2- moiety that may resemble CPA substrates in structure.

Two different mechanisms have been proposed to account for the inactivation of metalloenzymes by chelating agents (Coombs et al., 1962, Frieden, 1974; Riordan and Vallee, 1974; Chlebowski and Coleman, 1976; Vallee and Galdes, 1984). In the first, the chelating agent removes the metal ion from the active site of the enzyme, leaving an inactive apoenzyme. In this mechanism, restoration of activity occurs on addition of the metal ion. In the second, the chelating agent forms an inactive mixed complex with the enzyme. Prevention of inhibition by added metal ions can confirm that the metal-binding site of a chelating agent, rather than some other property, is responsible for observed inhibition. Further studies are desirable to define the role of metal ions in the inhibition process.

Our conclusions about the mode of inhibition of carboxypeptidase are consistent with those proposed by Charlier et al. (1984), for the active-site-directed inactivation of the zinc-containing D-alanyl-D-alanine-cleaving carboxypeptidase of Streptomyces albus G. They report that, depending on the structure of the inhibitor, enzyme inhibition may be competitive (binding occurs at the active site), or noncompetitive (binding causes disruption of the protein crystal lattice).

Another possibility is that since CPA's primary specificity is for aromatic amino-acid side chains such Phe, Tyr, and Trp, the phenyl group of either the D- or L-isomer of D, L-PEAA binds specifically to the active site, thus inactivating the enzyme. The alkaline phosphatase and polyphenol oxidase results suggest, however, that inhibition is not necessarily related to a structural resemblance of PEAA to CPA substrates (cf. Ghosh and Rao, 1984; Friedman and Liardon, 1985).

The fact that acetyl-PEAA does not inhibit CPA suggests that the free α-NH_2 of the alanine part of PEAA is required for inactivation of CPA, and that the amino group probably interacts with the active site of the enzyme during the inhibition process.

Since PEAA, LAL, and sodium phytate are probably too large to fit into the active site of cytochrome C oxidase to form bridged complexes illustrated earlier, inactivation probably takes place by a different mechanism. The following is an alternative possibility. Treatment of the enzyme with $NaBH_4$ reduces the Fe^{III} to Fe^{II}. As discussed in detail elsewhere (Friedman, 1962; Maldotti et al., 1985), the introduced electron is probably not concentrated completely in the d-orbital shell of the iron but is delocalized in the highly conjugated porphyrin ring to which the iron is attached. Such delocalization creates reactive sites at the periphery of the active site of the enzyme. The added PEAA, LAL, or sodium phytate could, in principle, chemically interact with such free-radical-type sites, thus suppressing enzymatic activity.

The visible spectra (Smith, 1957) of the reaction mixtures following assays were consistent with a higher fraction of reduced than oxidized cytochrome C when lysinoalanine or sodium phytate was present, corroborating the kinetic data.

Implications for Nutrition and Food Safety.

Inhibition of CPA by treated proteins may have nutritional implications. Inhibition would hinder further digestion of a food peptide by CPA or other carboxypeptidases so that action of aminopeptidase or endopeptidase would be required. Specifically, during in vivo digestion of food proteins, the protein is cleaved by endopeptidases (pepsin, trypsin, chymotrypsin, elastase), yielding large peptide fragments that are then further degraded by a variety of exopeptidases such as CPA and carboxypeptidase B (CPB), aminopeptidases, and dipeptidases (Floch, 1981; Schwimmer, 1981). CPA removes the C-terminal amino acid residue. Its primary specificity is for aromatic amino acids such as Phe, Tyr, and Trp. These considerations imply that if alkali-treated proteins inhibit CPA, the negative feedback mechanism which may account for the in vivo effects of naturally occurring carboxypeptidase (Hojima et al., 1979; Pearce et al., 1983) and soybean trypsin inhibitors (Gallaher and Schneeman, 1986) could also operate with the CPA inhibitors.

Although alkali treatment is widely used in food processing (Sternberg et al., 1975; Sternberg and Kim, 1977; Slump and Haagsma, 1978; Raymond, 1980; Friedman and Masters, 1982), such treatments may not always be as severe as those shown in Tables 2 and 3. The possible inhibition of metalloenzymes by commercially processed food proteins and by methylated histidine and lysine derivatives (Friedman and Gumbmann 1979, 1981) merit further study.

Our studies show no inhibition of trypsin by alkali-treated soy protein and slight (less than 5%) inhibition by alkali-treated casein. In related studies, Possompes and colleagues (Possompes and Berger, 1986; Vimont-Rispoli et al., 1980) reported that nondigestible peptides isolated from alkali-treated casein can bind and inactivate trypsin *in vitro*, and Percival and Schneeman (1979) demonstrated that poorly-digested casein-glucose browning products do not stimulate increased trypsin production by the rat pancreas. Additional studies are needed to better define the possible antinutritional significance of these findings.

The formation of PEAA in foods may also have nutritional, therapeutic, and toxicological consequences. First, formation of PEAA from its biogenic amine precursor may decrease the migraine-causing potential of phenylethylamine (Chaytor et al., 1975; Ingles et al., 1985, 1980; Luthy and Schlatter, 1983). This could be beneficial provided PEAA itself does not turn out to be deleterious. Second, inhibition of CPA by PEAA, LAL, and sodium phytate *in vivo* could affect the digestion of foods since CPA participates in the removal of C-terminal amino acid residues during the digestion of proteins (Schwimmer, 1981). These considerations also imply that *in vivo* chelation of the inhibitors to metalloenzymes and essential trace elements could affect mineral nutrition, provided the inhibitors are present at high enough concentrations in foods. This is especially true for sodium phytate, which is known to adversely affect the nutritional quality of foods (de Rham and Jost, 1979; Clydesdale and Camire, 1983; Graf and Eaton, 1984; Knuckles et al., 1985; Rodriguez et al., 1985). This aspect needs to be addressed separately.

The described inhibition of zinc- and copper-containing metalloenzymes by sodium phytate implies that dietary phytate can affect mineral nutrition by chelating to essential trace elements such as zinc, copper, and iron (Henkin, 1974; Lonnerdal et al., 1985). It can also alter the activity of important classes of enzymes needed for normal metabolic activity. The possible consequences of such inhibition for nutrition, food safety, and health await further study.

Since polyphenol oxidase catalyzes oxidative browning in many plant foods (Deshpande et al., 1984; Friedman and Smith, 1984a; Hurrell and Finot, 1984) efforts have been made to inactivate the enzyme in order to minimize possible undesirable deteriorative and antinutritional effects associated with such browning reactions (Golan-Goldhirsh and Whitaker, 1984). The results of the present study suggest that the tyrosinase inhibitors listed in Tables 12 and 16 should also be investigated for their ability to prevent oxidative browning in foods.

In summary, the described results show that lysinoalanine, lysinoalanine containing food proteins, phenylethylaminoalanine, sodium phytate, and cysteine have the ability to inactivate metalloenzymes such as carboxypeptidase A and tryosinase. Possible beneficial or adverse effects of such inactivations *in vivo* by both naturally occurring food ingredients (sodium phytate) and those formed during food processing (lysinoalanine, phenylethylaminoalanine) remain to be elucidated. It would also be of interest to find out whether such inactivations do actually take place in vivo, and if they do, whether they are related to the observed antinutritional and toxicological properties of these compounds. In the case of lysinoalanine, it would be useful to determine whether the adverse effects in vivo (Woodard et al., 1975; Gould and MacGregor, 1977; Fritsch et al., 1983; Finot, 1983; Abe et al., 1984; Friedman et al., 1984 a,b) can be avoided by consumption of alkali-treated proteins in which lysinoalanine formation was prevented (Finley et al., 1978; Friedman, 1978 a, 1982; Friedman et al., 1984 a,b) and whether these effects are due to binding of LAL or LAL-containing peptides to metalloproteins in kidney cells (Friedman, 1977; Hayashi, 1982; Friedman et al., 1986).

REFERENCES

Abe, K., Suehara, S., Homma, S., Fujimaki, M., and Arai, S. (1984). Metabolic transit of lysinoalanine in the rat body: observation of feeding tests with alkali-treated casein and gluten. Agric. Biol. Chem., 48. 579-584.

Bull, C. and Fee, J.A. (1985). Mechanism of superoxide dismutases: the manganese containing protein. Biochemistry, 24, 3378.

Charlier, P., Dideberg, O., Jamoulee, J.C., Frere, J.M., Ghuysen, J.M., Dive, G., and Lamotte-Brasseur, J. (1984). Active-site-directed inactivators of the Zn (II) containing D-alanyl-D-alanine-cleaving carboxypeptidase of Streptomyces albus G. Biochem. J., 219, 763-772.

Chaytor, J.P., Crathorne, B., and Saxby, M.J. (1975). The identification and significance of 2-phenylethylamine in foods. J. Sci. Agric., 26, 593-598.

Chlebowski, J.F. and Coleman, J.E. (1976). In: "Metal Ions in Biological Systems". H. Siegel, ed., Marcel Dekker, New York. Vol. 6, p. 1.

Clydesdale, F.M. and Camire, A.L. (1983). Effect of pH and heat on the binding of iron, calcium, magnesium, and zinc and the loss of phytic acid in soy flour. J. Food Sci., 48, 1272-1274.

Coombs, T.L., Felber, J.P., and Vallee, B.L. (1962). Metallocarboxypeptidase: mechanism of inhibition by chelating agents, mercaptans, and metal ions. Biochemistry, 1, 899-905.

Deshpande, S.S., Sathe, S.K., and Slunkhe, D.K. (1984). Chemistry and safety of plant polyphenols. In: "Nutritional and Toxicological Aspects of Food Safety", M. Friedman, ed., Plenum Press, New York, pp. 457-495.

Fee, J.A. and Bull, C. (1985). Mechanism of superoxide dismutases: the copper, zinc containing proteins. Biochemistry, 24, 3377.

Finley, J.W., Snow, J.T., Johnston, P.H., and Friedman, M. (1978). Inhibition of lysinoalanine formation in food proteins. J. Food Sci., 43, 619-621.

Finot, P.-A. (1983). Lysinoalanine in food proteins. Nutrition Abstracts and Reviews. Clin. Nutr., 53, 67-80.

Floch, M.H. "Nutrition and Diet Therapy in Gastrointestinal Diseases," Plenum Medical Book Co., New York, 1981.

Folk, J.E. and Schirmer, E.W. (1963). The porcine pancreatic carboxypeptidase A system. J. Biol. Chem., 238, 3884-3894.

Frieden, E. (1974). The evolution of metals as essential elements. In: "Protein-Metal Interactions", M. Friedman, ed., Plenum Press, New York, pp. 1-31.

Friedman, M. (1982). Lysinoalanine formation in soybean proteins: kinetics and mechanisms. In: "Food Protein Deterioration-Mechanisms and Funtionality", J. P. Cherry, ed., ACS Symposium Series, Washington, D.C. 206, 231-273.

Friedman, M. (1978). Inhibition of lysinoalanine synthesis by protein acylation. In: "Nutritional Improvement of Food and Feed Proteins", M. Friedman, ed., Plenum Press, New York, pp. 613-648.

Friedman, M. (1977). Crosslinking amino acids-stereochemistry and nomenclature. In "Protein Crosslinking: Nutritional and Medical Consequences", M. Friedman, ed., Plenum Press, New York, pp. 1-27.

Friedman, M. (1973). "The Chemistry and Biochemistry of the Sulfhydryl Group in Amino Acids, Peptides, and Proteins", Pergamon Press, Oxford, England, p. 40.

Friedman, M. (1962). "Oxidations and Reductions of Porphyrins and Metalloporphyrins", Ph. D. Thesis, University of Chicago, p. 76.

Friedman, M. and Gumbmann, M.R. (1981). Bioavailability of some lysine derivatives in mice. J. Nutrition, 111, 1362-1369.

Friedman, M. and Gumbmann, M.R. (1979). Biological availability of epsilon-N-methyl-L-lysine, 1-N-methyl-l-histidine, and 3-N-methyl-L-histidine in mice. Nutrition Reports International, 19, 437-443.

Friedman, M. and Liardon, R. (1985). Racemization kinetics of amino acid residues in alkali-treated soybean proteins. J. Agric. Food Chem., 33, 662-672.

Friedman, M. and Masters, P.M. (1982). Kinetics of racemization of amino acid residues in casein. J. Food Sci. 47, 760-764.

Friedman, M. and Noma, A.T. (1986). Mechanism of formation and analysis of phenylethylaminoalanine in food proteins. J. Agric. Food Chem., 34, in press.

Friedman, M. and Smith, G.A. (1984 a). Inactivation of quercetin mutagenicity. Food and Chem. Toxicology, 22, 535-539.

Friedman, M. and Smith, G.A. (1984 b). Factors which facilitate inactivation of quercetin mutagenicity. In: "Nutritional and Toxicological Aspects of Food Safety", M. Friedman, ed., Plenum Press, New York, pp. 527-544.

Friedman, M., Grosjean, O.K., and Zahnley, J.C. (1986). Inacativation of metalloenzymes by food constituents. J. Food Chem. Toxicol., 24, in press.

Friedman, M., Grosjean, O.K., and Zahnley, J.C. (1985). Carboxypeptidase inhibition by alkali-treated food proteins. J. Agric. Food Chem., 33, 208-213.

Friedman, M., Gumbmann, M.R., and Masters, P.M. (1984 a). Protein-alkali reactions: chemistry, toxicology, and nutritional consequences. In: "Nutritional and Toxicological Aspects of Food Safety", M. Friedman, ed., Plenum Press, New York, pp. 367-412.

Friedman, M., Gumbmann, M.R., and Savoie, L. (1982). The nutritional value of lysinoalanine as a source of lysine for mice. Nutrition Reports International, 26, 939-947.

Friedman, M., Levin, C.E., and Noma, A.T. (1984 b). Factors governing lysinoalanine formation in soy proteins. J. Food Sci., 49, 1282-1288.

Friedman, M., Zahnley, J.C., and Masters, P.M. (1981). Relationship between in vitro digestibility of casein and its content of lysinoalanine and D-amino acids. J. Food Sci., 46, 127-131 and 134.

Fritsch, R.J., Hoffman, H. and Klostermeyer, H. (1983). Formation of lysinoalanine during heat treatment of milk. Z. Lemensm. Unters. Forsch., 176, 341-345.

Gallaher, D. and Schneeman, B.O. (1986). Nutritional and metabolic response to plant inhibitors of digestive enzymes. This volume.

Ganote, C.E., Peterson, D.R., and Carone, F.A. (1974). The nature of D-serine induced nephrotoxicity. Am. J. Pathol., 77, 269-176.

Ghosh, I. and Rao, V.S.R. (1984). Effect of configuration of inhibitors on the mode of binding to the enzyme, thermolysin. J. Biomol. Structure and Dynamics, 2, 29-40.

Golan-Goldhirsh, A. and Whitaker, J.R. (1984). Relation between structure of polyphenol oxidase and prevention of browning. In: "Nutritional and Toxicological Aspects of Food Safety", M. Friedman, ed., Plenum Press, New York, pp. 437-456.

Gould, D.H. and MacGregor, J.T. (1977). Biological effects of alkali-treated protein and lysinoalanine: an overview. In: "Protein Crosslinking: Nutritional and Medical Consequences," M. Friedman, ed., Part B, Plenum Press, N.Y., pp. 29-48.

Graf, E. and Eaton, J.W. (1984). Effects of phytate on mineral bioavailability in mice. J. Nutr., 114, 1192-1198.

Hayashi, R. (1982). Lysinoalanine as a metal chelator. An implication for toxicity. J. Biol. Chem., 257, 13896-13898.

Henkin, R.I. (1974). Metal-albumin-amino acid interactions: chemical and physiological relationships. In: "Protein-Metal Interactions", M. Friedman, ed., Plenum Press, New York, pp. 299-328.

Hodges, T.K. and Leonard, R.D. (1974). Purification of a plasma membrane-bound adenosine triphosphate from plant roots. Meth. Enzymol., 32, 392-406.

Hojima, Y., Mariya, H., Moriwaki, C., and Ryan, C.Y. (1979). Metalloenzyme inhibitor from kidney beans. Plant Physiol., 63, 562-566.

Hurrell, R.F. and Finot, P.A. (1974). Nutritional consequences of the reactions between proteins and oxidized polyphenols. In: "Nutritional and Toxicological Aspects of Food Safety", M. Friedman, ed., Plenum Press, New York, pp. 423-435.

Ingles, D.L., Back, J.F., Allimore, D., Tindale, R., and Shaw, K.J. (1985). Estimation of biogenic amines in foods. J. Sci. Food Agric., 36, 402-406.

Ingles, D.L., Tindale, C.R., and Gallimore, D. (1980). Absorption of biogenic amines by food constituents and some related substances. Chem. Ind., 415-416.

Ingles, D.L., Tindale, C.R., and Gallimore, D. (1978). Recovery of biogenic amines in chocolate. Chem. Ind., 432-433.

Jones, G., Rivett, D.E., and Tucker, D.J. (1981). The reaction of biogenic amines with proteins. J. Sci. Food Agric., 32, 805-812.

Jung, M. J. (1985). Enzyme inhibition by amino acids and their derivatives. In: "Chemistry and Biochemistry of the Amino Acids", G. C. Barrett, ed., Chapman and Hall, London and New York, pp. 227-245.

Knuckles, B.E., Kuzmicky, D.D., and Betschart, A.A. (1985). Effect of phytate and partially hydrolyzed phytate on in vitro protein digestiblity. J. Food Sci., 50, 1080-1082.

Liardon, R. and Hurrell, R.F. (1983). Amino acid racemization in heated and alkali-treated proteins. J. Agric. Food Chem., 31, 432-437.

Lipscomb, W.N. (1982). Acceleration of reactions by enzymes. Accounts Chem. Res., 15, 232-238.

Lonnerdal, B., Keen, C.L., and Hurley, L.S. (1985). Zinc binding ligands and complexes in zinc metabolism. Adv. Nutr. Res., 6, 139-167.

Luthy, J. and Schlatter, C. (1983). Biogenic amines in foods: effects of histamine, tyramine, and phenylethylamine in humans. (German). Z. Lebensmitt. Unters. Forsch., 177, 439-443.

Maldotti, A., Bartocci, C., Chiorboli, C., Ferri, A., and Carassiti, V. (1985). The role of oxygen in the mechanism of the intramolecular photoredox reaction of Fe(III) protoporphyrin IX in alkaline aqueous ethanol. J. Chem. Soc. Chem. Commun., 881-882.

Masters, P.M. and Friedman, M. (1980). Amino acid racemization in alkali-treated food proteins--chemistry, toxicology, and nutritional consequences. In: "Chemical Deterioration of Proteins", J.R. Whitaker and M. Fujimaki, eds. ACS Symposium Series, Washington, D.C., 123, 165-194.

Masters, P.M. and Friedman, M. (1979). Racemization of amino acids in alkali-treated food proteins. J. Agric. Food Chem., 27, 507-511.

Nishino, T. and Murao, S. (1983). Isolation of some properties of an aspartame aminotransferase inhibitor, gostatin. Agric. Biol. Chem., 47, 1961-1966.

Pearce, G., McGinnis, J., and Ryan, C.A. (1983). Effects of feeding carboxypeptidase inhibitor from potatoes to newly hatched chicks. Proc. Soc. Exp. Biol. Med., 173, 447-453.

Percival, S. and Schneeman, B.O. (1979). The effect of heat treated proteins on pancreatic protein secretion in rats. Nutr. Repts. Int., 20, 475-483.

Pfaff, G. (1984). Lysinoalanine content of formula diets. Lancet, 1, 448.

Pfaff, G. (1985). Lysinoalanine content of infant formulas - consequences for postsurgical nutrition. International Congress of Pediatric Surgery, Lisbon, Portugal, July 10 - 13, 1985. (Abstract).

Possompes, B. and Berger, J. (1986). Effect of severely alkali-treated casein on gastrointestinal transit and selected intestinal enzyme activities. This volume.

Powers, L. and Chance, B. (1985). Multiple structures and functions of cytochrome oxidase. J. Inorg. Biochem., 23, 207-217.

Raymond, M.L. (1980). Studies concerning the determination of lysinoalanine in food proteins. J. Food Sci., 45, 56-59.

de Rham, O. and Jost, T. (1979). Phytate-protein interactions in soybean extracts and low-phytate soy protein products. J. Food Sci., 44, 596-600.

Riordan, J.F. and Vallee, B.L. (1974). The functional role of metals in metalloenzymes. In: "Protein-Metal Interactions", M. Friedman, ed., Plenum Press, New York, pp. 33-57.

Rodriguez, C.J., Morr, C.V., and Kunkel, M.E. (1985). Effect of partial phytate removal and heat upon iron bioavailability from soy protein-based diets. J. Food Sci., 50, 1072-1075.

Schwimmer, S. (1981). "Source Book of Food Enzymology", AVI, Westport, CT, Chapter 1.

Segel, I.H. (1975). "Enzyme Kinetics", John Wiley & Sons, New York, p. 100.

Slump, P. and Haagsma, N. (1978). Evaluation of lysinoalanine determinations in food proteins. Z. Lebensm. Unters. Forsch., 167, 238-244.

Smith, L. (1957). Spectrophotometric assay of cytochrome C oxidase. Methods Biochem. Analysis, 2, 427-434.

Sternberg, M. and Kim, C.Y. (1977). Lysinoalanine formation in protein food ingredients. In: "Protein Crosslinking: Nutritional and Medical Consequences", M. Friedman, ed., Plenum Press, New York, pp. 73-84.

Sternberg, M., Kim, C.Y., and Schwende, F.J. (1975). Lysinoalanine: presence in foods and food ingredients. Science, 190, 992-994.

Thompson, A.J., Greenwood, C., Gadsby, P.M.A., Peterson, J., Eglinton, D.G., Hill, B.C., and Nicholls, P. (1985). The structure of cytochrome a_3-Cu_B site of mammalian cytochrome C oxidase as probed by MCD and EPR spectroscopy. J. Inorg. Biochem.,23, 187-197.

Tucker, D.J., Jones G.P., and Rivett, D.E. (1983). Formation of β-phenylethylaminoalanine in protein foods heated in the presence of added amine. J. Sci. Food Agric., 34, 1427-1433.

Vallee, B.L. and Galdes, A. (1984). The metallobiochemistry of zinc enzymes. Adv. Enzymol. Relat. Areas Mol. Biol., 56, 283-431.

Vimont-Rispoli, S., Possompes, B., and Besancon, P. (1980). Mise en evidence de la formation d'inactivateurs de la trypsine au cours de traitment alcalins des proteines. C.R. Acad. Sci. Paris, Series, D., 291, 945-948.

Wharton, D.C. and Tzagoloff, A. (1967). Cytochrome oxidase from beef heart mitochondria. Methods Enzymol., 10, 245-250.

Woodard, J.C., Short, D.D., Alvarez, M.R., and Reyniers, J. (1975). Biologic effects of N-ε-(DL-2-amino-2-carboxyethyl)-L-lysine, lysinoalanine. In: "Protein Nutritional Quality of Foods and Feeds" M. Friedman, ed., Marcel Dekker, New York, pp. 595-616.

Worthington (Anonymous) "Enzymes and Related Biochemicals". Biochemical Products Division, Worthington Diagnostic Systems, Inc., Freehold, New Jersey, 1982, pp. 26, 149.

Zubay, G. (1983). "Biochemistry", Addison-Wesley, Reading, PA, p. 159.

Presented at the Symposium on "Nutritional and Toxicological Significance of Enzyme Inhibitors in Foods", American Institute of Nutrition (FASEB) Meeting, Anaheim, CA, April 24, 1985. Fed. Proc. 1985, 44, 1497 (Abstract).

Reference to a brand or firm name does not constitute endorsement by the U.S. Department of Agriculture over others not mentioned.

INDEX

AACN, atypical acinar cell foci and nodules, 81
Abdominal pain and pancreatic extract therapy, 150
Acacia seed inhibitors, 245
ACTH, 475
N-Acetyl-L-cysteine, 63
effect on antigenicity of soybean inhibitors, 457
inactivation of protease inhibitors 323, 364, 378,
inhibition of tyrosinase, 549
Acidofilic foci, 96, 103
Acinar adenoma of the pancreas, 57, 95, 130
Acinar ductular transformation, 112
Activation energies for inactivation of inhibitors, 324
Adaptation
intestine to alkali-treated proteins, 526
rats to meal feeding, 206
rats to raw soy flour, 195
Adenomatous proliferation, 48
Adipose tissue, 173
Adiposis of the pancreas, 190
Affinity chromatography
alkali-treated proteins, 526
marama bean, 435
millet inhibitors, 440
soybean inhibitors, 216
winged bean inhibitors, 418
Aflatoxins, 159, 334
Alanine aminotransferase, 552
Albino mice, 190
Alcohol
induced pancreatitis, 145
treatment to remove inhibitors, 337
Alfalfa inhibitors, 286
Alkali-treated proteins,
adaptation to, 519
and pancreatic hypertrophy, 525
nutritional significance, 517
renal effects, 517
Alkaline phosphatase, 523, 534
Allelic forms of soybean inhibitors, 396
Allergy
to cow's milk 87
use of inhibitors in 301
Ames assay, 117
Amino acid
absorption of D-isomers, 518
availability, 127
composition
amylase inhibitors, 485, 488
isoinhibitors, 397, 404
lentil inhibitors, 262
marama bean inhibitors, 430
millet inhibitors, 441
winged beans, 416, 423
depravation and tumor growth 155
endogenous losses, 168, 217
reactive sites of inhibitors, 11
sequences,
alfalfa inhibitors, 287
soybean inhibitors, 287, 305
supplementation and pancreatic growth, 129
1-Aminoacetonitrile inhibitor activity, 311
Amylase, 129, 204, 352
Amylase inhibitors
amino acid composition, 488
and glucose absorption, 174
and growth, 173
and starch hydrolysis, 173
as drugs, 176
distribution in plants, 483
double-headed, 491
families, 489
from Colocasia, 492
from peanuts, 492
from wheat, 489
from yams, 492
therapeutic uses, 499
toxicity, 499

Anionic and cationic human trypsins, 240, 394
Antibiotics and growth, 169
Antibodies to inhibitors, 308
Antibody-based analyses, 452
Anticarcinogencity of Bowman-Birk inhibtors, 474
Antienzymes, 299
Antigenic determinants of inhibitors, 287, 464
Antigenicity of soy flour, 462
Antigenicity of inhibitors, 377, 449
  and toxicity, 452
  effect of food processing, 452
Antinutritional effects of inhibitors
  and negative feedback control, 392
  animal species variation, 393
  millet inhibitors, 446
  soybean inhibitors, 232, 394
  winged bean inhibitors, 413
Antrum, 135
Apoptosis, 83
Aprotein, 144
Arascine, 109, 118
Arginine destruction, 358
Arginine modification, 266
Arylsulfenyl chlorides as photo-affinity labels, 475
Ascaris inhibitor, 8
Aspartate aminotransferase, 552
Aspartic acid inhibitors, 2
Assays
  cholecystokinin, 135
  gastrin, 135
  inhibitors, 306, 359, 415
Association constants of inhibitor complexes, 395
Atypical acinar cells, 81, 194
Autoclaving of inhibitors, 323, 533
Autoimmunity, 451
Azaserine,effect on the pancreas, 41, 70, 82, 86, 91, 95, 207, 300

Banana inhibitors, 322
Barley inhibitors, 8, 303, 357
Barnyard millet inhibitors, 445
Basal output, 110
Basophilic acinar foci, 96
BBI, 159, 244, 305
Benzamidine, 520
Benzoyl-DL-arginine-p-nitroanalide (BAPNA), 306, 359
Bile
  duct jaundice, 115
  pancreatic juice, 25
Biological function of inhibitors, 20
Black beauty bean inhibitors, 321
Black gram inhibitors, 246, 357
Blackeye bean inhibitors, 415
Blood
  chemistry of monkeys, 231, 277
  clots, removal by inhibitors, 154
  glucose levels, 173, 214
Bioassay of cholecystokinin, 124
Biopsy of monkey pancreata, 225
Biosynthesis of gastrin, 138
Biotinylated inhibitors, 454
Body weights of calves, 213
Bolus of food, 22
Bovine trypsin inhibition, 3, 6
Bowman-Birk inhibitors
  and X-ray cell transformation, 474
  as anticarcinogens, 474
  chemical modification, 472
  human antithrombin inactivation, 474
  molecular properties, 392, 400
  peptide fragments, 473
  reactive site, 394, 470
  ternary complex, 392, 470
  therapeutic significance, 153, 474
Bread inhibitor activity 322
Breast cancer, 153, 195
Broad bean inhibitors, 245, 357
Brush border cells, 518
Bush bean inhibitors, 245

Calorie-balance technique, 174
Calves
  blood chemistry, 214
  body weights, 213
  inhibitor intake and pancreatic weights, 215
Camera lucida, 95
Cancer and enzyme inhibitors, 153
Cancer of the pancreas, 35
Carboxypeptidase
  human, 253
  inhibition by
    alkali-treated proteins, 533
    lysinoalanine, 535
Carcinoma of the pancreas, 45, 112
Carob bean inhibitors, 246
Casein
  alkali-treated, 521, 539
  and plasma cholecystokinin, 124, 127, 128
  antitrypsin activity, 83
  hydrolysis, 158
  Maillard browning, 552
Caseinolytic activity of subtilisin, 394, 446
Cationic human trypsin, 240
Cebus monkeys, effect of inhibitors, 233
Cecum, 203

Celiac disease, 452
Cell density of the acinar pancreas, 52
Chemical modification of lentil inhibitors, 266
Chicken ovomucoid, 240
Chicken pancreatic enzymes, 394
Chickpea inhibitors, 247, 414, 469
reactive sites 474
Chloramphenicol acyltransferase, 285
Cholecystokinin (CCK), 35, 143, 170, 235
and feedback mechanism, 118
as cocarcinogen , 109, 111
assay, 124, 135
effect on pancreas, 110, 123, 134,
homeostasis, 65
plasma levels, 126, 134, 138
suppression of release by inhibitors, 87, 91, 128, 185
trophic effect, 110
tumor protection, 196
Cholesterol-binding pancreatic proteinase, 253
Chromatic protease staining, 416
Chronic pancreatitis, 144
Chymotrypsin
activity, 204
effect on pancreatic secretions, 146
inhibition, 3, 144, 227, 249, 361, 365, 440
photoreactive derivatives, 467
Chymotrypsinogen, 171
secretion, 27
synthesis, 129
Cigarette smoking and pancreatic cancer, 109, 154
Classes of inhibitors, 292
Classification of protease inhibitors, 1
Cocoonase, 2
Coffee consumption and pancreatic cancer, 160
Cognate enzymes, 2
Colocasin inhibitor, 492
Colon cancer, 153
Common pea inhibitors, 247
Competitive inhibition, 1, 4
Competitive inhibition assay, 459
Complex function of lentil inhibitors, 267
Complexes of amylase inhibitors, 493, 495
Composition
chickpeas, 414
cowpeas, 414
isoinhibitors, 401
peanuts, 414
soybeans, 414
winged beans, 414
Conglycinin antigenicity, 452
Cooking and inhibitor activity, 273
Copper diisopropyl salicylate, 155
Coprophagy, 169, 521
Corn, 357
alkali-treated, 539
inhibitors, 303, 322
raw soybean diets, 210
Cotyledonous mass, 21
Cow pancreatic enzymes, 394
Cowpea inhibitors, 247, 414
Cranberry inhibitors, 321
Cross-reactivity
inhibitors, 287
isoinhibitors, 406
L-Cysteine, 63
effect on antigenicity of soybean inhibitors, 457
effect on inhibitor activitv, 273
inactivation of inhibitors, 322, 364, 365
inhibition of tyrosinase, 549
photoaffinity labeling, 475
proteinase inhibitors, 2
residues of alfalfa inhibitors, 287
rich endogenous protein, 168
L-Cystine, 63
destruction, 358
in feather proteins, 216
fortification of soy flour, 373
Crystalline complexes of inhibitors, 307
Cytochrome C, 352
inhibition by alkali-treated food proteins, 532
inhibition by lysinoalanine, 532
Cytology of neoplastic tissues, 42
Cytoplasmic vacuolation in the pancreas, 190

Daily intake of inhibitors, 34
Dark red kidney bean inhibitors, 415
Defatted soy flour, 37
Deficient pancreatic growth, 129
Degradation
Bowman-Birk inhibitors in the intestine, 474
by ruminant microorganisms, 216
Deoxyribonucleosides, 156
Diabetes and amylase inhibitors, 176
Diafiltration of soy proteins, 354
Diagnostic uses of amylase inhibitors, 501
Diet(s)
and cancer, 91, 118, 153
and hyperstimulation of the pancreas, 88

and immunity, 450
composition, 124, 134, 191, 201, 230, 519
experimental, 99
Dietary
intake of inhibitors, 87, 204, 301
protein
and acinar adenoma, 130
and growth of the pancreas, 129
and cholecystokinin release, 128
Differential scanning calorimetry of soybean inhibitors, 370
Digestibility
alkali-treated proteins, 517
effect of tannins, 512
soybean meal, 168

Digestive
enzymes, 167
of insects, 153
reactions, 23
inhibition by polyphenols, 509
Di-(2-hydroxypropyl) amine, 189, 300
Dimethylhydrazine induced colon cancer, 154
Diphenylamine reaction, 111
Disease and ingestion of inhibitors, 19
Disulfide bonds
and inhibitor activity, 9, 273, 303
of Bowman-Birk inhibitors, 471
role in antigenicity, 454
Dithiobis (2-nitrobenzoic acid) (DTNB), 359
DNA, 69
effect of inhibitors, 83, 86, 114, 200, 229
protein crosslinks, 117
repair, 117, 156
sequence of tomato inhibitors, 284
synthesis and lysinoalanine, 532
Dogs, 200
Dose-responses to dietary inhibitors, 57
Double-headed inhibitors, 306
amylase, 491
lima beans, 491
Dry bean inhibitors, 357
Dry roasting of navy beans, 337
D-serine toxicity, 518, 532
Ductular morphology of the pancreas, 111, 112, 116
Duodenal infusion, 135, 144

Effect of inhibitors in monkeys, 186
Egg white inhibitors, 24, 26, 356
Eggplant inhibitors, 303
Eggs, inhibitor content 322
Eglin, 7
Elastase, 3, 403
human, 144, 253
inhibition
by Bowman-Birk inhibitors, 474
by marama beans, 434
by millet beans, 445
by soybeans, 434
Electrophilic radicals, 117
Electrophoresis
amylase inhibitors, 489
isoinhibitors, 400
marama inhibitors, 431
soybean inhibitors, 292
winged bean inhibitors, 415, 419
Elimination of inhibitors through plant breeding, 281
ELISA method for detection of inhibitors, 287
Emphysema, effect of inhibitors, 154, 301
Endogenous loss of amino acids, 216
Energetics of complex formation, 6
Enterokinase, 227
Enteropeptidase, 254
Enzyme
inhibitor complex, 3
and competitive inhibition, 4
profile of proteases, 171
role of tannins, 511
substrate complexes, 170
Epidemiology of pancreatic cancer, 71, 109, 118, 153
Ethylenediamine tetraacetic acid, 319
inhibition of metalloenzymes, 531, 543
Exocrine secretion of the pancreas, 112, 114, 143
Experimental models of pancreatic carcinogenesis, 92
Extrusion cooking, effect on inhibitors, 335

Fabricated foods, 349
Families of inhibitors, 8, 487
Fat in the diet and pancreatic cancer, 109
Feather proteins, 216
Fecal nitrogen excretion 168
Feedback mechanism, 70
Fermentation, 333

Fibrinogen, 154
Fibrinopeptide, 154
Fibrosis, 190
Field bean inhibitors, 247
Finger millet inhibitors, 445
Focal acinar lesions, 71
Focal nodular hyperplasia of the pancreas, 46
Food
  antigens, 450
  intake and body weight, 205
  processing
    and antigenicity, 452
    effect of alkali, 531
  safety
    and inhibitors in the diet, 65, 533
    and pancreatic cancer, 71
Fox enzymes, 394
Free radical traps, 379
Function of endogenous inhibitors, 28
Functional properties of soy proteins, 350

Gallotannins, 510
Garden bean inhibitors, 246, 357, 415
Gastric detoxification, 22
Gastrin, 234
  and pancreatic hypertrophy, 139
  assay, 135
  biosynthesis, 137
  immune-reactivity, 135
  plasma levels, 137
Gastrointestinal
  organ weights, 211
  transit of alkali-treated proteins, 520
Geese and chicks
  activity of digestive enzymes, 209
  diets, 208
  organ weights, 208
  sensitivity to raw soybeans, 208
Gene transfer of inhibitors, 300
Genetic
  basis of variability of isoinhibitors, 407
  breeding of soybeans lacking Kunitz inhibitors, 291, 304
Genotypes of soybean inhibitors, 294
Germination and inhibitor activity 331, 444
Globulins, 352
Glutathione inhibition of tyrosinase, 549
Glycinin antigenicity, 452
Gostatin, 536
Growth
  depression and inhibitor content of diets, 169, 185, 217, 426
  of mice, 192
  of rats, 41, 202
Groundnut inhibitors, 357
Guanine-cytosine pairing, 117
Guar inhibitors, 247
Guinea pea inhibitors, 248
Guinea pigs, 216

Half-cystine assay, 362
Half inhibition of trypsin, 309
Hamster nitrosamine model of pancreatic cancer, 92
Heat
  stability of inhibitors, 417, 443
  treatment of soy products, 234
    effect on antigenicity, 463
Hemagglutinin
  inactivation, 378
  toxicity, 377
Hematology of Cebus monkeys, 231, 233
Hirudin family of inhibitors, 443
Histology of the pancreas, 95, 111, 115
Historical discoveries of inhibitors, 239, 299
Homeostasis and cholecystokinin, 65
Hormonal stimulation of the pancreas, 65, 69
HSF, heated soy flour, 82, 86
HSM, heated soybean meal, 129
Human
  carboxypeptidase A, 253
  diets and meal feeding, 218
  experiments with inhibitors, 177
  group A blood, 359
  trypsin inhibitors, 21
    homogeneity, 240
    substrates, 240
Hydroxyethylglycine, 359
Hyperplasia of the pancreas, 170
Hyperplastic nodules, 234
Hypersecretic pancreas, 23, 170, 204
Hyperstimulation of the pancreas, 88
Hypertrophy and hyperplasia of the pancreas, 115, 170
Hypervariability of enzyme-inhibitor contacts, 9

Idiopathic chronic pancreatitis, 145
Image analysis of the pancreas, 97
Immune complexes, 450
Immune responses to inhibitors, 450
Immunoassays of inhibitors, 453
Immunochemistry of soybean inhibitors, 377

Immunological methods for inhibitor
assays, 308
Immunopharmacological activity, 449
Immunoreactivity of cholecystokinin,
135
Immunotoxicology, 450
Inactivation of inhibitors
activation energies, 324
by cysteine, 322
by dry roasting, 337
by extrusion cooking,335
by heating, 333
by thiols
effect of pH, 366
effect of temperature, 372
effect of sulfite, 374
role of sulfur amino acids, 366
kinetic aspects, 324
moisture effects, 325
Inactivation of hemagglutinins, 377
Incidence of pancreatic cell lesions,
92
Infant(s)
dietary inhibitor levels, 87,
118, 406
formulas, 118
susceptibility to inhibitors,
87
Inflammation, use of inhibitors,
301
Ingobsin lumen enzyme, 253
Inhibition
of cancer by inhibitors, 301
of digestive enzymes, 501
of human chymotrypsin, 249, 251, 252
of human elastase, 253
of human plant proteases, 300
of plant proteases, 300
of tumor promotion, 154
Inhibitors
as free radical traps, 379
in human foods, 288
of digestive enzymes, 240, 282,
257
peptide digestion, role in, 3
reactive sites, 11
substrates, 6
three-dimensional structure, 7
toxicity, 22
units, 308
Inhibitory domains, 1
Interaction of inhibitors with food
components, 320
Intestinal
adaptation, 519
mucosa, 451
trypsin activity, 126
Intraepithelial lymphocytes, 450
Intraluminal protease activity,
127, 130
Introduction of inhibitor genes,
288
Introns of tomato inhibitor genes,
285
Isoelectric precipitation, 350
Isoinhibitors, 287
amino acid composition, 397
amino acid sequence, 403
amylases, 489
cross-reactivity, 406
immunology, 398
reactive sites, 403
Isoline development of soybeans,
294
Italian millet inhibitors

Kallikrein, human, 253
Kazal family of inhibitors, 8
Kidney beans, 173
amylase inhibitors, 492
inhibitor activity, 321
Kinboki beans
inhibitor activity, 271
susceptibility to pepsin, 271
Kodo millet inhibitor, 445
Kunitz, Moses, 3
Kunitz soybean inhibitor
amino acid composition, 397
and newborn infants, 406
assays, 455
association constants, 395
covalent structure, 471
family of inhibitors, 8
function in plants, 453
immunoassays, 453
immunochemistry, 459
isoinhibitors, 396
molecular properties, 470
molecular weight, 392
photoreactive derivatives, 476
reactive site, 392
single gene locus coding, 396
structure, 305

Lactalbumin, alkali-treated, 539
Lactic casein, 37
Lacto-ovovegetarianism, 302
Laparotomy, 110
Large intestine, 303
Latency period of pancreatic tumor
development, 115
Lathrogen inhibitor activity, 311
LBI, lima bean inhibitor, 245
Legume inhibitor activity, 317
Lentil inhibitors
activities, 260-65
against human enzymes, 248, 256
composition, 262
enzyme complexes, 267
reactive sites, 267

Leucine
aminopeptidase, 524
beta-naphthylamine, 524
restricted diets, 155
Leupeptin, 3, 154
Lifetimes of inhibitor-enzyme complexes, 392
Lima bean inhibitors, 357, 470
hemagglutinin inactivation, 377
susceptibility to pepsin, 271
Linkage of soybean gene loci, 293
Lipase
activity in geese and chicks, 209
inhibition by tannins, 512
inhibitors, 175
effect of cellulose on activities, 175
effect on pancreatic weights, 176
occurrence, 175
Lipoxygenases, 352
Liver cancer protection against by inhibitors, 195
Liver tumors, 95
Lobular carcinoma of the pancreas, 116
Long-term effects of pancreatic acinar adenoma, 59-62
Long-term feeding of soy-based diets, 189, 223
Lysine
as sites for photoaffinity labeling, 475
deficiency, 525
destruction, 266
modification with maleic anhydride, 266
Lysinoalanine
and pancreatic hypertrophy, 518
complexing with metalloenzymes, 518
formation in foods, 536
mechanism of inhibition of metalloenzymes, 532
renal effects, 532
Macroglobulin-enzyme complexes, 3
Maillard reaction, 526
Maleylation of soybean inhibitors, 473
Malnutrition and immunity, 450
Maltase, 523
Mammary adenocarcinoma, 155
Man's lifespan, 153
Marama bean inhibitors, 357
affinity chromatography, 435
amino acid composition, 430
electrophoresis, 431
potential as a food source, 437
Meal-feeding
chicks and geese, 210
effect on food consumption, 211
effect on pancreatic weights, 211
rats, 201
Mechanism of action
protease inhibitors, 1, 4
metalloenzyme inhibitors, 550
Medical uses of amylase inhibitors, 499
Membrane separation of soy protein inhibitors, 351, 354
Metabolic response to inhibitors, 167
Metabolism of introsomes, 117
Metalloenzyme inhibitors, 2
alkali-treated proteins, 531
in foods, 531
lysinoalanine, 553
mechanism of action, 550
phytates, 538
role of D-amino acids, 552
Metaplasia of the pancreas, 190
Methionine
availability and inhibitors, 23
fortification of soy flour, 373, 393
oxidation, 154
sulfoxide, 368
Methylgramine, 117
Mice, 200
Microcomplement fixation assay, 453
Microheterogeneity in amino acid sequence of inhibitors, 304
Migraine and phenylethylaminoalanine, 554
Milk replacers, 213
Millets inhibitors
affinity chromatography, 440
amino acid sequence, 441
effect of germination, 444
scientific names, 440
Mineral nutrition, 554
Miso inhibitors, 314
Mitotic figures of the pancreas, 48
Mixed disulfides, 367
Moisture and inhibitor inactivation, 325
Molecular weights of inhibitors, 303, 421
Monkey pancreatic enzymes, 394
Monoclonal antibodies against soybean inhibitors, 456
classification, 460
epitope mapping, 460
3-(N-Morpholino) propane sulfonic acid (MOPS), 359

Multiple alleles, 292
Mungbean inhibitors, 248
Musotrypsin, human, 353
Mustard inhibitors,357
Mutagenicity, 156
Mutagenicity and carcinogenicity, 117
Navy bean inhibitors, 246, 357
Negative feedback control, 35, 123, 300, 392, 553
  and protection against cancer, 196
  in humans, 144, 186
  in monkeys, 236
  in pigs, 170, 171
Nitrogen solubility index, 327
2-Nitro-4-azidosulfenyl chloride (2,4-NAPS-Cl), 469
N-Nitro -bis-(2,4-oxopropyl)amine (BOP), 111
N-Nitro-2,6-dimethyl morpholine, 117
Nitrogen digestibilities, 40, 323
NMR, nuclear magnetic resonance of inhibitors, 6, 15
Noncompetitive inhibition, 4
Nondigestible peptides, 552
Nodular scores in rats, 207
Nodules in the pancreas, 207
Nonspecific inhibitor activity of tannins, 310
Nontoxicity of soybeans, 200
Nucleic acid synthesis, 393
Nutrient absorption, effect of inhibitors, 218
Nutrition
  and cholecystokinin release, 123
  and disease prevention, 153
  and gastrin biosynthesis, 138
  and pancreatic cancer, 97
  and unsaturated fat, 98
Nutritional
  effects of inhibitors, 300, 499, 544
  effects of tannins, 514
  improvement, 357-379
  quality and inhibitors, 391-407
    of heated soy proteins, 39
  response to inhibitors, 167
  significance of inhibitors, 473
    in humans, 274
  status and inhibitors, 171
  value of sulfur amino acids in inhibitors, 407

Obesity and amylase inhibitors, 176
Oilseed inhibitor activity, 312
Oligonucleotide probes, 283
Oral cancer, 153
Organ weights of rats, 202, 424
Ornithine decarboxylyase, 69
Overtoasted soy flour, 37
Ovogastric administration of trypsin, 123
Ovomucoid, 150
  amino acid sequence, 13
  inhibition of
    chymotrypsin, 14
    elastase, 14
    subtilisin, 14
  variants, 14
Oxygen radicals
  and DNA damage, 156
  and tumor promotion, 155
  induction of pol(ADP-ribose) formation, 157

Pancreas weights
  and lipase activity, 176
  and sulfur amino acids, 63

Pancreatic
  acinar adenoma, 95, 130, 193
  adaptation, 127, 186
  carcinogenesis, 46, 71, 82, 97, 159, 193
    promotion by azaserine, 35, 98
  disease, 144
  ducts, 110
  effects of inhibitors, 35, 63,170, 300, 473
  effects of lysinoalanine, 518
  effects of sulfur amino acids, 63
  enzymes, 394
  exocrine secretions, 114, 143, 186
  growth parameters, 85
  histology, 83, 111, 136, 214, 393
  hypertrophy, 84, 136, 139
  lobular carcinoma, 116
  nodules, 192
  pathology, 190
  role in amylase synthesis, 129
  weights and amino acid availability, 129
  wet weights, 83, 101, 114
Pancreaticotrophic effect, 110
Pancreatitis, use of inhibitors, 30
Pancreatopeptidase E, 514
Panlobular ductular carcinoma of the pancreas, 114
Paralogous inhibitors, 12
Pathologic sequellae, 19
Peanut(s)
  amylase inhibitors, 492
  composition, 414
  inhibitor content, 315
Pearl millet inhibitors, 445
Pepsin inhibitors from legumes, 269, 272
Pepsinogen, 22

Peptide mapping of antigenic sites, 452
PER of soy flour, 375, 377
PER of winged beans, 424
Percutaneous transhepatic cholangiography, 144
pH Stat assay of proteolytic digestion, 363
Phenolic antioxidants and pancreatic cancer, 95
PHSD, partially heated soy flour,213
Phenylethylaminoalanine
    and migraine, 532
    inhibition of metalloenzymes, 541
Photoaffinity labeling
    as probes of inhibitor structures, 477
    hormones, 475
    inhibitor-enzyme complexes, 477
Photoreactive
    chymotrypsin inhibitors, 477
    reagents, 475
Physiological role of enzyme inhibitors, 21
Phytate inhibition of metalloenzymes, 542
Phytohemagglutinins, 323
Placebo-controlled double-blind studies, 145
Plants
    as food sources, 167
    role of protease inhibitors in, 20, 281, 300
Plasma levels of cholecystokinin, 134
Plasminogen activator, 154
Pleomorphism in the pancreas, 190
Polyacrylamide gel screening of inhibitors, 292
Polyadenylation, 157, 285
Polymorphonuclear leucocytes, 154
Polyphenol oxidase inhibition by alkali-treated proteins, 534
Polyphenols
    enzyme inhibition, 509
    function in plants, 509
    structure, 509
Polysaccharide inhibition, 311
Polyvinylpyrrolidone (PVP), 310, 513
Positive adaptive mechanism, 216
Postinitiation of pancreatic carcinogenesis, 93, 98
Potato inhibitors, 8, 282, 303, 322, 357
Potentiation of nitrosamine action by inhibitors, 195
Preneoplastic lesions of the pancreas, 71, 115
Probit analysis, 57
Problems with inhibition assays,309
Processing
    effects on soy antigenicity, 464
    effects on trypsin inhibitor activity, 273
    of soybeans lacking the Kunitz inhibitor, 294
Properties of amylase inhibitors, 494
Prostate cancer, 153
Protease E, human, 253
Protease inhibitors
    biological action, 153
    effect on release of cholecystokinin, 127
    historical perspective, 299
    inhibition of tumor promotion, 154
    synthetic, 154
Protective effect of inhibitors against cancer, 153, 195
Protective effect of inhibitors against insects, 300
Protein
    analysis, 363
    and cholecystokinin release, 123
    complexes with tannins, 514
    content of diets and pancreatic carcinogenesis, 82, 83
    effects on cholecystokinin 65, 123, 127
    effects on pancreatic neoplasia, 58
    efficiency ratios (PER), 40, 363
    endogenous losses, 216
    enzyme complexes, 170, 299
    in diets and pancreatic cancer, 129
    synthesis, 69, 127, 393
    turnover, 2
Proteinase inhibitor(s)
    and proteolysis, 2
    classification, 2
    distribution in plants, 2, 469
    complexes, 264
    genes, 281
    inducing factor, 283
Proteolysis, 1, 362
    effect of inhibitors, 168
Proteolytic digestion assay, 361
Putative gastrointestinal hormones, 27
Putrescine, 69
PVP, 310

Quercetin, 509

Rabbit antigenicity against inhibitors, 456

Racemization of amino acids, 531, 552
Radioimmunoassay of cholecystokinin
  and gastrin, 135
Rat(s), 200
  antral mucosa, 136
  azaserine model of pancreatic
    carcinogenesis, 93, 98, 118
  body and organ weights, 202
  pancreatic enzymes, 394
Raw soy flour and pancreatic
  carcinogenesis, 37, 82
Raw soy protein isolates and
  carcinogenesis, 95, 100
Raw soybean milk, 406
Reactive sites of inhibitors, 1, 5,
  303, 306
  amino acid residues, 11
  of chickpea inhibitors, 474
  of lentil inhibitors, 267
  of millet inhibitors, 267
  of soybean inhibitors, 292, 403
Reactive sites of isoinhibitors,
  403
Red gram inhibitors, 248
Red kidney bean inhibitors, 258
Red sandal wood inhibitors, 248
Red wood inhibitors, 358
Regulation of
  cholecystokinin release, 128
  digestive reactions, 23
Removal of naturally occurring
  toxicants, 167
Residual levels of inhibitors in
  foods, 69, 358
Retinoids, 95, 155
Reversibility of pancreatic
  hypertrophy, 84, 200
Rhamnoside larval feeding, 509
Rice inhibitors, 322
RNA of pancreas and inhibitor acti-
  vity, 83
RNA transcription, 69, 83
mRNA and inhibitor biosynthesis,
  283
Role of amylase inhibitors in
  plants, 497
RSD, raw soy diets, 199
RSF, raw soy flour, 81, 189, 200
RSM, raw soybean meal, 129
Rutenoside larval feeding, 509
Rye inhibitors, 358

Safety of inhibitors, 33, 172,
  186
SBTI, soybean trypsin inhibitor,
  124, 134
Scarlet runner bean inhibitor, 248
Select genomic clones of tomato
  inhibitors, 285
Senilac inhibitors, 321
Sensitivity of mice to inhibitors,
  195
Separation of winged bean
  inhibitors, 418
Serine proteinase inhibitors, 2
Serpin family of inhibitors, 7
Serum chemistry of monkeys, 231
Shock, use of inhibitors in, 301
Short-term exposure to soy flour,
  84
Signal peptide removal, 2
Single-headed inhibitor, 306
Small
  intestine, 203
  red bean inhibitors, 321
  whole bean inhibitors, 321
Sodium phytate inhibition of
  metalloenzymes, 540
Solubility of soy proteins, 351
Sorghum inhibitors, 358, 441
Soy
  based diets
    infants, 118
    monkeys, 223
  flour
    and pancreatic cancer, 117
    antigenicity, 462
    composition, 82, 350
    effect of heat, 118, 190
    effect of long-term feeding
      118
  meat analog,inhibitor content,
    313
  milk,inhibitor content, 314
  protein
    amino acid analysis, 533
    as milk replacers, 452
    as antigens, 452
    autoclaving, 533
    effect of alkali-treatment,
      539
    extractability, 351
    interaction with fat, 98
    isolate, 349
    isoelectric properties, 350
    role in pancreatic adenoma,
      53
    trypsin inhibitor content,
      54, 350
Soybean(s)
  aflatoxin content, 334
  breeding, 291
  composition, 414
  cultivars, 394
  diets, 138
  digestibility, 158
  germplasm, 292
  heat treatment, 169
  isolines, 294
  trypsin inhibitors, 8, 270,
    299, 326, 370

removal of inhibitors, 169
urease activity, 335
Soybean-specific food allergy, 453
Species variation in susceptibilities to soybean inhibitors, 87, 169, 185, 196, 199
Specificity of amylase inhibitors, 490
Spiral-wound microfiltration module, 351
Squash inhibitors, 8, 358
Stages of pancreatic carcinogenesis, 92
Starch
blockers, 174
free diets, 173
hydrolysis, 173
Starfish trypsin inhibitor, 273
Stereology of acinar foci, 95
Stomach, 203
as detoxifying antichamber, 22
Structure
inhibitors, 304, 474
tannins, 510
Substrates for human trypsin, 240
Subtilisin inhibition, 3, 6, 446
Succinylation of soybean inhibitors, 473
Sulfhydryl(s)
assay of soy proteins, 363, 367
disulfide interchange, 367
of lima bean and soybean flours, 367
oxidation, 368
Sulfite inactivation of inhibitors, 374
Sulfur amino acids
and cholecystokinin release, 65
effect on pancreatic weights, 63
effect on PER, 63
of inhibitors, 168, 281, 291, 407
requirements by chicks and geese, 216
role in inhibitors, 362, 373, 393
Sunflower inhibitors, 315
Superoxide dismutase, 155, 550
Sweet potato inhibitors, 358
Synthetic protease inhibitors, 154
Syrian hamsters, 109, 115

T cell responses, 452
TAME, 245, 309
Tannic acid, 514
Tannin inhibitors,
activity, 310
and digestibilities, 511
and trypsin inhibitors, 514
protein complexes, 514
removal from food sources, 311
Taro inhibitors, 322
TCA-precipitable nitrogen, 168
Temporal induction of inhibitor mRNA's, 283
TIA, trypsin inhibitor activity, 309
Time-course of pancreatic carcinoma, 83
Time-paired controls, 124
Therapeutic benefits of inhibitors, 301, 500, 554
Therapy with pancreatic extracts, 148
Thin-layer chromatography of isoinhibitors, 402
Three-dimensional
structure of inhibitors, 7
inhibitor-enzyme complexes, 477
visualization of pancreatic tissue, 95
Threshold levels of inhibitors, 83
Thrombin, 154
Thymidine incorporation into pancreatic DNA, 69, 87
TIU, trypsin inhibitor units, 309
Tofu inhibitor content, 313
p-Toluenesulfonyl arginine methyl ester (TAME), 309
Tomato inhibitors, 282, 285
Tosyl-L-lysine chloromethyl ketone, (TLCK), 154
Tosyl-phenylalanine chloromethyl ketone, 154
Toxicological
consequences of alkali-treatment, 554
disease, 19
effects of amylase inhibitors, 499
Tracheostomy, 110
Triacetylglycerol, 175
Trimeric complexes of lentil inhibitors, 269
Triticale inhibitors, 322
Trophic effects of cholecystokinin, 87, 110
Trout pancreatic enzymes, 394
Trypsin, human forms, 172
Trypsin inhbitor(s)
analysis, 126, 130,190,204, 225,227, 351, 359, 415, 535
antigenicity, 449, 455
biosynthesis, 281
chromatography, 216
degradation in the rumen, 215
effect of heat, 234
effect on rat pancreas, 136
growth depression, 358
isolines, 295

methionine availability, 23
protection against digestion, 28, 168
protection against liver cancer, 195
removal and inactivation, 224, 234, 323
role in humans, 240
safety for humans, 172
species variation in susceptibility to, 169
susceptibility of infants, 87
units, 309
Trypsinogen, 17, 23, 27
Tryptophan as sites for photoaffinity labeling, 475
Tumor burden, 98
Tumorigenicity, 103
TUI, trypsin units inhibited, 309
Turkeys, 200
Tyrosinase inhibition, 546
Tyrosine modification, 266

Uncontrolled digestion, 28
Unsaturated fats and pancreatic cancer, 95, 98
Urease activity, 335

Vegetarian diets and inhibitor consumption, 302
Viva pink bean inhibitors, 321

Weight gain of rats fed inhibitors, 422
Wet weight of the pancreas, 114
Wheat inhibitors, 322, 357
Winged bean inhibitors, 248, 314, 358
amino acid analysis, 416, 423
antinutritional properties, 413, 424
autoclaving, 422
composition, 414
diets, 416
thermal stability, 417
Wound-induced inhibitors
amino acid sequence, 287
biochemistry, 283
biological role, 288
Bowman-Birk-type, 288
characterization, 283
DNA sequence, 284
genes, 285

X-ray crystallography of enzyme-inhibitor complexes, 6, 15
X,Y-digitizer, 95

Yam inhibitors, 492

Zein alkali treatment, 539
Zein inhibitors, 129
Zymogen activation, 2, 470

MIX
Papier aus verantwortungsvollen Quellen
Paper from responsible sources
FSC® C105338

If you have any concerns about our products,
you can contact us on
ProductSafety@springernature.com

In case Publisher is established outside the EU,
the EU authorized representative is:
Springer Nature Customer Service Center GmbH
Europaplatz 3, 69115 Heidelberg, Germany

Printed by Libri Plureos GmbH
in Hamburg, Germany